AGF-2374
TA1800
.C47
1995

DATE DUE

ELEMENTS OF OPTOELECTRONICS
AND FIBER OPTICS

ELEMENTS OF OPTOELECTRONICS AND FIBER OPTICS

Chin-Lin Chen
School of Electrical and Computer Engineering
Purdue University

IRWIN
Chicago Bogotá Boston Buenos Aires Caracas
London Madrid Mexico City Sydney Toronto

© **Richard D. Irwin,** a Times Mirror Higher Education Group, Inc. company, 1996

All rights reserved. No part of this publication may be reproduced, stored in a retrieval system, or transmitted, in any form or by any means, electronic, mechanical, photocopying, recording, or otherwise, without the prior written permission of the publisher.

Irwin Book Team

Sponsoring editor: Scott Isenberg
Editorial assistant: Tricia Howland
Marketing manager: Brian Kibby
Project editor: Paula M. Buschman
Production manager: Bob Lange
Cover designer: Tim Goldman
Art studio: Wellington Studios
Compositor: Graphic Composition, Inc.
Typeface: 10/12 Times Roman
Printer: R. R. Donnelley & Sons Company

**Times Mirror
Higher Education Group**

Library of Congress Cataloging-in-Publication Data

Chen, Chin-Lin.
 Elements of optoelectronics and fiber optics / Chin-Lin Chen.
   p.   cm.
 Includes index.
 ISBN 0-256-14182-7
 1. Fiber optics. 2. Optoelectronics. I. Title.
TA1800.C47   1996
621.36—dc20   95–14931

Printed in the United States of America
1 2 3 4 5 6 7 8 9 0 DO 2 1 0 9 8 7 6 5

FOREWORD

In view of recent advances in photonic devices and optical fiber communications systems, it is clear to me that electrical engineering students should have considerable exposure to optoelectronics and fiber optics in their undergraduate educations, even if their areas of specialization are not photonics. Since photonics comprises many different disciplines, a text on optoelectronics and fiber optics, even at the introductory level, must cover a multitude of topics. It should provide a broad overview of many of those topics and an in-depth treatment of selected subjects in lightwave technology. An overview is necessary for the reader to be able to discern subtle differences in existing photonic devices and appreciate the significance of new developments in photonic technology. An in-depth discussion should help the reader prepare for future graduate studies and professional updates. With this in mind, I have organized the materials in this text into 10 self-contained chapters. They include geometrical optics and gaussian beams, the generation and detection of light, the modulation and deflection of light beams, integrated optic waveguides and optical fibers, and guided-wave components. Each subject is discussed by starting with the basic principles and with emphasis on the physical concepts. Ample sketches and illustrations are included. Often, simplifying assumptions are made and special cases are considered, so that meaningful results can be deduced without relying on advanced mathematics. Extensions or generalizations are then presented. Whenever possible, final results are couched in terms of normalized parameters. Figures of merit of several photonic devices are also introduced to facilitate comparisons.

More than enough material is presented in the text for a typical one-semester, three-credit-hour course for electrical engineering seniors. Instructors can then select a subset of topics to suit their needs. The minimum mathematics requirements are calculus and elementary differential equations. Prerequisites also include introductory courses on modern physics and electromagnetics. Some exposure to basic semiconductor physics and devices would be helpful. However, a knowledge of quantum mechanics is neither assumed nor required.

I have presented the materials contained in this text at regular classes for senior electrical and computer engineering students at Purdue University and as short courses elsewhere. I owe much to the students in those classes and short courses. They made numerous suggestions for improvement and discovered several errors and omissions in the lecture notes on which this book is based. In a sense, they are the co-authors of this book. I am indebted to my colleagues at Purdue University for their active interest and free advice throughout the years. In addition, I would like to express my sincere appreciation to Professor Gregory J. Sonek of the University of California, Irvine, Professor James J. Burke of the University of Arizona, Professor John Buck of the Georgia Institute of Technology, and Professor Monish Chatterjee of the State University of New York at Binghamton for their thorough reading and constructive critiques of the manuscript.

Finally, I am grateful to my wife, Ching-Fong, for her constant smile, everlasting patience, and continuing encouragement, and for maintaining a pleasant, comfortable, and worry-free home in which to work and to watch our children laugh, grow, and mature.

Chin-Lin Chen

CONTENTS

Chapter 1
Introduction to Optoelectronics and Fiber Optics   1

- 1.1 Introduction   1
- 1.2 Compact Disk Players   2
- 1.3 Integrated Optic Temperature Sensors   5
- 1.4 An Optical Frequency Division Multiplexing Distribution System   8
- 1.5 Optical Fiber Gyroscopes   13
- 1.6 Overview and General Outline   17

Chapter 2
Optical Beams and Their Transformations   23

- 2.1 Introduction   24
- 2.2 Ray Vectors, Ray Transfer Matrices, and ABCD Matrices   25
- 2.3 Optical Systems Synthesis   44
- 2.4 Fundamental Gaussian Beams   49
- 2.5 Optical Components and Gaussian Beams   55
- 2.6 Higher-Order Modes   61
- 2.7 ABCD Law   64
- 2.8 Gaussian Beam Transformations by Thin Lenses   68
- 2.9 Optical Cavities   74

Chapter 3
Elements of Lasers   93

- 3.1 Introduction   94
- 3.2 Boltzmann's Distribution, Photons, and Blackbody Radiation   99
- 3.3 Lasers: A Qualitative Description   100
- 3.4 Absorption, Stimulated Emission, and Spontaneous Emission   102
- 3.5 Line Broadening   109

- 3.6 Rate Equations for Lasers with Optical Pumping 110
- 3.7 Q-Switched Lasers 125
- 3.8 Mode Locking Operation 133
- 3.9 Description of Gas Laser Systems 140
- 3.10 Description of Dye Lasers 149
- 3.11 Description of Solid-State Lasers 153

Appendix A
Identification of Atomic Energy Levels 169

Chapter 4
Semiconductor Injection Lasers, Light Emitting Diodes, and Superluminescent Diodes 181

- 4.1 Introduction 182
- 4.2 Intrinsic and Extrinsic Semiconductors 183
- 4.3 The Interaction of Semiconductors with Electromagnetic Waves 190
- 4.4 Ternary and Quaternary Semiconductors 197
- 4.5 Homojunctions and Heterojunctions 202
- 4.6 Basic Semiconductor Luminescent Diode Structures 206
- 4.7 Light Emitting Diodes 217
- 4.8 Basic Parameters: Semiconductor Injection Lasers 230
- 4.9 Dynamic Characteristics: Semiconductor Injection Lasers 239
- 4.10 Superluminescent Diodes 250
- 4.11 Comparison of ILDs, LEDs, and SLDs 251

Chapter 5
Optical Detection and Detectors 259

- 5.1 Introduction 260
- 5.2 Thermal Detectors 263
- 5.3 Photon Detectors 265
- 5.4 Noise and Noise Equivalent Power 285
- 5.5 Coherent and Incoherent Detection 290
- 5.6 Circuit Topology 298

Chapter 6
Modulation and Deflection of Optical Beams 305

- 6.1 Introduction 306
- 6.2 State of Polarization 308
- 6.3 Simplified Acoustooptic, Electrooptic, and Magnetooptic Effects 315
- 6.4 Faraday Rotation, and Magnetooptic Modulators and Isolators 322

6.5 Index Ellipsoids 327
6.6 Linear Electrooptic Effects 337
6.7 Electrooptic Modulators 341
6.8 Elasticity and Acoustooptic Effects 351
6.9 Acoustooptic Modulators and Deflectors 357

Appendix B
Elasticity for Three-Dimensional Objects 377

Chapter 7
Integrated Optics 391

7.1 Introduction 392
7.2 Guided Waves: A Physical Picture 392
7.3 Guided Wave Phase and Group Velocities 397
7.4 Reflection by Planar Dielectric Boundaries 400
7.5 Step Index Thin-Film Waveguides 403
7.6 Graded Index Thin-Film Waveguides 418
7.7 Prism Couplers 421
7.8 Film Index and Thickness Measurement: An Application 424
7.9 Optical Directional Couplers 426

Chapter 8
Optical Fibers 437

8.1 Introduction 438
8.2 Simple Characteristics of Step Index Fibers 440
8.3 Linearly Polarized Modes Guided by Weakly Guided Step Index Fibers 444
8.4 Information Capacity 457
8.5 Fiber Fabrication Processes 466
8.6 Fiber Losses 472

Appendix C
Bessel and Modified Bessel Functions 481

Appendix D
Weakly Guided Fibers with a Parabolic Index Profile 485

Chapter 9
Multimode Fiber Components and Systems 489

9.1 Introduction 490
9.2 Taps and Star Couplers 491

9.3   Multimode Fiber Switches   502
9.4   Power Coupled into Fibers   508
9.5   Connector Loss   517
9.6   Risetime, Pulse Width, and Bandwidth   523
9.7   Power Budget and Risetime Budget Example   526

Chapter 10
Polarization Effects: Single-Mode Fibers and Components   533

10.1   Introduction   534
10.2   Birefringent Media and Jones Calculus   535
10.3   Single-Mode Fiber Birefringence   550
10.4   Single-Mode Fiber Polarization Components   557
10.5   Single-Mode Fiber Directional Couplers   562
10.6   Erbium-Doped Fiber Amplifiers   564

Appendix E
Jones Matrix for a Linearly Birefringent Medium in Strong Magnetic Fields   575

Appendix F
Jones Matrix for a Linearly Birefringent Medium with a Continuously Rotating Axis   579

Index   585

INTRODUCTION TO OPTOELECTRONICS AND FIBER OPTICS

CHAPTER 1

CHAPTER OUTLINE

1.1 Introduction
1.2 Compact Disk Players
1.3 Integrated Optic Temperature Sensors
1.4 An Optical Frequency Division Multiplexing Distribution System
    1.4.1 Multiplexing
    1.4.2 Demultiplexing
1.5 Optical Fiber Gyroscopes
    1.5.1 Sagnac Interferometers
    1.5.2 Fiber Gyroscopes
1.6 Overview and General Outline
References
Problems

1.1 INTRODUCTION

In the not-too-distant past, optics and electrical engineering were viewed as two separate disciplines, each flourishing independently of the other. However, with the invention of ruby lasers in 1960, GaAs injection lasers in 1962, and the realization of low-loss optical fibers in 1970, the two technologies quickly merged into one. Several words or terms have since been coined to identify this emerging technology. They include, for example, photonics, optoelectronics, electrooptics, etc. These terms are often used yet rarely defined. There is no generally accepted definition for any of them.

    An explanation for "photonics" has been offered in a report issued by the U.S. National Research Council [1], which states:

> Photonics is concerned with the use of photons to work with or to replace electrons in certain communications, computer, or control applications traditionally carried out by electronics.

In the same report, photonics is also cited as "one of the key technologies of the information age." A more succinct definition for photonics was given by I. M. Ross in his address at the IEE's Michael Faraday Bicentennial Conference in 1991 [2], in which he stated that photonics is "the science and application of the photon." We may view the short definition of photonics as a paraphrase of the definition for electronics, "the science and application of the electron." Ross also identified two perspectives of optoelectronics:

> One is that optoelectronics is basically concerned with an integration of electronics and optics at a level where new physical phenomena are observed, and new functionality is created—functionality not possible with electrons or electromagnetic waves separated. The other perspective is that optoelectronics is concerned with optical implementation of tasks that were previously done electronically.

The use of dielectric cylinders as transmission media for electromagnetic waves is not a new concept. The propagation of electromagnetic waves along dielectric waveguides was studied by Hondros and Debye in 1910 [3]. However, low-loss (10 dB/km or less) silica glass fibers were not realized until the early 1970s. Since then, optical fibers have been used extensively in various communications systems. Lately, optical fibers have become almost synonymous with long distance telecommunications. In addition, optical fibers are used in local networks connecting offices and computers, as signal processing components, and as sensors to monitor physical, chemical, and biological variables. A new and major thrust for fiber technology is the prospect of incorporating fibers into cable television systems for bidirectional voice, video, and data transmission. This could lead to the massive deployment of fibers in cities and urban areas in the near future.

Clearly, there exist many optoelectronic devices and systems which make use of electrons and photons, and use optical fibers as the medium for transmitting electromagnetic waves. Existing devices or systems have been improved by replacing some electronic components with their optical counterparts. New devices have been conceived to take advantage of optic and electronic technologies. These new devices simply could not function with one technology alone. To design or improve such devices, an engineer must be conversant in both optoelectronic and fiber technologies. The purpose of this book is to introduce these subjects to the reader, and to provide the foundations for understanding and the tools necessary for designing these components and devices. As indicated by the title, the book covers two major topics: optoelectronics and fiber optics. Before the contents of the major topics are introduced, four specific examples that illustrate the concepts and the use of the two technologies are described.

1.2 COMPACT DISK PLAYERS

Many readers probably own, or at least have listened to, a compact disk (CD) player. The basic configuration of a CD player, so far as its reading function is concerned, is essentially the same as the playback unit of a video disk player, CD-ROM reader, or other optical data retrieval device. This is not surprising, since CD players were originally intended for use as video disk playback units for television. The early products were referred to as video long-play (VLP) systems [4].

The same basic setup also exists in equipment used for archiving optical data, laser direct patterning of semiconductor wafers, laser machining, welding, engraving, and other applications. However, the power levels involved in these systems are much higher.

Figure 1.1 is a schematic of the original VLP system playback unit. Audio and video information is encoded in the form of etched pits on the video disk. Pits of width 0.8 μm and depth 0.16 μm are etched on the flat reflecting surface of the video disk and protected by a transparent layer. The depth of the pits is approximately a quarter wavelength of the probing light in the protective transparent layer. Coherent and monochromatic light is used as the probing light, so that it can be focused to a small spot on each pit. In the original VLP design, HeNe lasers were used as the light source. Light with a wavelength of 0.633 μm from a low-power (~ 1 mW) HeNe laser is routed by prisms and focused by an objective onto the reflecting surface of the disk. Light falling on the flat reflecting surface without pits is reflected through the objective to the photodetector. Light hitting a pit is diffracted by it. Because of the divergence

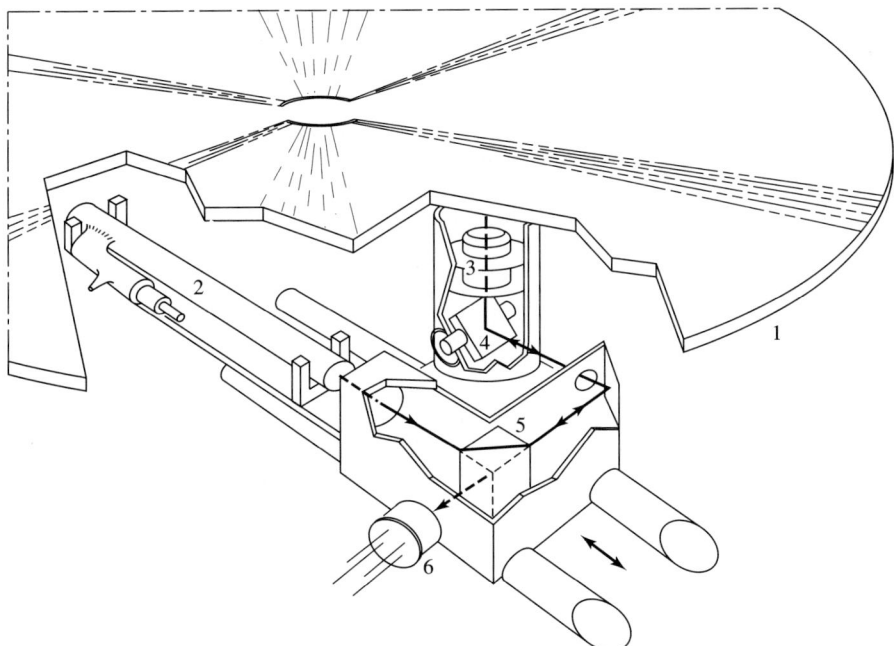

Figure 1.1 Schematic diagram of VLP playback unit (1, Optical disk; 2, HeNe laser; 3, objective; 4, pivoting mirror; 5, prism; 6, photodetector.) [4]

of diffracted light, a substantial fraction of diffracted light is diverted away from the light detector. Thus, there is a decrease in the light intensity reaching the photodetector. The phase of the light reflected by the pits is also delayed, relative to the light returning from the flat surface, by approximately a half wavelength. When two reflected beams are superimposed at the detector, the total light intensity is reduced due to destructive interference. The light intensity is converted to an electrical signal by the photodetector. The electrical signals thus generated are proportional to the light intensity incident upon the detector surface and are therefore a good replica of the information encoded on the disk.

Video disks have a diameter of 30.5 cm. Although the precise overall dimensions of the playback unit shown in Figure 1.1 are not known, a rough estimate is possible from the size of the HeNe laser. For HeNe lasers of 1970 vintage, the plasma tube is approximately 30 cm long and 5 cm in diameter. In modern CD audio players, units are smaller and HeNe lasers have been replaced by GaAlAs injection lasers with emissions near 0.82 μm. Figure 1.2 is an artist's impression of the optical pickup unit of a commercially available CD player. The pickup device is about 4.5 cm in length and 1.2 cm in diameter [5]. The essential optical components are shown schematically in Figure 1.3. The discrete optical components are fabricated individually and then assembled in a robust and compact manner. Two additional features should be noted. First, the reflected light is directed to four photodetectors. Electrical outputs from the photodetectors are combined to provide the audio information, as well as the electrical signals for

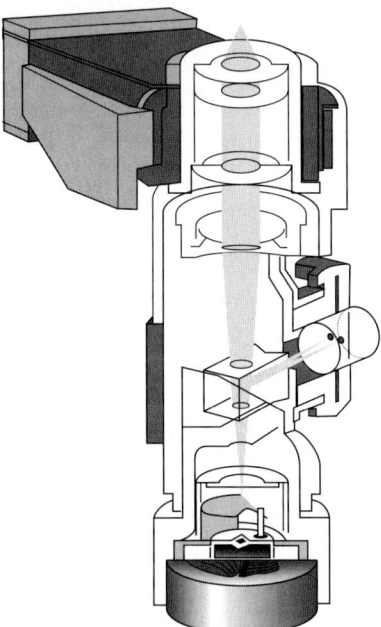

Figure 1.2 Artist's impression of the pickup unit of a CD player [5]

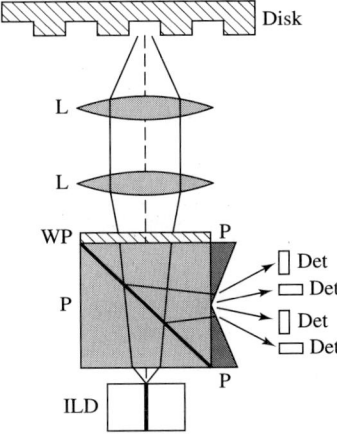

Figure 1.3 Essential optical elements of an optical pickup unit

Det: Detector
ILD: Injection laser diode
L: Lens
P: Prism
WP: Quarterwave plate

tracking and error correction. Second, a quarterwave plate is inserted between the beam splitter and the lenses. Because of the quarterwave plate and the polarizing beam splitter, the reflected beam is diverted to the photodetectors, rather than to the laser diode. As a result, instabilities due to optical feedback are minimized and system performance is greatly enhanced.

An integrated optic pickup unit has been reported in the literature (Figure 1.4). In this unit, optical beams are guided by thin-film waveguides, rather than propagated in an unguided manner in free space. The discrete or "bulk" optic components, like lenses and beam splitters, are replaced by their integrated optic counterparts. For example, gratings are used to both focus and split the guided beams. In addition, photodetectors are built on the same substrate used to form the thin-film waveguides, grating reflectors, and lenses [6, 7]. Note, however, that the pickup unit shown in Figure 1.4 is not yet fully "integrated," since the laser diode is attached separately to the pickup chip. In the future, it can be expected that the injection laser diode will be built monolithically onto the pickup unit. The resulting fully integrated pickup devices will be truly compact and very rugged.

1.3 INTEGRATED OPTIC TEMPERATURE SENSORS

Not all integrated optic devices are as complicated as the CD pickup unit shown in Figure 1.4. In this section, the operation of a relatively simple and yet sensitive temperature sensor is described. It is an integrated optic Michelson interfer-

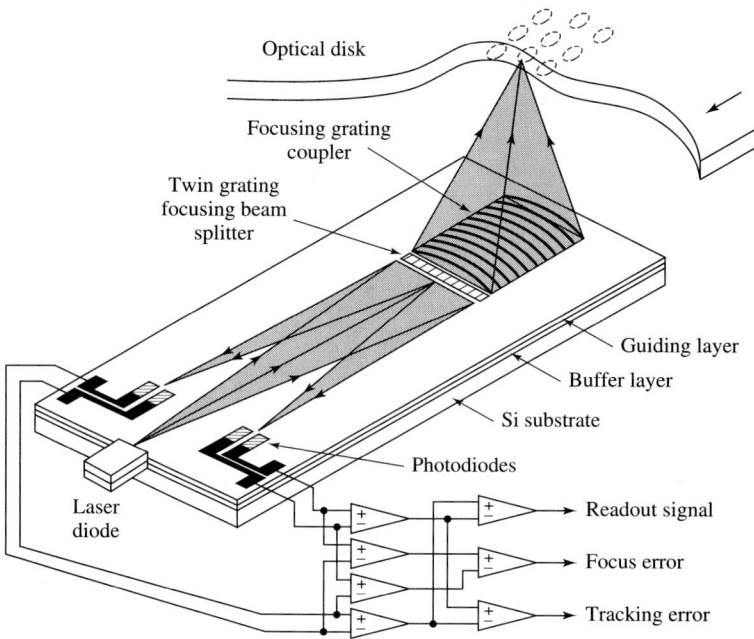

Figure 1.4 Schematic diagram of integrated optical disk pickup device with gratings for beam focusing and splitting ([7] © 1986 IEEE)

ometer. The essential components of the temperature sensor are shown schematically in Figure 1.5. In particular, the waveguide junction serves as both an optical beam splitter and a beam recombiner. The waveguides to the right of the junction have different lengths, L_1 and L_2. We will show shortly that the temperature sensitivity of the interferometric sensor increases with the path length difference, $\Delta L = L_1 - L_2$. To make the device robust, the reflectors terminating the waveguides are deposited directly on the waveguide ends.

Light of a known wavelength λ from a coherent source is fed to the input arm to the left of the junction. Because of the junction, the input is split between the two arms on the right. Let the fields emerging from the junction be E_1 and E_2. These fields traverse from the junction to the mirrors and then back to the junction. The round-trip phase delays in the two arms are $2kNL_1$ and $2kNL_2$, where N is the effective index of refraction of the thin-film waveguides and $k = 2\pi/\lambda$. Also, k can be expressed in terms of frequency f, i.e., $k = 2\pi f/c$. It will be shown in Chapter 7, "Integrated Optics," that N depends on the material properties and the waveguide geometry. We assume that the reflection coefficients introduced by the two reflectors are the same and they are set to 1 for simplicity. The fields returning to the junction are $E_1 e^{-j2kNL_1}$ and $E_2 e^{-j2kNL_2}$, respectively. The reflected fields are combined by the junction. The fields in the detector arm are:

$$E = E_1 e^{-j(2kNL_1 + \phi)} + E_2 e^{-j2kNL_2}$$

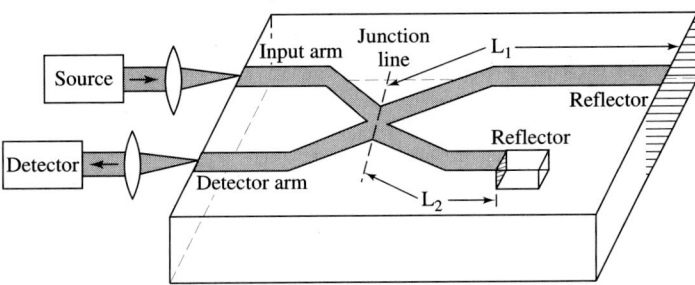

Figure 1.5 Integrated optic temperature sensor

where ϕ is the phase introduced by the waveguide junction. In Chapter 5, "Optical Detection and Detectors," it will be demonstrated that the detector output is proportional to $|E|^2$. A little manipulation will show that

$$|E|^2 = E_1^2 + E_2^2 + 2E_1E_2 \cos[2kN(L_1 - L_2) + \phi]$$

Thus, the detector output is of the form

$$V = 1 + m\cos(2kN\Delta L + \phi) \quad (1.1)$$

where $m = 2E_1E_2/(E_1^2 + E_2^2)$. Clearly, the detector output is a periodic function of $2kN\Delta L + \phi$ and varies between $1 + m$ and $1 - m$. Both N and ΔL are temperature dependent. As temperature changes, N, ΔL, and therefore V change as well. By monitoring the change in V, we infer the temperature change. This is the operating principle of the Michelson interferometric temperature sensor. To estimate the temperature sensitivity, we write $2kN_0\Delta L_0 + \phi = \Phi_0$ where $\Delta L_0 = \Delta L(T_0)$, $N_0 = N(T_0)$, and T_0 is the reference temperature. At temperature T,

$$2kN\Delta L + \phi \sim \Phi_0 + 2k\left[\frac{d}{dT}(N\Delta L)\right](T - T_0).$$

The derivative term at $T = T_0$ is:

$$\frac{d}{dT}(N\Delta L) = \frac{dN}{dT}\Delta L_0 + N_0\frac{d\Delta L}{dT} = \Delta L_0\left(\frac{dN}{dT} + \frac{N_0}{\Delta L_0}\frac{d\Delta L}{dT}\right) \quad (1.2)$$

The term $\frac{dN}{dT}$ is the rate of index change as temperature changes, and $\left(\frac{1}{\Delta L_0}\frac{d\Delta L}{dT}\right)$ is the linear thermal expansion coefficient of the waveguide. These terms depend on the material properties and the geometry of the waveguide cross section. Thus, $\frac{d}{dT}(N\Delta L)$ increases linearly with ΔL_0, which is also proportional to ΔL.

An integrated optic temperature sensor, as depicted in Figure 1.5, has been reported by Izutsu, et al. [8]. Their sensor was built on an $LiNbO_3$ substrate. By using published values of material properties, such as the thermal expansion coefficient of $LiNbO_3$, the specific waveguide dimensions (ΔL = 9.6 mm), and the operating condition (λ = 0.633 µm), they estimated that $\left[2k \dfrac{d}{dT}(N\Delta L) \right]$ is about 7. Thus, for a temperature increment of 0.44°C, $2kN\Delta L + \phi$ changes by π. In other words, a temperature change of 0.44°C would cause V given in (1.1) to change from its peak value to a valley, or vice versa. Such a variation in V can easily be monitored. The measured value is 0.32°C. As indicated in (1.2), $\dfrac{d}{dT}(N\Delta L)$ increases linearly with ΔL. By increasing the path length difference, the temperature increment needed to change $2kN\Delta L + \phi$ by π decreases; in other words, the temperature sensitivity increases. Izutsu, et al., predicted that this type of temperature sensor might be used to discern temperature variations as small as 0.01°C.

1.4 AN OPTICAL FREQUENCY DIVISION MULTIPLEXING DISTRIBUTION SYSTEM

The principal use of optical fibers is to transmit optical signals, with little attenuation and distortion. There are many types of fiber communications systems with various degrees of complexity and sophistication. The simplest fiber communications system consists of a transmitter at one end of a fiber and an optical detector at the other end. In such a simple configuration, the fiber transmits a single signal channel. In actuality, the fiber bandwidth is quite wide, and fibers can carry many signal channels. Therefore, a more economical use of fibers would be to transmit a large number of signal channels from several transmitters on one end to a multitude of receivers on the other end. Various signal multiplexing and demultiplexing schemes have been conceived for transmitting many signal channels over a single fiber. In this section, we will describe an optical frequency division multiplexing distribution system, originally presented by Toba, et al. [9].

In an optical frequency division multiplexing distribution system, optical signals with several carrier frequencies, separated by a small frequency separation, are transmitted in the same fiber. Each carrier is individually modulated and carries different information. For example, suppose that the optical signals have a wavelength near 1.5 µ, which in terms of frequency is about 200 THz, and that the frequency separation between neighboring channels is 10 GHz. The carrier frequencies, then, can be 200.00 THz, 200.01 THz, and 200.02 THz, etc. A block diagram of such an optical frequency division multiplexing distribution system is shown in Figure 1.6a. The optical signals are combined in a multiplexer and then fed to the fiber. If necessary, the optical signals are amplified along the way by one or more semiconductor amplifiers or, more likely, by doped fiber amplifiers. At the other end of the fiber, signals are branched out to several receivers. All receivers receive all the signal channels. The desired signal

1.4 An Optical Frequency Division Multiplexing Distribution System

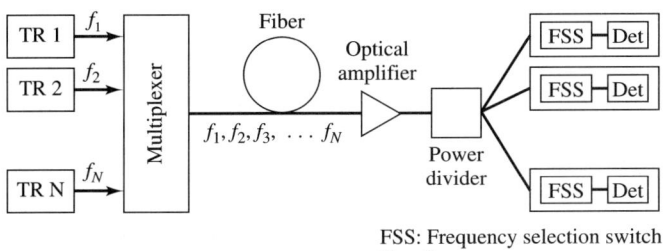

(a) Block diagram

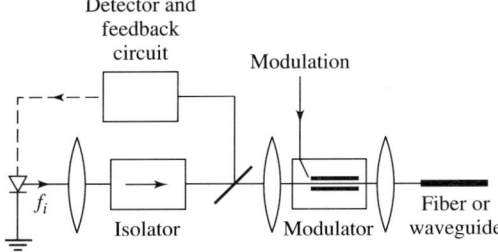

(b) Schematic diagram of a transmitter

Figure 1.6 Optical frequency division multiplexing distribution system ([9] © 1986 IEEE)

channel is then optically selected by each receiver individually, and the selected signal is detected and processed electronically. Obviously, there are many components to such a system. The function of some components is rather evident. Here, we will concentrate on two optical components—the optical multiplexer, and the frequency selecting switches.

1.4.1 Multiplexing

Each optical carrier is generated by a semiconductor laser diode. The frequency separation of neighboring channels is very narrow in terms of the carrier frequency, or center frequency, of the optical signal. In the numerical example previously given, the frequency separation (10 GHz) is a 50-millionth of the carrier frequency (about 200 THz). Therefore, each laser diode must be carefully stabilized. For this purpose, a small portion of the output is diverted to a detector, which produces a voltage for the electrical feedback control circuit (Figure 1.6b). An optical isolator is also inserted in the optical path, to reduce possible optical reflection back to the laser diode. In the presence of an optical isolator, the stability of the laser output is greatly improved. Modulation is imposed on each optical carrier via an external modulator. The optical isolator and external modulator are discussed in Chapter 6, "Modulation and Deflection of Optical Beams."

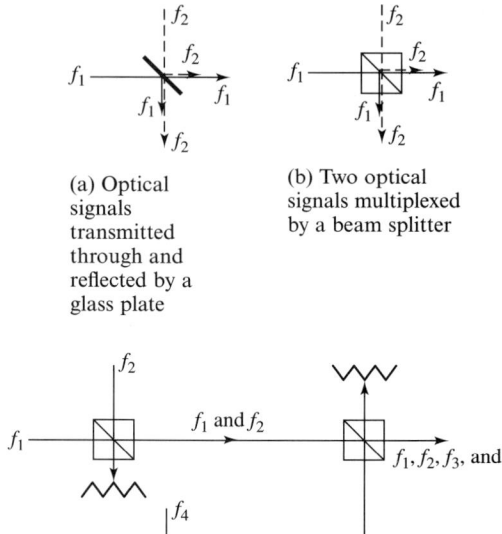

(a) Optical signals transmitted through and reflected by a glass plate

(b) Two optical signals multiplexed by a beam splitter

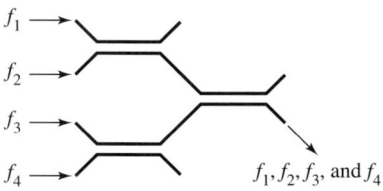

(c) Four signal channels multiplexed by three beam splitters

(d) Four signal channels multiplexed with beam splitters or directional couplers

Figure 1.7 Optical signal multiplexing with beam splitters or directional couplers

The multiplexing scheme is easy to understand. Conceptually, multiplexing can be done with glass plates. Figure 1.7a shows two optical beams incident upon a glass plate. Each beam is partially transmitted through, and partially reflected by, the glass plate. When a glass plate is used for beam splitting, the transmitted power is much stronger than the reflected power. Nevertheless, the two beams are superimposed in the output ports. A good beam splitter performs both of the functions; that is, beam splitting and superposition (Figure 1.7b). The power split ratio of the beam splitter can be controlled to any desired value. Usually, a 50–50 power split is desired. With three beam splitters, four signal channels can be multiplexed. This is shown in Figure 1.7c. Integrated optic or fiber optic directional couplers are also useful for beam splitting and combining. A multiplexer based on integrated optic or optical fiber directional couplers is shown in Figure 1.7d. The operation of directional couplers is discussed in Chapter 7, "Integrated Optics," and Chapter 10, "Polarization Effects in Single-Mode Fibers and Components."

1.4.2 Demultiplexing

In each receiver, there is a frequency selection switch, which is a series combination of Mach–Zehnder interferometers. To understand the operation of a frequency selection switch, let's examine the interference in a single Mach–Zehnder interferometer with unequal path lengths. As depicted in Figure 1.8a, the input beam is split into two beams by the first beam splitter. The two beams are then recombined by the second beam splitter. Depending on the frequency of the optical beam and the path length difference $\Delta L = L_1 - L_2$, the combined beam may emerge from port a or port b. There are three mirrors in each path. For simplicity, assume that the mirrors are identical. Let the total phase change due to the three mirrors be ϕ_m. The two beam splitters are also assumed to be lossless and to have symmetric structures. Incident power is divided equally between the transmitted and reflected beams by the beam splitter. In the process, phase delays ϕ_t and ϕ_r are introduced by the beam splitter to the transmitted and reflected beams, respectively. Of particular importance to the operation of a Mach–Zehnder interferometer is the difference between the phase delays of

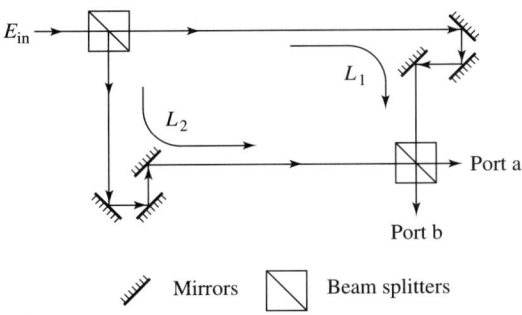

(a) Mach-Zehnder interferometer with unequal path lengths

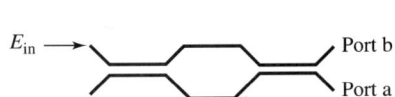

(b) Mach-Zehnder interferometer with unequal path lengths based on integrated optic or fiber optic directional couplers

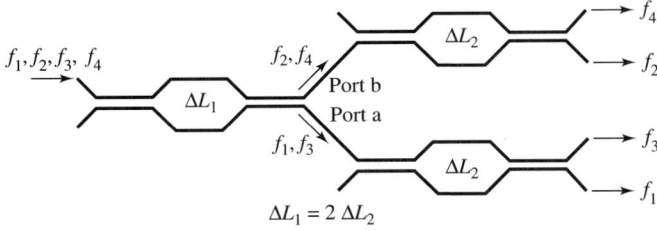

(c) Frequency selection switch with three directional couplers

Figure 1.8 Frequency selection switch based on Mach-Zehnder interferometers with unequal path lengths

the reflected and transmitted beams [10]. For a symmetric beam splitter, the phase delay difference is

$$\Delta\phi = \phi_r - \phi_t = \pm \pi/2 \tag{1.3}$$

Suppose that waves of amplitude E_{in} and frequency f are incident upon the first beam splitter. The transmitted and reflected waves are

$$\frac{E_{in}}{\sqrt{2}} e^{-j\phi_t} \quad \text{and} \quad \frac{E_{in}}{\sqrt{2}} e^{-j\phi_r}$$

Waves arriving at the second beam splitter are

$$\frac{E_{in}}{\sqrt{2}} e^{-j(\phi_t + \phi_m)} e^{-jkL_1} \quad \text{and} \quad \frac{E_{in}}{\sqrt{2}} e^{-j(\phi_r + \phi_m)} e^{-jkL_2}$$

where $k = 2\pi f/c$. The amplitude of the waves emerging from port a is

$$E_a = \frac{E_{in}}{\sqrt{2}} e^{-j(\phi_t + \phi_m)} e^{-jkL_1} \frac{1}{\sqrt{2}} e^{-j\phi_r} + \frac{E_{in}}{\sqrt{2}} e^{-j(\phi_r + \phi_m)} e^{-jkL_2} \frac{1}{\sqrt{2}} e^{-j\phi_t}$$

$$= \frac{E_{in}}{2} e^{-j(\phi_r + \phi_m + \phi_t)} e^{-jk(L_1 + L_2)/2} [e^{-jk\Delta L/2} + e^{jk\Delta L/2}]$$

The output at port b is

$$E_b = \frac{E_{in}}{\sqrt{2}} e^{-j(\phi_t + \phi_m)} e^{-jkL_1} \frac{1}{\sqrt{2}} e^{-j\phi_t} + \frac{E_{in}}{\sqrt{2}} e^{-j(\phi_r + \phi_m)} e^{-jkL_2} \frac{1}{\sqrt{2}} e^{-j\phi_r}$$

$$= \frac{E_{in}}{2} e^{-j(2\phi_t + \phi_m)} e^{-jk(L_1 + L_2)/2} [e^{-jk\Delta L/2} + e^{-j2(\phi_t - \phi_r)} e^{jk\Delta L/2}]$$

In view of (1.3), the above equation becomes

$$E_b = \frac{E_{in}}{2} e^{-j(2\phi_t + \phi_m)} e^{-jk(L_1 + L_2)/2} [e^{-jk\Delta L/2} - e^{jk\Delta L/2}]$$

Thus, we obtain

$$|E_a| = E_{in} |\cos\frac{k\Delta L}{2}|$$

and

$$|E_b| = E_{in} |\sin\frac{k\Delta L}{2}|$$

Clearly, if $k\Delta L$ is an even multiple of π, all input power emerges from port a and none comes from port b. On the other hand, when $k\Delta L$ is an odd multiple of π, all power goes to port b, none to port a. Thus, by choosing $k\Delta L$, we can direct the signal to the desired port.

To examine the use of the Mach–Zehnder interferometer with unequal path lengths as a frequency selection switch, suppose the input signal has two frequency components with frequencies f_1 and f_2. The frequency separation is $\Delta f = |f_2 - f_1|$. Choose L_1 and L_2 such that the signal channel with frequency f_1 goes to port a, while signals with frequency f_2 emerge from port b. This can be done by choosing ΔL such that

$$\frac{2\pi}{c} f_1 \Delta L = 2m\pi$$

$$\frac{2\pi}{c} f_2 \Delta L = (2m + 1)\pi$$

where m is an integer. From these two equations, we obtain

$$\Delta L = \frac{c}{2\Delta f}$$

The frequency selection switches can also be thin-film waveguides or optical fibers and optical directional couplers (Figure 1.8b). In the thin-film or fiber implementation, no mirror is needed. However, it is necessary to account for the effective index of refraction N of the waveguide. The path length difference is then

$$\Delta L = \frac{c}{2N\Delta f} \qquad (1.4)$$

As an example, consider two signal channels with a frequency separation of 10 GHz. Also, suppose that the Mach–Zehnder interferometer is based on fibers or integrated optic waveguides with an effective refraction index of 1.47. From (1.4), we have $\Delta L = 1.02$ cm. Next, modify the frequency selection scheme to accommodate four channels, with frequencies f_1, f_2, f_3, and f_4. For simplicity, we keep the frequency separation the same as before, that is, 10 GHz. Figure 1.8c shows the schematic diagram of such a design. The first Mach–Zehnder interferometer has a path length difference of 1.02 cm. Thus, the signals with f_1 and f_3 emerge from port a, while the signals with frequencies f_2 and f_4 emerge from the other port. To separate f_1 from f_3, and f_2 from f_4, we need Mach–Zehnder interferometers with a path length difference of 0.51 cm, since $|f_1 - f_3| = |f_2 - f_4| = 20$ GHz. Clearly, the scheme can be extended to systems with 2^n channels.

1.5 OPTICAL FIBER GYROSCOPES

Although the interest in silica-based glass fibers is mainly for information transmission applications, the use of fibers has permeated nearly all aspects of engineering. We note, for example, the use of optical fibers as sensors, signal processing elements, and media for transmitting microwave signals [11–14]. In the following example, we will concentrate on a particular application of fibers, namely, the use of fiber interferometers as rotation sensors. There are four basic interferometric configurations: Michelson, Mach–Zehnder, Sagnac, and Fabry–

Perot. The uses of a Michelson interferometer for temperature sensing and Mach–Zehnder interferometers for frequency discrimination were discussed in the last two sections. In this section, therefore, we will study the use of Sagnac interferometers.

1.5.1 Sagnac Interferometers

A basic Sagnac interferometer consists of an optical source, an optical beam splitter, several mirrors, and a photodetector. Light with a known wavelength λ is split by the beam splitter into two beams of equal intensity. With the help of mirrors, the beams are made to travel along the same path, but in opposite directions, as shown in Figure 1.9a. One beam propagates in the clockwise sense, and the other propagates in the counterclockwise direction. Upon completion of one round trip of travel, the two beams are recombined by the beam splitter. The resulting interference leads to a change in the light intensity. At the photodetector, the optical intensity is converted to electrical signals. In the following analysis, we relate the electrical signals to the rate of rotation of the interferometer. For simplicity, consider a Sagnac interferometer with a circular optical path of radius R (Figure 1.10). In a stationary Sagnac interferometer, the path lengths traveled by the two beams are identical, i.e., $2\pi R$. The path lengths are different when the interferometer is under rotation. Suppose the interferometer is rotated clockwise with an angular velocity Ω. It takes a finite time for light to travel a path of finite length. At the end of time interval t, the beam splitter has moved to a new position by a distance equal to $\Omega R t$. Thus, light traveling clockwise must traverse an extra length $\Omega R t$ before exiting from the interferometer. Let t_{cw} be the time of travel of the beam moving clockwise. The total path length of the clockwise beam is

$$L_{cw} = 2\pi R + \Omega R t_{cw} = \frac{c t_{cw}}{n}$$

where c is the speed of light in vacuum and n is the index of refraction of the medium. We solve for t_{cw} and obtain

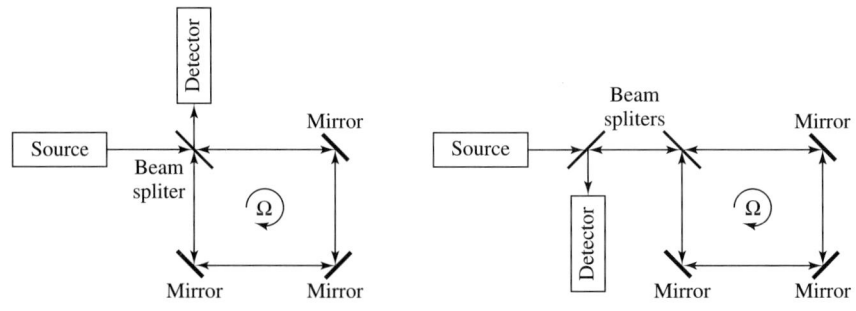

(a) Basic Sagnac interferometer (b) Improved setup for Sagnac interferometer

Figure 1.9 Schematic diagram of Sagnac interferometer

1.5 Optical Fiber Gyroscopes

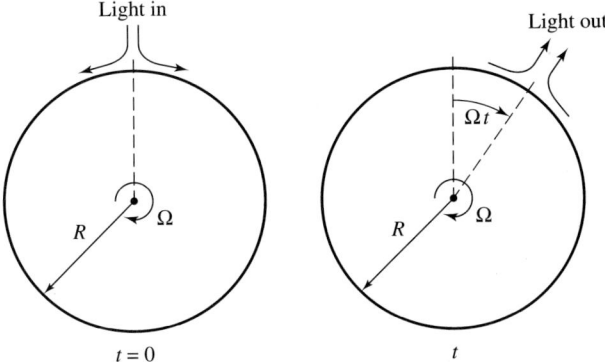

Figure 1.10 Operating principle of circular Sagnac interferometer

$$t_{cw} = \frac{2\pi R}{\frac{c}{n} - \Omega R} = \frac{2\pi n R}{c - n\Omega R}$$

For the beam moving in a counterclockwise direction, the path length is shortened by $\Omega R t_{ccw}$, where t_{ccw} is the time of travel of the counterclockwise beam. Thus, the path length is

$$L_{ccw} = 2\pi R - \Omega R t_{ccw} = \frac{c t_{ccw}}{n}$$

and

$$t_{ccw} = \frac{2\pi n R}{c + n\Omega R}$$

The time difference is

$$\Delta t = t_{cw} - t_{ccw} = \frac{4\pi n^2 R^2}{c^2 - n^2 \Omega^2 R^2} \Omega$$

Since c^2 is much greater than $(n\Omega R)^2$, Δt can be approximated by

$$\Delta t \approx \frac{4\pi n^2 R^2}{c^2} \Omega$$

The corresponding phase difference between the two beams is

$$\Delta\phi = \Delta t\, 2\pi f \approx \frac{8\pi^2 f n^2 R^2}{c^2} \Omega = \frac{8\pi^2 n^2 R^2}{c\lambda} \Omega \qquad (1.5)$$

The intensity of the interference beam is a function of the phase difference, which is linearly proportional to the rotation rate. By monitoring $\Delta\phi$, we can infer Ω. This is the operating principle of Sagnac interferometers as rotation sensors. It is interesting to note that the basic concept of using Sagnac interfer-

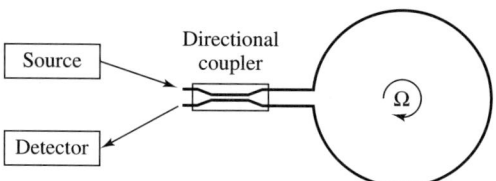

(a) Basic setup for rudimentary fiber gyroscope

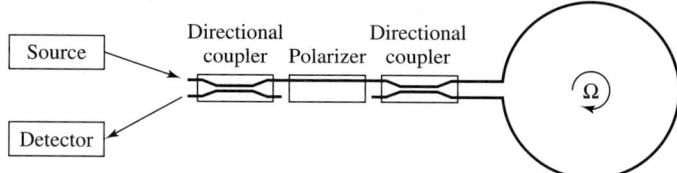

(b) Improved setup for fiber gyroscope

Figure 1.11 Schematic diagram of fiber gyroscope

ometers as rotation sensors was confirmed by Michelson, Gale, and Pearson in 1925. Their experiments were performed in evacuated 12″ pipes forming a 2010′ × 1113′ rectangle [15]. A photograph of the 12″ pipes and five workers can be found in [16].

1.5.2 Fiber Gyroscopes

Sagnac interferometers can be implemented with low-loss optical fibers, and this leads to optical fiber gyroscopes [17]. In fiber gyroscopes, light is guided by fibers, all mirrors are eliminated, and the beam splitter is replaced by a fiber directional coupler that splits the input beam into two parts. Since fibers are thin and reasonably flexible, they can be wound into coils with many turns. Thus, the path length can be increased without appreciably changing the gyroscope size. With the increase in path length, the gyroscope sensitivity is also greatly increased. For a fiber gyroscope coil with N_t turns, the total phase difference is

$$\Delta\phi \approx \frac{8\pi^2 N_t N^2 R^2}{c\lambda}\Omega = \frac{4\pi L N^2 R}{c\lambda}\Omega \qquad (1.6)$$

where $L = 2\pi N_t R$ is the total fiber length. In (1.6), we use the effective refractive index N of the fiber in lieu of the refractive index n of the medium. Figure 1.11a shows a schematic of a rudimentary fiber gyroscope setup.

For navigation applications, the gyroscope sensitivity and stability requirements are quite stringent. If all optical components are perfect and external disturbances are absent, the setups shown in Figures 1.9a or 1.11a would be adequate. In reality, no component is perfect, and disturbances are unavoidable.

For example, if the optical beam splitter or the fiber directional coupler is not perfect, or if the temperature fluctuates, gyroscope stability becomes a problem. Studies show that gyroscope stability can be improved by using a different topology, such as the setup with two beam splitters, shown in Figure 1.9b, or the corresponding fiber implementation with two fiber directional couplers as shown in Figure 1.11b. As a further improvement, a fiber polarizer is also added in Figure 1.11b. For each beam splitter shown in Figure 1.9b and each fiber directional coupler shown in Figure 1.11b, an output port is left open and power exiting from this port is lost. Although the net power incident on the photodetector is reduced, the gyroscope reading is more reliable since the gyroscope stability is greatly improved. A phase modulator can be added to enhance the gyroscope stability and to provide signal processing capabilities. Further improvement can be realized by using a superluminescent diode, in lieu of the laser diode, and by attaching fibers directly to the light source and photodetector. Figure 1.12 is a photograph of an actual gyroscope [18]. The fiber coil, "pigtailed" superluminescent diode and photodiode, and other optical components are clearly identified.

1.6 OVERVIEW AND GENERAL OUTLINE

The examples given previously in this chapter clearly demonstrate two important aspects of modern optical engineering, aspects that are not necessarily implied in the definition of "photonics" or "optoelectronics." First, contempo-

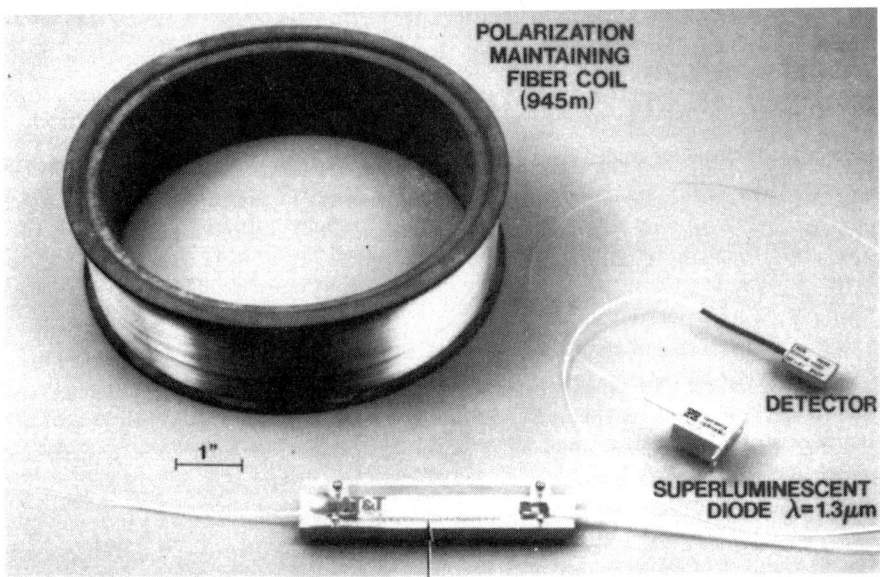

Figure 1.12 Photograph of all-fiber gyroscope, showing the fiber and optical components [18]

rary optical devices or systems contain many optical components in addition to the lasers and fibers. Such components include lenses, mirrors, beam splitters, and waveguide junctions, which play important roles in CD players and temperature sensors. These passive elements also perform important functions in fiber optic systems, such as the fiber optic frequency division multiplexing distribution system, although their presence may be concealed by the packaging. Second, electronic and optic components play equally significant roles in these systems. Therefore, the interfacing of optical components with electrical components must be an important consideration in the design of any optoelectronic system.

In the next chapter, we will study the effects of simple, passive optical elements on light beams. Coherent light beams are treated first as uniform plane waves and then as Gaussian beams. ABCD matrices and the ABCD law are introduced to quantify the effects of passive components on uniform plane waves and Gaussian beams, respectively. The relationships between optical resonators and Gaussian beams are also detailed, as are effects of simple lenses. In addition to analysis problems, simple synthesis problems are also studied.

Chapters 3 and 4 are concerned with the generation of coherent light. After a brief introduction to the absorption, and the spontaneous and stimulated emissions, of photons, rate equations are presented for three- and four-level lasers with optical pumping. In most instances, the continuous and transient operations of lasers are described in terms of such rate equations. Examples of gas lasers and solid-state lasers are given at the end of Chapter 3. Chapter 4 is devoted entirely to semiconductor light emitting diodes, injection laser diodes, and superluminescent diodes. Rate equations with a current injection term are used to examine the power–current and modulation characteristics of semiconductor injection laser diodes. The equivalent electrical circuits for these diodes are also presented.

Optical detection and photodetectors are the subjects of Chapter 5. Although our emphasis with respect to optical detectors is on photon detectors, including PIN diodes and avalanche photodiodes, the basic operation of thermal detectors is also briefly mentioned. Following the discussion on optical detectors and a brief study of noise, the methods of incoherent and coherent detection are studied and compared.

To fully utilize the huge bandwidth potential of optical beams and optical fibers, high-speed deflection and modulation techniques are required. Direct modulation of gas and solid-state laser outputs is very difficult, if not impossible. Although the emission from semiconductor diodes can be amplitude-modulated by varying the diode current, the amplitude modulation is accompanied by frequency chirping and mode hopping, which may be detrimental to many optical systems. External modulators, discussed in Chapter 6, are useful in these situations. In external modulators, electrooptic, magnetooptic and acoustooptic effects are used to create modulation or deflection. To understand the physics involved, we begin by presenting simplified pictures of electrooptic, magnetooptic, and acoustooptic effects. The concept of the index ellipsoid is then brought in to analyze the linear electrooptic effect quantitatively. Our attention is then turned to the basic configurations of electrooptic and acoustooptic modulators and the figures of merit of electrooptic and acoustooptic materials.

When intrinsic atmospheric absorption and the effects of air turbulence on the propagation of visible and near infrared (IR) light are considered, the need for waveguide structures at optical frequencies becomes obvious. Since all electrical conductors experience high losses at optical frequencies, waveguide structures at these frequencies must be based on low-loss dielectric materials. Thin-film waveguides and optical fibers are two basic forms of optical waveguides. Thin-film waveguides are the building blocks of optoelectronic integrated circuits (OEIC), which may be found in consumer products, such as CD players, simple temperature sensors, and multiplexing and demultiplexing components, and in complicated transmitters, repeaters, and receivers of optical communication systems. Integrated optics is therefore an important subject in its own right. In addition, the basic waveguide principle of planar dielectric waveguides is the same as that of circular fibers, and the mathematics is much simpler. We can therefore use planar dielectric waveguides as a vehicle to introduce various topics in guided wave optics. The main objectives of Chapter 7, then, are to introduce the following concepts: modes, phase and group velocities of guided modes, and, more importantly, the generalized parameters useful in characterizing optical waveguides. The operation of directional couplers is also studied in Chapter 7.

Optical fibers are discussed in Chapter 8. We begin by introducing important nomenclature, such as meridional and skew rays, step and graded index fibers, numerical aperture, and traditional mode designation. Although the dispersion relation of circular fibers is presented without derivation, considerable insight is gained by studying plots of the characteristic equation. (The derivation of that equation for linearly polarized (LP) modes guided by weakly guiding fibers is given in Appendix B.) The significance of weakly guiding fibers and LP mode designation are then discussed. Throughout the discussions on weakly guiding fibers, the use of generalized parameters is stressed. We note in particular the similarities of the b and V parameters and the bV plots of circular fibers with those of thin-film waveguides. Discussions on the intermodal dispersion of multimode fibers and the intramodal dispersion of single-mode fibers are also based on the bV diagrams. The discussion concludes with a comparison of the bandwidth–length product of various fibers. While many material combinations can be and have been used to form optical fibers, the chapter concentrates mainly on silica-based glass fibers. The causes of fiber losses are identified, and important techniques for fabricating low-loss silica fibers are described.

Having discussed the light sources, photodetectors, and light transmission media, we then combine these components into a complete, albeit simple, system in Chapter 9. In this chapter, we consider simple multimode fiber systems with light emitting diodes (LEDs) and PIN diodes. For a given output level from a transmitter and the minimum power required by a receiver, we determine the information capacity and the link length of a fiber system. Factors considered include: the source–to–fiber and fiber–to–detector coupling, fiber attenuation, the insertion loss of connectors or splices, the rise time of electronic components, and the pulse broadening of the fiber. Thus, the concepts of power budget and rise time budget are introduced.

In Chapter 10, we concentrate on single-mode fibers. Most single-mode fibers support two nearly-degenerate polarization modes. Since the propagation constants of the two polarization modes are not identical, single-mode fibers are birefringent. The origin of that birefringence is discussed. We use Jones calculus to characterize the birefringent components in general and single-mode fibers in particular. As noted in Chapter 8, the main advantage of single-mode fibers over multimode fibers is the bandwidth–length product. However, if two polarization modes are present, the advantages of single-mode fibers are seriously compromised. A coherent detection scheme is therefore needed to handle the large bandwidth available in single-mode fiber systems. Also, to optimize the sensitivity of the heterodyne detection system, the state of polarization of the local oscillator beam must be matched to that of the incoming signals, as discussed in Chapter 5. Therefore, polarization elements are needed at the front ends of the coherent receivers to select, or to maintain, the states of polarization. In this connection, we discuss fiber polarizers and polarization transformers. Jones calculus is also used to describe quantitatively the operation of these polarization components.

REFERENCES

1. *Photonics: Maintaining Competitiveness in the Information Era.* Washington D.C.: National Academy Press, 1988.
2. Ross, I. M. "Telecommunications in the era of photonics." *Sol. State Technol.* 35, (April, 1992), pp. 36–43.
3. Hondros, A.; and P. Debye. "Electromagnetic waves in dielectric waveguides." *Ann. Phys.* 32, (1910), pp. 465–476.
4. Compaan, K.; and P. Kramer. "The Philips 'VLP' system." *Philips Tech. Rev.* 33, (1973), pp. 178–180.
5. Carasso, M. G.; J. B. H. Peek; and J. P. Sinjou. "The compact disc digital audio system." *Philips Tech. Rev.* 40, (1982), pp. 151–155.
6. Suhara, T.; and H. Nishihara. "Integrated optic components and devices using periodic structure." *IEEE J. Quantum Electron.* QE-22, (1986), pp. 845–867.
7. Ura, S.; T. Suhara; H. Nishihara; and J. Koyama. "An integrated-optic disk pickup device." *IEEE J. Lightwave Technol.* LT-4, (1986), pp. 913–918.
8. Izutsu, M.; A. Enokihara; and T. Sueta. "Integrated optic temperature and humidity sensors." *IEEE J. Lightwave Technol.* LT-4, (1986), pp. 833–836.
9. Toba, H.; K. Inoue; and K. Nosu. "A conceptional design on optical frequency-division-multiplexing distribution systems with optical tunable filters." *IEEE J. on Selected Areas in Communications* SAC-4, (1986), pp. 1458–1467.
10. Pi, F.; and G. Orriols. "Energy balance in the superposition of light waves with lossless beam splitters." *Am. J. Phys.* 53, (1985), pp. 667–670.

11. Culshaw, B. *Optical fiber sensing and signal processing.* London: Peregrinus Ltd., 1984.
12. Udd, E. *Fiber optic sensors: an introduction for engineers and scientists.* New York, N.Y.: John Wiley and Sons, Inc., 1991.
13. Special issue on applications of lightwave technology to microwave devices, circuits, and systems. *IEEE Trans. Microwave Theory and Technol.* 38, ed. P. R. Herczfeld. (1990), pp. 467–688.
14. Special issue on optical/microwave interaction devices, circuits, and systems. *IEICE Transactions on Electronics* E76-C, ed. T. Yoneyama, and H. Ogawa. (1993), pp. 175–317.
15. Michelson, A. A.; H. G. Gale; and F. Pearson. "The effect of the earth's rotation on the velocity of light, part II." *Astrophys. J.* 61, (1925), pp. 140–145.
16. Shankland, R. S. "Michelson and his interferometer." In *History of Physics,* ed. S. R. Weart, and M. Phillips. New York, N.Y.: American Institute of Physics, (1985).
17. Vali, V.; and R. W. Shorthill. "Fiber ring interferometer." *Appl. Opt.* 15, (1976), pp. 1099–1100.
18. W. J. Minford, F. T. Stone, B. R. Youmans and R. K. Bartman, "Fiber optic gyroscope using an eight-component $LiNbO_3$ integrated optics circuit," SPIE Proceedings, *Fiber Optics and Laser Sensors VII* 1169, ed. E. Udd and R. P. DePaula, Boston, MA (September 5–7, 1989).

PROBLEMS

1. Show that the total phase difference of a fiber gyroscope with an N_t-turn coil is given by equation (1.6).
2. Consider a 1000-turn gyroscope coil with a 10-cm coil radius. The fiber gyroscope is used to sense the earth's rotation rate. Estimate the phase difference if the gyroscope is located at the north pole with its axis normal to the earth's surface. The effective index of refraction of the fiber is 1.485 and the emission wavelength from an optical source is 0.82 μm.
3. If the fiber gyroscope is located at the equator and its axis is normal to the earth's surface, estimate the total phase difference produced by the earth's rotation.

OPTICAL BEAMS AND THEIR TRANSFORMATIONS

CHAPTER 2

CHAPTER OUTLINE

2.1 Introduction
2.2 Ray Vectors, Ray Transfer Matrices, and ABCD Matrices
    2.2.1 Propagation through a Free-Space Region of Thickness d
    2.2.2 Propagation through a Homogeneous Dielectric Layer of Thickness d
    2.2.3 Refraction by a Spherical Surface with Radius R
    2.2.4 Thick Lenses
    2.2.5 Thin Lenses
    2.2.6 Reflection by Spherical and Planar Reflecting Mirrors
    2.2.7 Physical Meaning of ABCD Matrix Elements
    2.2.8 Cascading of ABCD Matrices
    2.2.9 Application of ABCD Matrices: Optical Cavities
2.3 Optical Systems Synthesis
    2.3.1 Factorization Methods
    2.3.2 Lens-Like and Propagation-Like Matrices
    2.3.3 Example
2.4 Fundamental Gaussian Beams
    2.4.1 Mathematical Expression
    2.4.2 The Constant C'
    2.4.3 Beam Radius $w(z)$
    2.4.4 Phase Variation $\Psi(z)$
    2.4.5 Radius of Curvature $R(z)$
    2.4.6 Rayleigh Range z_R
    2.4.7 Complex Radius of Curvature q
2.5 Optical Components and Gaussian Beams
    2.5.1 Sizes of Pinholes, Mirrors, and Lenses
    2.5.2 Determination of Beam Radius
    2.5.3 Example

2.6 Higher-Order Modes
2.7 ABCD Law
    2.7.1 Free-Space Layer of Thickness d
    2.7.2 Spherical Mirror with Radius R_m
    2.7.3 Thin Lens with Focal Length f
2.8 Gaussian Beam Transformations by Thin Lenses
    2.8.1 Given: $w_{o\,in}$, z_{in}, and f
    2.8.2 Given: $w_{o\,in}$, z_{in}, and $w_{o\,out}$
    2.8.3 Given: $w_{o\,in}$ and $w_{o\,out}$
    2.8.4 Example: Perfect Beam Expanders
    2.8.5 Comparison: Geometrical Optics
2.9 Optical Cavities
    2.9.1 Symmetric Optical Cavities
    2.9.2 Symmetric Confocal Resonators
    2.9.3 Parallel Planar Mirror Cavities
    2.9.4 Concentric Resonators
    2.9.5 Planar Mirror/Spherical Mirror Resonators
    2.9.6 Resonance Frequency: Spherical Mirror Cavities
References
Additional Reading
Problems

2.1 INTRODUCTION

In the use of coherent light beams in various engineering applications, two questions arise: how are light beams generated, and how can they be manipulated after generation. The generation of light will be addressed in the next two chapters. The manipulation and transformation of optical beams are discussed in this chapter.

In elementary physics or optics, optical beams are approximated by **uniform plane waves.** However, it is more accurate to describe laser beams as **Gaussian beams,** a term to be defined later. In this chapter, we will analyze the effects of lenses, mirrors, and other passive optical components on uniform plane waves and Gaussian beams. We will also discuss ways to design simple optical systems to transform beams.

There are three levels of sophistication in analyzing the effects of optical elements on optical beams. The simplest level treats beams as uniform plane waves and follows the waves as they propagate through various optical components. Uniform plane waves can also be viewed as **rays,** which are specified by **ray vectors** and can be traced as they transit through various optical elements. A **ray transfer matrix (RTM)** quantitatively describes the effects of an optical element on the rays. By cascading the RTMs of various components, we can obtain the overall effect of the entire optical system. RTMs are 4×4 matrixes and, for some optical elements, many matrix elements are redundant. With redundant matrix elements eliminated, 4×4 RTMs can be reduced to 2×2 matrixes. These 2×2 matrixes are commonly known as **ABCD matrixes.** The ABCD matrixes for various optical components are used to describe the effects of those components on optical beams.

The second level of sophistication involves the concept of **modes.** The fundamental mode of unguided, freely propagating beams can be described in terms of a Gaussian function. Higher-order modes can be expressed as products of a Gaussian function and Hermite polynomials, if a rectangular coordinate system is used, or as products of a Gaussian function and Laguerre polynomials when a cylindrical coordinate system is used. We will demonstrate that, for a given optical component, through the application of the **ABCD law,** the ABCD matrix elements are useful in describing the transformation of the modes. The third level of sophistication is the use of Huygen's principle to calculate the reflection, transmission, and refraction of the beams [1]. This level involves considerable mathematics and will not be attempted in this book.

2.2 RAY VECTORS, RAY TRANSFER MATRICES, AND ABCD MATRICES

Waves are specified by the **frequency** f (or **free-space** or **vacuum wavelength** λ), **amplitude, phase, polarization** and **direction of propagation.** If the constant-phase surface or wavefront of a wave is a plane, the wave is known as a **plane wave.** If the field amplitude is also constant on this constant-phase plane, the wave is known as a uniform plane wave. In many applications, fields may be approximated by uniform plane waves. However, it is important to recognize that not all waves are uniform plane waves.

For future reference, we note that the free-space or vacuum permittivity ε_0 and permeability μ_0 are

$$\varepsilon_0 \approx \frac{1}{36\pi} \times 10^{-9} \text{ F/m} \qquad \mu_0 = 4\pi \times 10^{-7} \text{ H/m}$$

In free space or vacuum, the speed c of uniform plane waves is

$$c = \frac{1}{\sqrt{\mu_0 \varepsilon_0}} \approx 3 \times 10^8 \text{ m/s}$$

In a homogeneous medium, the permittivity ε and the permeability μ are constants independent of position, and uniform plane waves propagate with a velocity of $v = 1/\sqrt{\mu\varepsilon}$. The **index of refraction** n is defined as the ratio of c/v. By definition, the index of refraction of vacuum is 1. It is customary to write $\varepsilon = \varepsilon_r \varepsilon_0$ and $\mu = \mu_r \mu_0$, where ε_r and μ_r are the **relative permittivity** and the **relative permeability.** Also, ε_r is known as the **relative dielectric constant,** or simply the **dielectric constant.** For nonmagnetic materials, $\mu_r \approx 1$ and n is $\sqrt{\varepsilon_r}$.

In studying wave motion, we can follow the motion of either a family of wavefronts or vectors normal to the wavefronts. For a family of wavefronts traveling through space, a sequence of normal vectors can be drawn. A curve connecting these vectors can be sketched as a visual aid. This is the **ray representation** of the waves. In a homogeneous and extended region, ε and μ are constants independent of position, and waves propagate along a straight line until a boundary is encountered, after which it follows a different straight line. Thus, ray trajectories in optical systems with many layers of homogeneous media consist of sequences of straight-line segments. If the position and the slope of a ray is known anywhere in the region, its position and slope elsewhere in the region can be determined. In a nonhomogeneous medium, μ and ε are functions of position. Hence, the rays bend continuously and follow curved paths. Graded index thin-film waveguides and fibers, to be discussed in Chapters 7 and 8, are good examples of optical components containing nonhomogeneous media. In this chapter, our interest is mainly on **homogeneous media.** At the boundary between two homogeneous media with different refractive indices, the direction of propagation changes abruptly and rays bend sharply. The bending angle depends on the boundary geometry and the ratio of refractive indices of the two neighboring media. The concept of rays is useful if the wavelength λ is much smaller than the radius of curvature of the boundary.

Many lenses are constructed from spherical or rotationally symmetric surfaces that share a common, aligned axis. Such an axis is taken as the optical axis of the lens. When several lenses are assembled in this manner, the line is denoted the **optical axis** of the system, and all angles are referenced to this axis. It is convenient and customary to choose the z axis as the optical axis. Consider a plane, referred to as the **reference plane (RP),** normal to the z axis. A ray can be specified by the x and y coordinates of the intercept of the ray with the reference plane, and by the slope of the ray, $x' = \dfrac{dx}{dz}$ and $y' = \dfrac{dy}{dz}$ with respect to the z axis. This information can be summarized in a 4×1 column matrix known as the **ray vector,** as follows:

$$\mathbf{R} = \begin{bmatrix} x \\ y \\ x' \\ y' \end{bmatrix} \tag{2.1}$$

Let $\mathbf{R}_{in}$ and $\mathbf{R}_{out}$ be the input and output ray vectors at reference planes RP_{in} and RP_{out}, respectively. We can trace $\mathbf{R}_{in}$ at RP_{in} to $\mathbf{R}_{out}$ at RP_{out} (Figure 2.1) to establish a relationship between $\mathbf{R}_{in}$ and $\mathbf{R}_{out}$. In general, such a relationship is too complicated to be useful unless some simplifying assumptions are made. For example, if under certain conditions $\mathbf{R}_{out}$ is a linear function of $\mathbf{R}_{in}$, the relationship may be written as

$$\begin{bmatrix} x_{out} \\ y_{out} \\ x'_{out} \\ y'_{out} \end{bmatrix} = \begin{bmatrix} m_{11} & m_{12} & m_{13} & m_{14} \\ m_{21} & m_{22} & m_{23} & m_{24} \\ m_{31} & m_{32} & m_{33} & m_{34} \\ m_{41} & m_{42} & m_{43} & m_{44} \end{bmatrix} \begin{bmatrix} x_{in} \\ y_{in} \\ x'_{in} \\ y'_{in} \end{bmatrix} \tag{2.2}$$

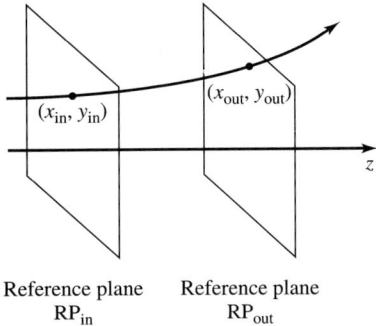

Figure 2.1 Input and output reference planes

This 4×4 matrix is the **ray transfer matrix (RTM).**

However, even a 4×4 matrix is too complicated for most applications and further reduction is desirable. Toward this objective, three simplifying assumptions are made. The first two assumptions reduce the ray vectors from 4×1 to 2×1 column matrices; the third simplifies the relationship between $\mathbf{R}_{in}$ and $\mathbf{R}_{out}$.

The first assumption involves the type of ray being studied. To explain, there are two types of rays. If a ray or its extension in either direction actually crosses the optical axis, a plane can be defined in terms of the ray and the optical axis. This plane is known as a **meridional plane** and the corresponding ray is a **meridional ray.** If a ray never crosses the optical axis, even when it is extended in either direction, it is a **skew ray.** In this chapter, we will deal exclusively with meridional rays.

The second simplifying assumption concerns the optical system. To explain, meridional rays on different meridional planes require different descriptions. Fortunately, many (although not all) optical systems are rotationally symmetrical with respect to the optical axis and, for these systems, all meridional planes are equivalent. Also, when a meridional ray is reflected or refracted in a rotationally symmetric optical system, it remains within the same meridional plane. Therefore, in tracing a meridional ray through a rotationally symmetric system, it is only necessary to deal with a single meridional plane.

Rotationally symmetric systems are the second assumption. On a meridional plane, a meridional ray is specified by its distance r from the optical axis and its slope $r' = \dfrac{dr}{dz} = \tan\theta$ with respect to the positive z direction. The corresponding ray vector may be reduced to a 2×1 column matrix

$$\mathbf{R} = \begin{bmatrix} r \\ r' \end{bmatrix} \qquad (2.3)$$

Here, r is positive if the intercept of the ray with the reference plane is above the z axis. The slope $r' = \tan\theta$ and the angle θ are taken to be positive if the direction of rotation from the z axis to the positive ray direction is counterclockwise. Thus, a positive r' and θ means that r increases as a function of z.

The third simplifying assumption is that the rays are very close to and almost parallel with the optical axis. "Close" means that $|r|$ is small in comparison with the cross section of the system, the radius of curvature of the boundary, and the linear dimension L of the system, that is, $|r/L| \ll 1$. An angle is considered to be small if it is less than, say, 5° (i.e., 0.0873 rad). An examination of any appropriate mathematical table shows that if $|\theta| \leq 0.0873$ rad,

$$\tan\theta \approx \sin\theta \approx \theta$$

Under this assumption, r' can be used directly as an approximation for θ.

The assumptions ($|r/L| \ll 1$ and $|r'| \ll 1$) are the basis for the **paraxial approximation.** Optics based on the paraxial approximation are referred to as **Gaussian optics,** honoring Carl Frederick Gauss who first introduced the concept in 1840 in his study of lens systems.

Combining all three assumptions, the outgoing ray vector $\mathbf{R}_{out}$ at RP_{out} is linearly related to the incoming ray vector $\mathbf{R}_{in}$ at RP_{in}, and the linear relationship can be expressed as a 2×2 ray transfer matrix $\mathbf{M}$

$$\mathbf{R}_{out} = \mathbf{M}\mathbf{R}_{in}$$

or, more explicitly,

$$\begin{bmatrix} r_{out} \\ r'_{out} \end{bmatrix} = \begin{bmatrix} A & B \\ C & D \end{bmatrix} \begin{bmatrix} r_{in} \\ r'_{in} \end{bmatrix} \tag{2.4}$$

The 2×2 RTM given in (2.4) is called the **ABCD matrix.** [2–5]

To develop the ABCD matrix for an arbitrary optical system, we will first study the ABCD matrices for a few basic elements which are the building blocks for most systems.

The three basic rules for the analysis are:

1. In a homogeneous medium, a ray follows a straight line.
2. The **incident, reflected,** and **refracted angles** are the angles between the incident, reflected, and refracted rays and the **normal to the surface,** respectively. Also, upon reflection of a ray from a smooth surface, the incident and reflected angles of the ray are the same.
3. At any smooth boundary between two media, the incident angle θ_1 and the refracted angle θ_2 are related to each other through **Snell's law:**

$$n_1 \sin\theta_1 = n_2 \sin\theta_2$$

where n_1 and n_2 are the refractive indices of the media on the incident and refracted sides, respectively.

2.2.1 Propagation through a Free-Space Region of Thickness d

The simplest possible building block is a free-space region of width d as depicted in Figure 2.2. Reference planes RP_{in} and RP_{out}, shown as dashed lines, are located an infinitesimal distance outside the region. The media inside and outside

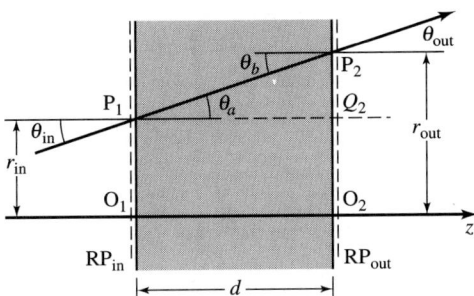

Figure 2.2 Propagation through free-space region of thickness d

the region are really the same; therefore, $\theta_{in} = \theta_a$ and $\theta_b = \theta_{out}$. Since a ray follows a straight line in a homogeneous region, $\theta_{in} = \theta_a = \theta_b = \theta_{out}$, and

$$r'_{out} = r'_{in} \tag{2.5}$$

With reference to Figure 2.2, we also note that

$$\overline{P_2O_2} = \overline{Q_2O_2} + \overline{P_2Q_2} = \overline{P_1O_1} + \overline{O_2O_1}\tan\theta_a \tag{2.6}$$

In terms of r_{in}, r'_{in}, and d, the above equation becomes

$$r_{out} = r_{in} + d\, r'_{in} \tag{2.6}$$

When arranged in matrix form, (2.5) and (2.6) become

$$\begin{bmatrix} r_{out} \\ r'_{out} \end{bmatrix} = \begin{bmatrix} 1 & d \\ 0 & 1 \end{bmatrix} \begin{bmatrix} r_{in} \\ r'_{in} \end{bmatrix}$$

Thus, the ABCD matrix for rays traveling in a homogeneous free-space region of thickness d is

$$\mathbf{M} = \begin{bmatrix} 1 & d \\ 0 & 1 \end{bmatrix} \tag{2.7}$$

It should be noted that (2.7) is exact; we have not used the paraxial approximation in its derivation. Also, for future reference, note that the determinant, AD–BC, of the ABCD matrix given in (2.7) is 1.

2.2.2 Propagation through a Homogeneous Dielectric Layer of Thickness d

To introduce the next building block, which is a homogeneous dielectric layer of thickness d, let the indices of the regions interior and exterior to the layer be n_{int} and n_{ext}, respectively. For this situation, the ABCD matrix is dependent on the locations of the reference planes. If RP_{in} and RP_{out} are *immediately inside* the dielectric layer, as shown by the dashed lines in Figure 2.3a, the ABCD

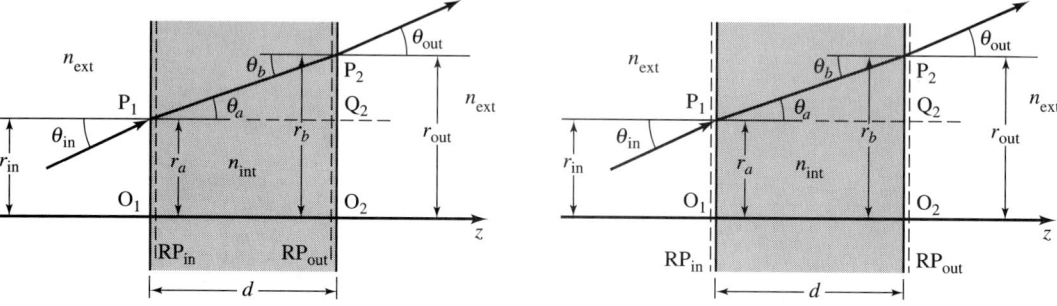

(a) Reference planes inside dielectric layer (b) Reference planes outside dielectric layer

Figure 2.3 Propagation through dielectric region of index n_{int} and thickness d

matrix is the *same* as that given by (2.7), and the indices of refraction are not involved. However, if RP_{in} and RP_{out} are *immediately outside* the dielectric region, as depicted in Figure 2.3b, the indices of refraction are involved, and the ABCD matrix is developed as follows. With the reference planes located immediately outside the layer, we have $r_{in} = r_a$ and $r_{out} = r_b$. In addition, θ_{in} and θ_{out} are related to θ_a and θ_b by Snell's law as follows:

$$n_{ext} \sin\theta_{in} = n_{int} \sin\theta_a \qquad n_{ext} \sin\theta_{out} = n_{int} \sin\theta_b \qquad (2.8)$$

Because a ray inside the homogeneous dielectric region follows a straight line with a constant slope, $\theta_a = \theta_b$; therefore, $\theta_{in} = \theta_{out}$. Hence,

$$r'_{in} = r'_{out} \qquad (2.9)$$

With reference to Figure 2.3b, we note that

$$\overline{P_2 O_2} = \overline{Q_2 O_2} + \overline{P_2 Q_2} = \overline{P_1 O_1} + \overline{O_2 O_1} \tan\theta_a$$

Making use of the paraxial approximation and (2.8), we have

$$\theta_a \approx \frac{n_{ext}}{n_{int}} \theta_{in} \approx \frac{n_{ext}}{n_{int}} r'_{in}$$

Combining these equations, we obtain

$$r_{out} \approx r_{in} + d \frac{n_{ext}}{n_{int}} r'_{in} \qquad (2.10)$$

From (2.9) and (2.10), we have

$$\begin{bmatrix} r_{out} \\ r'_{out} \end{bmatrix} = \begin{bmatrix} 1 & d\frac{n_{ext}}{n_{int}} \\ 0 & 1 \end{bmatrix} \begin{bmatrix} r_{in} \\ r'_{in} \end{bmatrix} \qquad (2.11)$$

Thus, for RP_{in} and RP_{out} situated *outside* the layer, the ABCD matrix describing the propagation of a ray through a layer of width d and an index n_{int} immersed in a medium with an index n_{ext} is

$$\mathbf{M} = \begin{bmatrix} 1 & d\dfrac{n_{ext}}{n_{int}} \\ 0 & 1 \end{bmatrix} \quad (2.12)$$

Comparing (2.7) with (2.12), we note that the thickness d is either reduced or increased by a factor of n_{ext}/n_{int}. The term dn_{ext}/n_{int} is commonly referred to as the **reduced thickness.** The determinant of the ABCD matrix given in (2.12) is also 1.

To summarize, if RP_{in} and RP_{out} are inside the dielectric layer, the ABCD matrix depends only on the layer thickness, as given in (2.7). If RP_{in} and RP_{out} are outside the dielectric layer, then the index ratio n_{ext}/n_{int} must be taken into account, and $\mathbf{M}$ is given by (2.12).

2.2.3 Refraction by a Spherical Surface with Radius R

Figure 2.4 depicts six types of simple lenses, which are the next building blocks of an optical system. Each lens is made of two spherical refractive surfaces separated by a distance d measured along the optical axis. The radii of curvature R_1 and R_2 may be positive or negative. To derive the ABCD matrices for these lenses, we must first consider the ABCD matrix for a spherical refractive surface.

Consider a convex spherical interface with a radius $|R|$ between two media with indices n_{ext} and n_{int} (Figure 2.5). Also, assume the reference planes, RP_{in} and RP_{out} are to the left and to the right of the boundary, respectively. For the paraxial approximation to be applicable, $|r|$ must be small in comparison with $|R|$. Thus, RP_{in} is immediately adjacent to RP_{out}, and

$$r_{in} = r_{out} \quad (2.13)$$

To relate the angles, we again apply Snell's law:

$$n_{ext} \sin\phi_{in} = n_{int} \sin\phi_{out}$$

where ϕ_{in} and ϕ_{out} are the angles of the incoming and outgoing rays with respect to the normal to the boundary. For a spherical boundary, the normal is a line along the radial direction. As shown in Figure 2.5, it is clear that $\phi_{in} = \phi_0 + \theta_{in}$ and $\phi_{out} = \phi_0 + \theta_{out}$. Using the paraxial approximation, these relations may be approximated as $\phi_{in} \approx \phi_0 + r'_{in}$ and $\phi_{out} \approx \phi_0 + r'_{out}$. We also note that $\phi_0 = \sin^{-1}(r_{in}/|R|) \approx r_{in}/|R|$. Combining these expressions, we have

$$r'_{out} = \frac{r_{in}}{|R|} \frac{n_{ext} - n_{int}}{n_{int}} + r'_{in} \frac{n_{ext}}{n_{int}} \quad (2.14)$$

From (2.13) and (2.14), we have the ABCD matrix describing the refraction by a convex spherical surface:

$$\mathbf{M} = \begin{bmatrix} 1 & 0 \\ \dfrac{1}{|R|} \dfrac{n_{ext} - n_{int}}{n_{int}} & \dfrac{n_{ext}}{n_{int}} \end{bmatrix} \quad \text{(convex surface)} \quad (2.15a)$$

The refraction by a concave spherical surface of radius $|R|$ can be considered in the same fashion. The details are left to the reader as an exercise (Problem 6). The result is

$$\mathbf{M} = \begin{bmatrix} 1 & 0 \\ -\dfrac{1}{|R|} \dfrac{n_{ext} - n_{int}}{n_{int}} & \dfrac{n_{ext}}{n_{int}} \end{bmatrix} \qquad \text{(concave surface)} \qquad (2.15b)$$

The equations (2.15a) and (2.15b) can be consolidated into one equation by allowing the radius R to assume a positive or negative value. Specifically, R is positive if the surface is convex as seen by the incident beam, and R is negative if the surface is concave. Stated another way, for incident rays coming from the left of the boundary, R is positive if the center of curvature of the boundary is

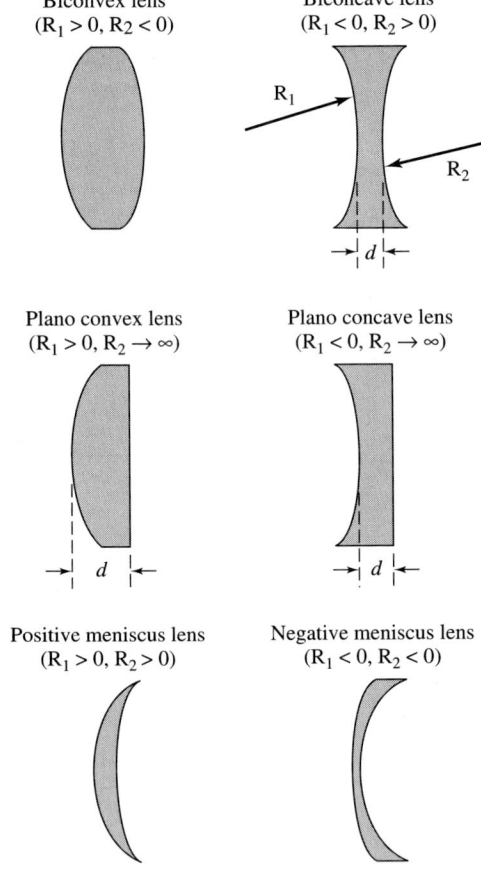

Figure 2.4 Six simple lenses

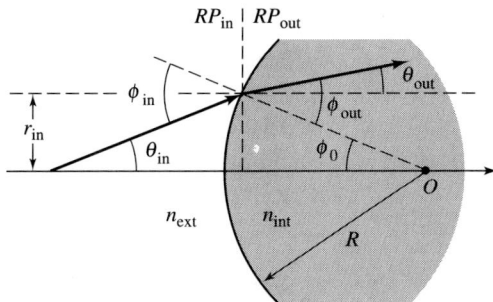

Figure 2.5 Refraction by a convex surface

to the right of the boundary, as shown in Figure 2.5, and R is negative if the center of curvature is to the left of the boundary. With this sign convention, (2.15a) and (2.15b) can be combined:

$$\mathbf{M} = \begin{bmatrix} 1 & 0 \\ \dfrac{1}{R} \dfrac{n_{\text{ext}} - n_{\text{int}}}{n_{\text{int}}} & \dfrac{n_{\text{ext}}}{n_{\text{int}}} \end{bmatrix} \quad (2.15)$$

A planar boundary corresponds to the limit $|R| \to \infty$. In this limit, (2.15) becomes

$$\mathbf{M} = \begin{bmatrix} 1 & 0 \\ 0 & \dfrac{n_{\text{ext}}}{n_{\text{int}}} \end{bmatrix} \quad (2.16)$$

The determinants of the matrices in (2.15a), (2.15b), (2.15), and (2.16) are $n_{\text{ext}}/n_{\text{int}}$, that is, the ratio of the index of the exterior region to that of the interior region.

2.2.4 Thick Lenses

Many passive optical systems can be constructed from refractive surfaces separated by free-space or homogeneous dielectric layers. An example is the thick, positive meniscus lens depicted in Figure 2.6. According to the sign convention for refractive surfaces, discussed in the previous section, both radii of curvature are positive. To develop the ABCD matrix for this lens, examine a ray incident upon the first boundary. Let $\mathbf{R}_{\text{in}}$ be the incident ray vector at RP_1. The refraction by the first convex spherical surface with a radius of R_1 is given by (2.15), and the ray leaving RP_2 is

$$\mathbf{R}_2 = \mathbf{M}_L \mathbf{R}_{\text{in}} \quad (2.17)$$

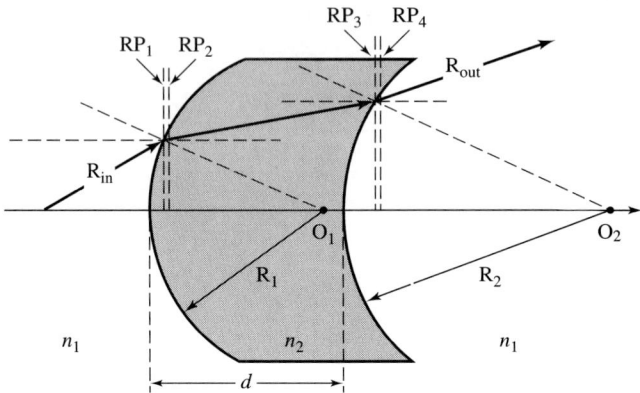

Figure 2.6 Thick convex lens

where

$$\mathbf{M}_L = \begin{bmatrix} 1 & 0 \\ \dfrac{1}{R_1} \dfrac{n_1 - n_2}{n_2} & \dfrac{n_1}{n_2} \end{bmatrix} \quad (2.18)$$

From RP_2 to RP_3, the ray propagates in a dielectric layer of thickness d. Since reference planes RP_2 and RP_3 are inside the lens medium, the ray arriving at RP_3 is represented by a ray vector

$$\mathbf{R}_3 = \mathbf{M}_D \mathbf{R}_2 \quad (2.19)$$

where

$$\mathbf{M}_D = \begin{bmatrix} 1 & d \\ 0 & 1 \end{bmatrix} \quad (2.20)$$

At RP_3, a second convex refractive surface with a radius of curvature R_2 is encountered. The output ray vector $\mathbf{R}_{\text{out}}$ is related to $\mathbf{R}_3$ by

$$\mathbf{R}_{\text{out}} = \mathbf{M}_R \mathbf{R}_3 \quad (2.21)$$

where

$$\mathbf{M}_R = \begin{bmatrix} 1 & 0 \\ \dfrac{1}{R_2} \dfrac{n_2 - n_1}{n_1} & \dfrac{n_2}{n_1} \end{bmatrix} \quad (2.22)$$

Note that the roles of n_1 and n_2 in (2.18) and (2.22) have been interchanged.
Combining (2.17)–(2.22), we have

$$\mathbf{R}_{\text{out}} = \mathbf{M}_R \mathbf{R}_3 = \mathbf{M}_R \mathbf{M}_D \mathbf{R}_2 = \mathbf{M}_R \mathbf{M}_D \mathbf{M}_L \mathbf{R}_{\text{in}} \quad (2.23)$$

If we let the overall ABCD matrix of the thick lens be **M**, then

$$\mathbf{R}_{out} = \mathbf{M}\,\mathbf{R}_{in}$$

and

$$\mathbf{M} = \mathbf{M}_R\,\mathbf{M}_D\,\mathbf{M}_L,$$

or

$$\mathbf{M} = \begin{bmatrix} 1 + \dfrac{d}{R_1}\dfrac{n_1 - n_2}{n_2} & d\dfrac{n_1}{n_2} \\ (\dfrac{1}{R_1} - \dfrac{1}{R_2})\dfrac{n_1 - n_2}{n_1} - \dfrac{d}{R_1 R_2}\dfrac{(n_1 - n_2)^2}{n_1 n_2} & 1 - \dfrac{d}{R_2}\dfrac{n_1 - n_2}{n_2} \end{bmatrix} \quad (2.24)$$

M is the ABCD matrix of a thick lens with an index n_2 situated in a medium with an index n_1. This remains valid even if one or both of the spherical surfaces is concave. Then R_1 and/or R_2 becomes negative. Note that the reference planes RP_{in} and RP_{out} are immediately outside the lens and in the region with the index n_1. Also, based on observations in the preceeding subsections, we would expect the determinant of the ABCD matrix in (2.24) to be 1. When $|\mathbf{M}|$ is evaluated, this is indeed the case.

2.2.5 Thin Lenses

Of all optical components, lenses are used most often and it is convenient to approximate lenses as thin lenses. The ABCD matrix of thin lenses can be deduced from (2.24) by taking the limit of $d \to 0$:

$$\mathbf{M} = \begin{bmatrix} 1 & 0 \\ (\dfrac{1}{R_1} - \dfrac{1}{R_2})\dfrac{n_1 - n_2}{n_1} & 1 \end{bmatrix} \quad (2.25)$$

It is instructive to relate the matrix elements of (2.25) with the focusing properties of the lens. From elementary optics, incident rays parallel with the optical axis are focused by a lens to its focal point. For such incident rays,

$$\mathbf{R}_{in} = \begin{bmatrix} r_{in} \\ 0 \end{bmatrix}$$

and the output ray vector becomes

$$\mathbf{R}_{out} = \begin{bmatrix} r_{in} \\ r_{in}(\dfrac{1}{R_1} - \dfrac{1}{R_2})\dfrac{n_1 - n_2}{n_1} \end{bmatrix}$$

If $(\dfrac{1}{R_1} - \dfrac{1}{R_2})\dfrac{n_1 - n_2}{n_1}$ is negative, then the output ray has a negative slope if $r_{in} > 0$ and a positive slope if $r_{in} < 0$. In both cases, the output rays bend

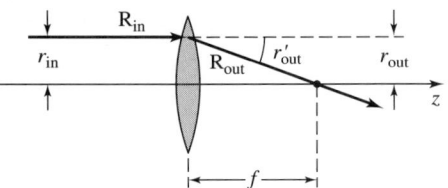

Figure 2.7 Focal length of a thin lens

toward the optical axis. Simple geometrical considerations (Figure 2.7) show that, regardless of the value of r_{in}, all output rays converge to a point on the optical axis, and that point is given by

$$f = \frac{r_{out}}{-r'_{out}} = \frac{1}{(\frac{1}{R_1} - \frac{1}{R_2})\frac{n_2 - n_1}{n_1}} \tag{2.26}$$

This is the **focal length** f of a thin lens. Therefore, (2.25) can be expressed in terms of the focal length f:

$$\mathbf{M} = \begin{bmatrix} 1 & 0 \\ -\frac{1}{f} & 1 \end{bmatrix} \tag{2.27}$$

Simple algebra also shows that convex lenses have a positive focal length and they are known as **positive lenses.** Positive lenses are also known as **converging lenses,** since incident rays parallel to the optical axis are bent to the focal point on the output side. Concave lenses have negative focal lengths and are known as **negative** or **diverging lenses.** A lens can be positive or negative depending on the values of R_1, R_2, n_1, and n_2. In fact, (2.26) explicitly displays the dependence of the focal length on the radii of curvature of the refractive surfaces, the refractive index of the lens material, and the index of the outside medium. Clearly, f is positive if $n_1 < n_2$ and $R_1 < R_2$. In most cases of practical interest, such as glass lenses in air, $n_1 < n_2$. It is also interesting to note that as $n_2 \to n_1$, the focusing properties of the lens disappear. This corresponds to the case in which a lens is immersed in an index-matching liquid, that is, a liquid for which $n_1 = n_2$. Since the lens and the surrounding region have the same index, the presence of the lens is not "detected" by the incoming rays and they are not refracted. Also note that, when $R_1 \to \infty$ and $R_2 \to \infty$, a lens reduces to a planar dielectric slab with an index of n_2, and the lens properties vanish as expected.

2.2.6 Reflection by Spherical and Planar Reflecting Mirrors

Planar or spherical reflecting mirrors are also used in many applications, such as laser cavities. There are two distinct differences between reflecting mirrors and the refractive surfaces previously discussed. First, for refractive surfaces, the incident and refracted angles are related through Snell's law; for mirrors, the angle of reflection *equals* the angle of incidence. Second, for refractive elements, the incoming and outgoing rays are roughly in the same direction; for mirrors, however, the direction of propagation changes abruptly upon reflection. Because it is difficult to account for these sudden changes in direction, we examine the *image* of the reflected ray rather than the reflected ray itself. In Figure 2.8a, the image of the reflected ray is depicted as a dashed line. As shown, the slope of the incident ray differs slightly from that of the image of the reflected ray. The ABCD matrix for a mirror describes the relationship between the incident ray and the image of the reflected ray in the *forward direction*.

Figure 2.8a shows a convex spherical mirror of radius $|R_m|$. Since RP_{in} coincides with RP_{out},

$$r_{out} = r_{in}$$

For a spherical surface, the normal is along the radial direction. The angle between the incident ray and the normal to the spherical surface is $\phi_o + \theta_{in}$, and the angle between the reflected ray and the $-z$ axis is $\phi_o + (\phi_o + \theta_{in})$. The angle of the image of the reflected ray with respect to the $+z$ axis is also $\phi_o + (\phi_o + \theta_{in})$. Under paraxial approximations, $\theta_{in} \approx r'_{in}$, $\theta_{out} \approx r'_{out}$, $|r| \ll |R_m|$, and $\phi_o \approx r_{in}/|R_m|$, and we have

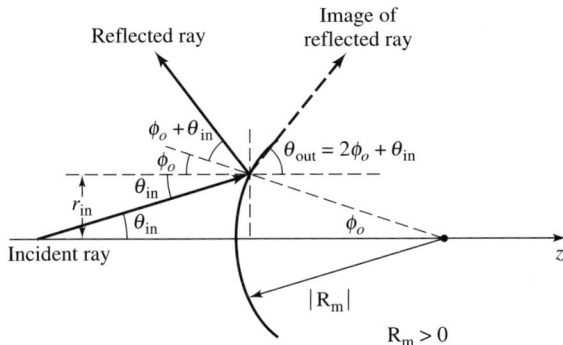

(a) Convex spherical mirror

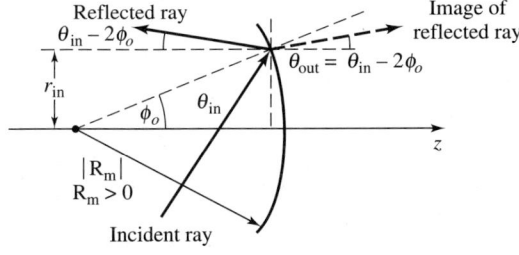

(b) Concave spherical mirror

Figure 2.8 Reflection by a spherical mirror

$$r'_{out} \approx 2\phi_o + r'_{in} \approx 2\frac{r_{in}}{|R_m|} + r'_{in}$$

Combining these equations, we have the ABCD matrix for a convex spherical mirror:

$$\mathbf{M} = \begin{bmatrix} 1 & 0 \\ \dfrac{2}{|R_m|} & 1 \end{bmatrix} \qquad (2.28a)$$

Similarly, for concave mirrors with a radius $|R_m|$ (Figure 2.8b), we can show that

$$\mathbf{M} = \begin{bmatrix} 1 & 0 \\ -\dfrac{2}{|R_m|} & 1 \end{bmatrix} \qquad (2.28b)$$

For refractive surfaces, the sign convention states that the radius of curvature is positive for a convex refractive surface, as seen by the incoming rays, and is negative for a concave surface. This has been noted in section 2.2.3. Adapting the same sign convention, (2.28a) and (2.28b) can be combined to yield

$$\mathbf{M} = \begin{bmatrix} 1 & 0 \\ \dfrac{2}{R_m} & 1 \end{bmatrix} \qquad (2.29)$$

which is applicable to convex as well as concave mirrors. Comparing (2.29) with (2.27), we see that a convex (concave) spherical mirror of radius R_m is equivalent to a diverging (converging) lens with a focal length of $-R_m/2$, so far as the incoming rays and images of the reflected rays in the forward direction are concerned.

The ABCD matrix for a **planar mirror** can be obtained by taking the limit of $R_m \to \infty$. In the limit, (2.29) becomes simply an identity matrix

$$\mathbf{M} = \begin{bmatrix} 1 & 0 \\ 0 & 1 \end{bmatrix}$$

This means that, for planar mirrors, the incoming rays and the images of the outgoing rays are in the same direction. This is expected since, for a planar mirror, the image of a reflected ray is the exact continuation of the incident ray.

The ABCD matrices for the various optical elements discussed thus far are summarized in Table 2.1. Matrices (d) and (e) in the table are the limiting cases of (b) and (c) as the radius of curvature approaches infinity. In addition, matrix (g) for dielectric slabs can be derived by making repeated use of (a) and (d). Specifically, matrix (g) is the product of

Table 2.1 ABCD matrices of elementary optical components

Optical element	ABCD matrix	Optical element	ABCD matrix
(a) Free-space layer	$\begin{bmatrix} 1 & d \\ 0 & 1 \end{bmatrix}$	(d) Plannar boundary	$\begin{bmatrix} 1 & 0 \\ 0 & \dfrac{n_{ext}}{n_{int}} \end{bmatrix}$
(b) Concave boundary ($R<0$) / Convex boundary ($R>0$)	$\begin{bmatrix} 1 & 0 \\ \dfrac{n_{ext}-n_{int}}{R\, n_{int}} & \dfrac{n_{ext}}{n_{int}} \end{bmatrix}$	(e) Planar mirror	$\begin{bmatrix} 1 & 0 \\ 0 & 1 \end{bmatrix}$
		(f)	$\begin{bmatrix} 1 & d \\ 0 & 1 \end{bmatrix}$
(c) Concave mirror ($R_m<0$) / Convex mirror ($R_m>0$)	$\begin{bmatrix} 1 & 0 \\ \dfrac{2}{R_m} & 1 \end{bmatrix}$	(g)	$\begin{bmatrix} 1 & \dfrac{n_{ext}d}{n_{int}} \\ 0 & 1 \end{bmatrix}$
		(h) Thin lens (focal length f)	$\begin{bmatrix} 1 & 0 \\ -\dfrac{1}{f} & 1 \end{bmatrix}$

$$\begin{bmatrix} 1 & 0 \\ 0 & \dfrac{n_{int}}{n_{ext}} \end{bmatrix} \begin{bmatrix} 1 & d \\ 0 & 1 \end{bmatrix} \begin{bmatrix} 1 & 0 \\ 0 & \dfrac{n_{ext}}{n_{int}} \end{bmatrix}$$

Thus, the matrices listed in the left column of Table 2.1 are basic matrices; all others, including those given in the right column, are the limiting cases of, or can be derived from, the basic matrices.

2.2.7 Physical Meaning of ABCD Matrix Elements

The matrix elements A and D are dimensionless, and B and C have the dimension of length and length^{-1}, respectively. This is true for all matrix elements of (2.7), (2.12), (2.15), (2.16), (2.24), (2.25), (2.26), (2.27), and (2.29). As far as the ABCD matrices themselves are concerned, it is immaterial which length unit is adopted, as long as it is used consistently. To avoid confusion, international (SI) units are recommended.

To understand the physical meaning of the matrix elements, as presented in (2.4), we consider input rays which are either parallel to the optical axis, $r'_{in} = 0$, or impinging on the axis at the input, $r_{in} = 0$. For rays parallel to the optical axis at the input, output ray vectors are

$$\mathbf{R}_{out} = \begin{bmatrix} Ar_{in} \\ Cr_{in} \end{bmatrix}$$

Thus, output rays are laterally shifted to location Ar_{in}, and A is the **lateral magnification** for input rays parallel to the axis:

$$A = \frac{r_{out}}{r_{in}} \bigg|_{r'_{in} = 0}$$

Since $r_{out} = Ar_{in}$ and $r'_{out} = Cr_{in}$, all output rays, regardless of r_{in}, converge to a point on the axis at a distance $-A/C$ from the output reference plane. Thus, we can treat $-A/C$ as the **effective focal length** relative to RP_{out}.

For rays incident upon the input reference plane with $r_{in} = 0$, the ray vector at the output reference plane is

$$\mathbf{R}_{out} = \begin{bmatrix} Br'_{in} \\ Dr'_{in} \end{bmatrix}$$

Comparing the slope of the outgoing rays with that of the incoming rays, we note that D is the **angular magnification** for rays impinging on the axis at the input:

$$D = \frac{r'_{out}}{r'_{in}} \bigg|_{r_{in} = 0}$$

To understand the meaning of B, recall that, for a free-space layer of thickness d, if the input rays cross the optical axis at RP_{in} (i.e., $r_{in} = 0$), the output rays emerge from the free-space layer at $r'_{in}d$. This means that, for the matrix element under consideration,

$$B = \frac{r_{out}}{r'_{out}} \bigg|_{r_{in} = 0}$$

Thus, B may be viewed as the **effective thickness** of the optical component, for input rays with $r_{in} = 0$.

In deriving the various equations thus far presented, the paraxial approximation has been invoked repeatedly, and $\tan\theta$ or $\sin\theta$ has been approximated by θ.

Under the paraxial approximation, the input and output ray vectors, $\mathbf{R}_{in}$ and $\mathbf{R}_{out}$, are related in a linear fashion, as indicated in (2.4). If the paraxial approximation had not been used, the relationship between the input and output ray vectors would have been very complicated. To avoid such complications, we confine ourselves to cases where the paraxial approximation is applicable.

2.2.8 Cascading of ABCD Matrices

One important feature of the ABCD matrix is that, when the matrices of the constituent elements are known, the matrix of the whole system can be obtained. The process involves cascading the ABCD matrices of the individual elements.

Suppose for example that an arbitrary system (Figure 2.9) is formed by k refractive surfaces, labeled 1 through k and ordered from left to right, with the far left boundary being boundary 1 and the far right one being boundary k. The planes RP_{in} and RP_{out} coincide with surfaces 1 and k and they are located in regions with indexes n_{in} and n_{out}, respectively. The ABCD matrices of these refractive surfaces are $\mathbf{M}_1, \mathbf{M}_2, \ldots$, respectively. The dielectric layers beween boundaries 1 and 2, 2 and 3, etc., have indices n_2, n_3, etc., and the ABCD matrices corresponding to the propagation from boundary 1 to 2, 2 to 3, etc., are $\mathbf{M}_{2,1}, \mathbf{M}_{3,2}$, etc., respectively. The ABCD matrix representing the overall effect of these refractive boundaries and layers is obtained by tracing a ray through succeeding boundaries and layers. Let the incident ray vector be $\mathbf{R}_{in}$. After crossing boundary 1, the ray is represented by $\mathbf{M}_1 \mathbf{R}_{in}$. Next, the ray goes from boundary 1 to boundary 2. Upon arrival at boundary 2, it is represented by $\mathbf{M}_{2,1} \mathbf{M}_1 \mathbf{R}_{in}$. This process is repeated until the last boundary is reached, at which

$$\mathbf{R}_{out} = \mathbf{M}\, \mathbf{R}_{in}$$

where

$$\mathbf{M} = \mathbf{M}_k\, (\mathbf{M}_{k, k-1})\, \mathbf{M}_{k-1}\, (\mathbf{M}_{k-1, k-2}) \cdots \mathbf{M}_3\, (\mathbf{M}_{3, 2})\, \mathbf{M}_2\, (\mathbf{M}_{2, 1})\, \mathbf{M}_1$$

This is exactly the same procedure used previously in obtaining the ABCD matrix for thick lenses. Note in particular that the matrices are cascaded *in reverse order* of the actual arrangement of the optical elements.

Since $\mathbf{M}_{i,i-1}$ represents the propagation from boundary $i-1$ to boundary i in the same medium, the determinant of $\mathbf{M}_{i,i-1}$ is 1. On the other hand, M_j is the

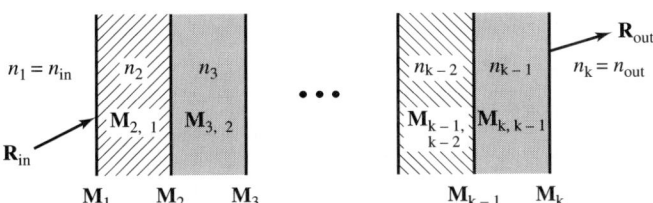

Figure 2.9 Arbitrary optical system with k refractive surfaces

ABCD matrix for the refraction at the boundary j between two media; hence, $|\mathbf{M}_j| = n_j/n_{j+1}$. Therefore, we have the determinant of the overall matrix $\mathbf{M}$,

$$|\mathbf{M}| = \frac{n_k}{n_{out}} \cdot 1 \cdot \frac{n_{k-1}}{n_k} \cdot 1 \cdots 1 \cdot \frac{n_2}{n_3} \cdot 1 \cdot \frac{n_{in}}{n_2} = \frac{n_{in}}{n_{out}} \qquad (2.30)$$

Thus, the determinant of an ABCD matrix depends only on the indices of the outermost regions. The indices of the intermediate layers cancel out completely.

2.2.9 Application of ABCD Matrices: Optical Cavities

Various lasers will be discussed in the following chapters. An essential part of any laser is the optical cavity formed by two mirrors. The operation of these cavities can be analyzed in terms of the ABCD matrices of the mirrors and the medium between the mirrors. Since the derivation is relatively lengthy, it will be presented in two parts. First, mirrors are replaced by an equivalent lens sequence. Thus a cavity is viewed as a system with a periodic lens sequence. Then, the criteria for the existence of stable periodic rays are established.

Lens Sequence Consider two concave spherical mirrors of radii R_{m1} and R_{m2}, as shown in Figure 2.10a. As noted previously, the radii of curvature of concave mirrors are negative. To bring out the minus signs explicitly, we write $R'_{m1} = -R_{m1}$ and $R'_{m2} = -R_{m2}$. Waves in the cavity are bounced back and forth by the mirrors. Physically, this is similar to the situation of sitting between the two

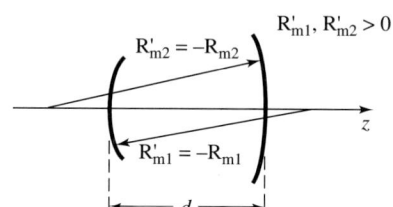

(a) Optical cavity with two concave mirrors

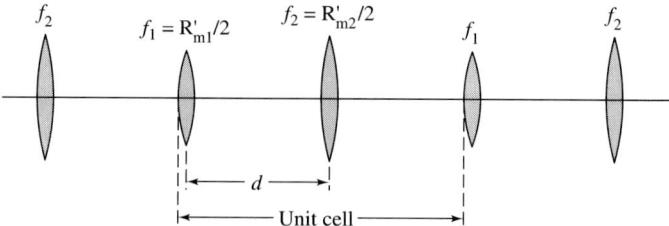

(b) Equivalent lens sequence

Figure 2.10 Optical cavity and its equivalent periodic lens sequence

mirrors. In looking into either mirror, we would see not just the image of ourselves, but also the image of the image, the image of the image of the image, etc. In discussing the images of the reflected rays, we replace the mirrors by equivalent thin lenses with focal lengths of $R'_{m1}/2$ and $R'_{m2}/2$, respectively, as discussed previously. To account for the multiple reflected rays, we also treat the images of the spherical mirrors as equivalent thin lenses. Thus, we have a sequence of thin lenses extending in both directions. This is shown in Figure 2.10b. Such a lens sequence can be viewed as a periodic structure. A **basic unit**, or **unit cell**, of the periodic structure consists of two lenses and two sections of free space between them. It is immaterial how a unit cell is defined, as long as it includes a section of length $2d$ and two equivalent lenses.

Consider the unit cell shown in Figure 2.10b. The ABCD matrix of the unit cell is

$$\begin{bmatrix} A & B \\ C & D \end{bmatrix} = \begin{bmatrix} 1 & 0 \\ -\dfrac{2}{R'_{m1}} & 1 \end{bmatrix} \begin{bmatrix} 1 & d \\ 0 & 1 \end{bmatrix} \begin{bmatrix} 1 & 0 \\ -\dfrac{2}{R'_{m2}} & 1 \end{bmatrix} \begin{bmatrix} 1 & d \\ 0 & 1 \end{bmatrix} \quad (2.31)$$

$$= \begin{bmatrix} 1 - \dfrac{2d}{R'_{m2}} & 2d - \dfrac{2d^2}{R'_{m2}} \\ -\dfrac{2}{R'_{m1}} - \dfrac{2}{R'_{m2}} + \dfrac{4d}{R'_{m1}R'_{m2}} & -\dfrac{2d}{R'_{m1}} + (1 - \dfrac{2d}{R'_{m1}})(1 - \dfrac{2d}{R'_{m2}}) \end{bmatrix}$$

There are infinite many unit cells. Consider an arbitrary unit cell and label it as cell k. Our input reference plane RP_{in} was chosen just to the right of the mirror R'_{m1} in cell k. If we let the ray vector in cell k be

$$\mathbf{R}_k = \begin{bmatrix} r_k \\ r'_k \end{bmatrix}$$

the ray vector $\mathbf{R}_{k+1}$ in the next unit cell, i.e., the $(k+1)$ cell, is related to $\mathbf{R}_k$, as follows

$$\begin{bmatrix} r_{k+1} \\ r'_{k+1} \end{bmatrix} = \begin{bmatrix} A & B \\ C & D \end{bmatrix} \begin{bmatrix} r_k \\ r'_k \end{bmatrix}$$

Similarly, ray vectors $\mathbf{R}_{k+1}$ and $\mathbf{R}_{k+2}$ are also related:

$$\begin{bmatrix} r_{k+2} \\ r'_{k+2} \end{bmatrix} = \begin{bmatrix} A & B \\ C & D \end{bmatrix} \begin{bmatrix} r_{k+1} \\ r'_{k+1} \end{bmatrix}$$

By eliminating r'_{k+1} and r'_k from these equations and recalling $AD - BC = 1$, we have a difference equation,

$$r_{k+2} - (A + D)r_{k+1} + r_k = 0 \quad (2.32)$$

Stable and unstable conditions The solution of (2.32) for r_k may assume different forms under different conditions. Depending on the value of $A + D$, $|r_k|$

may grow indefinitely or may vary periodically between limits. If an optical cavity is useful in supporting a mode, the mode must be confined spatially to the cavity; that is, $|r_k|$ must be smaller than the radii of the mirrors. In other words, $|r_k|$ cannot grow indefinitely. Let us write

$$r_k = r_0 e^{jk\zeta}$$

where ζ is an unknown variable. Substituting the above expression in (2.32), we have

$$e^{j2\zeta} - (A + D)e^{j\zeta} + 1 = 0$$

We can solve $e^{j\zeta}$ immediately and obtain

$$e^{j\zeta} = \frac{A + D \pm \sqrt{(A + D)^2 - 4}}{2}$$

If ζ is real, then $|r_k|$ is a constant $|r_0|$ for all values of k. The variable ζ is real if $\sqrt{(A + D)^2 - 4}$ is a complex quantity. The condition for real ζ is $-2 < A + D < 2$, which can be written as

$$0 < \frac{A + D + 2}{4} < 1 \tag{2.33}$$

When the expressions for matrix elements A and D given in (2.31) are used, (2.33) becomes

$$0 < (1 - \frac{d}{R'_{m1}})(1 - \frac{d}{R'_{m2}}) < 1 \tag{2.34}$$

When (2.34) is satisfied, $|r_k|$ remains a constant and rays are confined in the optical cavity by the mirrors. Physically, this means that the cavity loss is sufficiently small and laser oscillations can be easily sustained. These cavities are referred to as **stable cavities.** Oscillations in **unstable cavities** are difficult to maintain since these cavities incur tremendous losses. A **stability diagram** has been constructed by Kogelnik and Li [2] to identify the stable and unstable operating regions of cavities. In Figure 2.11, the unshaded region corresponds to the stable cavity region. Further discussion of this topic is contained in Section 2.9.

2.3 OPTICAL SYSTEMS SYNTHESIS

Having analyzed the passive optical elements in terms of ABCD matrices, we turn to the problem of synthesizing an optical system. We seek a systematic procedure for designing an optical system from a known ABCD matrix. As an example, a system must be designed such that it is capable of magnifying an incoming beam by a factor of 3 and reducing the angle of divergence to a third of its original value. The ABCD matrix satisfying this specification is

2.3 Optical Systems Synthesis

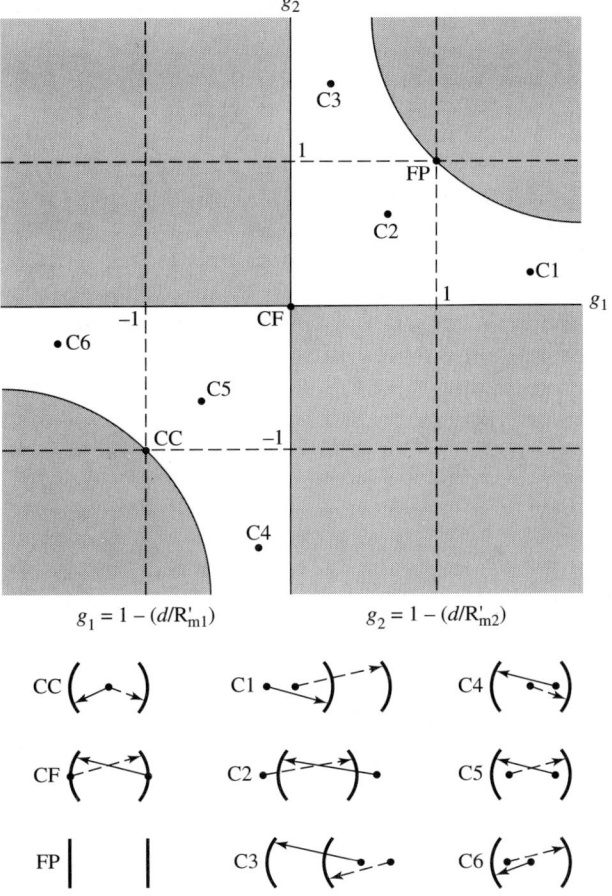

Unstable cavities lie in shaded regions.
FP = Fabry-Perot cavity
CC = Concentric cavity
CF = Confocal cavity

Figure 2.11 Stability diagram for optical cavities (after [2] © 1966 IEEE)

$$\begin{bmatrix} 3 & 0 \\ 0 & \dfrac{1}{3} \end{bmatrix}$$

The ABCD matrices of free-space or dielectric layers, lenses, and mirrors were derived in the last section and are listed in Table 2.1. If an arbitrary ABCD matrix can be factored as products of these matrices, the desired optical system can be formed with the corresponding elements. In other words, elements listed in Table 2.1 can be used as the building blocks to form the optical system. A general synthesis procedure for a complex ABCD matrix with a **real positive determinant** $AD - BC$ was discussed by Casperson [6]. A simplified version, also developed by Casperson, will be discussed here. Although it is a special, simplified version of the general procedure, it is probably the most practical and useful one. Specifically, we restrict ourselves to those cases where all matrix elements are real and the determinant $AD - BC$ is 1. In other words, RP_{in} and RP_{out} are in regions with the same refractive indices. Optical systems in which the input and output reference planes are in air are included in this category.

It is a simple matter to demonstrate that a 2×2 matrix can have at most two nonzero matrix elements if the matrix determinant is not zero. In addition, the nonzero elements are either the diagonal or off-diagonal elements. In other words, the zero matrix elements cannot be in the same column or the same row. Thus, matrices that need to be considered are of the forms

$$\begin{bmatrix} A & 0 \\ 0 & D \end{bmatrix}, \begin{bmatrix} 0 & B \\ C & 0 \end{bmatrix} \tag{2.35}$$

if two matrix elements are zeros, and are of the form

$$\begin{bmatrix} 0 & B \\ C & D \end{bmatrix}, \begin{bmatrix} A & 0 \\ C & D \end{bmatrix}, \begin{bmatrix} A & B \\ 0 & D \end{bmatrix}, \begin{bmatrix} A & B \\ C & 0 \end{bmatrix} \tag{2.36}$$

if one of the matrix elements vanishes. If none of the matrix elements is zero, we have

$$\begin{bmatrix} A & B \\ C & D \end{bmatrix} \tag{2.37}$$

It is only necessary to show that, by applying the factorization methods discussed in the next section, the matrices listed in (2.35)–(2.37) can be expressed as the products of the ABCD matrices of thin lenses and free-space layers. Thus, of all the optical components listed in Table 2.1, only thin lenses and free-space layers are used as the building blocks in the optical systems being considered.

2.3.1 Factorization Methods

Straightforward matrix multiplications will show that the following factorization methods are correct.

Factorization Method 1. If $AD - BC = 1$ and $C \neq 0$, an ABCD matrix can be written as the product of three matrices:

$$\begin{bmatrix} A & B \\ C & D \end{bmatrix} = \begin{bmatrix} 1 & \dfrac{A-1}{C} \\ 0 & 1 \end{bmatrix} \begin{bmatrix} 1 & 0 \\ C & 1 \end{bmatrix} \begin{bmatrix} 1 & \dfrac{D-1}{C} \\ 0 & 1 \end{bmatrix} \tag{2.38}$$

Factorization Method 2. If $AD - BC = 1$ and $B \neq 0$, a 2×2 matrix can be factored as

$$\begin{bmatrix} A & B \\ C & D \end{bmatrix} = \begin{bmatrix} 1 & 0 \\ \frac{D-1}{B} & 1 \end{bmatrix} \begin{bmatrix} 1 & B \\ 0 & 1 \end{bmatrix} \begin{bmatrix} 1 & 0 \\ \frac{A-1}{B} & 1 \end{bmatrix} \qquad (2.39)$$

Factorization Method 3. On the other hand, if B and C are 0, then $AD = 1$ and

$$\begin{bmatrix} A & 0 \\ 0 & D \end{bmatrix} = \begin{bmatrix} 1 & \alpha \\ 0 & 1 \end{bmatrix} \begin{bmatrix} A & -\alpha D \\ 0 & D \end{bmatrix} = \begin{bmatrix} A & -\alpha A \\ 0 & D \end{bmatrix} \begin{bmatrix} 1 & \alpha \\ 0 & 1 \end{bmatrix} \qquad (2.40a)$$

or,

$$\begin{bmatrix} A & 0 \\ 0 & D \end{bmatrix} = \begin{bmatrix} 1 & 0 \\ \beta & 1 \end{bmatrix} \begin{bmatrix} A & 0 \\ -\beta D & D \end{bmatrix} = \begin{bmatrix} A & 0 \\ -\beta D & D \end{bmatrix} \begin{bmatrix} 1 & 0 \\ \beta & 1 \end{bmatrix} \qquad (2.40b)$$

where α and β are arbitrary real numbers. Factorization methods 1 or 2 may be further used to reduce

$$\begin{bmatrix} A & 0 \\ -\beta D & D \end{bmatrix}, \begin{bmatrix} A & -\alpha D \\ 0 & D \end{bmatrix}$$

etc., to the desired form.

2.3.2 Lens-Like and Propagation-Like Matrices

Matrices in (2.38)–(2.40) are either of the form

$$\begin{bmatrix} 1 & 0 \\ \zeta & 1 \end{bmatrix} \qquad (2.41)$$

for a lens-like matrix, or of the form

$$\begin{bmatrix} 1 & \zeta \\ 0 & 1 \end{bmatrix} \qquad (2.42)$$

for a propagation-like matrix, where ζ is a real number that can be positive or negative. The matrix in (2.41) has the form of an ABCD matrix of a thin lens with a focal length $-1/\zeta$. Therefore, matrices of the form of (2.41) are referred to as **lens-like matrices**. Since the focal length of a thin lens can be positive or negative, a lens-like matrix is easily realizable. The matrix (2.42) is of the same form as (2.7) which results from the propagation through a free-space or homogeneous dielectric layer. Therefore, (2.42) is named a **propagation-like matrix**. However, there is a significant difference between (2.42) and (2.7). In (2.7), the layer thickness d is positive; but ζ in (2.42) can be either positive or negative. A propagation-like matrix with a positive ζ can be readily realized. However, additional manipulations are necessary for propagation-like matrices with a negative ζ. A possible factorization has been provided by Casperson. He showed that

$$\begin{bmatrix} 1 & \zeta \\ 0 & 1 \end{bmatrix} = \begin{bmatrix} 1 & 0 \\ \dfrac{2}{\zeta} - \dfrac{1}{h} & 1 \end{bmatrix} \begin{bmatrix} 1 & h \\ 0 & 1 \end{bmatrix} \begin{bmatrix} 1 & 0 \\ \dfrac{-\zeta}{h}\left(\dfrac{2}{\zeta} - \dfrac{1}{h}\right) & 1 \end{bmatrix} \begin{bmatrix} 1 & h \\ 0 & 1 \end{bmatrix} \begin{bmatrix} 1 & 0 \\ \dfrac{2}{\zeta} - \dfrac{1}{h} & 1 \end{bmatrix} \quad (2.43)$$

where h is an arbitrary constant. If the chosen h is real and positive, (2.43) can be realized with three thin lenses with a lens-to-lens separation of h.

2.3.3 Example

We conclude this section by working out the details of the example proposed at the beginning of the section. For the matrix in question, $B = C = 0$ and $AD = 1$; therefore, (2.40a) is applicable. By arbitrarily choosing $\alpha = 3$, we obtain

$$\begin{bmatrix} 3 & 0 \\ 0 & \dfrac{1}{3} \end{bmatrix} = \begin{bmatrix} 1 & 3 \\ 0 & 1 \end{bmatrix} \begin{bmatrix} 3 & -1 \\ 0 & \dfrac{1}{3} \end{bmatrix}$$

The second matrix on the right-hand side has a determinant of 1 and $B \ne 0$. Therefore (2.39) is applicable. Hence,

$$\begin{bmatrix} 3 & 0 \\ 0 & \dfrac{1}{3} \end{bmatrix} = \begin{bmatrix} 1 & 3 \\ 0 & 1 \end{bmatrix} \begin{bmatrix} 1 & 0 \\ \dfrac{2}{3} & 1 \end{bmatrix} \begin{bmatrix} 1 & -1 \\ 0 & 1 \end{bmatrix} \begin{bmatrix} 1 & 0 \\ -2 & 1 \end{bmatrix} \quad (2.44)$$

The third matrix on the right-hand side of (2.44) is a propagation-like matrix with a *negative* ζ. To factor this matrix further, we use (2.43) and choose, again arbitrarily, $h = 1$ to obtain

$$\begin{bmatrix} 1 & -1 \\ 0 & 1 \end{bmatrix} = \begin{bmatrix} 1 & 0 \\ -3 & 1 \end{bmatrix} \begin{bmatrix} 1 & 1 \\ 0 & 1 \end{bmatrix} \begin{bmatrix} 1 & 0 \\ -3 & 1 \end{bmatrix} \begin{bmatrix} 1 & 1 \\ 0 & 1 \end{bmatrix} \begin{bmatrix} 1 & 0 \\ -3 & 1 \end{bmatrix}$$

Substituting this equation in (2.44) and combining some of the matrices, we finally obtain

$$\begin{bmatrix} 3 & 0 \\ 0 & \dfrac{1}{3} \end{bmatrix} = \begin{bmatrix} 1 & 3 \\ 0 & 1 \end{bmatrix} \begin{bmatrix} 1 & 0 \\ -\dfrac{7}{3} & 1 \end{bmatrix} \begin{bmatrix} 1 & 1 \\ 0 & 1 \end{bmatrix} \begin{bmatrix} 1 & 0 \\ -3 & 1 \end{bmatrix} \begin{bmatrix} 1 & 1 \\ 0 & 1 \end{bmatrix} \begin{bmatrix} 1 & 0 \\ -5 & 1 \end{bmatrix}$$

(2.45)

Thus far, the length unit has not been chosen. We choose the meter as the unit of length. Then, (2.45) can be recognized as three lenses with focal lengths, from left to right, of $1/5$, $1/3$, and $3/7$ m, as shown in Figure 2.12. It is worthwhile to repeat that the matrices are arranged in **reverse order** to the way the optical elements are arranged. As mentioned previously, the choice of $\alpha = 3$ and $h = 1$ is quite arbitrary. Many alternate and equally acceptable choices exist. Clearly, the solutions to the synthesis problems are not unique.

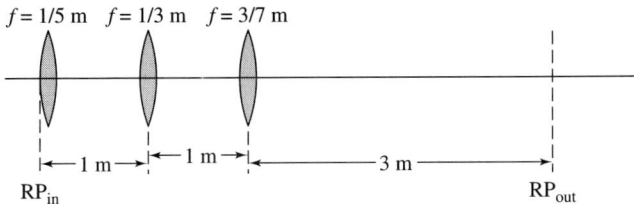

Figure 2.12 System synthesized from (2.45)

2.4 FUNDAMENTAL GAUSSIAN BEAMS

To discuss optical beams in a slightly more sophisticated manner, a more precise description of electromagnetic fields is required. If we let f be the frequency of the waves, then the free-space wave vector is $k = 2\pi f/c = \omega/c$. In terms of the free-space wavelength λ, $k = 2\pi/\lambda$. In a medium with an index of refraction n, the wave vector is nk. Waves in free space that are radiated by an infinitesimal source are **spherical waves** of the form e^{-jkr}/r. Waves radiated by an infinitely large source, or a source that is infinitely far away, are uniform plane waves.

Fields radiated by sources of finite size may be written as $f(\theta,\phi)e^{-jkr}/r$, where $f(\theta,\phi)$ is the **field pattern** of the source [7]. If the radiating source, such as a gas laser, is very large compared to the wavelength, an exact expression for the radiated fields in the far-field zone would be very complicated. A possible approximation expresses the fields as the superposition of fundamental and higher-order modes. Each mode may be expressed as the product of a Hermite or a Laguerre polynomial and a Gaussian function. The simplest, and lowest-order, mode is the **fundamental Gaussian beam,** which is frequently encountered in practice, and is also the most stable and hence the most preferred beam type in many applications. If higher-order modes are present, it is frequently necessary to eliminate them and retain the fundamental Gaussian beam. In this section, we will develop an expression for the fundamental Gaussian beam [2,8].

2.4.1 Mathematical Expression

The wave equation in a free-space region is

$$(\nabla^2 + k^2)E = 0 \qquad (2.46)$$

where E stands for a Cartesian component of the electric field. For convenience, we will use the cylindrical coordinate system (r,ϕ,z) and choose the z axis as the direction of propagation. In the limiting case where the radiating source is infinitely large, the resulting expression should reduce to that of uniform plane waves. Toward this end, we consider a Cartesian field component transverse to the z axis and express it as the product of a rapidly varying part, e^{-jkz}, and a slowly varying part, Φ,

$$E(r,\phi,z) = \Phi(r,\phi,z)e^{-jkz} \tag{2.47}$$

By "slowly varying," we mean that

$$\left|\frac{\partial^2 \Phi}{\partial z^2}\right| \ll k \left|\frac{\partial \Phi}{\partial z}\right| \tag{2.48}$$

Substituting (2.47) into (2.46), we have

$$(\nabla^2 + k^2)E = \left[\frac{1}{r}\frac{\partial}{\partial r}\left(r\frac{\partial \Phi}{\partial r}\right) + \frac{1}{r^2}\frac{\partial^2 \Phi}{\partial \phi^2} + \frac{\partial^2 \Phi}{\partial z^2} - j2k\frac{\partial \Phi}{\partial z}\right]e^{-jkz} = 0$$

In view of (2.48), $\partial^2\Phi/\partial z^2$ can be neglected, as an approximation. Also, in many cases of practical interest, the beam is circularly symmetric, that is, $\partial\Phi/\partial\phi = 0$, and the expression in brackets can be further simplified to

$$\frac{1}{r}\frac{\partial}{\partial r}\left(r\frac{\partial \Phi}{\partial r}\right) - j2k\frac{\partial \Phi}{\partial z} = 0 \tag{2.49}$$

The solution can therefore be written in the form

$$\Phi(r,\phi,z) = C'e^{-j\left[P(z) + \frac{kr^2}{2q(z)}\right]} \tag{2.50}$$

where C' is a constant, and $P(z)$ and $q(z)$ are functions to be determined. Equation (2.50) is the basic equation for Gaussian beams. To determine $P(z)$ and $q(z)$, we substitute (2.50) into (2.49) and obtain

$$\left[-2k\left(\frac{j}{q} + \frac{dP}{dz}\right) - \frac{k^2 r^2}{q^2}\left(1 - \frac{dq}{dz}\right)\right]\Phi = 0$$

The above equation consists of two parts. One is independent of r^2 and the other varies with r^2. If the equation is to hold for all values of r and z, terms dependent on and those independent of r^2 must vanish separately. Thus, we have two differential equations

$$\frac{dq}{dz} = 1$$

and

$$\frac{dP}{dz} = -\frac{j}{q}$$

From these differential equations, we can solve for $q(z)$ and $P(z)$, as follows:

$$q(z) = z + j\frac{\pi w_o^2}{\lambda} \tag{2.51}$$

$$P(z) = -j\left[\frac{1}{2}\ln\left[z^2 + \left(\frac{\pi w_o^2}{\lambda}\right)^2\right] - j\tan^{-1}\frac{z\lambda}{\pi w_o^2}\right] \tag{2.52}$$

where w_o is an undetermined constant and $\pi w_o^2/2$ is a constant of integration. The physical meaning of w_o will be explained shortly, as will $q(z)$, the **complex radius of curvature** of the Gaussian beam.

In solving the differential equation for $P(z)$, we have chosen a particular complex constant as the constant of integration. As a result, we are able to cast $P(z)$ in a convenient form given in (2.52). If desired, other constants may be selected or combined with the constant C'. Combining (2.50)–(2.52) with (2.47) and observing that $e^{-\ln x} = 1/x$, we have

$$E(r,\phi,z) = \frac{C'}{\sqrt{z^2 + (\pi w_o^2/\lambda)^2}} e^{-j[kz-\tan^{-1}\lambda z/(\pi w_o^2)]} e^{-j\left[\frac{kr^2}{2(z+j\,\pi w_o^2/\lambda)}\right]} \quad (2.53)$$

$$= \frac{C'}{\sqrt{z^2 + (\pi w_o^2/\lambda)^2}} e^{-j\left[kz-\tan^{-1}\frac{\lambda z}{\pi w_o^2}\right]} e^{-j\frac{kr^2 z}{2[z^2+(\pi w_o^2/\lambda)^2]}} e^{-\frac{kr^2 \pi w_o^2/\lambda}{2[z^2+(\pi w_o^2/\lambda)^2]}}$$

Clearly (2.53) is too unwieldy to handle. To simplify it, we define $w(z)$, $R(z)$, and $\psi(z)$ as follows:

$$w^2(z) = w_o^2\left[1 + \left(\frac{\lambda z}{\pi w_o^2}\right)^2\right] \quad (2.54)$$

$$R(z) = z\left[1 + \left(\frac{\pi w_o^2}{\lambda z}\right)^2\right] \quad (2.55)$$

$$\psi(z) = \tan^{-1}\frac{\lambda z}{\pi w_o^2} \quad (2.56)$$

Also note that $w_o = w(0)$. Equation (2.53) now becomes

$$E(r,\phi,z) = \frac{C'\lambda}{\pi w_o\, w(z)} e^{-j[kz-\psi(z)]} e^{-j\frac{kr^2}{2R(z)}} e^{-\frac{r^2}{w^2(z)}} \quad (2.57)$$

Consistent with the notation to be adopted in section 2.6 of this chapter, and recalling that a Cartesian field component transverse to z is being discussed here, the field described by (2.53) and (2.57) is commonly known as the TEM_{00} mode propagating in the z direction in an unbounded free space. Determination of the constant C' and the physical meaning of w, R, and Ψ are discussed in the following subsections.

2.4.2 The Constant C'

In this subsection, we will discuss three ways of specifying C'. Depending on the application, one form may be more convenient than the others. In the first way, choose C' such that $E = E_o$ at $z = 0$ on the axis, and $r = 0$. Therefore,

$$C' = E_o\, \pi w_o^2/\lambda$$

Then, (2.57) becomes

$$E(r,\phi,z) = E_o \frac{w_o}{w(z)} e^{-j[kz-\Psi(z)]} e^{-j\frac{kr^2}{2R(z)}} e^{-\frac{r^2}{w^2(z)}} \tag{2.58}$$

This is an expression describing a fundamental Gaussian beam with a **peak field intensity** E_o at $r = 0$ and $z = 0$.

The **irradiance**, I, defined as the power per unit area, is $|E|^2/(2\eta)$, where η is the intrinsic impedance of the medium. The constant C' may be chosen such that the beam's peak irradiance is I_o. Then,

$$I(r,z) = I_o \frac{w_o^2}{w^2(z)} e^{-2r^2/w^2(z)} \tag{2.59}$$

Clearly, $I_o = |E_o|^2/(2\eta)$. This is the second way of selecting C'.

The third way is to express C' in terms of the **total optical power** P_∞. At an arbitrary plane located at point z, the total power transported by the beam is found by integrating (2.59) over an entire cross section as follows:

$$P_\infty = \int_0^\infty \int_0^{2\pi} I(r,z)\, rdrd\phi = \frac{w_o^2}{w^2(z)} 2\pi I_o \int_0^\infty e^{-2r^2/w^2(z)}\, rdr = \frac{\pi w_o^2}{2} I_o$$

Thus, in terms of the total optical power, $I_o = \dfrac{2P_\infty}{\pi w_o^2}$ and (2.59) may be written as

$$I(r,z) = \frac{2P_\infty}{\pi w^2(z)} e^{-2r^2/w^2(z)} \tag{2.60}$$

From (2.59) and (2.60) it is obvious why these beams are generally known as **Gaussian beams.** There is an interesting interpretation for the coefficient $2P_\infty/(\pi w^2)$ in (2.60). In the next subsection, $w(z)$ is defined as the radius of the Gaussian beam at z. If the total power P_∞ is distributed uniformly over a circle of radius $w(z)$, then the irradiance is $P_\infty/(\pi w^2)$. However, the irradiance distribution is not uniform; the extra factor of 2 indicates that the beam is concentrated more near the axis.

2.4.3 Beam Radius w(z)

To understand the physical meaning of each term in (2.58), suppose that z is fixed while r varies. As r increases, $|E|$ decreases as $e^{-r^2/w^2(z)}$. At $r = w(z)$, $|E|$ reduces to $1/e$, that is, 36.8 percent of its peak on-axis value. Hence, $w(z)$ is viewed as the **beam radius,** or the **spot size,** of the beam at z. As shown in Figure 2.13, $w(z)$ has a minimum value w_o at the point $z = 0$, which is known as the **beam waist.** The constant w_o introduced in (2.51) is the spot size, or the radius of the beam waist.

2.4.4 Phase Variation ψ(z)

To understand the presence of the phase term $e^{j\psi(z)}$ in (2.53), (2.57), and (2.58), consider that, if the fields under consideration were uniform plane waves, the phase term e^{-jkz} alone would be sufficient to signify the propagation of waves.

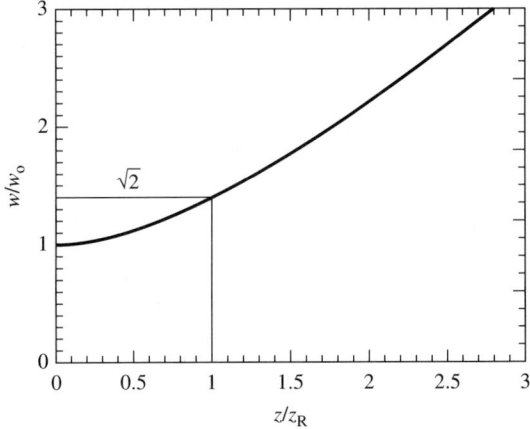

Figure 2.13 Normalized spot size of a Gaussian beam as a function z/z_R

However, the fields are not uniform plane waves, and the propagation of Gaussian beams differs from simple plane wave propagation. The difference in phase is the term $e^{j\psi(z)}$ where $\psi(z)$ signifies a longitudinal phase variation for the propagating Gaussian field. We also note that, due to the presence of $w(z)$ in (2.58), the field amplitude decreases as a function of z. Because of $w(z)$ in (2.53), (2.57), and (2.58), the total power carried by the beam remains unchanged as the beam propagates in the z direction. The phase factor $e^{j\psi(z)}$ is needed to accompany the amplitude change so that the wave equation is satisfied approximately by (2.53), (2.57), and (2.58).

2.4.5 Radius of Curvature R(z)

To understand the meaning of the exponential term $e^{-jkr^2/(2R(z))}$, consider spherical waves originating from the origin, as shown in Figure 2.14. Let R be the radius of the spherical wavefront. Consider two points A and B on the plane at $z = R$. Point A is on the axis and point B is off the axis. From the geometry shown in Figure 2.14, it is clear that, if $R \gg r$,

$$\overline{OB} - \overline{OA} = \sqrt{R^2 + r^2} - R \approx R[1 + \frac{1}{2}\frac{r^2}{R^2}] - R = \frac{r^2}{2R}$$

Thus, the fields at points A and B differ by a phase factor $e^{-jkr^2/(2R)}$, which is precisely the second exponential term of (2.58). Therefore, $R(z)$ is the **radius of curvature** of the constant-phase surface at z.

Figure 2.15 depicts the variation of $R(z)$ as a function of z. It is clear from (2.55) that for $R(z) > z$, the center of the spherical wavefront going in the +z direction is located at a point to the left of $z = 0$. At $z = 0$ and $z \to \infty$, $R(z)$ approaches infinity, which means that the constant-phase surfaces at the beam

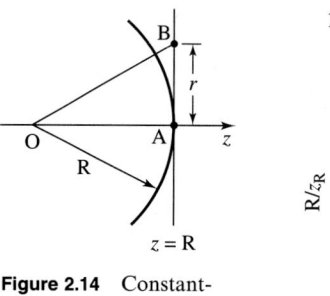

Figure 2.14 Constant-phase surface of a spherical wave

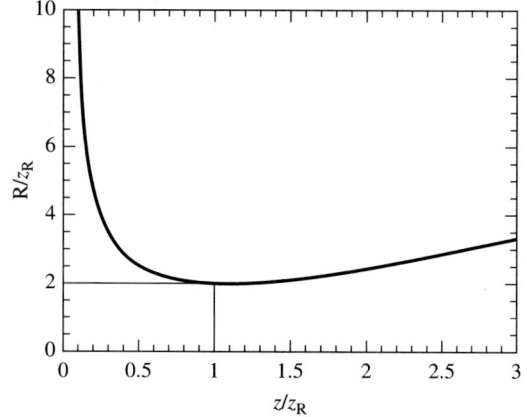

Figure 2.15 Normalized radius of curvature of a Gaussian beam as a function of z/z_R

waist and at infinity are planes. The minimum value of $R(z)$ is found by setting dR/dz to zero. The minimum radius is located at

$$z_R = \frac{\pi w_o^2}{\lambda} \tag{2.61}$$

and the minimum value is $R_{min} = 2z_R$. The term z_R is called the **Rayleigh range**.

It should be emphasized that (2.53), or its simplified form (2.57) or (2.58), is an approximate mathematical description for spherical diverging waves with a Gaussian amplitude distribution on the spherical wavefront. These expressions are derived based on the assumptions that the medium is an unbounded free-space region, and that the conditions for the paraxial approximations are satisfied.

2.4.6 Rayleigh Range z_R

The Rayleigh range given by (2.61) is an important beam parameter. At $z = z_R$, the beam radius is $\sqrt{2}\, w_o$, as shown in Figure 2.13, and the irradiance is half of the value at the beam waist. It is also customary to define the **depth of focus** or **collimated range** as the distance between $\pm z_R$; that is, the depth of focus is $2z_R$. Within the collimated range ($|z| < z_R$), $w(z)$ varies only slightly and may be approximated by w_o. This region is also known as the **near field region**. The region outside the collimated range ($|z| \gg z_R$) is the **far field region**. In the far field region, $w(z)$ increases linearly with $|z|$, $w(z) \approx \frac{\lambda}{\pi w_o} |z|$, and the beam diverges at a **half angle** of

$$\theta_h = \frac{\lambda}{\pi w_o} = \sqrt{\frac{\lambda}{\pi z_R}} \tag{2.62}$$

as shown schematically in Figure 2.16.

The range z_R can be viewed as a characteristic length for a Gaussian beam. Many distance or length variables related to Gaussian beams can be expressed in terms of z_R. In fact, when $w(z)$ and $R(z)$ were presented in Figures 2.13 and 2.15, these quantities were normalized with respect to z_R.

It is convenient to express $w(z)$, $R(z)$, and $\Psi(z)$ in terms of z_R, as well:

$$w(z) = \left(\frac{\lambda z_R}{\pi}\right)^{\frac{1}{2}} \left[1 + \left(\frac{z}{z_R}\right)^2\right]^{\frac{1}{2}} \tag{2.63}$$

$$R(z) = z + \frac{z_R^2}{z} \tag{2.64}$$

$$\Psi(z) = \tan^{-1}\left(\frac{z}{z_R}\right) \tag{2.65}$$

It is also convenient to use $z = z_R$ as the demarcation line between the near field region and the far field region.

2.4.7 Complex Radius of Curvature q

By using (2.54), (2.55), and (2.61), we can express $q(z)$ given in (2.51) as

$$\frac{1}{q(z)} = \frac{1}{R(z)} - j\frac{\lambda}{\pi w^2(z)} = \frac{1}{z + jz_R}$$

Because the real part of $1/q(z)$ is $1/R(z)$, and $R(z)$ is the radius of curvature of the wavefront, $q(z)$ is known as the **complex radius of curvature** of the Gaussian beam at z. The radius of curvature $R(z)$ and the beam radius $w(z)$ are specified completely once $q(z)$ is known. The transformation of Gaussian beams, to be discussed in section 2.7, amounts to a transformation of q.

2.5 OPTICAL COMPONENTS AND GAUSSIAN BEAMS

Two practical topics related to Gaussian beams are: (1) the size of optical components required to reflect, intercept, or focus Gaussian beams; and (2) the determination of the beam radius. Both are discussed in this section.

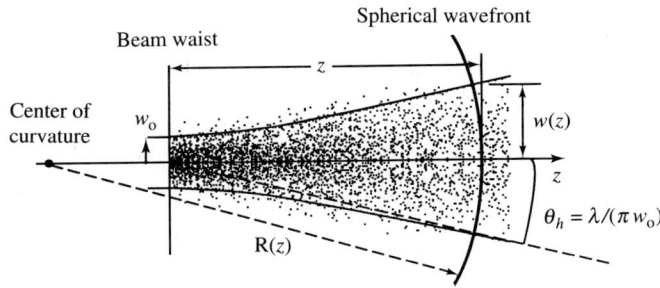

Figure 2.16 Various parameters of Gaussian beams

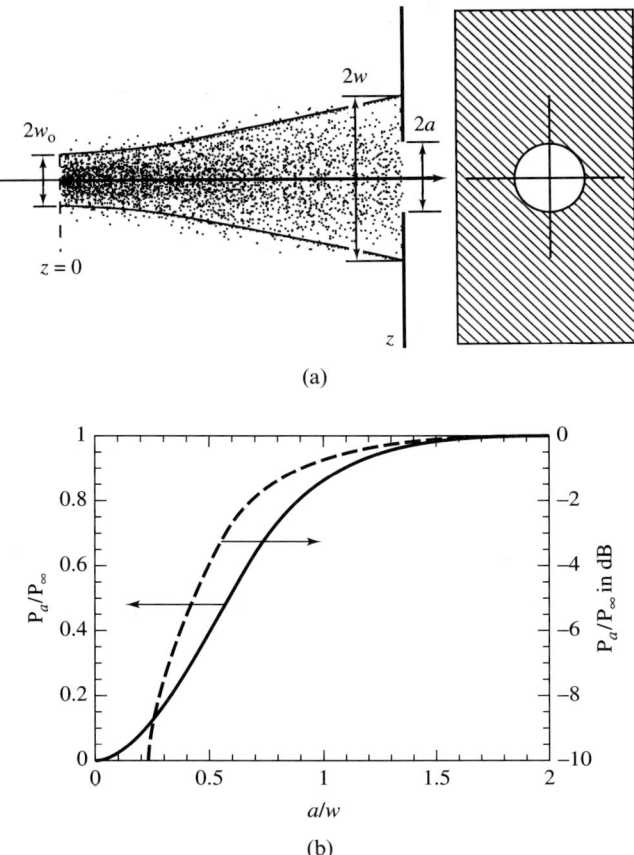

Figure 2.17 Portion of a Gaussian beam passing through a circular aperture on an opaque screen

2.5.1 Sizes of Pinholes, Mirrors, and Lenses

In designing optical systems, it may be necessary to select the radius of a pinhole, mirror, or lens for a given application. If diffraction-related effects are ignored, the component size can be selected from the power consideration.

Assume that the location ($z = 0$), the radius w_o of the beam waist, and the total power P_∞ of the Gaussian beam are all known. An opaque screen with a circular opening of radius a is placed at position z, as shown in Figure 2.17a. The power P_a passing through the aperture can be computed by integrating (2.60):

$$P_a = \int_0^{2\pi} \int_0^a I(r,z) \, r dr d\phi = \frac{2P_\infty}{\pi w^2(z)} 2\pi \int_0^a e^{-2r^2/w^2(z)} \, r dr = P_\infty [1 - e^{-2a^2/w^2(z)}]$$

The fraction of power passing through the circular opening of radius a on a large opaque screen is

2.5 Optical Components and Gaussian Beams

$$\frac{P_a}{P_\infty} = 1 - e^{-2a^2/w^2(z)} \qquad (2.67)$$

If a mirror or lens of radius a is located at z, P_a/P_∞ is also the fractional power intercepted by the mirror or lens. Figure 2.17b is a plot of P_a/P_∞ as a function of $a/w(z)$. Also shown in Figure 2.17b is P_a/P_∞ in decibels (dB). Three key values are listed in Table 2.2 at the bottom of the page.

Clearly, if a mirror or lens is used to reflect or transform a Gaussian beam, an aperture radius greater than $1.5w(z)$ or $2.0w(z)$ is preferred, purely from a power consideration.

2.5.2 Determination of Beam Radius

The experimental determination of the beam radius $w(z)$ is an important aspect of engineering practice. Using a small pinhole to probe the irradiance distribution of a beam is one approach, but such a method is very time consuming. Other methods, based on knife edges, thin wires, or narrow slits, have been devised and are more convenient [11–14]. These methods may be automated and can be used for measuring the beam radius quickly. However, such methods are predicated on the assumption that the beam has a Gaussian distribution. The result would be erroneous if the beam is not Gaussian.

The basis of the **knife edge method** is simple. Suppose a straightedge (i.e., the knife) is used to block a portion of a Gaussian beam (Figure 2.18a). The power P_1 not blocked by the knife is recorded as a function of the position x of the knife edge. From the measured data, the positions $x_{0.1}$ and $x_{0.9}$ of the knife edge, corresponding to P_1 at 10 percent and 90 percent of P_∞, respectively, may be obtained. In the following, we will show that $w(z)$ is proportional to the separation between the two positions, $(x_{0.1} - x_{0.9})$. An alternative method is to calculate the slope of $P_1(x)$ and to relate w to $-P_\infty/(dP_1/dx)$ at $x = 0$.

Referring to Figure 2.18a, the power not blocked by the knife is

$$P_1(x) = \frac{2P_\infty}{\pi w^2} \int_{-\infty}^{\infty} dy \int_{x}^{\infty} e^{-2(x^2+y^2)/w^2} \, dx = \frac{2P_\infty}{\pi w^2} \int_{-\infty}^{\infty} e^{-2y^2/w^2} \, dy \int_{x}^{\infty} e^{-2x^2/w^2} \, dx$$

$$= \frac{P_\infty}{w} \sqrt{\frac{2}{\pi}} \int_{x}^{\infty} e^{-2x^2/w^2} \, dx = \frac{P_\infty}{2} \operatorname{erfc}\left(\frac{2^{1/2}x}{w}\right) \qquad (2.68)$$

Table 2.2 Key values of P_a/P_∞

a	P_a/P_∞	P_a/P_∞
1.0w	86.5%	−0.63 dB
1.5w	98.9%	−0.48 dB
2.0w	99.97%	−0.001 dB

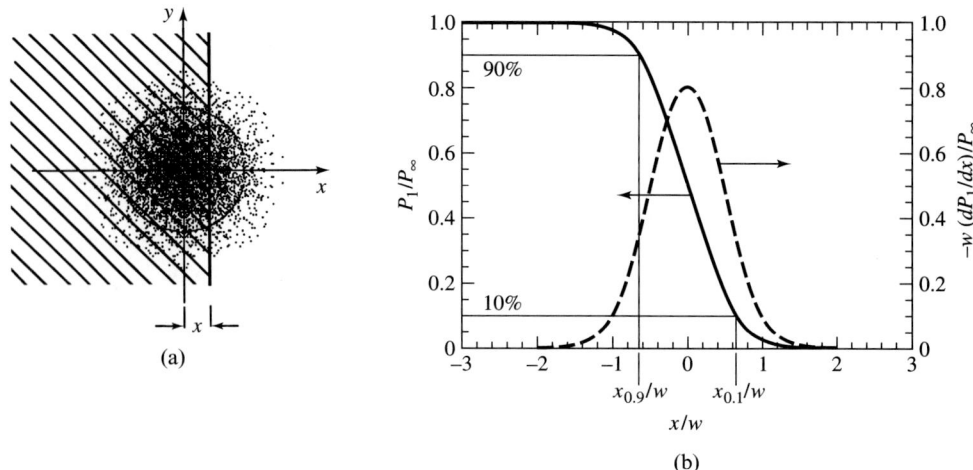

Figure 2.18 Gaussian beam partially blocked by a straight edge

Equation (2.68) uses the identity

$$\int_{-\infty}^{\infty} e^{-t^2}\, dt = \pi^{1/2} \qquad (2.69)$$

as well as the definitions of the error function and complimentary error function:

$$erf(x) = \frac{2}{\pi^{1/2}} \int_0^x e^{-t^2}\, dt \qquad (2.70)$$

$$erfc(x) = 1 - erf(x) = \frac{2}{\pi^{1/2}} \int_x^{\infty} e^{-t^2}\, dt \qquad (2.71)$$

Numerical values of error and complimentary error functions are given in Table 2.3.

In Figure 2.18b, $P_1(x)$ is shown as the solid curve. Let $x_{0.9}$ and $x_{0.1}$ be the positions of the knife edge when P_1 is 90 percent or 10 percent of P_∞, respectively. Then,

$$0.9 P_\infty = \left(\frac{P_\infty}{2}\right) erfc(2^{1/2} x_{0.9}/w)$$

$$0.1 P_\infty = \left(\frac{P_\infty}{2}\right) erfc(2^{1/2} x_{0.1}/w)$$

From Table 2.3, we may extrapolate that $erfc(+0.903) = 0.2$ and $erfc(-0.903) = 1.8$[1]. A simple equation relating w to $(x_{0.1} - x_{0.9})$ is

$$w = 0.7830\,(x_{0.1} - x_{0.9}) \qquad (2.72)$$

---

(1) Note that $erf(-x) = erf(x)$ and $erfc(-x) = 2 - erfc(x)$ for $x > 0$.

Table 2.3 Error and complimentary error functions

x	erf(x)	erfc(x)	x	erf(x)	erfc(x)	x	erf(x)	erfc(x)
0.00	0.00000	1.00000	0.50	0.52050	0.47950	1.00	0.84270	0.15730
0.01	0.01128	0.98872	0.51	0.52924	0.47076	1.01	0.84681	0.15319
0.02	0.02256	0.97744	0.52	0.53790	0.46210	1.02	0.85084	0.14916
0.03	0.03384	0.96616	0.53	0.54646	0.45354	1.03	0.85478	0.14522
0.04	0.04511	0.95489	0.54	0.55494	0.44506	1.04	0.85865	0.14135
0.05	0.05637	0.94363	0.55	0.56332	0.43668	1.05	0.86244	0.13756
0.06	0.06762	0.93238	0.56	0.57162	0.42838	1.06	0.86614	0.13386
0.07	0.07886	0.92114	0.57	0.57982	0.42018	1.07	0.86977	0.13023
0.08	0.09008	0.90992	0.58	0.58792	0.41208	1.08	0.87333	0.12667
0.09	0.10128	0.89872	0.59	0.59594	0.40406	1.09	0.87680	0.12320
0.10	0.11246	0.88754	0.60	0.60386	0.39614	1.10	0.88021	0.11979
0.11	0.12362	0.87638	0.61	0.61168	0.38832	1.11	0.88353	0.11647
0.12	0.13476	0.86524	0.62	0.61941	0.38059	1.12	0.88679	0.11321
0.13	0.14587	0.85413	0.63	0.62705	0.37295	1.13	0.88997	0.11003
0.14	0.15695	0.84305	0.64	0.63459	0.36541	1.14	0.89308	0.10692
0.15	0.16800	0.83200	0.65	0.64203	0.35797	1.15	0.89612	0.10388
0.16	0.17901	0.82099	0.66	0.64938	0.35062	1.16	0.89910	0.10090
0.17	0.18999	0.81001	0.67	0.65663	0.34337	1.17	0.90200	0.09800
0.18	0.20094	0.79906	0.68	0.66378	0.33622	1.18	0.90484	0.09516
0.19	0.21184	0.78816	0.69	0.67084	0.32916	1.19	0.90761	0.09239
0.20	0.22270	0.77730	0.70	0.67780	0.32220	1.20	0.91031	0.08969
0.21	0.23352	0.76648	0.71	0.68467	0.31533	1.21	0.91296	0.08704
0.22	0.24430	0.75570	0.72	0.69143	0.30857	1.22	0.91553	0.08447
0.23	0.25502	0.74498	0.73	0.69810	0.30190	1.23	0.91805	0.08195
0.24	0.26570	0.73430	0.74	0.70468	0.29532	1.24	0.92051	0.07949
0.25	0.27633	0.72367	0.75	0.71116	0.28884	1.25	0.92290	0.07710
0.26	0.28690	0.71310	0.76	0.71754	0.28246	1.26	0.92524	0.07476
0.27	0.29742	0.70258	0.77	0.72382	0.27618	1.27	0.92751	0.07249
0.28	0.30788	0.69212	0.78	0.73001	0.26999	1.28	0.92973	0.07027
0.29	0.31828	0.68172	0.79	0.73610	0.26390	1.29	0.93190	0.06810
0.30	0.32863	0.67137	0.80	0.74210	0.25790	1.30	0.93401	0.06599
0.31	0.33891	0.66109	0.81	0.74800	0.25200	1.31	0.93606	0.06394
0.32	0.34913	0.65087	0.82	0.75381	0.24619	1.32	0.93807	0.06193
0.33	0.35928	0.64072	0.83	0.75952	0.24048	1.33	0.94002	0.05998
0.34	0.36936	0.63064	0.84	0.76514	0.23486	1.34	0.94191	0.05809
0.35	0.37938	0.62062	0.85	0.77067	0.22933	1.35	0.94376	0.05624
0.36	0.38933	0.61067	0.86	0.77610	0.22390	1.36	0.94556	0.05444
0.37	0.39921	0.60079	0.87	0.78144	0.21856	1.37	0.94731	0.05269
0.38	0.40901	0.59099	0.88	0.78669	0.21331	1.38	0.94902	0.05098
0.39	0.41874	0.58126	0.89	0.79184	0.20816	1.39	0.95067	0.04933
0.40	0.42839	0.57161	0.90	0.79691	0.20309	1.40	0.95229	0.04771
0.41	0.43797	0.56203	0.91	0.80188	0.19812	1.41	0.95385	0.04615
0.42	0.44747	0.55253	0.92	0.80677	0.19323	1.42	0.95538	0.04462
0.43	0.45689	0.54311	0.93	0.81156	0.18844	1.43	0.95686	0.04314
0.44	0.46623	0.53377	0.94	0.81627	0.18373	1.44	0.95830	0.04170
0.45	0.47548	0.52452	0.95	0.82089	0.17911	1.45	0.95970	0.04030
0.46	0.48466	0.51534	0.96	0.82542	0.17458	1.46	0.96105	0.03895
0.47	0.49375	0.50625	0.97	0.82987	0.17013	1.47	0.96237	0.03763
0.48	0.50275	0.49725	0.98	0.83423	0.16577	1.48	0.96365	0.03635
0.49	0.51167	0.48833	0.99	0.83851	0.16149	1.49	0.96490	0.03510
0.50	0.52050	0.47950	1.00	0.84270	0.15730	1.50	0.96611	0.03389

In the above discussion, the choice of 10 percent and 90 percent, while convenient, is arbitrary. Any two convenient values may be used, with the understanding that, for different values, the numerical constant will also be different.

It is also possible to relate w to the slope of P_1. Simple differentiation of (2.68) yields

$$\frac{dP_1}{dx} = -\sqrt{\frac{2}{\pi}} \frac{P_\infty}{w} e^{-2x^2/w^2} \qquad (2.73)$$

As shown by the dashed line in Figure 2.18b, $-\dfrac{dP_1}{dx}$ reaches a peak value at $x = 0$, and the peak value is

$$-\frac{dP_1}{dx}\bigg|_{x=0} = \sqrt{\frac{2}{\pi}} \frac{P_\infty}{w} \qquad (2.74)$$

If the maximum slope is determined, by any means, such as a graph, we obtain

$$w = \sqrt{\frac{2}{\pi}} \frac{P_\infty}{-\dfrac{dP_1}{dx}\bigg|_{x=0}} \qquad (2.75)$$

In the experiment, the knife edge can be moved by a motor-driven micrometer and the derivative can be obtained electronically. After calibration, P_∞ and $-\dfrac{dP_1}{dx}$ can then be read directly from an oscilloscope.

Thin wires or **narrow ribbons** can also be used to determine w. By moving a wire to intercept the beam, we can block a portion of the optical power. (Figure 2.19a). Let P_2 be the power not blocked by the wire. Here, P_2 has a maximum value of P_∞ when the wire is absent, and is a minimum when the wire axis coincides with the beam center. Let the minimum be $P_{2,\min}$. We will show that w can be determined if the ratio $P_{2,\min}/P_\infty$ and the wire radius are known.

Let the wire radius be r_w and the wire center be at x relative to the beam center. The power not intercepted by the wire is

$$P_2(x) = \frac{2P_\infty}{\pi w^2} \left[\int_{-\infty}^{\infty} dy \int_{-\infty}^{x-r_w} dx\, e^{-2(x^2+y^2)/w^2} + \int_{-\infty}^{\infty} dy \int_{x+r_w}^{\infty} dx\, e^{-2(x^2+y^2)/w^2} \right] \qquad (2.76)$$

When the wire is located precisely at the center of the beam, the power unblocked by the wire would be the minimum, or

$$P_{2,\min} = \frac{2P_\infty}{\pi w^2} \left[\int_{-\infty}^{\infty} dy \int_{-\infty}^{-r_w} dx\, e^{-2(x^2+y^2)/w^2} + \int_{-\infty}^{\infty} dy \int_{r_w}^{\infty} dx\, e^{-2(x^2+y^2)/w^2} \right] \qquad (2.77)$$

The integrals can be expressed in terms of the complementary error function

$$\frac{P_{2,\min}}{P_\infty} = \mathrm{erfc}\left(\frac{2^{1/2} r_w}{w}\right) \qquad (2.78)$$

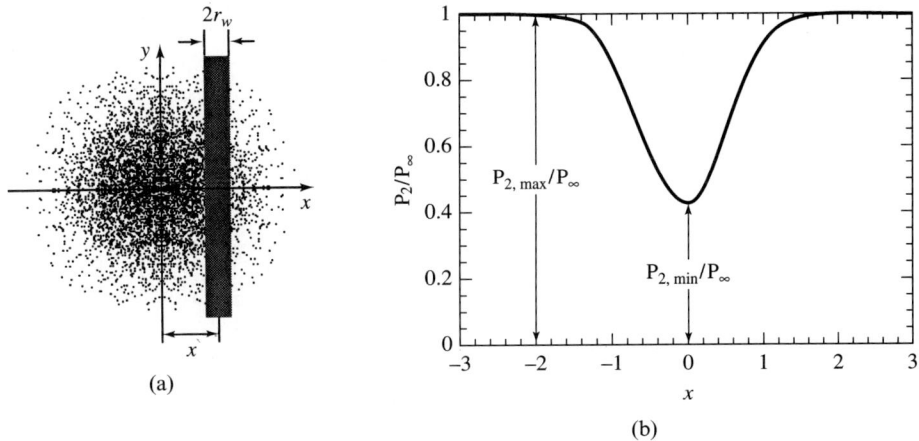

Figure 2.19 Gaussian beam partially blocked by a thin wire or strip

By measuring $P_{2,\min}$ and P_∞, we can deduce the beam radius w, provided the wire radius r_w is known. For best results, the wire radius r_w should be on the order of one-fourth to three-fourths of the beam radius w [11]. If the beam radius is very small, on the order of a few microns for example, **gratings** may be used in lieu of wires, slits, or knife edges [15,16].

Experience shows that circular wires and flat ribbons have the same effect [11,13,14]. This is understandable, since the effects of diffraction, have been ignored in our discussions.

2.5.3 Example

Suppose a #10 copper wire is used in the experiment and the results are $P_\infty = 1$ mW and $P_{2,\min} = 0.4$ mW. Then, from (2.78), we have $erfc(2^{1/2}r_w/w) = 0.4$. From Table 2.3 or any mathematical tables, we obtain $erfc(0.595) = 0.4$. The radius of a #10 wire is 0.129 cm (i.e., 50.9 mils). Thus,

$$w = 2^{1/2} \times 0.129/0.595 = 0.307 \text{ cm}$$

2.6 HIGHER-ORDER MODES

The Gaussian beams discussed in the preceding sections are the lowest-order modes propagating in an unbounded, isotropic, homogeneous medium. There are also higher order modes. In a lossless medium, the total power of a beam, that is the sum of all power in all modes, does not change as the beam propagates. However, the *distribution* of power in individual modes, including particularly higher-order modes, may be changed by the presence of scatterers or air turbulence. Therefore, lower order modes, and fundamental modes in particular, are preferred in most applications.

In the following discussion, we present the expressions for higher-order modes in both rectangular and cylindrical coordinates. These expressions can be derived in the same fashion as that given in section 2.4. Details of the derivation are omitted (see [2], [17], and [18]), and only the final results are presented here.

As a note, many reflectors used to form laser cavities have circular cross sections. Light beams emitted by these laser cavities are expected to have a circular cross section, as well, in which case cylindrical coordinates would be most convenient for describing these beams. However, it has been reported that, in the presence of **Brewster angle windows** or misaligned mirrors, the higher-order modes have a Hermite distribution in rectangular coordinates, rather than a circular Laguerre profile, even though the optical cavity has a circular aperture [8].

In rectangular coordinates, the electric field component of TEM_{mn} modes may be written

$$E(x,y,z) = C'' H_m(\sqrt{2}\frac{x}{w(z)}) H_n(\sqrt{2}\frac{y}{w(z)}) e^{-j[P_{mn}(z) + \frac{k(x^2+y^2)}{2q(z)}]} e^{-jkz} \qquad (2.79)$$

where C'' is a constant, and

$$P_{mn}(z) = -j\left[\frac{1}{2} \ln(z^2 + \frac{\pi w_o^2}{\lambda})^2) - j(m + n + 1)\Psi(z)\right]$$

The terms $q(z)$, $\psi(z)$, and $w(z)$ were defined in (2.51), (2.53), and (2.54). The term H_m is a Hermite polynomial of order m where $m = 0, 1, 2\cdots$. For $m \leq 3$,

$$H_0(x) = 1$$
$$H_1(x) = 2x$$
$$H_2(x) = 4x^2 - 2$$
$$H_3(x) = 8x^3 - 12x$$

The polynomial H_m has m roots; therefore, the electric field components of a TEM_{mn} mode have m nulls in the x direction and n nulls in the y direction, as shown in Figure 2.20.

In cylindrical coordinates, the Cartesian component of the electric field modes may be expressed as

$$E(r,\phi,z) = C''' \left[(\sqrt{2}\frac{r}{w(z)})^q L_p^q(r)\right] \begin{Bmatrix} \sin q\phi \\ \cos q\phi \end{Bmatrix} e^{-j\left[P_{pq}(z) + \frac{kr^2}{2q(z)}\right]} e^{-jkz} \qquad (2.80)$$

where C''' is a constant and

$$P_{pq}(z) = -j\left[\frac{1}{2} \ln\left[z^2 + (\frac{\pi w_o^2}{\lambda})^2\right] - j(2p + q + 1)\Psi(z)\right]$$

The terms L_p^q are associated Laguerre polynomials of order p and degree q. For $p \leq 2$, the polynomials are

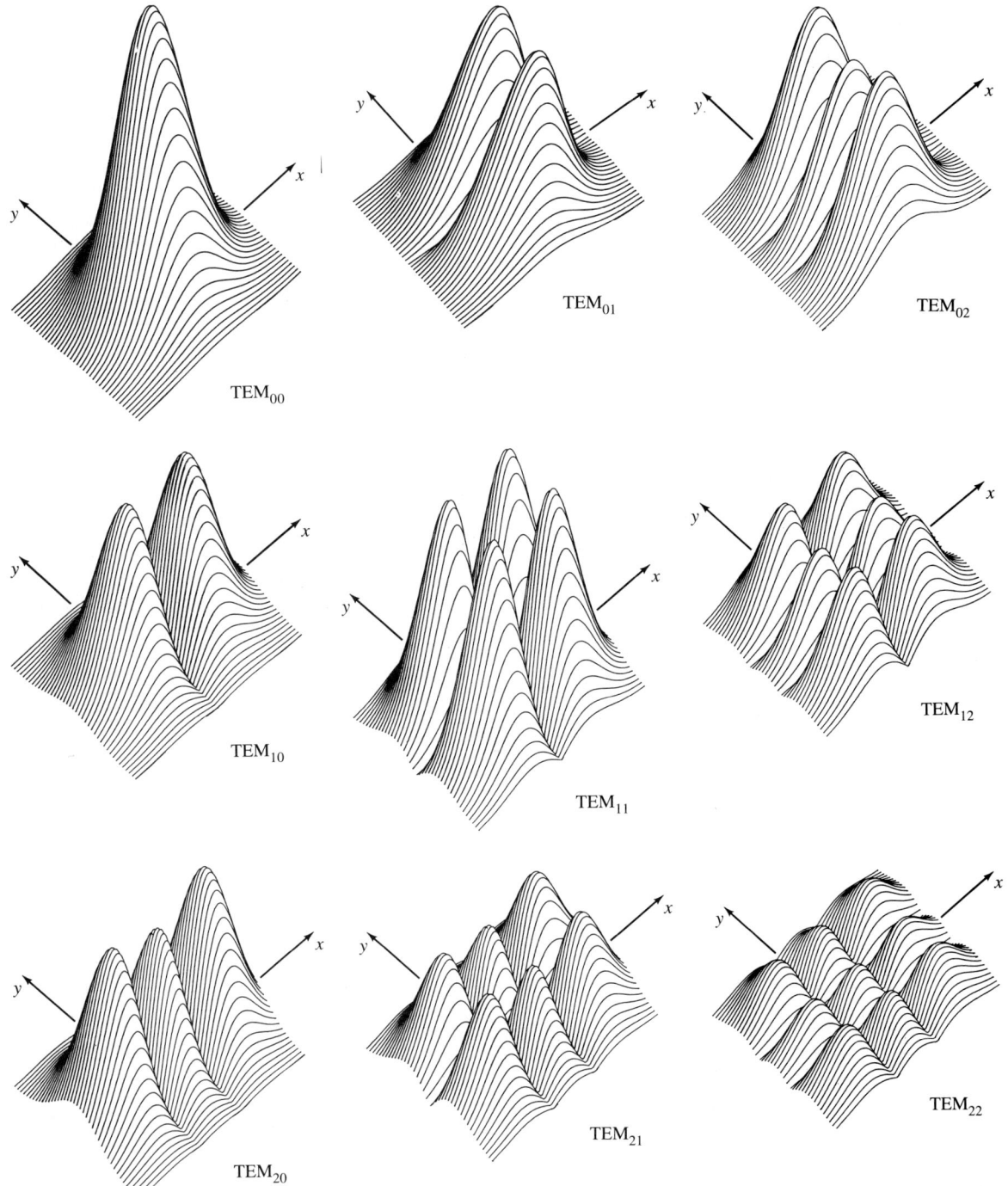

Figure 2.20 TEM$_{mn}$ Hermite-Gaussian modes in Cartesian coordinates

$$L_0^q(r) = 1$$
$$L_1^q(r) = (q + 1) - r$$
$$L_2^q(r) = \frac{(q + 1)(q + 2)}{2} - (q + 2)r + \frac{r^2}{2}$$

The profiles of higher-order modes in cylindrical coordinates are sketched in Figure 2.21. These modes are labeled TEM$_{pq}$ modes. The subscripts p and q signify that there are p nulls in the radial direction, and that the angular variation of the fields is sin qϕ or cos qϕ.

2.7 ABCD LAW

As stated previously, a Gaussian beam is specified if the complex radius of curvature q is given and the transformation of Gaussian beams is simply a transformation of q. If we let q_{in} and q_{out} be the complex radii of curvature of the original and transformed Gaussian beams, respectively, then the **ABCD law** is

$$q_{out} = \frac{Aq_{in} + B}{Cq_{in} + D} \qquad (2.81)$$

where A, B, C, and D are the elements of the ABCD matrix discussed in Section 2.2. Instead of establishing the validity in general, we will consider three specific cases. In each case, we will start from basic principles and show that (2.81) holds for that case.

2.7.1 Free-Space Layer of Thickness *d*

A free-space layer of thickness d is the first case to be considered. Let the input and output reference planes be located at z and $z + d$. The complex radii of curvature of a Gaussian beam at the reference planes are then given by

$$q_{in} = q(z) = z + j\frac{\pi w_o^2}{\lambda}$$
$$q_{out} = q(z + d) = z + d + j\frac{\pi w_o^2}{\lambda}$$

Comparing these equations, we see that $q_{out} = q_{in} + d$, which may be written in the form of the ABCD law as

$$q_{out} = \frac{1 \times q_{in} + d}{0 \times q_{in} + 1}$$

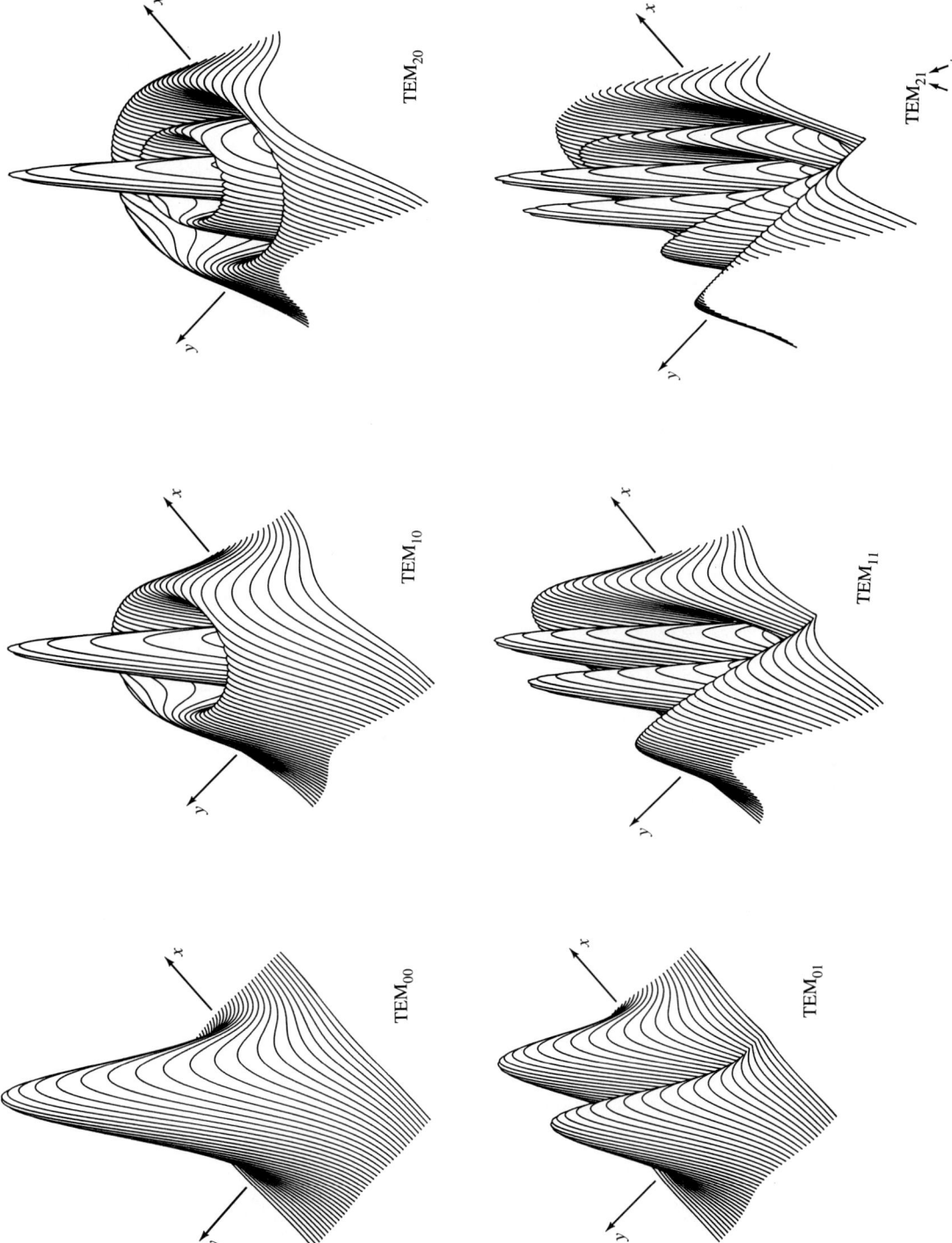

Figure 2.21 TEM$_{pq}$ Laguerre-Gaussian modes in cylindrical coordinates

2.7.2 Spherical Mirror with Radius R_m

The effects of spherical mirrors on Gaussian beams is treated next. We begin by studying the effects of a spherical mirror on incoming uniform plane waves. Consider uniform plane waves impinging on a convex spherical mirror that has a radius of curvature $|R_m|$ (Figure 2.22a). According to Huygen's principle [1], every point on a wavefront can be viewed as a source for the succeeding wavelets. By summing all wavelets thus emitted, we can construct succeeding wavefronts. Consider the wavelets originating from points on an incoming wavefront, as shown by the dashed line in Figure 2.22a. Wavelets originating from a point on the z axis are reflected by the mirror immediately. Wavelets originating from a point S at a distance r above the axis must travel an extra distance $\overline{ST}$ before reaching the mirror, and reflected wavelets must go an additional segment $\overline{ST}$ before returning to the dashed line. A phase delay, $2k\overline{ST}$, is introduced to account for these distances. When the paraxial condition $|r| \ll |R_m|$ is met, $\overline{ST}$ may be approximated as

$$\overline{ST} = |R_m| - \sqrt{|R_m|^2 - r^2} \approx \frac{r^2}{2|R_m|}$$

If the spherical mirror is concave (Figure 2.22b), the phase delay becomes a phase advance of $-2k\overline{ST}$. Both convex and concave mirrors are accounted for by writing $\overline{ST} \approx r^2/(2R_m)$ and allowing R_m to be positive or negative. Thus, if the incoming uniform plane wave arriving at the dashed line is represented by e^{-jkz}, the image of the reflected wave is given by $e^{-jkz} e^{-j2kr^2/(2R_m)}$.

The effects of spherical mirrors on Gaussian beams are the same. For an incoming Gaussian beam given by (2.58), which is rewritten here for convenience,

$$E_o \frac{w_o}{w(z)} e^{-j(kz - \Psi(z))} e^{-j\frac{kr^2}{2}[\frac{1}{R(z)} - j\frac{\lambda}{\pi w^2(z)}]}$$

the complex radius of curvature q_{in} is given by

$$\frac{1}{q_{\text{in}}} = \frac{1}{R(z)} - j\frac{\lambda}{\pi w^2(z)} \tag{2.82a}$$

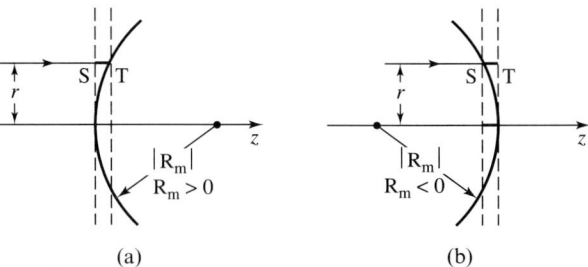

(a)   (b)

Figure 2.22 Spherical reflecting mirror with radius of curvature R'_m

By inserting an exponential term representing the phase delay or advance, we have the expression for the image of the reflected Gaussian beam

$$\frac{E_o w_o}{w(z)} e^{-j(kz - \Psi(z))} e^{-j\frac{kr^2}{2}[\frac{1}{R(z)} - j\frac{\lambda}{\pi w^2(z)}]} e^{-j\frac{kr^2}{2}\frac{2}{R_m}}$$

The complex radius of curvature q_{out} is given by

$$\frac{1}{q_{\text{out}}} = \frac{1}{R(z)} + \frac{2}{R_m} - j\frac{\lambda}{\pi w^2(z)} \qquad (2.82b)$$

From (2.82a) and (2.82b), we have

$$\frac{1}{q_{\text{out}}} = \frac{1}{q_{\text{in}}} + \frac{2}{R_m}$$

Upon rearrangement, this becomes

$$q_{\text{out}} = \frac{1 \times q_{\text{in}} + 0}{\frac{2}{R_m} \times q_{\text{in}} + 1}$$

A comparison of the above equation with the ABCD matrix for spherical mirrors given in (2.28) and in Table 2.1 reveals that the ABCD law (2.81) is indeed valid for spherical mirrors.

2.7.3 Thin Lens with Focal Length *f*

Finally, we consider the effects of thin lenses, beginning with their effects on spherical waves. We know that diverging spherical waves originating from the focal point of a positive lens are converted to plane waves by the lens. For convenience, the focal point is taken as the origin of the coordinates, and the spherical waves are considered to be originating from that point, as shown in Figure 2.23. In the region near the lens, that is, $z \sim f$, and under the paraxial approximation $|r| \ll z$, the incoming spherical waves are given by

$$e^{-jk\sqrt{z^2 + r^2}} \approx e^{-jkz} e^{-jkr^2/(2z)}$$

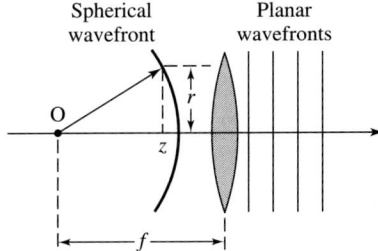

Figure 2.23 Thin lens with focal length *f*

After passing through the lens, the spherical waves are converted to plane waves given by e^{-jkz}. Thus, the effects of a thin lens that has a focal length f and is located at $z = f$ are represented mathematically by multiplying the spherical wave description by a phase term $e^{+jkr^2/(2f)}$ [19].

The effects of a thin lens on Gaussian beams is the same. Specifically, a lens applies a phase term $e^{+jkr^2/(2f)}$ to the incoming waves. If we let the incoming Gaussian beam be given by (2.58), then the outgoing Gaussian beam becomes

$$\frac{E_o w_o}{w(z)} e^{-j(kz-\Psi(z))} e^{-j\frac{kr^2}{2}\left[\frac{1}{R(z)} - j\frac{\lambda}{\pi w^2(z)}\right]} e^{+j\frac{kr^2}{2f}}$$

The complex radius of curvature q_{out} of the outgoing beams may be obtained from

$$\frac{1}{q_{out}} = \frac{1}{R(z)} - \frac{1}{f} - j\frac{\lambda}{\pi w^2(z)}$$

Therefore, the effects of a thin lens with a focal length f are simply

$$\frac{1}{q_{out}} = \frac{1}{q_{in}} - \frac{1}{f} \tag{2.83}$$

or equivalently,

$$q_{out} = \frac{1 \times q_{in} + 0}{-\frac{1}{f} \times q_{in} + 1}$$

The effects on Gaussian beams of a planar interface between dielectrics can also be studied from basic principles. This is left as an exercise for the reader (Problem 11). Additional examples demonstrating the validity of the ABCD law can be found in references 2, 8, 19–22.

2.8 GAUSSIAN BEAMS TRANSFORMATIONS BY THIN LENSES

Since lenses are the basic building blocks of many optical systems, it is worthwhile to devote additional study to their effects on Gaussian beams. Also, since the effects of thick lenses are too complicated to be considered in an introductory text [23], we will only examine the effects of thin lenses. All quantities associated with the incoming beam are identified by the subscript "in," and those associated with outgoing beams by the subscript "out."

The beam waist $w_{o\,in}$ of the input beam is located at a distance z_{in} from the thin lens, as shown in Figure 2.24, and the Rayleigh range is $z_{R\,in}$. The beam radius w_{in}, the radius of curvature R_{in}, and the complex radius of curvature q_{in} of the incoming beam at the thin lens can all be expressed in terms of $w_{o\,in}$ and z_{in}, as follows:

2.8 Gaussian Beams Transformations by Thin Lenses

$$\frac{1}{q_{in}} = \frac{1}{R_{in}} - j\frac{\lambda}{\pi w_{in}^2} = \frac{1}{z_{in} + j(\pi w_{o\,in}^2/\lambda)} \quad (2.84)$$

Corresponding terms and expressions may be written for the outgoing beam. Making use of (2.83) or the ABCD law (2.81) and equating real and imaginary parts of both sides of the equation separately, we have

$$w_{out} = w_{in} \quad (2.85)$$

$$\frac{1}{R_{out}} = \frac{1}{R_{in}} - \frac{1}{f} \quad (2.86)$$

It is important to distinguish the radius $w_{o\,in}$ of the input beam waist from the input beam radius w_{in} at the lens. The corresponding terms of the output beam are $w_{o\,out}$ and w_{out}. Equation (2.85) means that the beam radii w_{in} and w_{out} on both sides of the thin lens are the same. However, the radii of curvature are modified by the presence of the lens, as indicated in (2.86).

Several pairs of equations can be derived from (2.85) and (2.86). In the following sections, four pairs are presented: [(2.87) and (2.88)], [(2.89) and (2.90)], [(2.91) and (2.92)], and [(2.93) and (2.94)]. Although these pairs appear different, they are in fact equivalent. Depending on the specific application, one pair may be more convenient to use than the others.

2.8.1 Given: $w_{o\,in}$, z_{in}, and f

With the location and radius of the input beam waist and the focal length of a thin lens known, the radius and location of the output beam waist, relative to the thin lens, can be calculated from

$$z_{out} = f + \frac{(z_{in} - f)f^2}{(z_{in} - f)^2 + (\pi w_{o\,in}^2/\lambda)^2} = f + \frac{(z_{in} - f)f^2}{(z_{in} - f)^2 + z_{R\,in}^2} \quad (2.87)$$

$$\frac{1}{w_{o\,out}^2} = \frac{1}{w_{o\,in}^2}\left(1 - \frac{z_{in}}{f}\right)^2 + \frac{1}{f^2}\left(\frac{\pi w_{o\,in}}{\lambda}\right)^2 = \frac{1}{w_{o\,in}^2}\left[\left(1 - \frac{z_{in}}{f}\right)^2 + \left(\frac{z_{R\,in}}{f}\right)^2\right] \quad (2.88)$$

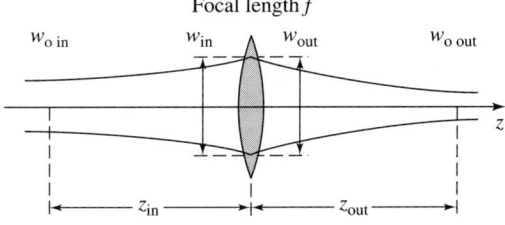

Figure 2.24 Effect of a convex lens on a Gaussian beam

These equations can be deduced from (2.85) and (2.86). Simple manipulation also shows that these equations are equivalent to

$$\frac{1}{z_{out}} + \frac{1}{z_{in} + \left[\frac{z_{R\,in}^2}{z_{in} - f}\right]} = \frac{1}{f} \qquad (2.89)$$

and

$$\frac{w_{o\,out}}{w_{o\,in}} = \frac{1}{\left[\left(1 - \frac{z_{in}}{f}\right)^2 + \frac{z_{R\,in}^2}{f^2}\right]^{1/2}} \qquad (2.90)$$

Equations [(2.87) and (2.88)] or [(2.89) and (2.90)] are useful in applications where $w_{o\,in}$, z_{in}, and the focal length f of the lens are given, and the parameters of the output beam are to be calculated.

2.8.2 Given: $w_{o\,in}$, z_{in}, and $w_{o\,out}$

There are situations in which a lens is used to transform an input beam with $w_{o\,in}$ and z_{in} to an output beam with a specific $w_{o\,out}$. However, the focal length f and z_{out} are unknown. For these applications, the focal length f is derived from (2.88):

$$\frac{w_{o\,in}^2}{w_{o\,out}^2} = \frac{(f - z_{in})^2 + (\pi w_{o\,in}^2/\lambda)^2}{f^2} = \left(1 - \frac{z_{in}}{f}\right)^2 + \left(\frac{z_{R\,in}}{f}\right)^2 \qquad (2.91)$$

and z_{out} is deduced from (2.87) and (2.91)

$$\frac{(z_{out} - f)}{(z_{in} - f)} = \left[\frac{w_{o\,out}}{w_{o\,in}}\right]^2 \qquad (2.92)$$

2.8.3 Given: $w_{o\,in}$ and $w_{o\,out}$

A given lens with a focal length f may be used to transform a beam with $w_{o\,in}$ to a beam with $w_{o\,out}$ without any restriction on z_{in}. For this purpose, z_{in} and z_{out} are solved for from (2.91) and (2.92)

$$\frac{z_{in}}{f} = 1 \pm \frac{w_{o\,in}}{w_{o\,out}} \left[1 - \left(\frac{f_{ml}}{f}\right)^2\right]^{1/2} \qquad (2.93)$$

$$\frac{z_{out}}{f} = 1 \pm \frac{w_{o\,out}}{w_{o\,in}} \left[1 - \left(\frac{f_{ml}}{f}\right)^2\right]^{1/2} \qquad (2.94)$$

In (2.93) and (2.94),

$$f_{ml} = \frac{\pi w_{o\,in} w_{o\,out}}{\lambda} \qquad (2.95)$$

is the characteristic **matching length** [20–22]. In view of (2.93) and (2.94), we conclude that a lens with a focal length f longer than the matching length f_{ml} may be used to transform $w_{o\,in}$ to $w_{o\,out}$. When plus signs of (2.93) and (2.94) are used together, they lead to a configuration with z_{in} and $z_{out} > f$. The minus signs of (2.93) and (2.94) used together correspond to an arrangement in which z_{in} and $z_{out} < f$.

2.8.4 Example: Perfect Beam Expanders

As an example, consider a perfectly adjusted beam expander [22] or telescope consisting of two thin lenses with focal lengths f_1 and f_2 separated by a distance $d = f_1 + f_2$. As shown in Figure 2.25, the incoming and outgoing beam waists $w_{o\,in}$ and $w_{o\,out}$ are located at distances z_{in} and z_{out} from the first and second lenses, respectively. Equations (2.93) and (2.94) are used to evaluate the effect of the first lens,

$$\frac{z_{in}}{f_1} = 1 \pm \frac{w_{o\,in}}{w_{o\,x}}\left[1 - \left(\frac{f_{ml1}}{f_1}\right)^2\right]^{1/2} \qquad (2.96)$$

$$\frac{z_x}{f_1} = 1 \pm \frac{w_{o\,x}}{w_{o\,in}}\left[1 - \left(\frac{f_{ml1}}{f_1}\right)^2\right]^{1/2} \qquad (2.97)$$

where z_x and $w_{o\,x}$ are the location and radius of the beam waist for the beam emerging from the first lens, and $f_{ml1} = \pi w_{o\,x} w_{o\,in}/\lambda$.

For the second lens, the incoming beam has a beam waist $w_{o\,x}$ at a distance $d - z_x$ from the second lens. Again (2.93) and (2.94) are used and we obtain

$$\frac{d - z_x}{f_2} = 1 \mp \frac{w_{o\,x}}{w_{o\,out}}\left[1 - \left(\frac{f_{ml2}}{f_2}\right)^2\right]^{1/2} \qquad (2.98)$$

$$\frac{z_{out}}{f_2} = 1 \mp \frac{w_{o\,out}}{w_{o\,x}}\left[1 - \left(\frac{f_{ml2}}{f_2}\right)^2\right]^{1/2} \qquad (2.99)$$

where $f_{ml2} = \pi w_{o\,x} w_{o\,out}/\lambda$. It is clear from Figure 2.25 that if z_x is greater (or smaller) than f_1, then $d - z_x$ is smaller (or greater) than f_2. This observation has

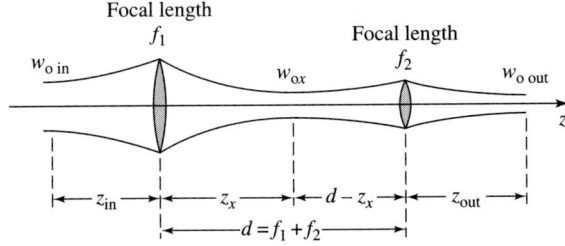

Figure 2.25 Perfect optical beam expander

been taken into account in selecting the signs for (2.96) to (2.99). Combining (2.97) with (2.98), and recalling that $d = f_1 + f_2$, we obtain

$$\frac{w_{o\,in}}{w_{o\,out}} = \frac{f_1}{f_2} \tag{2.100}$$

Similarly, from (2.96), (2.99), and (2.100), we obtain

$$\frac{(z_{in} - f_1)}{(z_{out} - f_2)} = -\frac{f_1^2}{f_2^2} \tag{2.101}$$

Thus, by using a beam expander with perfect alignment, the radius of the output beam waist is changed by a factor of f_1/f_2, and the distance from the output beam waist to the focal point is scaled as $-(f_1/f_2)^2$.

If the beam expander is misaligned,

$$d = f_1 + f_2 + \Delta d \tag{2.102}$$

then the exact expressions relating the output to the input are very complicated. Approximate relations similar to (2.100) and (2.101) have been determined by Kogelnik [22], as follows:

$$\frac{w_{o\,out}}{w_{o\,in}} \approx \frac{f_2}{f_1}\left[1 - \frac{\Delta d}{f_1}\left(1 - \frac{z_{in}}{f_1}\right)\right] \tag{2.103}$$

$$\frac{z_{in} - f_1}{f_1^2} + \frac{z_{out} - f_2}{f_2^2} \approx -\frac{\Delta d}{f_1^4}\left[\left(z_{in} - f_1\right)^2 - \left(\frac{\pi w_{o\,in}^2}{\lambda}\right)^2\right] \tag{2.104}$$

and those equations are accurate to the first order of Δd.

2.8.5 Comparison: Geometrical Optics

Transforming an object to its image, using thin lenses, is an important topic of **geometrical optics.** The focal length f, the object distance z_{in}, and the image distance z_{out} are all related by the Gauss lens equation

$$\frac{1}{z_{out}} + \frac{1}{z_{in}} = \frac{1}{f} \tag{2.105}$$

Lateral magnification M is the ratio of the image height h_{out} to the object height h_{in}. For the present purpose, we are mainly concerned with the absolute value of M, that is,

$$|M| = \left|\frac{h_{out}}{h_{in}}\right| = \frac{1}{\left|1 - \frac{z_{in}}{f}\right|} = \left|1 - \frac{z_{out}}{f}\right| \tag{2.106}$$

These equations can be easily established by simple geometric construction as shown in Figure 2.26. The object-to-image relationship can best be presented by plotting z_{out}/f and $|M|$ as functions of z_{in}/f. These curves are shown as solid lines in Figures 2.27a and 2.27b.

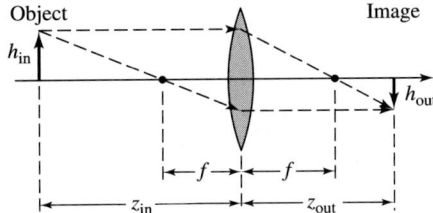

Figure 2.26 Object and image of a thin lens

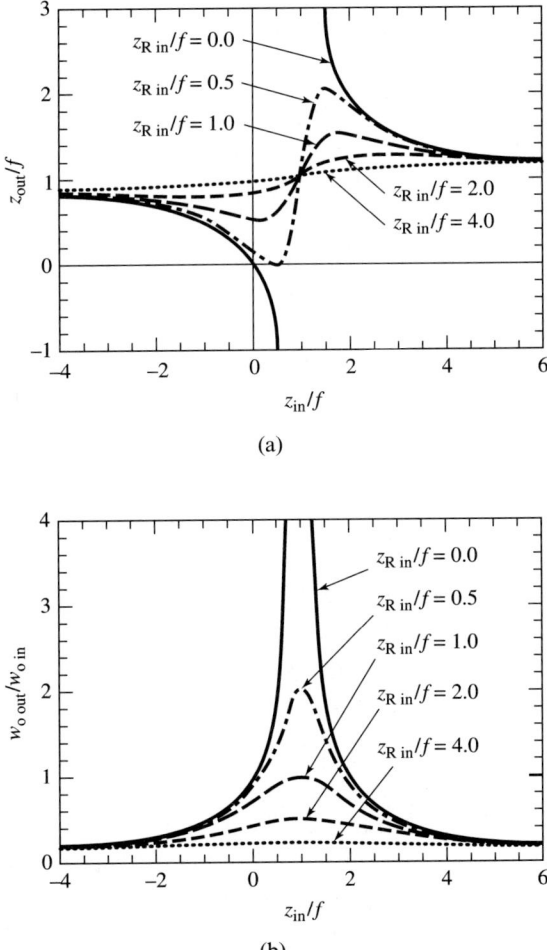

Figure 2.27 Position and beam radius of output beam waist as functions of input beam waist (after [24])

For Gaussian beams, the emphasis is on the transformation of the input beam waist $w_{o\,in}$ to the output beam waist $w_{o\,out}$. It is instructive to compare the relationship between the two beam waists with that between the object and the image [24, 25]. For this purpose, (2.105) and (2.106) are compared with (2.89) and (2.90). Clearly, if $z_{R\,in} = 0$, the object–to–image relationship is precisely the same as that between the input and output beam waists. To stress this point, the curves of z_{out}/f vs. z_{in}/f and $w_{o\,out}/w_{o\,in}$ vs. z_{in}/f are plotted in Figures 2.27a and 2.27b for five values of $z_{R\,in}/f$. If z_{in} is greater than f, so is z_{out}; if z_{in} is smaller than f, so is z_{out}. When $|z_{in}|$ is much greater than the focal length, $|z_{out}|$ approaches f. Therefore, collimated beams are focused to a point near the focal point. On the other hand, if the input beam waist is precisely located at the front focal point, and if $z_{R\,in} \neq 0$, the output beam waist is located at the back focal point. As $z_{R\,in}$ becomes smaller, the curves swing over a wider range. For $z_{R\,in} = 0$, $z_{out} \to \pm \infty$ as $z_{in} \to f$, which is exactly the object–image relationship given in (2.105).

The peak value of $w_{o\,out}/w_{o\,in}$ is $f/z_{R\,in}$ at $z_{in} = f$, as plotted in Figure 2.27b. The maximum value of $w_{o\,out}$ is obtained by setting $z_{in} = f$; then,

$$w_{o\,out} = \frac{f}{z_{R\,in}} w_{o\,in} = \frac{f\lambda}{w_{o\,in}\pi} = f\theta_{h\,in} \qquad (2.107)$$

For the special case of $z_{R\,in} = 0$, the peak value approaches ∞ and we have an infinite magnification, as indicated by (2.106).

A slowly diverging beam, that is, a beam for which $z_R \gg |f - z_{o\,in}|$ and hence $\theta_{h\,in} \ll 1$, is known as a **collimated beam**. For a collimated input beam, the output beam waist, from (2.90), is approximately

$$w_{o\,out} \approx \frac{f\lambda}{\pi w_{o\,in}} \qquad (2.108)$$

2.9 OPTICAL CAVITIES

Optical cavities were discussed briefly in section 2.3 as an application of ABCD matrices. In this section, optical cavities are studied from a different point of view. Our objective is to relate various beam parameters to various cavity parameters. The discussion focuses mainly on optical cavities for fundamental Gaussian modes.

As mentioned earlier, an optical cavity is formed by two mirrors. A cavity may be **stable** or **unstable,** depending upon the curvature of, and spacing between, the mirrors (Figure 2.28). A cavity is stable if beams are confined spatially in the cavity, despite repeated reflection by the mirrors. This means that the loss in stable cavities is low. Loss is minimized if the radii of curvature of the reflectors perfectly match the radii of curvature of the beam at the reflectors.

Suppose a fundamental Gaussian mode is known or given. Also, assume that a mirror with a radius of curvature matching the radius of curvature of the mode at that location is used to reflect the beam. The result is that the incoming

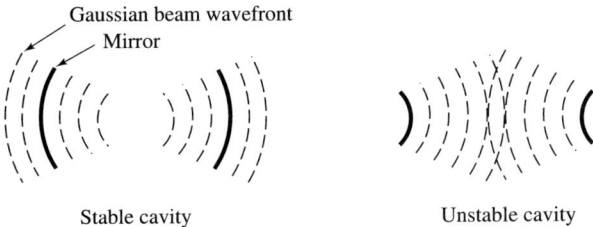

Figure 2.28 Stable and unstable optical cavities (Dashed curves correspond to the spherical wavefronts of Gaussian beams. Solid curves represent the mirrors.)

Gaussian beam is reflected without being converted to higher-order modes, and the reflection loss is small. If, on the other hand, the radii of curvature are not matched, a fraction of the power in the incoming mode is converted to a Gaussian beam with a different radius of curvature and higher-order modes. This constitutes loss for the incoming Gaussian mode. In other words, the Gaussian mode sustainable by an optical cavity is the mode for which the radii of curvature at the reflectors match those of the reflectors.

Stable cavities usually have two concave spherical mirrors, as shown schematically in Figure 2.29. Let d be the distance separating the two mirrors. Remembering that the radii of curvature R_{m1} and R_{m2} of concave mirrors are negative, we define $R'_{m1} = -R_{m1}$ and $R'_{m2} = -R_{m2}$. Suppose that the beam waist of the mode in question is located at $z = 0$, and that the reflectors 1 and 2 are located at $z = z_1$ and $z = z_2$, respectively. Hence, z_2, R'_{m1}, and R'_{m2} are positive, while z_1 is negative. We can treat either the beam radius at the waist w_o or the Rayleigh range $z_R = \pi w_o^2/\lambda$ as an unknown parameter. In terms of z_R, R'_{m1} and R'_{m2} may be written from (2.64) as

$$R'_{m1} = -\left(z_1 + \frac{z_R^2}{z_1}\right)$$

$$R'_{m2} = z_2 + \frac{z_R^2}{z_2}$$

which may be used to solve for z_1 and z_2, as follows:

$$z_1 = -\frac{1}{2}\left[R'_{m1} \mp \sqrt{R'^2_{m1} - 4z_R^2}\right]$$

$$z_2 = -\frac{1}{2}\left[R'_2 \pm \sqrt{R'^2_{m2} - 4z_R^2}\right]$$

An examination of Figure 2.15 shows that, for a given value of $R(z)$, there are two possible choices for z. These choices correspond to the plus and minus signs in the above equations. The plus signs correspond to mirrors located further away from the beam waist. The corresponding beam radii at the reflectors are also larger, according to Figure 2.13. This means that larger mirrors would

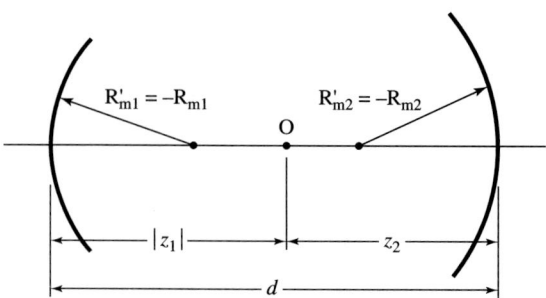

Figure 2.29 Optical cavity with two spherical reflectors

be required for the reflectors. The minus signs, corresponding to mirrors closer to the beam waist and thus requiring smaller reflectors, are more desirable. Choosing the minus signs, and referring to Figure 2.29, we obtain

$$2d = 2(z_2 - z_1) = R'_{m1} + R'_{m2} - \sqrt{R'^2_{m1} - 4z_R^2} - \sqrt{R'^2_{m2} - 4z_R^2}$$

which may be used to solve for z_R

$$z_R^2 = \frac{R'^2_{m1} R'^2_{m2} - [2d^2 - 2d(R'_{m1} + R'_{m2}) + R'_{m1} R'_{m2}]^2}{4[4d^2 - 4d(R'_{m1} + R'_{m2}) + (R'_{m1} + R'_{m2})^2]}$$

$$= -\frac{d(d - R'_{m1} - R'_{m2})(d - R'_{m1})(d - R'_{m2})}{(2d - R'_{m1} - R'_{m2})^2}$$

Introducing two **normalized cavity parameters** g_1 and g_2,

$$g_1 = 1 - \frac{d}{R'_{m1}} \qquad (2.109)$$

$$g_2 = 1 - \frac{d}{R'_{m2}} \qquad (2.110)$$

the term z_R^2 may be written as

$$z_R^2 = d^2 \frac{g_1 g_2 (1 - g_1 g_2)}{(g_1 + g_2 - 2 g_1 g_2)^2} \qquad (2.111)$$

Relative to the beam waist, the reflectors are located at

$$z_1 = -\frac{g_2 (1 - g_1)}{g_1 + g_2 - 2 g_1 g_2} d \qquad (2.112)$$

and

$$z_2 = +\frac{g_1 (1 - g_2)}{g_1 + g_2 - 2 g_1 g_2} d = z_1 + d \qquad (2.113)$$

Beam radii w_1 and w_2 at the reflectors are

$$w_1^2 = w_1^2(z_1) = \frac{d\lambda}{\pi}\left[\frac{g_2}{g_1(1-g_1g_2)}\right]^{\frac{1}{2}} \qquad (2.114)$$

and

$$w_2^2 = w_2^2(z_2) = \frac{d\lambda}{\pi}\left[\frac{g_1}{g_2(1-g_1g_2)}\right]^{\frac{1}{2}} \qquad (2.115)$$

Note that if $g_1g_2 < 0$ or $g_1g_2 > 1$, w_1 and w_2 become imaginary, which is obviously not acceptable nor physically realizable. Therefore, the **stability condition** for an optical cavity is $0 < g_1g_2 < 1$.

The stabilities of various optical cavities are shown in the stability diagram in Figure 2.11[2]. Of these, five cavity configurations deserve special discussion and are covered in the following subsections.

2.9.1 Symmetric Optical Cavities

If two reflectors are identical, that is, $R'_{m1} = R'_{m2}$ and $g_1 = g_2 = g$, we have a **symmetric cavity** (Figure 2.30a). For symmetric cavities, we have, from (2.112) and (2.113),

$$-z_1 = z_2 = \frac{d}{2} \qquad (2.116)$$

Thus, the beam waist is precisely midway between the two mirrors. This is expected, from a simple physical consideration. The beam radii at the mirrors are also the same:

$$w^2 = w_1^2 = w_2^2 = \frac{d\lambda}{\pi}\sqrt{\frac{1}{1-g^2}} \qquad (2.117)$$

For this case, the beam waist w_o and the Rayleigh range z_R, in terms of g, are

$$w_o = \sqrt{\frac{d\lambda}{2\pi}}\left[\frac{1+g}{1-g}\right]^{1/4} \qquad (2.118)$$

$$z_R = \frac{d}{2}\sqrt{\frac{1+g}{1-g}} \qquad (2.119)$$

2.9.2 Symmetric Confocal Resonators

In section 2.2, it was noted that a spherical mirror could be represented by an equivalent lens. For each lens, there is a front and a back focal point. If the front focal point of one equivalent lens coincides with the back focal point of another such lens, the cavity is known as a **confocal resonator**. Cavities with $R'_{m1} = R'_{m2} = d$ are symmetrical, as well as confocal. Thus, they are known as **symmetric confocal resonators**. Note that the center of one spherical

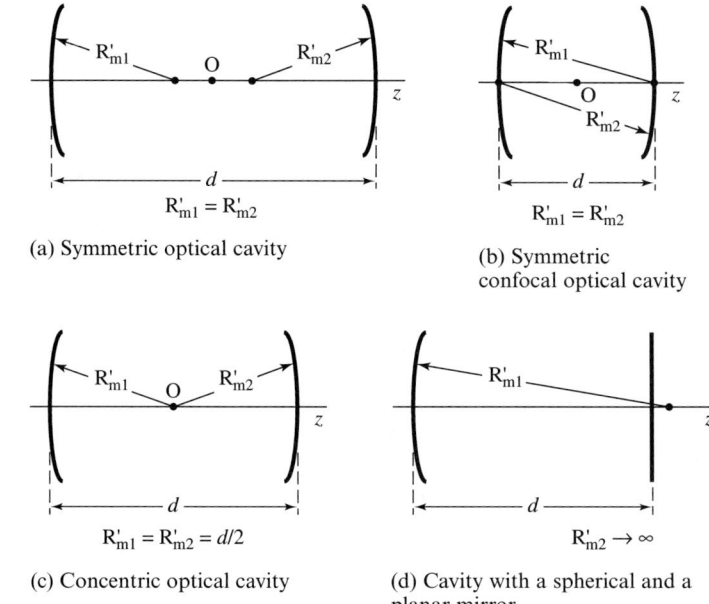

Figure 2.30 Four optical cavities

mirror is located precisely at the other mirror and vice versa, while the focal point is at the center O of the cavity (Figure 2.30b). Hence, the focal points of the two equivalent lenses coincide. For symmetric confocal resonators, $g_1 = g_2 = 0$. On the stability diagram (Figure 2.11), symmetric confocal cavities correspond to the origin of the diagram.

By setting $g = 0$, we have, from (2.117) and (2.118),

$$w = w_1 = w_2 = \sqrt{\frac{d\lambda}{\pi}} \qquad (2.120)$$

and the beam waist radius is

$$w_o = \sqrt{\frac{d\lambda}{2\pi}} \qquad (2.121)$$

The Rayleigh range z_R is

$$z_R = \frac{d}{2} = \frac{R'_{m1}}{2} = \frac{R'_{m2}}{2} \qquad (2.122)$$

Note that, for a given mirror separation d, the beam radii w_1 and w_2 at the mirrors of symmetric confocal cavities, as given by (2.120), are the *smallest* values realizable for any cavity with two spherical mirrors.

2.9.3 Parallel Planar Mirror Cavities

Optical cavities with two parallel planar mirrors are probably the simplest cavity type known. They are commonly known as the **Fabry–Perot cavities** or **interferometers**. While Fabry–Perot cavities are simple conceptually, they are difficult to align. For planar mirrors, R'_{m1} and $R'_{m2} \to \infty$, then $g_1 = g_2 = 1$ and w_1 and $w_2 \to \infty$. From section 2.5, we know that mirrors with radii equal to or larger than w_1 or w_2 are required to reflect a substantial portion of the beam power, say 86.5 percent or more. This is not possible, however, if w_1 and w_2 approach ∞, which is the case for Fabry–Perot cavities. Physically, this means that the diffraction loss of finite planar mirrors is large, and a precise alignment of the mirrors is required. Consequently, optical cavities with two planar mirrors are rarely used in gas lasers or dye lasers. Semiconductor injection lasers are the exception, because the laser cavities are quite short, which means that the diffraction losses are small despite the use of the planar mirrors. Furthermore, the optical gain in semiconductor junctions is quite large, easily compensating for the diffraction losses.

2.9.4 Concentric Resonators

For optical cavities defined by two concentric mirrors (Figure 2.30c), $R'_{m1} = R'_{m2} = d/2$, then $g_1 = g_2 = -1$, and w_1 and w_2 again become infinite. A completely enclosed reflecting sphere is needed to form a spherical resonator that confines the light beams inside the sphere.

2.9.5 Planar Mirror/Spherical Mirror Resonators

Resonators with a planar mirror and a spherical mirror (Figure 2.30d) are practical. If we let mirror 2 be a planar mirror, $R'_{m2} \to \infty$, then $g_2 = 1$. From (2.113), we obtain $z_2 = 0$. This means that the beam waist is located at the planar mirror. Suppose that the mirror separation d is slightly shorter than the radius of curvature of the spherical reflector (mirror 1), that is,

$$d = R'_{m1} - \varepsilon \tag{2.123}$$

and ε is smaller than d or R'_{m1}. Then

$$g_1 = \frac{\varepsilon}{R'_{m1}} \tag{2.124}$$

Using these values of g_1 and g_2, we have, from (2.114) and (2.115),

$$w_1^2 = \frac{\lambda}{\pi} \sqrt{\frac{R'^2_{m1} d}{R'_{m1} - d}} = \frac{\lambda}{\pi} \sqrt{\frac{R'^2_{m1} d}{\varepsilon}} \tag{2.125}$$

$$w_2^2 = w_o^2 = \frac{\lambda}{\pi} \sqrt{d(R'_{m1} - d)} = \frac{\lambda}{\pi} \sqrt{\varepsilon d} \tag{2.126}$$

Since ε is less than d, the beam radius at the planar mirror is small and the beam radius at the spherical mirror is large. By adjusting the distance d between the mirrors, we can vary the cavity loss and change the cavity from stable to unstable. In other words, we can use the distance d as a means to control the modes. This type of arrangement is used in many modern gas lasers.

2.9.6 Resonance Frequency: Spherical Mirror Cavities

In our study of the field distribution, beam divergence, spot size, and other characteristics of the Gaussian beam, we have assumed that R'_{m1}, R'_{m2}, d, and λ are known. In some applications, the emission wavelength λ is not known. To determine the value of λ, it is necessary to calculate the cavity resonant wavelength for a given set of R'_{m1}, R'_{m2}, and d.

Since the physical processes involved in Fabry–Perot cavities are relatively simple, we will use such a cavity with two planar mirrors to begin our discussion. Assume that the cavity length is d. The resonant wavelength must be such that the round-trip phase delay is an integer multiple of 2π; therefore, the one-way mirror-to-mirror phase delay is an integer multiple of π, that is, $kd = \ell\pi$, where ℓ is an integer called the **longitudinal mode number,** which indicates that the cavity is ℓ half-wavelengths long. In terms of the wavelength λ and frequency f, we have $\lambda_\ell = 2d/\ell$ and $f_\ell = \ell(c/2d)$. For most optical cavities, d is much longer than λ; therefore, ℓ is usually a huge number. Also, for a different ℓ, we have a different resonant frequency, which corresponds to a different longitudinal mode.

Extending this simple derivation to cavities with spherical mirrors, we must first evaluate the mirror-to-mirror phase delay. We could start from anywhere in the cavity, but a convenient starting point is the beam waist of the cavity mode. For a symmetric cavity, the beam waist is at the cavity center. Let the cavity center be at $z = 0$. At the mirror on the right, $z = d/2$, the phase delay relative to the cavity center is $kd/2 - \Psi(d/2)$, from (2.57) or (2.58). The phase difference between the other mirror and the center is the same. Thus, the total mirror-to-mirror phase delay is $kd - 2\Psi(d/2) = \ell\pi$, and the resonant frequency is

$$f_\ell = \frac{c}{2d}\left[\ell + \frac{2}{\pi}\Psi\left(\frac{d}{2}\right)\right] \qquad (2.127)$$

To relate the Gaussian beam parameters to the cavity parameters, we note that the phase term Ψ is given by

$$\tan\Psi\left(\frac{d}{2}\right) = \frac{\lambda d}{2\pi w_o^2}$$

For symmetric cavities ($R'_{m1} = R'_{m2} = R'_m$), we obtain the following from (2.55):

$$\frac{2R'_m}{d} = 1 + \left[\frac{2\pi w_o^2}{\lambda d}\right]^2$$

2.9 Optical Cavities

Thus, we have a simple relationship

$$\frac{2R'_m}{d} = 1 + \cot^2\Psi\left(\frac{d}{2}\right) = \csc^2\Psi\left(\frac{d}{2}\right)$$

When combined with (2.109) and (2.110), we obtain

$$g = 1 - \frac{d}{R'_m} = 1 - 2\sin^2\Psi\left(\frac{d}{2}\right) = \cos 2\Psi\left(\frac{d}{2}\right)$$

In other words,

$$2\Psi\left(\frac{d}{2}\right) = \cos^{-1}(1 - \frac{d}{R'_m}) = \cos^{-1}g$$

Substituting this result in (2.127), we have

$$f_\ell = \frac{c}{2d}[\ell + \frac{1}{\pi}\cos^{-1}g] \qquad (2.128)$$

This is the resonant frequency of the fundamental mode of a symmetric cavity. The fundamental mode is often labeled $TEM_{00\ell}$. Thus, f_ℓ in the above equation is commonly written $f_{00\ell}$.

For asymmetric cavities, the two mirrors are not identical. The normalized parameter g in (2.128) is replaced by $\sqrt{g_1 g_2}$ [2,18]. Thus, the resonant frequency is

$$f_{00\ell} = \frac{c}{2d}\left[\ell + \frac{1}{\pi}\cos^{-1}\sqrt{g_1 g_2}\right] \qquad (2.129)$$

For higher-order modes, there are two additional mode numbers, as noted in section 2.8. Expressed in rectangular coordinates, the **transverse mode numbers** are m and n, and the resonant frequency is

$$f_{mn\ell} = \frac{c}{2d}\left[\ell + \frac{m+n+1}{\pi}\cos^{-1}\sqrt{g_1 g_2}\right] \qquad (2.130)$$

In terms of cylindrical coordinates, the transverse mode numbers are p and q, and the resonant frequency is

$$f_{pq\ell} = \frac{c}{2d}\left[\ell + \frac{2p+q+1}{\pi}\cos^{-1}\sqrt{g_1 g_2}\right] \qquad (2.131)$$

The angular mode number is q, and it signifies the $\cos q\phi$ and $\sin q\phi$ type field variations. The radial mode number is p. In (2.130) and (2.131), m, n, p and q are integers.

The **longitudinal mode separation** between modes with the same transverse mode numbers is

$$\Delta f = f_{mn\ell+1} - f_{mn\ell} = \frac{c}{2d} \qquad (2.132)$$

The **transverse mode separation** between modes with the same longitudinal mode numbers is $\frac{c}{2d}\frac{1}{\pi}cos^{-1}\sqrt{g_1 g_2}$. Since the inverse cosine function has a value between 0 and $\pi/2$, $\frac{1}{\pi}cos^{-1}\sqrt{g_1 g_2}$ is smaller than $1/2$. In other words, the frequency separation of transverse modes is smaller than that of longitudinal modes.

REFERENCES

1. Born, M., and E. Wolf, *Principles of Optics.* 6th ed., Oxford, Pergamon Press, (1980).
2. Kogelnik, H.; and T. Li. "Laser beams and resonators." Proc. IEEE 54, (1966), pp. 1312–1329. Also: Kogelnik, H. "Propagation of laser beams." *Applied Optics and Optical Engineering.* VII, Chapter 6. ed. R. R. Shannon; and J. C. Wyant. New York, NY: Academic Press, 1979, pp. 136–190.
3. Brouwer, W. *Matrix Methods in Optical Instrument Design.* New York, NY: W. A. Benjamin, Inc., 1964.
4. Gerrard, A.; and J. M. Burch. *Introduction to Matrix Methods in Optics.* London, New York, NY: John Wiley & Sons, Inc., 1973.
5. Halbach, K. "Matrix representation of Gaussian optics." Am. J. Phys. 32, (1964), pp. 90–108.
6. Casperson, L. W. "Synthesis of Gaussian beam optical systems." Appl. Opt. 20, (1981), pp. 2243–2249.
7. Hayt, W. H., Jr. *Engineering Electromagnetics* 5th ed. New York, NY: McGraw-Hill Book Company, 1988.
8. Siegman, A. E. *Introduction to Lasers and Masers.* New York, NY: McGraw-Hill Book Company, 1971. Also: Siegman, A. E. *Lasers.* Mill Valley, CA: University Science Books, 1986.
9. Arnaud, J. A.; W. M. Hubbard; G. D. Mandeville; B. de la Claviere; E. A. Franke; and J. M. Franke. "Technique for measurement of Gaussian laser beam parameters." Appl. Opt. 10, (1971), pp. 2775–2776.
10. Suzaki, Y.; and A. Tachibana. "Measurement of the μm sized radius of Gaussian laser beam using the scanning knife edge." Appl. Opt. 14, (1975), pp. 2809–2810.
11. Tridimas, Y.; M. J. Lalor; and N. H. Woolly. "Beam waist location and measurement in dual-beam laser Doppler anemometer." J. Phys., E., Sci. Instrum. 11, (1978), pp. 203–206.
12. Khosrofian, J. M.; and B. A. Garetz. "Measurement of a Gaussian laser beam diameter through the direct inversion of knife-edge data." Appl. Opt. 22, (1983), pp. 3406–3410.
13. Yoshida, A.; and T. Asakura. "A simple technique for quickly measuring the spot size of Gaussian beams." Opt. Laser Technol. 8, (December 1976), pp. 273–274.
14. E. Stijns, "Measuring the spot size of a Gaussian beam with an oscillating wire." IEEE J. Quantum Electron. QE–16, (1980), pp. 1298–1299.
15. Dickson, L. D. "Ronchi ruling method for measuring Gaussian beam diameter." Opt. Eng. 18 (1979), pp. 70–75.

16. Karim, M. A.; A. A. S. Awwal; A. M. Nasiruddin; A. Basit; D. S. Vedak; C. C. Smith; and G. D. Miller. "Gaussian laser beam parameter measurement using sinusoidal and triangular rulings." Opt. Lett. 12, (1987), pp. 93–95.
17. Goubau, G.; and F. Schwering. "On the guided propagation of electromagnetic wave beams." IRE Trans. Antennas and Propagation, AP–10, (1961), pp. 248–256.
18. Milonni, P. W.; and J. H. Eberly. *Lasers.* New York, NY: John Wiley & Sons, 1988.
19. Goodman, J. W. *Introduction to Fourier Optics.* New York, NY: McGraw-Hill Book Company, 1968.
20. Kogelnik, H. "Matching of optical modes." Bell Sys. Tech. J. 42, (1964), pp. 335–336.
21. Kogelnik, H. "Imaging of optical modes—resonators with internal lenses." Bell Sys. Tech. J. 43, (1965), pp. 455–496.
22. Kogelnik, H. "Modes in optical resonators." In *Advances in Lasers,* ed. A. K. Levine. New York, NY: Dekker Publishers, 1965.
23. Nemoto, S. "Transformation of waist parameters of a Gaussian beam by a thick lens." Appl. Opt. 29, (1990), pp. 809–4446.
24. Self, S. A. "Focusing of spherical Gaussian beams." Appl. Opt. 22, (1983), pp. 658–661.
25. Herman, R. M.; J. Pardo; and T. A. Wiggins. "Diffraction and focusing of Gaussian beams." Appl. Opt. 24, (1985), pp. 1346–1354.

ADDITIONAL READING

1. Banerjee, P. P.; and T. C. Poon. *Principles of Applied Optics.* Homewood, IL and Boston, MA: R. D. Irwin, Inc. and Aksen Associates, Inc. (1991).
2. Guenther, B. D. *Modern Optics.* New York, NY: John Wiley & Sons, 1990.
3. Haus, H. A. *Waves and Fields in Optoelectronics.* Englewood Cliffs, NJ: Prentice Hall, Inc., 1984.
4. Hecht, E.; and A. Zajac. *Optics.* 2nd ed. Addison Wesley Publishing Co, 1987.
5. Jenkins, F. A.; and H. E. White. *Fundamentals of Optics.* 4th ed. New York, NY: McGraw-Hill Book Company, 1976.
6. Karim, M. A. *Electro-optical Devices and Systems.* Boston, MA: PWS–Kent Publishing Co., 1990.
7. Nussbaum, A.; and R. A. Phillips. *Contemporary Optics for Scientists and Engineers.* Englewood Cliffs, NJ: Prentice Hall, Inc., 1976.
8. Ramo, S.; J. R. Whinnery; and T. Van Duzer. *Fields and Waves in Communication Electronics.* 2nd ed. New York, NY: John Wiley & Sons, Inc., 1984.
9. Pedrotti, F. L.; and L. S. Pedrotti. *Introduction to Optics.* Englewood Cliffs, NJ: Prentice-Hall, Inc., 1992.
10. Saleh, B. E. A.; and M. C. Teich. *Fundamentals of Photonics.* New York, NY: John Wiley & Sons, Inc., 1991.
11. Siegman, A. E. *Lasers.* Mill Valley, CA: University Science Books, 1986.
12. Verdeyn, *Laser Electronics* 2nd ed. Englewood Cliffs, NJ: Prentice Hall, Inc., 1989.
13. Yariv, A. *Introduction to Optical Electronics* 3rd ed. New York, N.Y.: Holt, Rinehart & Winston, 1984.

PROBLEMS

1. A thin lens with a focal length of 10 cm is located 20 cm to the right of a glass plate. The glass plate has a thickness of 3 cm and an index of 1.5. Using the reference planes RP_{in} and RP_{out}, derive the ABCD matrix of the plate–lens combination.

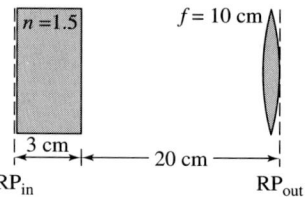

2. A glass hemisphere of radius R and index n is situated in air ($n = 1$). The reference planes RP_{in} and RP_{out} are located just outside the glass.
   a. Derive the ABCD matrix for the hemisphere.
   b. Calculate the focal length f of the hemisphere. f is the distance from RP_{out} to the focal point.

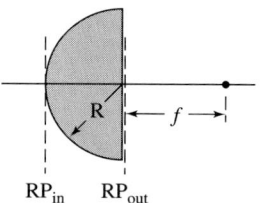

3. A glass rod with a spherical tip and an index of 1.5 is situated in water with $n \approx 1.3$. The spherical tip has a radius of 2.0 cm, and the total length of the rod is 5.0 cm.
   a. The reference planes RP_{in} and RP_{out} are just outside the rod. Derive the ABCD matrix for the rod.
   b. Calculate the focal length f of the glass rod in water. As shown in the figure, f is the distance from the flat surface of the rod.

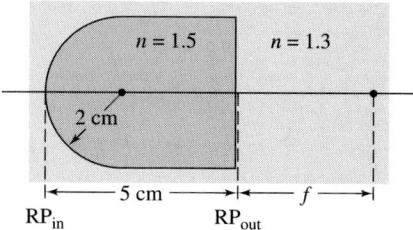

4. A glass hemisphere with an index of 1.5 is followed by a water-filled cylinder with an index of 1.3. As shown in the figure on the next page, reference planes RP_{in} and RP_{out} are outside the hemisphere and water cylinder, respectively.
   a. Derive the ABCD matrix of the hemisphere–water cylinder combination.
   b. Calculate the focal length f of the combination.

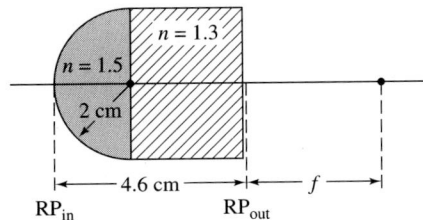

5. Derive the ABCD matrix for a thick biconcave lens as shown in the figure. Compare your results with equation (2.24).

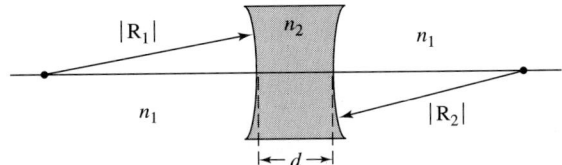

6. Consider rays incident upon a concave spherical surface separating two media with indices n_{ext} and n_{int}. Let $|R|$ be the radius of the spherical surface. Show that the ABCD matrix of the concave refractive spherical surface is

$$\begin{bmatrix} 1 & 0 \\ \dfrac{-1}{|R|}\dfrac{n_{ext}-n_{int}}{n_{int}} & \dfrac{n_{ext}}{n_{int}} \end{bmatrix}$$

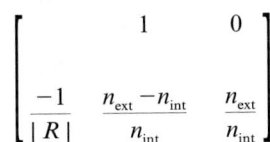

7. The figure (next page) shows a dielectric sphere of radius $|R|$ and index n situated in air. Show that the ABCD matrix is

$$\begin{bmatrix} \dfrac{2}{n}-1 & 2\dfrac{|R|}{n} \\ \dfrac{2(1-n)}{n|R|} & \dfrac{2}{n}-1 \end{bmatrix}$$

Also calculate the focal length of the sphere relative to RP_{out}.

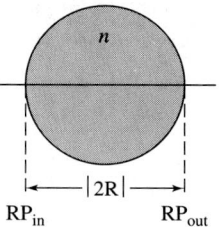

8. a. Consider a system consisting of three dielectric layers of thickness d_1, d_2, and d_3 and indices n_1, n_2, and n_3. Show that the ABCD matrix of the system is

$$\begin{bmatrix} 1 & \dfrac{d_1}{n_1} + \dfrac{d_2}{n_2} + \dfrac{d_3}{n_3} \\ 0 & 1 \end{bmatrix}$$

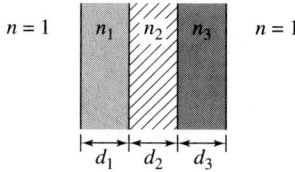

b. Generalize the result in (a) to a system with k layers (d_1, d_2 ... d_k, n_1, n_2, ... n_k). What is the ABCD matrix for the system with k layers?

c. As a further generalization, consider a nonhomogeneous dielectric layer where the index is a function of z, that is, $n(z)$. Let the overall thickness of the dielectric layer be d. What is the ABCD matrix for the nonhomogeneous dielectric layer?

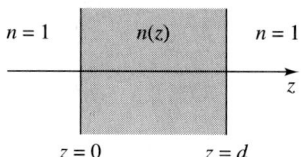

9. Given the ABCD matrix

$$\begin{bmatrix} 2 & 1 \\ 1 & 1 \end{bmatrix}$$

use factorization techniques to design the optical system and sketch the layout. Use the meter as the length unit. Is your design unique?

10. An optical system with the following ABCD matrix is desired:

$$\begin{bmatrix} \dfrac{1}{3} & 2 \\ -\dfrac{1}{2} & 0 \end{bmatrix}$$

Design the optical system and sketch the layout. Use the meter as the length unit. Is your design unique?

11. Consider a planar air/dielectric interface for which RP_{in} and RP_{out} are outside and inside the dielectric medium, respectively.
   a. State the ABCD matrix relating the ray vectors at RP_{out} to the ray vectors at RP_{in}.
   b. A Gaussian beam with a complex radius of curvature of q_{in} impinges upon the interface from the air side. Derive an expression for the complex radius of curvature q_{out} for the beam just inside the dielectric medium, which has an index of n. Show that the ABCD law is indeed satisfied for this case. (Note that waves with wavelength λ in air have a wavelength λ/n in the dielectric medium.)

12. Spatial filters consisting of a lens and a pinhole with a small aperture are often used to reduce higher-order modes. (See figure) Suppose the input is a Gaussian beam with wavelength λ that diverges with a half angle θ_h.
   a. The distance z_a between the lens and the beam waist of the incoming beam is larger than the focal length f of the lens. Where should the pinhole be placed if the power passing through the pinhole is to be maximized?
   b. Assuming that the pinhole of radius a is positioned precisely at the point where it is supposed to be, derive an expression for the fractional power passing through the pinhole as a function of z_a, λ, f, θ_h, and a.
   c. Assume $\lambda = 0.633$ μm, $\theta_h = 1.5$ mrad, and $z_a = 20$ cm. As the focal length changes from 2 mm to 40 mm, plot the fractional power passing through a pinhole with radius of 2.5 μm. Repeat for pinholes with radii of 5 and 10 μm.

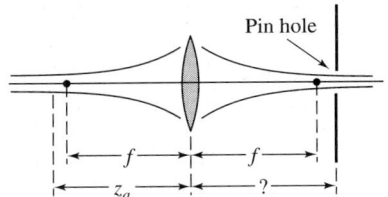

13. An opaque screen is used to intercept a Gaussian beam. At the screen, the beam radius is w. A ring-shaped aperture with an inner radius $a = w$ and an outer radius $b = 1.5\,w$ is cut onto the screen. Calculate the percentage of power transmitted through the aperture.

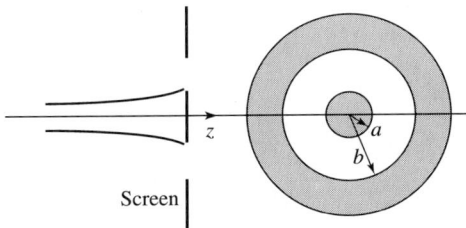

14. An opaque screen with a long, narrow slit of width d is used to block a Gaussian beam. At the screen the beam radius is w. Calculate the fractional power transmitted through the slit.

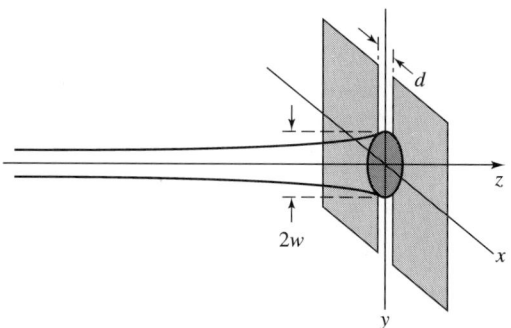

15. An opaque screen with a long, adjustable slit is used to measure the parameters of a Gaussian beam. In the measurement, the slit width d is adjusted until half of the beam power passes through the slit. Derive an expression relating the beam radius w to d.

16. A Gaussian beam with $\lambda = 0.543$ μm passes from air into a glass plate ($n \approx 1.5$), of thickness 30 cm, and back into air again (see figure below). At the left boundary, RP_{in}, the input Gaussian beam has a beam radius of 0.12 cm and a radius of curvature of 40 cm. Find the beam radius and the radius of curvature of the Gaussian beam emerging from the glass plate.

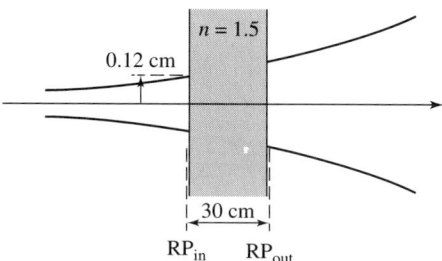

17. A Gaussian beam ($\lambda \approx 1.5708$ μm) with a radius of curvature of 8.0 m and a beam radius of 2 mm impinges upon the concave boundary of a dielectric material.
    a. Calculate the radius of the beam waist of the incident Gaussian beam.
    b. Calculate the location of the beam waist of the incident beam relative to the curved boundary of the dielectric.
    c. Calculate the complex radius of curvature of the beam just outside the dielectric and to the right of the straight boundary.

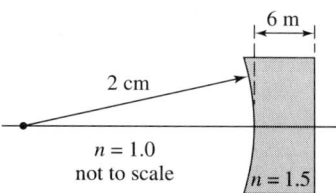

18. A Gaussian beam ($\lambda \approx 1.5708$ μm) with a beam waist radius of 0.25 mm is to be transformed to a beam with a beam waist radius of 0.10 mm. The separation of the

two beam waists is 12 cm. Determine the focal length f and the position x of the thin lens relative to the input beam waist.

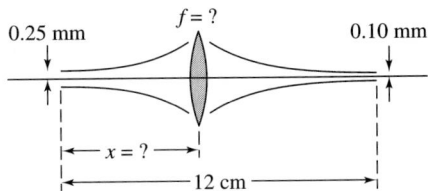

19. A lens of unknown radius a and focal length f is used to transform a Gaussian beam with a wavelength of 0.9 μm. The original beam has a waist radius of 0.5 mm and the thin lens is 2 m from the beam waist.
   a. What is the minimum lens radius a if 98 percent or more of the power is to be intercepted by the lens?
   b. What is the focal length f if the transformed beam has a waist radius of 1.25 mm?

20. An optical cavity has two spherical mirrors with radii of $R'_{m1} = R'_{m2} = 50$ cm, separated by a distance $d = 98$ cm. Assume that the optical output of the cavity is a fundamental Gaussian beam with 1 mW of power and a wavelength of 0.633 μm.
   a. What is the half-angle beam divergence in the far field region?
   b. What is the beam radius at a point 150 cm from the center of the cavity?
   c. What is the irradiance I of the beam at $r = 0$, $z = 150$ cm?

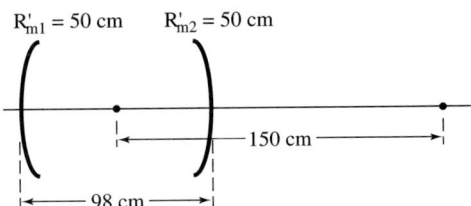

21. The figure below depicts a laser cavity ($d = 1$ m, $R'_{m1} = R'_{m2} = 2$ m) and an interferometer cavity ($d = 5$ cm, $R'_{m3} = R'_{m4} = 10$ cm). Mirrors 2 and 3 are 50 cm apart. A lens is used to match the output from the laser cavity to the interferometer cavity. What is the focal length f of the lens and where should the lens be placed? [Note: The answers are independent of λ. But if you need a numerical value for λ in your calculation, pick a reasonable value.]

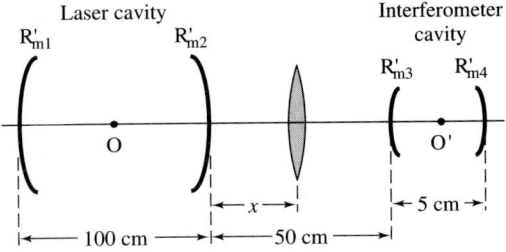

22. A laser cavity has a flat mirror and a spherical mirror with radius of curvature of 40 cm. The mirrors are separated by 20 cm. The flat mirror is partially transparent, while the spherical mirror is perfectly reflecting. An output of 2 mW at a wavelength of 0.543 μm (green) is emitted by the laser.
   a. Demonstrate that the cavity is stable.
   b. What is the half angle of the beam divergence (in radians)?
   c. An opaque screen with a circular hole of radius a is placed at a distance of 40 cm from the flat mirror. A total power of 1 mW passes through the hole. Calculate the radius a of the hole. Assume that the center of the hole coincides perfectly with the optical beam axis.

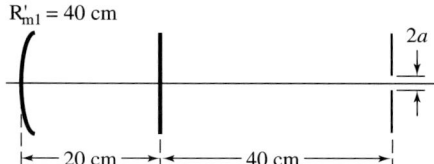

23. The figure below depicts a laser cavity with a flat mirror and a spherical mirror that has a radius of curvature $R'_m = 4.0$ cm. The spacing between the two mirrors is 2.0 cm. Waves with $\lambda = 1.05$ μm are emitted by the laser.
   a. What is the half angle of the beam divergence (in radians)?
   b. The diameters of the mirrors are 2.0 mm. Are the mirrors large enough? Why or why not?
   c. A huge water tank is placed in front of the laser cavity. What is the half angle of the beam divergence (in radians) in the water if the index of water is about 1.3?

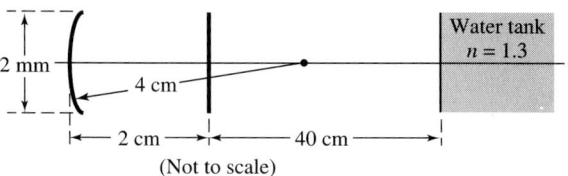

(Not to scale)

24. An optical cavity is formed by two identical spherical mirrors. The radius of curvature of the two mirrors and the separation between the mirrors is the same and is equal to 20 cm. A Gaussian beam with a wavelength of 1.5708 μm is emitted and propagates to the right. For convenience, choose the center of the cavity as $z = 0$.
   a. Determine the half angle of the beam divergence, in radians.
   b. Where is the beam waist relative to the center of the cavity?
   c. An opaque circular disk is located 70 cm from the nearest mirror and blocks 25 percent of the incident power. The center of the disk coincides perfectly with the optical beam axis. Calculate the radius of the disk.

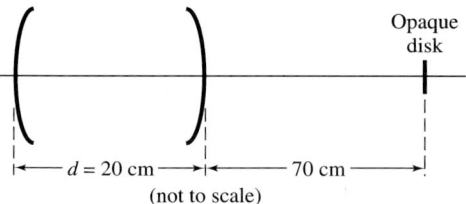

(not to scale)

25. A laser cavity has a perfectly reflecting spherical mirror and a partially transparent planar mirror, as shown in the figure. The emission wavelength is 0.9 μm.
   a. What is the waist radius, and where is the waist located relative to the flat mirror?
   b. A lens with a focal length of 10 cm is placed in front of the laser cavity. What is the waist radius of the transformed beam? Where is the beam waist relative to the lens?

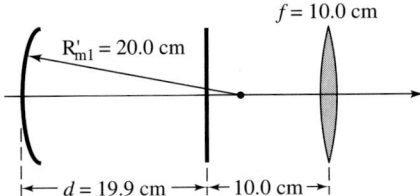

26. Consider a cavity with two identical, concave spherical mirrors, as shown in the figure. Light ($\lambda = 3.1416$ μm) emitted by the cavity diverges with a half angle of 2 mrad.
   a. If the mirror separation d is 4 cm, calculate the radii of curvature of the mirrors.
   b. A glass plate with an index of 1.5 and a thickness of 4 cm is placed in front of the cavity as shown. Calculate the complex radius of curvature of the beam as the beam enters the glass plate.
   c. Calculate the beam radius as the beam emerges from the glass plate.

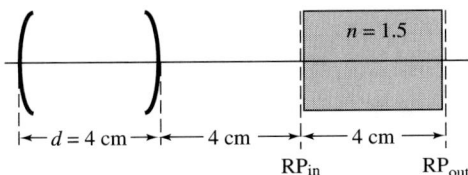

27. A laser cavity is formed by a flat mirror and a spherical mirror with a radius of curvature of 40 cm. The laser emission has a wavelength of 0.9 μm. In the far field, the optical beam diverges with a half angle of 1.3 mrad. Calculate the spacing d between the two mirrors.

28. A laser cavity is formed by a flat mirror and a spherical mirror that has a radius of curvature of 40 cm. The laser emission has a wavelength of 0.9 μm. An opaque screen with a circular hole of radius 1.5 mm is placed at a distance of 40 cm from the flat mirror, and the hole aligns perfectly with the optical beam axis. Measurements show that 50 percent of the laser output passes through the hole. Calculate the spacing between the two mirrors.

29. Starting from (2.85) and (2.86), derive (2.87) and (2.88).
30. Starting from (2.87) and (2.88), derive (2.89) and (2.90).
31. Starting from (2.87) and (2.88), derive (2.91) and (2.92).
32. Starting from (2.91) and (2.92), derive (2.93) and (2.94).

ELEMENTS OF LASERS

CHAPTER 3

CHAPTER OUTLINE

- 3.1 Introduction
    - 3.1.1 Units
    - 3.1.2 Energy Levels
- 3.2 Boltzmann's Distribution, Photons, and Blackbody Radiation
- 3.3 Lasers: A Qualitative Description
- 3.4 Absorption, Stimulated Emission, and Spontaneous Emission
    - 3.4.1 Absorption
    - 3.4.2 Stimulated Emission
    - 3.4.3 Spontaneous Emission
    - 3.4.4 Einstein Relations
    - 3.4.5 Wave Attenuation
    - 3.4.6 Two-Level System Rate Equation
- 3.5 Line Broadening
- 3.6 Rate Equations for Lasers with Optical Pumping
    - 3.6.1 Three-Level Systems with Optical Pumping
    - 3.6.2 Four-Level Systems with Optical Pumping
    - 3.6.3 System Comparisons
    - 3.6.4 Rate of Change of Photon Density
    - 3.6.5 Critical Pumping and Critical Population Inversion
    - 3.6.6 Steady-State Output Power
    - 3.6.7 Steady-State Gain of Three-Level Active Medium
- 3.7 Q-Switched Lasers
    - 3.7.1 Pulse Shape
    - 3.7.2 Peak Power
    - 3.7.3 Total Energy per Pulse
    - 3.7.4 Approximate Pulse Width
- 3.8 Mode Locking Operation
    - 3.8.1 Electrical Source Phase Locking
    - 3.8.2 Laser Mode Locking

3.9   Description of Gas Laser Systems
    3.9.1   Neutral Gas Lasers
    3.9.2   Ionized Gas Lasers
    3.9.3   Molecular Gas Lasers

3.10 Description of Dye Lasers

3.11 Description of Solid-State Lasers
    3.11.1 Energy Levels of Active Ions in Solid Host Materials
    3.11.2 Neodymium Lasers
    3.11.3 Ruby Lasers
    3.11.4 Alexandrite Lasers and Emerald Lasers

References

Additional Reading

Problems

PHYSICAL CONSTANTS

Boltzmann's constant	$k = 1.381 \times 10^{-23}$ J/°K $= 8.617 \times 10^{-5}$ eV/°K
Electron charge	$e = 1.6021 \times 10^{-19}$ C
Electron–volt	1 eV $= 1.6021 \times 10^{-12}$ erg $= 1.6021 \times 10^{-19}$ J
Planck's constant	$h = 6.6256 \times 10^{-34}$ J–s $= 4.1356 \times 10^{-15}$ eV–s
Speed of light in vacuum	$c = 2.9979 \times 10^{+8}$ m/s $\approx 3 \times 10^{+8}$ m/s

3.1 INTRODUCTION

For optical communications and signal and image processing applications, coherent light sources are often preferred, if not required. By a coherent light source, we mean a source which generates light with a constant frequency and a fixed phase relationship with respect to a certain reference point over a long duration. The first coherent light source was a ruby laser [1]. The word LASER stands for **l**ight **a**mplification by **s**timulated **e**mission of **r**adiation. It is derived from maser, which is an acronym for **m**icrowave **a**mplification by **s**timulated **e**mission of **r**adiation. The letter "a" in maser is appropriate because masers are used as low-noise amplifiers. While optically active media or devices may be used for light amplification, many optically active devices operate as light oscillators. However accurate it might be, it would not be attractive to replace the letter a in laser by the letter o.

    Amplifiers can be converted to oscillators by the inclusion of a **feedback mechanism**. Also, to sustain the oscillations, energy must be supplied to the system. The three basic components of an optical oscillator are:

- *An active medium* The energy needed to initiate and sustain laser action is stored in the **active medium,** which may be in the form of gases, liquids, glasses, crystalline solids, or semiconducting junctions.
- *A feedback mechanism* Optical feedback is provided by the optical cavity, which may be defined by two reflecting mirrors, as discussed in Chapter 2, by gratings which function as distributed feedback reflectors or by a recircu-

lating ring. The spectral purity, spatial distribution, and directionality of the emitted light are determined or controlled mainly by the optical cavity or recirculating ring.
- *Pumping* Energy is supplied to the laser by electrical, optical, chemical, or other means. This is called **pumping.**

One of the dominant factors determining the frequency (or equivalently, the free-space wavelength) of optical radiation is the energy difference between the energy levels involved, which may be either the discrete energy levels of isolated atoms or molecules, or energy bands, such as the conduction and valence bands of semiconductors. The notations used to describe the atomic energy levels are discussed briefly in this section. Detailed discussions are included in Appendix A.

In the next section, we will study the concepts and laws pertinent to the interaction of electromagnetic fields with materials. Qualitative and quantitative descriptions of such interactions are presented in sections 3 and 4. With these topics as the foundation, rate equations describing the operation of three- and four-level lasers are then introduced. The continuous-wave (cw), as well as the transient, operation of lasers can be understood in terms of such rate equations. The generation of optical pulses by Q-switching and mode locking is examined in sections 7 and 8. Finally, several types of gas, liquid, and solid-state lasers are described qualitatively in sections 9, 10, and 11. Semiconductor injection lasers, light-emitting diodes and superluminescent diodes are discussed in Chapter 4.

3.1.1 Units

Six fundamental units of the SI system are: *meter* for length, *kilogram* for mass, *second* for time, *kelvin degree* for temperature, *candela* for luminous intensity, and *ampere* for electric current (or *coulomb* for electric charge). In the field of lightwave technology, other units are also used, particularly for terms associated with basic device physics. For example, **angstroms** and **electron volts** are units for length and energy, respectively. Some of the most commonly used units and conversion factors are described in the following paragraphs.

The wavelength of electromagnetic waves is expressed in terms of micrometers (microns), nanometers, or angstroms (Å):

$$1 \ \mu m = 10^{-6} \ m = 10^{-4} \ cm = 10^{+3} \ nm = 10^{+4} \ \text{Å}$$
$$1 \ nm = 10^{-9} \ m = 10^{-7} \ cm = 10^{-3} \ \mu m = 10 \ \text{Å}$$
$$1 \ \text{Å} = 10^{-10} \ m = 10^{-8} \ cm = 10^{-4} \ \mu m = 10^{-1} \ nm$$

In the SI system, the unit for energy is joules (J). However, the photon energy and atomic energy levels relative to a zero energy reference are customarily expressed in terms of **electron volts** (eV). By definition, 1 eV is the energy required to change the potential of an electron by 1 volt. Numerically,

$$1 \text{ eV} = 1.6021 \times 10^{-12} \text{ erg} = 1.6021 \times 10^{-19} \text{ J}$$

The photon energy E is related to the frequency f or the vacuum wavelength λ of electromagnetic waves through a simple relation

$$E = hf = hc/\lambda$$

where $h = 6.6256 \times 10^{-34}$ J–s, or 4.1356×10^{-15} eV–s, is **Planck's constant,** and $c = 2.9979 \times 10^8$ m/s is the speed of light in vacuum. Therefore,

$$\lambda = \frac{hc}{E} = \frac{4.1356 \times 10^{-15} \text{ (eV-s)} \times 2.9979 \times 10^{+8} \text{ (m/s)}}{E_{\text{in eV}} \text{ (eV)}} \approx \frac{1.2398}{E_{\text{in eV}}} \mu\text{m} = \frac{1.2398}{E_{\text{in eV}}} \times 10^{-4} \text{ cm} = \frac{12398}{E_{\text{in eV}}} \text{ Å} \quad (3.2)$$

The wavelength can also be expressed in terms of a **wavenumber** ($1/\lambda$). A wavenumber, when expressed in units of cm^{-1}, is the number of wavelengths per cm. If λ is expressed in micrometers and the wavenumber is in cm^{-1}, then

$$[\lambda]_{\text{in } \mu\text{m}}[\text{wavenumber}]_{\text{in cm}^{-1}} = 10^4 \quad (3.3)$$

Energy can also be expressed in units of cm^{-1}. To convert energy in eV to energy in cm^{-1}, we note from (3.1) that $1/\lambda = E/(hc)$, and

$$\left[\frac{1}{\lambda}\right]_{\text{in cm}^{-1}} = \frac{E}{hc} = \frac{E_{\text{in eV}}}{4.1356 \times 10^{-15} \text{ (eV-s)} \times 2.9979 \times 10^{10} \text{ (cm/s)}} \approx 8066 \, E_{\text{in eV}} \quad (3.4)$$

Wavenumber is particularly useful in expressing the spectral width or line width of optical sources and components. The **spectral width** of light is typically defined in terms of two wavelengths, λ_1 and λ_2, or frequencies, f_1 and f_2, at the half-power points (Figure 3.1), and it is labeled as the **full width between half power points** (FWHP). The bandwidth of radio or microwave frequency signals is defined in the same manner. The frequency bandwidth $\Delta f = |f_2 - f_1|$ and spectral width $\Delta \lambda = |\lambda_2 - \lambda_1|$, or $\Delta(\frac{1}{\lambda}) = \left|\frac{1}{\lambda_1} - \frac{1}{\lambda_2}\right|$, are related. The relationship between Δf in Hz and $\Delta(\frac{1}{\lambda})$ in cm^{-1} is simple and exact:

$$\Delta f = |f_1 - f_2| = c \left|\frac{1}{\lambda_1} - \frac{1}{\lambda_2}\right| = c\Delta(\frac{1}{\lambda}) \quad (3.5)$$

Also shown in Figure 3.1 are the frequency f and wavelength λ where the frequency and spectral characteristics reach their maximum. For signals with a narrow line width, $\lambda_1 \approx \lambda_2 \approx \lambda$ and we obtain the following approximate relations:

$$\Delta f = c \left|\frac{1}{\lambda_1} - \frac{1}{\lambda_2}\right| = \frac{c|\lambda_2 - \lambda_1|}{\lambda_1 \lambda_2} \approx c\frac{\Delta \lambda}{\lambda^2} \quad (3.6)$$

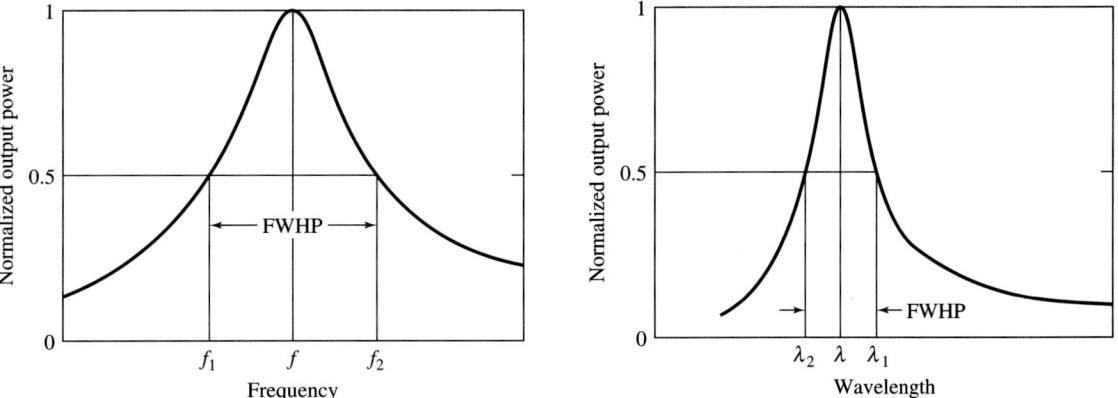

Figure 3.1 Definitions of the frequency bandwidth and spectral linewidth

$$\Delta(\frac{1}{\lambda}) = \left|\frac{1}{\lambda_1} - \frac{1}{\lambda_2}\right| = \frac{|\lambda_2 - \lambda_1|}{\lambda_1 \lambda_2} \approx \frac{\Delta\lambda}{\lambda^2} \qquad (3.7)$$

The relationships between energy in eV, frequency in Hz, energy or wavenumber in cm^{-1}, wavelength in nm, frequency bandwidth in Hz, and spectral width in cm^{-1} or Å, are summarized in the nomograph in Figure 3.2.

3.1.2 Energy Levels

The wavelength of a laser emission is determined primarily by the energy difference between the levels or bands involved in the laser transitions and the resonant wavelengths of the optical cavity or recirculating ring providing the optical feedback. The energy levels or bands involved may, in the case of neutral and ionized gas lasers, be the discrete levels of isolated atoms or ions, or, in the case of molecular gas lasers and liquid dye lasers, the vibrational and rotational states of molecules. Lasing in semiconductor injection lasers occurs between the conduction and valence bands of the semiconducting crystals. For solid-state lasers based on trivalent rare earth or actinide ions, the levels are essentially those of the isolated trivalent ions. On the other hand, for solid-state lasers based on trivalent transition metal ions, the levels are strongly influenced by the fields of the host crystal.

Since different lasers have different energy levels, it is important to identify the specific energy levels involved. They are identified in terms of spectroscopic notations, which are discussed in detail in Appendix A. For example, the states of isolated hydrogen atoms, which only have one electron, are specified by five quantum numbers: the principal quantum number, the orbital angular momentum quantum number, the magnetic quantum number, the electron spin quantum number, and the electron spin angular momentum quantum number [2–6]. However, the states of more complex many-electron atoms are specified by the

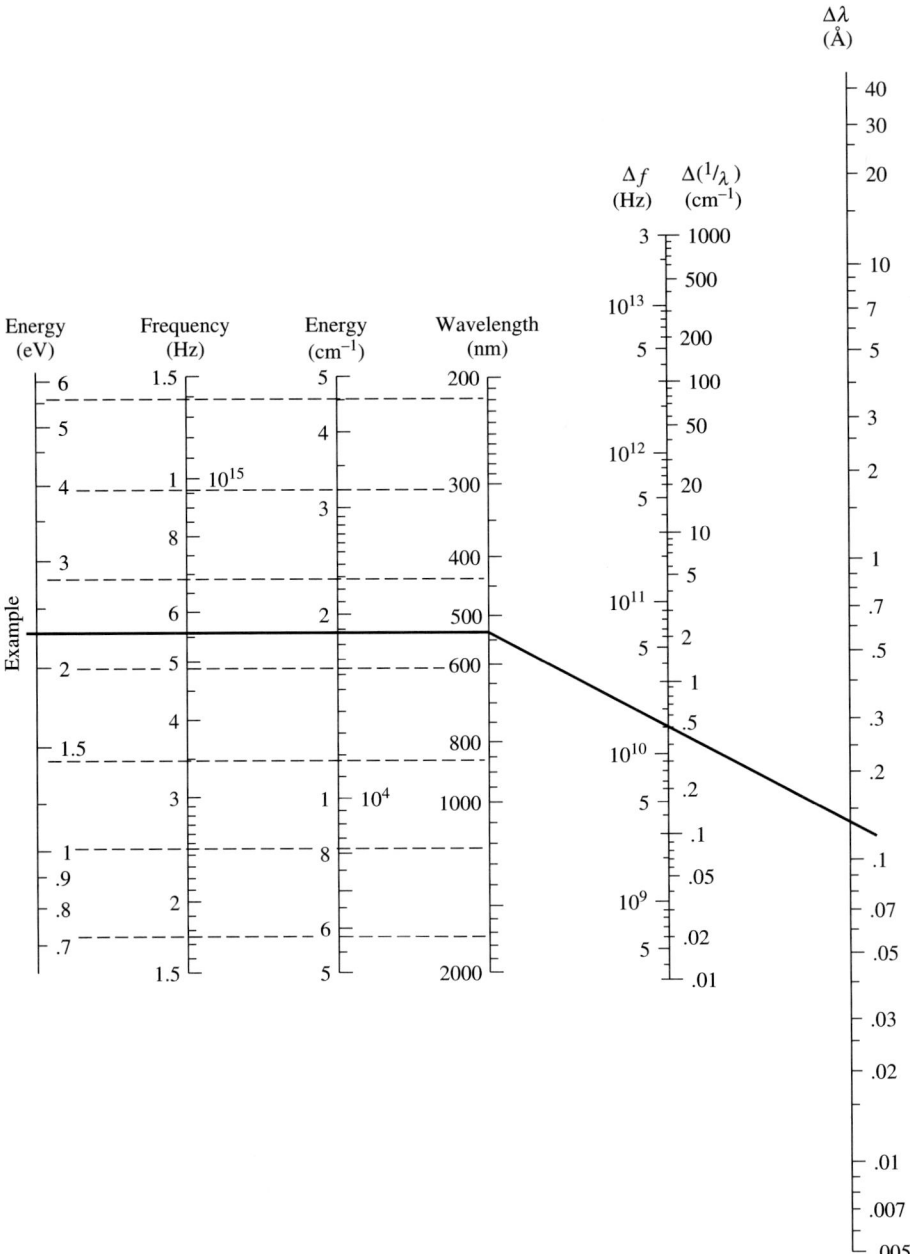

Figure 3.2 Energy, frequency, and wavelength nomograph (*Laser Focus Buyer's Guide,* 1980, © Penn Well Publishing Co., used by permission.)

This nomograph provides fast, convenient conversions among the various ways of describing light waves in energy, frequency and wavelength.

As an example, the left side of the nomograph shows that the 532-nanometer wavelength of a frequency-doubled neodymium-yag laser corresponds to an optical carrier frequency of 5.6×10^{14} hertz, and a photon energy of 2.3 electronvolts or 18,800 wavenumbers (cm^{-1}).

The right side of the nomograph shows that the laser's spectral width of 0.5 cm^{-1} is equivalent to a width of 15 gigahertz or 0.14 angstrom.

electron configuration, indicating the number of electrons in each shell, and a term symbol, signifying the vector combination of the orbital and spin angular momenta of these electrons. In all cases, different states have different energy levels. The state having the lowest energy level is referred to as the ground state of the atoms or molecules. For example, the ground state of Ne atoms is labeled $1s^2 2s^2 2p^6\ {}^1S_0$. The electron configuration $1s^2 2s^2 2p^6$ indicates the presence of two electrons in the 1s shell, two electrons in the 2s shell, and six electrons in the 2p shell; ${}^1S_0$ is the term symbol. Further discussion on spectroscopic notations is given in Appendix A.

3.2 BOLTZMANN'S DISTRIBUTION, PHOTONS, AND BLACKBODY RADIATION

Next, we consider the *probability* that any given state is occupied under a given condition. Consider a system with two energy levels, E_1 and E_2. Let the densities of occupied states, that is, the number of occupied states per unit volume, at these two energy levels be N_1 and N_2. The unit for N_1 and N_2 is m^{-3}. In **thermal equilibrium** at a temperature T, the ratio N_2/N_1 is

$$\frac{N_2}{N_1} = e^{-(E_2-E_1)/kT} \tag{3.8}$$

where $k = 1.381 \times 10^{-23}$ J/°K $= 8.617 \times 10^{-5}$ eV/°K is **Boltzmann's constant.** Equation (3.8) is the **Boltzmann equation** describing the distribution of occupied states in thermal equilibrium [2–5].

In discussions of the interaction of electromagnetic fields with materials, it is convenient to treat electromagnetic fields as *waves,* as well as *particles.* In turn, electromagnetic waves of frequency f can be viewed as **photons** or **light quanta** with an energy $E = hf$ and a momentum

$$k = 2\pi/\lambda = 2\pi f/c \tag{3.9}$$

Although material particles, such as electrons, protons, and ions, move with a velocity much slower than the speed of light, photons always "move" with a velocity c. The density of photon states in the frequency range between f and $f + \delta f$ is $8\pi f^2 \delta f/c^3$. Also, not all photon states are occupied. The probability that photon states are occupied is $[e^{hf/kT} - 1]^{-1}$. Thus, the photon density $\Phi(f)\delta f$ in the frequency range between f and $f + \delta f$ is

$$\Phi(f)\,\delta f = \frac{8\pi f^2 \delta f}{c^3} \frac{1}{e^{hf/kT} - 1} \tag{3.10}$$

The term $\Phi(f)\delta f$ is usually written simply as $\Phi \delta f$ and the unit for this term is m^{-3}. Taking the photon energy into account, we have **Planck's equation for blackbody radiation** [2–5]:

$$\rho(f)\delta f = \frac{8\pi}{c^3} \frac{hf^3\,\delta f}{e^{hf/kT} - 1} \tag{3.11}$$

where $\rho(f)\delta f$ is the energy density of the electromagnetic radiation in the frequency range between f and $f + \delta f$.

3.3 LASERS: A QUALITATIVE DESCRIPTION

Before we attempt a quantitative discussion on lasers it is instructive to have a qualitative picture of the physical processes involved. Lasers can be classified as **three-level** or **four-level** systems. In a three-level laser system, the energy levels are labeled **ground**, **metastable** and **excited**, **levels** or simply E_1, E_2, and E_3. For simplicity, we consider level 3 to be sharply defined, even though it is often a broad energy band, as depicted in Figure 3.3. The densities of occupied states in these levels are N_1, N_2, and N_3, respectively. In thermal equilibrium, $N_1 > N_2 > N_3$. In such a system, the sequence of events leading to lasing by optical pumping is as follows:

1. Intense radiation from an external light source with $f_{31} \approx (E_3 - E_1)/h$ illuminates the active medium. Electrons in the ground level E_1, after absorbing energy supplied by an external source, are raised to the upper level E_3. Photons are absorbed in the transitions. This process is the optical pumping. The external source may be a broad-band incoherent light emitter or a narrow-band coherent light source with a shorter wavelength. For efficient absorbtion of incoherent light, E_3 should be a broad energy band. If a coherent light source is used, the frequency of the pumping light must match f_{31}.

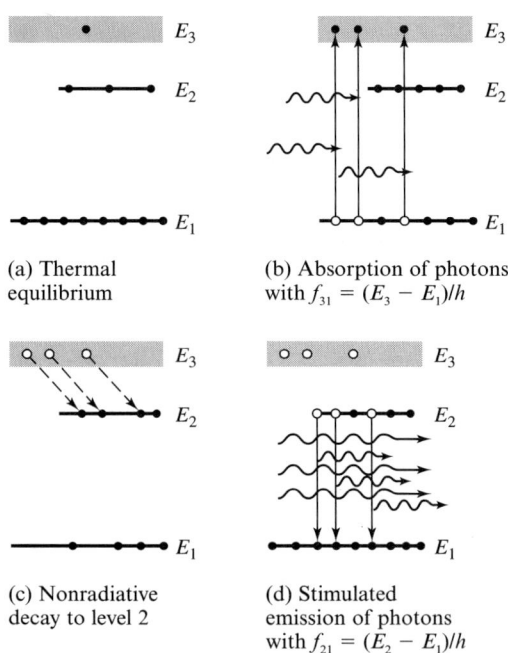

(a) Thermal equilibrium

(b) Absorption of photons with $f_{31} = (E_3 - E_1)/h$

(c) Nonradiative decay to level 2

(d) Stimulated emission of photons with $f_{21} = (E_2 - E_1)/h$

Figure 3.3 Emission processes in a three-level laser

2. Ideally, the time spent by the electrons in the excited level (i.e., the lifetime at the excited level) is relatively short. After being pumped to E_3, the electrons quickly decay either to the metastable or ground level. In an efficient lasing medium, a large number of electrons drop from E_3 to E_2 and a small fraction return to E_1 directly. If photons are radiated in a transition process, the process is a **radiative transition.** Transitions in which the energy released is in forms other than electromagnetic waves are known as **nonradiative transitions.**

3. Also ideally, the electron lifetime in level E_2 is relatively long. The population in E_2 builds up quickly as a result of electrons being pumped from the ground level to the excited level, as well as subsequent nonradiative decay from E_3 to E_2. If pumping is sufficiently strong, N_2 may become larger than N_1. When this occurs, it means that the population distributions in E_1 and E_2 have undergone **inversion,** and the pumped medium becomes an *active, gain,* or *amplifying medium.*

4. Some of the electrons in E_2 may drop to E_1 without external perturbation. In the process, photons are emitted. However, such emissions occur randomly, and the emitted photons are not correlated. In other words, incoherent waves are radiated. This photon emission process is known as **spontaneous emission.** The photons thus generated may be viewed as noise. However, it is important to note that this spontaneous emission process provides the "first few" photons that initiate the laser oscillation in the laser cavity.

5. Often, photons with frequency $f_{21} = (E_2 - E_1)/h$ exist in the cavity. These photons may be due to noise in the system, or may have been emitted by prior lasing transitions and been trapped by the cavity mirrors. In any event, if a medium is disturbed by perturbing photons of frequency f_{21}, electrons in level E_2 decay to E_1. In the process, additional photons with the same frequency f_{21} as the perturbing photons are emitted. This is the **stimulated emission process.** The newly emitted photons are *in time phase* with the perturbing photons and are radiated in a *confined spatial direction,* that is, as the laser output.

In four-level lasers, there is a level E_a in addition to E_1, E_2, and E_3. The energy diagrams of three- and four-level optical materials are shown in Figures 3.4a and b for comparison. In a three-level system, lasing takes place between levels E_2 and E_1. In a four-level system, the laser action occurs between levels E_2 and E_a. The main difference between three- and four-level lasers is the energy difference $E_a - E_1$, rather than the absence or the presence of level E_a. If $E_a - E_1 \gg kT$, then level E_a is sparsely populated, lasing can be achieved efficiently, and the system is considered a four-level system. However, if $E_a - E_1$ is comparable to or less than kT, the system is a three-level one, despite the presence of E_a.

Historically, the first optical source made to emit coherently was a three-level ruby laser. Today, a large number of lasers currently in use are four-level systems. Examples of four-level sources are: HeNe lasers emitting at 0.6328 μm, Nd:YAG lasers emitting at 1.06 μm, and semiconductor injection laser diodes. There are also active media which, with one set of cavity mirrors, can operate

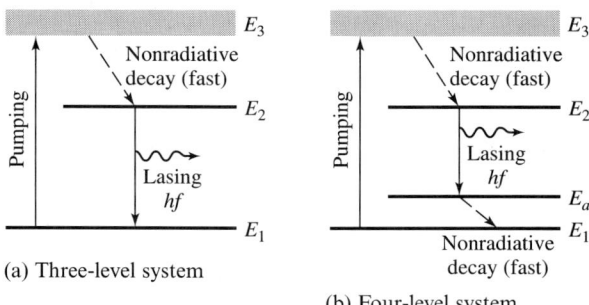

Figure 3.4 Energy diagrams of three- and four-level systems

as a three-level system or, with another set of mirrors, can operate as a four-level system. Of course, the resulting lasers have different wavelengths and emission characteristics. A good example is the injection laser diode-pumped Nd:YAG laser. (The term Nd:YAG will be defined in section 3.11).

3.4 ABSORPTION, STIMULATED EMISSION, AND SPONTANEOUS EMISSION

To understand three- and four-level laser systems, it is necessary to characterize the wave-material interaction quantitatively [7–12]. Although there are several levels in an active medium, each transition involves only two levels; therefore, we will focus our attention on levels E_1 and E_2. Let the density of occupied states in these levels be N_1 and N_2, respectively.

3.4.1 Absorption

In the presence of electromagnetic fields of frequency $f = (E_2 - E_1)/h$, there is a finite probability that electrons in the lower level will absorb photons and jump to the upper level. The time rate of change of N_1 is proportional to N_1, and the photon density $\Phi\delta f$ in the frequency range of f to $f + \delta f$. If we let the proportionality constant be B_{12}, then the time rate of change of N_1 may be written as

$$\left.\frac{dN_1}{dt}\right|_{ab} = -B_{12}(\Phi\delta f)N_1 \qquad (3.12)$$

The terms B_{12} in (3.12) and B_{21}, to be introduced in (3.13), are **Einstein's B-coefficients** and they have a dimension of m³/s. Equation (3.12) simply states that, in time interval dt, dN_1 states are raised from E_1 to E_2. In the process, dN_1 photons are absorbed by the medium. The minus sign in (3.12) indicates the decrease of the original population N_1 in E_1 by the amount dN_1. The absorption process is an *upward process,* as depicted in Figure 3.5a. The bandwidth δf in

(3.12) is written explicitly to emphasize the fact that the spectral width of photon absorption is finite. Unless indicated otherwise, a bandwidth of 1 Hz is assumed.

3.4.2 Stimulated Emission

Upon illumination by incoming photons of frequency $f = (E_2 - E_1)/h$, electrons in the upper level may emit additional photons with the same frequency and may drop to the lower level. In this case, the probability of transition is proportional to N_2 and $\Phi \delta f$

$$\left.\frac{dN_2}{dt}\right|_{st} = -B_{21}(\Phi \delta f)N_2 \qquad (3.13)$$

where B_{21} is a proportionality constant. The result is stimulated (or induced) emission, which is dependent upon the presence of photons of frequency $f = (E_2 - E_1)/h$. The newly emitted photons are in time phase with the existing photons and are radiated in a highly confined direction. These properties are distinct from the spontaneous emission to be discussed in the next subsection. As indicated in Figure 3.5b, stimulated emission is a *downward process*.

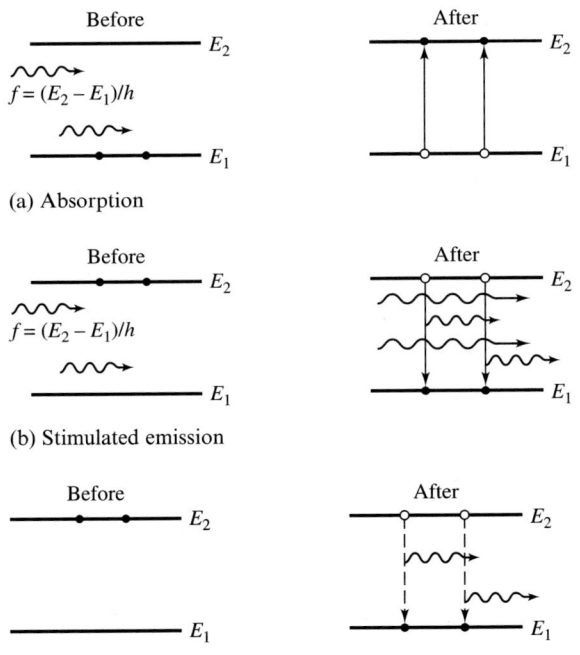

(a) Absorption

(b) Stimulated emission

(c) Spontaneous emission

Figure 3.5 Absorption, stimulated emission, and spontaneous emission of photons

3.4.3 Spontaneous Emission

There is always a finite probability that electrons in an upper level may drop to a lower level without provocation. The equation describing this process is

$$\left.\frac{dN_2}{dt}\right|_{sp} = -AN_2 \qquad (3.14)$$

where A is the proportionality constant and is **Einstein's A-coefficient** for *spontaneous emission* (Figure 3.5c), which has a dimension of s^{-1}. Such a *downward process* occurs randomly and does not require the presence of disturbing photons. The newly released photons are not in time phase with other photons (since there are no other photons with which to be in phase), and they are incoherent and lack any directionality.

3.4.4 Einstein Relations

Detailed quantum mechanics considerations show that B_{12} and B_{21} are determined primarily by the electric dipole moments of the interacting atoms or molecules [10–11]. While such detailed calculations *for these coefficients* are beyond the scope of our discussion, simple relationships between the two coefficients can be obtained. To establish these relationships, we use the Boltzmann equation (3.8) for the distribution of density of occupied states, and Planck's equation for blackbody radiation (3.10) for the distribution of the photon density in thermal equilibrium. Consider the steady state of an *isolated, nondegenerate* two-level system in thermal equilibrium. As there are no other levels in the system, the increase of population in the lower level must correspond to the population decrease in the upper level, and *vice versa*. Thus, the net rate of change of N_1 is

$$\frac{dN_1}{dt} = \left.\frac{dN_1}{dt}\right|_{ab} - \left.\frac{dN_2}{dt}\right|_{st} - \left.\frac{dN_2}{dt}\right|_{sp} = -B_{12}\Phi\delta f N_1 + B_{21}\Phi\delta f N_2 + AN_2$$

In the steady state, $dN_1/dt = 0$ and we have

$$\Phi\delta f = \frac{AN_2}{B_{12}N_1 - B_{21}N_2} = \frac{\dfrac{A}{B_{21}}}{\dfrac{B_{12}}{B_{21}}\dfrac{N_1}{N_2} - 1}$$

Making use of (3.8), we obtain

$$\Phi\delta f = \frac{\dfrac{A}{B_{21}}}{\dfrac{B_{12}}{B_{21}}e^{(E_2 - E_1)/kT} - 1}$$

Comparing this equation with (3.10), we conclude that the two Einstein coefficients are the same, and we can therefore label them simply as B

$$B_{12} = B_{21} = B \tag{3.15}$$

In addition, the ratio A/B is given by

$$\frac{A}{B} = \frac{8\pi f^2 \delta f}{c^3} = \frac{8\pi \delta f}{\lambda^3 f} = \frac{8\pi \delta f}{\lambda^2 c} \tag{3.16}$$

These relationships are known as the **Einstein relations**.

For degenerate systems, with degeneracy factors of g_1 and g_2 in E_1 and E_2, respectively, (3.15) should be modified as follows:

$$g_1 B_{12} = g_2 B_{21} \tag{3.17}$$

Equation (3.15) means that, for nondegenerate systems, stimulated emission and absorption processes are equally probable if $N_1 = N_2$. Actually, N_1 is much greater than N_2, in thermal equilibrium. Thus, more electrons are involved in the upward processes than the downward processes. That is, the incoming waves are attenuated by the two-level system.

As noted in (3.16), the Einstein A-coefficient increases with frequency for a given Einstein B-coefficient. The conclusion is that spontaneous emission is important in optical devices that operate at terahertz frequencies, but is relatively unimportant in low-frequency devices.

3.4.5 Wave Attenuation

The Einstein coefficients can be related to experimentally measurable quantities. Since the attenuation of waves in a medium is readily measurable, we can establish a relationship between B and the power attenuation constant. In a medium with an index of refraction n, photons "move" with a velocity c/n. The photon flux intensity S is

$$S = \frac{c}{n} \Phi \delta f \tag{3.18}$$

which is proportional to the power density of the electromagnetic waves. If the power attenuation constant is α, then in the presence of attenuation, S decays as a function of z, and the decay of $S(z)$ can be written as

$$S(z) = S(0) e^{-\alpha z} \tag{3.19}$$

where $S(0)$ is the photon flux intensity at $z = 0$. As a result of stimulated emission and absorption processes, the net change in the photon density in the time interval Δt is

$$\Delta(\Phi \delta f) \approx [B(N_2 - N_1)\Phi \delta f]\Delta t.$$

By letting Δt approach zero, we obtain a differential equation

$$\frac{d(\Phi \delta f)}{dt} = B(N_2 - N_1)\Phi \delta f \tag{3.20}$$

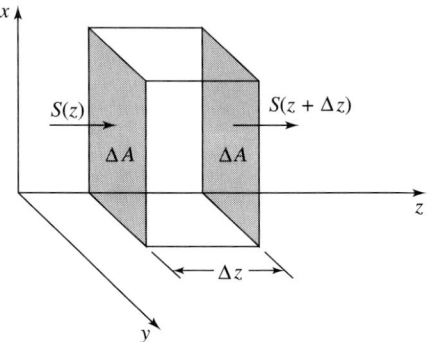

Figure 3.6 Attenuation of waves in a material medium

In this process, contributions due to spontaneous emission are ignored, since spontaneously emitted photons are not in time phase with the original photon flux, and they "move" in random directions.

Consider an elementary volume with a cross section ΔA and length Δz, as shown in Figure 3.6. The number of photons entering and leaving the elementary volume are $S(z)\Delta A$ and $S(z + \Delta z)\Delta A$, respectively. The change in the photon density in the elementary volume is

$$\frac{S(z + \Delta z)\Delta A - S(z)\Delta A}{\Delta A \Delta z} = \frac{S(z + \Delta z) - S(z)}{\Delta z}$$

In the limit as $\Delta z \to 0$, the ratio approaches $\frac{dS}{dz}$. Making use of (3.18) and (3.20), we can rewrite the above expression as

$$\frac{dS}{dz} = \frac{dS}{dt}\frac{dt}{dz} = \frac{n}{c}\frac{dS}{dt} = \frac{n}{c}\frac{d}{dt}(\frac{c}{n}\Phi\delta f) = -\frac{n}{c}B(N_1 - N_2)S$$

which may be solved to yield

$$S(z) = S(0)\, e^{-[B(N_1 - N_2)n/c]z} \tag{3.21}$$

A relation between α and B is readily obtained when (3.21) is compared with (3.19):

$$\alpha = B(N_1 - N_2)\frac{n}{c} \tag{3.22}$$

As noted previously, α, N_1, and N_2 are experimentally measurable quantities. By measuring α, N_1, and N_2, we can calculate the B-coefficient. With the help of the Einstein relations, the A-coefficient can also be determined.

Under normal circumstances, $N_1 > N_2$, α is positive, and the medium is lossy. However, if the density of occupied states in level E_2 is increased by some means to the extent that $N_2 > N_1$, loss in the medium becomes negative. The photon flux increases rather than decreases as waves propagate in the medium, and the medium becomes an amplifier. The gain coefficient g is

$$g = -\alpha = B(N_2 - N_1)\frac{n}{c} \qquad (3.23)$$

The condition of $N_2 > N_1$ is known as **population inversion.**

3.4.6 Two-Level System Rate Equation

According to the Boltzmann equation (3.8), the distribution of occupied states in thermal equilibrium is $N_2 < N_1$ if $E_2 > E_1$. It is informative to determine if N_2 can be made larger than N_1 by optical pumping. For this purpose, we consider a nondegenerate, isolated two-level system illuminated by an external light source that maintains a constant photon density $\Phi\delta f$ in the region. The system is isolated from "the rest of the world," except for the illumination of light. Let $N_1(t)$ and $N_2(t)$ be the density of occupied states in levels E_1 and E_2, respectively, at time t. The total density of occupied states $N_1(t) + N_2(t)$ is a constant N_t. For the two-level system, the population inversion is $\Delta N(t) = N_2(t) - N_1(t)$.

To study the changes in $N_1(t)$ and $N_2(t)$, we note that $N_2(t)$ may be decreased due to spontaneous and stimulated emission, and increased by the absorption of photons. The net rate of change of $N_2(t)$ is

$$\frac{dN_2(t)}{dt} = -AN_2(t) - B\Phi\delta f\,[N_2(t) - N_1(t)] \qquad (3.24)$$

In the following discussion, we will consider three cases. The simplest case involves an external light source that is turned off for $t \geq 0$. Setting $\Phi\delta f = 0$, the above equation becomes

$$\frac{dN_2(t)}{dt} = -AN_2(t)$$

Thus, the decay of $N_2(t)$ is given by

$$N_2(t) = N_2(0)e^{-At}$$

where $N_2(0)$ is the initial value of N_2 at $t = 0$. Clearly, in the absence of light illumination, $N_2(t)$ decays exponentially from $N_2(0)$. Specifically, N_2 reduces to 36.8 percent of its initial value at a time $t = 1/A$. In Figure 3.7, $N_2(t)/N_t$ is plotted as a function of time. Also in Figure 3.7, the increase of $N_1(t)/N_t$ and the evolution of $\Delta N(t)/N_t$ are shown as functions of time.

The time constant $\tau = 1/A$ is the **spontaneous emission lifetime** of the unpumped system. Since the physical meaning of τ is much more obvious than that of the Einstein A-coefficient, we will use τ in lieu of A in the following discussions.

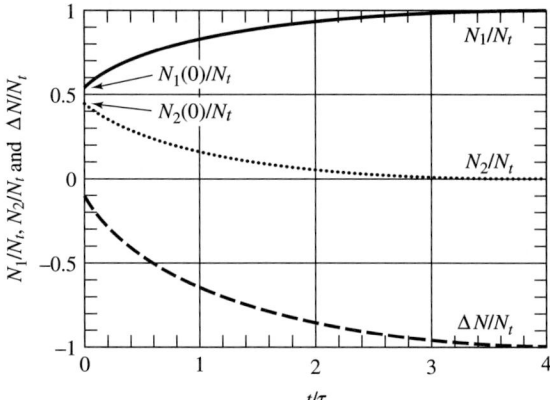

Figure 3.7 Evolution of N_1, N_2, and ΔN of a two-level system with no light illumination

The next case is slightly more complicated. Suppose that a steady state has been reached at $t = 0$, and the steady-state values are $N_1(0)$ and $N_2(0)$, respectively. Also suppose that an external light source of constant intensity is turned on at $t = 0$. We then wish to study the evolution of N_1, N_2, and ΔN as functions of time.

In terms of N_t and $\Delta N(t)$, we have $N_1(t) = (N_t - \Delta N(t))/2$ and $N_2(t) = (N_t + \Delta N(t))/2$. Therefore, (3.24) becomes

$$\frac{d}{dt}[\Delta N(t)] = -(2B\Phi\delta f + \frac{1}{\tau})\Delta N(t) - \frac{N_t}{\tau} \tag{3.25}$$

Solving the differential equation, we obtain

$$\Delta N(t) = [\Delta N(0) + \frac{N_t}{1 + 2B\tau\Phi\delta f}]e^{-(2B\Phi\delta f + \frac{1}{\tau})t} - \frac{N_t}{1 + 2B\tau\Phi\delta f} \tag{3.26}$$

By combining $\Delta N(t)$ with N_t, we obtain expressions for $N_1(t)$ and $N_2(t)$.

Figure 3.8 depicts the variations of $N_1(t)/N_t$, $N_2(t)/N_t$, and $\Delta N(t)/N_t$ as functions of t. As long as $\Delta N(0)$ is negative, $\Delta N(t)$ remains negative for all time and $N_2(t)$ is always less than $N_1(t)$. For large t,

$$N_1(t) \to N_t \frac{1 + B\tau\Phi\delta f}{1 + 2B\tau\Phi\delta f}$$

$$N_2(t) \to N_t \frac{B\tau\Phi\delta f}{1 + 2B\tau\Phi\delta f}$$

$$\Delta N(t) \to -N_t \frac{1}{1 + 2B\tau\Phi\delta f}$$

It is important to note that $N_2(t)$ is always less than $N_1(t)$ for all values of $\Phi\delta f$. Even if the light illumination is extremely intense, $\Delta N(t)$ remains negative as $t \to \infty$.

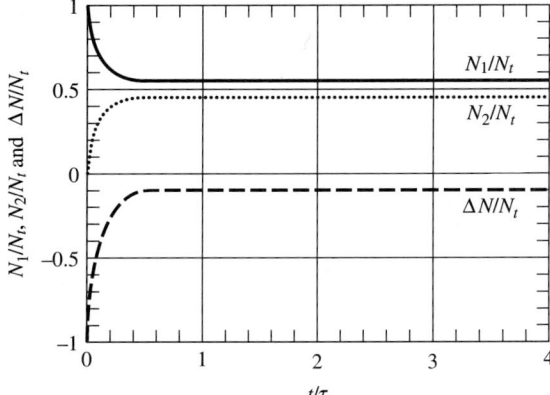

Figure 3.8 Evolution of N_1, N_2, and ΔN of a two-level system under constant light illumination ($B\Phi\delta f\tau = 4.5$).

Finally, we consider the case in which the external light source is turned on and then off. Suppose the light source is turned on for $0 \leq t \leq t_1$ and off for $t > t_1$. In the time interval of $0 \leq t \leq t_1$, $\Delta N(t)$ is given by (3.26). For $t > t_1$, ΔN is given by

$$\Delta N(t) = \Delta N(t_1) e^{-(t-t_1)/\tau} \qquad (3.27)$$

As t increases beyond t_1, $\Delta N(t)$ decays to its final unpumped value with a time constant τ. Similarly, with the pump turned off, $N_1(t)$ and $N_2(t)$ gradually return to their equilibrium values $N_1(0)$ and $N_2(0)$. This is shown in Figure 3.9.

Based on the results of all three cases, we conclude that it is *impossible* to achieve population inversion (i.e., $N_2 > N_1$) in an isolated two-level system by optical pumping. The most optical pumping can do in such a system is keep a little more than half of the electrons in the lower level and a little less than half of the electrons in the upper level.

3.5 LINE BROADENING

The linewidth of the emission or absorption process is an important parameter, and is represented by a complicated line shape function. To simplify the discussion, a linewidth of δf is used in lieu of the shape function. Because of the various physical processes involved, and possible material inhomogeneity, transitions between energy levels always have a finite spectral width. This is also known as **line broadening,** which can be classified as **homogeneous** or **inhomogeneous** line broadening. If the physical process is the same for all atoms, ions, or molecules everywhere in the medium, the process is homogeneous line broadening process. Examples include the finite lifetime of energy levels, and the collisions between atoms or ions with the crystal lattices in perfect crystals. If, however, the effects are different for atoms, ions, or molecules at different locations,

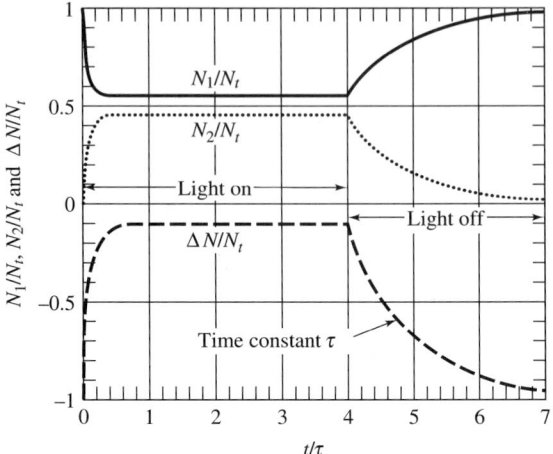

Figure 3.9 Evolution of N_1, N_2, and ΔN of a two-level system with $B\Phi\delta\tau = 4.5$, and light occurring during the interval $0 < t < 4\tau$

the process is inhomogeneous. For example, in real crystals, strain may be induced by defects, and the energy levels can be perturbed by the strain. Similarly, the fields and temperature disturbance may be different for different parts of the medium, resulting in different perturbations at the various locations, in turn resulting in varying energy differences in the transition at different locations. For gaseous media, for example, the thermal energy of the atoms, ions, and molecules at non-zero temperature, causes these particles to move randomly relative to a fixed laboratory reference frame. The emitted photons are therefore frequency shifted by the **Doppler effect.** Since different particles move with different thermal velocities, both in magnitude and in direction, the frequency shifts vary for different molecules. Therefore, for most gaseous media, inhomogeneous line broadening is the dominant effect. As an example, the line width of the HeNe laser emission at 0.633 μm is about 1500 MHz, with the Doppler shift being the main contributing factor.

3.6 RATE EQUATIONS FOR LASERS WITH OPTICAL PUMPING

Having demonstrated the impossibility of population inversion in isolated two-level media via optical pumping, we can now demonstrate that the population may be inverted by optical pumping in some three- and four-level systems [7–12]. We begin by setting up rate equations describing the rate of change of the density of occupied states in each level, and the photon density in the system. Based on the rate equations, we then calculate the pumping power required to initiate and sustain the laser operation, and the optical power generated by the laser. In addition, we will study the steady-state and transient characteristics of lasers in terms of rate equations, and will demonstrate that four-level lasers can be more efficient than three-level lasers.

3.6.1 Three-Level Systems with Optical Pumping

Consider an optical cavity containing an active medium, as shown in Figure 3.10. Suppose that the active medium is a three-level material system (Figure 3.4a), with energy levels E_1, E_2, and E_3, and that the densities of occupied states in these levels are N_1, N_2, and N_3, respectively. The medium is pumped optically by waves of frequency $f_{31} = (E_3 - E_1)/h$, which leads to transitions between E_1 and E_3. For simplicity, assume that the cavity mirrors are transparent at frequency f_{31}, and that they have reflectivities Γ_1 and Γ_2 at frequency $f_{21} = (E_2 - E_1)/h$. In other words, optical pumping is between E_3 and E_1, while lasing occurs between E_2 and E_1. As electrons drop from E_3 to E_2, the energy difference $E_3 - E_2$ is converted to acoustic waves, which are eventually dissipated in the host material as heat. Such a transition is known as a **nonradiative transition**. The time rates of change of N_1, N_2, and N_3 are (Figure 3.11):

$$\frac{dN_1}{dt} = A_P(N_3 - N_1) - A_L(N_1 - N_2) + \frac{N_3}{\tau_{31}} + \frac{N_2}{\tau_{21}} \quad (3.28)$$

$$\frac{dN_2}{dt} = \quad\quad\quad\quad - A_L(N_2 - N_1) + \frac{N_3}{\tau_{32}} - \frac{N_2}{\tau_{21}} \quad (3.29)$$

$$\frac{dN_3}{dt} = -A_P(N_3 - N_1) - \left(\frac{1}{\tau_{31}} + \frac{1}{\tau_{32}}\right)N_3 \quad (3.30)$$

In these equations, $\Phi_P \delta f$ is the density of photons with frequency f_{31}, $\Phi \delta f$ is the photon density with frequency f_{21}, and $A_P = B_{13}(\Phi_P \delta f)$ and $A_L = B_{12}(\Phi \delta f)$ are the **transition probabilities**. Although the pumping transition probability A_P is proportional to the photon density $\Phi_P \delta f$, it is often more convenient to think in terms of A_P rather than $\Phi_P \delta f$. Similarly, it may be easier to work with A_L instead of $\Phi \delta f$. **Spontaneous emission lifetimes** τ_{ij} are used in

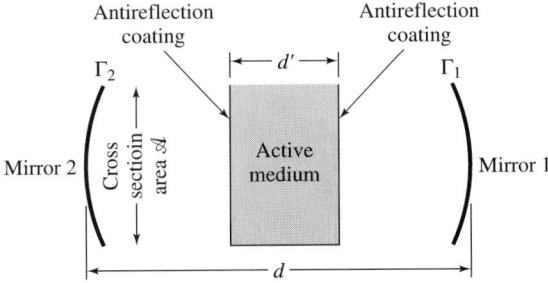

Figure 3.10 An active medium inside an optical cavity with a cross section $\mathcal{A}$

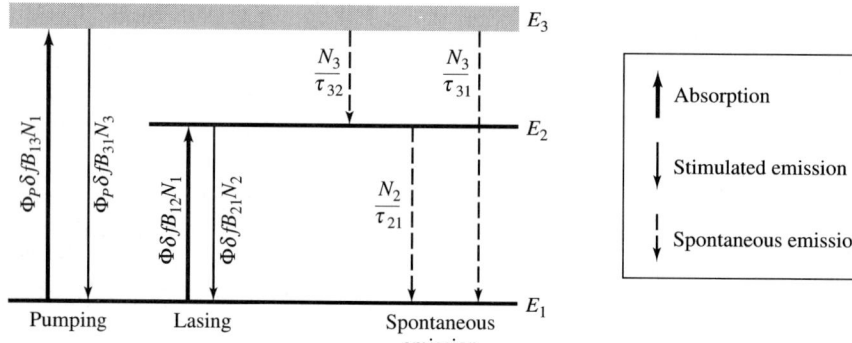

Figure 3.11 Transitions in a three-level system

lieu of the Einstein coefficients A_{ij} and $\tau_{ij} = 1/A_{ij}$. The subscripts identify the energy levels involved in the transitions. For convenience, we also write the lifetime τ_3 for level 3 as

$$\frac{1}{\tau_3} = \frac{1}{\tau_{31}} + \frac{1}{\tau_{32}}$$

The description for the three-level system also requires an equation describing the rate of change of $\Phi\delta f$. That equation is given in Section 3.6.4. For now, let us extract as much information as possible without the use of that equation.

Since we have assumed an isolated system, the total density of occupied states N_t is a constant

$$N_1 + N_2 + N_3 = N_t \tag{3.31}$$

The transient characteristics of three-level lasers will be treated in sections 7 and 8 of this chapter. The remainder of this section is devoted to the steady-state performance of three-level lasers. We determine N_1, N_2, N_3, and $\Phi\delta f$ under various conditions, assuming A_P is known. In the steady state, the derivatives with respect to time are set equal to zero. From (3.29) and (3.30), we can solve for N_2 and N_3:

$$N_3 = \frac{A_P N_1}{A_P + \dfrac{1}{\tau_3}} \tag{3.32}$$

$$N_2 = \frac{N_1 A_L + \dfrac{N_3}{\tau_{32}}}{A_L + \dfrac{1}{\tau_{21}}} = \frac{A_L + \dfrac{A_P}{\tau_{32}(A_P + \dfrac{1}{\tau_3})}}{A_L + \dfrac{1}{\tau_{21}}} N_1 \tag{3.33}$$

The term N_1 may be obtained by combining (3.31), (3.32), and (3.33). Finally, we have an expression for the population inversion $N_2 - N_1$. It is convenient to normalize $N_2 - N_1$ with respect to N_t as follows:

3.6 Rate Equations for Lasers with Optical Pumping

$$\frac{N_2 - N_1}{N_t} = \frac{A_P(\frac{1}{\tau_{32}} - \frac{1}{\tau_{21}}) - \frac{1}{\tau_{21}}\frac{1}{\tau_3}}{(A_P + \frac{1}{\tau_3})(2A_L + \frac{1}{\tau_{21}}) + A_P(A_L + \frac{1}{\tau_{21}} + \frac{1}{\tau_{32}})} \qquad (3.34)$$

Equations (32)–(34) are the basis for subsequent discussions.

Necessary Condition for Population Inversion In thermal equilibrium, the ratio $N_1:N_2:N_3$ is given by Boltzmann's equation (3.8), and is depicted schematically in Figure 3.12a. A pumped system (i.e., $A_P > 0$) is not in thermal equilibrium, and the ratio $N_1:N_2:N_3$ computed from (3.32) and (3.33) should, and does, differ from the thermal equilibrium value. For a weakly pumped system N_1, N_2, and N_3 deviate only slightly from the thermal equilibrium values, as shown schematically in Figure 3.12b. As A_P increases further, the distribution of occupied states may become quite different from the thermal equilibrium distribution. Also, if τ_{21} is less than τ_{32}, N_2 is always less than N_1, as indicated by (3.34). Thus, if $\tau_{21} < \tau_{32}$, population inversion is not possible with optical pumping. We conclude that the population in a three-level medium can be inverted by optical pumping only if

$$\tau_{21} > \tau_{32} \qquad (3.35)$$

This is the **necessary condition** for population inversion in three-level systems.

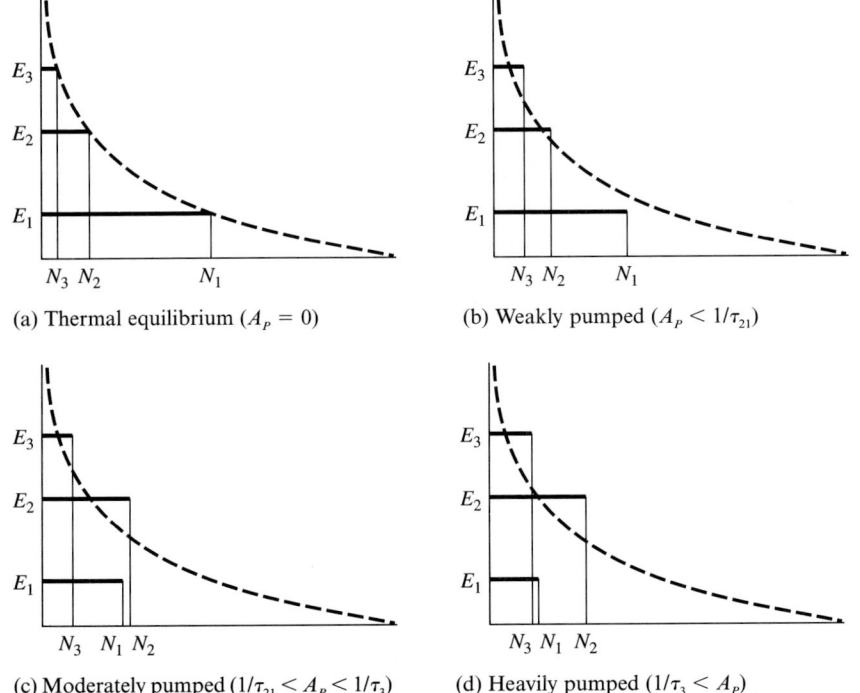

(a) Thermal equilibrium ($A_P = 0$)

(b) Weakly pumped ($A_P < 1/\tau_{21}$)

(c) Moderately pumped ($1/\tau_{21} < A_P < 1/\tau_3$)

(d) Heavily pumped ($1/\tau_3 < A_P$)

Figure 3.12 Schematic representation of the density of states as a function of pumping level

Minimum Pumping Transition Probability Suppose the necessary condition (3.35) is satisfied. The population is still not inverted unless the pumping is sufficiently intense that

$$A_P > \frac{\frac{1}{\tau_{21}}\frac{1}{\tau_3}}{\frac{1}{\tau_{32}}-\frac{1}{\tau_{21}}} = \frac{\frac{1}{\tau_{21}}(\frac{1}{\tau_{31}}+\frac{1}{\tau_{32}})}{\frac{1}{\tau_{32}}-\frac{1}{\tau_{21}}} \qquad (3.36)$$

The **minimum pumping transition probability** required for the population inversion is

$$A_{Pm} = \frac{\frac{1}{\tau_{21}}(\frac{1}{\tau_{31}}+\frac{1}{\tau_{32}})}{\frac{1}{\tau_{32}}-\frac{1}{\tau_{21}}} \qquad (3.37)$$

If τ_{32} is much smaller than τ_{31} and τ_{21}, then A_{Pm} may be approximated by

$$A_{Pm} \approx 1/\tau_{21}$$

Pumping is considered *weak* if $A_P < 1/\tau_{21}$, and *moderate* if $1/\tau_{21} < A_P < 1/\tau_3$. Under the condition of moderate pumping, lasing is very weak, if it occurs at all. In this case, we set A_L to zero and obtain, from (3.32) and (3.33),

$$N_3 \approx (A_P\tau_3)N_1 \qquad (3.38)$$

$$N_2 \approx (A_P\tau_{21})N_1 \qquad (3.39)$$

Thus, for moderate pumping, $N_2 > N_1 \gg N_3$, as depicted in Figure 3.12c.

A system is *heavily pumped* if $1/\tau_{21} < 1/\tau_3 < A_P$, and we obtain $N_2 > N_1 \approx N_3$ from (3.32) and (3.33) (Figure 3.12d). It can also be seen from (3.34) that, as A_P becomes very intense, $N_2 - N_1$ approaches a saturation value.

In three-level systems, we define the population inversion as $\Delta N = N_2 - N_1$. The change of ΔN as a function of A_P is presented schematically in Figure 3.13. As A_P increases, ΔN also increases. Although we have not discussed four-level systems, the change in ΔN for four-level systems is displayed as the dashed curve in Figure 3.13, for comparison.

Pumping Efficiency Factor We have now established the necessary condition for, and the minimum pumping transition probability required by, a population inversion in three-level media. The next consideration is the factor that dictates which three-level system is more efficient for lasing operation. Suppose that the electrons have been pumped to E_3. Most electrons drop to E_2 or E_1; only a few remain in E_3. The probabilities of electron decay from E_3 to E_1 and E_2 via spontaneous emission are proportional to $1/\tau_{31}$ and $1/\tau_{32}$, respectively. The fraction of electrons decaying from E_3 to E_1 is

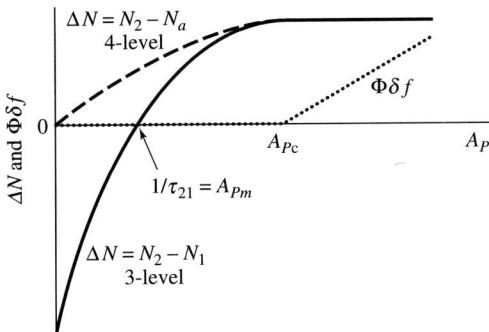

Figure 3.13 Population inversion and photon density as functions of pumping transition probability

$$\frac{\dfrac{1}{\tau_{31}}}{\dfrac{1}{\tau_{31}} + \dfrac{1}{\tau_{32}}} = \frac{\tau_{32}}{\tau_{31} + \tau_{32}}$$

and from E_3 to E_2 is $\tau_{31}/(\tau_{31} + \tau_{32})$. Transitions directly from E_3 to E_1 are really wasted as far as lasing is concerned; only the transitions from E_3 to E_2 are useful. Therefore, $\tau_{31}/(\tau_{31} + \tau_{32})$ can be viewed as the **pumping efficiency factor** of the three-level system, and the criterion for an *efficient* three-level system is simply $\tau_{32} \ll \tau_{31}$.

Ruby Lasers as a Specific Laser System In order to fully appreciate the discussion on laser operations, we must understand the order of magnitude of the various parameters. Consider a ruby laser as an example. Details of ruby lasers will be discussed in section 3.11. For the present purpose, it is only necessary to note that ruby lasers are three-level systems. The lifetimes of various levels, as reported by Maiman [1], are $\tau_{31} \approx (1/3) \times 10^{-5}$s, $\tau_{32} \approx 5 \times 10^{-8}$s, and $\tau_{21} \approx 3 \times 10^{-3}$s. Obviously, $\tau_{32} \ll \tau_{31}$ and $\tau_{32} \ll \tau_{21}$. The lifetime τ_3 of level E_3 is

$$\frac{1}{\tau_3} = \frac{1}{\tau_{31}} + \frac{1}{\tau_{32}} \approx \frac{1}{\tau_{32}} = 2 \times 10^7 \text{ s}^{-1}$$

and is determined by the E_3–to–E_2 transitions. In particular, the inequality

$$\frac{1}{\tau_3} \approx \frac{1}{\tau_{32}} \gg \frac{1}{\tau_{31}} \gg \frac{1}{\tau_{21}}$$

is well satisfied. Clearly, population inversion is achieved if $A_P > 3.33 \times 10^2 \text{ s}^{-1}$. The minimum pumping transition probability required is $A_{Pm} = 3.33 \times 10^2 \text{ s}^{-1}$.

Therefore, by choosing a pumping transition probability of $A_P \approx 3.33 \times 10^3 \text{ s}^{-1}$, which is 10 times larger than A_{Pm}, we obtain $N_3 \approx (1.667 \times 10^{-4})N_1$ and $N_2 \approx 10 N_1$, from (3.38). Note that N_3 is negligible in comparison to N_1. Calculations also show that 1.5 percent of the electrons drop directly from E_3 to E_1.

Rate Equation Approximation The next step is to combine several equations into a simplified and approximated expression. We combine (3.28) and (3.29) to yield

$$\frac{d}{dt}(N_2 - N_1) = -2A_L(N_2 - N_1) - \frac{2N_2}{\tau_{21}} + A_P(N_1 - N_3) + N_3\left(\frac{1}{\tau_{32}} - \frac{1}{\tau_{31}}\right)$$

Making use of the approximation (3.38) and the inequality $\tau_{32} \ll \tau_{31}$, the above equation becomes

$$\frac{d}{dt}(N_2 - N_1) \approx -2A_L(N_2 - N_1) - \frac{2N_2}{\tau_{21}} + 2A_P N_1 \qquad (3.40)$$

As demonstrated in the numerical example given in the previous paragraph, N_3 is small compared to N_1 and N_2. Then, $N_1 + N_2 \approx N_t$. Using the definition for $\Delta N = N_2 - N_1$, we can approximate N_1 and N_2 as

$$N_1 = \frac{N_1 + N_2 + N_1 - N_2}{2} \approx \frac{N_t - \Delta N}{2}$$

$$N_2 = \frac{N_1 + N_2 - N_1 + N_2}{2} \approx \frac{N_t + \Delta N}{2}$$

Thus we rewrite (3.40) as

$$\frac{d}{dt}(\Delta N) = -2A_L \Delta N - \frac{N_t + \Delta N}{\tau_{21}} + A_P(N_t - \Delta N) \qquad (3.41)$$

This is the desired approximate expression describing the *rate of change of the population inversion* in three-level media. The first term on the right-hand side of (3.41) represents the rate of change of the population inversion induced by stimulated emission. The second term is the rate of reduction of ΔN due to spontaneous emissions. The last term is the net rate of increase of the population inversion forced by optical pumping.

3.6.2 Four-Level Systems with Optical Pumping

We will now consider lasers with four-level active media. Like the three-level system discussed in the last section, optical pumping is between levels E_1 and E_3. However, lasing is between E_2 and E_a. The mirrors shown in Figure 3.10 now reflect waves of frequency $f_{2a} = (E_2 - E_a)/h$. Transitions between E_3 and E_2 and between E_a and E_1 are nonradiative. The time rates of change of N_a, N_2, and N_3 are (Figure 3.14)

3.6 Rate Equations for Lasers with Optical Pumping

$$\frac{dN_a}{dt} = -A_L(N_a - N_2) - \frac{N_a}{\tau_{1a}} + \frac{N_2}{\tau_{2a}} + \frac{N_3}{\tau_{3a}} \quad (3.42)$$

$$\frac{dN_2}{dt} = -A_L(N_2 - N_a) - \frac{N_2}{\tau_{2a}} - \frac{N_2}{\tau_{21}} + \frac{N_3}{\tau_{32}} \quad (3.43)$$

$$\frac{dN_3}{dt} = -A_P(N_3 - N_1) - N_3\left(\frac{1}{\tau_{32}} + \frac{1}{\tau_{3a}} + \frac{1}{\tau_{31}}\right) \quad (3.44)$$

where the transition probabilities and lifetimes are given by $A_P = B_{13}\Phi_P\delta f$, $A_L = B_{a2}\Phi\delta f$, $\tau_{32} = 1/A_{32}$, etc. An equation describing the rate of change of N_1 can be written in similar fashion, but it is really not necessary since the total number of occupied states $N_1 + N_2 + N_3 + N_a = N_t$ is a constant and therefore

$$\frac{dN_1}{dt} = -\left[\frac{dN_3}{dt} + \frac{dN_2}{dt} + \frac{dN_a}{dt}\right] \quad (3.45)$$

We could solve for the exact expressions for steady-state values of N_a, N_2, and N_3 from (3.42)–(3.45). However, the resulting expressions would be very complicated. To stress the essential physical concepts, we can look for approximate expressions instead. The steady-state expression for N_3 is obtained from (3.44)

$$N_3 = \frac{A_P N_1}{A_P + \dfrac{1}{\tau_3}} \quad (3.46)$$

where the total lifetime τ_3 of states in E_3 in a four-level system is given by

$$\frac{1}{\tau_3} = \frac{1}{\tau_{32}} + \frac{1}{\tau_{3a}} + \frac{1}{\tau_{31}}$$

As noted in (3.46), N_3 relates directly to N_1. Thus, the spontaneous emission terms N_3/τ_{3a} and N_3/τ_{32} of (3.42) and (3.43) can be written as

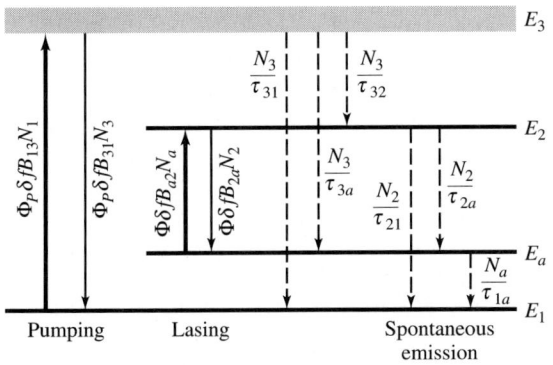
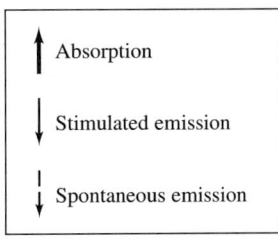

Figure 3.14 Transitions in a four-level system

$$\mathcal{R}_a = \frac{N_3}{\tau_{3a}} = \frac{A_p N_1}{\tau_{3a}(A_p + \frac{1}{\tau_3})} \qquad (3.47)$$

$$\mathcal{R}_2 = \frac{N_3}{\tau_{32}} = \frac{A_p N_1}{\tau_{32}(A_p + \frac{1}{\tau_3})} \qquad (3.48)$$

In four-level media only a small fraction of states in E_1 are pumped to higher levels. N_1 remains essentially unchanged by the pumping. In view of (3.46)–(3.48), N_3, $\mathcal{R}_2$, and $\mathcal{R}_a$ are also approximately constants. Thus, $\mathcal{R}_2$ and $\mathcal{R}_a$ may be interpreted as follows. In the actual process, electrons are raised from E_1 to E_3 and they then decay spontaneously to E_2 or E_a. If their short stay in E_3 is ignored, the pumping process may be viewed as if states are pumped "directly" from E_1 to E_2 or E_a. Thus, $\mathcal{R}_2$ and $\mathcal{R}_a$ can be taken to represent the density of states per second pumped "directly" from E_1 to E_2 and E_a, respectively. In terms of $\mathcal{R}_a$ and $\mathcal{R}_2$, (3.42) and (3.43) become

$$\frac{dN_a}{dt} \approx -A_L(N_a - N_2) - \frac{N_a}{\tau_{1a}} + \frac{N_2}{\tau_{2a}} + \mathcal{R}_a \qquad (3.49)$$

$$\frac{dN_2}{dt} \approx -A_L(N_2 - N_a) - \frac{N_2}{\tau_{2a}} - \frac{N_2}{\tau_{21}} + \mathcal{R}_2 \qquad (3.50)$$

Approximate steady-state values of N_a and N_2 may be obtained from (3.49) and (3.50). In particular, the steady-state population inversion in a four-level system is

$$\Delta N = N_2 - N_a \approx \frac{\mathcal{R}_2 \left[1 - \frac{\tau_{1a}}{\tau_{2a}}(1 + \frac{\mathcal{R}_a}{\mathcal{R}_2}[1 + \frac{\tau_{2a}}{\tau_{21}}])\right]}{A_L + \frac{1}{\tau_{2a}} + \frac{1}{\tau_{21}}(1 + A_L \tau_{1a})} \qquad (3.51)$$

As noted previously, states raised to E_3 will drop to the lower levels by spontaneous transitions. The fraction of states dropping from E_3 to E_2 and E_a are, respectively,

$$\frac{\frac{1}{\tau_{32}}}{\frac{1}{\tau_{32}} + \frac{1}{\tau_{3a}} + \frac{1}{\tau_{31}}} = \frac{\tau_{3a}\tau_{31}}{\tau_{3a}\tau_{31} + \tau_{32}\tau_{31} + \tau_{32}\tau_{3a}}$$

$$\frac{\frac{1}{\tau_{3a}}}{\frac{1}{\tau_{32}} + \frac{1}{\tau_{3a}} + \frac{1}{\tau_{31}}} = \frac{\tau_{32}\tau_{31}}{\tau_{3a}\tau_{31} + \tau_{32}\tau_{31} + \tau_{32}\tau_{3a}}$$

Therefore, the *pumping efficiency factor* of four-level media is

$$\frac{\tau_{3a}\tau_{31}}{\tau_{3a}\tau_{31} + \tau_{32}\tau_{31} + \tau_{32}\tau_{3a}}$$

Obviously, for efficient four-level lasers, τ_{32} should be much shorter than τ_{3a} and τ_{31}. As a result,

$$\frac{\mathfrak{R}_a}{\mathfrak{R}_2} = \frac{\tau_{32}}{\tau_{3a}} \ll 1 \qquad (3.52)$$

In other words, the density of states pumped "directly" from E_1 to E_a is much less than that pumped "directly" from E_1 to E_2.

In view of (3.52), Equation (3.51) may be approximated further as

$$\Delta N \approx \frac{\mathfrak{R}_2\left[1 - \dfrac{\tau_{1a}}{\tau_{2a}}\right]}{A_L + \dfrac{1}{\tau_{2a}} + \dfrac{1}{\tau_{21}}(1 + A_L \tau_{1a})} \qquad (3.53)$$

From the numerator of (3.51), we conclude that the population inversion can be achieved in a four-level system if

$$\tau_{2a} > \tau_{1a}\left[1 + \frac{\mathfrak{R}_a}{\mathfrak{R}_2}1 + \frac{\tau_{2a}}{\tau_{21}}\right].$$

The condition may be further simplified as $\tau_{2a} > \tau_{1a}$. This is the *necessary condition* for the population inversion in a four-level medium. Note also that the *minimum pumping transition probability* A_{Pm} required for maintaining population inversion in four-level systems is 0. This should be compared with the A_{Pm} given in (3.37) for three-level systems. The population inversion increases with the pumping transition probability until either a saturation value is reached, or the presence of the lasing transition probability A_L is felt. This is shown as the dashed curve in Figure 3.13.

In summary, the condition for population inversion in a four-level system is $\tau_{1a} < \tau_{2a}$, and a four-level system is efficient if $\tau_{32} \ll \tau_{3a}$ and τ_{31}. Usually, N_3 and N_a are negligibly small when these conditions are satisfied. It can also be shown that $A_P \tau_3 < 1$. Thus, $\mathfrak{R}_2 \approx A_P N_1$. We define $\Delta N = N_2 - N_a$ for four-level systems. Since N_3 and N_a are small, $\Delta N \approx N_2$ and $N_t \approx N_1 + N_2 \approx N_1 + \Delta N$. Thus, we have from (3.50)

$$\frac{d}{dt}\Delta N \approx -A_L \Delta N - \frac{\Delta N}{\tau_2} + A_P(N_t - \Delta N) \qquad (3.54)$$

where $\dfrac{1}{\tau_2} = \dfrac{1}{\tau_{21}} + \dfrac{1}{\tau_{2a}}$. The first two terms on the right-hand side of (3.54) represent the rate of depletion of the population inversion by stimulated and spontaneous emission processes. The third term is the net change of the population inversion by optical pumping.

3.6.3 System Comparisons

It is instructive to compare (3.54) for four-level media with (3.25) for two-level media and (3.41) for three-level media. Comparing (3.41) with (3.54), we note that, if τ_{21} and τ_2 are comparable, the spontaneous term $(N_t + \Delta N)/\tau_{21}$ in three-level systems is much larger than $\Delta N/\tau_2$ of four-level systems. As an application of the approximate rate equations (3.41) and (3.54), we can estimate the steady-state pumping transition probability required to maintain a given population inversion ΔN for three- and four-level systems prior to lasing. Since the system is not lasing, we set A_L equal to zero. Under the steady-state condition, we have, from (3.41) and (3.54), the following relationships between A_P and ΔN

$$\left(A_P\right)_{\text{3-level}} \approx \frac{1}{\tau_{21}} \frac{N_t + \Delta N}{N_t - \Delta N} \qquad \left(\frac{\Delta N}{N_t}\right)_{\text{3-level}} \approx \frac{A_P \tau_{21} - 1}{A_P \tau_{21} + 1} \qquad (3.55)$$

$$\left(A_P\right)_{\text{4-level}} \approx \frac{1}{\tau_2} \frac{\Delta N}{N_t - \Delta N} \qquad \left(\frac{\Delta N}{N_t}\right)_{\text{4-level}} \approx \frac{A_P \tau_2}{A_P \tau_2 + 1} \qquad (3.56)$$

Since $N_t \gg \Delta N$, a large pumping transition probability A_P is required to maintain a given ΔN in three-level systems. Assuming that τ_2 and τ_{21} are roughly the same, less pumping is required to maintain a given $\Delta N/N_t$ in a four-level system than in a three-level system. The minimum pumping transition probability A_{Pm} discussed previously corresponds to the pumping requirement for $\Delta N = 0$.

3.6.4 Rate of Change of Photon Density

To complete the description of optically pumped lasers, we need an expression for the rate of change of the photon density in the optical cavity. Figure 3.10, displayed earlier, shows an optical cavity with two mirrors spaced a distance d apart. The mirrors have power reflection coefficients Γ_1 and Γ_2 at frequency f_{21} for three-level systems and frequency f_{2a} for four-level systems. The cavity is partially filled with an active medium of length d' and index n. For simplicity, we assume that the cross section $\mathcal{A}$ of the cavity is the same as that of the active medium. We also assume that the ends of the active medium are coated with antireflection layers, so that reflections from the end surfaces are negligible.

First, we consider an *unpumped cavity*. Without pumping, there is no optical gain in the medium. The losses due to the diffraction by mirrors of finite size, and the scattering and absorption by mirrors and various scatterers in the cavity, including the unpumped medium, are lumped into an effective attenuation constant α. Begin at an arbitrary point in the cavity and follow the photons as they "travel" in the cavity. Assume the initial photon density is $\Phi \delta f$. Upon completing one round trip of travel in the unpumped cavity, the photon density becomes $\Phi \delta f \Gamma_1 e^{-\alpha d} \Gamma_2 e^{-\alpha d}$, which means that the net change in photon density is $[\Gamma_1 e^{-\alpha d} \Gamma_2 e^{-\alpha d} - 1] \Phi \delta f$.

Next, we take into consideration the pumping process and the resulting optical gain. Let the gain constant due to population inversion be g, which is

3.6 Rate Equations for Lasers with Optical Pumping

$B_{21}n(N_2 - N_1)/c$ for three-level media and $B_{2a}n(N_2 - N_a)/c$ for four-level media. Since the length of the gain medium is d', the net change in photon density upon one round trip of travel in a pumped cavity is

$$\Delta(\Phi\delta f) = [\Gamma_1 \Gamma_2 \, e^{2(gd' - \alpha d)} - 1]\Phi\delta f$$

To simplify this expression further, we define γ_1, γ_2 in terms of Γ_1 and Γ_2, as follows:

$$\Gamma_1 = e^{-\gamma_1} \qquad \Gamma_2 = e^{-\gamma_2}$$

Usually, the net change in photon density per round trip of travel in the pumped cavity is small, that is $\gamma_1 + \gamma_2 + 2\alpha d - 2gd'$ is small. Consequently, an approximation for the net change[1] is

$$\Delta(\Phi\delta f) = [e^{-(\gamma_1 + \gamma_2 + 2\alpha d - 2gd')} - 1]\,\Phi\delta f \approx [2(gd' - \alpha d) - (\gamma_1 + \gamma_2)]\Phi\delta f$$

The time required for photons to complete a round-trip excursion in the cavity is

$$\Delta t = \frac{2[(d - d') + nd']}{c}$$

We define the **effective cavity length** as $d_e = d + (n - 1)d'$. Thus, the time rate of change of the photon density is

$$\frac{d(\Phi\delta f)}{dt} \approx \frac{\Delta(\Phi\delta f)}{\Delta t} \approx \frac{c}{d_e}\left[(gd' - \alpha d) - \frac{\gamma_1 + \gamma_2}{2}\right]\Phi\delta f \tag{3.57}$$

The terms on the right-hand side may be grouped into two categories, one representing power loss in the cavity and the other the gain in the medium. To do so, we again consider the cavity with the pumping source turned off. Setting g equal to zero, (3.57) can be solved as

$$\Phi(t)\delta f = \Phi(0)\delta f \, e^{-t/\tau_{ph}}$$

where τ_{ph} is given by

$$\frac{1}{\tau_{ph}} = \frac{c}{d_e}\left(\alpha d + \frac{\gamma_1 + \gamma_2}{2}\right) = \frac{cd}{d_e}\left(\alpha + \frac{1}{2d}\ln\frac{1}{\Gamma_1\Gamma_2}\right) \tag{3.58}$$

and τ_{ph} is the time constant or lifetime of the unpumped cavity. This constant relates to the loss caused by scattering, absorption, and diffraction in the cavity, as represented by αd, and by the transmission $(1 - \Gamma_1)$ and $(1 - \Gamma_2)$ through the mirrors.

If the cross section of the active medium is different from that of the cavity, an additional factor must be included to account for this difference (see Problem 16). As shown in Figure 3.10, the cavity is partially filled with an active medium of volume $\mathcal{A}d'$ and index n, and the total population inversion in the active medium is $\mathcal{A}d' \cdot \Delta N$. The *total* volume of the cavity is $\mathcal{A}d$. To account for the difference, we introduce an **effective active volume** V_a, where

---

[1] Note that $e^{-x} = 1 - x + (x^2/2) + \ldots$

$$V_a = \frac{nd'}{d_e}$$

The effective active volume is a dimensionless term introduced to account for the fact that the cavity is only partially filled by the active medium. Using the effective active volume, we can treat a partially filled cavity as if it were completely filled. In terms of τ_{ph} and V_a, (3.57) becomes

$$\frac{d(\Phi\delta f)}{dt} = [BV_a \Delta N - \frac{1}{\tau_{ph}}] \Phi\delta f \quad (3.59)$$

Now, we are ready to examine three- and four-level laser systems. In particular, we note that the essential features of three-level systems are described by (3.41) and (3.59) with $\Delta N = N_2 - N_1$ and $B = B_{21}$, and four-level lasers are described by (3.54) and (3.59) with $\Delta N = N_2 - N_a$ and $B = B_{2a}$.

3.6.5 Critical Pumping and Critical Population Inversion

When the lasing action is just initiated and output power increases from zero, $\Phi\delta f$ is small and increasing, and A_L is small and increasing. For $\Phi\delta f$ to increase, the population inversion must be greater than $1/(BV_a\tau_{ph})$, as indicated by (3.59). We define the **critical population inversion** as the minimum ΔN required to guarantee the increase of the photon density in the cavity

$$\Delta N_c = \frac{1}{BV_a\tau_{ph}} \quad (3.60)$$

By substituting ΔN_c in (3.55) and (3.56), we obtain the **critical pumping transition probability** A_{Pc} needed to maintain the critical population inversion for three-level systems, as follows:

$$\left(A_{Pc}\right)_{\text{3-level}} = \frac{1}{\tau_{21}} \frac{N_t + \Delta N_c}{N_t - \Delta N_c} \quad (3.61)$$

Noting that $N_t \gg \Delta N_c$, we have

$$\left(A_{Pc}\right)_{\text{3-level}} \approx \frac{1}{\tau_{21}}$$

The corresponding term for four-level systems is

$$\left(A_{Pc}\right)_{\text{4-level}} = \frac{1}{\tau_2} \frac{\Delta N_c}{N_t - \Delta N_c} \quad (3.62)$$

which is very small because N_t is much greater than ΔN_c. A comparison of equations (3.61) and (3.62) clearly shows the advantage of four-level systems over three-level systems.

3.6.6 Steady-State Output Power

The steady-state photon density and steady-state population inversion are needed to calculate the steady-state output power. In the following, we use a subscript 0 to signify the steady-state values. We set the time derivatives to zero and obtain from (3.59),

$$\Delta N_0 = (N_2 - N_1)_0 = \frac{1}{BV_a \tau_{ph}} \quad (3.63)$$

Note that ΔN_0 is equal to ΔN_c. This means that when the pumping transition probability increases beyond A_{Pc}, the steady-state population inversion ΔN_0 does not increase beyond ΔN_c. All additional pumping results in an increase in the photon emission. From (3.41) we can solve for the steady-state population inversion as a function of A_P and A_{L0}:

$$\left(\frac{\Delta N_0}{N_t}\right)_{3\text{-level}} \approx \frac{A_P - \frac{1}{\tau_{21}}}{A_P + 2A_{L0} + \frac{1}{\tau_{21}}}$$

Making use of (3.61) and noting that $A_{L0} = B\Phi_0 \delta f$, we solve for $\Phi_0 \delta f$ as follows:

$$(\Phi_0 \delta f)_{3\text{-level}} = \frac{N_t + \Delta N_0}{2B\Delta N_0 \tau_{21}} \left(\frac{A_P}{A_{Pc}} - 1\right) = V_a \frac{\tau_{ph}}{\tau_{21}} \frac{N_t + \Delta N_0}{2} \left(\frac{A_P}{A_{Pc}} - 1\right) \quad (3.64)$$

As A_P increases beyond A_{Pc}, $\Phi_0 \delta f$ increases linearly and ΔN_0 approaches a constant value. This is shown schematically in Figure 3.13.

For four-level systems, we have from (3.54)

$$\left(\frac{\Delta N_0}{N_t}\right)_{4\text{-level}} \approx \frac{A_P}{A_P + A_{L0} + \frac{1}{\tau_2}}$$

and

$$(\Phi_0 \delta f)_{4\text{-level}} = \frac{1}{B\tau_2} \left(\frac{A_P}{A_{Pc}} - 1\right) = V_a \frac{\tau_{ph}}{\tau_2} \Delta N_0 \left(\frac{A_P}{A_{Pc}} - 1\right) \quad (3.65)$$

To calculate the steady-state power output, we recall that there are $\mathcal{A}d(\Phi_0 \delta f)$ photons in the cavity and the photon energy is hf. Since the cavity lifetime is τ_{ph}, the time rate of decrease of the total energy in the cavity is $\mathcal{A}d(\Phi_0 \delta f)hf/\tau_{ph}$. The power loss is due to scattering, absorption, and diffraction in the cavity and to radiation through the mirrors. The fraction of power decrease in the cavity due to the power transmitted through mirror "i" is $\gamma_i/(2\alpha d + \gamma_1 + \gamma_2)$. Thus, the steady-state power radiated through mirror "i" is

$$P_{i0} = \frac{\mathcal{A}d(\Phi_0\delta f)hf}{\tau_{ph}} \frac{\gamma_i}{2\alpha d + \gamma_1 + \gamma_2} = \frac{c}{2}\frac{\mathcal{A}d}{d_e}\Phi_0\delta f hf \gamma_i$$

where $i = 1$ or 2. Since the mirror reflectivity Γ_i is very close to 1, $\gamma_i \approx 1 - \Gamma_i$ and P_{i0} can be approximated as

$$P_{i0} \approx \frac{c}{2}\frac{\mathcal{A}d}{d_e}\Phi_0\delta f hf(1 - \Gamma_i) \tag{3.66}$$

A physical interpretation or alternative derivation of (3.66) is possible. In an optical cavity, photons can "move" in either direction. Half of the photons move toward mirror "i", which means that, in a time interval of d_e/c, $(1/2)\mathcal{A}d(\Phi_0\delta f)$ photons "move" toward mirror "i" and the power incident upon mirror "i" is $[(1/2)\mathcal{A}d(\Phi_0\delta f)hf]/(d_e/c)$. Since the power reflection coefficient of mirror "i" is Γ_i, the optical power transmitted through mirror "i" is

$$P_{i0} = \frac{1}{2}\frac{\mathcal{A}d(\Phi_0\delta f)hf}{d_e/c}(1 - \Gamma_i)$$

and (3.66) is recovered. Using (3.64), we have

$$\left(P_{i0}\right)_{3\text{-level}} \approx (1 - \Gamma_i)\frac{\mathcal{A}d}{d_e}\frac{c(N_t + \Delta N_0)}{4B\tau_{21}\Delta N_0}hf\left(\frac{A_P}{A_{Pc}} - 1\right)$$

$$= (1 - \Gamma_i)\frac{\mathcal{A}d}{d_e}\frac{c}{4}\left(\frac{\tau_{ph}}{\tau_{21}}\right)V_a(N_t + \Delta N_0)hf\left(\frac{A_P}{A_{Pc}} - 1\right) \tag{3.67}$$

Equations (3.66) and (3.67) give the steady-state power output of three-level lasers. Following the same procedure, we can show (see Problem 9) that the steady-state power output of four-level systems is

$$\left(P_{i0}\right)_{4\text{-level}} \approx (1 - \Gamma_i)\frac{\mathcal{A}d}{d_e}\frac{c}{2}\left(\frac{\tau_{ph}}{\tau_2}\right)V_a \Delta N_0 hf\left(\frac{A_P}{A_{Pc}} - 1\right) \tag{3.68}$$

Recall that ΔN_0 and A_{Pc} are functions of Γ_1, Γ_2, and α, etc. To increase laser power output, we must reduce Γ_1 and/or Γ_2. In the process, however, more photons would leak from the cavity and more pumping power would be required to sustain the laser oscillation. By varying Γ_1 and Γ_2, we can optimize laser output.

3.6.7 Steady-State Gain of a Three-Level Active Medium

If the optical feedback via the mirrors is removed, a laser is converted back to an optical amplifier. The steady-state gain of the optical amplifier is, from (3.23),

$$g_0 = \frac{Bn}{c}\Delta N \approx \frac{Bn}{c}N_t\frac{A_P - \dfrac{1}{\tau_{21}}}{A_P + \dfrac{1}{\tau_{21}} + 2B\Phi_0\delta f}$$

The optical gain is larger when $\Phi_0 \delta f$ is smaller. By ignoring $\Phi_0 \delta f$, we have the **small-signal gain**

$$g_{ss0} \approx \frac{Bn}{c} N_t \frac{A_p - \frac{1}{\tau_{21}}}{A_p + \frac{1}{\tau_{21}}}$$

Then, the steady-state gain of the amplifying device with a finite $\Phi_0 \delta f$ can be expressed in terms of the small-signal gain

$$g_0 = \frac{g_{ss0}}{1 + \dfrac{B\phi_0 \delta f}{A_p + (1/\tau_{21})}} \quad (3.69)$$

3.7 Q-SWITCHED LASERS

The previous discussions centered on the steady-state continuous-wave (cw) operation of lasers. The following discussions concern the transient or pulsed operation of lasers. **Q-switching** and **mode locking** are two techniques for generating narrow and possibly very intense optical pulses [9–12]. Pulses generated by Q-switching are on the order of a few nanoseconds, while those produced by mode locking may be as narrow as, or even narrower than, a few tenths of a picosecond. The rate equations (3.41) and (3.59) can be used for our discussion on Q-switched three-level lasers, and (3.54) and (3.59) can be used for the four-level lasers.

To review briefly, two distinct physical processes are involved in lasers. A medium is made unstable by raising states from E_1 to E_2 via E_3. When a pumped medium is stimulated by photons, stimulated transition occurs and states in level E_2 drop to level E_1 in three-level lasers (or E_a in four-level lasers). For cw operations, the pumping and lasing processes take place continuously and simultaneously, and eventually a steady state is reached. For Q-switched lasers, the two processes occur *sequentially*. To produce extremely intense pulses, steps are taken to insure that as many states as possible are stored in level E_2. In this period, the stimulation is kept to a minimum to suppress lasing. This is done by "turning off" the optical feedback. At the instant an optical pulse is desired, the optical feedback is "switched on." With the optical feedback turned on, photons are trapped in the cavity and stimulated emission is initiated. This is the basic operation of Q-switching, also known as **Q-spoiling.**

Any fast mechanism for varying the cavity loss may be used for Q-switching. Eight Q-switching methods are depicted schematically in Figure 3.15. These methods can be grouped into four generic schemes. Figure 3.15a shows the simplest scheme, with a rotating mirror. The rotating mirror may be replaced by a rotating prism (Figure 3.15b), a rotating polygon, or a chopper (Figure 3.15c). When the mirrors or prisms are not aligned, or the beam is interrupted

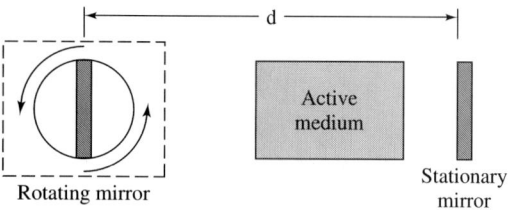

(a) A cavity with a rotating mirror

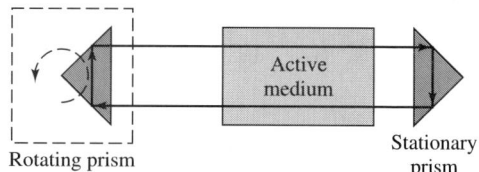

(b) A cavity with a rotating prism

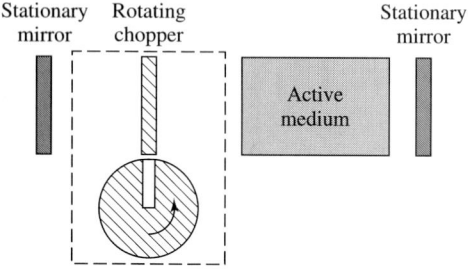

(c) A cavity with a rotating chopper

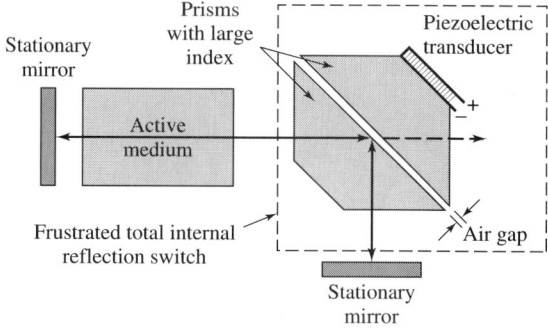

(d) A cavity with a frustrated total internal reflection switch

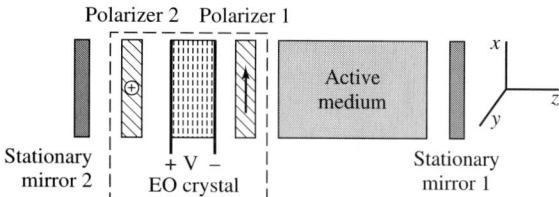

(e) An electrically controlled gate with a half-wave EO crystal and two polarizers with crossed transmission axes

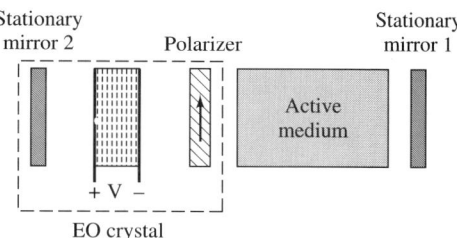

(f) A polarizer, a quarter-wave EO crystal, and mirror 2 acting as an electrically controlled switch

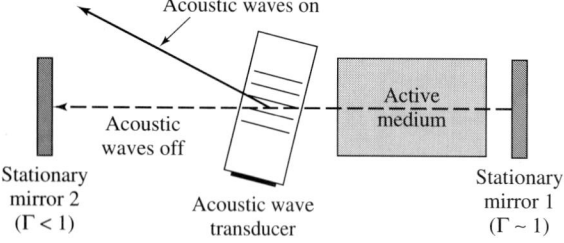

(g) Acoustooptic Q-switch

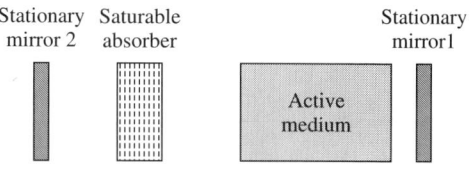

(h) A cavity with a saturable absorber

Figure 3.15 Various methods used to achieve Q-switching

by the chopper, there is little or no optical feedback and lasing is suppressed. Optical pulses are generated only when the mirrors or prisms are aligned or the optical path is unblocked by the chopper. The main limitation of these mechanical devices is the speed of rotation, which is on the order of 10,000 to 30,000 rpm or less.

In Figure 3.15d, two high-index prisms and one low-index air gap form an optical switch. When the air gap is wide, that is, the prisms are far apart, optical beams are reflected toward the stationary mirrors by total internal reflection at the air gap. When the two prisms are pushed closer together (less than 1 μm), the total internal reflection at the air gap is frustrated by the presence of the second prism. Instead of the light being bent, it propagates straight through the prisms, as indicated by the dashed line in the figure, and the light misses the reflecting mirrors. Thus, the cavity Q is changed by changing the spacing between the two prisms.

Instead of using mechanical rotation or translation, we could use an electrically controlled optical gate in the form of a combination of either a pair of crossed polarizers and a half-wave electrooptic (EO) crystal (Figure 3.15e), or a polarizer, a quarter-wave EO crystal, and a mirror (mirror 2 in Figure 3.15f). The electrooptic crystals may be coated with antireflection layers to suppress the reflections due to the air/crystal boundaries. In the absence of an applied voltage, the EO crystals may be considered absent, except for an extra, fixed phase delay introduced by the crystals. The effects of EO crystals in the presence of applied electric fields and polarizers is discussed in detail in Chapter 6. For the present discussion, it is sufficient to note that the half-wave EO crystal is arranged in such a way that linearly polarized optical field vectors are rotated spatially by 90° by the EO crystal in the presence of an applied voltage.

A polarizer is an optical element that passes waves oriented in a specific direction and blocks waves oriented in the orthogonal direction. The direction of transmitted polarization is called the **transmission axis.** For the crossed polarizers shown in Figure 3.15e, the transmission axes of the two polarizers are mutually perpendicular. If the EO crystal is absent, or if there is no voltage applied across the crystal, no light can be transmitted through the crossed polarizers. In the presence of an applied voltage across the half-wave EO crystal, the electric field vectors are rotated by 90°, waves polarized along the x axis pass freely through polarizer 1, and the transmitted light is converted to y-polarized waves by the half-wave EO crystal. The y-polarized waves then pass freely through the second polarizer. Thus, the combination of crossed polarizers and the half-wave EO crystal acts as an optical switch. The optical cavity Q is very high when the voltage is applied and very low when the voltage is turned off.

The arrangement shown in Figure 3.15f also functions as an electrically controlled gate, except that the cavity Q is *low* when the voltage is on and *high* when the voltage is off. Since there are no moving parts in the schemes involving EO crystals and polarizers, the switching speed is limited primarily by the RC time constant of the electric circuit.

In the **acoustooptic** Q-switching arrangement (Figure 3.15g), the optical beam is steered away from mirror 2 by acoustic waves. The **saturable absorber** shown in Figure 3.15h is a nonlinear material that is lossy when the light intensity is low, and highly transparent when the light intensity is high. Thus, the cavity Q is controlled by the optical nonlinearity of the saturable absorber.

Of all the methods discussed thus far, the scheme based on a stationary mirror and a rotating mirror (Figure 3.15a) is the simplest. We will use this scheme as the vehicle for explaining Q-switching. When the two mirrors are not in parallel, the cavity Q is low, and the system can't lase because there is insufficient optical feedback. Narrow pulses are only generated during the very short time interval when the mirrors are in parallel. The operation of a Q-switched laser can therefore be divided into two distinct stages: a long **pumping stage** when the cavity Q is low, and a short **lasing stage** when the Q is high (Figure 3.16). As an example, suppose the mirror is rotating at a speed of 30,000 rpm (500 rps). If the rotating mirror is single-sided, the pumping stage is about 2 ms; for a double-sided mirror, the pumping stage is 1 ms. The lasing stage is about 10 to 30 ns and is determined by the characteristics of the active medium and pumping.

For the pumping stage, $t < 0$, the optical feedback is too weak to sustain stimulated emission. We can therefore ignore A_L and use (3.55) or (3.56) to determine the steady-state population inversion. Since the pumping stage is long, we may assume that the steady-state condition has been reached at $t = 0$. The steady-state population inversion thus obtained is used as the *initial value* of ΔN for the lasing stage. The initial value ΔN_i is

$$\left(\Delta N_i\right)_{\text{3-level}} = \frac{A_P\tau_{21} - 1}{A_P\tau_{21} + 1} N_t \qquad \left(\Delta N_i\right)_{\text{4-level}} = \frac{A_P\tau_2}{A_P\tau_2 + 1} N_t \qquad (3.70)$$

If $A_P \gg 1/\tau_{21}$ for three-level systems, or $A_P \gg 1/\tau_2$ for four-level systems, $\Delta N_i \approx N_t$. This means that, when a three- or four-level laser system is pumped hard, most of the states are raised to level E_2.

In the short time interval near $t \approx 0$, the two mirrors are either exactly or approximately in parallel, the cavity Q is high, and the laser is in the lasing stage. During the short lasing stage, the contributions to $\dfrac{d\Delta N}{dt}$ due to optical pumping and spontaneous emission are negligible compared to that of the stimulated emission. Thus (3.41) may be simplified to

$$\frac{d}{dt}[\Delta N(t)] \approx -2B[\Phi(t)\delta f]\Delta N(t) \qquad (3.71)$$

The transient behavior of Q-switched lasers in the lasing stage can be understood in the framework of (3.59) and (3.71), including the pulse shape, peak instantaneous power, total energy per pulse, and approximate pulse width.

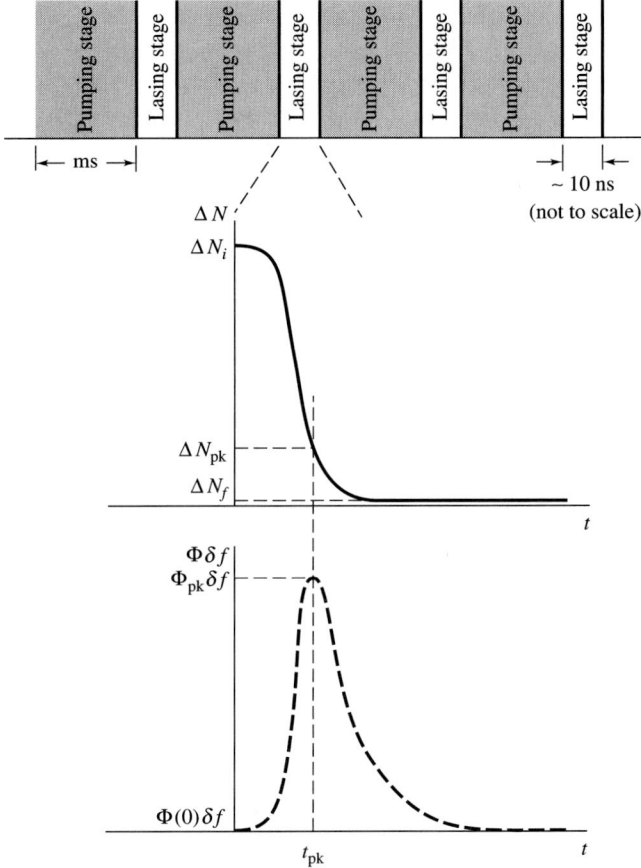

Figure 3.16 Pumping and lasing stages of Q-switching

3.7.1 Pulse Shape

A *qualitative* description of the pulse shape is relatively simple. From (3.71) we see that $\Delta N(t)$ decreases monotonically from the initial value ΔN_i to a yet unspecified final value ΔN_f (Figure 3.16). Also, from (3.59), we see that $\Phi(t)\delta f$ increases initially, reaches a peak value $\Phi_{pk}\delta f$, then drops to a final value when t is large.

If a *quantitative* description of pulse shape is desired, it would be necessary to solve (3.59) and (3.71), subject to the initial conditions $\Phi(0)\delta f$ and $\Delta N(0) = \Delta N_i$. However, no analytical closed form solution is known for these seemingly simple equations [13]. Figure 3.17 shows the results of numerical calculations under a specific set of initial conditions. The pulse shape is *asymmetric* in that the rise time is shorter than the fall time. For a smaller ΔN_i, the peak instantaneous photon density $\Phi_{pk}\delta f$ is smaller and the peak occurs at a

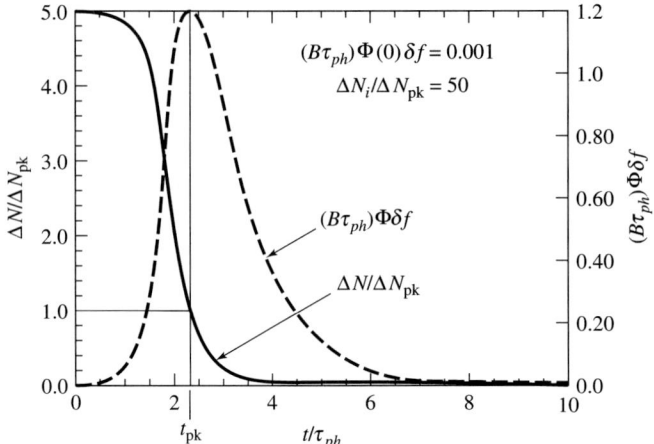

Figure 3.17 Numerical solution of $\Delta N(t)$ and $\Phi(t)\delta f$ of a Q-switched laser

later time t_{pk}. These properties are the general features of the pulses generated by Q-switched lasers.

3.7.2 Peak Power

The time rate of change of the photon density in the cavity is given by (3.59). The photon loss due to scattering, absorption, and diffraction in the cavity and to transmission through the mirrors is $\mathcal{A}d\Phi(t)\delta f/\tau_{ph}$. The fraction of the loss attributable to mirror "i" is $\gamma_i/(2\alpha d + \gamma_1 + \gamma_2)$, where $i = 1$ or 2. This can be understood from the definition of τ_{ph} given in (3.58). The instantaneous power radiated through mirror "i" is

$$P_i(t) = \frac{\gamma_i}{2\alpha d + \gamma_1 + \gamma_2} \mathcal{A}d \frac{\Phi(t)\delta f}{\tau_{ph}} hf$$

To calculate the peak instantaneous power, we need to know the peak value of $\Phi(t)\delta f$. When $\Phi(t)\delta f$ reaches its peak value $\Phi_{pk}\delta f$ at $t = t_{pk}$, $\frac{d(\Phi\delta f)}{dt}$ vanishes. Let the population inversion at this instant be ΔN_{pk}. From (3.59) we obtain

$$\Delta N_{pk} = \frac{1}{V_a B \tau_{ph}} \tag{3.72}$$

Note that ΔN_{pk} is *precisely the same* as the critical population inversion ΔN_c given in (3.60) and the steady-state population inversion ΔN_0 given in (3.62). To evaluate $\Phi_{pk}\delta f$, we combine (3.59) with (3.71) and rearrange the resulting equation to get

$$d(\Phi\delta f) = \frac{1}{2B}\left[\frac{1}{\tau_{ph}\Delta N} - V_a B\right]d(\Delta N) = \frac{1}{2B\tau_{ph}}\left[\frac{1}{\Delta N} - \frac{1}{\Delta N_{pk}}\right]d(\Delta N)$$

We then integrate both sides of the equation to obtain

$$\Phi(t)\delta f - \Phi(0)\delta f = \frac{1}{2B\tau_{ph}}\left[\ln\frac{\Delta N(t)}{\Delta N_i} - \frac{\Delta N(t) - \Delta N_i}{\Delta N_{pk}}\right] \quad (3.73)$$

At $t = t_{pk}$, $\Delta N(t)$ is ΔN_{pk}. Thus, we can solve for the peak value of $\Phi_{pk}\delta f$, as follows:

$$\Phi_{pk}\delta f - \Phi(0)\delta f = \frac{1}{2}\Delta N_{pk} V_a \left[\frac{\Delta N_i}{\Delta N_{pk}} - 1 + \ln\frac{\Delta N_{pk}}{\Delta N_i}\right] \quad (3.74)$$

It is reasonable to expect that the initial photon density $\Phi(0)\delta f$ is negligibly small and can be set equal to zero as an approximation. Since ΔN_i and ΔN_{pk} are known, $\Phi_{pk}\delta f$ can be determined from (3.74). Therefore, in terms of $\Phi_{pk}\delta f$, the peak instantaneous power radiated through mirror "i" is

$$(P_i)_{pk} = \frac{\gamma_i}{2\alpha d + \gamma_1 + \gamma_2}\mathcal{A}d\frac{\Phi_{pk}\delta f}{\tau_{ph}}hf \quad (3.75)$$

3.7.3 Total Energy per Pulse

The total energy radiated per pulse through the two mirrors is

$$\int_0^\infty [P_1(t) + P_2(t)]dt = \frac{hf\mathcal{A}d}{\tau_{ph}}\frac{\gamma_1 + \gamma_2}{2\alpha d + \gamma_1 + \gamma_2}\int_0^\infty \Phi(t)\delta f\, dt \quad (3.76)$$

To evaluate the integral, we combine (3.59) with (3.71) to obtain

$$\frac{d}{dt}\left[\Phi(t)\delta f + \frac{V_a}{2}\Delta N(t)\right] = -\frac{\Phi(t)\delta f}{\tau_{ph}}$$

By integrating both sides of the equation and noting that the initial and final values of $\Phi\delta f$ are small, we obtain

$$\int_0^\infty \Phi(t)\delta f\, dt = -\tau_{ph}\left[\Phi(t)\delta f + \frac{V_a}{2}\Delta N(t)\right]\Big|_0^\infty = \tau_{ph}V_a\frac{\Delta N_i - \Delta N_f}{2}$$

Thus the total energy per pulse is

$$\int_0^\infty [P_1(t) + P_2(t)]dt = \frac{\gamma_1 + \gamma_2}{2\alpha d + \gamma_1 + \gamma_2}\mathcal{A}dV_a\, hf\frac{\Delta N_i - \Delta N_f}{2} \quad (3.77)$$

It is instructive to examine various terms in (3.77) closely. The physical meaning of $(\gamma_1 + \gamma_2)/(2\alpha d + \gamma_1 + \gamma_2)$ has been explained previously, and hf is the photon energy. In the optical cavity of volume $\mathcal{A}d$, there are $\mathcal{A}dV_a\Delta N_i$ states in level E_2 at the beginning of a pulse, and there are $\mathcal{A}dV_a \Delta N_f$ states left at the end of the pulse. In a three-level system, for each photon emitted, an electron drops from E_2 to E_1. Thus, for each photon emitted, N_2 decreases by 1, N_1 increases by 1, and ΔN changes by 2. This accounts for the factor 1/2 in the equation.

However, ΔN_f in (3.77) is yet undetermined. To solve for ΔN_f, we note that as $t \to \infty$, $\Phi(t)\delta f$ approaches 0. From (3.73) we have an expression for the final value ΔN_f

$$\frac{\Delta N_i - \Delta N_f}{\Delta N_i} = \frac{\Delta N_{pk}}{\Delta N_i} \ln \frac{\Delta N_i}{\Delta N_f}$$

which can be written as

$$\frac{\Delta N_f}{\Delta N_i} = e^{-(\Delta N_i - \Delta N_f)/\Delta N_{pk}} \tag{3.78}$$

Again, $\Phi(0)\delta f$ has been set equal to 0 as an approximation. Since ΔN_i and ΔN_{pk} are known, ΔN_f can be solved numerically from (3.78). An approximate expression for ΔN_f can be obtained if we assume $\Delta N_i \gg \Delta N_{pk}$. When ΔN_i is much greater than ΔN_{pk}, ΔN_i must be much greater than ΔN_f, as well. By neglecting ΔN_f in the exponent of (3.78), we have an approximate expression for ΔN_f

$$\Delta N_f \approx \Delta N_i \, e^{-\Delta N_i/\Delta N_{pk}} \tag{3.79}$$

This completes our discussion on the total energy contained in a Q-switched laser pulse.

3.7.4 Approximate Pulse Width

As noted previously, an accurate determination of pulse width would require detailed numerical calculations. However, an approximate estimate for the pulse width is possible, based on the expressions for the peak instantaneous power and the total energy per pulse. For a rectangular pulse with a peak instantaneous power $P_{pk} = (P_1)_{pk} + (P_2)_{pk}$ and a pulse width of Δt_{rec}, the total energy per pulse is simply $P_{pk}\Delta t_{rec}$. Thus, we have as an approximation,

$$\Delta t_{rec} = \frac{\int_0^\infty [P_1(t) + P_2(t)]dt}{(P_1)_{pk} + (P_2)_{pk}} \tag{3.80}$$

Using (3.74), (3.75), and (3.77), we have

$$\Delta t_{rec} = \tau_{ph} \frac{\Delta N_i - \Delta N_f}{\Delta N_i - \Delta N_{pk} + \Delta N_{pk} \ln \frac{\Delta N_{pk}}{\Delta N_i}} \tag{3.81}$$

For pulses with other pulse shapes, it would be necessary to introduce a numerical factor S_p such that the total energy per pulse is given by $P_{pk}\Delta t/S_p$. Table 3.1 lists the numerical values of S_p for five pulse shapes [14].

Table 3.1 Pulse shape function

Pulse shape	S_p
Rectangular	1.00
Gaussian	0.94
sech	0.88
sech^2	0.84
Lorentzian	0.64

3.8 MODE LOCKING OPERATION

Mode locking is the second method for generating narrow pulses [15–18]. As mentioned briefly in section 5 of this chapter, an active medium has a useful gain over a finite bandwidth B_g, which may be wide enough to support several longitudinal modes. The frequency separation between the longitudinal modes is determined by cavity length, as discussed in Chapter 2. In some lasers, the optical cavities are long enough that many longitudinal modes can be supported within the gain bandwidth of the active medium. By controlling the phases of these longitudinal modes, the laser system can produce narrow and intense pulses. This is the basis of mode locking.

3.8.1 Electrical Source Phase Locking

We can use the locking of electronic oscillators to illustrate the essential features of mode locking in lasers. To monitor narrow optical pulses with electronic instruments, it is necessary to use a photodetector to convert optical pulses to electrical pulses. The responses of photodetectors are much slower than the optical frequencies. The effect of slow detectors has to be accounted for. Therefore a slow detector is included in our circuit model for the mode locking of signals. Consider electrical signals fed to a detector. All detectors are nonlinear devices. The functions of detectors may be modeled by a square-law characteristic, representing the nonlinear behavior of the detectors, and a low-pass filter characteristic, signifying the slowness of the detector response. The equivalent circuit of a detector therefore consists of an ideal square-law element and an ideal low-pass filter, as shown in Figure 3.18a.

Suppose a signal of angular frequency ω is applied to the detector input. Then,

$$V_1(t) = V_o \cos\omega t$$

After passing through the square-law element, the signal becomes

$$[V_1(t)]^2 = \frac{V_o^2}{2}(1 + \cos 2\omega t)$$

Because of the low-pass filter, the direct current (dc) portion of the squared signal is the only frequency component preserved. The detector output is then

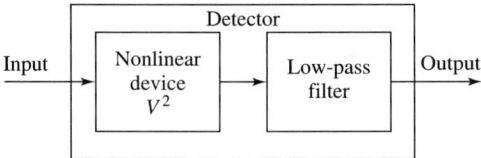

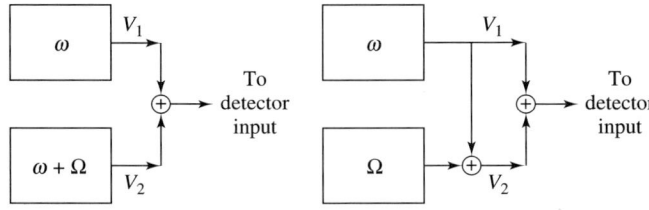

(a) Equivalent circuit for a typical detector

(b) Two isolated sources

(c) Two sources with fixed phase relationship

Figure 3.18 Phase locking of two electrical sources

$$[V_{out}(t)]_1 = \Re \frac{V_o^2}{2}$$

where $\Re$ is a constant representing the detector sensitivity. The output can also be viewed as a term proportional to the time average of $[V_1(t)]^2$. Averaging over a time longer than $2\pi/\omega$, the $\cos 2\omega t$ term drops out and the dc term remains.

Suppose there are two signals, $V_1(t)$ and $V_2(t)$, and the frequency of $V_2(t)$ is slightly different from $V_1(t)$. That is,

$$V_2(t) = V_o \cos[(\omega + \Omega)t + \phi]$$

and $\omega \gg \Omega$. In addition to the small frequency difference, there is also a possible phase difference ϕ. If $\phi = 0$, then $V_1(t)$ and $V_2(t)$ peak simultaneously at $t = 0$.

Suppose the two signals are superimposed and fed into the detector. Because of the ideal square-law element, we have

$$[V_1(t) + V_2(t)]^2 = V_o^2 \left[1 + \cos(\Omega t + \phi) + \frac{1}{2}\cos 2\omega t \right. \\ \left. + \frac{1}{2}\cos 2[(\omega + \Omega)t + \phi] + \cos(2\omega t + \Omega t + \phi) \right]$$

The frequency components 2ω, $2\omega + \Omega$, and $2(\omega + \Omega)$ are rejected by the low-pass filter or removed by averaging over a long period. Thus, the detector output can be represented by

$$[V_{out}(t)]_2 = \Re V_o^2 [1 + \cos(\Omega t + \phi)]$$

3.8 Mode Locking Operation

which is a function of Ω and ϕ. Note that $[V_{out}(t)]_2$ peaks at $(2m\pi - \phi)/\Omega$, where m is an integer. The peak value of $[V_{out}(t)]_2$ is

$$[V_{out}(t)]_{2,pk} = \Re \, 4 \frac{V_o^2}{2}$$

which is four times the peak value from V_1 or V_2 alone. Also note that the separation between major peaks is $2\pi/\Omega$

From this discussion, two interesting features of locking emerge:

1. The peak detector output is quadrupled when two sources of equal amplitude are applied to the detector.
2. The occurrence of major peaks is determined by ϕ, while the separation between major peaks depends on Ω.

Figures 3.19a–d show respectively $V_1(t)$, $V_2(t)$, $[V_1(t) + V_2(t)]$, and $[V_1(t) + V_2(t)]^2$ for $\phi = 0$. The dashed curve shown in Figure 3.19d is the envelop of $[V_1(t) + V_2(t)]^2$.

Next, we consider the implementation of locking. Suppose two sources isolated from each other, as shown schematically in Figure 3.18b, are used as sources for $V_1(t)$ and $V_2(t)$. Although ω and Ω are fixed, ϕ can be any value.

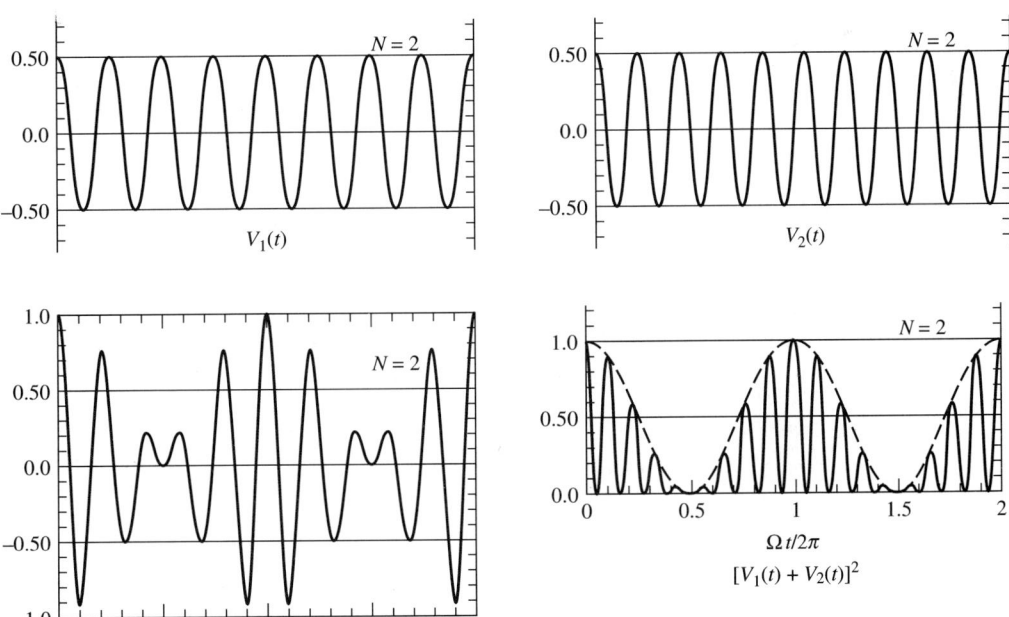

Figure 3.19 Locking of two voltage sources

For example, the sources may be turned on and off repeatedly. Each time the two sources are turned on, ϕ may have a different value, $[V_1(t) + V_2(t)]$ will be different, and the peaks of $[V_{out}(t)]_2$ will occur at different times. In other words, the result of superimposing two outputs is not predictable if the phase ϕ is not controlled.

If measures are taken to ensure a fixed phase relationship, that is, a fixed ϕ, then $[V_{out}(t)]_2$ is always the same and the peaks occur at a fixed time. One possible way to guarantee a fixed phase relationship is to use two oscillators with angular frequencies of ω and Ω, as shown in Figure 3.18c. In this scheme, $V_1(t)$ is generated by one signal source, while $V_2(t)$ is the sum frequency signal obtained by mixing the outputs of the two sources, as shown in Figure 3.18c. With the phase relationship fixed, the superposition of two sources leads to predictable results. This is known as the **coherent superposition of sources**, or simply **locking.**

Let us now consider the superposition of N sources:

$$V_k(t) = V_o \cos[(\omega + [k-1]\Omega)t + \phi_k], \qquad k = 1, 2, \ldots n \quad (3.82)$$

When these signals are combined and fed into the detector, the filtered detector output is proportional to the time average of $[\sum_{k=1}^{N} V_k(t)]^2$. From (3.82), we see that the detector output is dependent on ϕ_k. If all ϕ_k are not controlled, the output is unpredictable. If the ϕ_k are random variables, the average detector output is $\Re N V_o^2/2$. For phase-locked sources, the ϕ_k are fixed and the situation is different. For simplicity, we consider the case of $\phi_k = 0$, that is, all voltages peak simultaneously at $t = 0$ and at regular time intervals $2\pi/\Omega$ thereafter. Then,

$$[\sum_{k=1}^{N} V_k(t)]^2 = V_o^2 \cos^2\left(\omega t + \frac{N-1}{2}\Omega t\right)\left[\frac{\sin(N\Omega t/2)}{\sin(\Omega t/2)}\right]^2$$

The detector output following the low-pass filtering is

$$[V_{out}(t)]_N = \Re \frac{V_o^2}{2}\left[\frac{\sin(N\Omega t/2)}{\sin(\Omega t/2)}\right]^2 \quad (3.83)$$

The features of coherent superposition of two, four, eight, and sixteen sources are shown in Figure 3.20a. The solid lines depict the detailed variations of $[\sum_{k=1}^{N} V_k(t)]^2$. The envelopes, shown as the dashed curves, are the pulse shapes monitored by a slow detector and observed on an oscilloscope. In the following discussion, we are mainly concerned with these envelopes.

For $N > 2$, minor maxima appear, in addition to the major peaks. The major peaks have a value of $\Re N^2 V_o^2/2$, since

$$\left[\frac{\sin(N\Omega t/2)}{\sin(\Omega t/2)}\right]^2 \to N^2$$

as $\Omega t \to 0$ or $2m\pi$. The time separation between major peaks is derived from (3.83):

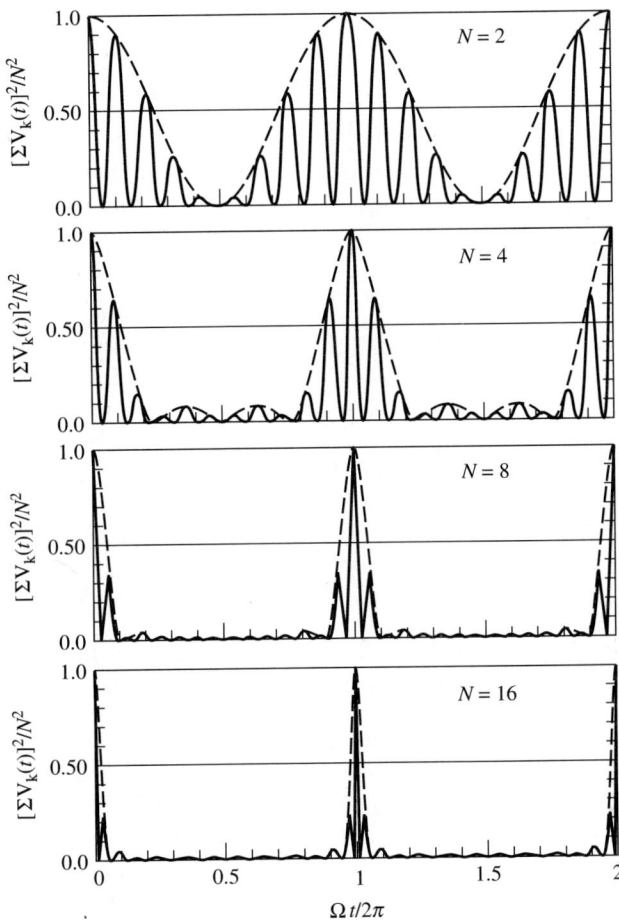

(a) Coherent superposition or locking of 2, 4, 8, and 16 sources

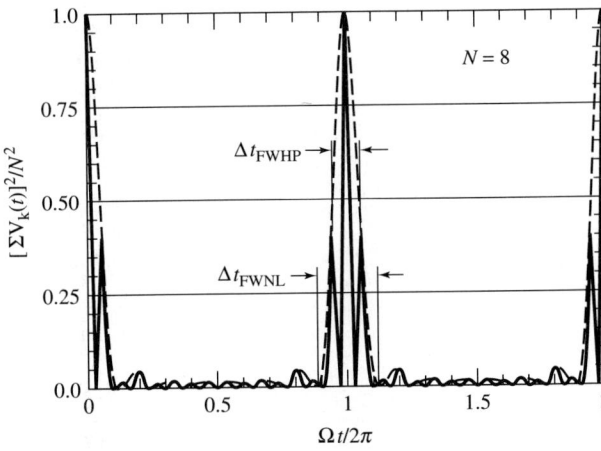

(b) Definition for pulse width Δt_{FWHP} and Δt_{FWNL}

Figure 3.20 Locking of several sources

$$T = 2\pi/\Omega \tag{3.84}$$

The width of the major lobes may be defined as the time separation between the *nearest nulls* on each side of a major peak (Figure 3.20b). The pulse width thus defined is the **full width between nulls** (FWNL). From (3.83) we have

$$\Delta t_{\text{FWNL}} = \frac{4\pi}{N\Omega} \tag{3.85}$$

Pulse width may also be defined as the time separation between the *half-power or 3 dB points*. The pulse width thus defined is the **full width between half-power** points (FWHP), as indicated in Figure 3.20b. Table 3.2 lists the $N\Omega\Delta t_{\text{FWHP}}$ as a function of N.

An *approximate* expression for the pulse width for N greater than 5 is possible. For large N, the pulses are narrow, and $[V_{\text{out}}(t)]_N$ reduces to one-half of its peak value when $\Omega t/2$ is still small. Therefore, the denominator of (3.83) may be approximated as $\Omega t/2$, which means that (3.83) behaves like $\sin^2 x/x^2$. Numerical calculations show that $\sin x/x$ is $1/\sqrt{2}$ when x is 1.3916. Thus, for N greater than 5,

$$\Delta t_{\text{FWHP}} \approx \frac{4 \times 1.3916}{N\Omega} \approx \frac{5.6}{N\Omega} \tag{3.86}$$

For sources with different amplitudes or phase relationships, $[V_{\text{out}}(t)]_N$ has a different pulse shape, but the basic features remain the same; namely, the pulse width is on the order of $a/(N\Omega)$, where a is a numerical constant depending on the definition of the pulse width, the amplitude distribution, and the phase difference between various signal sources.

3.8.2 Laser Mode-Locking

As noted in Chapter 2, a large number of longitudinal modes can be supported by a long optical cavity. For simplicity, consider an optical cavity with two planar mirrors separated by a distance d. The wavelength of a mode is such that the round-trip phase delay in the cavity is an integer multiple of 2π, that is,

$$2kd = 2\frac{2\pi}{\lambda}d = 2\pi m$$

where m is an integer. The free-space wavelength λ_m and frequency f_m of longitudinal mode m are

$$\lambda_m = \frac{2d}{m} \tag{3.87}$$

$$f_m = \frac{c}{2d}m \tag{3.88}$$

The frequency separation between two adjacent longitudinal modes is

3.8 Mode Locking Operation

Table 3.2 Exact and approximate values of $N\Omega\Delta t_{FWHP}$

N	Exact value	Approx. value based on (86)
2	6.28	5.56
3	5.85	5.56
4	5.72	5.56
5	5.67	5.56
6	5.63	5.56
7	5.62	5.56
8	5.61	5.56
9	5.60	5.56
10	5.59	5.56
20	5.58	5.56
30	5.58	5.56
40	5.56	5.56
50	5.56	5.56

$$\Delta f = f_{m+1} - f_m = \frac{c}{2d} \tag{3.89}$$

For gas, liquid, solid-state, or semiconductor injection lasers with external cavities, m can be quite large. For example, for a typical HeNe laser with a cavity length of 0.2 m, emitting at 0.633 µm, the integer m is on the order of 6×10^5 and Δf is approximately 750 MHz.

The locking of several longitudinal modes can be understood in the framework developed in section 3.8.1. In the terminology introduced there, the angular frequency Ω is $2\pi\Delta f$. When N longitudinal modes are superimposed coherently and $N > 5$, we have, from (3.85), (3.86), and (3.87),

$$T = \frac{2d}{c} \tag{3.90}$$

$$\Delta t_{FWNL} = \frac{4d}{Nc} = \frac{2T}{N} \tag{3.91}$$

$$\Delta t_{FWHP} \approx \frac{5.6\, d}{\pi Nc} = \frac{0.89\, T}{N} \tag{3.92}$$

The resulting pulses are very narrow, if a large number of modes are locked together. It is also interesting to observe that the T given in (3.90) is precisely the time required for light to complete one round trip of travel in the cavity at the speed of light.

The number of longitudinal modes of a laser is determined by the gain bandwidth B_g of the active medium, the frequency separation Δf between adjacent longitudinal modes, and the bandwidth of the mirror reflectivity. If the dominant limiting factors are B_g and Δf, and

$$N = \frac{B_g}{\Delta f} = \frac{2d\, B_g}{c} \tag{3.93}$$

then the pulse widths for $N > 5$ are

$$\Delta t_{\text{FWNL}} = \frac{2}{B_g} \qquad (3.94)$$

$$\Delta t_{\text{FWHP}} \approx \frac{0.89}{B_g} \qquad (3.95)$$

Note that the pulse widths are now independent of the cavity length.

Actual mode locking can be accomplished by inserting an amplitude or phase modulator in the optical path and driving the modulator with a frequency Δf [17,18], as shown in Figure 3.21a. Such a scheme is known as **active mode locking**, since electric power is required to drive the amplitude or phase modulator. Locking can also be achieved by incorporating a saturable absorber in the optical cavity (Figure 3.21b). Such schemes are depicted in Figure 3.21b and are referred to as **passive mode locking** since no external power source is needed for locking.

3.9 DESCRIPTION OF GAS LASER SYSTEMS

There are various types of lasers, involving different physical processes in a variety of materials, as well as numerous cavity configurations. The important types are neutral gas lasers, ionized gas lasers, molecular gas lasers, liquid dye lasers, solid-state lasers, semiconductor injection lasers, and free-electron lasers. A few

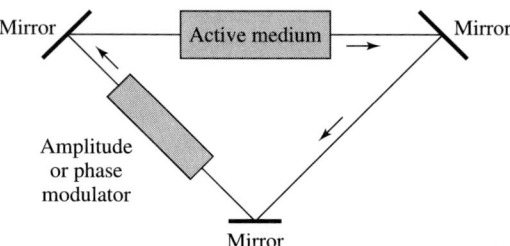

(a) Active mode locking

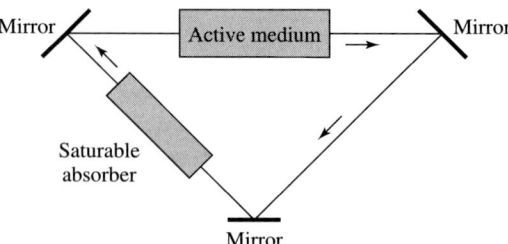

(b) Passive mode locking

Figure 3.21 Active and passive mode locking

Table 3.3 Coherence length of typical optical sources

Source	Coherence length
Sun	10^{-7} m
High-pressure Hg arc lamp	10^{-5} m
Low-pressure Hg arc lamp	10^{-4} m
Argon laser without mode selection etalon	10^{-2} to 10^{-1} m
Argon laser with mode selection etalon	10^{1} m
HeNe laser	10^{-1} m
Single-frequency, stablized HeNe laser	10^{2} m
Semiconductor injection laser without external cavity	10^{-2} m

of the more important gas, dye, and solid-state laser systems are described briefly in this and the following sections. Semiconductor injection lasers and related light sources are discussed in the next chapter.

Generally speaking, gas lasers have relatively long and large optical cavities. Their emission is characterized by a narrow spectral width and a narrow beam divergence angle. In particular, continuous-wave gas lasers have exceptional temporal and spatial coherence properties and are useful for various scientific and engineering applications. The coherence length for a number of typical optical sources, including gas and semiconductor injection lasers, is given in Table 3.3.

Gas lasers are usually pumped by electrical discharges, through which electrical energy is converted into electron or ion kinetic energy. Due to particle collisions, the kinetic energy of the charged particles is transferred to the constituent lasing particles, thereby raising their energy levels from the ground state to excited and metastable states. Laser action involving **ionized atoms** usually results in radiation in the visible or ultraviolet (UV) region (0.25 to 0.6 μm), **neutral atoms** in the near infrared (IR) region (1.2 to 12 μm), and **molecular gases** in the middle IR region (10 to 100 μm). If the emission is due to transitions between **molecular rotational and vibrational states,** the radiation is in the far IR and submillimeter wavelength regions [19].

3.9.1 Neutral Gas Lasers

Coherent emissions in 29 neutral atomic elements have been reported. These include noble gases, mixtures of noble gases, metal and halogen vapors[2], and others. Of all neutral gas lasers in existence, HeNe lasers operating at 0.6328 μm are the best known. In a HeNe laser tube, a dc or radio-frequency discharge is established in the mixture of He and Ne. Because of this electric discharge,

---

[2] Halogens are atoms that have one fewer electron than the inert gases. They are fluorine (F, atomic number 9), chlorine (Cl, 17), bromine (Br, 35), iodine (I, 53), and astatine (At, 85).

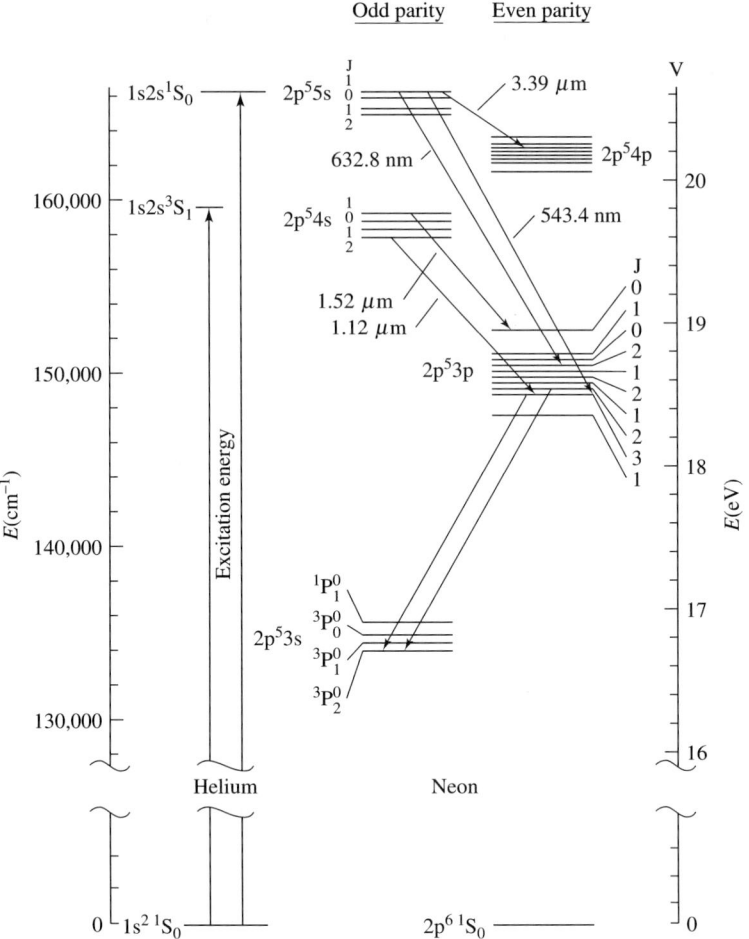

Figure 3.22 Energy level diagram of helium and neon in HeNe lasers [6]

He atoms are excited to long-lived metastable states by impact with energetic electrons. The energy is then exchanged between the metastable states of He and Ne, which is the **active lasing constituent.** The laser action occurs between the energy levels of Ne atoms. More than 100 different energy lines of Ne can be made to lase, most of which are quite weak. Five emission lines are particularly strong and have been developed into commercial products. They are:

$1s^22s^22p^55s\ ^1P_1 \to 1s^22s^22p^53p\ ^3S_1$, $\lambda_{vacuo} = 0.5435$ μm, $\lambda_{air} = 0.5433$ μm (green)
$1s^22s^22p^55s\ ^1P_1 \to 1s^22s^22p^53p\ ^3P_2$, $\lambda_{vacuo} = 0.6330$ μm, $\lambda_{air} = 0.6328$ μm (red)
$1s^22s^22p^54s\ ^1P_1 \to 1s^22s^22p^53p\ ^3P_2$, $\lambda_{vacuo} = 1.1526$ μm, $\lambda_{air} = 1.1523$ μm
$1s^22s^22p^54s\ ^1P_1 \to 1s^22s^22p^53p\ ^1S_0$, $\lambda_{vacuo} = 1.5235$ μm, $\lambda_{air} = 1.5231$ μm
$1s^22s^22p^55s\ ^1P_1 \to 1s^22s^22p^54p\ ^3P_2$, $\lambda_{vacuo} = 3.3922$ μm, $\lambda_{air} = 3.3913$ μm

3.9 Description of Gas Laser Systems

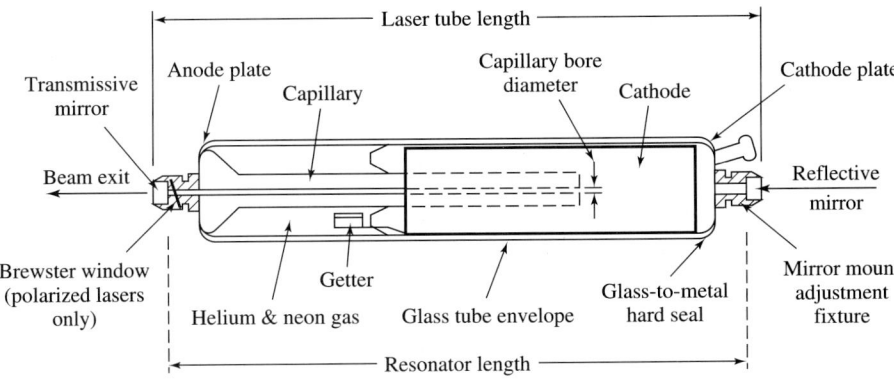

Figure 3.23 Construction of a modern HeNe laser plasma tube ([25], © Laurin Publishing Co., Inc.)

The energy levels involved in these commercial HeNe lasers are shown in Figure 3.22 [6], and their detailed numerical values are listed in Table 3.4 [20]. Note that the emission wavelength in air is slightly shorter than that in vacuum. Also, under normal circumstances, the gain at 3.3913 μm, for example, is much stronger than that at 0.6328 μm. For maximum power output at 0.6328 μm, emission at 3.3913 μm must be suppressed either by wavelength selection schemes or by magnetic fields. The laser output power is a function of gas pressure, the mixture ratio of He and Ne, tube diameter, discharge current, and gas temperature [21–24]. Typically, approximately 90 percent He is mixed with 10 percent Ne to produce a total pressure of 1 to 2 Torr[3]. A typical HeNe laser gas discharge tube has an inner diameter of a few mm. To maintain the electric discharge for an output power of a few milliwatts, a dc source supplying 2000 to 3000 V and 10 to 20 mA is required. The technology for manufacturing HeNe lasers is quite mature. Figure 3.23 shows the detailed construction of a commercial HeNe laser tube. The expected lifetime of modern HeNe lasers is 20,000 to 100,000 hours (2.3 to 11.6 years) [25]. Substitution of ^{4}He atoms by the ^{3}He isotope can typically increase the laser output by 25 percent, because the mass of the ^{3}He isotope is smaller than that of the ^{4}He, thereby resulting in higher electron temperatures.

3.9.2 Ionized Gas Lasers

Neutral gas lasers in general, and HeNe lasers in particular, are relatively simple. Colloquially, HeNe lasers may be considered the "glass receiving tubes" of the laser world. By comparison, ionized gas lasers are so complex that they would be viewed as the "klystrons" or "traveling wave tubes" of the laser world.[4] How-

---

[3] Torricelli (Torr) = 1 mm Hg at 0°C and 1 ATM = 760 mm Hg at 0°C.
[4] Free-electron lasers based on linear accelerators are probably the real "optical traveling wave tubes" [26].

ever, except for HeNe lasers operating at 0.5433 μm and 0.6328 μm, few cw neutral atomic or molecular gas lasers emit in the visible and UV wavelength ranges, and *no* neutral atomic gas lasers have wavelengths shorter than 0.5433 μm. On the other hand, more than 400 lines originating from 29 elements, ranging in wavelength from the UV to the IR, have been observed with ionized gas lasers.

In ionized gas lasers, or simply **ion lasers,** stimulated emission takes place between the energy levels of ionized atoms. Although many transitions are allowed in ionized elements, only a few lines have been developed into practical laser systems. **Metal ion lasers** use either Cd, Se, Zn, Pb, or Sn ions as the active media. In this group, HeCd lasers, with outputs at 0.4416 μm (blue) and 0.3250 μm (UV) are the most representative. In the **noble gas ion laser** family, lasers using Ar II [0.4880 μm (blue) and 0.5145 μm (green)], Kr II [0.6471 μm (red), 0.5682 μm (yellow), and 0.5309 μm], Ne III (UV), Ar III (UV), and Kr III (UV) are the important ones.

The most useful of the ion laser family are probably the Ar II lasers. In these lasers, several lines are emitted simultaneously, the two dominant lines being 0.4880 μm and 0.5145 μm. In addition, Ar and Kr lines can be emitted simultaneously from the same laser tube [27]. The emission lines of argon lasers, krypton lasers, and mixed argon and krypton lasers are listed in Table 3.5.

The excitation mechanism of Ar II lasers is quite complicated. For Ar lasers with a multimode, multiwavelength output of a few watts in the blue–green region, the output power increases *quadratically* with the *current density* [28]. Thus, to increase output power, a large current with a small bore diameter is desirable. An electrical discharge can be established in Ar at low pressure (0.1 to 1 Torr) in a plasma tube with a bore diameter of a few millimeters. The power efficiency is less than 1 percent, and most of the electrical power is dissipated thermally.

The plasma tubes are usually water cooled. In the early days, fused silica tubes were used as the plasma tubes. Due to ion sputtering, and the gas cleanup resulting from the sputtering and decomposition of the bore, the lifetime of an argon laser used to be very short. Even with water cooling, the useful lifetime of cw argon lasers was only a few hundred hours. However, the introduction of segmented disc plasma tubes with bore segments of graphite or tungsten was a milestone in the advancement of argon laser technology. Modern argon lasers use boron nitride, alumina, or beryllium oxide as the bore material. Because of its high thermal conductivity, good thermal shock resistance, high electrical resistivity, small thermal expansion coefficient, and low release of contaminating gases, BeO ceramic is well suited for the construction of laser tubes. With BeO plasma tubes, a lifetime of 10,000 hours may be expected. Recently, low-power (few milliwatts), air-cooled, cw Ar lasers have also become available commercially (Figure 3.24).

Table 3.4 The lower energy levels of He and Ne [20]

He			
Electron configuration	Term symbol	Paschen notation	Energy cm^{-1}
1s^2	$^1S_0$		0
1s 2s	$^3S_1$		159850.32
1s 2s	$^1S_0$		166271.70

Ne 1s^2 2s^2 −			
Electron configuration	Term symbol †	Paschen notation	Energy cm^{-1}
2p^6	$^1S_0$		0
2p^5 3s	$^3P_2$	1s$_5$	134043.79
2p^5 3s	$^3P_1$	1s$_4$	134461.24
2p^5 3s	$^3P_0$	1s$_3$	134820.59
2p^5 3s	$^1P_1$	1s$_2$	135890.67
2p^5 3p	$^3S_1$	2p$_{10}$	148259.75
2p^5 3p	$^3D_3$	2p$_9$	149659.00
2p^5 3p	$^3D_2$	2p$_8$	149826.18
2p^5 3p	$^3D_1$	2p$_7$	150123.55
2p^5 3p	$^1D_2$	2p$_6$	150317.82
2p^5 3p	$^3P_0$	2p$_3$*	150919.39
2p^5 3p	$^1P_1$	2p$_5$	150774.07
2p^5 3p	$^3P_2$	2p$_4$	150860.47
2p^5 3p	$^3P_1$	2p$_2$	151040.41
2p^5 3p	$^1S_0$	2p$_1$	152972.70
2p^5 4s	$^3P_2$	2s$_5$	158603.07
2p^5 4s	$^3P_1$	2s$_4$	158797.95
2p^5 4s	$^3P_0$	2s$_3$	159381.94
2p^5 4s	$^1P_1$	2s$_2$	159536.57
2p^5 4p	$^3S_1$	3p$_{10}$	162519.85
2p^5 4p	$^3D_3$	3p$_9$	162832.68
2p^5 4p	$^3D_2$	3p$_8$	162901.09
2p^5 4p	$^3D_1$	3p$_7$	163014.60
2p^5 4p	$^1D_2$	3p$_6$	163040.33
2p^5 4p	$^3P_0$	3p$_3$*	163403.28
2p^5 4p	$^1P_1$	3p$_5$	163659.25
2p^5 4p	$^3P_2$	3p$_4$	163710.58
2p^5 4p	$^3P_1$	3p$_2$	163709.70
2p^5 4p	$^1S_0$	3p$_1$	164287.86
2p^5 5s	$^3P_2$	3s$_5$	165830.14
2p^5 5s	$^3P_1$	3s$_4$	165914.76
2p^5 5s	$^3P_0$	3s$_3$	166608.31
2p^5 5s	$^1P_1$	3s$_2$	166658.48

† Term symbols are based on LS coupling scheme.
* The Paschen notations for 2p$_3$ and 3p$_3$ are listed correctly.

Table 3.5 Prominent emission lines in Ar and Kr ion lasers [27]

Ar lasers wavelength (μm)	Kr lasers wavelength (μm)	Ar/Kr lasers wavelength (μm)
0.3511	0.3507	
0.3638	0.3564	
0.4545	0.4619	0.4579
0.4579	0.4680	0.4765
0.4658	0.4762	0.4825
0.4727	0.4825	0.4880*
0.4765*	0.5208	0.4965
0.4880*	0.5309*	0.5017
0.4965	0.5682*	0.5145*
0.5017	0.6471*	0.5208
0.5145*	0.6764	0.5309
	0.7525	0.5682
	0.7931	0.6471*
	0.7993	0.6764

Three strongest lines are indicated by *.

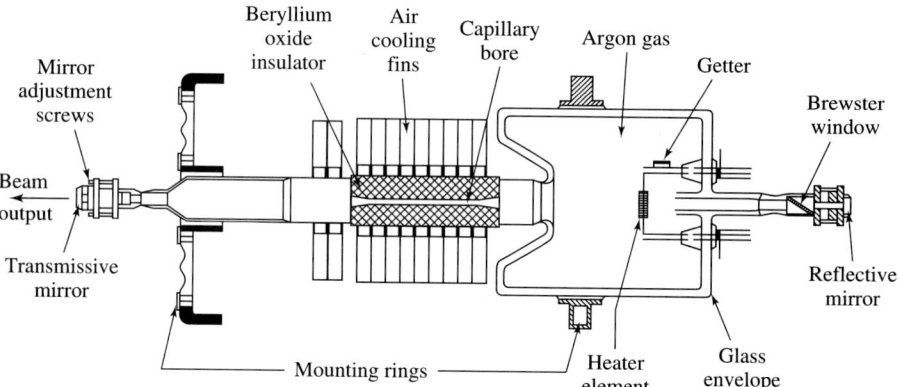

Figure 3.24 Construction of a modern low-power air-cooled argon laser tube ([29], © Laurin Publishing Co., Inc.)

3.9.3 Molecular Gas Lasers

In molecular gas lasers, lasing is due to transitions between the vibrational and rotational energy levels of gas molecules, which include: **diatomic molecules** (CN, CO, HBr, DBr, HCl, DCl, HF, DF, H_2, HD, D_2, NO, and N_2), **triatomic molecules** (CO_2, CO_2^{18}, CS_2, HCN, HCN^{15}, DCN, H_2O, H_2O^{18}, D_2O, H_2S, N_2O, and SO_2), and **polyatomic molecules** (CH_3F, CH_3OH, $H_2C{:}CHCL$, and NH_3). The spectral range of molecular gas lasers extends from 0.15 μm (H_2 lasers) in the UV to 773.5 μm (HCN lasers) in the far IR.

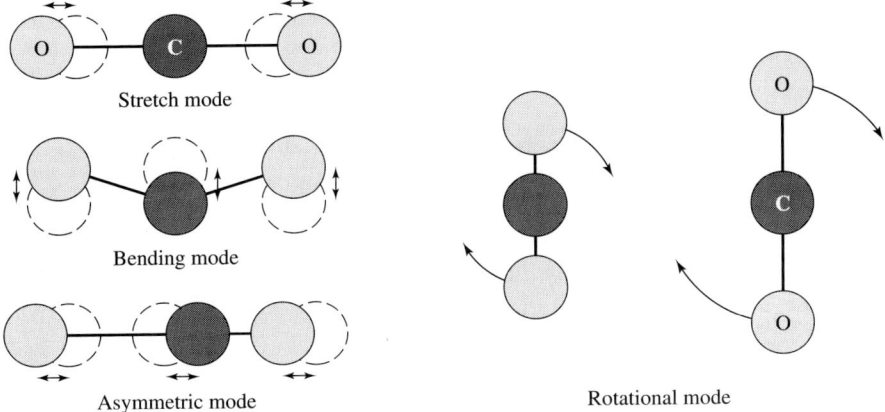

Figure 3.25 Vibrational and rotational modes of CO_2 molecules [7]

Of all the molecular gas lasers mentioned, CO_2 lasers are probably the best known and are certainly the most powerful lasers. CO_2 gas is an efficient, high-gain medium with high power capability. Gasdynamic CO_2 lasers with output powers in excess of 100 kW for a few seconds have been used in many industrial applications. CO_2 molecules have three normal vibration modes: **symmetric stretch, double degenerate bending,** and **asymmetric stretch** (Figure 3.25). The vibration–rotation energy levels are specified by three quantum numbers p, f, and r. A superscript ℓ on the bending quantum number represents the angular momentum of bending with respect to the molecular axis. Thus, the vibrational–rotational state of a CO_2 molecule is denoted by a symbol $(pf^\ell r)$.

Figure 3.26 illustrates a greatly simplified energy level diagram of CO_2 molecules, in which the fine structure associated with the rotational levels has been omitted. Among the multitude of possible transitions, two of the stronger groups are the (00^01) to (10^00) and the (00^01) to (02^00) transitions, with emission bands near 10.4 μm and 9.4 μm, respectively. Many CO_2 lasers emit at the (00^01)–(10^00) transition, with a vacuum wavelength of 10.59 μm. This emission line is particularly significant because the earth's atmosphere has a transmission window around 10 μm.

An important improvement in CO_2 lasers came with the discovery of the resonant energy transfer of vibrational energy from the metastable state of N_2 to the 00^01 level of CO_2. The output power of CO_2 lasers is increased by four orders of magnitude when a mixture of CO_2, N_2, and He is used instead of pure CO_2. CO_2 lasers containing He can also be driven with higher discharge currents to yield a larger population inversion density.

Conventional CO_2 Lasers Conventional CO_2 lasers can use either a continuously flowing CO_2–N_2–He mixture along a discharge tube, or a sealed-off discharge tube containing a CO_2–N_2–He–H_2–Xe mixture. The output from a sealed-off CO_2 laser can be as much as 60 percent higher than that of a corresponding flowing system. Lasing is maintained by the application of a dc or radio-frequency discharge, and heat dissipation is provided by water cooling in the laser tube walls.

The gain of a CO_2 medium is a function of the discharge current, the gas mixture ratio, the tube diameter, the wall temperature, and the flow rate. At low pressure, the emission linewidth is mainly determined by the Doppler effect. As the gas pressure increases, the collisions play an increasingly important role in determining that linewidth. Figure 3.27 shows the transition from **Doppler broadening** at low pressure to **collision broadening,** also known as **pressure broadening,** at higher pressures [12].

Waveguide CO_2 Lasers As the inner diameter of a discharge tube decreases, the spectral linewidth, the saturation intensity for cw operation, and the peak energy density for pulse operation all increase, while the efficiency, gain, and output power per unit length remain constant. If the optimum operating conditions are to be maintained as the tube diameter is narrowed, then the total gas pressure should be increased, the discharge current should be decreased, and the voltage across the discharge tube should be increased, according to the scaling laws of gas discharge with high particle densities [30]. Optical attenuation

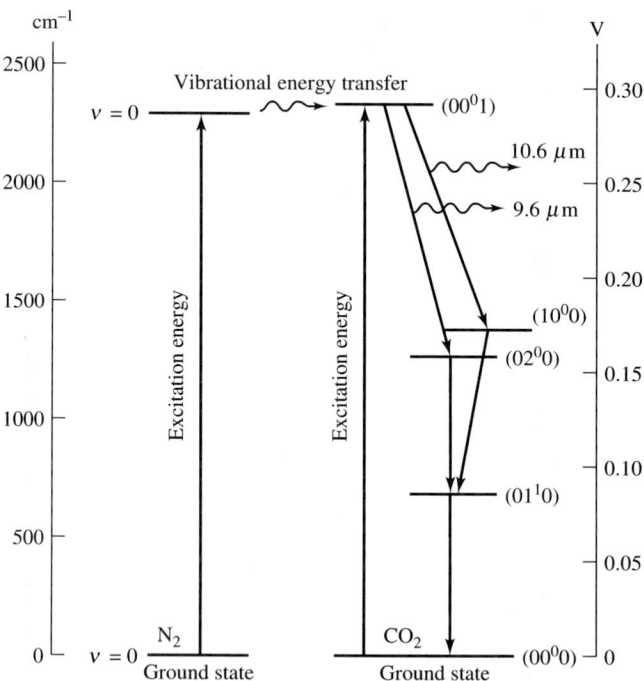

Figure 3.26 Energy levels of N_2 and CO_2 molecules [7]

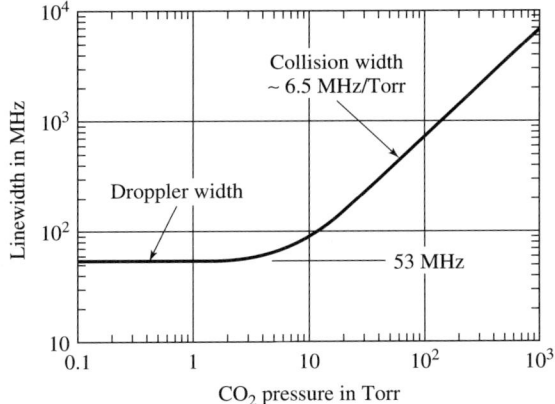

Figure 3.27 Linewidth of a CO_2 laser at 10.6 μm as a function of CO_2 pressure [12]

also increases when discharge tube sizes become smaller. Nevertheless, the loss may be overcome by the gain as long as the bore diameter is larger than 10λ. Consequently, by using discharge tubes with small bore diameters and high gas pressures, we can produce compact **waveguide lasers.** In conventional lasers, the beams are not guided by a waveguide structure. In waveguide lasers, the waves are guided by the discharge tube. This is the basic difference between conventional lasers and waveguide lasers.

Typical waveguide CO_2 lasers have hollow glass or quartz capillary tubes with bore diameters of a few millimeters. They can be gas-flowing or sealed-off systems. The total gas pressure is in the 100–to–700 Torr range. Like conventional CO_2 lasers, the small-signal gains and cw power outputs of waveguide CO_2 lasers are also functions of bore diameter, discharge current, gas pressure, wall or coolant temperature, and, in the case of gas-flowing systems, the gas flow rate. To reduce the wall temperature, high thermal conductivity ceramics like beryllium oxide (BeO), boron nitride (BN), and alumina (Al_2O_3) may be used in lieu of glass or quartz capillaries. The most unique feature of waveguide CO_2 lasers is that their pressure-broadened linewidths are very broad (Figure 3.27), and hence the lasers may be tuned over a range of gigahertz.

3.10 DESCRIPTION OF DYE LASERS

Lasing can also be obtained by dissolving organic dyes in solvents and pumping the dye-solvent solution with an optical source. These lasers are called **dye lasers.** More than 500 different dyes have been tested as the active material. Also, many liquids, including water, alcohol, ethylene glycol (antifreeze), ethanol, cyclohexane, toluene, and others can be used as the solvents. Flash lamps, pulsed N_2 or ruby lasers, continuous-wave HeNe or Ar lasers, or semiconductor injection laser diodes can be used as the source for optical pumping.

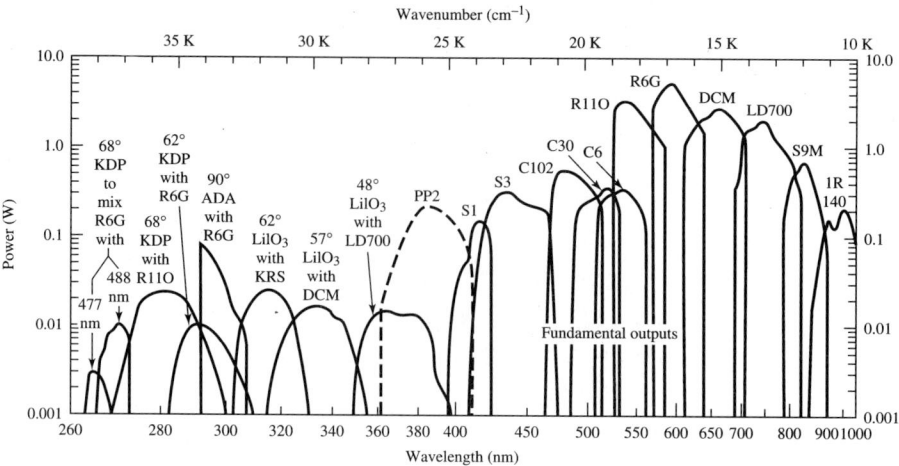

Figure 3.28 Power output versus wavelength of single-frequency dye laser with various dyes [31]

The most distinguishing feature of dye lasers is their tunability. To date, emissions from cw dye lasers extend continuously from near UV (395 nm) to near IR (1000 nm) regions. By harmonic generation, frequency mixing, or other nonlinear optical processes, the range can be further increased in the UV region, as shown in Figure 3.28 [31]. Of all dye lasers known, those based on rhodamine 6G, or R6G for short, pumped by cw Ar lasers are probably the most popular.

Organic dye molecules have large, complicated molecular structures, and there are many vibrational and rotational modes. Since the energy differences between the rotational modes are very small, the energy diagrams of organic dye compounds appear as bands instead of discrete levels [32]. Figure 3.29 is a simplified energy diagram of a typical dye–solvent system. In the diagram, heavy lines represent the levels of vibrational modes, and thin lines represent the levels of rotational modes. Of interest to the present discussion are the ground-state band S_0 and the excited-state band S_1. These levels are singlet states, identified by the letter S. The absorption spectra of the dye solution is very broad, since light with the wavelength corresponding to the energy difference between a level in the S_0 band and a level in the S_1 band can be absorbed. For example, the absorption band of R6G is about 30 nm wide, centering at 525 nm [32]. Because of the broad absorption spectra, broadband incoherent light sources, as well as narrow-line coherent sources, can be used efficiently as pumping sources. In addition to the singlet states, there are also triplet states, such as the T_0 band. The presence of T_0 and other triplet bands often leads to degradation of the dye laser's performance [32, 33].

In a dye laser, molecules are elevated by optical pumping from the bottom of the S_0 band to a level, such as level E_b, in the S_1 band. The excited molecules in the S_1 band redistribute themselves and relax thermally and nonradiatively to the bottom of the S_1 band. The thermal relaxation processes are very fast, and the molecules accumulate quickly at the bottom of the S_1 band. In the presence of stimulation, molecules decay from the bottom of the S_1 band to a level, such as level E_a, and photons are emitted. Finally, again by the thermal relaxation, molecules drop from level E_a to the bottom of the S_0 band.

As noted previously, the stimulated emission process occurs between the bottom of the S_1 band and a level in the S_0 band. Therefore, the tuning range of a dye–solvent medium is quite wide. For example, R6G dye lasers with cw Ar lasers as pumping sources can be tuned over 100 nm. The precise emission wavelength is determined by the stimulation, that is, the photons existing in the optical cavity. The linewidth is affected to a great extent by the cavity design. For cw R6G dye lasers with a broad-band cavity configuration, the linewidth is about 3 nm. On the other hand, if a frequency selection element, such as a prism or a grating, is incorporated into the cavity, the linewidth can be greatly reduced. The linewidth of a modern single-frequency dye laser can be as narrow as 10^{-6} nm, corresponding to a frequency bandwidth of 1 MHz.

When dye lasers were first introduced in 1966, the dye–solvent liquid was contained in a dye cell. However, there were difficulties caused by nonhomogeneities due to local heating and possible optical damage of the cell windows. To alleviate these problems, most modern dye lasers have a **flowing dye jet** arrangement in lieu of a dye cell [31]. These features are particularly important for high power operations.

There are two basic dye laser configurations, shown in Figure 3.30a and b [31]. In the **three-mirror folded cavity configuration** (Figure 3.30a), the dye laser cavity is defined by mirrors M_1, M_2, and M_3. Mirror M_3 also serves as the output mirror. Pumping light is focused to the dye jet either by a mirror M_p or a lens. To reduce the linewidth, one or more frequency selection elements, prisms, gratings, birefringent filters, optical interferometers, or other dispersive elements may be inserted in the cavity. In the configuration shown in Figure 3.30a, optical beams travel in both directions, and standing optical fields are formed in the cavity. The spatial distribution of the standing-wave field pattern in the optical cavity, including in particular the dye jet region, has peaks and valleys. However,

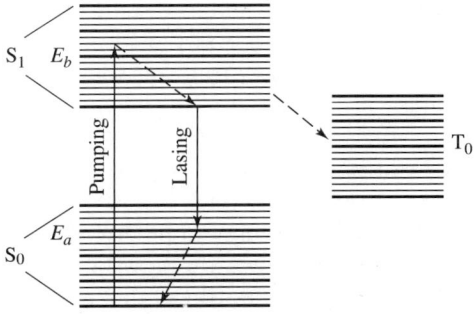

Figure 3.29 Simplified energy diagram of a dye laser

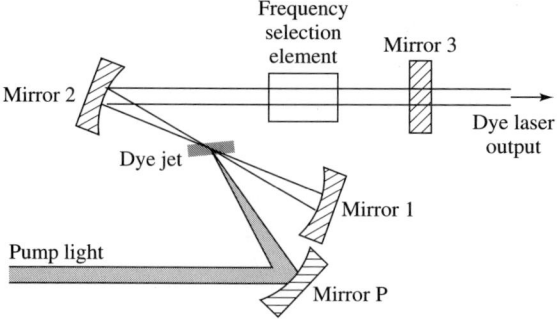

(a) Three-mirror folded configuration

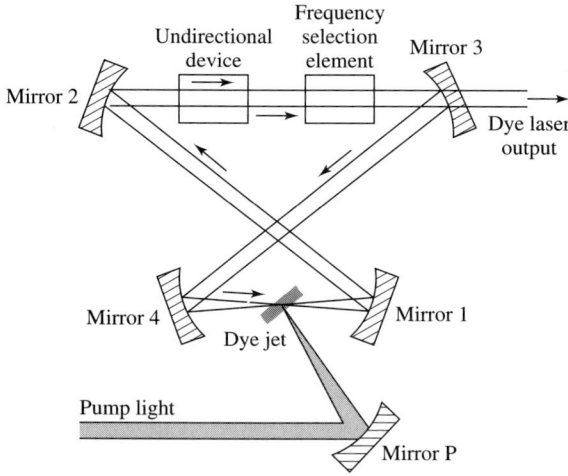

(b) Four-mirror unidirectional ring configuration

Figure 3.30 Two basic dye laser configurations

the pumping beam is a tightly focused optical beam. Consequently the overlap between the standing wave laser beam and the traveling wave pumping beam may become unstable. This leads to output instabilities. In the **unidirectional ring arrangement,** the light beam is routed in the cavity by mirrors. In Figure 3.30b, four mirrors, M_1, M_2, M_3, and M_4, are shown. More importantly, there is a *unidirectional device,* which is a Faraday-effect component subjected to a strong dc magnetic field. As discussed in Chapter 6, a Faraday-effect device introduces excessive loss for waves going in one direction relative to the strong magnetic field, and has a minimal effect for waves circulating in the opposite direction. Because of the unidirectional device, the laser beam circulates in the cavity in one direction only, and no standing wave is formed. Thus, the interaction between traveling laser fields and the traveling pump beam is much more stable. In fact, the introduction of the ring laser configuration (Figure 3.30b) is a major development in dye laser technology.

3.11 DESCRIPTION OF SOLID-STATE LASERS

The active media of solid-state lasers are solid host materials "doped" with rare earth, actinide, or transition metal ions. Lasing action is due to radiative transitions in these ions, maintained by optical pumping. Examples include ruby lasers, neodymium doped yttrium–aluminum garnet, neodymium- or erbium-doped glass lasers, and neodymium- or erbium-doped fiber lasers. Some of the solid-state lasers, such as the alexandrite lasers and Ti:sapphire lasers, can be tuned over a wide range.

We know that electrons in closed (i.e., filled) shells are optically inactive. Optically active electrons are those in partially filled shells. Therefore, in our discussions on solid-state lasers, we are mainly interested in electrons residing in partially filled shells. We also know that electrons in the inner shells are shielded from external perturbations by the electrons in the outer shells. Conversely, electrons in the outermost shells can be easily perturbed by external fields. Therefore, depending on the position of the unfilled shell, solid-state lasers can be classified into two distinctively different groups: those with rare earth or actinide ions, and those with transition metal ions.

3.11.1 Energy Levels of Active Ions in Solid Host Materials

In trivalent **rare earth ions** (Nd^{+3}, Pr^{+3}, Er^{+3}, Tm^{+3}, and Yb^{+3}) and **actinide ions** (U^{+3}), all shells except one are completely filled. The exception is an inner shell that is only partially filled. Take trivalent neodymium ions (Nd^{+3}) as an example. The electron configuration of *neutral* Nd atoms is $1s^2 2s^2 2p^6 3s^2 3p^6 3d^{10} 4s^2 4p^6 4d^{10} 4f^4 5s^2 5p^6 6s^2$. Except for the $4f$ and $6s$ electrons, the electron configuration of neutral Nd atoms is exactly the same as that of Xe, whose shells are filled completely. Therefore, the electron configuration of neutral Nd atoms can be written as $[Xe]4f^4 6s^2$, where [Xe] denotes the electron configuration of Xe. Trivalent neodymium ions are the result of removing one 4f electron and two 6s electrons from neutral Nd atoms; thus, these ions have an electron configuration of $[Xe]4f^3$. Similarly, the electron configuration of Er^{+3} can be written as $[Xe] 4f^{11}$. For all trivalent rare earth ions, the 4f level, an inner shell, is partially filled and all other shells, including in particular the shells residing outside the 4f shell, are completely filled. The electron configuration of actinide ions is similar, except that the outermost (i.e., completely filled) shells are 6s and 6p and the partially filled shell is 5f.

When incorporated into crystals or glasses, the ions are surrounded by a large number of host atoms and are under the influence of the electric fields due to these host atoms. These electric fields are called crystal fields, even though the host materials may be glass. The effects of the crystal fields are twofold. The numerous states of isolated ions are degenerate and the corresponding distribu-

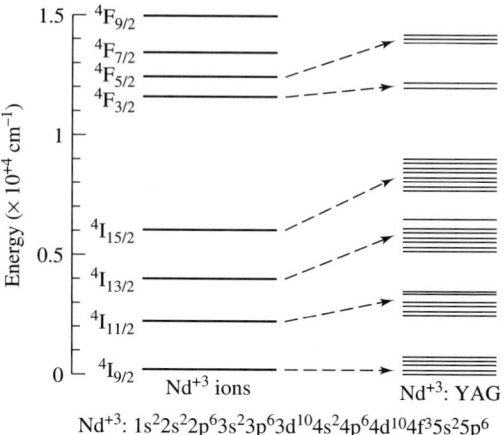

Figure 3.31 Energy levels of Nd^{+3} ions and Nd^{+3}:YAG (energy levels of multiplets are not to scale)

tions of the "electron clouds" are equivalent. In the presence of crystal fields, the "shape" of the electron clouds is distorted, states are no longer equivalent, and the energy levels are split into closely spaced multiplets. The transition probability between levels also changes due to the crystal field perturbation. As a result, the selection rule, described in Appendix A, is no longer applicable.

As an example, consider the energy levels of Nd^{+3}:YAG, where YAG stands for yttrium–aluminum garnet ($Y_3Al_5O_{12}$). The levels of isolated Nd^{+3} ions are shown on the left-hand side of Figure 3.31. The terms symbols $^4I_{9/2}$ and $^4I_{11/2}$, etc., reflect the interactions of three 4f electrons. When Nd^{+3} ions are incorporated randomly into a YAG crystal, the ion levels split under the influence of the crystal fields. The level splitting is shown schematically on the right-hand side of Figure 3.31. As noted previously, 4f electrons of Nd^{+3} are shielded from external fields by 5s and 5p electrons on the outermost shells. Thus, the effects of crystal field perturbations are weak, and the splitting is small. For the same reason, the energy level designations associated with isolated Nd^{+3} ions remain appropriate for the levels of Nd^{+3}:YAG. This is true for all solid-state lasers based on trivalent rare earth and actinide ions. The energy level diagram of Er^{+3} in glass shown in Figure 3.32 is another example. Note that Er^{+3} lasers can be either three-level or four-level systems, as indicated in Figure 3.32.

The situation for transition metal ions (Cr^{+3}, Ni^{+2}, and Co^{+2}) is different. For example, the electron configuration of Cr^{+3} ions is [Ar] $3d^3$. All shells of Ar atoms are filled, but the outermost shell of Cr^{+3}, the 3d shell, is only partially filled. Since the *outermost* shell is the one that is only partially filled, the effects due to crystal field perturbations are so strong that the symmetry of the host crystal is the dominant factor influencing the energy levels. This means that the

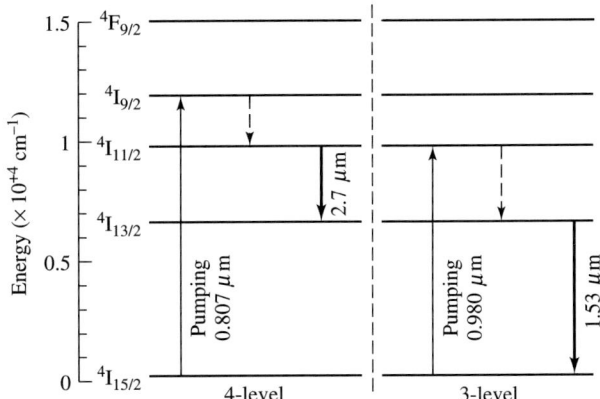

Figure 3.32 Simplified energy diagram of Er^{+3}:glass lasers

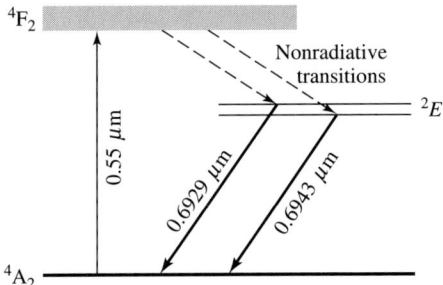

Figure 3.33 Energy diagram of ruby lasers

energy level designation of isolated Cr^{+3} ions is no longer meaningful in describing the energy levels of Cr^{+3}:Al$_2$O$_3$. Designations such as $^4A_2$, ^2E, etc. for ruby lasers, shown in Figure 3.33, are related to the symmetry of the crystalline host and are more appropriate for transition metal ions [1, 34, 35].

3.11.2 Neodymium Lasers

Neodymium ions can be imbedded in many host materials, including YAG, glass, or silica-based fibers. Nd:YAG lasers are mainly used for continuous-wave operation, while Nd^{+3} doped glass lasers are used in pulsed applications. In Nd^{+3}:YAG rods, some of the Y^{+3} ions of Y$_3$Al$_5$O$_{12}$ are replaced by Nd^{+3} ions. The most prominent room temperature emission is the 1.064 μm line (Figure 3.34), which involves the transition between levels $^4F_{3/2}$ and $^4I_{11/2}$. The absorption bands near 0.75 μm, 0.81 μm, and 0.88 μm may be used for optical pumping.

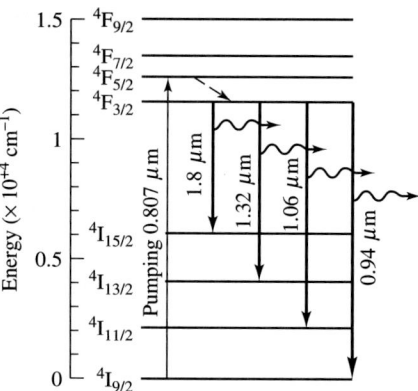

Figure 3.34 Energy level diagram of an Nd^{+3}: YAG laser

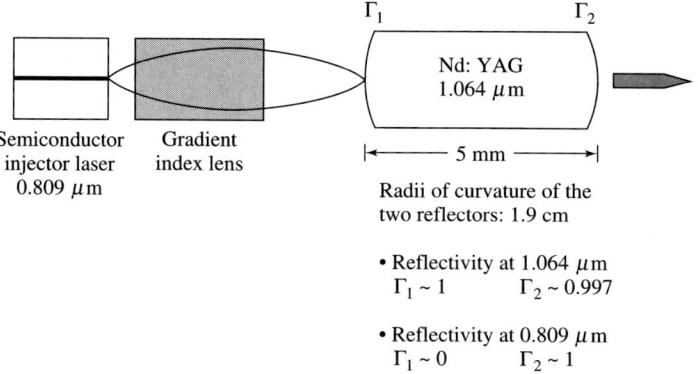

Figure 3.35 Nd: YAG rod pumped by a semiconductor injection laser diode, with a gradient index lens used to focus the light onto the rod

In many Nd^{+3} lasers, particularly early ones, arc lamps and flash lamps are used as pumping sources. However, these lamps are bulky, inefficient, and short lived. Recent advances in high-power semiconductor injection lasers, light emitting diodes, and superluminescent diodes have made these semiconductor diodes powerful enough to be used as the pumping sources. GaAlAs injection laser diodes with emissions near 0.81 μm or 0.88 μm are particularly useful in such applications, since these emission lines closely match the absorption band of the YAG rods. The energy level diagram in Figure 3.34 shows a pumping transition corresponding to 0.81 μm radiation from an injection laser diode. Depending on the mirrors and the pumping condition, the laser may emit at 0.94, 1.06, 1.32, or 1.8 μm. When photons with wavelengths of 0.94 μm are radiated, the system is a three-level system, with $^4I_{9/2}$, $^4F_{3/2}$, and $^4F_{5/2}$ corresponding to E_1, E_2 and E_3 respectively. Nd:YAG lasers emitting at 1.06 μm are four-level systems, with $^4I_{9/2}$, $^4I_{11/2}$, $^4F_{3/2}$, and $^4F_{5/2}$ corresponding to E_1, E_a, E_2, and E_3 levels, respectively. Note that the energy difference between $^4I_{11/2}$ and $^4I_{9/2}$ is approximately 0.25 eV, which is about $10kT$ at room temperature. For the four-level Nd:YAG laser with outputs at 1.32 μm or 1.8 μm, $^4I_{13/2}$ or $^4I_{15/2}$ is the level E_a. The advantages of diode-pumped Nd:YAG lasers include narrow spectral linewidths ($\Delta\lambda \approx 0.1$ nm), long lifetimes, compactness, and high power efficiency [36, 37, and 38]. A schematic diagram of a diode-pumped Nd:YAG laser is shown in Figure 3.35. Diode pumped light entering the Nd:YAG rod from one end is trapped by total internal reflection until it passes the entire rod length. This is an **end pumping** scheme. An alternate pumping scheme is the **side pumping** method shown in Figure 3.36 [39]. Side pumping is used mainly for high power applications (greater than 10 W).

3.11.3 Ruby Lasers

As mentioned earlier, a ruby laser was the first laser ever invented. The active medium of a ruby laser is an aluminum oxide (Al_2O_3) crystal lightly doped with trivalent chromium ions. Typically, the Cr^{+3} impurity is about 0.05 percent by weight. Because of doping, some of the Al^{+3} ions are replaced by Cr^{+3} ions. The laser action at 0.6934 μm is due to transitions between the energy levels of the $Cr^{+3}:Al_2O_3$ ions, as shown in Figure 3.33 [1].

3.11.4 Alexandrite Lasers and Emerald Lasers

In addition to ruby and neodymium lasers, other dopant–host combinations can be made to lase. Examples include: Ni^{+2} in MgF_2 or MgO; Co^{+2} in MgF_2; Ce^{+3} in $YLiF_4$ (YAL); Ti^{+3} in Al_2O_3 (Ti:sapphire lasers); and Cr^{+3} ions in various host material. Particularly noteworthy are **chromium doped alexandrite** ($Cr^{+3}:BeAl_2O_4$) and **chromium doped emerald** ($Cr^{+3}:Be_3Al_2Si_6O_{18}$) lasers, which are also known as **gem lasers**. In these four-level systems, one of the energy levels is coupled to the lattice vibration. Consequently, the emission wavelength can be tuned over a wide range. For example, alexandrite lasers can be tuned in

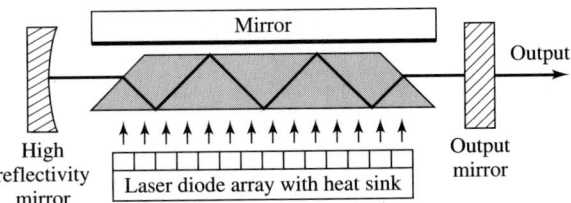

Figure 3.36 Schematic of solid-state laser pumped by a laser diode array in a side-pumped configuration [39]

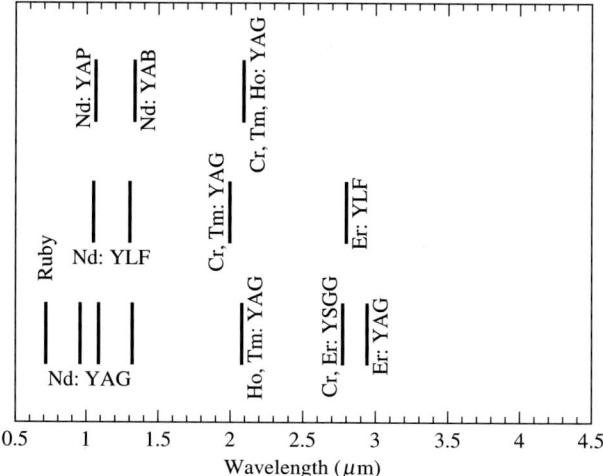

(a) Emission spectra

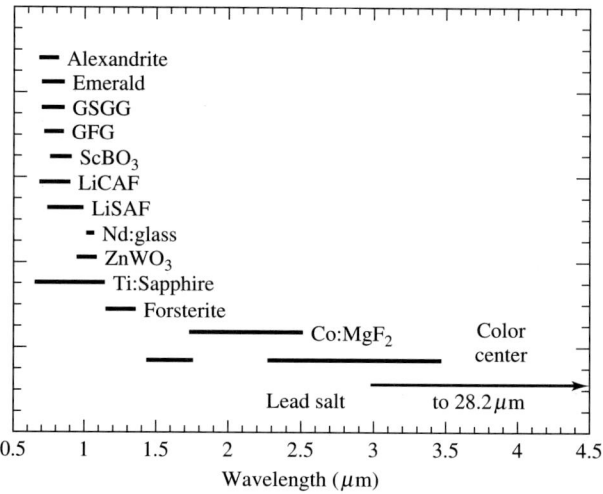

(b) Tuning range

Figure 3.37 Emission spectra and tuning range of selected solid-state lasers

the range of 0.70–0.82 μm, while emerald lasers are tunable between 0.70 μm and 0.85 μm. Figure 3.37 shows the discrete emission lines and tuning ranges of selected tunable solid-state lasers [40, 41].

An abbreviated summary of commercially available lasers is listed in Table 3.6 [42].

Table 3.6 Commercially available lasers (Excerpted from [42])

Type	Wavelength	Power	Nature of output	Lifetime	Beam diameter	Beam divergence
Excimer:						
Fluoride Argon	193 nm	Up to 50 W average	5–25 ns pulses, 500 mJ @ 1–1000 Hz	10^4–5×10^6 shots per gas fill	2×4 mm –25×30 mm	2–6 mrad oval beam
Krypton fluoride	248 nm	Up to 100 W average	2–50 ns pulses, to 1 J @ 1–500 Hz	10^4–10^7 shots per gas fill	Similar to ArF	Similar to ArF
Xenon chloride	308 nm	To 150 W average	1–80 ns pulses, to 1.5 J @ 1–500 Hz	10^5–2×10^7 shots per gas fill	Similar to ArF	Similar to ArF
Xenon fluoride	351 nm	To 30W average	1–30 ns pulses, to 500 mJ @ 1–500 Hz	10^4–10^7 shots per gas fill	Similar to ArF	Similar to ArF
Dye laser pumped by:						
Nitrogen, excimer, Nd: YAG	300–1000 nm tunable	0.05–15W average	3 to 50-ns pulses @ 1–10 KHz	Dye limited hrs. to two months	2–10 mm	0.36–4 mrad
Flashlamp	340–940 nm tunable	0.25–50 W average	0.05–50 J pulses, 0.2–4 μs @ 0.03–50 Hz	10^4–10^6 shots flashlamp	5–20 mm	0.5–5 mrad
Ion laser	400–1000 nm tunable	To 2 W	cw or ps pulses from mode-locked systems	Dye-limited hrs. to a week	0.6–1.0 mm	1–2 mrad
Nitrogen	337 nm	1–330 mW average	0.3 to 10 ns pulses @ 1–1000 Hz; (0.001–10 mJ)	Kilohours; clean after 10^6 shots	2×3 mm– 6×30 mm	0.3 to 3×7 mrad

Table 3.6 (continued)

Type	Wavelength	Power	Nature of output	Lifetime	Beam diameter	Beam divergence
Ion: Argon	Several lines, 351 to 528 nm (main lines 488 nm and 514.5 nm)	2 mW–20 W	cw (can be mode locked)	10^3 hrs. to 5 years	0.6–2 mm	0.40–1.0 mrad
Krypton	Several lines 350–800 nm (strongest at 647.1 nm)	5 mW–6 W (10%–20% of argon in same tube)	cw (can be mode locked)	10^3–10^4 hrs.	0.6–2 mm	0.4–1.5 mrad
Argon-Krypton	Several lines 450–670 nm	0.5–6 W	cw	1×10^3 plus hrs.	2 mm	2 mrad
Helium-Cadmium	442 or 325 nm	2–50 mW @ 442 nm 1.5–10 mW @ 325 nm	cw	4000 hrs., visible 2000 hrs., UV	0.3–1.2 mm	0.4–1.9 mrad
Helium-Neon	543, 594, 604, 633, 1152, 1523, or 3391 nm	0.1–50 mW @ 633 nm to 15 mw @ 1152 or 3391 nm, ~1 mW other lines,	cw	5×10^3–10^5 hrs.	0.3–3 mm	0.6–6.0 mrad
Ruby	694 nm	Pulses of 0.03–100 J, 10 ns–10 ms	Pulsed @ 0.01–4 Hz	10^6 shots per flashlamp	1.5–25 mm	0.2–10 mrad
Semiconductor diode:						
GaAs/GaAlAs	780 to 905 nm composition dependent	1–40 mW average or cw	cw or pulsed	10^4–10^7 hrs.	Not meaningful	$10° \times 35°$
Phase-coupled GaAlAs arrays	790 to 850 nm composition dependent	0.1–1 W cw, 1 to 10 W peak 1 ns to 200 μs in pulse mode	cw or 1 ns –200 μs pulses	10^4–7×10^7 hrs.	Not meaningful	$5° \times 10°$ $19° \times 35°$
InGaAsP	1100–1600 nm composition dependent	1–10 mW	cw or pulsed	10^5 hrs	Not meaningful	$10° \times 30°$ to $20° \times 40°$

Laser	Wavelength	Power	Operation	Lifetime	Beam diameter	Divergence
Nd:YAG (pulsed)	1.064 μm (low power @ 1.32 μm)	to 600 W average	Pulses 0.01–150 J @ 50 kHz	10^6 shots	1–10 mm	0.3–20 mrad
Diode-pumped Nd:YAG	1.064 μm (lower power @ 1.32 μm)	0.5–10 mW	cw or pulsed, can be mode locked or Q-switched	10^4 hrs. depending on diodes	1–2 mm	0.5–2.0 mrad
Nd:YAG (cw)	1.064 μm (low power @ 1.32 μm)	0.04–600W	cw	Arc lamp lasts 200 hrs.	0.7–8 mm	2–25 mrad
Nd doped glass	1.06 μm	Pulses of 0.1–100 J	Pulsed @ 0.1–2 Hz	10^6 shots per flashlamp	3–25 mm (some rect.)	3–10 mrad
F-center	1.43–1.58 μm, 2.3–3.5 μm	1–100 mw	cw or mode locked pulses of a few ps	10^3 hrs.	1.35 mm	1.6 mrad
Hydrogen fluoride (chemical)	2.6–3 μm (many lines)	0.01–150 W cw or 2–600 mJ pulses	cw or 50–200 ns pulses @ 0.5–20 Hz	Maintainance every 50–100 hrs.	2–40 mm	1–15 mrad
Deuterium fluoride (chemical)	3.6–4 μm	0.01–100 W cw or 2–600 mJ pulses	cw or 50–200 ns pulses @ 0.5–20 Hz	Maintainance every 50–100 hrs.	2–40 mm	1–14 mrad
Carbon dioxide:						
Axial flowing-gas	9–11 μm or 10.6 μm	20W–5kW	cw or long pulses	10^3 hrs.	3–25 mm	1–3 mrad
Transvere flowing-gas	9–11 μm or 10.6 μm	500W–15kW	cw or long pulses	10^3 hrs.	10–50 mm	1–3 mrad
Sealed-tube	9–11 μm or 10.6 μm	3–100 W	Generally cw	10^4 hrs.	3–4 mm	1–2 mrad
Pulsed, TEA	9–11 μm or 10.6 μm	0.03–to 150 J pulses	50 ns–100 μs pulses @ 0.1–1000 Hz	10^6 shots	5–100 mm	0.5–10 mrad
Waveguide	9–11 μm or 10.6 μm	0.1–50 W	cw or pulsed	to 10^4 hrs.	1–10 mm	4–10 mrad

REFERENCES

1. Maiman, T. H. "Stimulated Optical Radiation in Ruby." *Nature* 187, (1960), p. 493.
2. Born, M. *Atomic Physics* 8th ed. London: Blackie, 1969.
3. Richtmyer, F. K.; E. M. Kennard; and T. Lauritsen. *Introduction to Modern Physics* 5th ed. New York, NY: McGraw-Hill Book Company, 1955.
4. Ashby, N.; and S. C. Miller. *Principles of Modern Physics*. San Francisco, CA: Holden-Day Inc., 1970.
5. Herzberg, G. *Atomic Spectra and Atomic Structure*. New York, NY: Dover Publications, Inc. 1944.
6. Radziemski, L. J. "Spectroscopic notation for the energy levels of helium and neon." *Optics News* 15, no. 1, (1989), pp. 15–16.
7. Jenkins, F. A.; and H. E. White. *Fundamentals of Optics*, 4th ed. New York, NY: McGraw-Hill Book Company, 1976.
8. Pedrotti, F. L.; and L. S. Pedrotti. *Introduction to Optics*, 2nd ed. Englewood Cliffs, NJ: Prentice Hall, 1992.
9. Svelto, O. *Principles of Lasers*. New York, NY: Plenum Press, 1976.
10. Yariv, A. *Introduction to Optical Electronics*. Holt, Rinehart & Winston, New York, (1971).
11. Yariv, A. *Quantum Electronics*. 3rd ed. New York, NY: John Wiley & Sons, Inc. 1987.
12. Siegman, A. E. *Lasers*. Mill Valley, CA: University Science Books, 1986.
13. Reardon, A. C. "Exact solution of laser rate equations." *J. Modern Optics* 38, (1991), pp. 857–864.
14. Zayhowski, J. J.; and P. L. Kelly. "Optimization of Q-switched lasers." *IEEE J. Quantum Electron.* QE-27, (1991), pp. 2220–2225.
15. Weber, H.; and G. Herziger. *Laser, Grundlagen und Ahwendungen*. Berlin: Physik Verlag, 1972.
16. Smith, P. W. "Mode-locking of lasers." *Proc. IEEE* 58, (1970), pp. 1324–1357.
17. Kuizenga, D. J.; and A. E. Siegman. "FM and AM mode locking of homogeneous lasers-part I: Theory." *IEEE J. Quantum Electron.* QE-6, (1970), pp. 694–708.
18. Kuizenga, D. J.; and A. E. Siegman. "FM and AM mode locking of homogeneous lasers-part II: Experimental results in Nd:YAG laser with internal FM modulation." *IEEE J. Quantum Electron.* QE-6, (1970), pp. 709–715.
19. Harvey, A. F. *Coherent Light*. Chapter 10, New York, NY: Wiley-Interscience, 1970.
20. Moore, C. E. "Atomic Energy Levels." Gaithersburg, MD: National Bureau of Standards I, no. 467 (1949).
21. Arrathoon, R. "Helium-Neon Lasers and Positive Column." *Lasers*, Chapter 3, ed. K. Levine; and J. DeMaria. New York, NY: M. Dekker Inc., 1976.
22. Smith, P. W. "The Output Power of a 6328 Å He-Ne Laser." *IEEE J. Quantum Electron.* QE-2, (1966), pp. 62–68.
23. Smith, P. W. "On the Optimum Geometry of a 6328 Å Laser Oscillator." *IEEE J. Quantum Electron.* QE-2, (1966), pp. 77–79.
24. Field, R. L., Jr. "Operating Characteristics of dc Excited HeNe Gas Lasers." *Rev. Sci. Instrum.* 38, (1967), pp. 1720–1722.

25. Patel, B. "The Helium-Neon Laser." *Photonics Spectra* Parts 1–4, January, February, March, and April 1983.
26. Brau, C. A. *Free-Electron Lasers*. Boston, MA: Academic Press, 1990.
27. Davis, C. C.; and T. A. King. "Gaseous Lasers." *Advances in Quantum Electronics*, 3, ed. D. W. Goodwin. Boston, MA: Academic Press, 1975.
28. Bridges, W. B.; and A. N. Chester. "Ionized Gas Lasers." *Handbook of Lasers with Selected Data on Optical Technology* Chapter 7, ed. R. J. Pressley. Cleveland, OH: The Chemical Rubber Co., 1971.
29. Patel, B. "Designing the Argon laser into your system." *Photonics Spectra,* August and September 1984.
30. Degnan, J. J. "The Waveguide Laser: A Review." *Appl. Phys.* 11, (1976), pp. 1–33.
31. Hollberg, L. "CW dye lasers." *Dye Laser Principles with Applications* Chapter 2, Boston, MA: Academic Press, Inc., 1990.
32. Snavely, B. B. "Flashlamp-excited organic dye lasers." *Proc. IEEE* 57, (1969), pp. 1374–1390.
33. Shank, C. V. "Physics of dye laser." *Rev. Mod. Phys.* 47, (1975), pp. 649–657.
34. Yariv, A.; and J. P. Gordon. "The laser." *Proc. IEEE* 51, (1963), pp. 4–29.
35. Koechner, W. *Solid-State Laser Engineering* 3rd ed. Berlin, Germany: Springer-Verlag, 1992.
36. Zhou, B.; T. J. Kane; G. J. Dixon; and R. L. Byer. "Efficient, frequency stable laser-diode-pumped Nd:YAG laser." *Opt. Lett.* 10, (1985), pp. 62–64.
37. Fan, T. Y.; and R. L. Byer. "Diode laser pumped solid-state lasers." *IEEE J. Quantum Electron.* 24, (1988), pp. 895–912.
38. Malcolm, G. P. A.; and A. I. Ferguson. "Diode-pumped solid-state lasers." *Contempo. Phys.* 32, (1991), pp. 305–319.
39. Reed, M. K.; W. J. Kozlovsky; R. L. Byer; G. L. Harnagel; and P. S. Cross. "Diode-laser-array-pumped neodymium slab oscillators." *Opt. Lett.* 13, (1988), pp. 204–206.
40. Caird, J. A.; S. A. Payne; P. R. Staver; A. J. Ramponi; L. L. Chase; and W. F. Krupke. "Quantum electronic properties of the $Na_3Ga_2Li_3F_{12}:Cr^{3+}$ laser." *IEEE J. Quantum Electron.* 24, (1988), pp. 1077–1099.
41. Petricevic, V.; S. K. Gayen; and R. R. Alfano. "Continuous-wave laser operation of chromium-doped forsterite." *Opt. Lett.* 14, (1989), pp. 612–614.
42. Hecht, J. "Laser application matrix." *1984 Lasers & Applications Designer's Handbook and Production Directory,* (1984), pp. 146–149.
43. Harris, S. E. "Proposal for a 207-Å laser in lithium." *Opt. Lett.* 5, (1980), pp. 1–3.

ADDITIONAL READING

1. Banerjee, P. P.; and T. C. Poon. *Principles of Applied Optics.* Homewood, IL and Boston, MA, R. D. Irwin, Inc., and Aksen Associates, Inc., 1991.
2. Jenkins, F. A.; and H. E. White. *Fundamentals of Optics* 4th ed. New York, NY: McGraw-Hill Book Company, 1976.
3. Jones, K. A. *Introduction to Optical Electronics.* Harper & Row, 1987.
4. Koechner, W. *Solid-State Laser Engineering,* 3rd ed. Berlin, Germany: Springer-Verlag, 1992.

5. Nussbaum, A.; and R. A. Phillips. *Contemporary Optics for Scientists and Engineers.* Englewood Cliffs, NJ: Prentice Hall, 1976.
6. Pedrotti, F. L.; and L. S. Pedrotti. *Introduction to Optics,* 2nd ed. Prentice Hall, 1992.
7. Saleh, B. E. A.; and M. C. Teich. *Fundamentals of Photonics.* Wiley, 1991.
8. Senior, J. M. *Optical Fiber Communications, Principles and Practice.* 2nd ed. Prentice Hall, Inc. 1992.
9. Siegman, A. E. *Lasers.* University Science Books, Mill Valley, CA 1986.
10. Svelto, O. *Principles of Lasers.* New York: Plenum Press, 1976.
11. Verdeyn. *Laser Electronics* 2nd ed. Englewood Cliffs, NJ: Prentice Hall, 1989.
12. Wilson, J.; and J. F. B. Hawkes. *Optoelectronics, An Introduction.* 2nd ed. Englewood Cliffs, NJ: Prentice Hall, 1989.
13. Yariv, A. *Introduction to Optical Electronics* 3rd ed. New York: Holt, Rinehart & Winston, New York, 1984.
14. Yariv, A. *Quantum Electronics,* 3rd ed. New York: Wiley & Sons, 1987.

PROBLEMS

1. Consider a system with $E_1 = 166658.484$ cm^{-1} and $E_2 = 148259.749$ cm^{-1}.
   a. What is the wavelength (in μm) of the photons emitted or absorbed?
   b. What is the frequency (in Hz) of the emission?
   c. If the output has a spectral width of 0.2Å, what is the bandwidth in Hz?

2. Consider a system with $E_1 = 1.0$ eV, $E_2 = 1.1$ eV, and $A = 2.0 \times 10^6$ s^{-1}. Let the density of occupied states in the levels be N_1 and N_2, respectively, and let $N_t = N_1 + N_2 = 2 \times 10^{16}$ cm^{-3}.
   a. What is the wavelength (in μm) of the E_2-to-E_1 transition?
   b. What is N_2 at 300 °K?
   c. The system is now pumped to a steady state with $N_1/N_2 = 3$. How much pump power (in W/cm^3) is absorbed?
   d. How much power (in W/cm^3) is radiated by the spontaneous emission process in the steady state mentioned in (c)?

3. For a certain two-level system at thermal equilibrium at 600 °K, the densities of occupied states are $N_1 = 10^{+24}$ cm^{-3} and $N_2 = 10^{+14}$ cm^{-3}. What is the wavelength (in μm) of the photons absorbed or emitted by the system?

4. Find the ratio N_2/N_1 of a two-level system at thermal equilibrium at room temperature (300 °K), if the transition wavelength falls in:
   a. Near IR region, say 1.0 μm.
   b. IR region, say 10.0 μm.
   c. Millimeter region, say 1.0 mm.

5. The ratio N_2/N_1 of a two-level system in thermal equilibrium at room temperature is $1/e$. Calculate the transition frequency of the system.

6. Given a two-level system at room temperature. If the number of photons radiated due to the stimulated emission process is the same as that due to the spontaneous emission process, what is the wavelength of the radiation? If the wavelength of radiation is 10.6 μm, what is the operating temperature of the system?

7. Consider a two-level system with $E_1 = 150856$ cm^{-1} and $E_2 = 159534$ cm^{-1}. At $T = 300$ °K, $N_1 = 1.0 \times 10^{20}$ cm^{-3}.

a. Calculate the wavelength (in μm) of the photons absorbed or radiated by the system.
b. Calculate N_2.

8. Consider a three-level system in which radiative decay takes place between the E_3-to-E_2 transition (*not* the E_2-to-E_1 transition), and pumping occurs from level E_1 to level E_3.
   a. Write the rate equations for this system. Identify each term.
   b. Can population inversion be achieved in the system? If not, why not? If it can, establish the condition for population inversion.

9. Derive expressions for the following parameters of a four-level system (Figure 3.4b).
   a. The critical population inversion, ΔN_c.
   b. The steady-state population inversion ΔN_0.
   c. The critical pumping transition probability, A_{Pc}.
   d. The steady-state optical power output, P_1 and P_2, through mirrors 1 and 2, respectively.

10. Consider a usual three-level system (Figure 3.4a). What is the pumping transition probability A_P required to achieve $N_2 = 0.1 N_1$ in the steady state?

11. Consider the three-level system shown in Figure 3.4a, with the following energy levels and Einstein A coefficients:

    $E_1 = 0$ eV, $E_2 = 1.2$ eV,   and $E_3 = 2.0$ eV
    $A_{32} = 5.0 \times 10^7$ s^{-1}, $A_{31} = 2.5 \times 10^7$ s^{-1},   and $A_{21} = 1.0 \times 10^6$ s^{-1}

    a. What are the wavelengths (in μm) of the E_3-to-E_2, E_2-to-E_1, and E_3-to-E_1 transitions?
    b. At thermal equilibrium at 300 °K, what is the ratio $N_1:N_2:N_3$?
    c. At what temperature would $N_1:N_2 = 2:1$? What is the ratio $N_2:N_3$ at this temperature?
    d. Calculate B_{32}, B_{31}, and B_{21}. Take $\delta f = 1$ Hz.
    e. What is the lifetime of level 3?
    f. What percentage of states in level 3 would drop from level 3 to level 2?

12. The three-level system discussed in problem 11 is pumped by some means such that a steady-state value of $N_3 = 1 \times 10^{14}$ states/cm³ is maintained. Assume that $N_t \approx 1 \times 10^{19}$ states/cm³ and pumping is very weak.
    a. What is the steady-state value of N_2?
    b. How much pumping power (in W/cm³) is absorbed?
    c. How much power is radiated spontaneously in the 3–to–2 transition?

13. A four-level system is shown in the figure. Pumping is done with two sources, a and b, which maintain photon densities $\Phi_{Pa}\delta f$ and $\Phi_{Pb}\delta f$, with wavelengths corresponding to the E_1-to-E_3 and E_3-to-E_4 transitions, respectively. Lasing occurs between levels E_4 and E_2, with a photon density of $\Phi \delta f$. Write the rate equations for levels E_2 and E_4.

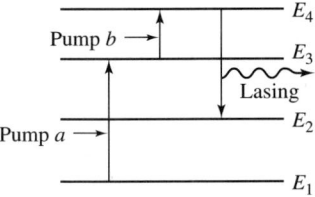

14. Consider a ring cavity with three mirrors as shown in the figure. The lengths of the three sections are d_i, where $i = 1, 2,$ and 3. The reflectivities of the mirrors are Γ_1, Γ_2, and Γ_3. The net losses of sections 1 and 2 are represented by attenuation constants α_1 and α_2. The net gain in section 3 is given by a gain constant g_3. The indices of these sections are approximately 1. Derive an expression for the rate of change of the photon density in the ring cavity for waves circulating in the clockwise direction.

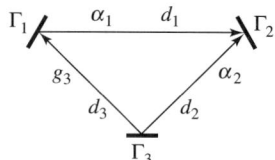

15. The optical cavity parameters of a three-level laser are:
    $\Gamma_1 = 0.95,$ $\quad \Gamma_2 \approx 0.99,$ $\quad \alpha = 0.01 \text{cm}^{-1},$
    $d' = 10.0$ cm, $\quad d = 50.0$ cm, $\quad \mathcal{A} = 0.8$ cm², $\quad V_a = \pi \times 10^{-2}.$
    At the lasing wavelength of 0.693 μm, the parameters of the lasing medium are:
    $B_{21} = 1.9 \times 10^{-9}$ m³s⁻¹, $\quad N_t = 1.6 \times 10^{19}$ cm⁻³, $\quad n = 1.76,$
    $\tau_{21} = 3.0 \times 10^{-3}$ s, $\quad \tau_{31} = 3.33 \times 10^{-6}$ s, $\quad \tau_{32} = 5.0 \times 10^{-8}$ s.
    a. Calculate the critical population inversion ΔN_c.
    b. Calculate the critical pumping transition probability A_{Pc}.
    c. What is the steady-state population inversion ΔN_0?
    d. Calculate the power output from the mirrors when the medium is pumped to $4A_{Pc}$.

16. Consider a Q-switched ruby laser ($\lambda = 0.6934$ μm). The active medium is a ruby rod doped with Cr^{+3} ions, resulting in a total population of $N_t = 1.6 \times 10^{19}$ cm⁻³. The index of refraction of ruby is 1.76, $\alpha = 1.$ m⁻¹, $d = 0.50$ m, and $d' = 0.015$ m. Additional material parameters can be found in section 6 of this chapter. The cross-sectional area of the cavity is 1.2 cm². The radii of curvature of the mirrors are chosen such that a beam waist w_o of 5.0×10^{-4} m is produced. To account for the fact that πw_o^2 is much smaller than the cross-sectional area $\mathcal{A}$ of the cavity, a factor of $\pi w_o^2 / \mathcal{A}$ should be added to the expression for the effective mode volume. Therefore, the effective mode volume V_a becomes $V_a = \pi w_o^2 d' n/(d_e \mathcal{A})$, where w_o, d', and d are given in meters.

    First, consider the *pumping stage*. During the pumping stage, the effective mirror reflectivities are $\Gamma_1 = 0.9$ and $\Gamma_2 = 0.1$.
    a. Calculate B_{12}.
    b. If the system is pumped at 10 times the critical pumping transition probability A_{Pc}, what is ΔN?
    Next consider the *lasing stage*. In the lasing stage, $\Gamma_1 = 0.9$ and $\Gamma_2 = 0.99$. The ΔN obtained in part (b) is the initial population inversion for the lasing stage.
    c. Calculate the peak power radiated through mirror 1.
    d. Calculate the peak power radiated through mirror 2.
    e. Calculate the total energy per pulse radiated through the two mirrors.

17. The gain bandwidth of an HeNe laser operating at 0.633 μm is about 1500 MHz. The laser cavity is 2.0 m long.

a. Estimate the number of longitudinal modes supported by the laser.
b. If all longitudinal modes are locked to generate narrow pulses, what is the pulse width Δt_{FWHP} (in s)?
c. What is the time interval (in s) between pulses?

18. A CO_2 laser ($\lambda \approx 10.6 \mu m$) has an optical cavity length of 1.5 m and the gain medium has a gain bandwidth of 800 MHz. When an individual longitudinal mode is excited, the power output is 0.1 W.
    a. How many longitudinal modes can be sustained by the laser medium?
    b. If all longitudinal modes are excited and mode locked, estimate the peak power output.
    c. Estimate the full width between half-power points of the pulses.

19. Pulses from a mode locked laser have a pulse separation of 10 ns and a pulse width (Δt_{FWHP}) of 0.5 ns. Assume that all longitudinal modes have the same amplitudes.
    a. How many longitudinal modes are locked?
    b. What is the frequency difference between two neighboring longitudinal modes?

20. A CO_2 laser ($\lambda \approx 10.6 \mu m$) has a long optical cavity and a sufficient gain bandwidth to support several longitudinal modes. When *each* longitudinal mode is excited individually, the power output is 0.1 W. When *all* longitudinal modes are excited and locked, the output pulses have a repetition rate of 100 MHz, and each pulse has a pulse width (Δt_{FWHP}) of 0.5 ns.
    a. Calculate the cavity length (in m).
    b. Calculate the gain bandwidth (in MHz) of the active medium.
    c. Estimate the peak power of the mode locked pulses.

21. The output of a mode locked laser has a pulse separation of 12 ns, a pulse width (Δt_{FWNL}) of 2.0 ns, and a peak power of 0.8 W. Assume that all longitudinal modes have the same amplitude.
    a. How many longitudinal modes are locked?
    b. Calculate the frequency difference between two neighboring modes.
    c. Suppose the cavity length is increased by 25 percent. Factors or parameters which might affect the bandwidth of the gain medium, or the mirrors, are not changed. Estimate the peak power radiated by the laser with the lengthened cavity.

IDENTIFICATION OF ATOMIC ENERGY LEVELS

APPENDIX A

APPENDIX OUTLINE

A.1 Hydrogen Atoms
A.2 Hydrogen-like Atoms
A.3 Electron Configuration
A.4 Term Symbol
A.5 Selection Rule
References
Problems

The main objective of this appendix is to introduce the spectroscopic notations used in the energy diagrams of neutral and ionized gas lasers, and in discussions of solid-state lasers based on trivalent rare earth or actinide ions.

Five quantum numbers are required to completely specify an atomic energy level. Three of these, n, ℓ, and m_ℓ, are related to the **electron orbital motion,** and the other two, s and m_s, are related to the **electron spin.** Historically, *old quantum theory* was mainly based on theories of Bohr and Sommerfeld and was conceptionally simple; however, quantum theory has many drawbacks. For example, it cannot explain the fine structure of spectral lines of many-electron atoms, nor can it describe transitions between energy levels. In contrast, the *new quantum mechanics* provides a firm foundation for the understanding of atomic spectral lines, as well as for computing the probability of transitions between energy levels [1–3]. However, the details of quantum mechanics are much too complicated to be presented here. Instead, we will present a pictorial description.

Appendix A Identification of Atomic Energy Levels

A.1 HYDROGEN ATOMS

Hydrogen atoms, which have only one electron and one proton each, are the simplest kind. In the **Bohr theory** for the hydrogen atom, an electron can move in circular orbits around the proton, just as the earth orbits around the sun. Consider an electron with charge $-e$ and mass m_e revolving in a circular path of radius r around a nucleus of charge $+e$. For simplicity, the nucleus is assumed to be stationary; therefore, its mass is immaterial for the present discussion. The Coulomb force between the electron and the nucleus is $e^2/(4\pi\varepsilon_o r^2)$. For an electron revolving with a velocity v in a circular orbit, the centripetal force acting on the electron is $m_e v^2/r$, and the angular momentum of the electron is $m_e vr$. According to classical electromagnetic theory, a charged particle moving in a circular path experiences constant acceleration and radiates energy continuously. As the total energy decreases, the orbit shrinks continuously, and the emission from such an atom has a continuous, rather than a discrete, spectrum. However, this is contrary to experimental observations. To avoid this difficulty, Bohr postulated that:

1. An electron orbit is *stable and permissible* only if the electron angular momentum is an integer multiple of $h/(2\pi)$, that is,

$$m_e vr = n\frac{h}{2\pi} \qquad n = 1, 2, 3, \ldots \qquad (A1)$$

2. No energy is radiated by electrons moving in one of the stable and permissible orbits.

This is known as **Bohr's quantization rule.** Because of this rule, the radius of a stable circular orbit, the electron velocity, and the total energy are restricted to certain discrete values. In this sense, the energy and angular momentum of an orbital electron are **quantized.** For the nth stationary orbit, the radius of the orbit, electron velocity, and energy are

$$r_n = n^2 \frac{\varepsilon_o h^2}{m_e \pi e^2} \qquad (A2)$$

$$v_n = \frac{1}{n}\frac{e^2}{2\varepsilon_o h} \qquad (A3)$$

$$E_n = -\frac{hc\, R_H}{n^2} \qquad (A4)$$

where $n = 1, 2, 3 \ldots$, and

$$R_H = \frac{1}{2} m_e \frac{c}{h}\left(\frac{e^2}{2\varepsilon_o hc}\right)^2 \qquad (A5)$$

is the **Rydberg constant** for a stationary H nucleus.

In reality, a nucleus has a finite mass M and is not stationary. If the motion of the nucleus is accounted for, then R_H becomes

$$R_H = \frac{1}{2} \frac{m_e M}{m_e + M} \frac{c}{h} \left(\frac{e^2}{2\varepsilon_o hc}\right)^2 \tag{A6}$$

When the numerical values for e, m_e, and M are substituted in (A6), the Rydberg constant for hydrogen atoms in 10967758.1 m^{-1}. In the unit of eV, (A4) may be written as

$$E_n = \left[\frac{-13.6}{n^2}\right]_{\text{in eV}} \tag{A7}$$

The numerical values for R_H and E_n in (A6) and (A7) are for H atoms. For other one-electron systems, such as ionized helium, the Rydberg constant and the energy levels have different values because the values for the charges and masses are different.

The integer n given in (A2)–(A4) and (A7) is the **principal quantum number.** For one-electron atoms, the principal quantum number is the only quantum number required to determine the atomic energy levels. Different energy levels are associated with different electronic orbits. Therefore, n is also related to the radius of the electron orbit. The lowest energy level, E_1, is the **ground state.** The corresponding orbit has a radius r_1 and is closest to the nucleus. As n approaches ∞, $r_n \to \infty$, the electron is infinitely far away from the nucleus, and the energy E_n approaches 0. By definition, the zero energy level is the **ionization level** of the atom.

Bohr's theory for hydrogen atoms with circular electron orbits has been generalized by Sommerfeld to cover elliptical orbits. Sommerfeld had also modified Bohr's postulates on quantization. In Sommerfeld's theory, two quantum numbers are used to describe a stable elliptical orbit. In addition to the principal quantum number n, an **orbital angular momentum quantum number** ℓ is introduced. For a given value of n, ℓ can range from 0 to $(n - 1)$. An elliptical orbit is defined by semimajor and semiminor axes. The ratio of the semimajor axis to the semiminor axis is $n/(\ell + 1)$. Therefore, a thin elliptical orbit corresponds to $\ell = 0$, and a perfectly circular orbit corresponds to $\ell = n - 1$ ([2]).

In spectroscopic notation, states with $\ell = 0, 1$, and 2 are referred to as the s **(sharp)**, p **(principal)**, and d **(diffused)** states, respectively. For $\ell \geq 3$, the states are labeled alphabetically. This is summarized as follows:

$\ell = 0$	s (sharp)
$\ell = 1$	p (principal)
$\ell = 2$	d (diffused)
$\ell = 3$	f
$\ell = 4$	g
$\ell = 5$	h
$\ell = 6$	i
$\ell = 7$	k

The angular momentum vector $\vec{\ell}$ is perpendicular to the orbital plane and has a magnitude of $\ell h/(2\pi)$. The projection of $\vec{\ell}$ in *any direction* is also quantized, and an integer quantum number m_ℓ is assigned. This means that $\vec{\ell}$ is restricted to a few discrete angles relative to a given direction. From the perspective of electron orbits, we can visualize the orbits as being oriented in discrete angles relative to the given direction. The need for assigning integer values to m_ℓ was first recognized when the effects of strong magnetic fields on spectral lines were studied. As a result, m_ℓ is commonly referred to as the **magnetic quantum number.** For a given value of ℓ, m_ℓ can be any integer between $-\ell$ and $+\ell$ inclusive.

In addition to its orbital motion around the proton, an electron also spins with respect to its own axis. This is the **electron spin.** An **electron spin quantum number** s is associated with the **spin angular momentum.** The electron spin quantum number is always $1/2$,[1] and the magnitude of the vector $\vec{s}$ for the electron spin angular momentum is $h/(4\pi)$. The projection of the electron spin angular momentum vector onto any direction is also quantized, and is represented by a **magnetic spin quantum number** m_s. The quantum number m_s of electron spin projected onto an axis is either $+1/2$ or $-1/2$, and is commonly referred to as **spin up** or **spin down,** respectively.

To summarize, an atomic energy level or state is specified by five quantum numbers. The "size" and "shape" of an electron orbit depend on n and ℓ. The orientation of the orbital plane is specified by ℓ and m_ℓ. For a given n, ℓ can be any integer between 0 and $(n-1)$, inclusive, and for a given ℓ, m_ℓ can assume any integer value between $-\ell$ and ℓ, inclusive. The electron spin quantum number s is always $1/2$. However, m_s can be either $1/2$ or $-1/2$.

It is instructive to compute the number of possible states associated with a given principal quantum number n. For each value of ℓ, m_ℓ can assume the values of $-\ell, -\ell + 1, \ldots -1, 0, 1, \ldots \ell - 1, \ell$. Thus, for each value of ℓ, there are $(2\ell + 1)$ possible values of m_ℓ. Also, for a given value of n, ℓ ranges from 0, 1, ... to $(n-1)$. Thus, for a given n, the number of allowable values of ℓ and m_ℓ is

$$\sum_{\ell=0}^{n-1} (2\ell + 1) = \frac{[(n-1) + 1][1 + 2(n-1) + 1]}{2} = n^2 \qquad (A8)$$

Finally, taking into account the electron spin, the total number of states is $2n^2$.

By **Pauli's exclusion principle,** there can be at most one electron in any given state, including the spin state. Thus, the *n*th shell has $2n^2$ states and can accept at most $2n^2$ electrons.

While the Bohr-Sommerfeld theory for hydrogen atoms is intuitively satisfying, it cannot be extended to many-electron atoms. The problem may be understood from the perspective of quantum mechanics. One approach to quantum mechanics is the **Schrodinger equation**. By solving this equation, we can obtain the eigenfunctions ψ and associated eigenvalues E. The eigenfunctions are the electron wave functions, and the eigenvalues are the energy levels

---

[1]For electrons, protons, and neutrons, the spin quantum number is always 1/2, and they are known as **fermions**. Particles with 0 or integer spin quantum numbers are known as **bosons**. Photons, α particles, and deuterons are bosons.

of the states represented by the eigenfunctions. From the Schrodinger equation, five quantum numbers, n, ℓ, m_ℓ, s, and m_s, arise naturally, without the need for postulates. However, it should be noted that, according to quantum mechanics, the angular momentum vector associated with a quantum number ℓ has a magnitude of

$$|\vec{\ell}| = \sqrt{\ell(\ell+1)}\,\frac{h}{2\pi} \qquad \ell = 0, 1, 2, \ldots (n-1) \tag{A9}$$

which differs slightly from the value of $\ell h/(2\pi)$ given by the quantum theory. Similarly, according to quantum mechanics, the magnitude of a spin angular momentum vector $\vec{s}$ is

$$|\vec{s}| = \sqrt{s(s+1)}\,\frac{h}{2\pi} \tag{A10}$$

No attempt is made here to solve the Schrodinger equation. As a matter of fact, the equation is not given at all.[2] However, an interpretation of the electron wave function ψ is important to the understanding of quantum mechanics.

Consider an arbitrary point (x,y,z) in space and volume element $dv = dx\,dy\,dz$ surrounding that point. The *probability* of finding an electron in the volume element dv is $|\psi(x,y,z)|^2 dv$. Clearly, $|\psi(x,y,z)|^2 dv$ depends on the quantum numbers. Figure A1 displays the computer-generated probability density plots of the following states: $1s$, $2s$, $2p$ $(m_\ell = 0)$; $2p$ $(m_\ell = \pm 1)$; $3s$, $3p$ $(m_\ell = 0)$; $3p$ $(m_\ell = \pm 1)$; $3d$ $(m_\ell = 0)$; $3d$ $(m_\ell = \pm 1)$; and $3d$ $(m_\ell = \pm 2)$. The plots may be rotated about the vertical axis to depict a three-dimensional representation. The density of dots in a region corresponds to the probability of finding an electron in that region. In other words, these plots provide a pictorial representation of the probable distribution of the "electron cloud" around the proton.

A.2 HYDROGEN-LIKE ATOMS

For many-electron atoms, each electron is under the influence of the nucleus and other electrons. As a simple approximation, we can lump the nucleus and all except one of the electrons into a combined nucleus, and can then treat the "exception" electron as if it moves in the field of the combined nucleus. The electron and the combined nucleus are viewed as a **hydrogen-like atom,** which serves as the model for many-electron atoms. The nomenclature developed previously in this appendix for labeling electronic energy levels of hydrogen atoms is also used for hydrogen-like atoms. However, details of the energy levels of many-electron atoms depend not only on the number of electrons in a shell, but

---

[2]On the other hand, if the reader wishes to know, the time-independent Schrodinger equation of a particle of mass m in moving a potential $V(x,y,z)$ is

$$-\frac{h^2}{8\pi^2 m}\nabla^2 \psi(x,y,z) + V(x,y,z)\psi(x,y,z) = E\,\psi(x,y,z)$$

where E is the energy eigenvalue.

Appendix A Identification of Atomic Energy Levels

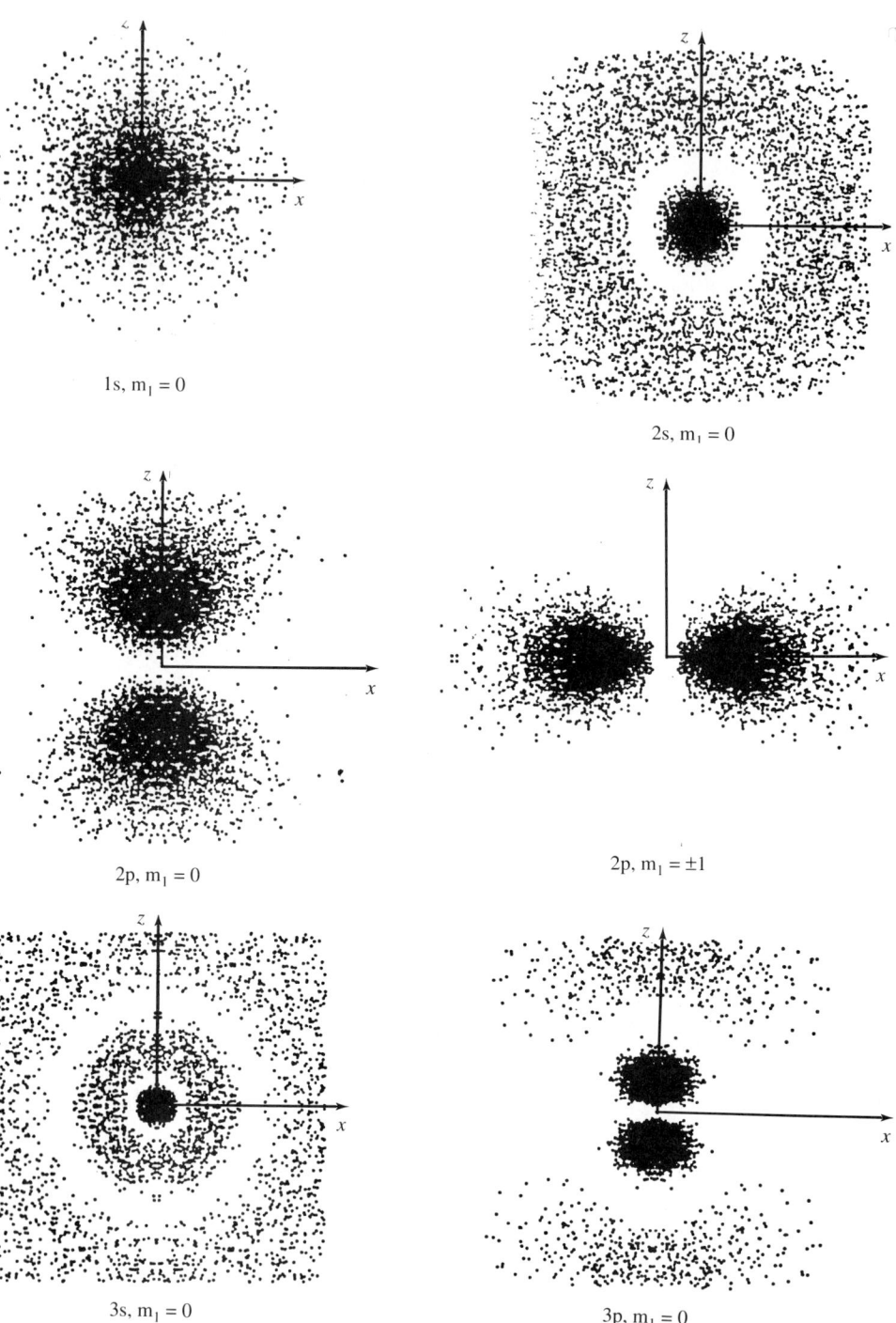

Figure A.1 Probability density plots of electron cloud distributions

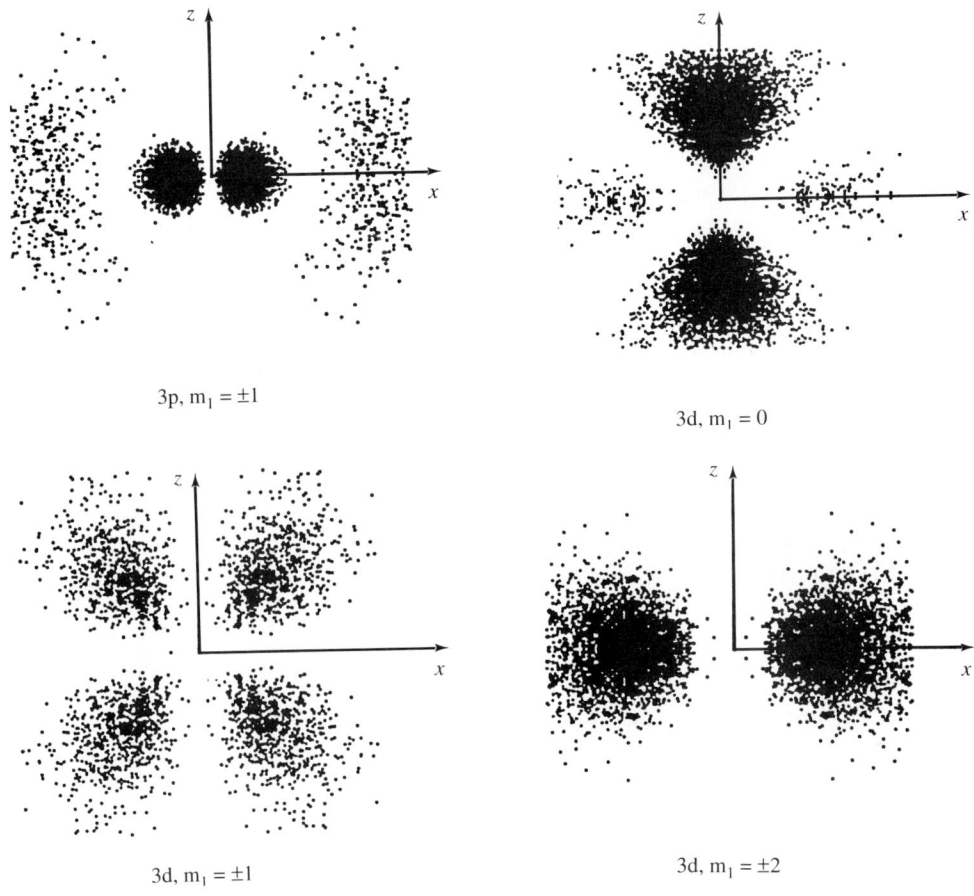

3p, m$_l$ = ±1

3d, m$_l$ = 0

3d, m$_l$ = ±1

3d, m$_l$ = ±2

Figure A.1 *(concluded)*

also on the interaction between those electrons. In other words, the energy levels of many-electron atoms are identified by two terms: the **electron configuration** and the **term symbol.** The electron configuration specifies the number of electrons in each shell, and the term symbol designates the vector sums of the orbital and spin angular momenta of all electrons and the sum of the total orbital and spin angular momenta.

A.3 ELECTRON CONFIGURATION

To introduce the electron configuration, we consider two simple many-electron atoms: helium and lithium.

Helium There are two electrons in a helium (He) atom. In the ground state of He, two electrons are in the lowest energy level, $n = 1$, $\ell = 0$, $m_\ell = 0$, and

$m_s = \pm 1/2$. In other words, the electron configuration for He in the ground state is 1s1s, or $1s^2$ for short.

Lithium A lithium (Li) atom has three electrons. However, by Pauli's exclusion principle, there can only be two electrons at most in the 1s shell. The third electron must go to the next lowest state, which is a 2s state. Thus, the electron configuration for an Li atom in the ground state is $1s^2 2s$.

A.4 TERM SYMBOL

To describe the states of many-electron atoms accurately, we must include the interactions between electrons. This can be done according to the **Russell-Saunders** (or LS) coupling scheme for light atoms, and JJ coupling or other schemes for heavy atoms. Details of such coupling schemes and the approximations involved are too complicated to be presented here [1–3]. For simplicity, however, the LS coupling scheme will be described briefly.

In **LS coupling,** the orbital angular momentum vectors $\vec{\ell}_i$ and electron spin angular momentum vectors $\vec{s}_i$ of each electron are identified. All $\vec{\ell}_i$ are summed vectorially, producing the total orbital angular momentum $\vec{L}$ of the atom

$$\vec{L} = \sum_i \vec{\ell}_i \qquad (A11)$$

The total orbital angular momentum quantum number L associated with the total orbital angular momentum vector $\vec{L}$ must be one of the allowed integers. As an example, consider a two-electron atom. Let ℓ_1 and ℓ_2 be the orbital angular momentum quantum numbers of the electrons; then, L can be any integer between $|\ell_1 - \ell_2|$ and $|\ell_1 + \ell_2|$, inclusive.

Similarly, by combining all electron spin angular momentum vectors, we can obtain the total spin angular momentum vector $\vec{S}$

$$\vec{S} = \sum_i \vec{s}_i \qquad (A12)$$

and $\vec{S}$ has a quantum number S. From $\vec{L}$ and $\vec{S}$, we obtain the total angular momentum vector

$$\vec{J} = \vec{L} + \vec{S} \qquad (A13)$$

Depending on the values of S and L, the value of J may be as follows:

For $L \geq S$, $J = |L - S|, |L - S + 1|, \ldots, |L + S - 1|, |L + S|$, with a total of $2S + 1$ values.

For $L \leq S$, $J = |S - L|, |S - L + 1|, \ldots, |S + L - 1|, |S + L|$, with a total of $2L + 1$ values.

The resulting electron interaction is designated by the **term symbol** $^{2S+1}L_J$. The superscript $2S + 1$ is the **multiplicity** of the state. A letter is used to designate the total orbital angular momentum quantum numbers L. Letters S, P, D,

F, etc. indicate an L of 0, 1, 2 and 3 etc. The numerical subscript is used to indicate the value of J.

As an example, consider He atoms. The orbital quantum numbers of the electrons in the ground state are $n = 1$, $\ell = 0$, $m_\ell = 0$, and the spin quantum numbers are $m_s = \pm 1/2$. The ground state *electron configuration* is $1s^2$. Under the LS coupling scheme, $L = 0 + 0 = 0$. Since two electron spins are oriented in opposite directions, the total electron spin quantum number is

$$S = |(1/2) + (-1/2)| = 0.$$

and the multiplicity is $2S + 1 = 1$. Thus,

$$J = L + S = 0 + 0 = 0.$$

The *term designation* for the ground state of He atoms is $^1S_0$. Hence, the complete description of the ground state of He is $1s^2\ ^1S_0$.

When one of the electrons is excited to the second shell, the electron configuration becomes $1s2s$. The quantum numbers of the electron in the first shell are $n = 1$, $\ell = 0$, $m_\ell = 0$, and $m_s = \pm 1/2$. The electron in the second shell has quantum numbers $n = 2$, $\ell = 0$ or 1, and $m_s = \pm 1/2$. It is beyond the scope of this book to calculate the energy associated with each state. We simply note that the state with $\ell = 1$ is at a higher energy level than that with $\ell = 0$. Therefore, for the first excited state, the quantum numbers of the second electron are $n = 2$, $\ell = 0$, and $m_\ell = 0$, which means that $L = 0$. Since two electrons are in different shells ($1s$ and $2s$), there is no restriction on their spin quantum numbers. Two m_s can have the same or opposite signs. If the spin angular momentum vectors are in opposite directions, then $S = |(1/2) + (-1/2)| = 0$, the term symbol is again $^1S_0$, and the complete description for the excited state is $1s2s\ ^1S_0$. If the spin angular momentum vectors are in the same direction, then $S = |(1/2) + (1/2)| = 1$ or $S = |(-1/2) + (-1/2)| = 1$, and the multiplicity is $2S + 1 = 3$. Since $J = |L + S| = |0 \pm 1| = 1$, the term symbol becomes $^3S_1$, which means the other excited state is labeled $1s2s\ ^3S_1$. These designations are used to label the energy levels of the He atoms of HeNe lasers (See Figure 3.22).

We can now consider the levels of Ne atoms. The ground state of Ne is $1s^2\ 2s^2\ 2p^6\ ^1S_0$. In fact, for any completely filled shell, the vector sum of all $\vec{\ell}$ is zero and all $\vec{s}$ cancel completely. Thus, the term symbol for a *closed* or *filled shell* is always $^1S_0$.

The first excited state of Ne has an electron configuration of $1s^2\ 2s^2\ 2p^5\ 3s$; one of the $2p$ electrons is raised to the $3s$ shell. The orbital angular momentum vectors for the electrons in the $1s$, $2s$, and $3s$ shells are all zero. It is therefore only necessary to consider five $2p$ electrons and a $3s$ electron. Recall that six electrons are needed to fill the $2p$ states completely, and the vector sum for $\vec{\ell}$ of six $2p$ electrons is zero. As a shortcut, we view five $2p$ electrons as six $2p$ electrons (with a vector sum of $\vec{\ell}$ of 0) minus a $2p$ electron (with $\ell = 1$). For the $3s$ electron, $\ell = 0$. Thus, the total orbital angular momentum quantum number L of the first excited state is 1. Next, we consider the total spin angular momentum. The spin angular momenta of the first eight electrons cancel completely.

The spin angular momenta of the last two electrons are either in the same direction, leading to S = 1, or in opposite directions, resulting in S = 0. Therefore, we conclude:

1. For L = 1 and S = 0, J is 1, and the term symbol is $^1P_1$.
2. For L = 1 and S = 1, J can be any integer between (1 + 1) and (1 − 1), that is, 2, 1, or 0. The multiplicity is 2S + 1 = 2 × 1 + 1 = 3. The term symbol can be $^3P_2$, $^3P_1$, $^3P_0$. These term symbols are shown in Figure 3.22.

Neon (Ne), Argon (Ar), and other inert gases are often used in various gas lasers. The electron configurations and term symbols for the inert gases in their ground states are

Helium	(He)	$1s^2\ ^1S_0$
Neon	(Ne)	$1s^22s^22p^6\ ^1S_0$
Argon	(Ar)	$1s^22s^22p^63s^23p^6\ ^1S_0$
Krypton	(Kr)	$1s^22s^22p^63s^23p^63d^{10}4s^24p^6\ ^1S_0$
Xenon	(Xe)	$1s^22s^22p^63s^23p^63d^{10}4s^24p^64d^{10}5s^25p^6\ ^1S_0$

Note that for the ground states of inert gases, all shells are filled completely and the term symbol is always $^1S_0$, as noted previously.

A.5 SELECTION RULE

When electromagnetic waves are absorbed or radiated, atoms are either raised to a higher or dropped to a lower energy level. However, not all transitions are permissible. Transitions can take place only if the levels involved are such that L changes by ± 1 and J by ± 1 or 0; that is,

$$\Delta L = \pm 1 \quad \text{and} \quad \Delta J = \pm 1 \text{ or } 0 \qquad (A14)$$

This is known as the **selection rule.** Consequently, transitions are allowed between the S and P states, and between the P and D states, of isolated atoms. However, no transition is allowed between the S and D states or the S and F states.

REFERENCES

1. Born, M. *Atomic Physics*, 8th ed. London: Blackie, 1969.
2. Richtmyer, F. K.; E. M. Kennard; and T. Lauritsen. *Introduction to Modern Physics*, 5th ed. New York, NY: McGraw-Hill, 1955.
3. Ashby, N.; and S. C. Miller. *Principles of Modern Physics*. San Francisco, CA: Holden-Day, 1970.
4. Herzberg, G. *Atomic Spectra and Atomic Structure*. New York, NY: Dover Publications, 1944.
5. Harris, S. E. "Proposal for a 207-Å laser in lithium." *Opt. Lett.* 5, (1980), pp. 1–3.

PROBLEMS

A1. An Na atom has 11 electrons. What is the electron configuration and the term symbol for the ground state of Na atoms? Explain how each term comes about.

A2. It has been proposed that neutral Li atoms be used to construct a soft X-ray laser [5]. Electron beams and tunable laser sources together would serve as the pump. The energy levels and the electron configuration for the relevant states of Li are shown in the figure.

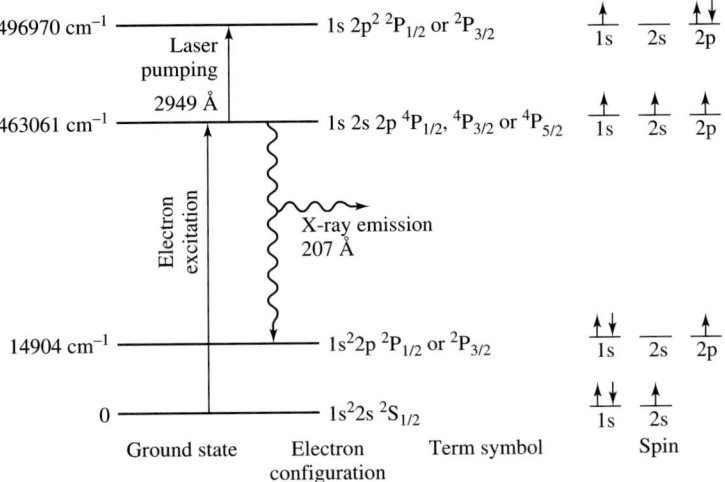

a. Identify the term symbol for each level.
b. Compute the energy (in eV) required for the electron beam excitation.
c. Compute the wavelength of the radiated X-ray.

SEMICONDUCTOR INJECTION LASERS, LIGHT EMITTING DIODES, AND SUPERLUMINESCENT DIODES

CHAPTER 4

CHAPTER OUTLINE

4.1 Introduction
4.2 Intrinsic and Extrinsic Semiconductors
 4.2.1 Energy Band Diagram
 4.2.2 Elemental Semiconductors
 4.2.3 Binary Semiconductors
4.3 The Interaction of Semiconductors with Electromagnetic Waves
 4.3.1 Direct and Indirect Bandgaps
 4.3.2 Absorption in Semiconductors
 4.3.3 Photon Emission in Semiconductors
4.4 Ternary and Quaternary Semiconductors
 4.4.1 III–V Semiconductors
 4.4.2 II–VI Semiconductors
 4.4.3 IV–VI Semiconductors
4.5 Homojunctions and Heterojunctions
 4.5.1 Homojunctions
 4.5.2 Heterojunctions
4.6 Basic Semiconductor Luminescent Diode Structures
 4.6.1 Broad Area Lasers
 4.6.2 Stripe Geometry Lasers
 4.6.3 Single-Frequency Single-Mode Injection Lasers
 4.6.4 Recent Developments
4.7 Light Emitting Diodes
 4.7.1 Two Basic LED Structures
 4.7.2 Power and Light Extraction Efficiency: Surface-Emitting LEDs
 4.7.3 Radiant Intensity Pattern: Surface-Emitting LEDs
 4.7.4 Radiant Intensity Pattern: Edge-Emitting LEDs
 4.7.5 PI and Spectral Characteristics
 4.7.6 Modulation Characteristics
 4.7.7 Coupling from LED–to–Multimode Fiber

4.8 Basic Parameters: Semiconductor Injection Lasers
    4.8.1 Wavelength of Radiation
    4.8.2 Threshold Current Density
    4.8.3 Internal and External Quantum Efficiencies
    4.8.4 Power Efficiency
    4.8.5 Far Field Radiation Pattern
    4.8.6 Spectral Linewidth
4.9 Dynamic Characteristics: Semiconductor Injection Lasers
    4.9.1 Rate Equations with a Current Injection Term
    4.9.2 Steady-State PI Characteristics
    4.9.3 Transient Characteristics
    4.9.4 Amplitude Modulation
4.10 Superluminescent Diodes
4.11 Comparison of ILDs, LEDs, and SLDs
References
Problems

4.1 INTRODUCTION

The lasers discussed in the last chapter are rather bulky, due to the size of the laser cavity and the pumping hardware and associated power supply. Although their emissions have superior spectral and spatial characteristics and they are capable of a large power output, their use in some applications may be impractical because of their size and weight. In addition, high-speed modulation of such lasers is possible only by indirect or external means. These undesirable features are eliminated in **semiconductor injection lasers (ILDs).** While the spectral and spatial characteristics and the power levels of ILDs are not comparable to those of gas, dye, or solid-state lasers, their compact sizes, coupled with the possibility of direct, high-speed modulation, makes ILDs attractive in many applications. Also, light coherence may be unnecessary, or even detrimental, in some applications. If **incoherent light** is needed, **light emitting diodes (LEDs)** and **superluminescent diodes (SLDs)** may be used. LED and SLD are actually preferred in many short-haul, moderate-bandwidth communications or sensory systems. Because ILDs, LEDs, and SLDs have the same basic device structure and are fabricated by the same basic technology, they are treated together in this chapter.

    Since ILDs, LEDs, and SLDs are based on semiconducting materials, we will begin by reviewing the basics of semiconductors. We will distinguish between **intrinsic** and **extrinsic semiconductors, direct** and **indirect bandgap** materials, and **homojunctions** versus **heterojunctions.** Also discussed are **elemental** and **binary compound semiconductors,** as well as **ternary** and **quaternary solid solutions.** These topics are presented in Sections 4.2 to 4.5. The basic structures of semiconductor luminescent diodes are discussed in Section 4.6, along with the various stripe geometry double-heterostructure diodes. Section 4.7 is devoted entirely to discussions on LEDs. Sections 4.8 and 4.9 are devoted to quantitative examinations of ILDs, including their amplitude modulation characteristics. Section 4.11 contains a brief discussion on SLDs, and the final section is a comparison of various ILDs, LEDs, and SLDs.

4.2 INTRINSIC AND EXTRINSIC SEMICONDUCTORS

4.2.1 Energy Band Diagram

In gaseous media, there are relatively few atoms and molecules, and they are far apart. As a result, interactions between them are very weak, which means that the energy levels in gas lasers are those of isolated atoms or molecules. In solid-state lasers, however, optically active ions are dispersed throughout the solid host materials. The effects of the host atoms and molecules on the levels of active ions depend strongly on the electron configuration of the active ions. The effects of the host material are minimal in some solid-state lasers and strong in others, as discussed in Chapter 3. In contrast, in semiconductor injection lasers, the semiconductor material is the active medium, and the interaction between the atoms or molecules in the semiconductor has a major influence on its energy levels.

The interactions between atoms or molecules are very similar to the coupling between electric circuits. Take as an example two identical LC circuits and assume that the two circuits have the same resonant frequency $f = 1/(2\pi\sqrt{LC})$ in the absence of coupling. The resonant frequencies become $f = 1/[2\pi\sqrt{C(L \pm M)}]$ in the presence of M, the mutual inductance. One resonant frequency is above and the other is below the original resonant frequency. When N identical LC circuits are coupled, there will be N resonant frequencies. This same frequency splitting also occurs when atoms interact. When a large number of atoms are involved, the energy levels split into a large number of levels as a result of coupling. Since the energy difference between these levels is small, the levels are closely spaced, and they may be viewed as bands.

The evolution from discrete energy levels to energy bands is depicted schematically in Figure 4.1. Beginning with the discrete levels of isolated atoms in Figure 4.1a, the energy diagram evolves into the energy band diagram of a crystalline solid, in Figure 4.1b. As an example, diamonds are carbon (C) atoms arranged in a cubic structure. Suppose that N carbon atoms are far apart initially and they are brought together slowly by some means. We can follow the evolution of the energy levels as the C atoms are moved closer. The ground state of isolated C atoms has an electron configuration of $1s^2 2s^2 2p^2$ and a term symbol of $^3P_0$. The first shell ($1s$) is completely filled and the second shell ($2s2p$) is partially filled. The electrons in the filled inner shells are customarily referred to as the **core electrons;** those in the outermost, and possibly unfilled, shell are the **valence electrons.** A C atom has two core electrons ($1s^2$) and four

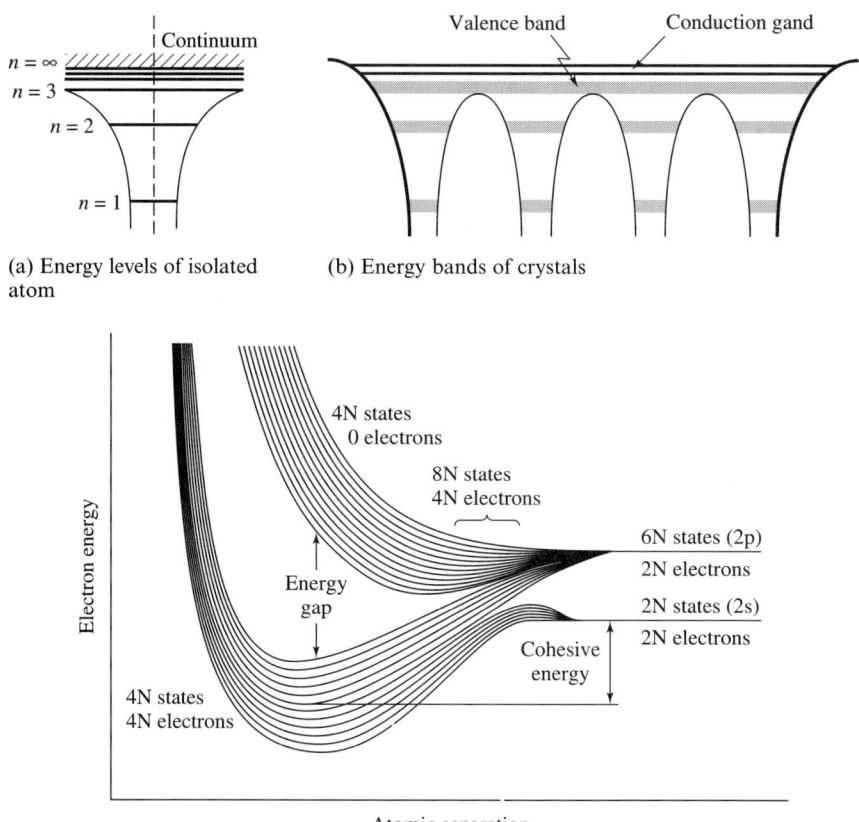

(a) Energy levels of isolated atom

(b) Energy bands of crystals

(c) Evolution from discrete levels of isolated C atoms to bands of diamond crystal

Figure 4.1 Evolution from discrete levels to energy bands

valence electrons ($2s^2 2p^2$). When several C atoms are brought together, neighboring C atoms exert minimal influence on the core electrons, unless the atoms are extremely close. Neighboring atoms mainly affect the valence electrons. Since a carbon atom has two $2s$ states and six $2p$ states, for N carbon atoms, there are $2N$ $2s$ states and $6N$ $2p$ states. All the $2s$ states are occupied, while a third of the $2p$ states are occupied. When the atoms are far apart, the energy levels of the $2s$ and $2p$ states are the same as those of isolated C atoms. They are represented by the two lines on the right-hand side of Figure 4.1c. The $2s$ line is $2N$-fold degenerate and the $2p$ line is $6N$-fold degenerate. By $2N$- or $6N$-degeneracy, we mean that $2N$ or $6N$ states have the same energy levels. When the atoms are brought together, however, the degeneracy is broken and each line becomes a group of lines. The lower group has $2N$ lines corresponding to $2N$ states, all of which are filled. The upper group has $6N$ states, a third of which are filled. As the atoms are pushed even closer, the two groups merge into a

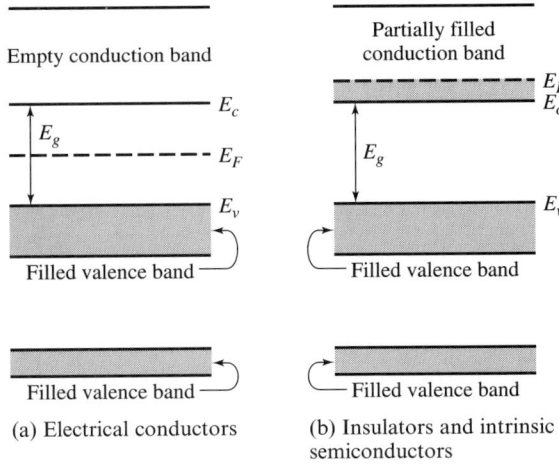

Figure 4.2 Energy band diagram for insulators, intrinsic semiconductors, and electrical conductors at 0 °K

continuum. As the distance is further reduced, two continua coalesce into one and then separate into two bands. Each band has a 4N-fold degeneracy. The lower band is completely filled with 4N electrons, while the upper band is completely empty. For energy levels in these bands, the old designations 2s and 2p are no longer appropriate. In the terminology to be introduced shortly, we refer to the lower and upper bands as the valence and conduction bands, respectively.

Energy bands in crystals can also be understood from a different point of view. Since the atoms or molecules are arranged in a perfect periodic array, the electric fields experienced by electrons are also periodic. The presence of the periodic electric fields causes the energy levels to break into interlaced *allowed* and *forbidden* bands. The allowed bands are the **energy bands** and the forbidden ones are the **bandgaps.**

The energy bands may or may not actually be occupied. Depending on temperature, some of them are likely to be filled, while others may be empty. At 0 °K, all states below a certain energy level, known as the **Fermi level,** are filled, while those above that level are empty.

Of particular interest is the highest filled band and the band immediately above it. These are known as the **valence band** and **conduction band,** respectively, and are in the immediate vicinity of the Fermi level. Figure 4.2 shows the energy band diagram of **intrinsic semiconductors**[1], insulators, and conductors at 0 °K, and their corresponding Fermi levels. The vertical axis represents the **electron energy.** At present, no physical significance is assigned to the horizontal axis. Horizontal lines are drawn merely as a visual aid. The top edge of the valence

---

[1] The term "intrinsic semiconductor" is discussed in the next section.

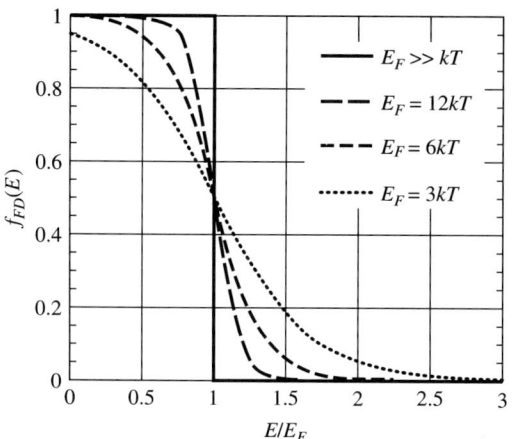

Figure 4.3 Fermi–Dirac distribution function

band is identified as E_v, and the bottom edge of the conduction band is E_c. The bandgap $E_g = E_c - E_v$ and the location of the Fermi level E_F relative to E_v and E_c are the dominant factors affecting the electrical and optical properties of the material.

It is instructive to compare the energy band diagram of insulating materials with that of conducting materials. A material with a partially filled band, in addition to completely filled or completely vacant bands, at 0 °K is a conductor (Figure 4.2a). The electrons in the partially filled band are free to move in the presence of electric fields, the partially filled band is a conduction band, and the material is electrically conducting. On the other hand, if the highest valence band is completely filled and the lowest conduction band is completely empty at 0 °K (Figure 4.2b), the electrons in the filled valence band cannot move, even in the presence of electric fields. There are no electrons in the empty conduction band. Therefore there are no currents in the material unless the electric fields are large enough to cause electric breakdown. Such a material is an insulator or intrinsic semiconductor. In other words, a material whose energy bands are either completely filled or completely empty at 0°K is an insulator or intrinsic semiconductor.

For $T > 0$ °K, the probability that a state with energy E is occupied is given by the **Fermi–Dirac distribution function**

$$f_{FD}(E) = \frac{1}{1 + e^{(E-E_F)/kT}} = \frac{1}{1 + e^{(E_F/kT)(E-E_F)/E_F}} \tag{4.1}$$

where k is the Boltzmann constant. Figure 4.3 is a plot of the Fermi–Dirac distribution as a function of E/E_F for four values of E_F/kT. Since the conduction band of a conductor is partially filled, even at 0 °K, its electrical characteristic is qualitatively the same whether T is precisely at 0 °K or above 0 °K. For insulators and intrinsic semiconductors, there is a finite probability that electrons can be thermally excited from the valence band to the conduction band at $T > 0\ °K$. Once that happens, excited electrons are free to move, as are the vacancies left behind in the valence band. Consequently, insulators and intrinsic semiconductors become electrically conducting when their temperature rises above $0\ °K$. This means that an insulator's electrical characteristics at $0\ °K$ are quite different from those at higher temperatures. The probability that electrons are elevated to the conduction band is greater if the bandgap E_g is narrower or the temperature is higher. Clearly, all insulators become highly conductive if the temperature is sufficiently elevated such that $T \gg E_g/k$.

4.2.2 Elemental Semiconductors

Si and Ge are the two best known **elemental semiconducting materials.** Although C is generally considered to be a good insulator at room temperature, it is also a good semiconducting material at high temperatures. Tin is also an elemental semiconductor, although it is less well known than Si, Ge, and C. An Si atom has a ground state of $1s^2 2s^2 2p^6 3s^2 3p^2\ ^3P_0$ and has four valence electrons ($3s^2 3p^2$). A Ge atom ($1s^2 2s^2 2p^6 3s^2 3p^6 3d^{10} 4s^2 4p^2\ ^3P_0$) also has four valence electrons ($4s^2 4p^2$), as does a C atom ($2s^2 2p^2$). These materials are referred to as **group IV materials.**

Pure semiconductors are known as **intrinsic semiconductors.** The energy band diagrams of intrinsic semiconductors differ from those of insulators only in the width of the bandgap. For example, diamond has a bandgap of 5 eV and is an insulating material at room temperature. The bandgaps of Ge and Si are approximately 0.67 and 1.12 eV, respectively. Although these bandgaps are narrower than those of insulators, they are still much greater than kT at room temperature. This means that, at room temperature, the conduction bands of Si and Ge are mostly empty and the valence bands are full, and intrinsic Si and Ge behave like insulating materials.[2]

As previously stated, as long as the bandgap is finite and the temperature is above $0\ °K$, there exists a finite probability that electrons can be excited from the valence band to the conduction band, where the electrons are free to move. The motion of an electron in a crystal is represented by a negative charge $-e$ and an **effective mass** m_e^*. The effects of periodic electric fields in the crystal are lumped into the effective mass m_e^*, which differs from the free-electron mass m_e. The vacancies left behind in the valence band also move under the influence of applied electric fields. The motion of a vacancy is equivalent to a charged particle with a positive charge $+e$ and an effective mass m_p^*. This is the concept of a **hole**. The Fermi level and the effective mass of an intrinsic semiconductor are related as shown by the expression

$$E_F = \frac{1}{2}(E_c + E_v) + \frac{3}{4} kT \ln\left(\frac{m_p^*}{m_e^*}\right)$$

---

[2] At room temperature, $kT \approx 0.0258$ eV $\approx 1/40$ eV.

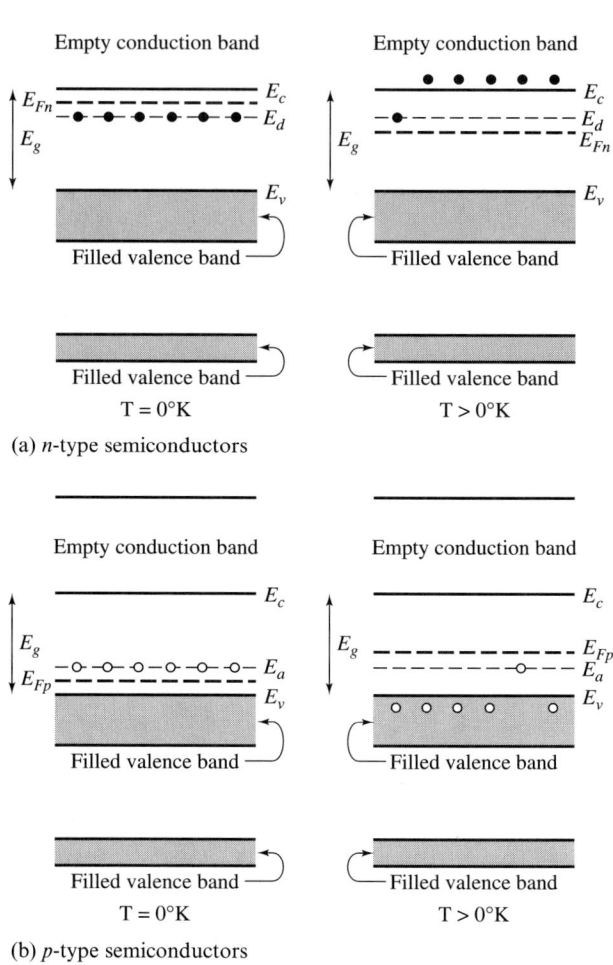

Figure 4.4 Energy band diagrams of n-type and p-type semiconductors

Thus, E_F is roughly midway between E_c and E_v and is slightly temperature dependent [1]. We also note that for intrinsic semiconductors, there are as many electrons in the conduction band as there are holes in the valence band.

If impurities are introduced into intrinsic semiconductors, the result is **extrinsic semiconductors.** The impurities are **dopants,** and the process of introducing impurities is known as **doping.** When a minute amount of **group V material,** such as Sb, P, or As, is added to Si or Ge, some of the group V atoms (the impurities) substitute for the group IV atoms. Since a group V atom has five valence electrons, an *excess* electron is donated to the semiconductor material. In other words, group V impurities are electron **donors.** The energy levels E_d of donor states are typically a few hundredths of an eV below E_c (Figure 4.4a). At 0 °K, donor electrons are bound to the donor states, that is, the donor levels are occupied. Since $E_c - E_d$ is on the order of kT at room temperature, the donor electrons are likely to be detached from the donor states at room temperature. Once detached, these electrons are free to move in the conduction band. The ionized donor sites, however, are not free to move. In donor-doped semiconductors, most of the electrons in the conduction band originate from the donor impurities rather than from the valence band; hence, there are more electrons than holes. Semiconductors doped with donor impurities are referred to as **n-type** semiconductors. The Fermi level E_{Fn} of an n-type semiconductor is located above the center of the bandgap, as shown in Figure 4.4a. As the donor concentration increases, E_{Fn} moves toward E_c.

If the impurity atoms are from **group III**, or atoms with three valence electrons, an opposite situation arises. For each host atom of group IV replaced by an impurity atom of group III, there is one *less* valence electron. Group III atoms, such as B, Al, Ga, or In, act as **acceptors** of electrons. The energy level E_a of acceptors is a few hundredths of an eV above the edge of the valence band E_v. At room temperature, electrons in the valence band are likely to be elevated to the acceptor states, leaving vacant sites in the valence band. These vacant sites are holes, which are free to move and contribute to electrical conduction. Because of the presence of acceptor impurities, there are more holes in the valence band than electrons in the conduction band. Semiconductors doped with group III impurities are **p-type** semiconductors. As shown schematically in Figure 4.4b, the Fermi level E_{Fp} of a *p*-type semiconductor is closer to the valence band edge than the center of the bandgap.

4.2.3 Binary Semiconductors

Compound semiconductors, such as AlAs, GaAs, GaSb, GaP, InP, InAs, and InSb, are often used in optical and electronic applications and are called **binary semiconductors.** As an example, Ga has a ground state electron configuration of $1s^2 2s^2 2p^6 3s^2 3p^6 3d^{10} 4s^2 4p^1$ $^2P_{1/2}$, with three valence electrons ($4s^2 4p^1$), while As ($1s^2 2s^2 2p^6 3s^2 3p^6 3d^{10} 4s^2 4p^3$ $^4S_{3/2}$) has five valence electrons ($4s^2 4p^3$). Since Al, Ga, and In are group III atoms and P, As, and Sb are in group V, compound semiconductors, such as GaAs, InSb, etc., are referred to as **III–V semiconductors.** These semiconductors can also be doped; group VI atoms, such as Te, Se, or S, may be used as donor impurities, and group II atoms (Zn, Cd, and Mn) as acceptor impurities. Depending on the host atoms they replace, Si or Ge atoms can be used as either donor or acceptor impurities for III–V semiconductors.

Elements from groups II and VI can form **II–VI semiconductors.** Examples are: CdS, CdSe, ZnO, and ZnS. A compilation of properties of selected elemental, binary semiconductors is given in Table 4.1 [2]. Semiconducting materials are further discussed in section 4.4.

Table 4.1 Selected semiconductor characteristics (excerpted from [2, 3])

Type	Semiconductor	E_g at 300 °K eV	Dir/ind bandgap	Index n	Relative dielectric constant ε_r	Lattice constant a nm	Electron mobility μ_e cm²/V s	Hole mobility μ_h cm²/V s	Electron affinity χ eV
IV	Si	1.11	ind 100	3.44	11.7	0.543	1350	480	4.01
	Ge	0.67	ind 111	4.00	16.3	0.566	3900	1900	4.13
III–V	AlAs	2.16	ind	2.9	12	0.566	1000	~100	2.62
	AlSb	1.6	ind 100	3.4	11	0.6135	50	400	3.6
	GaP	2.25	ind 100	3.37	10	0.5450	120	120	4.0
	GaAs	1.43	dir 000	3.4	12	0.5653	8600	400	4.07
	GaSb	0.69	dir 000	3.9	15	0.6095	4000	650	4.06
	InP	1.28	dir 000	3.37	12.1	0.5869	4000	650	4.40
	InAs	0.36	dir 000	3.42	12.5	0.6058	30000	240	4.90
	InSb	0.17	dir 000	3.75	18	0.6479	76000	5000†	4.59
II–VI	ZnO	3.2	dir 0000	2.02	7.9	a 0.3250 c 0.2065	180 180		
	ZnSe	2.58	dir 000	2.89	8.1	0.5667	100		4.09
	ZnTe	2.28	dir 000	3.56	9.7	0.6101		7	3.53
	CdS	2.53	dir 0000	2.5	8.9	a 0.4136 c 0.6713	210		4.79
	CdSe	1.74	dir 0000		10.6	a 0.4299 c 0.7010	500		4.95
	CdTe	1.50	dir 000	2.75	10.9	0.6477	600		4.28
IV–VI	PbS	0.37	dir 111	3.7	170	0.5936	550	600	3.3
	PbSe	0.26	dir 111		250	0.6124	1020	930	
	PbTe	0.29	dir 111	3.8	412	0.6460	1620	750	4.6

† at 78 °K.

4.3 THE INTERACTION OF SEMICONDUCTORS WITH ELECTROMAGNETIC WAVES

4.3.1 Direct and Indirect Bandgaps

Going back to the energy band diagrams introduced in the last section (Figures 4.2 and 4.4), we can now assign a physical meaning to the horizontal axis. To do so, examine the electron waves that propagate in crystal lattices. In isotropic media, all directions are equivalent, and it would be superfluous to distinguish

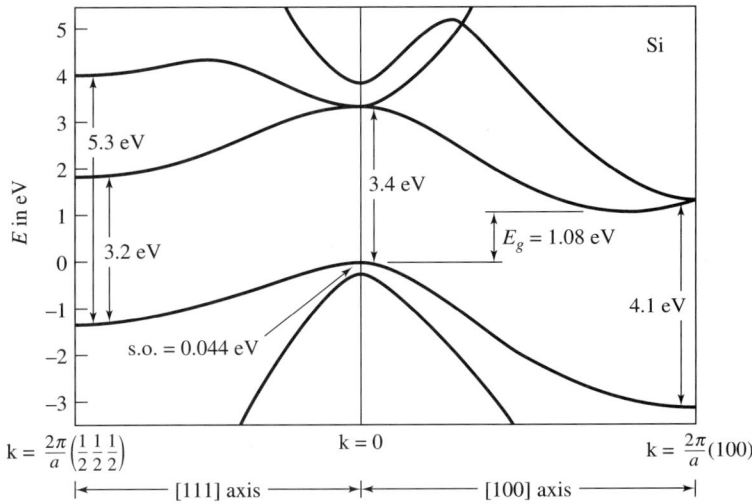

Figure 4.5 Energy band diagram of Si near the bandgap
Split of valence band at center is greatly exaggerated. [1]

these directions. However, many crystalline solids are not isotropic; that is, all directions are not equivalent. This means that, for crystals, the energy diagrams of electron wave functions propagating in different directions are different. It is customary to use the direction of the **wave vector k** to represent the propagation direction of these waves. The wave vectors of electron waves in crystals are referred to as the **crystal momenta,** and the horizontal axis of an energy diagram represents the wave vector of electron waves. Figures 4.5 and 4.6 display energy diagrams of Si and GaAs along *selected directions* [1,4]. If the main peak of the valence band and the lowest valley of the conduction band occur at different wave vectors, $\mathbf{k}_c \neq \mathbf{k}_v$, the semiconductor is referred to as an **indirect bandgap** semiconductor [5, 6]. The energy band diagram of Si shown in Figure 4.5 is a good example. In contrast, for GaAs, the main peak of the valence band and the lowest valley of the conduction band occur in the same direction and have the same wave vector, $\mathbf{k}_c = \mathbf{k}_v$, as shown in Figure 4.6. Semiconductors with this type of energy band diagram are **direct bandgap** semiconductors.

The optical properties of direct versus indirect bandgap semiconductors are fundamentally different. The difference can be understood in terms of the conservation of energy and the conservation of momentum. The corresponding conservation laws in classical mechanics are often used to evaluate particle collisions. Let m_i and v_i be the mass and the velocity of the particle i prior to collision, where $i = 1, 2$, etc. Assuming the potential energy of the particles remains unchanged, the velocities v'_i following the collision must satisfy the **laws of conservation of energy**

$$\sum_{i=1} (\tfrac{1}{2} m_i v_i^2) = \sum_{i=1} (\tfrac{1}{2} m_i v_i'^2)$$

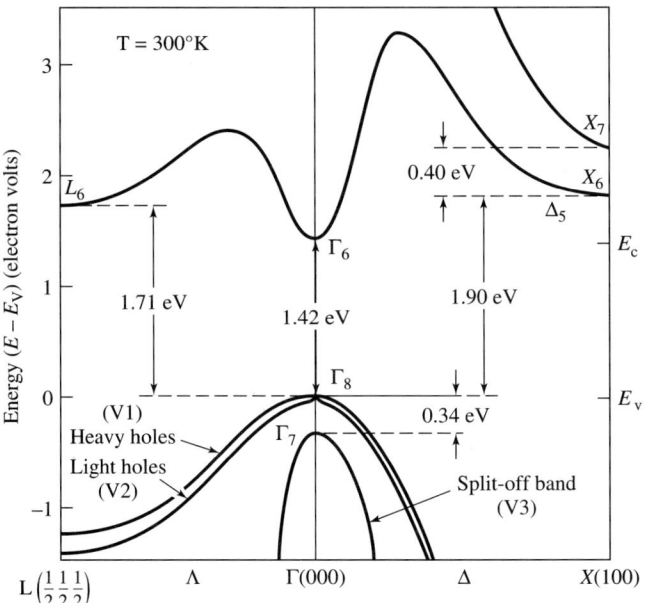

Figure 4.6 Energy band diagram of GaAs near the bandgap at 300 °K [4]

and the **laws of conservation of momentum**

$$\sum_{i=1} m_i \mathbf{v}_i = \sum_{i=1} m_i \mathbf{v}'_i$$

Analogous to these laws, in classical mechanics, we have the **conservation of energy** and the **conservation of wave vectors** in describing the interaction between waves and solids. Let E_i and $\mathbf{k}_i$ and E'_i and $\mathbf{k}'_i$ be the energy and wave vector of a wave i prior to and following the interaction. Wave interaction is *strong and probable if*

$$\sum_{i=1} E_i = \sum_{i=1} E'_i \qquad (4.2)$$

$$\sum_{i=1} \mathbf{k}_i = \sum_{i=1} \mathbf{k}'_i \qquad (4.3)$$

If either equality (4.2) or (4.3) is not satisfied, wave interaction is *weak* and *unlikely*.

In this chapter, we are interested in the interactions between electrons in solids, electromagnetic waves, and acoustic waves in crystalline solids. To focus on these interactions, the central portions of the energy diagrams in Figures 4.5 and 4.6 have been extracted and are depicted in Figures 4.7a and 4.7b. The energy and wave vector of the electrons near the lowest valley of the conduction band are E_c and $\mathbf{k}_c$. The corresponding terms near the main peak of the valence band are E_v and $\mathbf{k}_v$, respectively. As noted previously, $E_v - E_c$ is 1.11 eV for Si and 1.43 eV for GaAs. For photons of vacuum wavelength λ, the photon energy

4.3 The Interaction of Semiconductors with Electromagnetic Waves

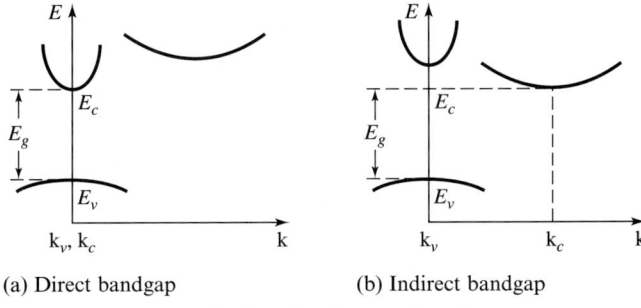

(a) Direct bandgap (b) Indirect bandgap

Figure 4.7 Direct and indirect bandgap semiconductors

is hc/λ and the wave vector has a magnitude of $|\mathbf{k}_{op}| = 2\pi/\lambda$. For electromagnetic waves in the visible or near infrared (IR) regions, λ is on the order of 1 µm, the photon energy is on the order of 1 eV, and $|\mathbf{k}_{op}|$ is on the order of 10^{+6} m^{-1}. In contrast, **acoustic waves**, known as **phonons**, in crystals have an energy E_{ac} on the order of a few tenths of an eV and a wave vector $|\mathbf{K}|$ on the order of $2\pi/a$, where a is the crystal lattice constant. Table 4.1 shows that the lattice constants are about 0.5 nm, which means that the phonon wave vector is on the order of 10^{+10} m^{-1}. Clearly, the wave vectors of photons in the visible or near IR spectra are four orders of magnitude *smaller* than those of phonons.

4.3.2 Absorption in Semiconductors

Absorption of electromagnetic waves by **direct bandgap** semiconductors is a relatively simple process. For electrons near the central portion of the energy diagram, $\mathbf{k}_v$ and $\mathbf{k}_c$ are the same, and the law of conservation of wave vectors is automatically satisfied. We need only consider the conservation of energy. Photons with lower energy, that is, $hf < E_g$, do not interact with, and are not absorbed by, the semiconductors since the law of conservation of energy could not be satisfied. Only photons that are sufficiently energetic, that is, for which $hf \geq E_g$, can interact with, and are likely absorbed by, the semiconductors. In the process, electrons are elevated from the valence band to the conduction band and electron–hole pairs are created (Figure 4.8a). As the photon energy increases, more photons and electrons are available for interaction, more photons are absorbed, and the electromagnetic waves are attenuated. Consequently, the attenuation characteristics of direct bandgap semiconductors have a sharp bend at $hf = E_g$. Beyond the bend, the attenuation constant increases abruptly. This is shown schematically in Figure 4.8b. Note that in direct bandgap semiconductors, no phonon is involved in the photon absorption process, and the process is known as a **direct transition** process.

We next consider **indirect bandgap** semiconductors. Photons with energies much less than E_g are simply not absorbed. For photons with energies greater than E_g and less than $E_g + E'_g$, the interaction is more complex. The law of conservation of energy can be satisfied by electrons near the peak of the valence band, electrons near the valley of the conduction band, and the photons just mentioned. However, the law of conservation of wave vectors cannot, because $|\mathbf{k}_c - \mathbf{k}_v|$ is at least four orders of magnitude larger than $|\mathbf{k}_{op}|$. Therefore, optical absorption by raising electrons from E_v to E_c, without the assistance of "other participants," is not likely. By "other participants" we mean phonons.

Now suppose that there are phonons involved in the interaction. If the photon energy is within a few tenths of an eV of E_g and is less than $E_g + E'_g$, the law of conservation of energy may be satisfied by adding or subtracting the phonon energy. It is also possible to satisfy the law of conservation of wave vectors, because $|\mathbf{k}_c - \mathbf{k}_v|$ and $|\mathbf{K}|$ are on the same order of magnitude. When both conservation laws are satisfied, the absorption of photons and the generation of electron–hole pairs become probable. Phonon-assisted absorption processes are also referred to as **indirect transitions.** In the photon absorption process, when photons with energies less than $E_g + E'_g$ are absorbed by indirect bandgap semiconductors, phonons can be *either emitted* or *absorbed*. The crucial point is the *participation of the phonons*.

Suppose phonons with an energy E_{ac} and a wave vector $\mathbf{K}$ are *absorbed* (Figure 4.9a). The conservation laws require that

$$E_v + E_{ac} + hf = E_c \tag{4.4a}$$

and

$$\mathbf{k}_v + \mathbf{K} + 0 = \mathbf{k}_c \tag{4.4b}$$

The "0" is included in (4.4b) to emphasize that $|\mathbf{k}_{op}|$ is negligibly small. The photon energy absorbed by the semiconductor is

$$E_c - E_v - E_{ac} = E_g - E_{ac}$$

The phonon energy E_{ac} is also absorbed by the indirect bandgap semiconductor. Since $\mathbf{k}_v$ is small, $\mathbf{K}$ and $\mathbf{k}_c$ are in the same direction.

Phonons may also be *emitted* by an indirect bandgap semiconductor in the photon absorption process (Figure 4.9b). The process occurs when the conservation laws

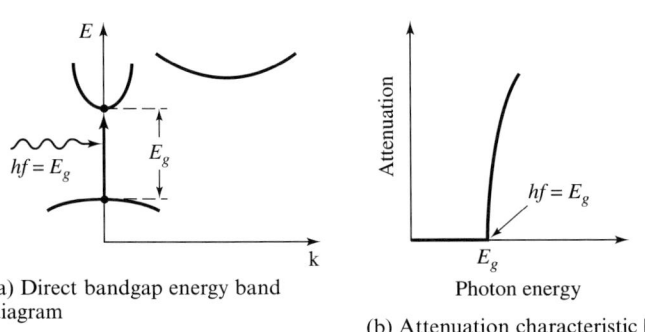

(a) Direct bandgap energy band diagram

(b) Attenuation characteristic [5]

Figure 4.8 Optical absorption in direct bandgap semiconductors

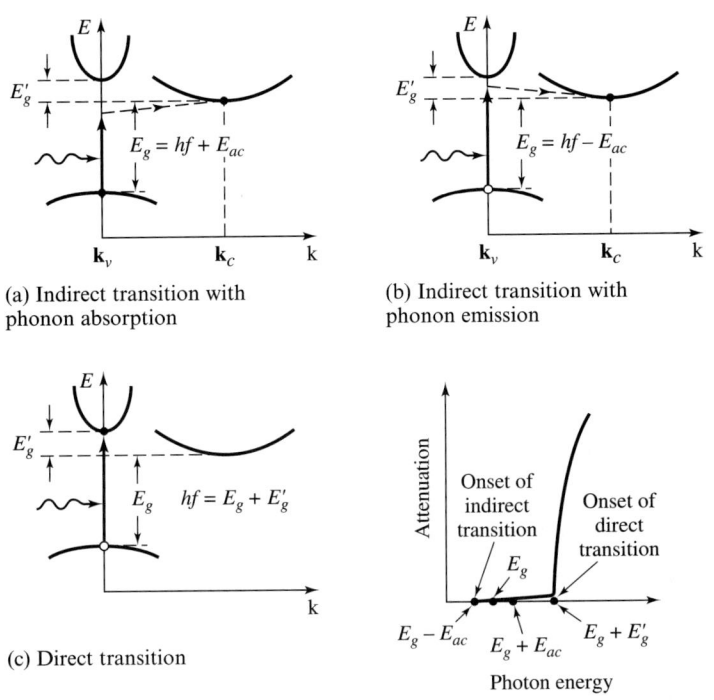

Figure 4.9 Optical absorption in indirect bandgap semiconductors

$$E_v + hf = E_c + E_{ac} \qquad (4.5a)$$

and

$$\mathbf{k}_v + 0 = \mathbf{k}_c + \mathbf{K} \qquad (4.5b)$$

are satisfied. Photons with energy $E_g + E_{ac}$ are absorbed, and a phonon is released. In this case, since $|\mathbf{k}_v|$ is very small, $\mathbf{k}_c$ and $\mathbf{K}$ are in opposite directions.

In (4.4a), (4.4b), (4.5a), and (4.5b), the photon absorption processes are weak and the participation of phonons is required. When the attenuation constant is plotted as a function of photon energy, the attenuation constant gradually increases as photon energy increases. The onset of phonon-assisted absorption process begins within a few tenths of an eV of E_g, since the phonon energy is also on the order of a few tenths of an eV. As the photon energy increases further and $hf > E_g + E'_g$, photons are sufficiently energetic to raise the electrons from the peak of the valence band directly to the conduction band, while keeping the same wave vector (Figures 4.9c). The conservation laws are satisfied by the electrons and photons alone, and no phonon participation is required. This *direct transition* interaction is strong and once it begins, the attenuation constant increases sharply as a function of photon energy. The *total* absorption is the sum of the direct and indirect transitions, as shown in Figure 4.9d.

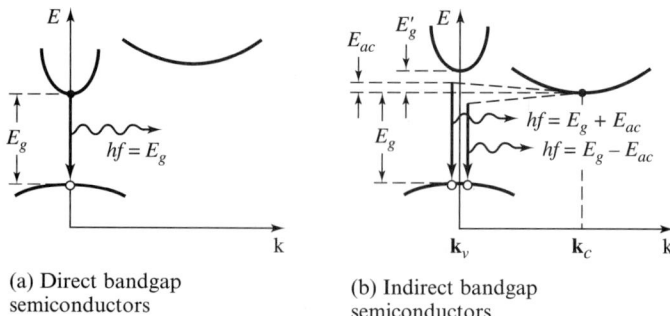

(a) Direct bandgap semiconductors

(b) Indirect bandgap semiconductors

Figure 4.10 Photon emission in direct and indirect bandgap semiconductors

4.3.3 Photon Emission in Semiconductors

In the electron–hole recombination process, electrons drop from the conduction band to the valence band. The energy difference can be released as photons, phonons, or both. Depending on the presence or absence of photon emission, the process is labeled a **radiative recombination** or a **nonradiative recombination.** If a radiative recombination occurs without external stimulation, it is a **spontaneous emission** process. Since energy and wave vectors must be conserved, the process is more likely to occur in direct bandgap semiconductors than in indirect bandgap materials. In direct bandgap semiconductors, the energy of the emitted photons is equal to the bandgap energy as shown in Figure 4.10a. In indirect bandgap materials, the photon emission is accompanied by the emission or absorption of phonons. The emitted photon has a energy of $E_g - E_{ac}$ if a phonon is also emitted, or $E_g + E_{ac}$ if a phonon is absorbed, as shown in Figure 4.10b.

To illustrate the difference between direct and indirect transitions, consider the generation of light by current injection in p-type semiconductors. Let p be the hole density (i.e., majority carrier concentration), in a p-type semiconductor and Δn be the excess electron density injected into the semiconductor. The **rate of radiative recombination** R, defined as the number of photons emitted per unit volume per second, is proportional to Δn and p as follows:

$$R = B_{rec} p \Delta n \quad (4.6)$$

From equation (4.6), we see that the radiative recombination rate increases with the minority carrier density, and the majority carrier concentration. The majority carrier concentration p may be increased by increasing the impurity concentration. The minority carrier density may be increased by injection of the charge carrier. The proportionality constant B_{rec} in (4.6) is the **recombination coefficient.** Table 4.2 [7] shows the recombination coefficients for several semiconductors. Note in particular that the B_{rec} of indirect bandgap semiconductors, such as Ge, Si, and GaP, is *smaller* than that of direct bandgap semiconductors by three to five orders of magnitude.

To reiterate, the interaction of electromagnetic waves with direct bandgap

Table 4.2 Recombination coefficients for several semiconductors [7]

Material	Bandgap type	B_{rec} in cm³/s
Si	Indirect	1.79×10^{-15}
Ge	Indirect	5.25×10^{-14}
GaP	Indirect	5.37×10^{-14}
InSb	Direct	4.58×10^{-11}
InAs	Direct	8.5×10^{-11}
GaSb	Direct	2.39×10^{-10}
GaAs	Direct	7.21×10^{-10}

semiconductors is strong and probable, while the interaction of electromagnetic waves with indirect bandgap materials is weak and unlikely.

4.4 TERNARY AND QUATERNARY SEMICONDUCTORS

In the last section, we learned that the energy bandgap plays an important role in determining the emission and absorption wavelengths of semiconductor diodes. Each elemental and binary semiconductor has a specific bandgap at a given temperature. If the choice of semiconductors were restricted to elemental and binary semiconductors only, then the available wavelengths would be rather limited. Fortunately, however, light is also emitted by solid solutions of semiconductors. When two or more binary semiconductors are mixed, **ternary** or **quaternary crystalline solid solutions** are created [3,8]. More importantly, the energy bandgap, refractive index, and lattice constant of such solid solutions can be adjusted by varying the composition of the contributing materials and the growth conditions of the solutions.

4.4.1 III–V Semiconductors

Ternary solid solutions are of the form of $A_x B_{1-x} C$, where the mole fraction x can range from 0 to 1. Solid solutions based on III–V semiconductors may be viewed as mixtures of a binary semiconductor AC with a binary semiconductor BC, where A and B stand for group III [or V] atoms, and C for group V [or III] atoms. In other words, the structure of a III–V solid solution consists of both group III and group V lattice sites. Each group III site is occupied by a group III atom and each group V site by a group V atom. Two or more types of group III atoms can be distributed randomly in group III lattice sites, and the same is true for group V atoms in group V sites. For example, by placing Al and Ga atoms randomly at the group III lattice sites and As atoms at the group V sites, a solid solution of $Ga_x Al_{1-x} As$ is formed. Since there are three group III atoms (Al, Ga, and In) and three group V atoms (P, As, and Sb), there are 18 possible ternary III–V solid solutions.

Some of these III–V solid solutions have direct bandgaps and others have indirect bandgaps. This means that not all ternary semiconductors are good

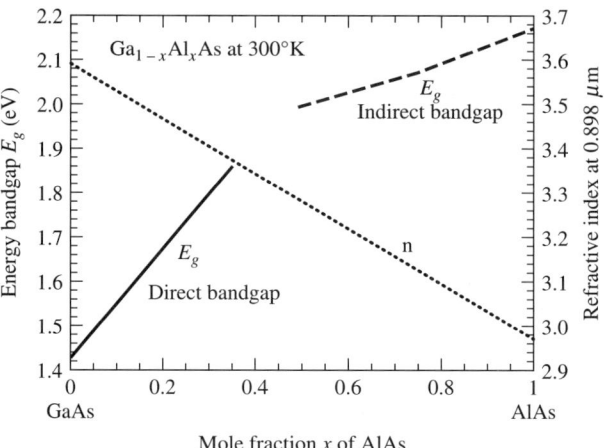

Figure 4.11 Bandgap energy and refractive index of $Ga_{1-x}Al_xAs$ as a function of Al mole fraction at 300 °K [9]

optical materials. Furthermore, solid solutions are good optical quality only if the lattice constants of the constituent binary semiconductors match very well, say within 0.1 percent. Without a good lattice match, excessive crystal defects appear and distribute themselves randomly throughout the material. As a result, the material becomes too lossy to be useful for optical applications. Therefore, a comprehensive knowledge of various semiconductor properties is extremely useful. In general, the lattice constant of a solid solution $A_xB_{1-x}C$ varies linearly from the lattice constant of the semiconductor AC to that of the semiconductor BC. This is known as **Vegard's law** [6]. On the other hand, the bandgap energy changes as a quadratic function of x,

$$E_g = a_E + b_E x + c_E x^2$$

where a_E, b_E, and c_E are constants. The index of refraction is also a function of the mole fraction. Since there is only one degree of freedom, the mole fraction x, the lattice constant, the energy bandgap, and the index of refraction of a ternary solid solution cannot vary independently. Again, take $Ga_xAl_{1-x}As$ as an example. GaAs and AlAs have the same lattice structure, with almost identical lattice constants. As x changes, the lattice constant of $Ga_xAl_{1-x}As$ varies from 5.653 Å of GaAs to 5.661 Å of AlAs, and the bandgap energy varies from 1.43 eV of GaAs to 2.16 eV of AlAs as x changes. $Ga_xAl_{1-x}As$ has a direct bandgap for $x < 0.37$ and an indirect bandgap for $x > 0.45$. A plot of E_g as a function of x is given in Figure 4.11 [9]. In the region of $0.37 < x < 0.45$, the band structure changes from a direct bandgap to an indirect bandgap. Since the precise crossover point is not known, the curve in the transition region is not plotted. The refractive index of $Ga_xAl_{1-x}As$ can be interpolated to a good approximation from that of GaAs ($n = 3.590$) and AlAs ($n = 2.971$). This is also shown in Figure 4.11.

4.4 Ternary and Quaternary Semiconductors

Quaternary solid solutions are of the form $A_xB_{1-x}D_yE_{1-y}$ or $(A_xB_{1-x})_yC_{1-y}D$, where A, B, and C are atoms of one group, and D and E atoms of the other group. As there are two degrees of freedom, the mole fractions x and y, the energy bandgaps, and the lattice constants of quaternary solid solutions can be varied independently by controlling the mole fractions.

The bandgaps and lattice constants of several III–V binary, ternary, and quaternary semiconductor materials are shown in Figures 4.12a and 4.12b [10,11].

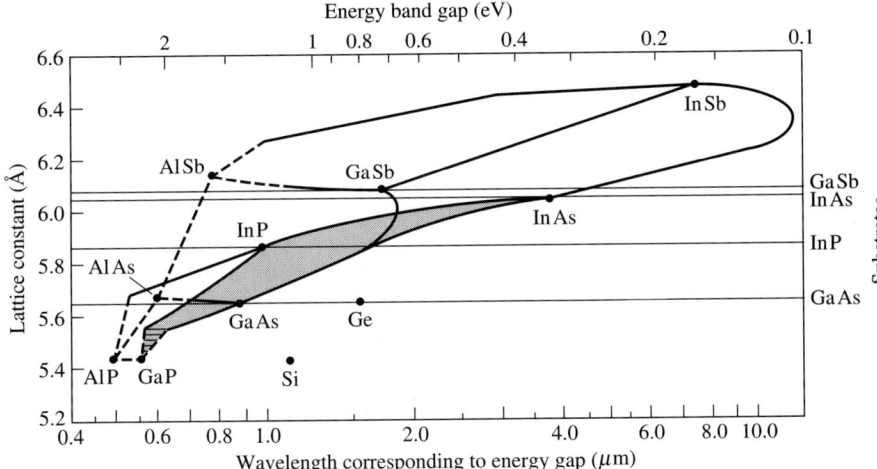

(a) $Ga_xIn_{1-x}P_yAs_{1-y}$ ([10], © 1980 IEEE)

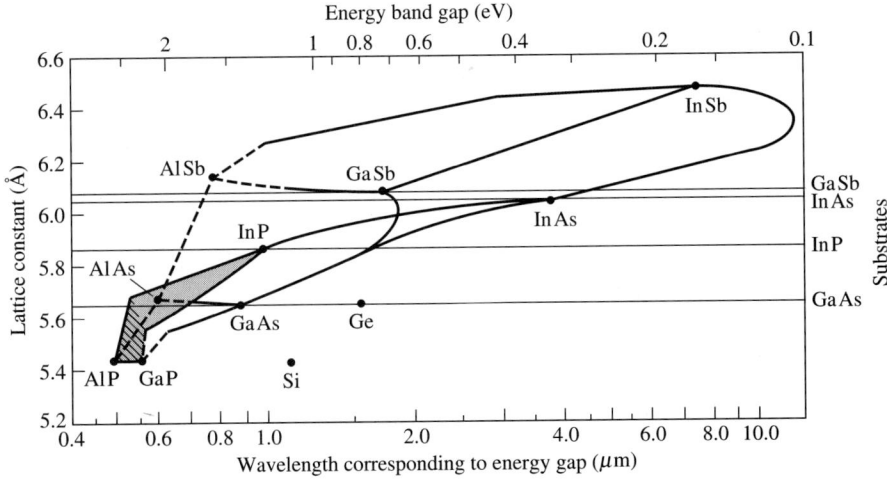

(b) $(Al_xGa_{1-x})_yIn_{1-y}P$ ([11])

Figure 4.12 Bandgap energy versus lattice constant for two ternary III–V compound semiconductors
Solid and dashed lines correspond to direct and indirect bandgaps, respectively. Dotted and crosshatched regions indicate the direct and indirect quaternary compounds $Ga_xIn_{1-x}P_yAs_{1-y}$ and $(Al_xGa_{1-x})_yIn_{1-y}P$

In these figures, the binary semiconductors are shown as small circles. Two circles corresponding to Si and Ge are also included for reference purposes. The lattice constants of GaSb, InAs, InP, and GaAs are indicated by the four horizontal lines. Note that the lattice constant of Ge matches very well with that of GaAs. Ternary solid solutions are represented by solid or dashed curves joining binary semiconductors. Solid curves signify direct bandgaps and dashed curves mean indirect bandgaps. For example, $Ga_xAl_{1-x}As$ is represented by a solid line and a dashed line connecting GaAs with AlAs. A quaternary solid solution is represented by an area defined by three or four binary III–V semiconductors. In Figure 4.12a, the dotted and crosshatched regions correspond to solid solutions $Ga_xIn_{1-x}P_yAs_{1-y}$ with direct and indirect bandgaps, respectively. The corresponding regions for $(Al_xGa_{1-x})_yIn_{1-y}P$ are shown in Figure 4.12b.

4.4.2 II–VI Semiconductors

The energy bandgaps and lattice constants of most II–VI semiconductors are shown in Figure 4.13 [12]. Note that E_g ranges from a very wide value (3.8 eV for ZnS) to a very narrow or even negative value (-0.28 eV for HgTe). As indicated in Table 4.1, all II–VI binary semiconductors have direct bandgaps. Ternary II–VI semiconductors are also of the form of $A_xB_{1-x}C$, where A and B are atoms of group II [or VI] and C is elements of group VI [or II]. The main drawback for II–VI materials is the difficulty in forming n-type and p-type II–VI semiconductors on the same substrate. It is also challenging to form good ohmic contacts. Because of these difficulties, II–VI ILDs and LEDs have only recently been realized [13, 14]. In the blue-green lasers based on ZnCdSe reported recently, quaternary II–VI compounds such as $Zn_{1-x}Mg_xS_{1-y}Se_y$ have been used as the cladding layers.

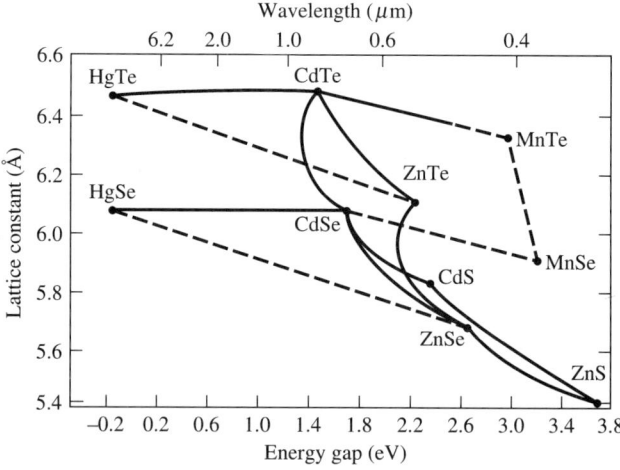

Figure 4.13 Bandgap energy versus lattice constant for ternary II–VI compound semiconductors [12]

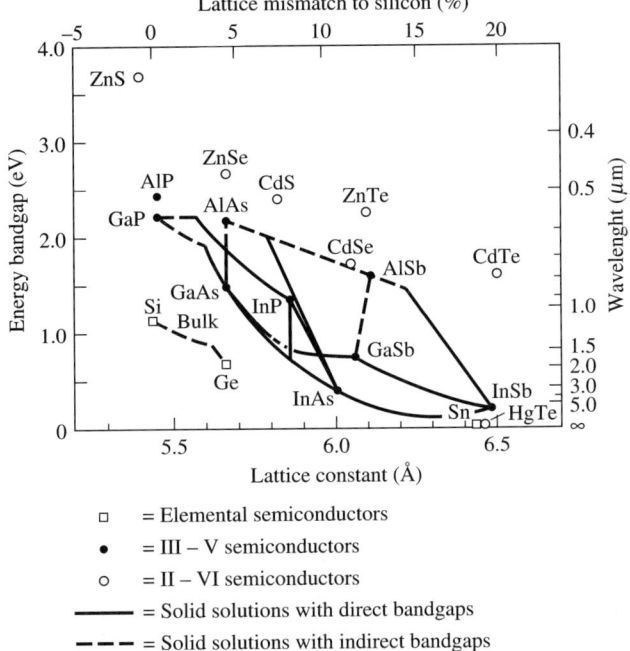

Figure 4.14 Energy bandgap versus lattice constant for common semiconductors
Squares correspond to elemental semiconductors, filled circles to III–V semiconductors, and unfilled circles to II–VI semiconductors. Solid and dashed lines are for solid solutions with direct and indirect bandgaps, respectively. ([15] © 1992 IEEE)

The energy bandgaps and lattice constants of common elemental III–V and II–VI semiconductors are combined in Figure 4.14 [15].

4.4.3 IV–VI Semiconductors

Stable IV–VI crystals, such as lead selenide (PbSe), lead sulphide (PbS), and lead telluride (PbTe), exist in various stoichiometric compositions. They have an interesting property: excess Pb atoms in PbSe act as electron donors, and excess Se atoms act as electron acceptors. By simply changing the proportional composition, we can change a IV–VI semiconductor from an n-type to a p-type.

The bandgaps of IV–VI semiconductors are very narrow and the emission wavelengths are long. Furthermore, E_g can be tuned over a wide range by controlling the temperature of, the pressure on, or the magnetic fields applied to, the semiconductors. Tunable semiconductor light sources based on these materials are useful for high-resolution spectroscopy applications. Most light-emitting IV–VI semiconductor devices operate at cryogenic temperatures, typically 50 °K. However, the continuous operation of $PbTe/Pb_{1-x}Eu_xSe_yTe_{1-y}$ heterostructure lasers (which will be explained in the next section) is possible at 174 °K, while pulsed operation can occur at 241 °K [16].

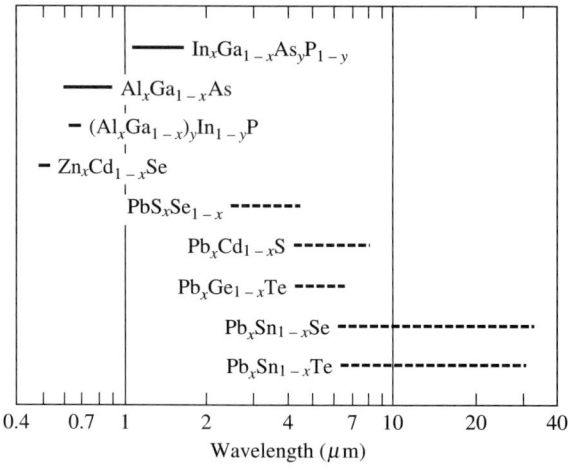

Figure 4.15 Emission wavelengths of selected semiconductor ILDs
Solid and dashed lines represent continuous operation at room temperature and cryogenic temperature, respectively. (after [17])

The emission wavelengths of various semiconducting materials are summarized in Figure 4.15 [17]. However, the detailed characteristics of compound semiconductors and solid solutions are well beyond the scope of this book.

4.5 HOMOJUNCTIONS AND HETEROJUNCTIONS

The two types of semiconductor junctions are **homojunctions** and **heterojunctions** [3, 8, 11]. If the materials on two sides of a junction have essentially the same energy bandgaps, the junction is known as a homojunction. For example, if a section of an intrinsic semiconductor is doped with donor impurities, resulting in an *n*-type semiconductor, and the other section is doped with acceptor impurities, forming a p-type semiconductor, the changes in the energy bandgap will be small. Then, the two sides of the junction will have roughly the same energy bandgap, and the junction is a homojunction. The *pn* diodes based on Si or Ge found in many electron devices are of this type junction. If, on the other hand, the two sides of the junction are based on different semiconducting materials with different bandgaps, the junction is a heterojunction. Junctions formed by joining Ge (E_g = 0.66 eV) with GaAs (E_g = 1.43 eV), or by joining GaAs with $Al_xGa_{1-x}As$, are examples of heterojunctions.

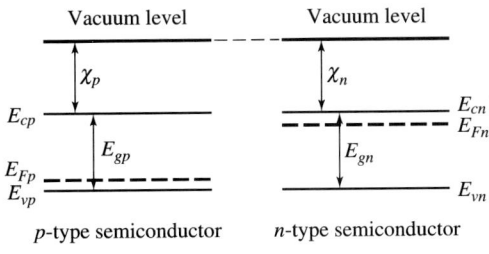

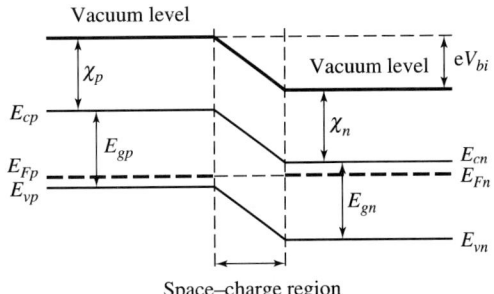

(a) Two separated semiconductors

(b) Two semiconductors in contact: zero bias

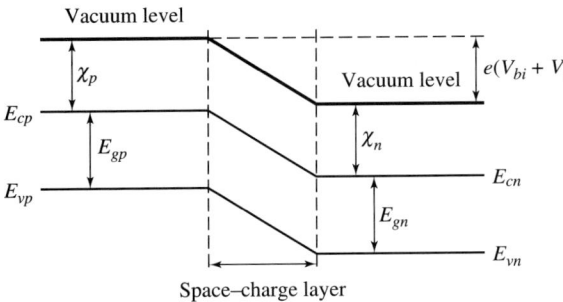

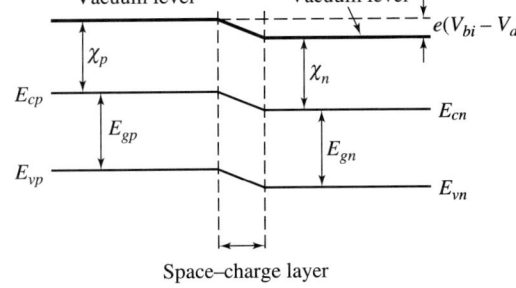

(c) Two semiconductors in contact: reverse bias

(d) Two semiconductors in contact: forward bias

Figure 4.16 Energy diagram of a homojunction under various bias conditions

4.5.1 Homojunctions

Four important parameters used in characterizing the energy diagrams of semiconductor junctions are E_v, E_c, E_F, and the electron affinity χ. The energy levels E_v, E_c, and E_F were introduced in Section 4.2. The **electron affinity** is the energy required to move an electron from the bottom of the conduction band E_c to infinity, also called the **vacuum level.** Table 4.1 lists the electron affinities of various semiconductors. To appreciate the significance of χ, let us consider a pn junction. Let the Fermi levels for the n- and p-type semiconductor sections be E_{Fn} and E_{Fp} and the electron affinities be χ_n and χ_p, respectively (Figure 4.16a). When the two sections are *not* in physical contact, the vacuum levels of both sides line up. Since the materials on both sides of a homojunction are similar, the electron affinities of both sides are the same, and the edges of the conduction and valence bands of the two sides line up. However, the Fermi levels are not aligned, since the doping impurities are different. When the sections are physically joined, the situation is changed. Since there are more electrons on the n-side than on the p-side, and more holes on the p-side than on the n-side, a concentration gradient exists near the junction. Electrons diffuse from the n-side to the p-side, and holes go in the opposite direction. As a result, charges of opposite polarity build up near the junction. Because of excess charges, an electric field is set up to impede the movement of charged particles across the boundary. The diffusion of charged carriers ceases when the effect of the concentration gradient is balanced by the electric field established by the excess charges. The region in the immediate neighborhood is known as the **space–charge region** or the **depletion layer.**

Establishing the space–charge layer can also be viewed in terms of energy. Recall that the ordinate in the energy band diagram refers to the electron energy. An energy barrier is formed when the energy levels of the *n*-type material are shifted downward with respect to those of the *p*-type. Equilibrium is reached when E_{Fn} lines up with E_{Fp}, as shown in Figure 4.16b. Note that E_{cn} and E_{vn} are now lower than E_{cp} and E_{vp}, respectively. This leads to a built-in potential barrier, V_{bi}, as shown in Figure 4.16b. If a voltage source V_a is used to further depress the energy levels in the *n*-type region relative to those of the *p*-type, the potential barrier is increased and the junction is **reverse biased** (Figure 4.16c). On the other hand, if the effect of the voltage source is to raise the energy level of the *n*-type material relative to that of the *p*-type, the energy barrier is reduced and the junction is **forward biased** (Figure 4.16d). Forward and reverse bias are also applicable to heterojunction devices. Some photodetectors operate with a reverse bias. ILDs, SLDs, and LEDs, however, are forward biased.

4.5.2 Heterojunctions

As noted previously, the materials on two sides of a heterojunction have different bandgaps. We will use a lowercase letter, *n* or *p,* to designate the impurity type of the *narrow-bandgap* semiconductor and a capital letter, *N* or *P,* for the *wide-bandgap* semiconductor. For example, a heterojunction formed by a *P*-type wide-bandgap semiconductor and an *n*-type narrow-bandgap semiconductor is a **Pn junction**. A **pN junction** is a junction with the wide bandgap material doped with donor impurities on one side and the narrow-bandgap material doped with acceptor impurities on the other side. In addition to *pN* and *Pn* junctions, there can also be **nN junctions** and **pP junctions.**

To explain the energy band diagram of heterojunctions, we will consider a *pN* junction. Since the materials on the two sides of a heterojunction are different, the electron affinities χ_p and χ_N, conduction band edges E_{cp} and E_{cN}, valence band edges E_{vp} and E_{vN}, and band gaps $E_{gp} = E_{cp} - E_{vp}$ and $E_{gN} = E_{cN} - E_{vN}$ are different. When the two sections are apart, the vacuum levels are aligned (Figure 4.17a). Since $\chi_p \neq \chi_N$, there may be a step or a gap between the conduction band edges, and between the valence band edges. We label the gaps or steps as $\Delta E_c = E_{cN} - E_{cp}$, or $\Delta E_v = E_{vN} - E_{vp}$, as shown in Figure 4.17a. Also, the Fermi levels E_{Fp} and E_{FN} are not necessarily aligned.

When the two semiconductors are physically joined, the energy band diagram may be constructed in the following manner.

1. In thermal equilibrium, the Fermi levels E_{Fp} and E_{FN} are aligned.
2. For each side, the vacuum level remains in parallel with E_{cp} and E_{vp} on the *p*-side, and with E_{cN} and E_{vN} on the *N*-side.

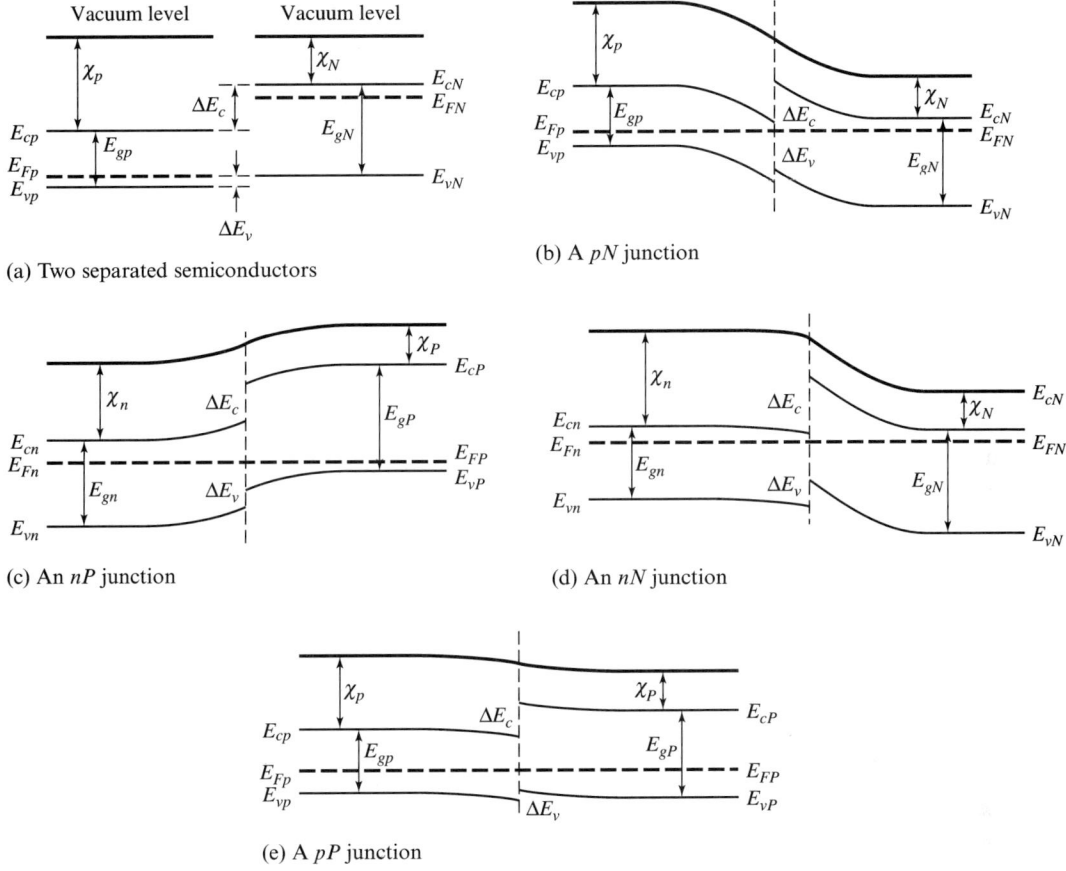

Figure 4.17 Energy diagrams for four types of heterojunctions

3. The vacuum levels are not aligned. Instead, they are joined smoothly and continuously.
4. The values of χ_p, χ_N, ΔE_c, and ΔE_v are not changed when the two sides are joined. There can be a step or a spike between E_{cp} and E_{cN}, or a discontinuity at E_{vp} and E_{vN}.

The resulting energy band diagram of an unbiased pN junction is shown in Figure 4.17b. Note in particular the presence of a gap between the edges of the conduction bands and a discontinuity between the tops of the valence bands. Because of the step and spike, the motion of electrons across the heterojunction can be quite different from that of the holes.

The energy band diagrams of nP junctions (Figure 4.17c), nN junctions (Figure 4.17d), and pP junctions (Figure 4.17e) are constructed in the same manner. Depending on the energy bandgaps and electron affinities, the energy diagrams of different heterojunctions can be very different in shape. The sketches presented in Figures 4.17b through e are only four examples.

4.6 BASIC SEMICONDUCTOR LUMINESCENT DIODE STRUCTURES

The heart of a semiconductor luminescent diode is an active semiconducting layer, which is sandwiched between two cladding layers. There are two junctions, one on each side of the active layer. Although the first ILDs and LEDs were homojunction diodes, most modern electroluminescent diodes have one or two heterojunctions. A **single heterostructure (SH)** diode has *two* different materials and has both a homojunction and a heterojunction. A **double heterostructure (DH)** diode is formed with *three* materials and has *two* heterojunctions.

It is customary to use the **threshold current density** J_{th}, a term to be defined shortly, as a parameter for quantitatively comparing the performances of various diodes. When the optical power output P of a diode is plotted as a function of the injection current I, we have the **PI characteristic,** which will be discussed in Section 4.8. The PI curves of ILDs have a sharp "knee," from which a **threshold current** can be extracted. Thus, the threshold current is an easily measurable quantity. Based on the threshold current and the junction area, the **threshold current density** can be determined. When the injection current is less than the threshold value, the radiation of the ILD is mainly a broadband, incoherent, spontaneous emission. For a current greater than the threshold current, the emission becomes mostly coherent and has a narrow spectral width.

Figure 4.18a–d shows the energy band diagram, the refractive index profile, and the optical intensity distribution of various diode structures. The composition and type of impurity in each semiconductor layer are also identified. If we consider diodes having emissions in the 0.8 to 09 μm range, the light generating layer, shown as the shaded region, is a GaAs layer.

Historically, the first-generation ILDs were fabricated by diffusion techniques and had homojunctions. In these diodes (Figure 4.18a), all layers were made of GaAs. Because of the diffusion techniques used in the fabrication, one side of the active region (i.e., the boundary between p-GaAs and GaAs) was not well defined. Thus, the injected charge carriers were not confined in the narrow active region. In addition, the index difference Δn between various layers was very small. Because of these deficiencies, the J_{th} of these ILDs was as high as 100 kA/cm².[3]

The second-generation laser diodes were made by liquid phase epitaxy techniques. The boundaries between various layers (n-GaAs, p-GaAs, and p^+-GaAs) were well defined, as shown in Figure 4.18b. Since all layers were based on GaAs, the energy bandgap difference was very small, and the junctions were homojunctions. However, the junction interfaces were sharp, and charge carriers were weakly confined in the active (p-GaAs) region. Due to the improvement in the charge carrier confinement, the threshold current density was reduced to 40 kA/cm². However, the index difference between layers was still quite small.

The use of heterostructures for charge carrier and optical beam confinement

---

[3]For comparison purposes, we note that the current rating for bare cooper AWG #10 wires is 16 A and 28 A if the temperature rise in the wire is restricted to 10°C and 35°C respectively. AWG #10 wires have a diameter of 101.9 mil. Thus, 10 A and 28 A correspond to current densities of 0.3 kA/cm² and 0.5 kA/cm², respectively.

4.6 Basic Semiconductor Luminescent Diode Structures

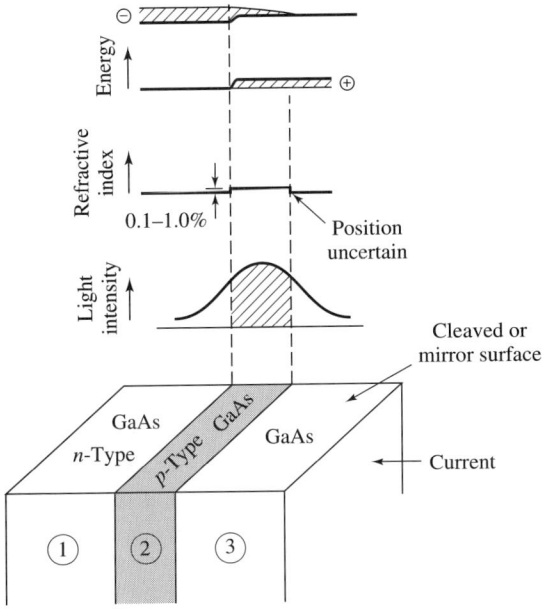

(a) Diffused homostructure diodes

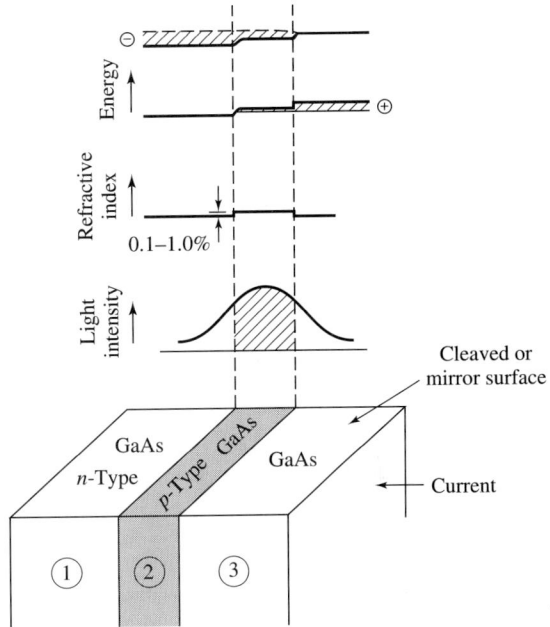

(b) Liquid phase epitaxial homostructure diodes

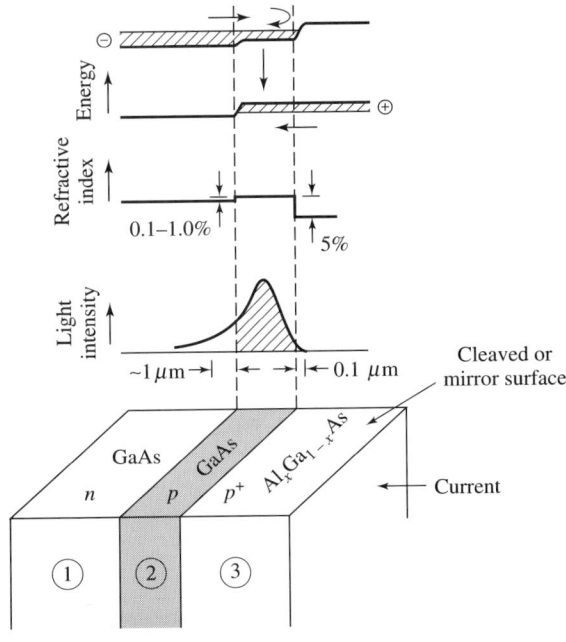

(c) Single heterostructure diodes

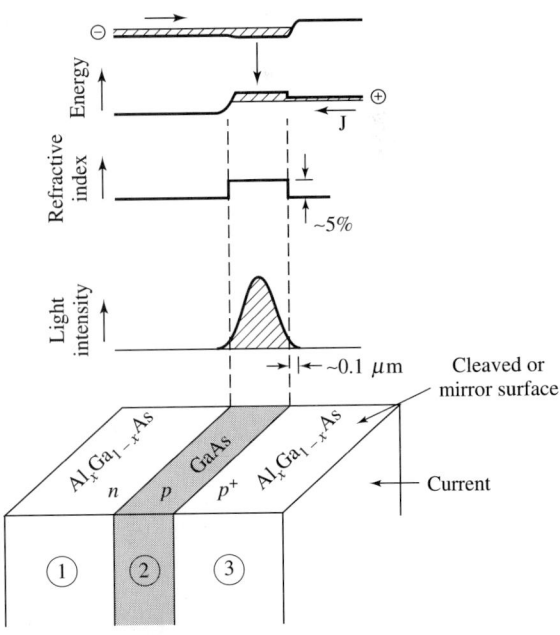

(d) Double heterostructure diodes

Figure 4.18 Schematic representation of energy band diagram, refractive index profile, and optical field distribution of homostructure, single heterostructure, and double heterostructure LEDs and ILDs ([38])

was proposed by Alferov and Kazarinov [18] and by Kroemer [19]. A comparison of Figure 4.18b and Figure 4.18c reveals that the p^+-GaAs layer of Figure 4.18b has been replaced by a p^+-$Al_xGa_{1-x}As$ layer. Since the bandgap of $Al_xGa_{1-x}As$ is wider than that of p-GaAs, the junction between p-GaAs and $p^+Al_xGa_{1-x}As$ is a single heterojunction, forming an SH diode. The p-GaAs/p^+-Al_xGa_xAs heterojunction serves two functions: it confines the optical power in the active layer and thus minimizes the optical loss in the cladding layer; and it provides a potential barrier at the heterojunction and thus confines the charge carriers in the active region. With the use of a GaAs/$Al_xGa_{1-x}As$ single heterostructure, the threshold current density further decreases to about 8 to 10 kA/cm^2.

Confinement of the optical power and the charge carriers is further improved by incorporating a second heterojunction, resulting in DH lasers (Figure 4.18d). In a DH laser diode, there is a potential barrier for holes at the pN junction and for electrons at the Pp junction. The index difference between the active layer and surrounding cladding layers also acts as a guide for the resulting optical waves. Because of these improvements, J_{th} can be reduced further, to about 0.5 kA/cm^2 at room temperature.

Further reduction of the threshold current density is possible by using very thin layers which are known as **quantum wells.** A brief discussion of quantum wells is presented in Section 4.6.4.

To summarize, the basic structure of modern ILDs, LEDs, and SLDs consists of an active layer and two cladding layers, surrounded by a substrate on one side and a contact layer on the other side. Each layer serves a specific purpose, as follows:

1. The **active layer** is the light generating medium of the diode. It should be a direct bandgap material, with E_g corresponding to the desired emission wavelength. The bandgap should be *narrower* and the refractive index *greater* than those of the surrounding cladding regions.
2. The **cladding layers** restrict the motion of the charge carriers and confine the optical power in the active layer. To prevent the charge carriers from diffusing away from the active layer, the cladding layers must have bandgaps *wider* than those of the active layer. To confine the optical power in the active region, they must also have an index of refraction that is *lower* than that of the active layer. Since no photons are emitted from the cladding layers, direct as well as indirect bandgap materials may be used.
3. The **substrate** provides mechanical support, heat dissipation capability, and electrical contacts. It should have the proper crystal symmetry and lattice constant, from which succeeding compound semiconductor layers or solid solutions can be grown.
4. The **contact layer** provides electrical contacts. In the case of surface-emitting diodes, this layer should also be transparent to the emission wavelength.

The lattice match between various layers is crucial. In Figure 4.19, two material combinations are listed [20]. As noted in Figures 4.12a and b, the lattice constant of $Al_xGa_{1-x}As$ matches well with that of GaAs for all values of

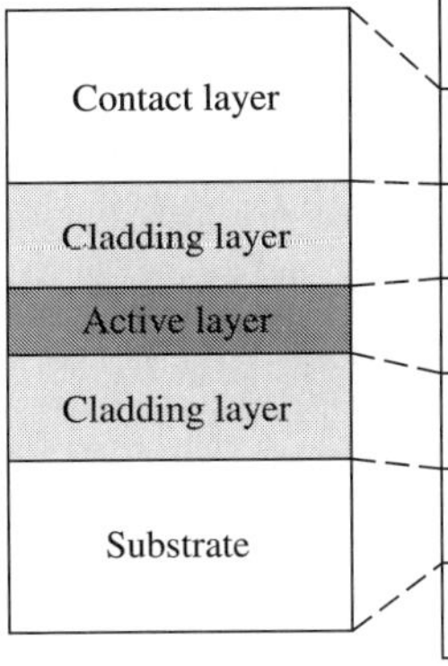

4.6 Basic Semiconductor Luminescent Diode Structures

Layer	Wavelength	
	0.7 to 0.87 μm	1.1 to 1.67 μm
Contact layer	GaAs	$In_{1-u}Ga_uAs_vP_{1-v}$
Cladding layer	$Ga_{1-y}Al_yAs$	InP
Active layer	$Ga_{1-x}Al_xAs$	$In_{1-x}Ga_xAs_yP_{1-y}$
Cladding layer	$Ga_{1-z}Al_zAs$	InP
Substrate	GaAs	InP
	$y > x, z > x$	$x > u, y > v$

Figure 4.19 Basic structure and material composition of LEDs and ILDs

x. Therefore, the $Al_xGa_{1-x}As-GaAs$ combination is often used in light generating diodes emitting in the 0.7 to 0.87 μm wavelength region at room temperature. $Ga_xIn_{1-x}As_yP_{1-y}$ layers and InP substrates are a good material combination for sources in the 1.1 to 1.67 μm region, which covers almost the entire spectrum of interest for fiber optical communications applications.

Depending on the geometric structures normal to the junction layers, ILDs can be further classified as broad area or stripe geometry lasers.

4.6.1 Broad Area Lasers

Broad area lasers are injection lasers of the most rudimentary form (Figure 4.20a). The active layer is typically a few tenths of a micrometers thick and is surrounded by relatively thick cladding layers. Electrical contacts are made with both top and bottom layers. The front and back facets are smooth and sharp, and are usually obtained by cleaving. The optical cavity length d, defined by the separation between the front and back facets, may be between 70 and 1000 μm, and has a typical value of about a few hundred micrometers. The lateral width s of a broad area diode is typically 200–300 μm wide in the y direction (Figure 4.20a). Since the width is much larger than the thickness, we can assume that the optical fields and injection current are distributed uniformly in the y direction. Again, since the diode is rather wide, many transverse modes can be supported by the structure, if they are not suppressed. To discourage the excitation of higher-order transverse modes, the side surfaces are made rough, introducing losses to the higher-order modes, which have fields extending to the regions near the boundaries. Thus, the right and left facets may be cut with wire saws.

The disadvantage of broad area lasers is that, since material nonhomogeneities and fabrication imperfections are inevitable, the current distribution is not necessarily uniform throughout the width in the y direction. As a result, the local current density may exceed the threshold level in some regions, and may be below that level elsewhere. There may also be excessive strain near imperfec-

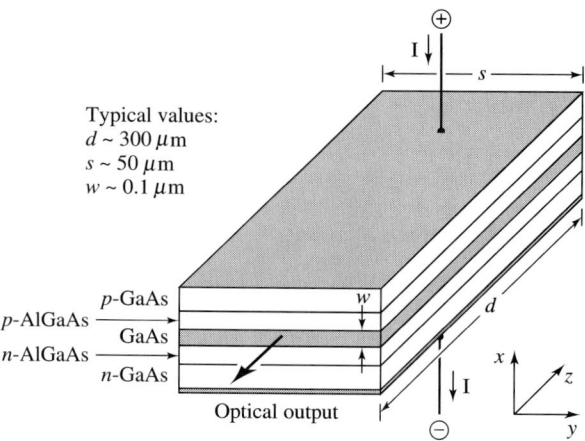

(a) Broad area injection laser

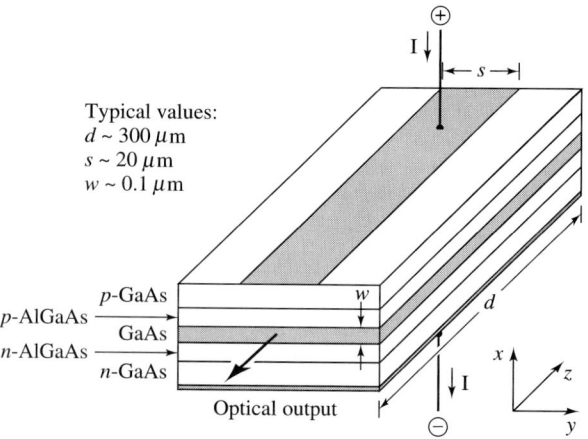

(b) Stripe geometry injection laser

Figure 4.20 Geometry of broad area injection lasers and stripe geometry injection lasers

tions, which leads to an increase in the refractive index, which in turn causes fields to concentrate in the high-index region. As a result, lasing is usually confined to narrow regions, and light appears in a filamentary form. These filaments are unstable and may vary with the injection current. As the current increases beyond the threshold level, several filaments may appear. When the PI characteristic is plotted, one or more *kinks* may appear in an otherwise smooth curve. While the injection current in all regions leads to junction heating, only the current in or near the lasing filament contributes to the stimulated emission process. As a result, broad area ILDs are inefficient, unstable, and obsolete. They are mentioned here purely for historical purposes.

4.6.2 Stripe Geometry Lasers

Experimental observations reveal that lasing filaments are typically 10 to 20 μm wide. Instead of trying to suppress the instabilities associated with filaments, it is better to forcibly control filament formation. This can be done by restricting the current to a narrow strip 10 to 20 μm wide. ILDs with this type of lateral current confinement are known as **stripe geometry injection lasers.** One way to achieve this objective is to use a narrow conducting strip 10 to 20 μm in width as one of the electrodes. Figure 4.20b depicts the basic configuration of stripe geometry ILDs. Because of the narrow metal contact, current is confined to a narrow region in the lateral dimension. With the current confined to the emission region, recombination of the injected electrons with holes is more efficient, and threshold current density is reduced. This is the basic concept of stripe geometry injection lasers. Methods have also been developed to introduce index variations in order to confine the optical beams within narrow stripes.

Gain-Guided Stripe Geometry Lasers Several schemes have been conceived to restrict current in a narrow region. Figure 4.21 shows the cross sections of three types of stripe geometry lasers. The laser with an **oxide insulating stripe** (Figure 4.21a) is probably the simplest to understand. An SiO_2 or other insulating layer is deposited on the GaAs surface, and narrow windows are opened lithographically before metal contacts are made. An alternative uses **ion implantation techniques** to implant protons (hydrogen ions) or oxygen ions in the areas outside the stripe region, thereby rendering the exterior regions resistive and preventing

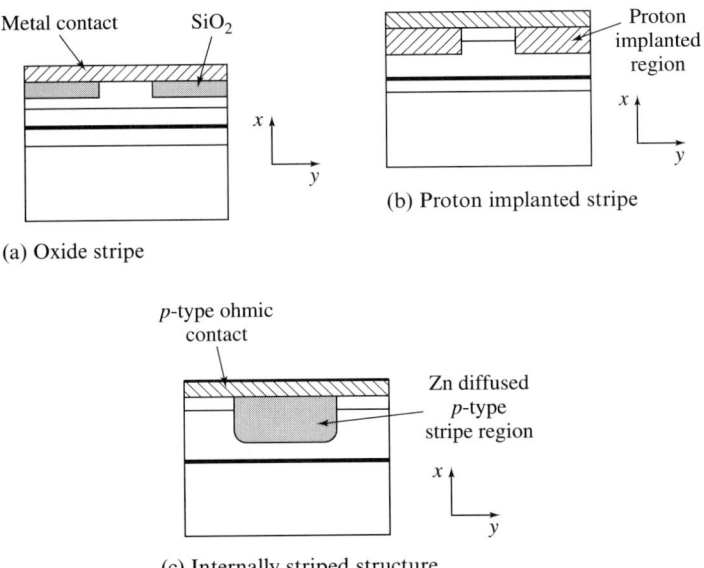

(a) Oxide stripe

(b) Proton implanted stripe

(c) Internally striped structure

Figure 4.21 Schematic cross section of three gain-guided stripe geometry ILDs

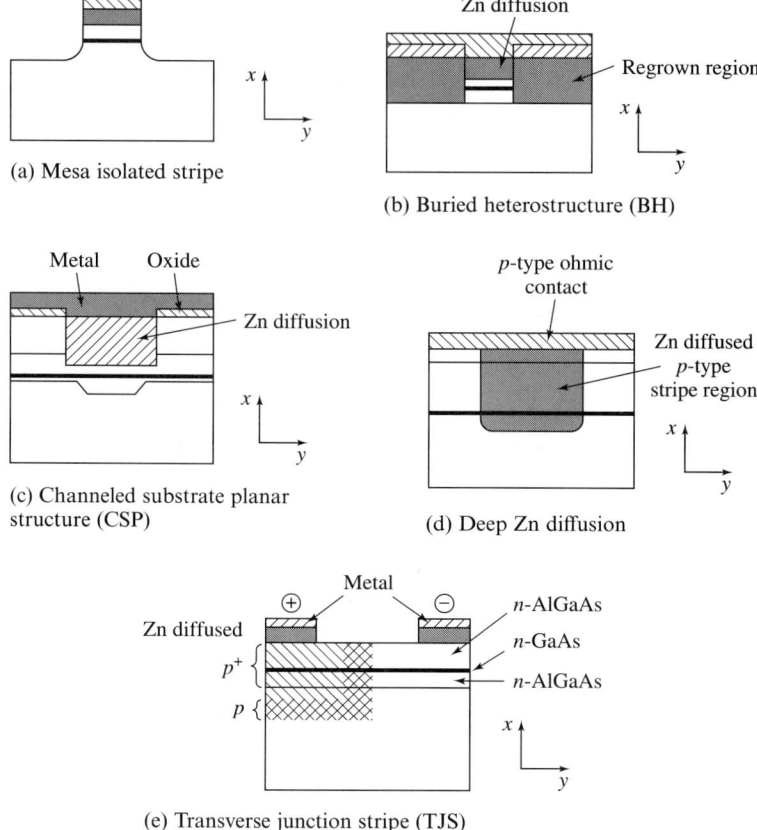

Figure 4.22 Schematic cross section of five index-guided stripe geometry ILDs

the current from spreading (Figure 4.21b). In the third approach, Zn is diffused into the cladding region to confine the current in the stripe region, but the Zn diffusion is not deep enough to reach the active region. This is known as an **internally striped structure** (Figure 4.21c). As mentioned previously, the stripe width is typically 10 to 20 μm. If the stripe is much narrower than 10 μm, the fraction of current confined in the stripe region will be too small to be effective. For stripes much narrower than 10 μm, the threshold current density increases due to **current spreading.** In the structures shown in Figure 4.21, lateral current confinement is purposely built into the diode structure. Since the current is confined to the stripe region, gain in the narrow region is high. These structures are known as **gain-guided structures.** Although no optical confinement in the lateral direction is intentionally built into these structures, a weak optical guidance does exist. The optical gain is related to the imaginary part of the refractive index. The increase in optical gain due to carrier injection, like the increase in refractive index, has the effect of confining the optical beams to the stripe region.

Index-Guided Stripe Geometry Lasers Figure 4.22 shows the cross sections of five stripe geometry lasers in which there is a large index change in the lateral dimension. Because of this, lateral optical confinement is built into the laser structure. The ILDs shown in Figure 4.22 are therefore known as **index-guided structures**. In the **mesa isolated stripe** structure (Figure 4.22a), the regions outside the stripe are removed completely, and the index for that region is reduced to that of air ($n = 1$). The disadvantage of this particular structure is that the index difference between the stripe region and the surrounding region is too large, and many higher-order transverse modes are excited unless the active stripe is very narrow. As a remedy, the etched regions are regrown with a low-index insulating material (Figure 4.22b). The result is the **buried heterostructure (BH)** laser. An alternative technique begins with a substrate that has a channel already etched before various layers are grown epitaxially. Due to the special properties of the liquid phase epitaxy technique, the indices of the epitaxially grown materials in the nearby regions are different. Lasers made with this technique known as **channeled substrate planar (CSP)** lasers (Figure 4.22c). A fourth option is ILDs with **deep Zn diffusion,** in which the optical confinement is realized by diffusing Zn deep into the active region, as shown in Figure 4.22d.

The structure of a **transverse junction stripe (TJS)** laser (Figure 4.22e) is quite different from other diode structures. In a TJS laser, impurity diffusion is performed twice. There are two homojunctions in the direction parallel to the junction: a p^+-GaAs/p-GaAs junction to the left of the active region, and a p-GaAs/n-GaAs to the right. Also, there are three layers in the direction perpendicular to the junction: n-AlGaAs, n-GaAs, and n-AlGaAs. These layers form two heterojunctions. Lateral confinement in the x direction is provided by the two heterojunctions, and in the y direction it is provided by the two homojunctions. In TJS lasers, current is injected *laterally* into the active region (e.g., horizontally in Figure 4.22e). This is quite different from other stripe geometry laser diodes, in which current flows *across* the active region (i.e., downward in other figures).

4.6.3 Single-Frequency Single-Mode Injection Lasers

The gain bandwidth of direct bandgap semiconductors is about a few nanometers. Many longitudinal and transverse modes can be supported within the gain bandwidth, and each mode oscillates at a slightly different wavelength. We can show that the separation of longitudinal modes of a Fabry–Perot cavity of length d is

$$\Delta\lambda \approx \frac{\lambda^2}{2d\left[n - \lambda \dfrac{dn}{d\lambda}\right]} \tag{4.7}$$

where n is the refractive index of the material and $\dfrac{dn}{d\lambda}$ is the **material dispersion.**

For GaAs, $\lambda\dfrac{dn}{d\lambda}$ is between -1.5 and -2. Thus, for a diode with length d of

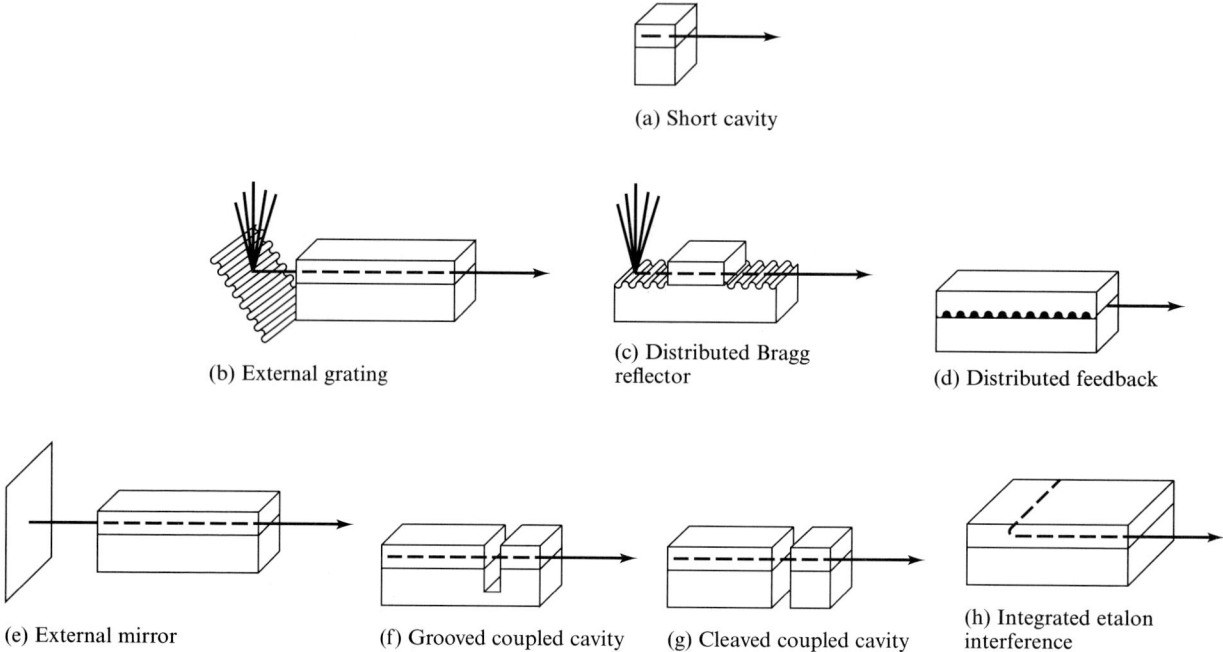

Figure 4.23 Methods for controlling longitudinal modes of ILDs ([21], © 1983 IEEE)

500 μm, the longitudinal mode separation is in the range of 0.1 to 0.3 nm (1 to 3 Å). The spectral separation of transverse modes is even narrower. However, higher-order transverse modes can be suppressed by introducing excessive losses to the higher-order modes. The index variation or gain stripes are specifically designed for this purpose.

The discrimination of **longitudinal modes** is much more difficult. Three basic schemes have been devised specifically to select or restrict longitudinal modes [21].

Geometry Control Since the longitudinal mode separation is inversely proportional to the cavity length d, as shown in (4.7), one and only one longitudinal mode can fit within the gain bandwidth if d is sufficiently small. Such a **short cavity laser** (Figure 4.23a) has been reported by Burrus et al. [22]. However, keeping the cavity short means that the optical output is also reduced.

Frequency-Selective Feedback In the usual Fabry–Perot cavities, broadband reflectors are used, and the frequency selectivity is mainly due to the large mirror separation. As an alternative, we could use gratings with a large number of grooves. For such gratings, the reflectivity is highly frequency dependent. Thus there are lasers that use *external gratings* as reflectors (Figure 4.23b). There are

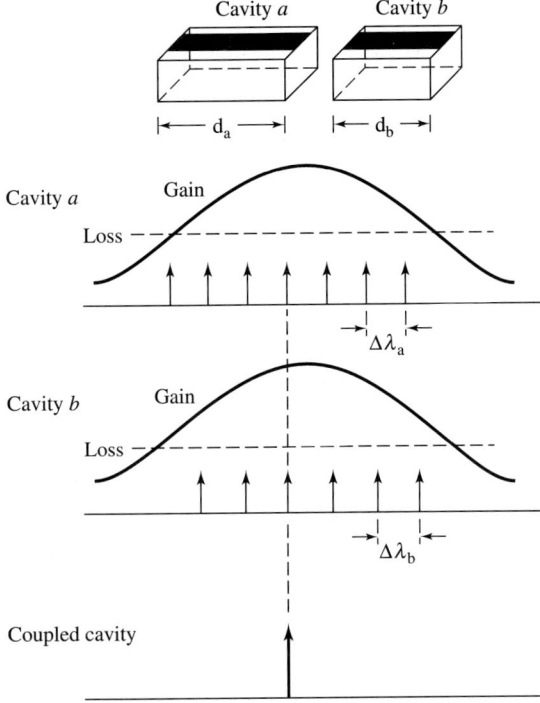

Figure 4.24 Principle of operation of coupled cavity ILDs

also lasers that use gratings on the same semiconductor substrate as distributed reflectors, which are known as **distributed Bragg reflectors (DBR)** (Figure 4.23c). If the gratings are incorporated into the active region, as shown in Figure 4.23d, to provide optical feedback, the lasers are known as **distributed feedback (DFB)** lasers.

Coupled Cavities Instead of having a single optical cavity, a laser diode can have two coupled cavities with slightly different cavity lengths. Let the cavity lengths be d_a and d_b, as shown in Figure 4.24. Each cavity has a set of longitudinal modes within the gain bandwidth of the active medium. Since $d_a \neq d_b$, two different sets of wavelengths are generally produced. If one set is tuned such that the two wavelength sets have one wavelength in common, the laser diode will lase at that particular wavelength. This is the operating principle of **coupled cavity lasers.** Several variations of this basic theme are shown in Figure 4.23e through h, including the **external mirror cavity, grooved coupled cavity, cleaved coupled cavity,** and **integrated etalon interference structure,** respectively.

The evolution of structures and geometries of laser diodes is summarized in Table 4.3. The objectives of laser diode development are to reduce the threshold current density in order to make continuous room-temperature operation possible.

Table 4.3 Evolution of laser diode structures and geometries

I. Layer structures
  1. Homojunctions [by diffusion or liquid phase epitaxial (LPE) techniques]
    Example: n-GaAs/p-GaAs/GaAs
  2. Single heterojunctions (by LPE techniques)
    Example: n-GaAs/p-GaAs/Al$_x$Ga$_{1-x}$As
  3. Double heterojunctions (by various epitaxial methods)
    Example: n–Al$_x$Ga$_{1-x}$As/GaAs/p–Al$_x$Ga$_{1-x}$As
  4. Quantum wells

II. Geometry in the transverse direction
  1. Broad area lasers: multiple longitudinal modes and multiple transverse modes
  2. Stripe geometry lasers (single-mode lasers): single transverse mode and multiple longitudinal modes
    i. Gain-guided stripe geometry lasers
    ii. Index-guided stripe geometry lasers

III. Optical cavity (single-mode single-frequency)
  1. Short cavity lasers
  2. Lasers with distributed Bragg reflectors, distributed feedback, or external grating reflectors
  3. Lasers with external mirrors, grooved couple cavity, cleaved coupled cavity, or integrated etalon interference

4.6.4 Recent Developments

Quantum Well Lasers The basic structure of a DH laser is a narrow-bandgap semiconductor layer sandwiched between two wide-bandgap materials, and the active layer thickness is on the order of 0.1 μm. When the layer thickness is 0.1 μm or greater, electrons are free to move in all three directions. However, when the layer thickness is reduced to 20 nm or less, the situation changes. Electron motion in the direction *normal* to the layer structure is allowed only if electron wave functions satisfy conditions dictated by quantum mechanics. This is because the electrons in the narrow-bandgap material are trapped by the potential barriers due to the wide-bandgap materials on both sides. This situation is similar to a well-known problem in quantum mechanics: the confinement of electrons in a "potential well." In a potential well, the kinetic energy associated with electron motion in the direction normal to the layer structure is quantized. The energy quantization is dependent upon the thickness of the thin active layer and the energy bandgaps of the active and cladding layers. The active region can also have multiple thin narrow-bandgap layers separated by thin barrier layers of wide-bandgap materials. Double heterostructure lasers with a *single* active layer 20 nm thick or less are known as **single quantum well lasers,** and those with *several* thin active layers are known as **multiple quantum well lasers** [23]. Since in these lasers the electron motion normal to the layered structure is restricted and the electrons are free to move only in the other two directions, the *density of states* describing the probability of occupation of electron states in the layered structure also changes. Because of this change in the density of states, the threshold current density is greatly reduced. For example, a threshold current density of only 80 A/cm^2 has been observed on a GaAs/AlGaAs quantum well laser having a very long cavity ($d \approx 3.3$ mm) [24].

The emission wavelength can also be varied by controlling the thin layer thickness. Reduction of the threshold current density and the possibility of tuning the emission wavelength by controlling the thin layer thickness are two key features of **quantum well lasers.**

Surface-Emitting Semiconductor Injection Lasers As we have discussed, for basic laser diode structures, an active layer medium is surrounded by cladding layers on either side. Reflecting mirrors normal to the layers are made by cleaving the semiconductors, and light is emitted from one or both of the cleaved edges. These ILDs are referred to as conventional or **edge-emitting ILDs.** Although used extensively in CD players and optical communications applications, etc., such ILDs have their shortcomings.

1. The radiated beam is elliptical and it diverges quickly. In particular, the divergence angle in the plane normal to the junction is quite large.
2. ILDs are large compared to other semiconductor diodes, transistors, and integrated circuit (IC) components. Because of this size, high electrical power is needed to drive them.
3. Since most cavity mirrors are formed by cleaving, these lasers are not amenable to mass production.
4. While it is possible to fabricate one-dimensional laser diode arrays based on the edge-emitting configuration, it is rather difficult to form two-dimensional laser arrays, particularly the densely packed ones that are needed in many optical image and signal processing applications. A key feature of a two-dimensional array is the ability to electronically control the emission from each emitter independently.

To avoid these shortcomings, a new type of ILD has been conceived: **surface-emitting ILDs**. An edge-emitting ILD can be converted to a surface-emitting configuration by the incorporation of deflecting mirrors, gratings, or intracavity bent waveguides, as shown in Figure 4.25a through c [25]. Such ILDs are therefore derivatives of edge-emitting ILDs. A true surface-emitting ILD is shown schematically in Figure 4.25d [26]. Note that the reflecting mirrors are in the form of thin layers. Also note that light emits in a direction normal to the layered structure. These lasers are quite small. Typically, each emitter is about 6 to 10 μm long and 10 μm in diameter. For this reason, they are often referred to as **microlasers** [26]. The configuration shown in Figure 4.25d is also known as the **vertical cavity surface-emitting lasers (VCSEL).**

4.7 LIGHT EMITTING DIODES

Light emitting diodes (LEDs) are *forward biased* semiconductor diodes that emit *incoherent* light when current passes through the semiconductor junction. As noted in section 4.5, there is a potential barrier at the unbiased junction impeding electrons and holes from diffusing across the depletion region. When a diode is forward biased, the potential barrier is reduced and charge carriers are in-

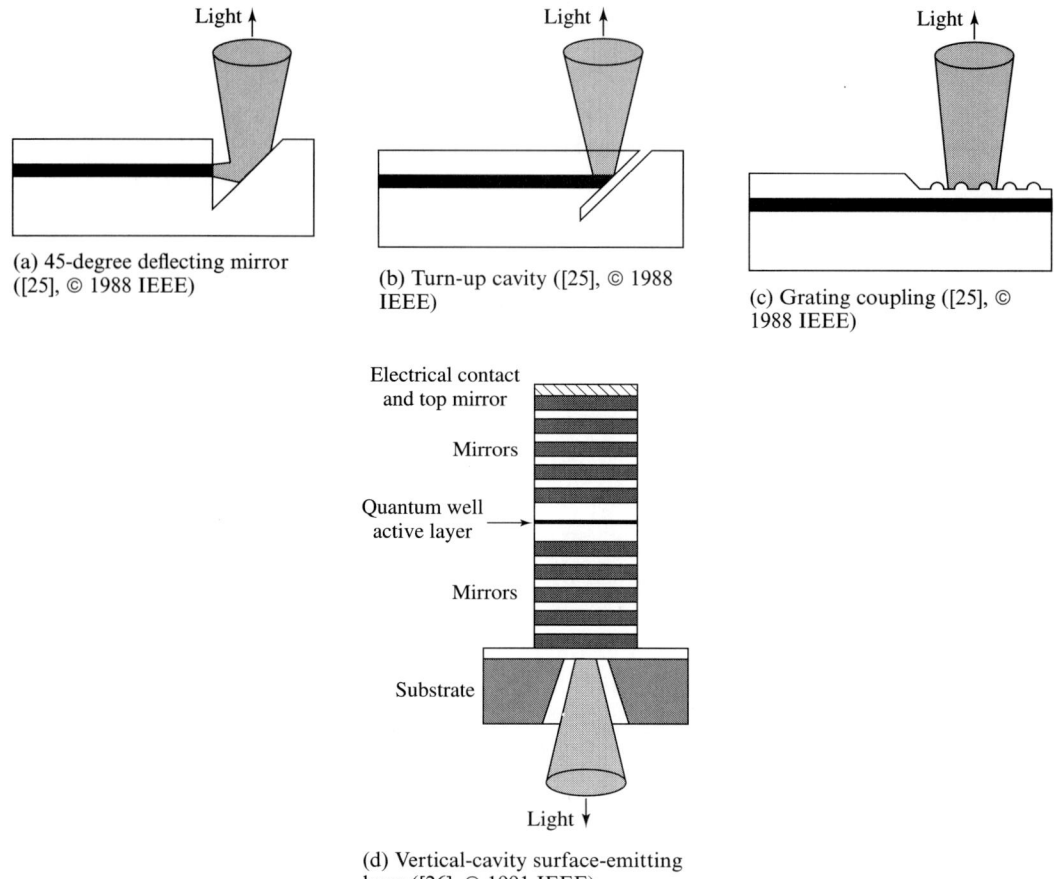

Figure 4.25 Types of surface-emitting injection lasers

jected across the junction. Having crossed the junction, the electrons and holes become the minority charge carriers on the other side of the junction. When the minority charge carriers recombine *radiatively* with the majority charge carriers, photons are emitted. This is the basic *light generation* process in semiconductors.

We can estimate the minority charge carrier density produced by current injection. In a typical display LED, the current density is on the order of 10 A/cm². This corresponds to an injection of $10/(1.6021 \times 10^{-19}) \approx 6 \times 10^{+19}$ carriers/(cm²s). Suppose that the injected charge carriers are confined to, and distributed uniformly in, a region that is 2 μm thick. The minority charge carrier density in the narrow region then increases at a *rate* of $6 \times 10^{19}/(2 \times 10^{-4}) \approx 3 \times 10^{+23}$ carriers/(cm³s). With the minority charge carrier lifetime on the order of 10^{-9}s, the minority charge carrier density in the narrow region is approximately $3 \times 10^{+23} \times 10^{-9} \approx 3 \times 10^{+14}$ carriers/cm³. Even for a moderate current injection, such as 10 A/cm², the number of *injected* minority charge carriers far exceeds that of the minority charge carriers in thermal equilibrium at room temperature. For LEDs designed for communications applications, the current density is typically 10^{+3} A/cm² or higher. This means that the injected minority charge carrier density in these LEDs is about two orders of magnitude greater than typical display LEDs.

4.7.1 Two Basic LED Structures

Two basic LED structures are surface-emitting and edge-emitting. Because of their structural differences, their optical characteristics are also quite different. Figure 4.26a and b shows these basic structures and the corresponding distributions of light in the far field. **Surface-emitting LEDs** are relatively simple. A narrow active layer is sandwiched between two cladding layers. Because of heterojunctions on either side of the active layer (Figure 4.26a), injected carriers are confined within the thin active region. A thick substrate is used to provide mechanical support and electrical contacts. An electrical contact layer is also deposited on the other side of the heterostructure. The diode is then mounted on a stud, which also serves as a heat sink. Because radiative recombination is

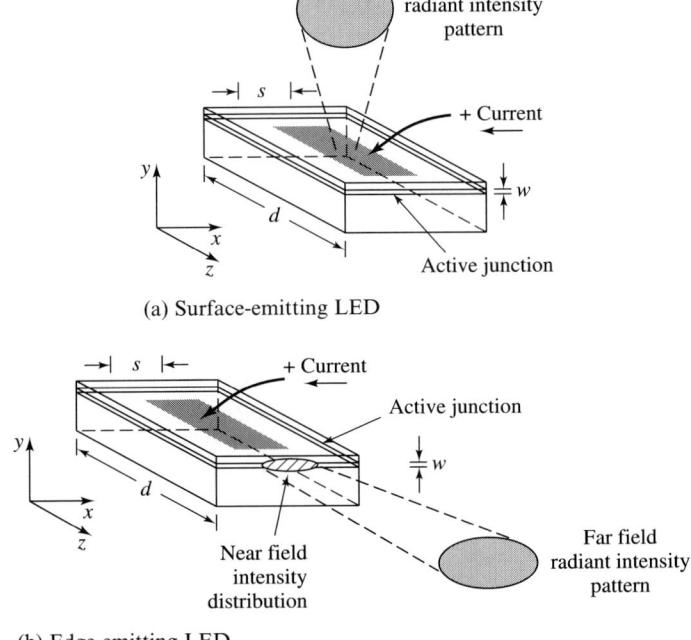

(a) Surface-emitting LED

(b) Edge-emitting LED

Figure 4.26 Geometries of surface-emitting and edge-emitting LEDs

more probable in direct-bandgap semiconductors than in indirect-bandgap semiconductors, direct-bandgap materials are preferred for the light emitting layers. Before reaching the air or a fiber, light may actually travel through several layers of different semiconductor material, and a portion of the emission is absorbed by these layers. For typical surface-emitting LEDs designed for optical fiber communication purposes, the active area, defined for example by an oxide stripe, has a width of 20 to 50 μm, which matches the core diameter of typical multimode fibers.

The structure of an **edge-emitting LED** (Figure 4.26b) is similar to that of ILDs. The cladding layers are chosen to provide *carrier confinement,* as well as *optical guidance*. Additional layers may also be used to provide optical confinement. Since the index of the cladding layers is slightly lower than that of the active layer, rays with a large incident angle are totally reflected internally by the boundaries between the two layers. This means that light is trapped inside the high-index light-emitting layer and emerges only from the diode edge. Edge-emitting LEDs have an elliptical pattern, as shown schematically in Figure 4.26b. Antireflection coating layers are often deposited on the two end facets to suppress reflection. In some edge-emitting LEDs, current injection in the region near the rear facet is reduced by an electrically resistive region, which means that the region near the rear facet absorbs rather than generates light.

4.7.2 Power and Light Extraction Efficiency: Surface-Emitting LEDs

Two terms are now needed in order for us to be able to quantify optical power. Figure 4.27 shows light generated by an *incoherent source* with a light-emitting area ΔS_s, and captured by a receiver with a receiving aperture of ΔS_r located at a distance r from the light source. **Irradiance** H is defined as the optical power ΔP per unit area received by the receiver

$$H = \frac{\Delta P}{\Delta S_r} \tag{4.8}$$

The irradiance H is a function of the orientation of ΔS_r relative to the radial direction, and H decreases as r^{-2} if the source-to-receiver separation is sufficiently large. For our purposes, a term *independent* of the orientation of the receiver area and the source-to-receiver separation is preferred. We therefore select the receiver area $\Delta S_{r\,n}$ normal to the radial direction, in lieu of an area in an arbitrary direction. We also work with $r^2 \Delta P$ in lieu of ΔP, and we define the **radiant intensity** J as

$$J = r^2 \frac{\Delta P}{\Delta S_{r\,n}} = \frac{\Delta P}{\frac{\Delta S_{r\,n}}{r^2}} = \frac{\Delta P}{\Delta \Omega} \tag{4.9}$$

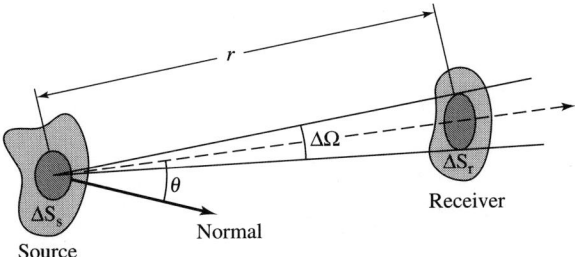

Figure 4.27 Geometry for defining irradiance, radiant intensity, and radiance

where $\Delta\Omega = \Delta S_{r,n}/r^2$ is the solid angle of the power impinging on the receiver surface element.

In view of (4.9), we can interpret J as the power ΔP contained in a solid angle $\Delta\Omega$. The dimension of J is watts per **spherical radian,** which is also watts per **steradian** [27].

Returning to our discussion on LEDs, consider a planar surface-emitting LED without a dome or other light focusing element (Figure 4.28a). This simple, flat LED can be modeled by a planar radiating surface, representing the active junction layer, located at a distance d inside the semiconductor that has an index of n_{sc} (Figure 4.28b). All parameters pertaining to the semiconductor region are identified with a subscript sc. The medium outside the semiconductor is air. It is reasonable to assume that the active junction emits light with equal radiant intensity in all directions; that is, $J_{sc}(\theta_{sc}, \phi_{sc}) = J_o$. If the semiconductor is sufficiently large, there is no reflection from any boundaries, then the total power radiated by the active semiconductor junction in all directions is

$$P_{sc} = \int_0^{2\pi} d\phi_{sc} \int_0^{\pi} J_{sc}(\theta_{sc}) \sin\theta_{sc} \, d\theta_{sc}$$
$$= J_o \int_0^{2\pi} d\phi_{sc} \int_0^{\pi} \sin\theta_{sc} \, d\theta_{sc} = 4\pi J_o \qquad (4.10)$$

However, the semiconductor size is finite, and reflection from the semiconductor/air boundary does exist. To study the total power P_{air} radiated into the air and the radiant intensity pattern $J(\theta, \phi)$ in air, we must take into account the presence of these boundaries. In air, the angular variables θ and ϕ are measured perpendicular to the light-emitting surface. Since J_{sc} is independent of ϕ_{sc}, J is also independent of ϕ. Therefore, we can ignore all references to ϕ_{sc} and ϕ.

We recognize that not all power radiated by the junction is transmitted into the air. Rays radiated away from the air/semiconductor boundary are heavily attenuated, since they follow long, zigzag paths in a lossy medium. In addition, the back surface is also not a good reflecting boundary. Therefore, rays such as R_1 in Figure 4.28b are severely attenuated and can be neglected.

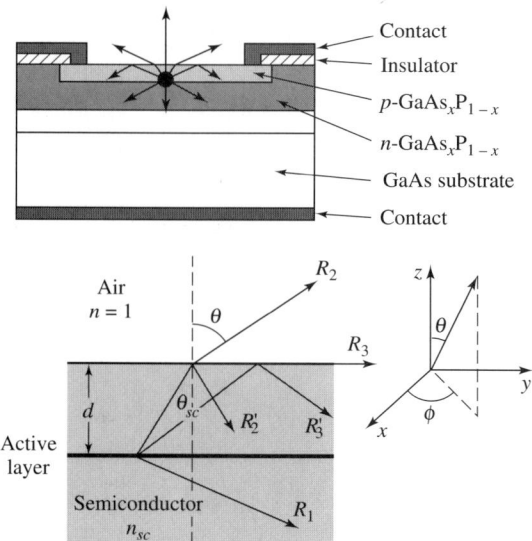

Figure 4.28 Model for undomed surface-emitting LED

A substantial portion of power radiated in the forward direction is also lost. This is due to the reflection and the total internal reflection at the semiconductor/air boundary, and to attenuation in the intervening layers. For rays impinging on the semiconductor/air boundary along directions close to the normal, a portion of the incident power is transmitted into the air and the rest is reflected by the boundary. These rays are represented by R_2 and R'_2 in Figure 4.28b. The fraction of power reflected by the semiconductor/air boundary can be calculated by evaluating the reflection coefficient of that boundary.

As the incident angle increases beyond a critical value, the situation changes. To appreciate this effect, we begin with Snell's law

$$n_{sc} \sin \theta_{sc} = 1 \sin \theta \tag{4.11}$$

Since n_{sc} is greater than 1, the angle θ of the transmitted rays is larger than θ_{sc}, and the transmitted rays are refracted away from the normal. As the angle of incidence increases, a ray occurs with an angle θ_{sc} such that $n_{sc} \sin \theta_{sc} = 1$. For this incident angle, the transmitted ray angle becomes $\theta = 90°$, and the refracted ray, identified as R_3 in Figure 4.28b, is precisely along the air/semiconductor boundary. This specific angle of incidence θ_{sc} is called the **critical angle** $\theta_{sc\,c}$

$$\theta_{sc\,c} = \sin^{-1}\left(\frac{1}{n_{sc}}\right) \tag{4.12}$$

For future reference, we note that at the GaAs/air boundary, the critical angle is 16.1 degrees. For rays with an angle of incidence greater than $\theta_{sc\,c}$, no real θ exists and no power is transmitted into the air. These rays are totally reflected by the semiconductor/air boundary. This is known as **total internal reflection.**

To study the power radiated into the air, it is only necessary to consider rays leaving the semiconductor junction with θ_{sc} less than $\theta_{sc\,c}$. The fractional power contained in a cone with an angle $\theta_{sc} < \theta_{sc\,c}$ is[4]

$$F = \frac{\int_0^{2\pi}\int_0^{\theta_{sc\,c}} J_{sc}(\theta_{sc}) \sin\theta_{sc}\, d\theta_{sc}\, d\phi_{sc}}{\int_0^{2\pi}\int_0^{\pi} J_{sc}(\theta_{sc}) \sin\theta_{sc}\, d\theta_{sc}\, d\phi_{sc}} = \frac{1 - \cos\theta_{sc\,c}}{2} \approx \frac{1}{4n_{sc}^2} \quad (4.13)$$

Rays within the cone of angle $\theta_{sc} \leq \theta_{sc\,c}$ also suffer loss due to reflection at the boundary and attenuation in the semiconductor. These rays may be approximated by rays perpendicular to the boundary. For such rays, the power reflection coefficient is $[(n_{sc} - 1)/(n_{sc} + 1)]^2$. Thus, the fraction of power transmitted across the semiconductor/air boundary is

$$T \approx 1 - \left(\frac{1 - n_{sc}}{1 + n_{sc}}\right)^2 = \frac{4n_{sc}}{(1 + n_{sc})^2}$$

Let the attenuation constant of the semiconductor be α. For perpendicular incident rays, the path length is d. For oblique rays with $\theta_{sc} < \theta_{sc\,c}$, the path length is slightly longer than d. The power reaching the semiconductor/air boundary is approximately $e^{-2\alpha d}$.

Combining all of the preceding loss terms, the total power transmitted to the air is

$$P_{air} \approx FTe^{-2\alpha d} P_{sc} \approx \frac{1}{4n_{sc}^2} \frac{4n_{sc}}{(n_{sc} + 1)^2} e^{-2\alpha d} P_{sc} = \frac{4\pi J_o e^{-2\alpha d}}{n_{sc}(n_{sc} + 1)^2} \quad (4.14)$$

We define the **light extraction efficiency** as

$$\eta = \frac{P_{air}}{P_{sc}} \approx \frac{e^{-2\alpha d}}{n_{sc}(n_{sc} + 1)^2} \quad (4.15)$$

As a numerical example, consider a GaAs LED with $d = 10$ μm, $n_{sc} \approx 3.6$, and $\alpha \approx 10$ cm^{-1} at 0.9 μm. Based on these numerical values, we obtain $F \approx 0.019$, $T \approx 0.68$, $e^{-2\alpha d} \approx 0.98$, and $\eta \approx 0.013$. In other words, less than 1.3 percent of the light generated by the semiconductor junction is extracted from the LED. Also note that 98 percent (i.e., $1 - F \approx 0.98$) of the power generated by the junction is lost due to total internal reflection. For most surface emitting LEDs, total internal reflection is in fact the dominant loss term. It is possible to cap an LED with a high-index dome to reduce the power loss due to the total internal reflection process. Domes can be **hemispherical, Weierstrass spherical,** or other shapes. Also, we can show that if the radius of the hemispherical dome is sufficiently large, no ray is totally reflected internally by

---

[4]Recall that $\cos^2\theta_{sc\,c} = 1 - \sin^2\theta_{sc\,c} = 1 - (1/n_{sc}^2)$ and note that $n_{sc}^2 \gg 1$

the dome [28], and the power extraction efficiency is greatly improved. This is left as an exercise for the reader (Problem 6).

4.7.3 Radiant Intensity Patterns: Surface-Emitting LEDs

The **radiant intensity pattern** $J(\theta,\phi)$ depicts the directional characteristic of the light-emitting properties of a source. We are particularly interested in the peaks, and in the valleys or nulls, if any, of the pattern. To quantify the beam width further, we identify the peak radiant intensity J_{max} of the main peak lobe, as well as the positions where $J(\theta,\phi)$ reduces to half to the peak value. The angular separation between these positions, one on each side of the major peak, is known as the **full width between half intensity (FWHI)** (Figure 4.29). The radiant intensity pattern of an LED depends not only on the characteristics of the semiconductor junction, but also on the diode geometry and the dome. Consider, for example, an undomed diode. To find the radiant intensity pattern in air, we must examine the power contained in a cone of angle $d\theta_{sc}$ impinging upon the semiconductor/air interface at an angle θ_{sc}. The power contained in $d\theta$ is proportional to that contained in $d\theta_{sc}$, with a proportionality constant K, as follows:

$$J(\theta)d\theta = K J_{sc}(\theta_{sc})d\theta_{sc}$$

The constant K will be determined later. As before, we treat $J_{sc}(\theta_{sc})$ as a constant J_o. To develop an explicit expression for $d\theta$, we use Snell's law and find that

$$n_{sc} \cos\theta_{sc} d\theta_{sc} = \cos\theta d\theta$$

Thus,

$$J(\theta) = \frac{K J_o \cos\theta}{\sqrt{n_{sc}^2 - \sin^2\theta}}$$

for $0 \le \theta \le \pi/2$. Since $n_{sc}^2 \gg 1$, the denominator can be approximated by n_{sc} and the radiant intensity pattern in the air can be approximately stated as

$$J(\theta) \approx \frac{K J_o}{n_{sc}} \cos\theta$$

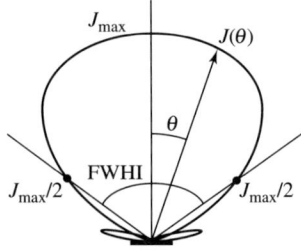

Figure 4.29 Radiant intensity pattern of incoherent source

4.7 Light Emitting Diodes

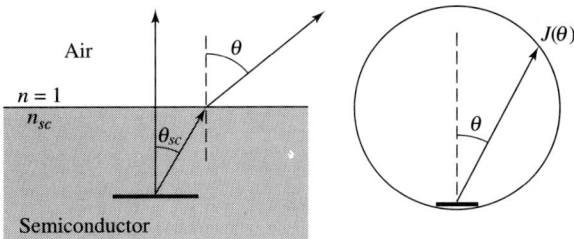

(a) Planar LED and its radiant intensity pattern

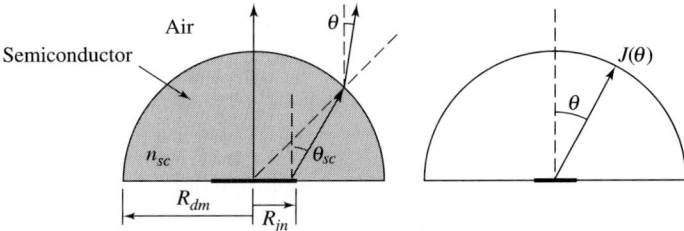

(b) LED with hemispherical dome and its radiant intensity pattern

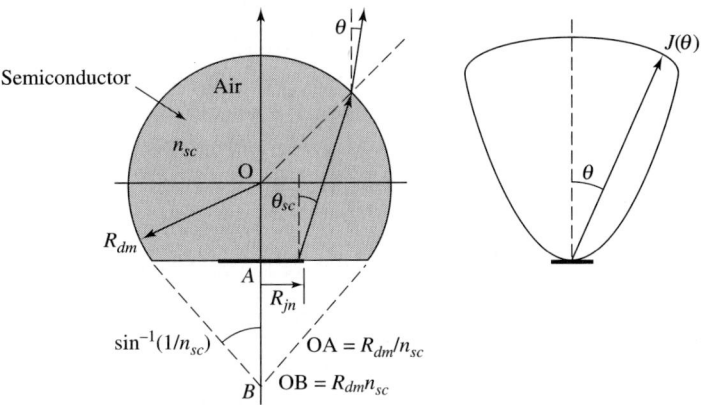

(c) LED with Weierstrass spherical dome and its radiant intensity pattern

Figure 4.30 Surface-emitting LEDs with and without high index domes [28, 29]

This shows that the radiation from a planar, undomed surface-emitting LED approximately follows **Lambert's cosine law** [27], which states that the radiant intensity pattern of any incoherent emitter is given by a cosine function if every point on the emitting surface radiates uniformly (i.e., equal intensity) in all directions. For an emitter with a **cosine radiant intensity pattern,** the half-power point is $\theta = 60°$, and the FWHI is 120 degrees. This is shown in Figure 4.30a.

In terms of K and J_o, the total power radiated into the air by an undomed LED is

$$\int_0^{2\pi} d\phi \int_0^{\pi/2} \frac{KJ_o}{n_{sc}} \cos\theta \sin\theta \, d\theta = \frac{\pi KJ_o}{n_{sc}}$$

By comparing this expression with (4.14), we obtain

$$K = \frac{4e^{-2\alpha d}}{(n_{sc}+1)^2}$$

and

$$J(\theta) = \frac{4J_o e^{-2\alpha d}}{n_{sc}(n_{sc}+1)^2} \cos\theta \qquad (4.16)$$

Many LEDs have a high-index dome. The function of a dome is to reduce the loss due to total internal reflection, and to modify the radiant intensity pattern. A **hemispherical dome** (Figure 4.30b) redistributes power uniformly in all directions and provides a broad beam angle over a large area. In contrast, in a **Weierstrass spherical dome**, light is distributed uniformly in a narrow cone (Figure 4.30c) ([28, 29]).

4.7.4 Radiant Intensity Pattern: Edge-Emitting LEDs

As noted previously, the radiant intensity pattern of an edge-emitting LED is quite different from that of a surface-emitting LED. An edge-emitting LED has an *elliptical* radiant intensity pattern, which cannot be described by a simple cosine function. In such LEDs, light follows zigzag trajectories in the thin active layer before reaching the open space. The radiation in the plane perpendicular to the junction is strongly influenced by the index difference between the active layer and the cladding layers, and is concentrated more in the forward direction. The beam angle (FWHI) in this plane $\theta_\perp$, is typically 30 degrees. The optical confinement in the junction plane itself is very weak or nonexistent, and therefore light is not guided in the lateral direction. The radiant intensity pattern in this plane is about the same as that of a surface-emitting LED. In other words, $\theta_\parallel$ is about 120 degrees. Figure 4.31 clearly shows the contrast in the radiant intensity patterns in the two planes, one parallel and the other perpendicular to the junction plane. To a good approximation, the radiant intensity pattern $J(\theta,\phi)$ may be expressed as

$$J(\theta,\phi) = \left[\frac{\cos^2\phi}{J_\perp(\theta)} + \frac{\sin^2\phi}{J_\parallel(\theta)}\right]^{-1} \qquad (4.17)$$

where $J_\perp(\theta)$ and $J_\parallel(\theta)$ are the radiant intensity patterns measured in the planes normal to and parallel to the junction, respectively.

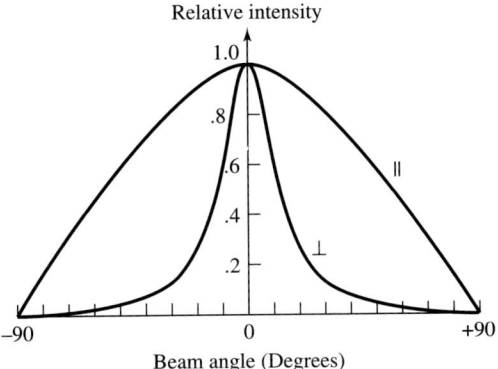

Figure 4.31 Radiant intensity patterns of an edge-emitting LED in the planes normal and perpendicular to the junction ([30], © 1976 IEEE).

4.7.5 PI and Spectral Characteristics

Figure 4.32 displays the optical power versus current (PI) characteristics of a typical surface-emitting LED, an edge-emitting LED, an SLD, and an ILD. Generally speaking, the optical power of an LED increases linearly with the input current, until saturation sets in. The saturation is due to junction heating. In contrast, the PI characteristic of an ILD has a sharp knee that corresponds to a threshold for stimulated emission. This difference provides a simple means for distinguishing between semiconductor diodes.

Another way to differentiate between various semiconductor sources is to compare the spectral widths of their outputs. For surface-emitting and edge-emitting LEDs with center wavelengths near 0.85 μm, the spectral width

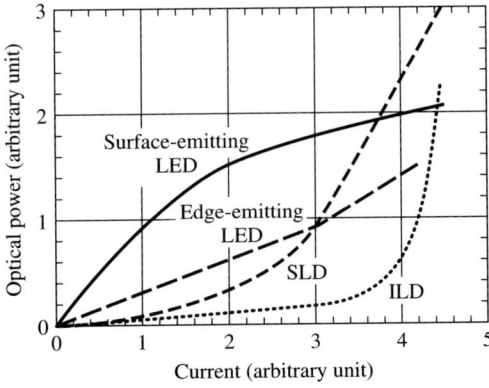

Figure 4.32 Comparison of the PI characteristics of surface-emitting and edge-emitting LEDs, ILDs, and SLDs

(FWHP) $\Delta\lambda$ is about 40 nm and 15 nm, respectively. However, the spectral width of SLDs is only 25 percent of that of surface-emitting LEDs. For ILDs, the spectral width is on the order of a few nanometers or less.

4.7.6 Modulation Characteristics

From the PI relationships shown in Figure 4.32, we see the possibility for direct amplitude modulation of semiconductor diode sources by changing the injection current to the diodes. This is an important feature of semiconductor luminescent diodes. Liu and Smith [31] showed that the modulated power output of LEDs can be expressed as

$$P(\omega_m) = \frac{P(0)}{\sqrt{1 + (\omega_m \tau)^2}} \quad (4.18)$$

where $P(0)$ is the LED power output with a dc input, and $f_m = \omega_m/(2\pi)$ is the modulation frequency. The time constant τ depends on the diode resistance, the junction capacitance, and the lifetime of the minority carriers. The frequency response (4.18) may be modeled by a simple RC parallel circuit. The capacitance C includes the diffusion capacitance and the space–charge capacitance for low injection current. At high injection levels, the space–charge capacitance has already been "charged," and the additional charges are mainly used to charge the diffusion capacitance [32]. Under this condition, the carrier diffusion is the key factor controlling the charge carrier motion in the junction region. As a result, the minority carrier lifetime τ_e becomes the main factor limiting the frequency response in amplitude modulation. The term τ_e can be expressed as

$$\frac{1}{\tau_e} = \frac{1}{\tau_r} + \frac{1}{\tau_{nr}} \quad (4.19)$$

where τ_r and τ_{nr} are the lifetimes of radiative and nonradiative recombination, respectively. To reduce the minority carrier lifetime, we can either increase the doping level or decrease the thickness of the active region, which has the effect of increasing the minority carrier concentration for the same amount of injection. In either case, the total LED output is reduced. It is interesting to note that the **power-modulation bandwidth product** of LEDs is a constant under various conditions [33]. In other words, the modulation frequency response of an LED may be increased at the expense of the optical power output. An amplitude modulation bandwidth of 450 MHz has been reported for a GaAs homojunction LED with a surface-emitting structure, and 200 MHz has been obtained for an AlGaAs double-heterojunction edge-emitting LED [34].

4.7.7 Coupling from LED–to–Multimode Fiber

Multimode fibers are discussed in Chapters 8 and 9. Of interest here is the power coupled from an LED into a multimode fiber. The only waves guided by the fiber are those incident upon the fiber end with an incident angle less than the fiber critical angle. As noted in Chapter 8, the critical angle of a fiber depends on its core and cladding indices, n_{co} and n_{cl}, and on the index profile. For a step index fiber, the critical angle is

$$\theta_{fb\,c} = \sin^{-1}\sqrt{n_{co}^2 - n_{cl}^2}$$

It is common practice to specify fibers in terms of the **numerical aperture (NA)**, to which the critical angle is directly related. For a step index fiber, the NA is

$$NA = \sqrt{n_{co}^2 - n_{cl}^2}$$

Thus, $\theta_{fb\,c}$ is simply $\sin^{-1}(NA)$. Detailed discussions on fibers are presented in Chapter 8.

To estimate the percentage of power coupled to a fiber from an LED with a radiant intensity pattern $J(\theta,\phi)$, we calculate the fractional power contained within a cone of angle $\theta_{fb\,c}$, as follows:

$$\eta_c = \frac{\int_0^{2\pi}\int_0^{\theta_{fb\,c}} J(\theta,\phi)\sin\theta\, d\theta\, d\phi}{\int_0^{2\pi}\int_0^{\pi/2} J(\theta,\phi)\sin\theta\, d\theta\, d\phi} \qquad (4.20)$$

For planar, undomed surface-emitting LEDs, the radiant intensity pattern is given by a cosine function, and the integrals in (4.20) can be evaluated in a closed form, to yield

$$\eta_c = \sin^2\theta_{fb\,c} = (NA)^2 \qquad (4.21)$$

As an example, consider a typical multimode fiber with an NA of 0.20. Only 4 percent of the power radiated by an undomed surface-emitting LED is coupled into such a fiber, even with perfect alignment. Coupling into a fiber is usually even less efficient due to Fresnel reflection loss at the air–fiber interface, imperfection of the fiber tip, misalignment, and area mismatch between the fiber and the LED. In any LED-to-fiber coupling, a large percentage of the power is lost.

It should be emphasized that (4.21) is valid *only* for LEDs with a cosine radiant intensity distribution. For emitters with a different $J(\theta,\phi)$, the integrals in (4.20) must be reevaluated individually. We stress in particular that (4.21) is not valid for edge-emitting LEDs, nor for domed surface-emitting LEDs. In the case of edge-emitting LEDs, the radiant intensity pattern in the plane normal to the junction is much narrower than that of a surface-emitting LED. Thus the coupling of light from edge-emitting LEDs to fibers can be much greater than that indicated by (4.21).

There are several ways to improve the coupling between surface-emitting LEDs and fibers. For example, a circular well can be etched into the substrate, and an optical fiber can be placed in the immediate vicinity of the active region. Epoxy resin or another index matching material is then used to hold the fiber

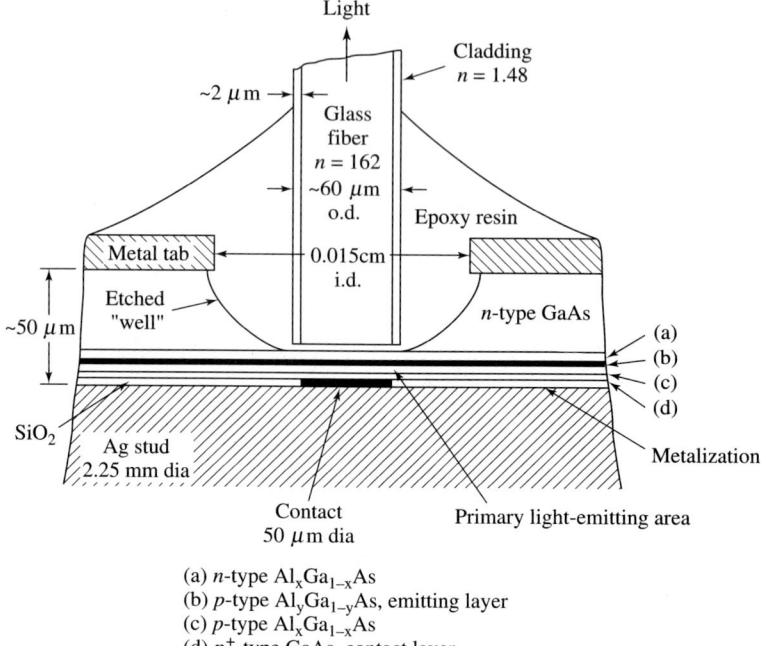

(a) n-type $Al_xGa_{1-x}As$
(b) p-type $Al_yGa_{1-y}As$, emitting layer
(c) p-type $Al_xGa_{1-x}As$
(d) p^+-type GaAs, contact layer

(a) Burrus' type LED

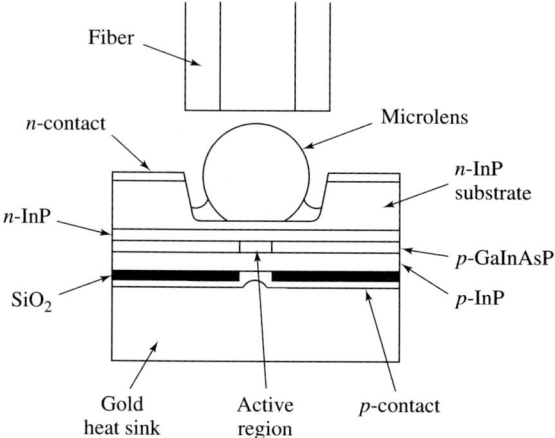

(b) Fiber coupling with a microlens

Figure 4.33 Surface-emitting LEDs with provisions for attaching fibers [35]

onto the substrate. Since the refractive index of the epoxy is greater than that of air, the presence of the epoxy reduces loss caused by the index mismatch and total internal reflection. In such an arrangement, the mechanical stability of the device is also greatly improved. This structure, shown in Figure 4.33a, is commonly referred to as a **Burrus LED** [35]. Alternatively, we can use small spheres, microlenses, or fibers with a spherical tip to improve the LED–to–fiber coupling (Figure 4.33b).

4.8 BASIC PARAMETERS: SEMICONDUCTOR INJECTION LASERS

As discussed in the last chapter, an amplifying medium, an optical feedback mechanism, and a pumping source are the three essential components in a laser system. In direct bandgap semiconductors, optical amplification can be realized with optical pumping, electron beam pumping, or current injection, the most convenient of which is current injection. Since the optical gain in direct bandgap semiconductors is very high, a modest amount of feedback is sufficient to make a semiconductor diode lase. Because the refractive index of GaAs and related semiconductors is about 3, the power reflection coefficient of a smooth semiconductor/air boundary is about 0.3, which is large enough to sustain semiconductor laser oscillations. In particular, cleaved semiconductor surfaces are sufficiently smooth for most ILDs. Therefore, the optical cavity of an ILD is made simply of two cleaved surfaces.

The operation of an ILD can be described as follows:

1. Electrons are injected from the n-side of a junction to the p-side, and holes are injected from the opposite direction.
2. In the depletion region, electrons are near the bottom of the conduction band, while holes are near the top of the valence band.
3. Photons are emitted when electrons and holes recombine.
4. Electric charges return through the external circuit.

Both electrons and holes are involved in the light emission process. However, for most compound semiconductors, electrons are much more mobile than holes. For example, the electron and hole mobilities of GaAs are 8600 and 400 cm^2/V-s, respectively (see Table 4.1). The motion of the electrons is the dominant component of the injection current. For simplicity, the motion of holes is usually ignored.

Lasers are classified as three-level or four-level systems. To determine how ILDs should be classified, consider a DH injection laser with an energy diagram as shown in Figure 4.34. Lasing occurs in the narrow bandgap semiconducting layer sandwiched between two wide bandgap semiconductors. In the terminology of three- or four-level schemes discussed in Chapter 3, the filled valence band in the P-region outside the active junction area may be considered level E_1. The conduction band on the N-region outside the active region corresponds

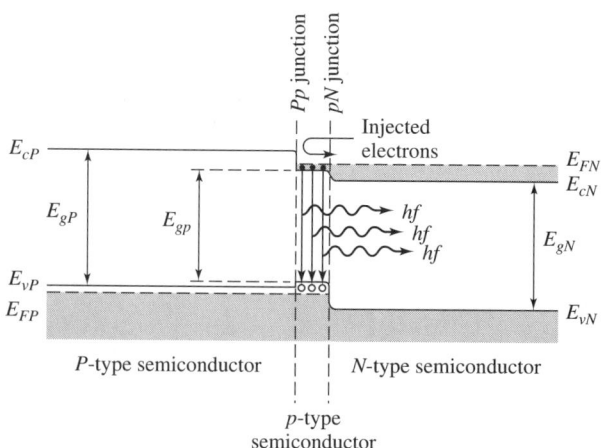

Figure 4.34 Energy diagram of heterostructure ILD

to level E_3. The conduction and valence bands in the narrow bandgap semiconductor are levels E_2 and E_a, respectively. Recall that the difference between three- and four-level systems involves whether level E_a is normally occupied. Since in an ILD, the top of the valence band in the active region is normally vacant, ILDs are considered *four-level systems* [36].

Due to the differences in the active medium and the pumping scheme, ILDs are different from gas, liquid dye, and solid-state lasers in several respects.

1. Lasing transitions in ILDs are between the valence and conduction bands of semiconductors rather than between the discrete energy levels of isolated atoms or ions in solid-state host materials.
2. The energy diagram of semiconductors depends on the orientation with respect to the crystal axis. The energy levels in gases or liquids are independent of direction or orientation.
3. Direct, high-speed amplitude modulation of ILDs is possible by controlling the current injected into the semiconductor junction. For other lasers, direct amplitude modulation is extremely difficult.
4. The overall size of ILDs, including the active medium and the optical cavity, is very small in comparison with other lasers. Because ILDs are small, the output is not as well confined spatially or spectrally, as in other lasers that are larger.

The characteristics of semiconductor injection lasers may be described by six parameters: radiation wavelength, threshold current density, external quantum efficiency, power efficiency, far-field radiation pattern, and spectral width. These topics are discussed in the following subsections. However, the discussion on amplitude modulation characteristics is postponed until the rate equations describing ILD operation are introduced in Section 4.9.

4.8.1 Wavelength of Radiation

For convenience, we will only discuss heterojunction lasers (Figure 4.34). Similar conclusions may be reached for homostructure lasers. When a diode is forward biased, electrons are injected from the N-side, while holes are injected from the P-side, into the p-side. As a result, there are a large number of electrons in the conduction band and a large number of holes in the valence band in the junction region, and population inversion occurs in the junction region. Spontaneous, stimulated emission occurs as a result of the radiative recombination of electrons and holes. An examination of Figure 4.34 shows that the photon energy is in the range of

$$E_{FN} - E_{FP} > hf > E_{cp} - E_{vp} = E_{gp} \quad (4.22)$$

where E_{gp} is the bandgap in the narrow-bandgap semiconductor region. In terms of wavelength λ, we have

$$\frac{hc}{E_{FN} - E_{FP}} < \lambda < \frac{hc}{E_{gp}} \quad (4.23)$$

Since the difference between E_{gp} and $(E_{FN} - E_{FP})$ is very small, λ is approximately hc/E_{gp}. The precise wavelength within the range specified by (4.23) is determined mainly by the cavity length, and, to a lesser degree, by the transverse dimensions of the laser cavity.

4.8.2 Threshold Current Density

The PI characteristics of various semiconductor luminescent diodes are depicted in Figure 4.32. Let us focus our attention on the PI curve of ILDs. Initially, P increases slowly with I. Near a certain critical value, the PI characteristic turns sharply upward. Beyond the critical value, the PI curve becomes a straight line with a steep slope. By extrapolating the linear segment of the PI characteristic in this region, we can obtain a **threshold current** I_{th}. Knowing I_{th} and the junction area sd, as shown in Figure 4.35, we can then calculate the **threshold current density** J_{th}. If the PI characteristic curve bends slowly, it is probably because different regions lase with different current densities.

The threshold current density J_{th} can be related to various device parameters. Consider an optical cavity of length d with an active region of thickness w and a lateral width s, as shown in Figure 4.35. Let the power reflection coefficients of the mirrors be Γ_1 and Γ_2, and let the attenuation constant and gain constant of the active region be α and g, respectively. Define the **threshold gain** g_{th} as the optical gain needed to balance the total power loss, due to various losses in the cavity, and the power transmission through the mirrors. Then,

$$\Gamma_1 \Gamma_2 e^{2d(g_{th} - \alpha)} = 1$$

Thus we have

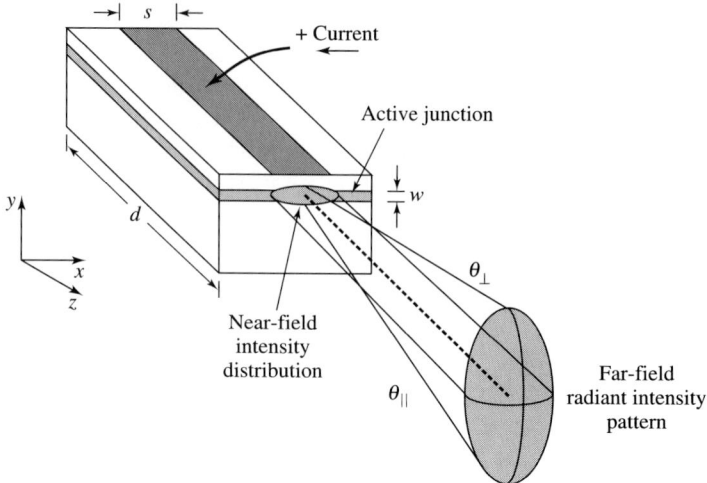

Figure 4.35 Schematic of semiconductor injection laser, and its near-field and far-field intensity distribution

$$g_{th} = \alpha + \frac{1}{2d} \ln \frac{1}{\Gamma_1 \Gamma_2} \qquad (4.24)$$

The first term of (4.24) represents the loss per unit length due to attenuation in the cavity. The second term accounts for the power radiated through the mirrors and is also expressed in terms of loss per unit length of the cavity. As a rough approximation, the gain g is proportional to the current density, that is, $g = \beta J$, where β is a proportionality constant. Let J_{th} be the threshold current density and the corresponding gain be g_{th}. Then,

$$J_{th} = \frac{1}{\beta}[\alpha + \frac{1}{2d} \ln \frac{1}{\Gamma_1 \Gamma_2}] \qquad (4.25)$$

By measuring J_{th}, α, d, Γ_1, and Γ_2, etc., we can determine β. Precise values of α and β would be dependent upon the materials and the junction structure. Typical values of α and β for GaAs lasers are listed in Table 4.4.

The emission spectrum is also a function of the injection current. For $I < I_{th}$, the emission spectrum is relatively broad ($\Delta\lambda > 10$ nm for λ in the 0.8 to 0.9 μm range). For $I > I_{th}$, the spectral line suddenly becomes very narrow (1 to 2 nm) (Figure 4.36) [38]. This narrowing of the spectral line is a good indication that the radiation is dominated by stimulated emission. As I increases, the linewidth becomes narrower, as shown in Figure 4.36.

4.8.3 Internal and External Quantum Efficiencies

The **quantum efficiency** is the average number of photons generated for each electron–hole pair injected into the semiconductor junction. In other words, it is a measure of the efficiency of the electron–to–photon conversion process. If photons are "counted" at the semiconductor junction region, the quantum

4.8 Basic Parameters: Semiconductor Injection Lasers

Table 4.4 Parameters of GaAs junctions at 300 °K [37]

	Homojunction	Single heterostructure	Double heterostructure
a (cm^{-1})	100	20	15
β (cm/A)	$(2-3) \times 10^{-3}$	$(3-6) \times 10^{-3}$	1.5×10^{-2}

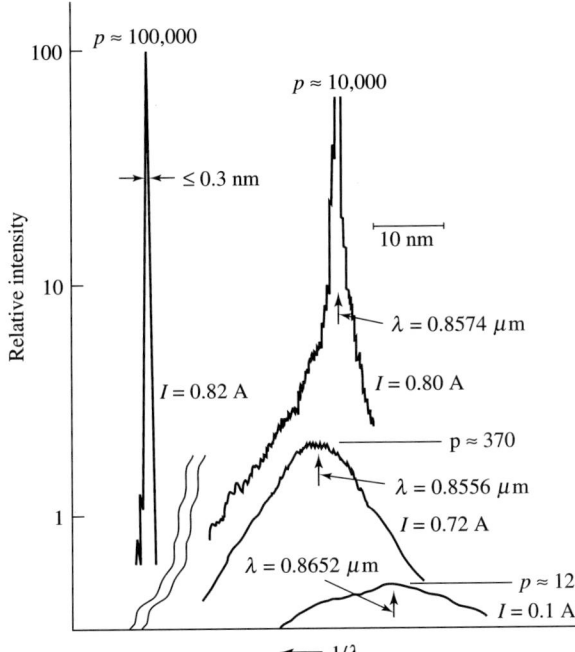

Figure 4.36 Spectra for a DH GaAs-AlGaAs injection laser at 295 °K at four current levels
The number p is the relative intensity. The peak intensity at 0.8 A, although incomplete, is included to illustrate the shape of the base. (adapted from [38])

efficiency is known as the **internal quantum efficiency** η_{int}, which depends on the materials of the active junction and the neighboring regions. For GaAs lasers, η_{int} ranges from 65 percent to almost 100 percent. For the **external quantum efficiency** η_{out}, the photons are counted outside the semiconductor diode.

Based on the physical meaning of the terms in (4.24), the ratio of the power radiated through the mirrors to the total power generated by the semiconductor junction is

$$\frac{\frac{1}{2d} \ln \frac{1}{\Gamma_1 \Gamma_2}}{\alpha + \frac{1}{2d} \ln \frac{1}{\Gamma_1 \Gamma_2}}$$

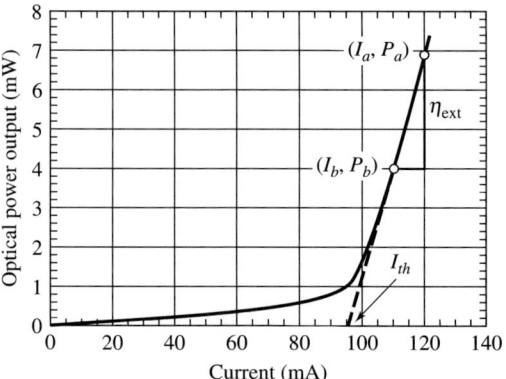

Figure 4.37 PI characteristics of an ILD

We therefore have the following expression relating η_{int} to η_{ext}:

$$\eta_{ext} = \eta_{int} \frac{\frac{1}{2d} \ln \frac{1}{\Gamma_1 \Gamma_2}}{\alpha + \frac{1}{2d} \ln \frac{1}{\Gamma_1 \Gamma_2}} \quad (4.26)$$

The external quantum efficiency can be determined experimentally from the PI characteristic. Consider two points (I_a, P_a) and (I_b, P_b) on the PI curve, with I_a and $I_b > I_{th}$ (Figure 4.37). For a given current I_a, the number of electrons injected into the active area *per second* is I_a/e. Also, since the photon energy is hf, the number of photons emitted *per second* is P_a/hf. From the definition of η_{ext}, we have

$$\frac{P_a}{hf} = \eta_{ext} \frac{I_a}{e}$$

From the second point (I_b, P_b) on the PI characteristic, we also obtain

$$\frac{P_b}{hf} = \eta_{ext} \frac{I_b}{e}$$

Combining these equations produces

$$\eta_{ext} = \frac{P_a - P_b}{(I_a - I_b)hf/e} = \frac{e}{hf} \frac{\Delta P}{\Delta I} \quad (4.27)$$

where $\Delta P = P_a - P_b$ is the increment of optical power output due to a current increment $\Delta I = I_a - I_b$. From this, we see that the *external quantum efficiency* is proportional to the *slope* of the PI characteristic in the region $I > I_{th}$. If we choose I_{th} as I_b, then $P_b \approx 0$ and (4.27) becomes approximately

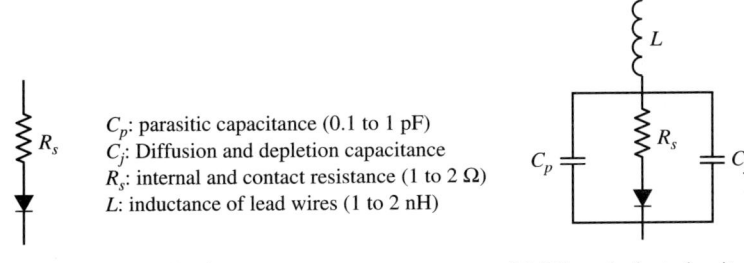

Figure 4.38 Equivalent circuit of a typical ILD

$$\eta_{ext} \approx \frac{e}{hf} \frac{P}{I - I_{th}} \qquad (4.28)$$

As noted previously, the photon energy and the energy bandgap of the junction material are approximately the same, that is, $hf \approx E_g$. When either hf or E_g is expressed in electron volts, then hf/e or E_g/e gives the voltage across the junction in volts.

4.8.4 Power Efficiency

At dc or low frequencies, the electrical characteristic of ILDs may be viewed as an ideal diode in series with a resistor R_s (Figure 4.38a). Here, R_s represents the resistance of the semiconductor and the electrical contacts. The voltage across the ideal diode is E_g/e. The **power efficiency** η_P is the ratio of the optical power output P to the dc electrical power input:

$$\eta_P = \frac{P}{I \frac{E_g}{e} + I^2 R_s} \qquad (4.29)$$

The optical power output can then be related directly to η_{ext} and $I - I_{th}$. When this is done, we can express η_P as

$$\eta_P \approx \frac{\eta_{ext}(I - I_{th})\frac{E_g}{e}}{I \frac{E_g}{e} + I^2 R_s} \qquad (4.30)$$

4.8.5 Far Field Radiation Pattern

Most ILDs support several longitudinal and transverse modes. The transverse mode structure depends on the thickness of the active region, the stripe width, and the refractive indices of the active and surrounding regions. All modes contribute to the total radiation. However, each transverse mode has a specific radiation pattern. For example, the radiation due to higher-order transverse modes diverges faster than that of lower-order transverse modes. For a different input current, a different set of longitudinal and transverse modes is excited. Thus, the radiation pattern of an ILD is different for a different input current. This is shown in Figure 4.39 [39]. For diodes with a narrow stripe ($s = 13$ μm) and a thin active region ($w = 0.9$ μm), the zeroth-order transverse mode is the dominant mode, even for an injection current substantially above the threshold value. For ILDs with a wide stripe ($s = 25$ μm), higher-order transverse modes appear in the plane parallel to the junction. If the active region is thick and the stripe is wide ($s = 25$ μm, $w = 1.8$ μm), higher-order transverse modes also appear in the plane perpendicular to the junction plane.

If the active region is not too thin, the radiation in the plane perpendicular to the junction plane can be approximated by a gaussian beam with a **half-power beamwidth** $\theta_\perp$ of

$$\theta_\perp = 2 \tan^{-1} \left[\frac{0.59 \lambda}{\pi w (0.31 + 3.15 V^{-3/2} + 2 V^{-6})} \right] \quad (4.31)$$

where

$$V = \frac{2\pi}{\lambda} w \sqrt{n_{ac}^2 - n_{cl}^2}$$

and n_{ac} and n_{cl} are the indices of the active and cladding layers, respectively [40]. Interestingly, this definition for V is exactly the same as that for thin-film optical waveguides and optical fibers. (See Chapters 7 and 8.) Equation (4.31) is valid for $1.5 < V < 6$. For $V < 1.5$, the zeroth-order beam pattern cannot be approximated by a gaussian function and (4.31) is not accurate. A detailed derivation and justification for (4.31) can be found in reference 40.

4.8.6 Spectral Linewidth

As noted previously, the gain profiles of GaAs at room temperature are on the order of a few nanometer. Many transverse and longitudinal modes can be sustained in the laser cavity. Figure 4.40 shows the spectrum of a typical DH laser with several transverse and longitudinal modes [41]. In the figure, each peak represents a resonant wavelength of the laser cavity. As discussed in Section 2.9.6, a laser cavity mode is identified by three integers, and they are the two transverse mode numbers and one longitudinal mode number. In each mode group, one of the transverse mode number changes by one in each consecutive resonance while the other two mode numbers remain unchanged.

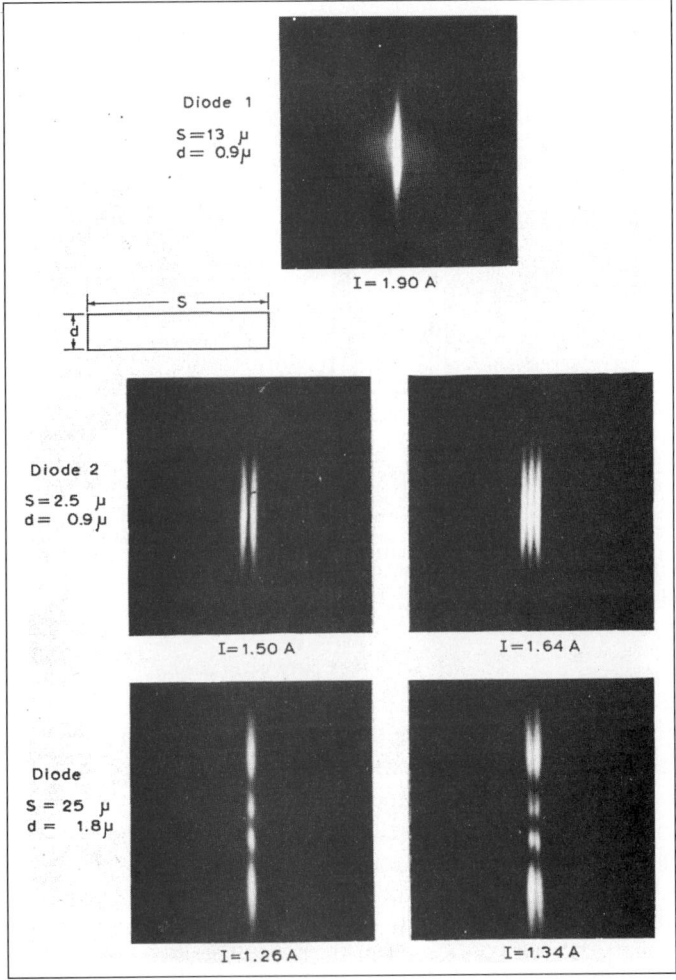

Figure 4.39 Radiation patterns of various transverse modes of strip geometry laser diodes [39]

4.9 DYNAMIC CHARACTERISTICS: SEMICONDUCTOR INJECTION LASERS

4.9.1 Rate Equations with a Current Injection Term

In this section we will address the modulation characteristics and transient response of ILDs. Our approach is based on the rate equations describing the time rate of change of charge carriers and photons in the optical cavity [4, 42–44]. In the process, we have the opportunity to examine the steady state response of ILDs again. As before, the contributions due to holes are ignored.

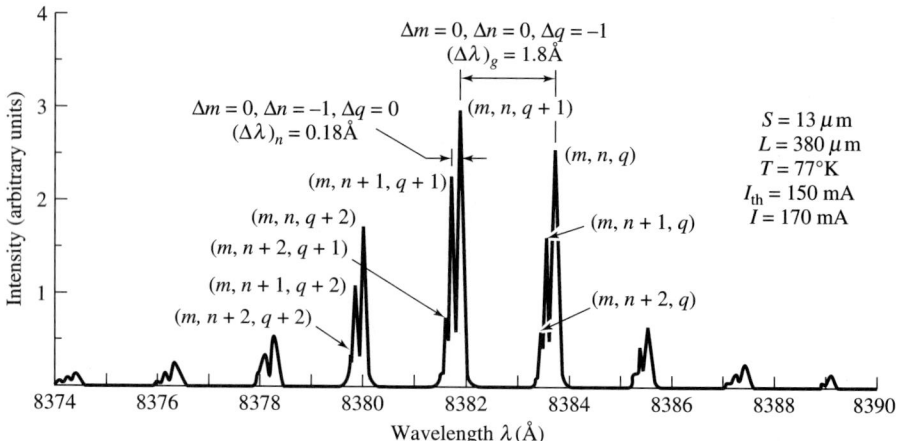

Figure 4.40 Frequency spectrum of a stripe geometry laser diode [41]

The equation describing the time rate of change of the electron density N in the thin active region in the absence of photons and current injection is

$$\frac{dN}{dt} = -\frac{N}{\tau_e} \qquad (4.32)$$

where τ_e is the **electron lifetime,** including the radiative and nonradiative recombination with holes. With (4.32) as the basis, we add several terms.

First, we consider the effects of the *current injection*. Let the injected current density be J. The number of electrons injected into the active region per second is Jsd/e. Since the active region is only a few tenths of 1 μm, we assume that injected electrons are distributed uniformly in the active region of width w. Therefore, the rate of increase of the electron density due to the current injection is

$$\frac{J\,ds}{e\,sd\,w} = \frac{J}{ew}$$

This is the *pumping term* for ILDs. If the thickness of the active region is comparable to the diffusion length of the charge carriers in the semiconductor, additional terms must be inserted to account for the electron diffusion in the active region.

Second, we consider the effects of photons that have wavelengths compatible with the laser emission. Photons are represented by cavity modes. For simplicity, we assume that there is a single longitudinal and transverse mode. Let the photon density corresponding to this mode be $\Phi\delta f$, where δf is the emission linewidth. The absorption and stimulated emission of photons per unit volume are proportional to $\Phi\delta f$, and are also a function of the electron density N. Therefore, we can express the net effect of *absorption* and *stimulated emission* as $g(N)\Phi\delta f$, where $g(N)$ is a function of the electron density N.

4.9 Dynamic Characteristics: Semiconductor Injection Lasers

With these terms included, the *time rate of change of electron density* in the active region becomes

$$\frac{dN}{dt} = \frac{J}{ew} - g(N)\Phi\delta f - \frac{N}{\tau_e} \quad (4.33)$$

In the absence of electrons, the time rate of change of photon density $\Phi\delta f$ in the passive cavity is

$$\frac{d\Phi\delta f}{dt} = -\frac{\Phi\delta f}{\tau_{ph}} \quad (4.34)$$

The cavity loss and the power transmitted through the cavity mirrors are lumped into the **photon lifetime** τ_{ph}. Based on (3.58) in Chapter 3 and (4.24) in this chapter, we have an expression for τ_{ph}, which is

$$\tau_{ph} = \frac{n}{c[\alpha + \frac{1}{2d}\ln\frac{1}{\Gamma_1\Gamma_2}]}$$

Rigorously speaking, τ_{ph} also changes with the electron density N, because α varies with N due to the free carrier absorption. However, for simplicity, the dependence of τ_{ph} on N is ignored.

To account for the absorption and stimulated emission in the presence of charge carriers, (4.34) must be amended by a term $g(N)\Phi\delta f$. Also, not all electron–hole recombinations are radiative. Photons emitted due to *spontaneous emission* may not be of the same wavelength as that of the laser emission. Only a small fraction of electron–hole pairs recombine to generate photons with the "right" wavelength. If we let the fraction be γ, then the increase of photons due to spontaneous emission can be written as $\gamma N/\tau_e$, where γ is generally small.

Thus, in the presence of electrons, the *time rate of change of photon density* in the cavity is

$$\frac{d\Phi\delta f}{dt} = \gamma\frac{N}{\tau_e} + g(N)\Phi\delta f - \frac{\Phi\delta f}{\tau_{ph}} \quad (4.35)$$

The gain function $g(N)$ in (4.33) and (4.35) depends on the material composition and junction structure, and may be approximated in various forms. Here, we choose the simplest possible form, $g(N) \approx g_1(N - N_{min})$, where g_1 is a constant with a dimension of $m^{-3}s^{-1}$. This simple representation is accurate for many diodes of interest [42], such as SH lasers on Si-doped GaAs substrate. We will see shortly that N_{min} is related to the minimum electron density required to achieve a positive gain, as well as to the threshold current density J_{th}. If $g(N)$ is written as as $g_1(N - N_{min})$, then (4.33) and (4.35) become

$$\frac{dN}{dt} = \frac{J}{ew} - g_1(N - N_{min})\Phi\delta f - \frac{N}{\tau_e} \quad (4.36)$$

$$\frac{d\Phi\delta f}{dt} = +g_1(N - N_{min})\Phi\delta f + \gamma\frac{N}{\tau_e} - \frac{\Phi\delta f}{\tau_{ph}} \quad (4.37)$$

It should be emphasized that these equations are valid *only* for ILDs with a single longitudinal and transverse mode and a particularly simple gain function. Nevertheless, many features of the steady-state and transient characteristics of ILDs can be explained in terms of (4.36) and (4.37).

4.9.2 Steady-State PI Characteristics

To study the steady-state PI characteristics, we set the time derivatives of (4.36) and (4.37) to zero and obtain

$$\frac{J_0}{ew} - \frac{N_0}{\tau_e} - g_1(N_0 - N_{min})\Phi_0\delta f = 0 \tag{4.38}$$

$$g_1(N_0 - N_{min})\Phi_0\delta f - \frac{\Phi_0\delta f}{\tau_{ph}} + \gamma\frac{N_0}{\tau_e} = 0 \tag{4.39}$$

From these equations, we can solve for the steady-state solutions N_0 and $\Phi_0\delta f$ for a given injection current density J_0. When expressed in terms of $\Phi_0\delta f$, N_0 is

$$N_0 = \frac{g_1 N_{min} + \dfrac{1}{\tau_{ph}}}{g_1 \Phi_0 \delta f + \dfrac{\gamma}{\tau_e}} \Phi_0\delta f \tag{4.40}$$

The product term $g_1(N_0 - N_{min})\Phi_0\delta f$ in (4.38) and (4.39) is eliminated by combining the two equations. The result is

$$\frac{J_0}{ew} - (1-\gamma)\frac{N_0}{\tau_e} - \frac{\Phi_0\delta f}{\tau_{ph}} = 0$$

By substituting (4.40) into the above equation, we obtain

$$\frac{J_0}{ew} = \left[\frac{g_1 N_{min} + \dfrac{1}{\tau_{ph}}}{g_1 \Phi_0 \delta f + \dfrac{\gamma}{\tau_e}}\frac{1-\gamma}{\tau_e} + \frac{1}{\tau_{ph}}\right]\Phi_0\delta f \tag{4.41}$$

It is possible to solve for N_0 and $\Phi_0\delta f$ in terms of J_0. Alternatively, we can treat $\Phi_0\delta f$ as an independent variable, and J_0 and N_0 as dependent variables. Of particular interest are the two limiting cases. If $\Phi_0\delta f \ll \gamma/(g_1\tau_e)$, equation (4.41) can be approximated as

$$\frac{J_0}{ew} \approx \left[(g_1 N_{min} + \frac{1}{\tau_{ph}})\frac{1-\gamma}{\gamma} + \frac{1}{\tau_{ph}}\right]\Phi_0\delta f \tag{4.42}$$

which represents a straight line passing through the origin of the $\Phi_0\delta f$–J_0 plane. On the other hand, if $\Phi_0\delta f$ is large, (4.41) becomes approximately

4.9 Dynamic Characteristics: Semiconductor Injection Lasers

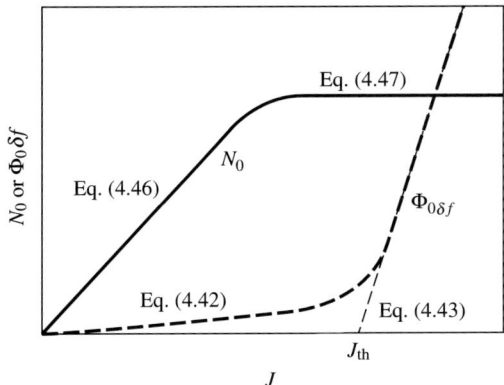

Figure 4.41 Theoretical response of an ILD, depicting the electron and photon densities as functions of injection current density

$$\frac{J_0}{ew} \approx \frac{g_1 N_{min} + \frac{1}{\tau_{ph}}}{g_1} \frac{1-\gamma}{\tau_e} + \frac{1}{\tau_{ph}} \Phi_0 \delta f \quad (4.43)$$

which is also a straight line. Therefore, in the limiting cases examined above, plots of $\Phi_0 \delta f$ versus J_0 are straight lines. Now, recall that the optical power emitted by an ILD is proportional to $\Phi_0 \delta f$, and the electric current injected into a diode is proportional to J_0. This means that a plot of $\Phi_0 \delta f$ as a function of J_0, shown as the dashed curve in Figure 4.41, has the same shape as the steady-state PI characteristic of the ILD. Consequently, the initial and final linear segments of the PI characteristic shown in Figures 4.32 and 4.37 can be understood in terms of the rate equations.

When extrapolated, the straight-line segment given by (4.43) intersects the J_0 axis at

$$\frac{J_{th}}{ew} = \frac{g_1 N_{min} + \frac{1}{\tau_{ph}}}{g_1} \frac{1-\gamma}{\tau_e} \approx \frac{1}{\tau_e}(N_{min} + \frac{1}{g_1 \tau_{ph}}) \quad (4.44)$$

which gives the threshold current density of the ILD. The last approximation was made on account of the fact that γ is small.

From (4.43) and (4.44), the steady-state photon density for $J_0 > J_{th}$ can be written as

$$\Phi_0 \delta f = \frac{\tau_{ph}}{ew}(J_0 - J_{th}) \quad (4.45)$$

We can also derive an expression relating the steady-state electron density directly to the steady-state photon density. From (4.40), we have

$$N_0 \approx \frac{\tau_e}{\gamma}(g_1 N_{min} + \frac{1}{\tau_{ph}})\Phi_0 \delta f \qquad \text{for } J < J_{th} \qquad (4.46)$$

$$N_0 = N_{min} + \frac{1}{g_1 \tau_{ph}} \approx \frac{\tau_e}{ew} J_{th} \qquad \text{for } J > J_{th} \qquad (4.47)$$

Combining (4.42) and (4.46), we see that N_0 increases linearly with J_0 for $J_0 < J_{th}$. For $J_0 > J_{th}$, N_0 is a constant. The dependence of N_0 as a function of J_0 is depicted in Figure 4.41. To summarize, the steady-state electron and photon densities increase linearly with the current density when J_0 is less than the threshold value J_{th}. When J_0 is greater than J_{th}, N_0 is a constant and $\Phi_0 \delta f$ increases linearly with the current density.

4.9.3 Transient Characteristics

Direct amplitude modulation of ILD output is a desirable feature for many applications, such as digital communications. To investigate the resulting transient characteristics, we will examine the light output waveform when the injection current is a rectangular pulse, i.e., when the ILD is turned on and off. Figure 4.42 schematically depicts the waveforms of the injection current and the resulting optical power. Three features are noted.

1. When the current pulse is turned on abruptly, there is a time delay t_{on} in the light output.
2. Depending on various conditions, there may be *damped, relaxation,* or *self-sustained oscillations* following the turn-on.
3. When the injection current is turned off abruptly, the light output again shows a time delay, that is, a slowed turn-off.

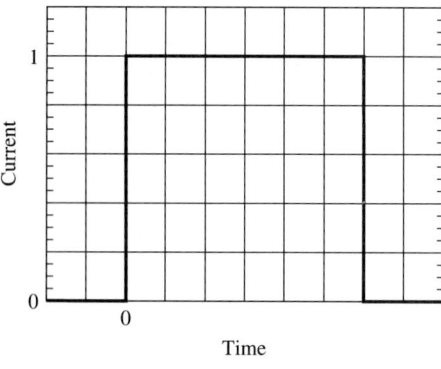

(a) Input current pulse

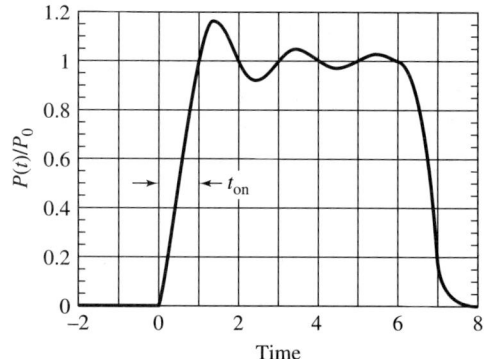

(b) Waveform of optical output

Figure 4.42 Transient response of an ILD

Experimental confirmation can be seen in Figure 4.43a through c [45]. We can understand these transient behaviors from the rate equations. Although (4.36) and (4.37) appear simple, they are actually nonlinear in N and $\Phi\delta f$. An exact analytic solution for the nonlinear coupled differential equations is not known. Approximate solutions for N and $\Phi\delta f$ have been reported by Boers, Vlaardingbrook, and Danielsen [46]. However, they are too complicated to be presented here. Instead, we duplicate the plots presented by them (Figure 44). In their study, the dependence of τ_{ph} on the electron density $N(t)$ has been taken into account. The effects of the spontaneous emissions, denoted by different values of γ, on the transient response are particularly noteworthy. Specifically, note in Figure 4.44 the drastic increase of oscillations or spikes as γ decreases.

If the current pulse is sufficiently long, $N(t)$ should approach the steady-state value given by (4.47) as t increases. We define the **turn-on time** (t_{on}) as the time required for $N(t)$ to reach the steady-state value for the first time. A simple estimate of t_{on} can be obtained as follows [46]. Prior to the ILD being turned on, there are very few photons in the active region. Therefore, we ignore $g_1(N(t) - N_{min})\Phi\delta f$, set $J(t) = J_0 u(t)$, and deduce from (4.36)

$$\frac{dN}{dt} = \frac{J_0}{ew} u(t) - \frac{N}{\tau_e}$$

where $u(t)$ is a unit step function. This simple differential equation can be solved, subject to the initial condition $N(0) \approx 0$. The result is

$$N(t) = \frac{\tau_e J_0}{ew}(1 - e^{-t/\tau_e})u(t)$$

Of practical interest is the case in which $J_0 > J_{th}$. Setting $N(t_{on}) \approx N_0$ and using (4.47), we have

$$N_0 \approx \frac{\tau_e}{ew} J_{th} \approx \frac{\tau_e J_0}{ew}(1 - e^{-t_{on}/\tau_e})$$

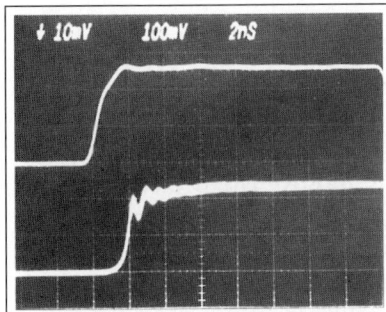

(a) Turn-on of AlGaAs laser, showing current pulse (top) and delayed optical response with damped relaxation oscillations (bottom)

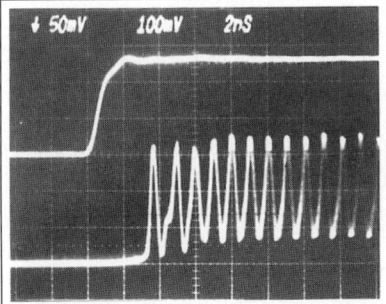

(b) AlGaAs laser, showing self-sustained output oscillations (bottom) from current pulse (top)

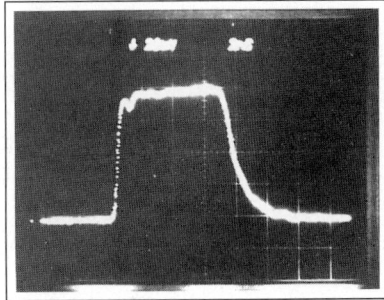

(c) Pulse response of InGaAsP laser, showing slowed turn-off

Figure 4.43 Measured transient response of an ILD with pulsed input [45]

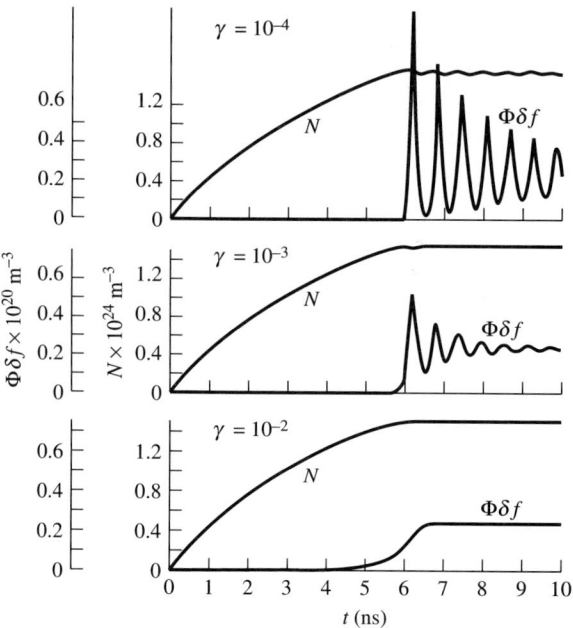

Figure 4.44 Calculated transient response of ILD with pulsed input [46].

From this equation, a simple expression for t_{on} is obtained:

$$t_{on} \approx \tau_e \ln\left(\frac{J_0}{J_0 - J_{th}}\right) \tag{4.48}$$

Figure 4.45 is a plot of t_{on}/τ_e as a function of J_0/J_{th}. It is possible to determine τ_e by measuring the time delay t_{on} as a function of J_0/J_{th}.

4.9.4 Amplitude Modulation

As shown in Figures 4.32 and 4.37, the PI characteristic of ILDs is not a straight line over the entire range of operation. Therefore, if an ILD is driven by a sinusoidal current, the optical output does not have a sinusoidal envelope. However, in the region where $J > J_{th}$, the PI characteristic is approximately linear. If the injection current density consists of a dc component J_0 and a time-varying component $J_1(t)$, corresponding to a dc current I_0 and a time-varying current $I_1(t)$, the light output will have a time-varying component that resembles $I_1(t)$, provided the ILD is not driven below the threshold value nor into saturation. If any part of the driving current is below the threshold value, the optical power output is truncated. The inputs $I_a(t)$ and $I_b(t)$ with resulting outputs $P_a(t)$ and $P_b(t)$ shown in Figure 4.46 are two examples of this relation. Note that the lower portion of $P_a(t)$ and $P_b(t)$ is distorted. On the other hand, the output is amplitude

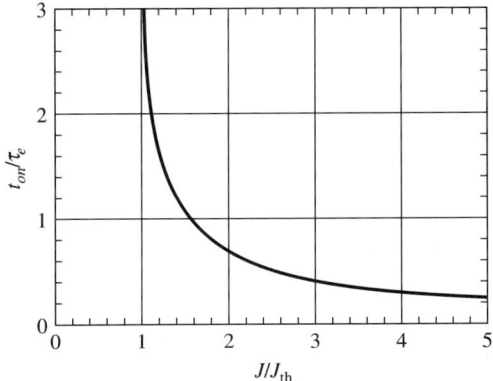

Figure 4.45 Turn-on time as a function of injection current

modulated with little distortion if the diode is biased properly. The input current $I_c(t)$ and optical output $P_c(t)$ shown in Figure 4.46 is an example. To examine the frequency response of the amplitude modulation further, we write the injection current density as

$$J(t) = J_0 + J_1 e^{+j\omega_m t} \tag{4.49}$$

where ω_m is the **angular modulation frequency** and $\omega_m = 2\pi f_m$. We also assume that J_0 is much larger than $|J_1|$, and that J_0 is greater than J_{th}. Under these conditions, $N(t)$ and $\Phi(t)\delta f$ can be expressed in the same form as $J(t)$. Thus, we write

$$N(t) \approx N_0 + N_1 e^{+j\omega_m t} \tag{4.50}$$

$$\Phi(t)\delta f \approx \Phi_0 \, \delta f + \Phi_1 \, \delta f \, e^{+j\omega_m t} \tag{4.51}$$

where N_1 and $\Phi_1 \delta f$ are complex quantities with small amplitudes, that is, $N_0 \gg |N_1|$ and $\Phi_0 \, \delta f \gg |\Phi_1 f|$. To seek an approximate expression, we treat the dc components J_0, N_0, and $\Phi_0 \delta f$ as terms on the order of 1, and the time harmonic terms $|J_1/J_0|$, $|N_1/N_0|$, and $|(\Phi_1 \delta f)/(\Phi_0 \delta f)|$ as terms on the order of $|J_1/J_0|$. Before substituting (4.49)–(4.51) into (4.36) and (4.37), we consider the product term

$$\begin{aligned}[N(t) - N_{\min}]\Phi(t)\delta f &= (N_0 + N_1 e^{+j\omega_m t} - N_{\min})(\Phi_0 \delta f + \Phi_1 \, \delta f \, e^{+j\omega_m t}) \\ &= (N_0 - N_{\min})\Phi_0 \delta f \\ &\quad + [(N_0 - N_{\min})\Phi_1 \, \delta f + \Phi_0 \, \delta f \, N_1] \, e^{+j\omega_m t} \\ &\quad + N_1 \Phi_1 \, \delta f \, e^{+j2\omega_m t}\end{aligned}$$

Since we are looking for an approximate expression, terms on the order of $|J_1/J_0|^2$ or smaller can be neglected. Therefore, we ignore $|(N_1 \Phi_1 \delta f)/(N_0 \Phi_0 \delta f)|$, which is on the order of $|J_1/J_0|^2$, and we obtain

$$(N(t) - N_{min})\Phi(t)\delta f \approx (N_0 - N_{min})\Phi_0 \delta f \qquad (4.52)$$
$$+ \left[(N_0 - N_{min})\Phi_1 \delta f + \Phi_0 \delta f N_1 \right] e^{+j\omega_m t}$$

Now, we substitute (4.49)–(4.52) into (4.36) and (4.37), and we group terms in (4.36) and (4.37) in time independent and time harmonic components. For the time independent terms of these equations, we obtain

$$0 = \frac{J_0}{ew} - g_1 (N_0 - N_{min})\Phi_0 \delta f - \frac{N_0}{\tau_e} \qquad (4.53)$$

$$0 = g_1 (N_0 - N_{min})\Phi_0 \delta f - \frac{\Phi_0 \delta f}{\tau_{ph}} + \gamma \frac{N_0}{\tau_e} \qquad (4.54)$$

Expressions for the time harmonic terms are obtained in the same manner. When the common exponential term $e^{+j\omega_m t}$ is factored out, we obtain

$$j\omega_m N_1 = \frac{J_1}{ew} - g_1 \left[(N_0 - N_{min})\Phi_1 \delta f + N_1 \Phi_0 \delta f \right] - \frac{N_1}{\tau_e} \qquad (4.55)$$

$$j\omega_m \Phi_1 \delta f = \quad + g_1 \left[(N_0 - N_{min})\Phi_1 \delta f + N_1 \Phi_0 \delta f \right] + \gamma \frac{N_1}{\tau_e} - \frac{\Phi_1 \delta f}{\tau_{ph}} \qquad (4.56)$$

Since (4.53) and (4.54) are identical to (4.38) and (4.39), N_0 and $\Phi_0 \delta f$ from (4.40) and (4.41) may be used immediately. Henceforth, we can treat N_0 and $\Phi_0 \delta f$ as known quantities. We then use (4.55) and (4.56) to solve for N_1 and $\Phi_1 \delta f$, as follows:

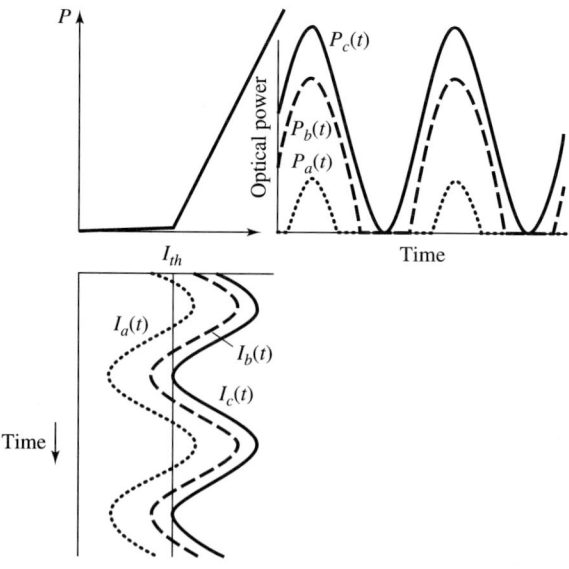

Figure 4.46 Direct amplitude modulation of an ILD

4.9 Dynamic Characteristics: Semiconductor Injection Lasers

$$\Phi_1 \delta f = \frac{J_1}{ew\Delta}(g_1\Phi_0\delta f + \frac{\gamma}{\tau_e}) \qquad (4.57)$$

where

$$\Delta = \left[j\omega_m + \frac{1}{\tau_e} + g_1\Phi_0\delta f\right]\left[j\omega_m + \frac{1}{\tau_{ph}} - g_1(N_0 - N_{min})\right] + g_1(N_0 - N_{min})(g_1\Phi_0\delta f + \frac{\gamma}{\tau_e}) \qquad (4.58)$$

The ratio $\Phi_1\delta f/J_1$ is the frequency response of the direct-amplitude modulation of the ILD output [47]. It is a measure of the effectiveness of the direct modulation by varying the current injection. To appreciate the meaning of (4.57) and (4.58), we examine the special case $\gamma = 0$. For the special case, (4.57) becomes:

$$\frac{\Phi_1\delta f}{\frac{J_1}{ew}} = \frac{\tau_{ph}\,\omega_{mo}^2}{(\omega_{mo}^2 - \omega_m^2) + j\omega_m(\frac{1}{\tau_e} + \tau_{ph}\,\omega_{mo}^2)} \qquad (4.59)$$

where

$$\omega_{mo}^2 = \frac{1}{\tau_e\tau_{ph}}(1 + g_1 N_{min}\tau_{ph})(\frac{J_0}{J_{th}} - 1) \qquad (4.60)$$

In Figure 4.47, the term $|\Phi_1\delta f/(J_1/ew)|$ is plotted as a function of $\omega_m\tau_e$, for $\gamma = 0$. The curve is relatively flat until ω_m approaches ω_{mo}, at which point a sharp peak appears. The curves for $\gamma \neq 0$ have the same general shape, although the peaks are less pronounced. Note that as J increases, or as the ratio τ_e/τ_{ph} increases, the resonance peak ω_m is moved to higher frequencies.

As indicated by (4.60), ω_{mo} depends on $\tau_e\tau_{ph}$ and J_0/J_{th}. Using (4.44) and (4.45), we can write (4.60) as

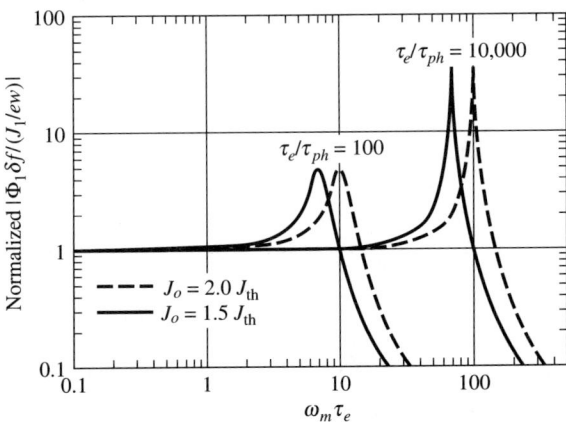

Figure 4.47 Frequency response of a direction modulated GaAs ILD (after [47])

$$\omega_{mo}^2 = \frac{g_1}{ew}(J_0 - J_{th}) = \frac{g_1 \Phi_0 \delta f}{\tau_{ph}} \qquad (4.61)$$

From (4.61), we see that ω_{mo}^2 varies linearly with $\Phi_0 \delta f$, which is proportional to the average optical power output of the ILD, which means that the frequency response of the direct-amplitude modulation of the ILD output is flat from dc to

$$f_{mo} = \frac{\omega_{mo}}{2\pi} = \frac{1}{2\pi}\sqrt{g_1 \frac{\Phi_0 \delta f}{\tau_{ph}}} \qquad (4.62)$$

The term f_{mo} may be viewed as the **modulation bandwidth.** Note that f_{mo} is proportional to the square root of the steady-state optical power output and $1/\tau_{ph}$. Since τ_e and τ_{ph} are on the order of nanoseconds and picoseconds, respectively, f_{mo} is on the order of GHz.

In designing electric circuits to drive ILDs at radio or microwave frequencies, we must look at both the optical and the microwave characteristics of the ILD. So far as the radio or microwave circuitry is concerned, an ILD can be represented by the equivalent circuit shown in Figure 4.38b. Note the presence of junction and parasitic capacitances, the lead inductance, and the contact and internal resistances in the circuits.

4.10 SUPERLUMINESCENT DIODES

The structure of a superluminescent diode (SLD) is very similar to that of an ILD or edge-emitting LED. All lasers have three basic components, an amplifying medium, an optical feedback mechanism, and a means to receive power from external sources. An SLD has no optical feedback, or cavity. This is the main, crucial difference between an SLD and an ILD. To ensure that optical feedback does not occur, two steps may be taken. The front surface is coated with an antireflection material, which may consist of several dielectric layers, such as silicon monoxide. In some SLDs, the back surface is also coated. Alternatively, a lossy section can be built into the diode structure near the back facet, to attenuate reflected waves in this region. Consequently, an SLD may have an amplifying segment and a lossy segment in the longitudinal direction. One way of producing the lossy segment is to remove the electrode near the back facet, as shown in Figure 4.48 [48]. Since there is no metal contact in the rear portion, no current is injected into the region. Alternatively, the region can be made highly resistive by the addition of protons [49].

Because of the anti-reflection coating on the back facet and/or the lossy segment near the rear end of an SLD, optical feedback and stimulated emission are greatly suppressed.

As current is injected into the front portion of an SLD, photons are amplified as they pass through the amplifying section. The radiation emitted by an SLD is therefore **amplified spontaneous emission.** Although the emission is not coherent, its spectral width is much narrower than that of an LED. The radiant intensity pattern of SLDs is comparable to that of edge-emitting LEDs and is narrower than that of surface-emitting LEDs.

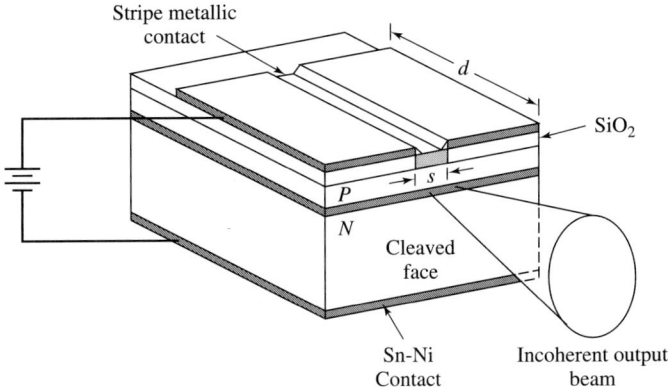

Figure 4.48 Geometry of a superluminescent diode ([48], © 1973 IEEE)

Since the optical amplification occurs only in a *single pass* in the *forward direction,* the realizable gain is limited. A large injected current is required to produce a substantial gain. In general, the current density required for SLD operation is much higher than that required for LEDs or ILDs. This has a significant impact on the reliability of SLDs. Figure 4.32 shows the PI characteristics of an SLD and other semiconductor diodes with the same basic structure. Note the absence of a sharply defined knee in the PI characteristic of an SLD. Nevertheless, SLD power output does vary with the injection current, implying that the optical output can also be modulated directly.

4.11 COMPARISON OF ILDS, LEDS, AND SLDS

An excellent comparison of three semiconductor devices has been given by Kaminow et al. [50]. Their comparison is quoted below:

> An idealized light-emitting diode (LED) emits incoherent spontaneous emission over a wide spectral range into a large solid angle. The unamplified light emerges in one pass from a depth limited by the material absorption. The output is unpolarized and increases linearly with input current. The modulation bandwidth is limited by the spontaneous lifetime.
>
> An idealized laser diode (LD) emits coherent stimulated emission (and negligible spontaneous emission) over a narrow spectral range and solid angle. The light emerges after many passes over an extended length with intermediate partial mirror reflections. The output is usually polarized and increases abruptly at a threshold current that provides just enough stimulated gain to overcome losses along the round-trip path and at the mirrors.
>
> In an idealized superluminescent diode (SLD), the spontaneous emission experiences stimulated gain over an extended path and, possibly, one mirror reflection, but no feedback is provided. The output is incoherent but the stimulated emission narrows the spectral width and solid angle and increases the modulation bandwidth. Thus, the SLD has optical properties, bounded by the LED and LD, that can be adjusted by changing the driving current. The output, which may be polarized, increases superlinearly with current with a knee that occurs near the current providing a significant net positive gain.

An actual SLD will have some feedback because it is not possible to make a perfect antireflection (AR) coating. An actual LED will have some stimulated gain (particularly an edge-emitter) if its length is greater than ~ *1* μm, the typical absorption length. An actual LD has relatively weak feedback from the cleaved mirrors and depends on the amplification of spontaneous emission continuously coupled into the laser mode. Thus, in practice these three devices form a continuum with no sharp boundaries.

For multimode fiber applications, the high output power and coupling efficiency of the LD are attractive but its high coherence can be a source of modal, partition, and feedback noise. On the other hand, the LED has too low output power, too low coupling efficiency, and too low modulation bandwidth for many applications. And in some cases its very wide spectral width may cause material dispersion to limit repeater spacing. By contrast, an SLD can be designed to give a combination of properties tailored to a given system.

It is also instructive to compare LEDs with ILDs from a practical point of view.

1. Since LEDs are relatively simple and the operating current density is relatively low, LEDs can tolerate variations in the fabrication processes. They can operate over a wide range of temperatures, and they require only a simple circuit to maintain a stable output. In contrast, ILDs must be fabricated precisely. Also, the PI characteristic of an ILD changes abruptly beyond the threshold level, and its spectral and spatial characteristics are critically dependent on the operating current and temperature. This means that provisions must be made to maintain a stable operating environment for ILDs.

2. Emissions from LEDs are broadband, unpolarized, incoherent light. The optical output can be amplitude modulated to a few hundred megahertz. In contrast, the output from ILDs is narrow band, polarized, and mainly coherent light. Direct-amplitude modulation of ILDs at 20 GHz has been achieved.

3. Radiation from surface-emitting LEDs is not very well confined spatially, and only a small fraction of the output can be coupled into a multimode fiber. Radiation from edge-emitting LEDs is better confined spatially and spectrally. By comparison, the directional characteristics of conventional ILDs are much better, and coupling from ILDs to fibers can be done more efficiently.

REFERENCES

1. Kittel, C. *Introduction to Solid State Physics* 3rd ed., New York, NY: John Wiley and Sons, Inc., 1968.
2. Pankove, J. I. *Optical Processes in Semiconductors.* New York, NY: Dover Publications, 1971.
3. Sharma, B. L.; and R. K. Purohit. *Semiconductor Heterojunctions.* New York, NY: Pergamon Press, 1980.
4. Blakemore, J. S. "Semiconducting and other major properties of gallium arsenide." *J. Appl. Phys.* 53, (1982), pp. R123–R181.

5. Dalven, R. *Introduction to Applied Solid State Physics.* New York, NY: Plenum Press, 1980.
6. Kressel, H.; and J. K. Butler. *Semiconductor Lasers and Heterojunction LEDs.* New York, NY: Academic Press, 1977.
7. Nuese, C. J.; H. Kressel; and I. Ladany. "The future for LEDs." *IEEE Spectrum* (May 1972), pp. 28–37.
8. Sze, S. M. *Physics of Semiconductor Devices.* 2nd ed. New York, NY: John Wiley and Sons, Inc., 1981.
9. Casey, H. C., Jr.; D. D. Sell; and M. B. Panish. "Refractive index of $Al_xGa_{1-x}As$ between 1.2 and 1.8 eV." *Appl. Phys. Lett.* 24, (1974), pp. 63–65.
10. Botez, D.; and G. J. Herskowitz. "Components for optical communications systems: a review." *Proc. IEEE* 68, (1980), pp. 689–731.
11. Casey, H. C., Jr.; and M. B. Panish. *Heterostructure Lasers.* Parts A and B. New York, NY: Academic Press, 1978.
12. Glass, A. M. "Optical materials." *Science* 235, (1987), pp. 1003–1009.
13. Hasse, M. A.; J. Qui; J. M. DePuydt; and H. Cheng. "Blue-green laser diodes." *Appl. Phys. Lett.* 59, (1991), pp. 1272–1274.
14. Jeon, H.; J. Ding; A. V. Nurmikko; W. Xie; M. Kobayashi; and R. L. Gunshor. "ZnSe based multilayer pn junction as efficient light emitting diodes for display applications." *Appl. Phys. Lett.* 60, (1992), pp. 892–894.
15. Bean, J. C. "Silicon-based semiconductor heterostructures: column IV bandgap engineering." *Proc. IEEE* 80, (1992), pp. 571–587.
16. Rosma, R.; A. Katzir; P. Norton; K. H. Bachem; and H. M. Preier. "On the performance of selenium rich lead-salt heterostructure lasers with remote p-n junction." *IEEE J. Quantum Electron.* QE-23, (1987), pp. 94–102.
17. Suematsu, Y.; and K. I. Iga. *Introduction to Optical Fiber Communications.* New York, NY: John Wiley & Sons, Inc., 1982.
18. Alferov, Z. I.; and R. F. Kazarinov. Author's certificate 1032155/26-25, USSR, (1963), as cited in *Sov. Phys. Solid-State* 9, (1967), pp. 208–210.
19. Kroemer, H. "A proposed class of heterojunction injection lasers." *Proc. IEEE* 51, (1963), pp. 1782–1783.
20. Suematsu, Y. "Long-wavelength optical fiber communication." *Proc. IEEE* 71, (1983), pp. 692–721.
21. Bell, T. E. "Single-frequency semiconductor lasers." *IEEE Spectrum,* (December 1983), pp. 38–45.
22. Burrus, C. A.; T. P. Lee; and A. G. Dentai. "Short-cavity single-mode 1.3 μm InGaAsP lasers with evaporated high-reflectivity mirrors." *Electron. Lett.* 17, (1981), pp. 964–956.
23. Zory, P. S. *Quantum well lasers.* New York, NY: Academic Press, 1993.
24. Chen, H. Z.; A. Ghaffari; H. Morkoc; and A. Yariv. "Effect of substrate tilting on molecular beam epitaxial grown AlGaAs/GaAs lasers having very low threshold current density." *Appl. Phys. Lett.* 51, (1987), pp. 2094–2096.
25. Iga, K.; F. Koyama; and S. Kinoshita. "Surface emitting semiconductor lasers." *IEEE J. Quantum Electron.* QE-24, (1988), pp. 1845–1855.
26. Jewell, J. L.; J. P. Harbison; A. Scherer; Y. H. Lee; and L. T. Florez; "Vertical-cavity

surface-emitting lasers: design, fabrication, characterization." *IEEE Quantum Electron.* 27, (1991), pp. 1332–1346.

27. Nussbaum, A.; and R. A. Phillips. *Contemporary Optics for Scientists and Engineers.* Englewood Cliffs, NJ: Prentice-Hall Inc., 1976.

28. Gooch, C. H. *Injection Electroluminescent Devices.* New York, NY: John Wiley & Sons, Inc., 1973.

29. Carr, W. N. "Photometric figures of merit for semiconductor luminescent sources operating in spontaneous mode." *Infrared Phys.* 6, (1966), pp.1–19.

30. Ettenberg, M.; H Kressel; and J. P. Wittke. "Very high radiance edge-emitting LED." *IEEE J. Quantum Electron.* QE-12, (1976), pp. 360–364.

31. Liu, Y. S.; and D. A. Smith. "The frequency response of an amplitude-modulated GaAs luminescence diode." *Proc. IEEE* 63, (1975), pp. 542–543.

32. Burrus, C. A.; T. P. Lee; and W. S. Holden. "Direct-modulation efficiency of LED's for optical fiber transmission applications." *Proc. IEEE* 63, (1975), pp. 329–330.

33. Lee, T. P.; and A. G. Dentai. "Power and modulation bandwidth of GaAs-AlGaAs high-radiance LED's for optical communication systems." *IEEE J. Quantum Electron.* QE-14, (1978), pp. 150–159.

34. Kressel, H.; M. Ettenberg; J. P. Wittke; and I. Ladany. "Laser diodes and LEDs for fiber optical communication." Chapter 2, *Semiconductor Devices for Optical Communication,* ed. H. Kressel. Berlin, Germany: Springer-Verlag, 1980.

35. Burrus, C. A.; and B. I. Miller. "Small-area, double-heterostructure Aluminum-Gallium Arsenide electroluminescent diode sources for optical-fiber transmission lines." *Opt. Comm.* 4, (1971), pp. 307–309.

36. Ross, D. *Lasers: Light Amplifiers and Oscillators.* Chapter 4. New York, NY: Academic Press, 1969.

37. Pankove, J. I. "Injection Lasers." In *Handbook of Lasers, with Selected Data on Optical Technology,"* ed. R. J. Pressely. CRC Press, (1971), pp. 365–370. Cleveland, OH.

38. Hayashi, I.; M. B. Panish; P. W. Foy; and S. Sumski. "Junction lasers which operate continuously at room temperature." *Appl. Phys. Lett.* 17, (1970), pp. 109–111. Also M. B. Panish and I. Hayashi, "Heterostructure junction lasers," *Applied Solid State Science: Advances in Materials and Device Research,* Vol. 4, edited by R. Wolfe, New York, NY, Academic Press, 1974.

39. D'Asaro, L. A. "Advances in GaAs junction lasers with stripe geometry." *J. Lumin.* 7, (1973), pp. 310–337.

40. Botez, D.; and M. Ettenberg. "Beamwidth approximations for the fundamental mode in symmetric double-heterojunction lasers." *IEEE J. Quantum Electron.* QE-14, (1978), pp. 827–830.

41. Zachos, T. H.; and J. E. Ripper. "Resonant modes of GaAs junction lasers." *IEEE J. Quantum Electron.* QE-5, (1969), pp. 29–37.

42. Adams, M. J. "Rate equations and transient phenomena in semiconductor lasers." *IEE Proc. J. Opto-electron.* 5, (1973), pp. 201–215.

43. Adams, M. J.; and M. Osinski. "Longitudinal mode competition in semiconductor lasers, rate equations revisited." *Proc. IEE* 129, pt. I, (1982), pp. 271–274.

44. Arnold, G.; and P. Russer. "Modulation behavior of semiconductor injection lasers." *Appl. Phys.* 14, (1977), pp. 255–268.

45. Chanin, D. J. "High data rate modulation of laser diodes." In *Fiber Optics for Communications and Control. SPIE Proc.* 224, (1980), pp. 128–132.
46. Boers, P. M.; M. T. Vlaardingerbroek; and M. Danielsen. "Dynamic behavior of semiconductor lasers." *Electron. Lett.* 11, (1975), pp. 206–208.
47. Paoli, T. L.; and J. E. Ripper. "Direct modulation of semiconductor lasers." *Proc. IEEE* 58, (1970), pp. 1457–1465.
48. Lee, T. P.; C. A. Burrus, Jr.; and B. I. Miller. "A stripe-geometry double-heterostructure amplified-spontaneous-emission (superluminescent) diode." *IEEE J. Quantum Electron.* QE-9, (1973), pp. 820–828.
49. Wang, C. S.; W. H. Cheng; C. J. Hwang; W. K. Burns; and R. P. Moeller. "High-power low-divergence superradiance diode." *Appl. Phys. Lett.* 41, (1982), pp. 587–589.
50. Kaminow, I. P.; G. Eisenstein; L. W. Stulz; and A. G. Dentai. "Lateral confinement InGaAsP superluminescent diode at 1.3 μm." *IEEE J. Quantum Electron.* QE-19, (1983), pp. 78–82.

PROBLEMS

1. Show that when the index dispersion of the semiconducting material is taken into account, the spectral spacing of the longitudinal modes of an injection laser is given by (4.7).

2. The spectral characteristic of a certain injection laser has peaks at 0.67970 μm, 0.67985 μm, 0.68000 μm, 0.68015 μm, and 0.68030 μm. These peaks correspond to the wavelengths of various longitudinal modes. Given that the index is 3.7 at the center wavelength of 0.68000 μm and that the laser cavity is 300 μm long, estimate the index dispersion, $dn/d\lambda$, of the semiconductor.

3. The output of a 1-mW source is

$$J(\theta, \phi) = \begin{cases} P_i \cos^i\theta & 0°\leq\theta\leq 90°, \quad 0°\leq\phi\leq 360° \\ 0 & 90°\leq\theta\leq 180°, \quad 0°\leq\phi\leq 360° \end{cases}$$

where $i = 0, 1, 2,$ and 3.
   a. Derive a general expression for P_i as a function of i.
   b. Derive a general expression for the full width between half-intensity points (FWHI). Then calculate the numerical value for the FWHI for the cases of $i = 0, 1, 2,$ and 3.
   c. The source is used to excite a multimode fiber with a numerical aperture of NA. Derive a general expression for the power P_{fb} coupled into the fiber. Then, plot P_{fb} vs. NA for $0 < NA < 0.5$ and $i = 0, 1, 2,$ and 3.

4. The radiant intensity distribution of an LED is

$$J(\theta, \phi) = \begin{cases} P' \cos^3\theta & 0°\leq\theta\leq 45°, \quad 0°\leq\phi\leq 360° \\ 0 & 45°<\theta\leq 180°, \quad 0°\leq\phi\leq 360° \end{cases}$$

where P' is a constant. A power of 0.2 mW is coupled into a multimode fiber with an NA of 0.15. When the fiber is replaced by another fiber with an unknown NA, a power of 0.4 mW is coupled to the unknown fiber. Calculate the NA of the unknown fiber.

5. The radiant intensity distribution of a source is of the form

$$J(\theta, \phi) = \begin{cases} P' \cos^2 \theta & 0° \leq \theta \leq 60° \quad 0° \leq \phi \leq 360° \\ 0 & 60° < \theta \leq 180° \quad 0° \leq \phi \leq 360° \end{cases}$$

a. What is the beam width (full angle between half-intensity points) of this source?
b. The source is used to excite a multimode fiber with an NA of 0.15, and a total power of 0.05 mW is coupled into the fiber. What is the total power radiated by the source?

6. Figure 4.30b depicts an LED with a hemispherical dome, which has a radius R_{dm} and refractive index n_{sc}. Radiation, with radiant intensity J_o, is emitted isotropically in all directions from all points on an active junction area of radius R_{jn}. Show that if $R_{dm} \geq n_{sc} R_{jn}$, then all radiation from the active junction area will emerge from the dome and none will be internally reflected.

7. A flat, undomed LED is immersed in a medium with an index of n and $n < n_{sc}$. Suppose $n_{sc} = 3.6$ and the radiant intensity distribution in the semiconductor is

$$J(\theta_{sc}, \phi_{sc}) = \begin{cases} J_0 & 0° \leq \theta_{sc} \leq 90°, \quad 0° \leq \phi_{sc} \leq 360° \\ 0 & 90° < \theta_{sc} \leq 180°, \quad 0° \leq \phi_{sc} \leq 360° \end{cases}$$

The attenuation in the semiconductor region is negligible. Estimate the minimum value of n if 50 percent or more of the total power is transmitted through the flat interface.

8. Two injection laser diodes (diodes A and B) are made of the same materials and have similar configurations. For both diodes, mirror 1 is perfectly reflecting and mirror 2 has a reflectivity of 0.3. The attenuation constant of the semiconductor is about 15 cm^{-1}. Diode A has an optical cavity volume of $500 \times 30 \times 0.2$ μm^3, while the dimensions of diode B are $400 \times 30 \times 0.2$ μm^3. The PI characteristic of diode A is shown in the figure. Estimate the threshold current (in mA) of diode B.

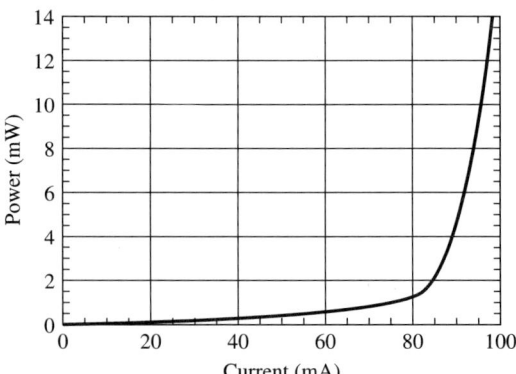

9. The PI characteristic of an injection laser is shown in the figure on the next page. The laser has an optical cavity volume of $500 \times 30 \times 0.2$ μm^3 and emits at a wavelength of 0.9 μm. Other parameters are: attenuation constant $\alpha = 15$ cm^{-1}, mirror reflectivity $\Gamma_1 = \Gamma_2 = 0.35$, and internal resistance 2 Ω. Calculate:

a. The threshold current.
b. The external quantum efficiency.
c. The internal quantum efficiency.
d. The power efficiency at $I = 30$ mA.

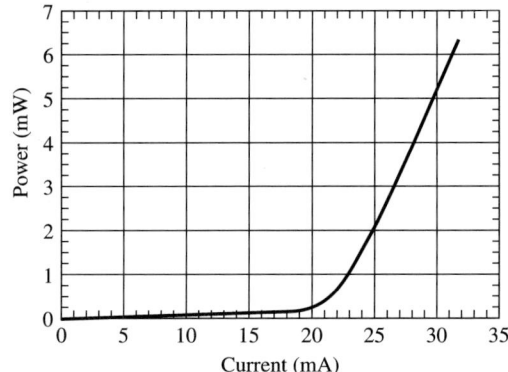

10. The PI characteristic of a fictitious injection laser is shown in the figure. The laser emits at a wavelength of 1.3 μm, and the optical cavity is 400 μm long. The semiconductor has an attenuation constant $\alpha = 12$ cm^{-1}. The internal resistance of the diode is 2.5 Ω. Mirror 1 is perfectly reflecting, that is, $\Gamma_1 = 1.0$, but the reflectivity of mirror 2 is not known. However, it is known that the internal quantum efficiency is 95 percent. Calculate:
    a. The threshold current.
    b. The external quantum efficiency.
    c. The reflectivity of mirror 2.

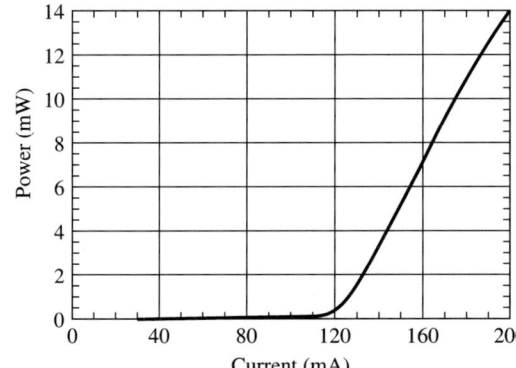

11. The PI characteristic of a fictitious injection laser is shown in the figure for problem 10. Other parameters known include: $d = 400$ μm, $\lambda = 1.2$ μm, $\Gamma_1 = 0.3$, $\Gamma_2 = 1.0$, and $\eta_{int} = 0.5$. Calculate the attenuation constant α of the semiconductor.

12. An injection laser diode has an active junction region of $300 \times 5 \times 0.2$ μm³, and $\tau_e = 3$ ns, $\tau_{ph} = 2$ ps, $g_1 \approx 5 \times 10^{-7}$ cm³ s^{-1}, $\gamma = 0$, and $N_{min} = 5 \times 10^{17}$ cm^{-3}.

a. Calculate the threshold current (in mA) of the diode.
b. The diode is driven with a dc current of 50 mA. Calculate the steady-state electron density and photon density in the pn junction.
c. An RF current of frequency f_m is superimposed onto the dc current of 30 mA. Sketch the response curve, that is, $|ew\Phi_1\delta f/J_1|$ vs. f_m.
d. The response curve has a peak at frequency f_{mo}. What is f_{mo} in GHz?
e. Would f_{mo} be different if $\gamma \approx 10^{-3}$? Why? How about 10^{-2}? Why?

13. The PI and the spectral characteristics of a visible injection laser diode are shown in the figure. The laser diode has a stripe geometry with a length d, width s, and thickness w. The index and the dispersion of the medium is such that

$$n - \lambda \frac{dn}{d\lambda} \approx 4.48$$ at the center wavelength of 683.5 nm. As shown the figure, in the wavelength range between 681 nm and 686 nm there are 25 longitudinal modes. The threshold current of 48 mA, as read from the PI characteristic, corresponds to a threshold current density of 3.2 kA/cm².

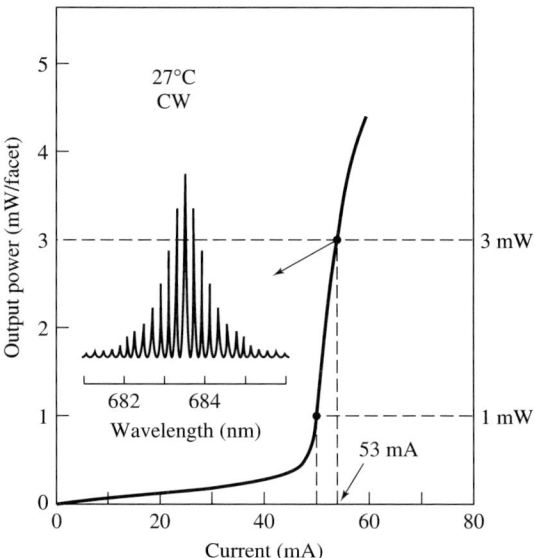

a. Calculate the length d (in μm) of the diode.
b. Calculate the width s (in μm) of the stripe.
c. Calculate the external quantum efficiency.

Clearly, the strengths of the longitudinal modes are different. To simplify the problem in parts d and e, assume that the six longitudinal modes near the center wavelength are of equal strength and each has an output of 0.2 mW.

d. Suppose that six longitudinal modes near the center wavelength are excited and mode locked and all other modes are suppressed. Calculate the peak power of the mode locked pulses.
e. Calculate the full width between half-power points of the pulses.

OPTICAL DETECTION AND DETECTORS

CHAPTER 5

CHAPTER OUTLINE

5.1 Introduction
5.2 Thermal Detectors
    5.2.1 Thermocouples and Thermopiles
    5.2.2 Bolometers and Thermistors
    5.2.3 Golay's Pneumatic Detectors
    5.2.4 Pyroelectric Detectors
5.3 Photon Detectors
    5.3.1 Photoemissive Detectors
    5.3.2 Quantum Efficiency of Semiconductor Detectors
    5.3.3 Photoconductive Detectors
    5.3.4 Photovoltaic Detectors
    5.3.5 Comparison of Optical Detectors
5.4 Noise and Noise Equivalent Power
    5.4.1 Noise Types
    5.4.2 Noise Equivalent Power
5.5 Coherent and Incoherent Detection
    5.5.1 Incoherent Detection
    5.5.2 Coherent Detection
    5.5.3 Detection Method Comparisons
    5.5.4 Stringent Coherent Detection Requirements
5.6 Circuit Topology
References
Additional Reading
Problems

5.1 INTRODUCTION

Optical detection is a process in which optical signals are converted to electrical signals, and photodetectors are the key components of such detection systems. In most applications, photodetectors are used to extract the information contained in the optical beams, as well as to monitor the presence of light. However, no existing detector in the visible and near infrared spectrum is fast enough to follow the instantaneous values of the optical carrier. Instead, the photodetector output is proportional to the mean square of the electric field averaged over several optical periods. Suppose that the input electric field is $\mathbf{E}_s(t)\cos(\omega t + \phi)$, where ϕ is the phase relative to an arbitrary reference frame, and $\mathbf{E}_s(t)$ varies slowly in comparison with the optical carrier $\cos\omega t$. The detector output is proportional to $\frac{1}{2}\mathbf{E}_s(t)\cdot\mathbf{E}_s(t)$ averaged over several optical periods. In other words, it is proportional to the time-average optical power falling onto the photodetector.

To carry information, optical beams are modulated in some manner. We follow the practice of using $f = \omega/(2\pi)$ and $f_m = \omega_m/(2\pi)$ to denote the optical carrier frequency and the modulation frequency, respectively. To detect weak optical signals and to produce an electrical replica of the modulation signals with minimal distortion, a detector must: (a) be *sensitive* to optical emissions of optical angular frequency ω; and (b) have sufficient bandwidth (*speed*) to follow the modulation signals at the modulation angular frequency ω_m. Many physical processes can be, and have been, used to accomplish the photon-to-electron conversion and the information extraction process. The results are multitudes of photodetectors. To characterize and compare various photodetectors, several terms or figures of merit are introduced. They include: the voltage or current responsivity R_v or R_i; the noise equivalent power (NEP); the detectivity D; the normalized detectivity D^* and D^{**}; and the speed of response or the modulation frequency bandwidth [1–6].

The **voltage** or **current responsivity** is the ratio of the electric voltage or current produced by a detector to the optical power P_{in} incident upon the detecting surface of area A,

$$R_v = \frac{V}{P_{in}} \tag{5.1}$$

$$R_i = \frac{I}{P_{in}} \tag{5.2}$$

The dimensions of R_v and R_i are V/W and A/W, respectively. For photodetectors designed for monochromatic radiation, the responsivity is specified in terms of the wavelength λ of the incoming radiation and the modulation frequency f_m. Thus, R_v and R_i can be expressed as $R_v(\lambda, f_m)$ or $R_i(\lambda, f_m)$. For detectors designed for broadband applications, it is more meaningful to use blackbody radiation in the detector specification, since blackbody radiation is a well characterized broadband emission. An examination of **Planck's equation** for blackbody radiation (see equation (3.11)) shows that once the temperature T is specified, the center wavelength and the bandwidth of blackbody radiation are specified. Therefore, the responsivity of broadband detectors may be specified in terms of temperature T of blackbody radiation and can be written as $R_v(T, f_m)$ or $R_i(T, f_m)$.

Depending on the physical processes involved, photodetectors may be classified as thermal detectors or photon detectors. In **thermal detectors,** heat is generated by the impinging radiation, causing any of several parameters to change: the voltage across dissimilar conductors or semiconductors; the resistivity of conductors; the electric polarization of dielectric materials; or the shape of thin membranes. These changes result in electrical signals. In **photon detectors,** electric charge carriers are ejected into a low-pressure or vacuum region, or are generated in semiconductors directly, by the incident photons. When these charge carriers interact with external circuits, electrical signals are generated. Although heat may be produced as a by-product by incoming radiation, it plays no direct role in the emission or generation of charge carriers. A few selected photodetectors are discussed briefly in sections 5.2 and 5.3. Our emphasis is on semiconductor photodetectors.

The two configurations or schemes that make use of photodetectors are the incoherent and coherent detection methods [7,8], shown schematically in Figure 5.1. The **incoherent detection method,** also known as the **direct detection method,** is relatively simple (Figure 5.1a). Monochromatic or broadband radiation falls on a photodetector, and the electric output is fed to an amplifier or other electronic device for further processing. In the **coherent detection method** (Figure 5.1b), monochromatic light of angular frequency ω is combined with fields of

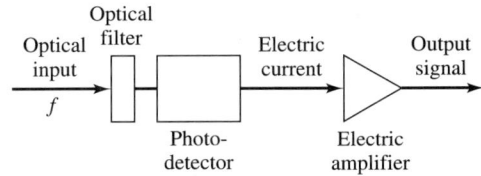

(a) Incoherent detection method

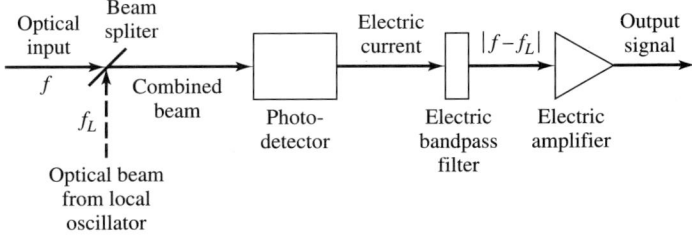

(b) Coherent detection method

Figure 5.1 Comparison of two detection methods

angular frequency ω_L from a coherent source, before reaching the detector. Because the configuration is identical to the setup used in superheterodyne radio or television receivers, the coherent detection method is also referred to as the **optical heterodyne detection method,** and the coherent source of frequency ω_L is known as the **local oscillator.** These terminologies are clearly borrowed from radio and microwave technology.

Photodetectors are square law devices in that they provide an electric replica of the mean square of the incident electric fields averaged over several optical periods. Optical fields of the incoming beam and the local oscillator beam are mixed in the detector, where sum and difference frequency signals are generated. In the photodetector output, signals with angular frequencies of ω, ω_L, 2ω, $2\omega_L$, and $\omega + \omega_L$ are rejected. The dc and difference frequency signals $|\omega - \omega_L|$ are fed into the next stage for further processing. In particular, the difference frequency component is selected by a narrow-band electric filter situated immediately following the photodetector. In this manner the angular frequency ω of the incoming beam is shifted downward to $|\omega - \omega_L|$. Thus, the combination of a photodetector in a coherent detection configuration and a narrow-band electric filter may be viewed as a **frequency translator** or **frequency down convertor.** An additional frequency translation may also take place before the information is extracted.

The ultimate limit of a photodetector's ability to extract information from optical beams is the noise encountered and introduced by the detection system. Noise may originate from the optical transmitters, the environment surrounding the photodetector, the photodetector itself, the amplifiers, or other electronic components used for post-detection processing. Since noise itself cannot be quantified in a deterministic manner we can only specify it in terms of mean noise power, mean square noise current, or mean square noise voltage.

Let the instantaneous signal and noise current at a load resistance R be $I_{sg}(t)$ and $i_n(t)$. The **signal-to-noise power ratio (S/N)** is the ratio of the average signal power to the average noise power at load R,

$$\frac{S}{N} = \frac{\overline{I_{sg}^2}R}{\overline{i_n^2}R} = \frac{\overline{I_{sq}^2}}{\overline{i_n^2}} \qquad (5.3)$$

where $\overline{I_{sg}^2}$ and $\overline{i_n^2}$, respectively, are the mean values of $I_{sq}^2(t)$ and $i_n^2(t)$ averaged over several periods of optical frequency. We know that $\overline{I_{sg}^2}$ is a function of the input optical power. A portion of $\overline{i_n^2}$ also arises from the current produced by the incoming optical signals, and is sometimes referred to as *noise in signal.* This means that part of $\overline{i_n^2}$ also depends on the input optical power. We can solve for the input optical power required to maintain a given *S/N*. However, a detector used in different detection systems will produce different noise. For a particular detection arrangement with a given photodetector, the minimum optical power required to maintain a signal-to-noise power ratio of 1 is defined as the **noise equivalent power (NEP)** which has the dimensions of W/Hz$^{1/2}$. In other words, the *NEP* is the input optical power required to produce the *same* average electric power at the load as the *average* electric power produced by *various noise sources* in the detection system. Furthermore, we specify the mean square noise current as that contained in a modulation frequency bandwidth Δf_m. A modulation frequency bandwidth of 1 Hz is usually understood.

The **detectivity** D is the inverse of the NEP. For many detectors, detectivity varies with detector area A. Therefore, it is convenient to define a term independent of detector area. For this purpose, the **normalized** or **specific detectivity** is introduced:

$$D^* \equiv \frac{\sqrt{A}}{NEP} \tag{5.4}$$

and it has the dimension of (cm Hz$^{1/2}$)/W.

Various types of noise are discussed in section 5.4. In some applications, the dominant noise term is shot noise due to background radiation. Since the background radiation reaching a detector is a function of the field of view of the detector, it is desirable to remove the dependence of D^* on that field of view. For this purpose, another normalization factor is introduced, leading to a different normalized detectivity. The newly defined normalized detectivity is commonly referred to as D^{**}, and is defined as $D^{**} = D^* \sin\Theta$, where Θ is the angle of the field of view. The dimensions for D^{**} is (cm Hz$^{1/2}$ ster$^{1/2}$)/W. For detectors with a hemispherical field of view, $\Theta = \pi/2$, and D^{**} and D^* are identical.

5.2 THERMAL DETECTORS

In thermal detectors, heat generation is an intermediate and necessary step of the detection process. Generally speaking, thermal detectors have a broad spectral range and slow response speed. While they are used extensively in infrared or far infrared detection, their use in high-speed optical communications or data processing applications is limited. There are many types of thermal detectors [1, 4 and 6]. In this section, we will discuss the physical mechanisms of four types. Although detailed characterization of these detectors are omitted, the D^{**} of several thermal detectors are compared with those of photon detectors in Figure 5.15 [4].

5.2.1 Thermocouples and Thermopiles

Whenever junctions of two or more dissimilar materials are kept at different temperatures, an electromotive force is induced. This is a thermodynamic effect known as the **Seebeck effect.** The materials used to form such **thermocouples** may be conductors, alloys, or semiconductors. In a typical arrangement (Figure 5.2a), a voltage is developed across the open-circuit terminals of the thermocouple. If the thermocouple is used as a photodetector, one junction, referred to as the **reference** or **cold junction,** is shielded, while the other, the **active** or **hot**

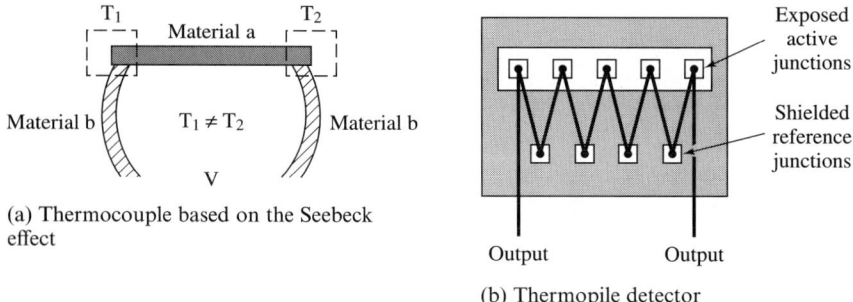

Figure 5.2 Thermocouple and thermopile detector

junction, is exposed to the incoming radiation. With the absorption of incoming radiation, a temperature difference between the two junctions is established, leading to an increase in the open-circuit voltage.

To increase sensitivity, two or more thermocouples may be connected in series such that the voltages due to the Seebeck effect are additive. A **thermopile** is such a series combination of several thermocouples, an example of which is shown in Figure 5.2b [1].

5.2.2 Bolometers and Thermistors

The electrical conductivity of all materials is temperature dependent. To explain, when incoming optical radiation is absorbed by a material, heat is generated, and the temperature rise leads to a variation in the electrical resistance. Various schemes are available to convert the electrical resistance variations to electrical signals. This is the basic operation of bolometers and thermistors. **Bolometers** are made of thin, conducting wires or filaments, typically, platinum or nickel, chosen for their large temperature coefficients. **Thermistors** are semiconducting bolometers.

5.2.3 Golay's Pneumatic Detectors

A Golay's detector is basically a gas-filled chamber defined by two thin membranes, one of which is coated with an absorbing layer and the other of which is flexible and reflecting [1, 9]. Incoming radiation is absorbed by the coated membrane, and heat is generated. As a result of this temperature change, the cell pressure is changed and the flexible membrane is distorted. A known optical source, S_1, shown in Figure 5.3, is on the other side of the flexible membrane. Light from S_1 is reflected by that membrane and diverted to a detector D_1. When the flexible membrane deforms, the amount of light impinging upon D_1 varies, and the change in light illumination is converted to electrical output from D_1. Golay cells were originally designed for IR detection applications. However, their spectral response is quite broad, ranging from ultraviolet to microwave

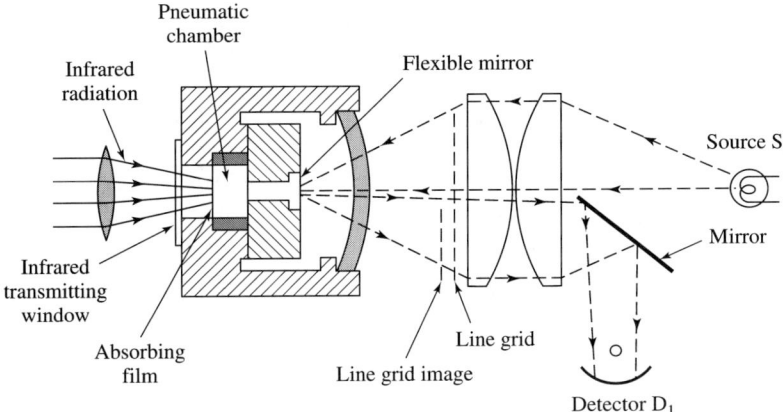

Figure 5.3 Golay's pneumatic detector [9]

frequencies. Therefore, Golay cells may find use in many other applications. Their slow speed of response is the main limitation.

5.2.4 Pyroelectric Detectors

The pyroelectric effect is a process in which a temperature change is converted to a variation of the electric polarization. In simple terms, a pyroelectric detector is simply electrodes deposited on an insulating pyroelectric crystal to form a capacitor. When the pyroelectric crystal is exposed to radiation, heat is produced, which changes the electric polarization of the crystal and thus the capacitance between the electrodes. Electronic methods are then used to monitor the capacitance change.

5.3 PHOTON DETECTORS

The photon-to-electron conversion processes in photon detectors may be classified as photoemissive, photoconductive, and photovoltaic. The photoemissive effect is also known as the **external photoelectric effect,** while the others are known as **internal photoelectric effects.**

5.3.1 Photoemissive Detectors

Vacuum photodiodes and **photomultiplier tubes** are **photoemissive detectors.** The heart of a vacuum tube is the photosensitive cathode. When a photosensitive cathode is illuminated by incoming radiation, electrons may be ejected into the vacuum. In a vacuum photodiode, ejected electrons are accelerated toward and collected by an anode (Figure 5.4). A complete electric path is formed through an external circuit. There is no internal gain.

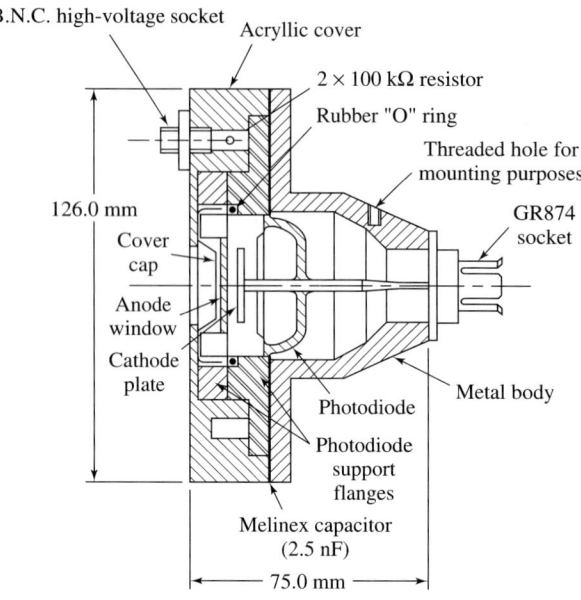
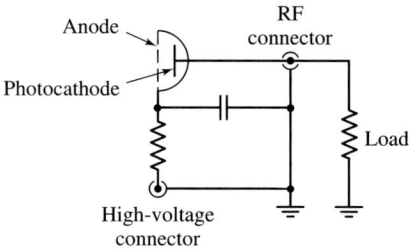

(a) Sectional view of a mounted photodiode

(b) Electrical circuit for a photodiode

Figure 5.4 Photodiode and electrical circuit (EG&G Inc.)

In a photomultiplier tube (PMT), electrons ejected by a photosensitive cathode are accelerated to secondary electrodes known as **dynodes** (Figure 5.5). When impacting electrons have sufficient kinetic energy, additional electrons are released from the dynodes. The newly released electrons are also accelerated by the applied fields to the next dynode, thus generating even more electrons. Finally, the electrons are collected and passed to the external circuits. Because of the electrons emitted by the dynode stages, there are more electrons reaching the external circuit than there are initial electrons emitted by the photocathode. This means there is an **internal current gain** in PMTs. The internal gain is dependent upon the *electron emission efficiency,* the *geometry of the dynodes,* and the *applied voltage.* The presence of this gain is the main advantage of PMTs over vacuum diodes.

The spectral characteristics of photodiodes and PMTs are determined by the spectral sensitivity of the photocathodes, which may be understood qualitatively. To "liberate" electrons from a photocathode, the photon energy must be greater than the work function of the material, where the work function is the energy required to raise an electron from the conduction band to the vacuum level. Therefore, a simple estimate of the spectral response of a photocathode material is possible once the work function of the material is known.

The sensitivity of photocathodes can be expressed in terms of the current responsivity (5.2) or the **quantum efficiency** η. Let the optical power incident upon a photocathode be P_{in}, and let the resulting electric current be I. The

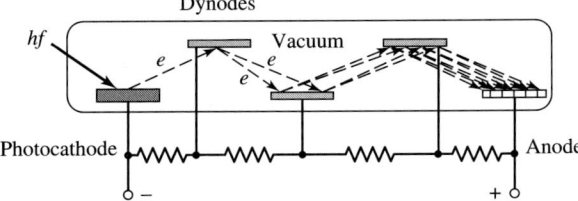

(a) Schematic diagram of a photomultiplier tube

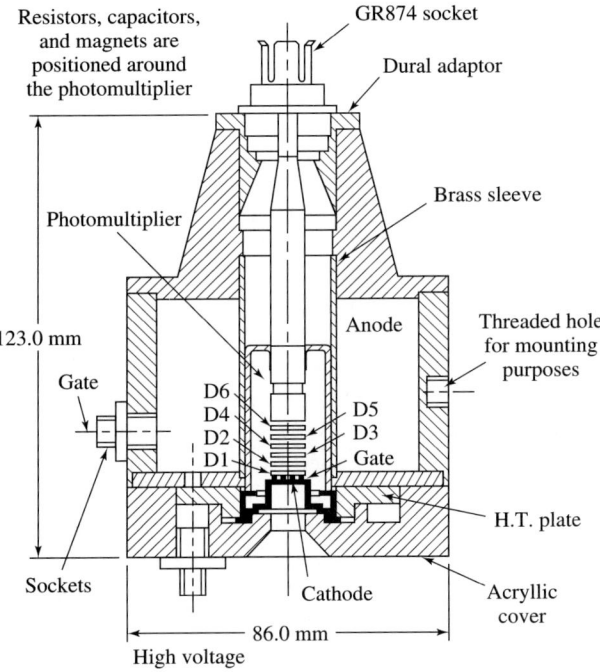

(b) Sectional view of a mounted photomultiplier tube (EG&E Inc.)

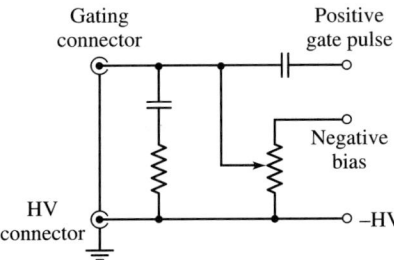

(c) Electrical circuit for a photomultiplier tube (EG&G Inc.)

Figure 5.5 Photomultiplier tube and electrical circuit

number of incoming photons per second is P_{in}/hf, and the number of electrons ejected in 1 second is I/e. In an ideal photocathode, an electron is ejected for each photon absorbed. In real photocathodes, however, fewer electrons are generated. The quantum efficiency then, is defined as the average number of electrons ejected by a photocathode for each photon impinging upon it, as follows:

$$\eta = \frac{\dfrac{I}{e}}{\dfrac{P_{in}}{hf}} = \frac{hf}{e}\frac{I}{P_{in}} \tag{5.5}$$

Comparing the definition of the quantum efficiency with that of the current responsivity introduced in (5.2), we have

$$R_i = \frac{\eta e}{hf}$$

Both R_i and η are dependent on the absorption and reflection of light by the photocathode, and on the probability that photoexcited electrons can escape from the photocathode. Figure 5.6 is a plot of R_i as a function of wavelength for various photocathode materials. Also shown are contours for the constant quantum efficiencies 0.1, 1.0, and 10 percent. These photocathode materials are very sensitive in the near UV and visible portions of the spectrum. Vacuum photodiodes and PMTs are most useful for detecting emissions with wavelengths in the visible region or shorter. Some vacuum diodes are sensitive to radiation as short as 0.1 μm; however, photoemissive detectors are not very useful beyond 1.1 μm. In fact, the quantum efficiencies of all photocathode materials shown in Figure 5.6 are less than 1 percent for $\lambda > 1.0$ μm.

The speed of response of vacuum diodes and PMTs is determined by the time of flight of electrons from the photocathode to the dynodes or the anode, and the spread of electron kinetic energy. Reducing this transit time increases the response speed. In general, the response is limited to a few hundred megahertz. To take advantage of the available bandwidth, the electrodes must be impedance-matched with the rest of the electrical circuitry. For the vacuum photodiodes and PMTs shown in Figures 5.4 and 5.5, the anodes are carefully shaped to impedance-match with the 50 Ω or 125 Ω circuits.

5.3.2 Quantum Efficiency of Semiconductor Detectors

Two essential processes in photoconductive and photovoltaic detectors are the generation and transport of charge carriers in semiconductors. We will review these processes briefly before discussing specific semiconductor detectors.

Although the quantum efficiency defined in (5.5) is presented in terms of electrons ejected by photocathodes, it is also applicable to semiconductor photodetectors, for which the quantum efficiency is also less than 1. Not all incoming optical power is absorbed by the semiconductors. A substantial fraction of

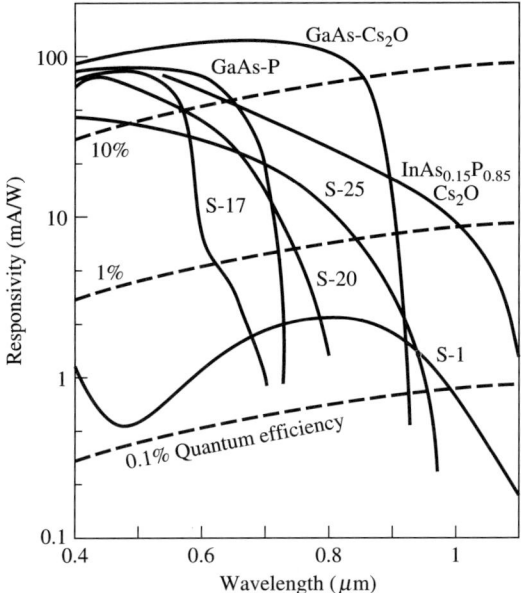

Figure 5.6 Current responsivity of various photocathodes ([19], © 1970 IEEE)

incident light is reflected by the semiconductor/air boundary, unless the sensing surface is antireflection coated. Also, not all photons entering a semiconductor result in the generation of charge carriers. Heat may be produced when radiation is absorbed by the semiconductor, and electron–hole pairs generated by the photons may recombine before reaching the electrodes. Thus, not all charge carriers generated in a semiconductor are collected by the external circuits. In this section, we discuss the factors affecting the quantum efficiency of various semiconductor detectors.

Reflection As shown in Table 4.1 of Chapter 4, the refractive index n_{sc} of semiconductors is much greater than 1.0, and the incident optical beam is partially reflected by the semiconductor/air boundary. As an estimate, consider an optical power P_{in} impinging perpendicular to the semiconductor surface. The power P_{sc} entering the semiconductors is

$$P_{sc} = \frac{4n_{sc}}{(n_{sc} + 1)^2} P_{in} \qquad (5.6)$$

where the factor $4n_{sc}/(n_{sc} + 1)^2$ is the power transmission coefficient. Si has an index of 3.4 in the visible and near IR regions. This means that 29.8 percent of the incident power is reflected by the air/Si boundary; hence, the power transmission coefficient is about 0.70. If the semiconductor surface is coated with a thin layer or layers of dielectric material with an index between 1 and n_{sc}, the power transmission coefficient can be greatly improved. All antireflecting coatings are wavelength dependent [10]. Thus, it is not possible to have a single antireflecting coating covering a very broad band range.

Attenuation Many physical mechanisms contribute to wave attenuation in semiconductors. Light is absorbed if photons are sufficiently energetic to elevate electrons from the valence band to the conduction band. In **band-to-band transitions,** electron–hole pairs are generated. The minimum photon energy required for band-to-band transitions, and the resulting generation of electron–hole pairs, is $hf = E_g$, where E_g is the semiconductor energy bandgap. In extrinsic semiconductors, mobile charge carriers are generated when the photons have sufficient energy to raise electrons from the donor impurity levels to the conduction band, or from the valence band to the acceptor levels. When the impurity levels are ionized, mobile electrons or holes are generated. The interaction of light with **free charge carriers** in semiconductors also contributes to the attenuation of light. However, no electron or hole is generated in the process. The **scattering** of light by defects or inhomogeneities in semiconductors is the fourth mechanism that causes light attenuation. Of the four processes mentioned, mobile charge carriers are generated only in *band-to-band transition* or *impurity ionization* processes. Let the attenuation coefficients due to band-to-band transitions, impurity ionization, interactions with free charge carriers, and scattering be α_b, α_i, α_f, and α_s, respectively. The total attenuation constant is then

$$\alpha = \alpha_b + \alpha_i + \alpha_f + \alpha_s$$

The fraction of photon absorption actually contributing to the generation of charge carriers is $(\alpha_b + \alpha_i)/\alpha$. In the visible and near IR regions, band-to-band transitions are the main photon absorption processes. For longer wavelengths, and in extrinsic semiconductors, the absorption due to impurity ionization becomes more important.

The attenuation α is strongly wavelength and temperature dependent, and increases very sharply as λ decreases, as shown in Figure 5.7 [11]. The absorption of optical power is nearly complete only if the semiconductor is sufficiently thick. If we let the power entering the semiconductor boundary at $x = 0$ be P_{sc}, the optical power elsewhere in the semiconductor is then given by

$$P(x) = P_{sc} e^{-\alpha x} \tag{5.7}$$

Carrier Transport Not all photogenerated charge carriers contribute to a current or voltage change in an external circuit, since many electrons and holes recombine before reaching the terminals. To reach the electrodes, electrons and holes must move under the influence of electric fields, density gradients, or a combination of the two. For simplicity, we consider fields and density variations in the longitudinal direction and ignore motion in the transverse direction. For electrons, the **drift velocity** is $-\mu_e E$ and the **diffusion velocity** is $-D_e \frac{dn}{dx}$, where $\frac{dn}{dx}$ is the electron concentration gradient. Similar expressions may be written

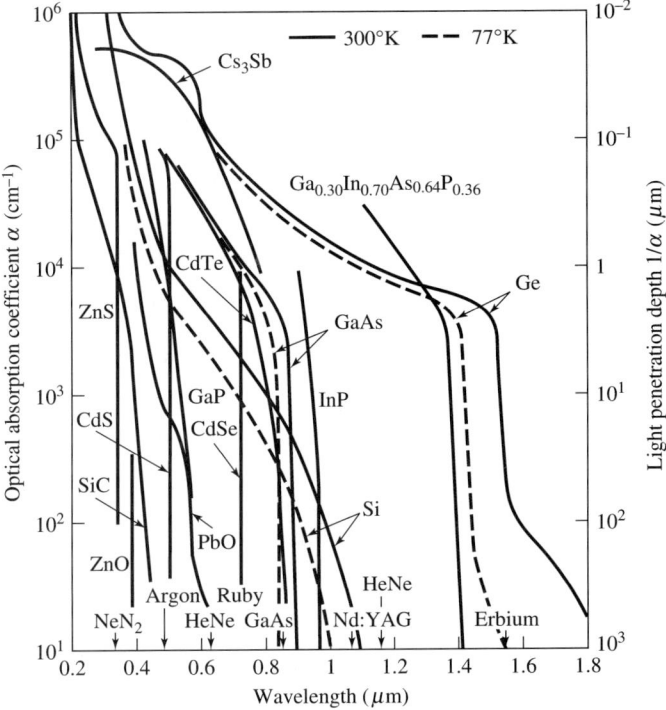

Figure 5.7 Light absorption of various semiconducting materials [11]

for holes. The total electron and hole current densities are composed of drift and diffusion components, as follows:

$$J_e = e(\mu_e n E + D_e \frac{dn}{dx}) \qquad (5.8)$$

$$J_h = e(\mu_h p E - D_h \frac{dp}{dx}) \qquad (5.9)$$

where n and p are the electron and hole densities, μ_e and μ_h are the electron and hole mobilities, and D_e and D_h are the corresponding diffusion coefficients. Numerical values of μ_e and μ_h for several semiconductors can be found in Table 4.1 of Chapter 4.

5.3.3 Photoconductive Detectors

Photoconductive (PC) detectors are simply *junctionless* semiconducting devices. Since the semiconductor conductivity σ varies linearly with the electron and hole densities,

$$\sigma = e(n\mu_e + p\mu_h) \qquad (5.10)$$

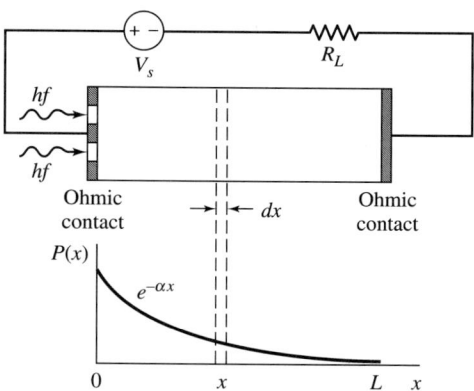

Figure 5.8 Photoconductive detector

σ is changed when n and p are changed by incident light. The conductivity change leads to variations in the conductance or resistance of the semiconductor, which in turn leads to current or voltage variations in the circuit containing the junctionless semiconductor. This is the basic operating principle of PC detectors.

Figure 5.8 shows a semiconductor of length L and cross section A. Consider a thin semiconductor layer of thickness dx located at x. Let the optical power entering and leaving the layer be $P(x)$ and $P(x + dx)$. Then, the optical power difference

$$P(x) - P(x + dx) \approx -\frac{dP}{dx}dx$$

gives the power absorbed by the layer. The number of photons absorbed per second in the semiconductor layer is $-\frac{1}{hf}\frac{dP}{dx}dx$. The total number of charge carriers generated per second in the semiconductor of length L is

$$\int_0^L \frac{\alpha_b + \alpha_i}{\alpha}\left(-\frac{1}{hf}\frac{dP}{dx}\right)dx = \frac{1}{hf}\frac{\alpha_b + \alpha_i}{\alpha}[P(0) - P(L)]$$

In semiconductors, the generation and recombination of electrons and holes takes place simultaneously. However, the lifetime τ_c of charge carriers in the semiconductor is less than 1 second. Thus, in the steady state, the net increase of electrons and holes in the semiconductor is reduced by a factor $\tau_c/1$, and the changes in electron and hole densities are, from (5.6) and (5.7),

$$\delta n = \delta p = \frac{1}{AL}\frac{P_{in}}{hf}\frac{4n_{sc}}{(n_{sc}+1)^2}\frac{\alpha_b + \alpha_i}{\alpha}[1 - e^{-\alpha L}]\tau_c \qquad (5.11)$$

Using (5.10) and (5.11), we obtain the following expression for the conductivity change

$$\delta\sigma = \frac{1}{AL}\frac{e\eta_{pc}}{hf}\tau_c(\mu_e + \mu_h)P_{in} \tag{5.12}$$

where

$$\eta_{pc} = \frac{4n_{sc}}{(n_{sc} + 1)^2}\frac{\alpha_b + \alpha_i}{\alpha}[1 - e^{-\alpha L}] \tag{5.13}$$

is the **quantum efficiency for PC detectors.** Due to the conductivity change, the change in the conductance is $\delta G = \delta\sigma A/L$.

Suppose a constant voltage V is maintained across the PC detector. The current variation due to the conductivity change is

$$\delta I = V\delta G = \frac{VA\delta\sigma}{L}.$$

From (5.12), we obtain

$$\delta I = \frac{V}{L^2}\frac{e\eta_{pc}}{hf}\tau_c(\mu_e + \mu_h)P_{in}$$

For the circuit containing a PC detector, the photon-to-electron conversion efficiency is

$$\frac{\delta I/e}{P_{in}/hf} = \frac{V}{L^2}\eta_{pc}\tau_c(\mu_e + \mu_h) \tag{5.14}$$

Since $\mu_e \gg \mu_h$ for most semiconductors, the contribution due to the hole motion may be ignored. In addition, we note that $E = V/L$, and the electron transit time required to drift through a semiconductor of length L is

$$\tau = \frac{L}{E\mu_e} = \frac{L^2}{V\mu_e}$$

Then, (5.14) can be expressed as

$$\frac{\delta I/e}{P_{in}/hf} = \eta_{pc}\frac{\tau_c}{\tau} \tag{5.15}$$

Thus, we see that the photon-to-electron conversion efficiency of a PC detector, as given in (5.14), is the product of the quantum efficiency η_{pc} and the **photoconductive gain** τ_c/τ. Since the electron transit time τ of PC detectors is very short in comparison with the carrier lifetime τ_c, the photoconductive gain can be substantial.

The current responsivity of a PC detector operating under a constant-voltage condition can be obtained from (5.15):

$$R_i = \frac{\delta I}{P_{in}} = \frac{e\eta_{pc}}{hf}\frac{\tau_c}{\tau} \tag{5.16}$$

For a PC detector driven by a constant-current source I, the change in voltage across the PC detector due to light illumination is

$$\delta V = I\delta R = IR\frac{\delta R}{R} = -V\frac{\delta \sigma}{\sigma}$$

where $R = L/(\sigma A)$ is the resistance of the junctionless semiconductor. From (5.11)–(5.13), we obtain

$$\frac{\delta V}{P_{in}} = -\frac{VR}{L^2}\frac{e\eta_{pc}}{hf}\tau_c(\mu_n + \mu_p)$$

Again, by ignoring the effects due to hole motion, we have the voltage responsivity of a PC detector

$$R_v = \left|\frac{\delta V}{P_{in}}\right| = \frac{e\eta_{pc}}{hf}\frac{\tau_c}{\tau}R \tag{5.17}$$

Note that $R_v = RR_i$.

Since PC detectors are junctionless semiconductors, intrinsic and extrinsic semiconducting materials may be used. For example, $Hg_{1-x}Cd_xTe$ is a good *intrinsic* PC material. It is a II–IV ternary compound semiconductor with a narrow bandgap. The bandgap is a function of the mole fraction x and the operating temperature T. Specifically, E_g is given by

$$E_g = -0.302 + 1.93x - 0.81\ x^2 + 0.832\ x^3 + 5.35 \times 10^{-4}T(1 - 2x)$$

where E_g is in eV, and T is in Kelvin degrees [12]. HgCdTe detectors are designed to operate at cryogenic temperatures (77 °K) and to monitor wavelengths between 3 and 30 μm. The mole fraction x is in the range of $0.18 \leq x \leq 0.4$. Figure 5.9 shows a practical circuit for a commercially available HgCdTe photoconductive detector designed for IR detection [13].

Doped Ge and Si are good *extrinsic* PC materials at long wavelengths. Typical symbols for extrinsic PC materials are Si:XX or Ge:XX, where XX stands for the chemical symbol for the added dopants. Extrinsic PC detectors are also designed to operate at cryogenic temperatures to minimize the thermal ionization of doping impurities. For example, Ge:Ga is useful in detecting 100 μm radiation at 4.2 °K [14].

A photoconductive gain of 50 to 100 and a response time of 1 or 2 ns may be realized by InGaAs PC detectors with a thin absorbing layer and interdigitated electrodes (Figure 5.10a). These PC detectors may be useful for optical communications systems with moderate speeds at the 1.3 or 1.65 μm wavelengths [15, 16].

5.3.4 Photovoltaic Detectors

The main difference between **photovoltaic (PV) detectors** and PC detectors is the presence of a semiconductor junction. The junction may be a homojunction, a heterojunction, a metal-semiconductor junction (such as a Schottky barrier), a point contact junction, or a PIN structure. In most detector applications, the

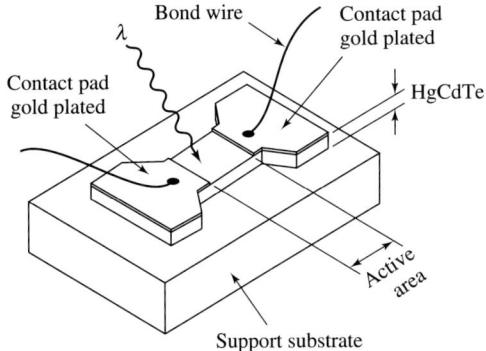

(a) Schematic diagram

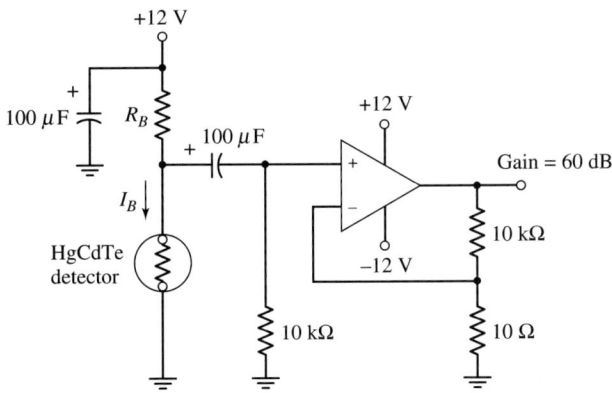

Note: Adjust R_B for proper bias current (I_B)

(b) Electrical circuit

Figure 5.9 Schematic diagram and electrical circuit for a HgCdTe PC detector [13]

semiconductor diode is reverse biased, and the junction current is modulated by the incident optical power. If the bias voltage is less than the breakdown value, there is no charge carrier multiplication, and current gain is not present. This is the usual mode of operation of many semiconductor junction diodes, including PIN diodes. However, **avalanche photodiodes (APD)** are designed to operate near the breakdown voltage. For APDs, the fields in the junction region are strong enough to cause avalanche multiplication, creating internal current gain.

PIN Diodes Figure 5.11 is a schematic diagram of an ideal PIN diode. The p^+ and n^+ regions are heavily doped end regions for making ohmic contact with the external circuit. The p^+ region is very thin, or is constructed with a material transparent to the incoming light. The intrinsic (i) layer is relatively thick. The

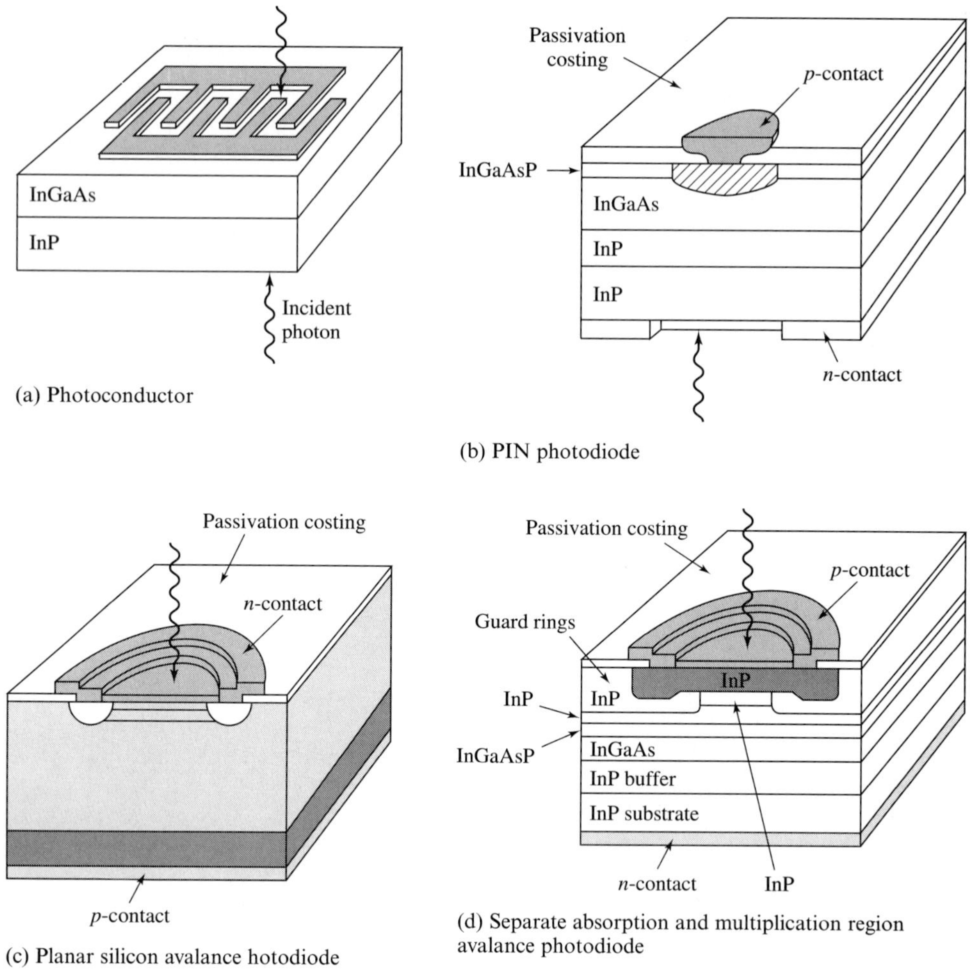

Figure 5.10 Planar, photoconductor, PIN, and APD detectors ([15], © 1986 IEEE)

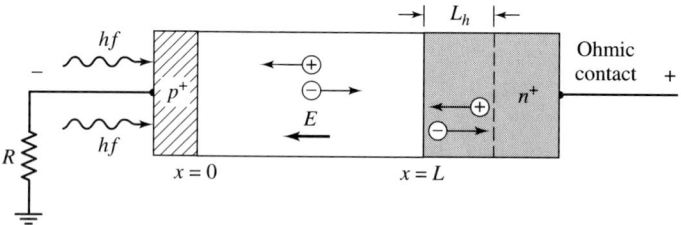

Figure 5.11 An ideal PIN diode

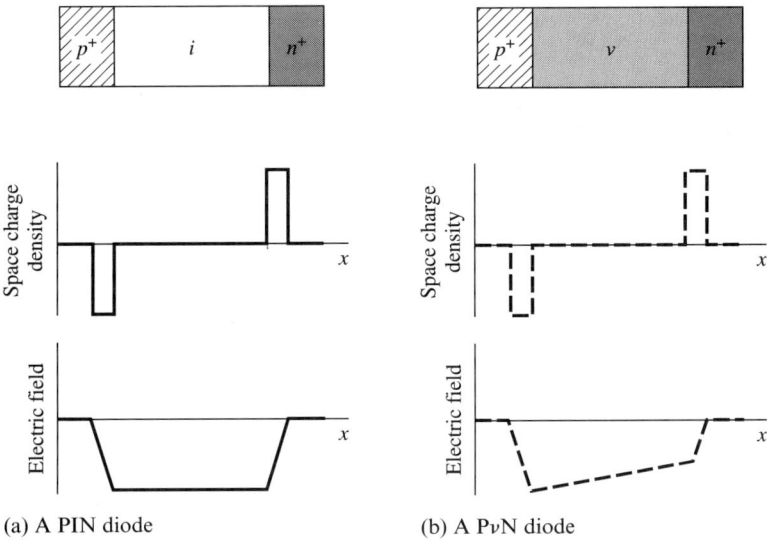

Figure 5.12 Distribution of space charges and electric fields in PIN diodes [17]

thickness of the n^+ region is not important for the elementary discussions presented here.

To discuss the operation of PIN diodes, we consider the ideal PIN diode under the *open-circuit condition*. Since the hole density in the p^+ region is much larger than that in the intrinsic and n^+ regions, holes diffuse from the p^+ region through the intrinsic region to the n^+ region. Electrons diffuse in the opposite direction. With the removal of mobile charge carriers, each region is left with stationary charges. A negative space-charge layer is formed near the p^+–i boundary and a positive space-charge layer is near the i–n^+ boundary. A uniform electric field distribution is established by the space-charge layer. This is shown in Figure 5.12a [17]. The hole and electron diffusion ceases when the electric fields due to the space-charge layers are strong enough to negate the effect of the density gradient. The region extending from the space-charge layer near the p^+–i boundary to that near the i–n^+ boundary is depleted of mobile charges and is therefore referred to as the **depletion region.** When the PIN diode is reverse biased, the depletion region becomes wider and the electric fields stronger.

In a real PIN diode, the central region is not purposely doped. However, impurities are introduced unintentionally during the fabrication process. As a result, the central region is lightly doped. The lightly doped n- or p-type region is labeled ν- or π-type, respectively. Suppose that the central region is lightly doped with n-type impurities (Figure 5.12b). A depletion region then begins at the p^+–ν boundary. Since the p^+ region is heavily doped, the depletion region is mainly in the ν region. When the reverse bias is sufficiently strong, the depletion region may extend from the p^+–ν interface to the ν–n^+ boundary. The corresponding bias voltage is known as the **punchthrough voltage.** Except for the difference in the distribution of fields in the depletion region, the essential characteristics of pνn diodes and ideal PIN diodes are similar.

Although a PIN diode can operate in the open-circuit (zero current), reverse bias, or short-circuit (zero voltage) modes[1], our discussion will focus mainly on the reverse bias mode. Since the p^+ region is thin, or is transparent to the incident light, the motion of charge carriers in the p^+ region can be ignored. In addition, electron-hole pairs generated in the p^+ region quickly recombine and have little effect on the external circuit. Therefore, the characteristics of PIN diodes can be determined mainly by the charge carriers generated in the depletion region and in a thin layer immediately following the depletion region.

In normal PIN diode operation, the electric field in the depletion region is on the order of 10^{+4} V/cm, which is strong enough to drive charge carriers at their saturation velocity, but not strong enough to cause avalanche multiplication. This means that the drift current is the dominant current component. Outside the depletion region, there is no electric field, and the diffusion current is the dominant current component.

To relate the characteristics of PIN diodes with various semiconductor parameters, we return to the ideal PIN diode shown in Figure 5.11. Since the p^+ region is thin, transparent to the incident light, or both, we ignore the attenuation in the p^+ region. Let the x coordinate origin be at the p^+–i boundary, and the thickness of the depletion region be L. The total optical power dissipated in the depletion region is $P_{sc}(1 - e^{-\alpha L})$. Suppose that all photons absorbed in the depletion region lead to the generation of mobile charge carriers; then, the number of electron–hole pairs generated per second in the depletion layer is $P_{sc}(1 - e^{-\alpha L})/hf$. Since the depletion region is only sparsely populated with mobile charge carriers, the probability of electron–hole recombination is very small. In addition, the field in the depletion region is strong, all charge carriers are quickly accelerated to their saturation velocity, and all photoexcited charge carriers are swept away from the depletion layer. As a numerical estimate, we note that the saturation velocity in Si or Ge is on the order of 10^{+5} m/s, and the depletion layer is about 10 μm or less [19]. Consequently, it takes 0.1 ns or less for an electron to traverse the depletion region. Since the lifetime of electrons in the depletion region is much longer than 0.1 ns, all electrons can reach the boundary of the depletion layer before recombining with holes.

At $x = L$, the electron drift current density is

$$J_{dr} = -\frac{e}{hf}\frac{P_{sc}}{A}(1 - e^{-\alpha L}) \tag{5.18}$$

Electrons and holes are also generated in the thin layer immediately adjacent to the depletion region. Since there is no electric field outside the depletion region, charges move under the influence of the density gradient. However, due to the

---

[1] In some literature, the short-circuit mode of operation is also referred to as the "photodiode mode," the open circuit mode as the "photovoltaic mode," and the reverse bias mode as the "photoconductive mode." Since this is confusing, as noted in [18], we will not use this terminology.

presence of a large number of majority charge carriers in the n⁺ region, the lifetime of photoexcited minority charge carriers in that region is extremely short. Only holes generated in a layer within a diffusion length L_h near the depletion region would last long enough to reach the depletion layer. In other words, only charges generated in the region $L < x < L + L_h$ contribute to the current in the external circuit.

Gartner [20] showed that the hole diffusion current density is

$$J_{df} = -\frac{eD_h}{L_h}p_{n0} - \frac{e}{hf}\frac{P_{sc}}{A}\frac{\alpha L_h}{1+\alpha L_h}e^{-\alpha L} \tag{5.19}$$

where p_{n0} is the hole density in the n⁺ semiconductor under thermal equilibrium conditions.

The total current density is the sum of the drift and diffusion current components:

$$J_{tot} = -\frac{eD_h}{L_h}p_{n0} - \frac{e}{hf}\frac{P_{sc}}{A}\left[1 - \frac{1}{1+\alpha L_h}e^{-\alpha L}\right] \tag{5.20}$$

The term $\frac{eD_h}{L_h}p_{n0}$ is independent of the incoming optical power. It is present even in the absence of any incoming radiation, and is known as the **dark current**. In normal detector operation, p_{n0} is usually small. Thus, the dark current term can be neglected. With p_{n0} ignored, the detector current becomes linearly proportional to the incident optical power. We therefore define the **quantum efficiency** for PIN diodes as

$$\eta_{PIN} \approx \frac{4n_{sc}}{(n_{sc}+1)^2}\left[1 - \frac{1}{1+\alpha L_h}e^{-\alpha L}\right] \tag{5.21}$$

The total current of a PIN diode with a cross sectional area A may be written, from (5.6), (5.20), and (5.21), as

$$I_{tot} \approx -\frac{e\eta_{PIN}}{hf}P_{in}$$

By definition, the current responsivity of a reverse biased PIN photodiode is

$$R_i = \left|\frac{I_{tot}}{P_{in}}\right| = \frac{e\eta_{PIN}}{hf} \tag{5.22}$$

The total current density J_{tot} is large if αL is large. This means that the intrinsic region in a PIN photodiode should be wide to increase light absorption in that region. However, L should not be too large, because a long depletion region also means a large transit time, which in turn reduces response speed.

The spectral response of a PIN diode has both a long- and short-wavelength cutoff. Photons are absorbed by a semiconductor if the photon energy is greater than the energy gap. In terms of wavelength, this condition is $\lambda < hc/E_g$. Thus,

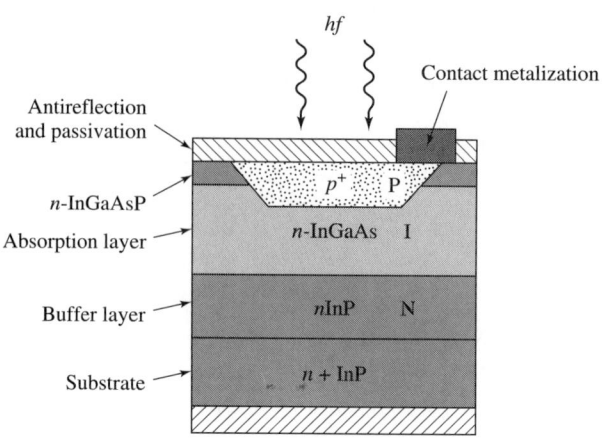

Figure 5.13 PIN diode designed for long-wavelength operation

the **long-wavelength cutoff** in the the spectral response of a PIN diode is at hc/E_g. As the wavelength gets shorter, the attenuation increases, and most of the light is absorbed by the p⁺ region. However, electron–hole pairs generated in p^+ region quickly recombine before reaching the external circuit and serve no useful purpose for light detection. This leads to the **short-wavelength cutoff**.

To consider the speed of response, we recall that the drift current component is important in the depletion region, and the diffusion current component dominants in the thin layer immediately following the depletion region. The speed of response of a PIN detector is affected by: (a) the transit time τ_{dr} of charge carriers drifting across the depletion region; and (b) the time delay τ_{df} associated with the diffusion of charge carriers in a thin layer of thickness L_h immediately outside the depletion region. Another factor is the time required to charge and discharge the junction capacitance. Let τ_{rc} be the RC time constant of the circuit consisting of various capacitors, including the depletion layer capacitance, and the resistors. The total response time of a PIN diode is

$$\tau = [\tau_{dr}^2 + \tau_{df}^2 + \tau_{rc}^2]^{1/2}$$

To reduce τ_{df}, the hole diffusion length L_h in the n^+ region must be kept to a minimum. If we wish to minimize τ_{dr}, many factors must be considered. For example, if we reduce the depletion layer thickness, the absorption of light would be incomplete, quantum efficiency would be reduced, and τ_{rc} would increase. There is a tradeoff between the speed or bandwidth of a PIN diode and its quantum efficiency. An optimal choice for the depletion layer thickness L would allow the carrier transit time in the depletion layer to be half the period of the modulation signal [19, 21].

Most PIN diodes designed for visible or near IR wavelengths are Si diodes. Si PIN diodes are known for their high speed and low noise. At longer wavelengths, such as 1.3 or 1.65 μm, InGaAs PIN diodes may be used (Figures 5.10b and 5.13).

Avalanche Photodiodes The basic structure of an avalanche photodiode (Figure 5.14a) is very similar to that of a PIN photodiode (Figure 5.11). However, APDs are designed to operate *near the reverse breakdown voltage*. The field in the depletion region of an APD is much stronger than that in a PIN diode. Typically, the field is on the order of 10^{+5} V/cm [5]. Because of the strong electric fields, electrons and holes acquire sufficient kinetic energy to ionize additional charge carriers when they collide with the crystal lattice. This process is known as the **impact ionization process**. The newly generated electrons and holes are also accelerated by the fields, and they in turn acquire sufficient kinetic energy to excite additional electrons and holes. This is **avalanche multiplication,** which is the current gain mechanism in APDs. Figure 5.14b schematically depicts the avalanche multiplication process in a semiconductor. Suppose an electron–hole pair is generated at the left boundary of a depletion region by an incoming photon. The electron, labeled e_1, is accelerated to the right by the electric field, while the hole, not shown in Figure 5.14b, is accelerated to the left. If e_1 is sufficiently energetic as it interacts with the crystal lattice, a new electron–hole pair is generated. The newly generated electron and hole are labeled e_2 and h_2. Following the collision, e_1 and e_2 are again accelerated to the right, and h_2 goes to the left. After gaining sufficient energy, each charge carrier again interacts with the lattices and additional charge carriers are produced (e_3 and h_3 by e_1; e_4 and h_4 by e_2; and, e_5 and h_5 by h_2; etc.) As depicted in Figure 5.14b, beginning with one photoexcited electron e_1, four electron–hole pairs have been generated

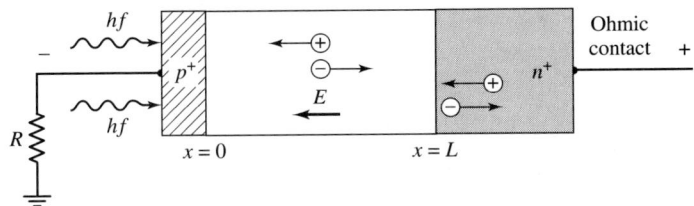

(a) An avalanche photodiode

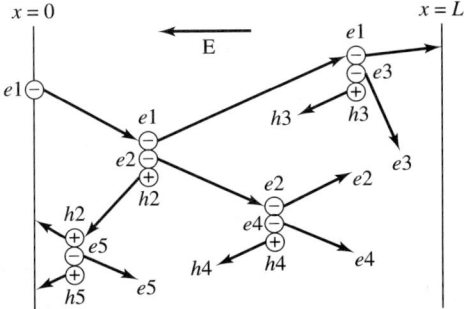

(b) Avalanche multiplication in semiconductors

Figure 5.14 Avalanche photodiode

by the impact ionization process. Since avalanche multiplication processes are not deterministic processes, the number of electron–hole pairs generated is represented by a random variable with certain statistical characteristics. The average value is known as the **avalanche multiplication gain factor, M**.

To estimate M, consider a thin semiconductor layer of thickness dx. The probability of collision and resulting impact ionization occurring in the layer is a function of the electron and hole densities within the layer. The function can be expressed in terms of the electron and hole current densities, which are related directly to external circuit parameters. As electrons move to the right, the electron current density J_e increases. Similarly, as holes move to the left, the hole current density J_h increases. This means that $J_e(x)$ is an increasing function of x and $J_h(x)$ is a decreasing function of x. The rate of increase of $J_e(x)$ is proportional to $J_e(x)$ and $J_h(x)$ [2, 18], as follows:

$$\frac{dJ_e(x)}{dx} = \alpha_e J_e(x) + \alpha_h J_h(x) \tag{5.24}$$

and the rate of decrease of the hole current density is

$$\frac{dJ_h(x)}{dx} = -[\alpha_e J_e(x) + \alpha_h J_h(x)] \tag{5.25}$$

The proportionality constants α_e and α_h are the ionization rates of electrons and holes, respectively. Experimental investigations show that α_e and α_h depend on the strength of the electric fields. As shown in Figure 5.12a, the field in an ideal PIN diode has a uniform distribution in the depletion region. This is also true for the field in an ideal APD. Thus, α_e and α_h are independent of x. Since

$$\frac{dJ_e(x)}{dx} + \frac{dJ_h(x)}{dx} = 0$$

we see that $[J_e(x) + J_h(x)]$ is a constant independent of x. Let the constant be J. In terms of J, (5.24) can be rewritten as

$$\frac{dJ_e(x)}{dx} = (\alpha_e - \alpha_h)J_e(x) + \alpha_h J \tag{5.26}$$

To determine the gain factor M, we solve (5.26), subject to the two initial conditions. First, we assume that electrons are photoexcited at the left boundary of the depletion layer, and we calculate the number of electrons arriving at the right boundary of the depletion region. Next, we assume that light is sufficiently attenuated at $x = L$, and that no holes are generated there by photons. Therefore, the initial conditions are $J_h(L) = 0$ and $J_e(0)$ is finite. However, it is convenient to express $J_e(0)$ and $J_h(L)$ in terms of J. Using the definition of the avalanche multiplication gain factor M, we have $J_e(L) = MJ_e(0)$. Thus at, $x = L$, we have

$$J = J_e(L) + J_h(L) = J_e(L) + 0 = MJ_e(0)$$

In terms of J and M, the two initial conditions are $J_e(0) = J/M$ and $J_e(L) = J$. The solution of (5.26) with constant α_e and α_h is

$$J_e(x) = C_1 e^{(\alpha_e - \alpha_h)x} - \frac{\alpha_h}{\alpha_e - \alpha_h} J \qquad (5.27)$$

The initial conditions mentioned are used to determine the unknowns C_1 and M. By simple substitution, we obtain

$$1 = \left[\left(\frac{1}{M} + \frac{\alpha_h}{\alpha_e - \alpha_h} \right) e^{(\alpha_e - \alpha_h)L} - \frac{\alpha_h}{\alpha_e - \alpha_h} \right] \qquad (5.28)$$

An explicit expression for M is obtained from (5.28)

$$M = \frac{\alpha_e - \alpha_h}{\alpha_e e^{-(\alpha_e - \alpha_h)L} - \alpha_h} \qquad (5.29)$$

If $\alpha_h \sim \alpha_e e^{-(\alpha_e - \alpha_h)L}$, M can be very large indeed. In terms of M and J, the growth of $J_e(x)$ can be written as

$$J_e(x) = J \left[\left(\frac{1}{M} + \frac{\alpha_h}{\alpha_e - \alpha_h} \right) e^{(\alpha_e - \alpha_h)x} - \frac{\alpha_h}{\alpha_e - \alpha_h} \right] \qquad (5.30)$$

As noted earlier, α_e and α_h, and therefore M, depend strongly on the electric field, which is directly related to the bias voltage. To emphasize this point, we consider the case of $\alpha_e = \alpha_h$. As α_h approaches α_e, (5.29) becomes

$$M = \frac{1}{1 - \alpha_e L} \qquad (5.31)$$

To a good approximation, $\alpha_e \sim KV^n$, and (5.31) becomes

$$M = \frac{1}{1 - \left(\dfrac{V}{V_B} \right)^n} \qquad (5.32)$$

From (5.32), we see that M increases sharply as V approaches the **breakdown voltage** V_B, which is given by $V_B = (KL)^{-1/n}$. This demonstrates that M is very sensitive to changes in the bias voltage. In general, M depends on the design of the APD, the bias voltage, and the operating temperature. Typically, M ranges between 10 and 10^{+3}, although multiplication gain factors greater than 10^{+4} have been realized with Si APDs [19].

The speed of response of an APD is limited by the transit time of charge carriers in the depletion layer, and the time needed to build up avalanche multiplication. However, because of the internal gain, the gain–bandwidth product of APDs is quite large. For example, in Si APDs (Figure 5.10c), a gain–bandwidth product of 100 GHz can be expected. For InGaAs APDs, a gain–bandwidth product of 70 GHz has been reported [22].

While the photoexcited current is amplified by the internal gain of an APD, excess noise is also introduced by the avalanche multiplication process. Consequently, the noise characteristics of APDs are more complicated than those of PINs. Fortunately, the signal amplification in APDs is large and a net increase in the signal-to-noise ratio is possible.

5.3.5 Comparison of Optical Detectors

As noted previously, there are many types of detectors with various spectral responses and response speeds. In Figure 5.15, D^{**} of a large number of photodetectors under various operating conditions in the spectral range of 1 to 1000

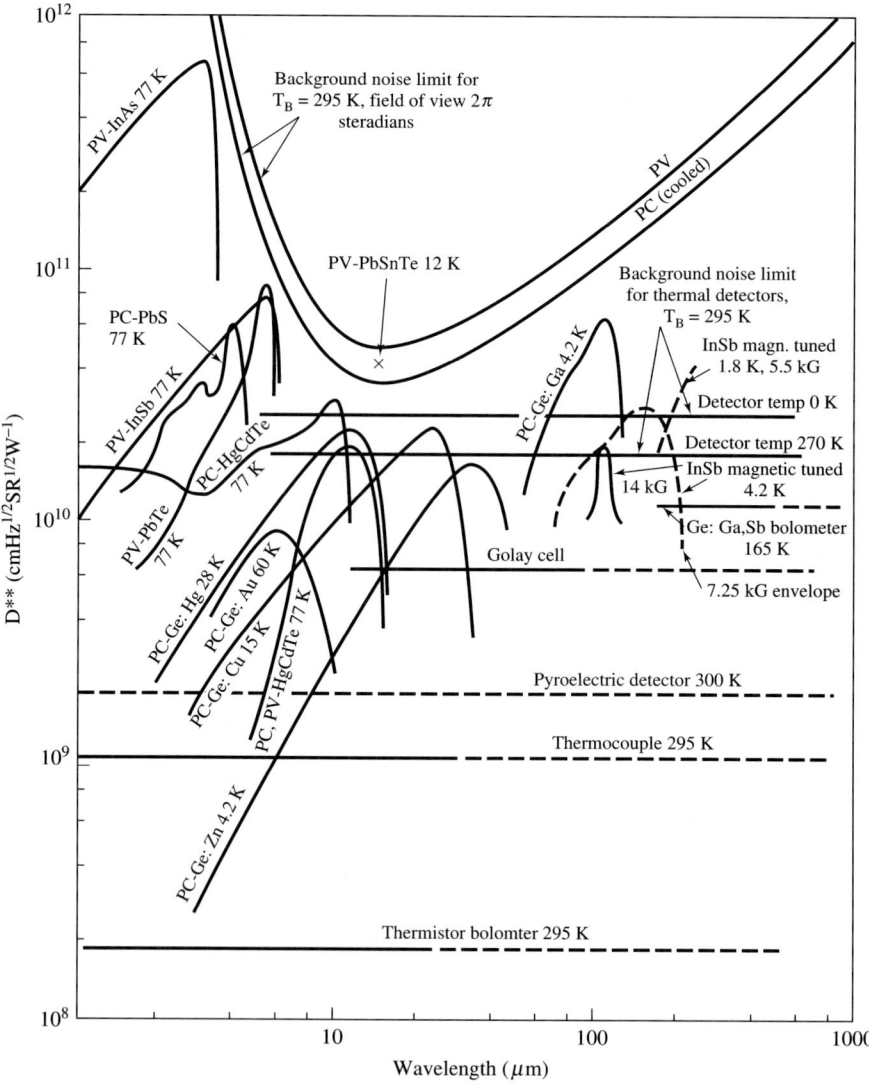

Figure 5.15 Spectral response of various photodetectors [4]

Table 5.1 Typical characteristics of semiconductor photodetectors (excerpt from [23] © 1988 IEEE)

	InGaAs PIN	Ge APD	InGaAs APD	Si APD
Quantum efficiency (%)	80	80	80	80
Bandwidth (GHz)	5	1	3	1
Capacitance (pF)	0.5	0.5	0.5	0.5
Dark current (nA)	10	100	100	10
Gain–bandwidth product (GHz)		10	20	>200

μm are compared [4]. Typical device characteristics of PINs and APDs are listed in Table 5.1 [23].

5.4 NOISE AND NOISE EQUIVALENT POWER

5.4.1 Noise Types

Noise from various origins may be introduced into a detection system. In this section, we will briefly review the types of noise affecting optical detection [1, 4].

Thermal Noise At a finite temperature, all particles, charged or uncharged, move with random speed in random directions. The random thermal motion of charge carriers is the most fundamental noise-generating mechanism in all resistive components, as long as the temperature is above 0 °K. This **thermal noise,** also known as **Johnson–Nyquist noise,** or simply **Johnson noise,** is the minimum noise generated by a resistor. Depending on the resistor's material and construction, additional noise may be introduced by other physical processes. At temperature T, the thermal noise generated by a resistance R in a modulation frequency bandwidth Δf_m may be represented by an open-circuit voltage source v_{th} and a series resistance R, as shown in Figure 5.16a, or by a short-circuit current source i_{th} in parallel with a resistance R (Figure 5.16b). The mean square noise voltage and current are, respectively,

$$\overline{v_{th}^2} = 4kTR\Delta f_m \tag{5.33}$$

$$\overline{i_{th}^2} = \frac{4kT\Delta f_m}{R} \tag{5.34}$$

where k is the Boltzmann constant. The noise power delivered to a load R_L depends on the load resistance. The noise power to the load is maximized when the load resistance matches the internal resistance R. Thus, the maximum mean thermal noise power in a modulation frequency bandwidth Δf_m from a resistor R is

$$\overline{p_{th}} = kT\,\Delta f_m \tag{5.35}$$

This is the **available thermal noise power** from a resistor R (Figure 5.16c) [24]. Note that the available thermal noise power is independent of the resistance value, although it varies linearly with T and Δf_m. It is also important to note that $\overline{p_{th}}$ is independent of the modulation frequency f_m. Noise for which the mean noise power is independent of the modulation frequency f_m is commonly referred to as **white noise,** which means that thermal noise is white noise.

Consider a 50-Ω resistor and a bandwidth of 1 kHz. At a temperature of 300 °K (room temperature), $\overline{i_{th}^2}$ is 3.31×10^{-19} A². The root mean square (rms) thermal current generated by a 50-Ω resistor in a 1-kHz bandwidth at room temperature is 0.57 nA. The rms thermal noise current can be reduced by keeping the resistor at a cryogenic temperature. For example, the rms thermal currents are 0.29 nA and 0.068 nA at 77 °K (the normal boiling point of liquid nitrogen) and 4.2 °K (the normal boiling point of liquid helium), respectively.

Shot Noise In photon detectors, charge carriers are generated by photon arrivals, which in a given time interval, are random and discrete events. Although the *average* number of photons arriving in a finite time interval is known, the *actual* number of photons is not known. The deviation of the actual value from the average is the **shot noise.** Since our interest is in the noise in electrical circuits, it is convenient to think in terms of electrons excited by the incoming photons, instead of the photons themselves. We treat the number of electrons generated in a time interval as a random variable. Assume that the finite time interval is divided into a large number of shorter time intervals $\delta\tau$, and that $\delta\tau$ is so short that there can be at most one electron generated in that time interval. Since an electron cannot be divided further, there may be only one, or no, electron generated in any given $\delta\tau$. Also, the generation of an electron in a given $\delta\tau$ is a random event independent of, and unrelated to, the generation or nongeneration of an electron at any other time interval. The probability of an electron generated in one particular $\delta\tau$ is exactly the same as that in any other $\delta\tau$. Equivalently, we can consider a large time interval ΔT, which is much greater than $\delta\tau$, and examine the number of electrons generated in ΔT. Let n be the number of

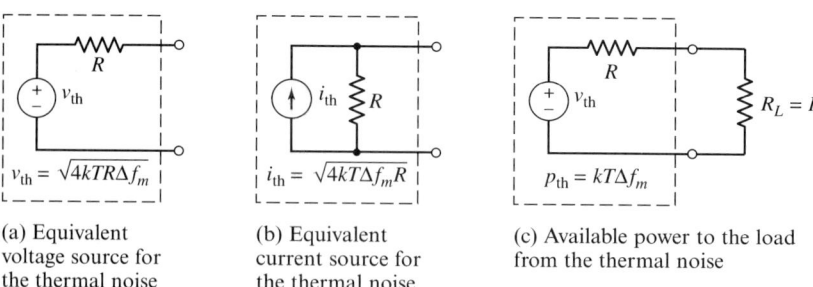

(a) Equivalent voltage source for the thermal noise

(b) Equivalent current source for the thermal noise

(c) Available power to the load from the thermal noise

Figure 5.16 Thermal noise due to a resistance R

photoexcited electrons generated in ΔT, and let $\bar{n}$ be the average number of electrons generated. The probability of n electrons generated in ΔT is characterized by a **Poisson distribution function** as

$$p_n = e^{-\bar{n}} \frac{(\bar{n})^n}{n!} \tag{5.36}$$

It is simple to show (problem 1) that, for random numbers characterized by a Poisson distribution function, the average values of n and n^2 are:

$$\bar{n} = \sum_{n=0}^{\infty} n\, p_n \tag{5.37}$$

$$\overline{n^2} = \sum_{n=0}^{\infty} n^2\, p_n = (\bar{n})^2 + \bar{n} \tag{5.38}$$

A particular feature of the Poisson distribution function is that, for random numbers characterized by (5.36), the mean square deviation of n from $\bar{n}$ equals the average number $\bar{n}$,

$$\overline{(n-\bar{n})^2} = \overline{n^2 - 2n\bar{n} + (\bar{n})^2} = \overline{n^2} - (\bar{n})^2 = \bar{n} \tag{5.39}$$

So far, we have discussed the occurrence of random events in terms of *time domain description*. These random events can also be described in the *frequency domain*. It can be shown that the spectral density of $n - \bar{n}$ is

$$S_n(f) = 2\,\bar{n} \tag{5.40}$$

for n given by a Poisson distribution function. This is known as **Schottky's theorem** and its derivation can be found in reference 25. Note the addition of a factor of 2 in (5.40).

We use (5.36)–(5.40) to determine the mean square noise current due to the random arrival of photons. The instantaneous current due to photoexcited charge carriers generated in a time interval ΔT is $i(t) = ne/\Delta T$, and the time average current is $I_{av} = \bar{n}e/\Delta T$. The shot noise current is the difference between the two, $i(t) - I_{av}$. In the time domain, the mean square shot noise current is

$$\overline{[i(t) - I_{av}]^2} = \overline{\left[\frac{(n-\bar{n})e}{\Delta T}\right]^2} = \frac{\bar{n}e^2}{\Delta T^2} = \frac{I_{av}e}{\Delta T} \tag{5.41}$$

In the frequency domain, the mean square shot noise current in a bandwidth Δf_m is, from (5.40),

$$\overline{i_{sn}^2} = 2e\, I_{av}\, \Delta f_m \tag{5.42}$$

Shot noise can also be expressed in terms of the mean square shot noise voltage across a resistor R, as follows:

$$\overline{v_{sh}^2} = 2eR^2\, I_{av}\, \Delta f_m$$

Also, since shot noise is independent of f_m, it too is white noise.

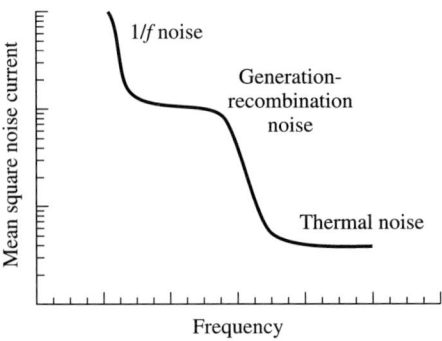

Figure 5.17 Noise in semiconductor detectors [6]

Dark Current In an ideal diode, there is no current when the diode is reverse biased. In a real pn diode, however, the reverse current is not zero. This is due to the leakage current in the reverse bias junction. PIN and APD diodes are no different from other semiconductor diodes in that the reverse current is not zero, even in the absence of incoming illumination. For example, the first term of (5.19) does not depend on P_{in}; there is a finite current in the PIN diode even with zero optical input. The reverse current is commonly referred to as the dark current. Surface states due to surface contamination also contribute to dark current. For photodiodes, the dark current is an important noise term. In APDs, the dark current is also amplified by the avalanche multiplication processes.

As noted earlier, shot noise is due to the occurrence of random discrete events. Dark current, due to the thermal generation of electron-hole pairs, is also a random process. Thus, the mean square noise current due to dark current has the same form as (5.42). Let the average dark current be I_{dk}. The mean square noise current due to the dark current is

$$\overline{i_{dk}^2} = 2e\, I_{dk}\, \Delta f_m \qquad (5.43)$$

Semiconductor Noise Noise is also produced in semiconductors by other processes. For example, noise may be caused by the thermal generation and recombination of charge carriers, as well as the motion of charge carriers. Experiments reveal that the mean square noise current in semiconductors in total darkness may be written as

$$\overline{i_{sc}^2} = \left[\frac{K_c I^\alpha}{f_m^\beta} + \frac{K_{rg} I^2}{1 + (f_m/f_{rg})^2} + \frac{4kT}{R} \right] \Delta f_m \qquad (5.44)$$

The constants K_c, K_{rg}, f_{rg}, α, and β, can be determined empirically. Numerically, α is about 2 and β ranges from 1.0 to 1.5. As depicted schematically in Figure 5.17, the first term of (5.44) is the dominant noise term at low modulation frequencies and is known as **current noise, contact noise, excess noise, modulation** noise, or simply **1/f noise**, even though β is not exactly 1. Noise arising from thermal generation and recombination of charge carriers is represented by the second term of (5.44). This is the major noise term at intermediate modulation frequencies. The constant f_{rg} depends on the host semiconducting material, type of impurities, and impurity concentration. The last term of (5.44) is the thermal noise in semiconductors. While thermal noise is present at all frequencies, it is the dominant noise component at high modulation frequencies. The first two terms of $\overline{i_{sc}^2}$ are not white noise, since they vary with f_m.

5.4.2 Noise Equivalent Power

In an optical detection system, the signal-to-noise power ratio depends not only on the photodetector, but also on the processing electronics, detection method, optical signal intensity, and modulation scheme. Consider a direct detection method based on a detector with a current responsivity R_i. Assume that the incident optical power is amplitude modulated

$$P_{in}(t) = P_0 + P_m \cos\omega_m t \tag{5.45}$$

and $P_0 \geq P_m$. The detector current is then $R_i P_0 + R_i P_m \cos\omega_m t$. The dc or time-average current is $R_i P_0$. The information-carrying part of the output is the sinusoidal current $R_i P_m \cos\omega_m t$. Let R be the effective or equivalent resistance of the detector load. The mean electric power to the load generated by the sinusoidal current is

$$\overline{I_{sg}^2} R = \frac{1}{2}(R_i P_m)^2 R \tag{5.46}$$

As discussed in section 5.4.1, noise is introduced via various mechanisms at various stages of the detection process (Figure 5.18). Let the total mean square noise current be $\overline{i_n^2}$, which includes shot noise, the dark current, thermal noise, and other noise generated by the detector or electronic circuits. For PMTs and APDs, noise arriving at or produced by the front portion of the detector is also amplified by the internal gain. However, for simplicity, assume that the detector has no internal gain. In the case where shot noise is the dominant noise term, the signal-to-noise power ratio is, from (5.3),

$$\left(\frac{S}{N}\right)_{ideal} = \frac{\frac{1}{2}(R_i P_m)^2}{2eR_i P_0 \Delta f_m} = \frac{R_i P_m^2}{4e \Delta f_m P_0} \tag{5.47}$$

The minimum detectable signal power P_m is the optical power required for a signal-to-noise ratio of 1. Thus the minimum P_m is $\sqrt{4e \Delta f_m P_0 / R_i}$. If we assume that $P_0 = P_m$, then the minimum P_m is $4e\Delta f_m/R_i$. By setting the bandwidth to 1 Hz, we have the **noise equivalent power** $NEP = 4e/R_i$.

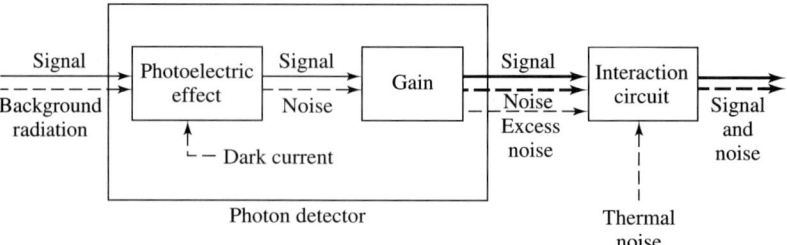

Figure 5.18 Generalized photodetection process with noise (after [2])

5.5 COHERENT AND INCOHERENT DETECTION

In this section, the two basic detection methods, coherent and incoherent, are discussed and compared [7, 8]. Again, we ignore the internal gain of the detector. The results would have to be modified for PMTs and APDs [26].

Assume that the electric field of the incident beam is $\mathbf{E}_s(t)\cos(\omega t + \phi)$, where the amplitude of the field vector $\mathbf{E}_s(t)$ and the phase term ϕ vary slowly in comparison with the optical carrier $\cos\omega t$. The time-average optical power incident upon a detector is proportional to $\frac{1}{2}\mathbf{E}_s(t)\cdot\mathbf{E}_s(t)$. The proportionality constant depends on the detector area and the medium in which the optical waves propagate. We also assume that the incoming optical beam is amplitude modulated, with an angular modulation frequency ω_m, as given in (5.45). Thus,

$$\frac{1}{2}\mathbf{E}_s(t)\cdot\mathbf{E}_s(t) \sim P_{in}(t) = P_0 + P_m\cos\omega_m t \qquad (5.48)$$

Since the detector area is quite large in comparison to the wavelength, $\mathbf{E}_t(t)$ and ϕ may be functions of position. In our present discussions, we will treat them as constants. However, the effects of position on ϕ will be examined in section 5.5.4.

5.5.1 Incoherent Detection

For the incoherent detection method (Figure 5.1a), the input electric field is squared by the detector, as follows:

$$[\mathbf{E}_s(t)\cos(\omega t + \phi)]^2 = \frac{1}{2}[\mathbf{E}_s(t)\cdot\mathbf{E}_s(t)][1 + \cos2(\omega t + \phi)]$$

When averaged over several optical periods, $2\pi/\omega$, this becomes $\frac{1}{2}\mathbf{E}_s(t)\cdot\mathbf{E}_s(t)$.

From (5.48), the detector current is

$$I(t) = R_i(P_0 + P_m\cos\omega_m t)$$

The average value or the dc component of $I(t)$ is $I_{av} = R_i P_0$. The modulation frequency component of the detector current is

5.5 Coherent and Incoherent Detection

$$I_{sg}(t) = R_i P_m \cos\omega_m t \qquad (5.49)$$

The mean square signal current at the modulation frequency is

$$\overline{I_{sg}^2} = \frac{1}{2} R_i^2 P_m^2 \qquad (5.50)$$

The mean square noise current $\overline{i_n^2}$ can contain four important components. They are:

1. Shot noise due to optical signals with an average current $R_i P_0$.

$$\overline{i_{sh\ sg}^2} = 2eR_i P_0 \Delta f_m$$

2. Shot noise due to background radiation P_{bg} impinging upon the detector.

$$\overline{i_{sh\ bg}^2} = 2eR_i P_{bg} \Delta f_m$$

3. Noise due to the detector dark current I_{dk}.

$$\overline{i_{dk}^2} = 2eI_{dk}\Delta f_m$$

4. Other noise generated in the semiconductor $\overline{i_{sc}^2}$.

Thus, the signal-to-noise power ratio is

$$\frac{S}{N} = \frac{\overline{I_{sg}^2}}{\overline{i_{sh\ sg}^2} + \overline{i_{sh\ bg}^2} + \overline{i_{dk}^2} + \overline{i_{sc}^2}} \qquad (5.51)$$

Since $\overline{I_{sg}^2}$ and $\overline{i_{sh\ sg}^2}$ are functions of $P_{in}(t)$, it is possible to solve for the input power required for a given S/N [21]. However, the result is very complicated. Instead, we will take an easier approach and treat three simple cases individually.

In the ideal situation, the incoming signals are sufficiently strong that the shot noise due to the incoming optical signals is much stronger than the other noise terms. By ignoring the other noise terms, we obtain

$$\left(\frac{S}{N}\right)_{dd} \approx \frac{\overline{I_{sg}^2}}{\overline{i_{sh\ sg}^2}} = \frac{\frac{1}{2}|R_i P_m|^2}{2eR_i P_0 \Delta f_m} = \frac{R_i P_m^2}{4e\Delta f_m P_0} \qquad (5.52)$$

This is the maximum signal-to-noise ratio achievable with the given detector in the incoherent detection arrangement, and is commonly known as the **shot noise limited signal-to-noise ratio** [5].

Usually, the input signals are rather weak and the dominant noise term is not $\overline{i_{sh\ sg}^2}$. In the case of low-frequency detection systems, the dominant noise terms may be the shot noise due to the background radiation and the dark current. For these detectors, the signal-to-noise ratio is

$$\left(\frac{S}{N}\right)_{dd\,lf} \approx \frac{\overline{I_{sg}^2}}{\overline{i_{sh\,bg}^2} + \overline{i_{dk}^2}} = \frac{R_i^2 P_m^2}{4e(R_i P_{bg} + I_{dk})\Delta f_m} \qquad (5.53)$$

For optical communications and other high-frequency applications, the thermal noise in semiconductors is the main noise term. Let R_{eq} be the equivalent resistance of the photodetector and its load. Then, $\overline{i_{sc}^2}$ can be written as $4kT\Delta f_m/R_{eq}$. The high frequency (hf) signal-to-noise power ratio can then be approximated as

$$\left(\frac{S}{N}\right)_{dd\,hf} \approx \frac{\overline{I_{sg}^2}}{\overline{i_{sc}^2}} \approx \frac{R_i^2 P_m^2 R_{eq}}{8kT\Delta f_m} \qquad (5.54)$$

As an example, consider a commercially available InGaAs PIN diode with a photosensitive area of 7.85×10^{-5} cm^2, a current responsivity of 0.90 A/W at 1.55 μm, and a dark current of 1.0 nA. Suppose the dark current is the dominate noise term and the signal bandwidths is 1 MHz. Then, the signal-to-noise ratio is, from (5.53),

$$\frac{S}{N} \approx \frac{R_i^2 P_m^2}{4eI_{dk}\Delta f_m} = \frac{(0.9\,P_m)^2}{4 \times 1.6021 \times 10^{-19} \times 1 \times 10^{-9} \times 1 \times 10^6} = 1.26 \times 10^{21}\,P_m^2$$

To achieve an S/N of 1, a power (P_m) of 2.81×10^{-11} W is required.

If the bandwidth is 1 Hz, the power required to achieve signal-to-noise ratio of 1 is

$$2.81 \times 10^{-11} / \sqrt{1 \times 10^6} = 2.81 \times 10^{-14}\,W$$

Thus, for the InGaAs PIN diode operating in the dark current limited condition, the *NEP* is 2.8×10^{-14} W/Hz$^{1/2}$. The detectivity D is

$$D = \frac{1}{NEP} = 3.55 \times 10^{13}\,Hz^{1/2}/W$$

The normalized detectivity D^* is

$$D^* = \frac{\sqrt{A}}{NEP} = 3.15 \times 10^{11}\,cm\,Hz^{1/2}/W$$

where the detector area A is in cm^2.

5.5.2 Coherent Detection

For the coherent detection method (Figure 5.1b), the incoming optical signals are combined with $\mathbf{E_L}\cos \omega_L t$ from a local oscillator before impinging upon the detector. Usually, $|\mathbf{E_L}|$ is much stronger than $|\mathbf{E_s}|$. The local oscillator frequency ω_L is slightly different from the signal frequency ω, and the difference frequency $|\omega - \omega_L|$ is much greater than ω_m; that is,

$$\omega_m << |\omega - \omega_L| << \omega,\,\omega_L. \qquad (5.55)$$

Typically, $|\omega - \omega_L|$ is in the radio and microwave frequency ranges.
The detector current is proportional to the time average of

$$[\mathbf{E}_s(t)\cos(\omega t + \phi) + \mathbf{E}_L \cos\omega_L t]^2$$
$$= \frac{1}{2}[\mathbf{E}_s(t)\cdot\mathbf{E}_s(t) + \mathbf{E}_L \cdot \mathbf{E}_L] + \mathbf{E}_s(t)\cdot\mathbf{E}_L \cos[(\omega - \omega_L)t + \phi]$$
$$+ \frac{1}{2}[\mathbf{E}_s(t)\cdot\mathbf{E}_s(t)\cos 2(\omega t + \phi) + \mathbf{E}_L^2 \cos 2\omega_L t] + \mathbf{E}_s(t)\cdot\mathbf{E}_L \cos[(\omega + \omega_L)t + \phi]$$

When averaged over long intervals compared to the period of the optical signals, second harmonics and the sum frequency terms drop out. The terms left are then

$$\frac{1}{2}[\mathbf{E}_s(t)\cdot\mathbf{E}_s(t) + \mathbf{E}_L \cdot \mathbf{E}_L] + \mathbf{E}_s(t)\cdot\mathbf{E}_L \cos[(\omega - \omega_L)t + \phi] \qquad (5.56)$$

The first term of (5.56) may be viewed as the dc or low-frequency component. The second term is the radio and microwave frequency component. Because of the presence of the scalar product of two field vectors, the second term in (5.56) is maximized if the two fields have the same state of polarization. For simplicity, assume that $\mathbf{E}_s$ and $\mathbf{E}_L$ are linearly polarized fields and θ is the angle between them. By writing $\mathbf{E}_s \cdot \mathbf{E}_L = |\mathbf{E}_s||\mathbf{E}_L|\cos\theta$, we obtain an expression for the detector current:

$$I(t) = R_i\{[P_{in}(t) + P_L] + 2\sqrt{P_{in}(t)\,P_L}\,\cos\theta\,\cos[(\omega - \omega_L)t + \phi]\} \qquad (5.57)$$

With $P_{in}(t)$ given by (5.45), the average detector current is $I_{av} = R_i[P_0 + P_L]$ and the radio and microwave frequency component is

$$2R_i\sqrt{P_{in}(t)\,P_L}\,\cos\theta\,\cos[(\omega - \omega_L)t + \phi]$$

Since P_L is much stronger than $P_{in}(t)$, the shot noise due to P_L is strong enough to overwhelm the other noise terms. The mean square shot noise current is then approximately

$$\overline{i_{sh}^2} = 2e\,I_{av}\,\Delta f_m \approx 2e\,R_i P_L\,\Delta f_m$$

The mean square of the radio and microwave frequency current component of (5.57) is

$$\frac{1}{2}\{R_i[2\sqrt{P_m P_L}\cos\theta]\}^2 = 2R_i^2\,(P_0 + P_m\cos\omega_m t)\,P_L\cos^2\theta$$

With the help of an electrical voltage source as the second local oscillator, the modulation frequency component $2R_i^2\,P_m P_L \cos^2\theta\,\cos\omega_m t$ can be extracted. This is the signal current, and the mean square signal current is

$$\overline{I_{sg}^2} = 2R_i^2\,P_m P_L\,\cos^2\theta \qquad (5.58)$$

The signal-to-noise power ratio is

$$\left(\frac{S}{N}\right)_{coh} \approx \frac{\overline{i_{sg}^2}}{\overline{i_{sh}^2}} \approx \frac{2R_i^2\, P_m P_L \cos^2\theta}{2e\, R_i P_L\, \Delta f_m} = \frac{R_i\, P_m \cos^2\theta}{e\Delta f_m} \qquad (5.59)$$

Note that in the coherent detection method, the signal-to-noise ratio is independent of local oscillator power P_L, as long as P_L is sufficiently strong. Precisely because the local oscillator power can be very strong, the shot noise limit is easily attainable. This is a special feature of the coherent detection arrangement.

The coherent detection method just described is also referred to as the **optical heterodyne detection method**. The special case in which $\omega_L = \omega$ is known as the **optical homodyne detection method.** It can be shown that the signal-to-noise ratio of optical homodyne detection may be better than that of optical heterodyne detection, by 3 dB [27].

5.5.3 Detection Method Comparisons

Several comparisons can be drawn between the two detection methods. While most comparisons are qualitative, two are quantitative: the signal power available to the load, and the signal-to-noise ratio.

The *setup* for the incoherent detection scheme is simple and requires only minimal alignment. In contrast, the coherent detection arrangement is very delicate and careful alignment is required to achieve optimal sensitivity. Coherent detection requires stable coherent sources to serve as the oscillators at the transmitter end and as the local oscillator at the receiving end. In addition, it requires high-speed photodetectors that can respond to signals of frequency $|\omega - \omega_L|$. In contrast, slower photodetectors may be used in the direct detection setup, since it need only respond to the modulation frequency ω_m, which may be much lower than $|\omega - \omega_L|$, as indicated in (5.55).

From (5.50) and (5.58), we see that the *electrical signal power* delivered to the load in a coherent detection arrangement is greater than that in an incoherent detection arrangement, by a factor of $4(P_L/P_m)\cos^2\theta$. Since P_L can be and usually is much stronger than P_m, the increase in electrical signal power can be substantial. Thus, the signal beam is effectively "amplified" by the local oscillator beam, even though the photodetector itself has no internal gain. The factor $4P_L/P_m$ may be viewed as an **effective gain.** Since no internal gain is required in the photodetector, detectors may be selected based mainly on bandwidth considerations.

As noted previously, if a strong local oscillator source is used, the *signal-to-noise ratio* of a coherent detection scheme can easily approach the shot noise limit. In contrast, when the incoming optical signals are very weak, the signal-to-noise ratio in an incoherent detection scheme is much less than the shot noise limit. Usually, the S/N of a direct detection scheme is determined by the dark current and other noise terms attributable to the electronic circuits and background radiation. However, even when the signal-to-noise ratio approaches the shot noise limit, the *S/N* of the direct detection method is smaller than that of the coherent detection method, by a factor of $4(P_0/P_m)\cos^2\theta$. In the limit of $P_0 = P_m$, the factor is $4\cos^2\theta$.

It is often desirable to use narrow-band filters to reduce background radiation, or to confine inputs to a narrow spectral range, so that a large number of communications channels can be used by the same transmission media. In coherent detection systems, electric filters may be used in the post-detection circuits. With a tunable ω_L, we have an effective, tunable, narrow-band electrical frequency discriminator. In direct detection systems, narrow-band optical filters can be inserted in front of the detector, as shown in Figure 5.1a. However, the bandwidth of a narrow-band optical filter is typically on the order of 100 GHz, which is several orders of magnitude broader than the bandwidth of electric filters. This means that more *information channels* are available in coherent detection systems than in incoherent detection systems. In addition, it is difficult to tune the pass band of optical fibers.

In incoherent detection systems, *phase information* and *polarization state* of the optical signals are lost in the detection process. In coherent detection systems, the polarization state of the incoming signals can be inferred by noting the $\cos^2\theta$ term in (5.58). The phase information may be extracted if the detector current given in (5.57) is mixed with signals from an electric source to generate an intermediate frequency component whose intermediate frequency voltage can then be processed electronically. It should be noted, however, that the presence of the phase and polarization terms also causes complications, as explained in the next subsection.

5.5.4 Stringent Coherent Detection Requirements

While the coherent detection method is advantageous in many respects, these advantages are realizable only if stringent requirements are met [7]. The requirements have to do with the active area of the photodetector, which is much larger than the optical wavelength, and the matching of the phase and polarization state of the signal and local oscillator beams. Since the detector area is large, the phase and polarization state of the beams may vary over the detection surface. Typically, the active aperture of a PIN diode or an APD is on the order of $100\,\lambda$ in diameter. The sensing areas of photodiodes or PMTs are even larger.

To estimate the effects of phase and polarization variations over the detector area, we envision that a detector is divided into small sensing elements and each sensing element is small enough that the phase and polarization state remain the same throughout the elemental sensing area. We also assume that the current responsivity is the same for all elemental sensing areas. Then, the total detector response is the superposition of the responses of all sensing elements.

Beam Diameters Figure 5.19 shows the cross sectional areas A_S and A_L of the signal and local oscillator beams, respectively. The shot noise, being proportional to the average current produced by the signal beam and the local oscillator beam, is determined by the *union* of A_S and A_L. On the other hand, the mean

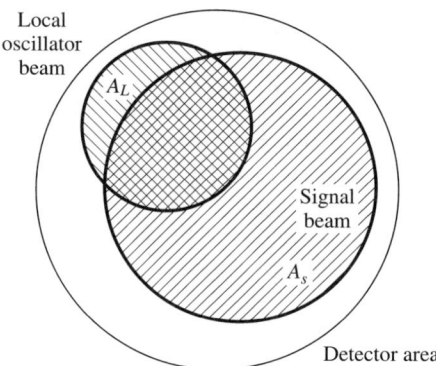

Figure 5.19 Superposition of signal and local oscillator beams

square signal current, being the result of two-beam mixing, is proportional to the *intersection* of A_S and A_L. In Figure 5.19, the overlapping area is shown as the cross shaded area. Maximizing the S/N requires optimizing the overlapped area without changing A_S and A_L. This is accomplished when A_S and A_L coincide completely at the active detector surface [7].

States of Polarization We recall the presence of the $\cos\theta$ term in (5.57) and the $\cos^2\theta$ term in (5.58). To optimize the sensitivity of the coherent detection method, θ must be zero. In other words, signal and local oscillator field vectors must be oriented in the same direction, if they are linearly polarized fields. In general, the two beams must have identical states of polarization.

Propagation Directions Thus far, we have assumed that the two beams are uniform plane waves incident normal to the detector surface. However, if the two beams propagate in different directions, it would be impossible to maintain a constant phase relationship throughout the entire sensing area. To estimate the effect of different propagation directions, we consider a photodetector with a rectangular sensitive area of width $2a$ and length $2b$. We then assume that the local oscillator beam is incident normal to the detector surface, while the signal beam impinges on the detector at a small angle ζ, as shown in Figure 5.20. We also assume that the two plane waves are in phase at the center of the detecting surface. At a distance x from the center, the phase difference is $\phi(x) = kx \sin\zeta$. Since ζ is small, $\phi(x)$ may be approximated as $k\zeta x$. Substituting $\phi(x)$ for ϕ in (5.56) or (5.57) and integrating over the detector area, we find that the total current generated by a detector with a rectangular aperture is proportional to

$$\int_{-b}^{b}\int_{-a}^{a} \cos[(\omega - \omega_L)t + \phi(x)]\, dx dy \approx 2b \int_{-a}^{a} \cos[(\omega - \omega_L)t + k\zeta x]\, dx$$
$$= [4ab \cos(\omega - \omega_L)t]\, \frac{\sin k\zeta a}{k\zeta a} \tag{5.60}$$

5.5 Coherent and Incoherent Detection

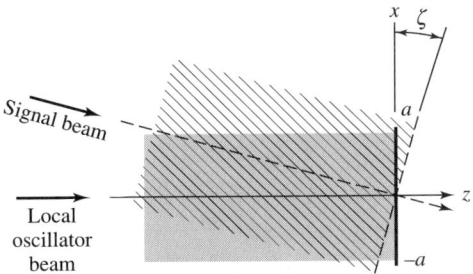

Figure 5.20 Two plane waves propagating in different directions

If the two plane beams propagate in the same direction and are in phase throughout the detector surface of area $4ab$, the detector current in the coherent detection arrangement is proportional to $4ab\cos(\omega - \omega_L)t$. However, if the two beams propagate in slightly different directions, the detection current is degraded by a factor of $\sin k\zeta a/(k\zeta a)$, due to the presence of the phase term in (5.56) and (5.57), [7]. Note in particular that the coherent detection response degrades as $\sin k\zeta a/(k\zeta a)$, and vanishes completely as $k\zeta a$ approaches multiples of π. Numerical calculations show that $\sin k\zeta a/(k\zeta a)$ reduces to 0.5 when $k\zeta a$ is about 1.89 rad. Thus, the detector current reduces to half its maximum value when the misalignment ζ is $0.294\lambda/a$, which is very small when the detector area is much larger than the wavelength.

As an example, consider a photodetector for monitoring radiation at 0.9 μm. Suppose that the detector width is 100 μm, that is, $a = 50$ μm. The misalignment ζ must be within 5.4×10^{-3} rad, i.e., 0.31 degrees, if the response is to be better than half of the maximum value. Similar sinc function type degradation can be expected for detectors with circular detector areas.

Mode Structures and Wavefront Curvature In the previous discussions, the two beams are treated as uniform plane waves. However, it would be more accurate to treat the two unguided beams as Gaussian beams. In addition to the fundamental Gaussian modes, there are higher-order Gaussian modes, as discussed in Chapter 2. We will show in Chapters 7 and 8 that there are also fundamental and higher-order modes for waves guided by thin-film waveguides and optical fibers. As an example, suppose that the signal beam is a fundamental Gaussian beam, while the local oscillator beam is a TEM_{10} or TEM_{01} mode in cartesian coordinates (See Figure 2.21 Chapter 2). In the left (or top) portion of the overlapping area, fields of the local oscillator beam would be in phase with fields of the signal beam, while in the right (or bottom) portion, the fields would be out of phase by 180 degrees. Therefore, the total detector response vanishes when summed over the entire detecting surface. In other words, maximum detector response is obtained only if the two beams have similar mode structures.

Also, if the two beams are treated as Gaussian beams, the effect of the wavefront curvature should be taken into account. If the wavefront curvatures are different, it would again not be possible to maintain a constant phase relationship across the detector surface. The result again leads to degradation in detection sensitivity [7].

5.6 CIRCUIT TOPOLOGY

Since the electrical signals generated by photodetectors are weak, further amplification is often necessary. Preamplifiers or "front ends" of optical receivers are designed to increase the electrical signal level, hopefully without introducing excessive distortion and noise. Since the photodetectors are connected directly to the preamplifiers, the design for the preamplifiers must take into account the circuit characteristics of the photodetectors. In this sense, the designs may be viewed as "unique." Once the signal level is increased to the desired level, the rest of the optical receiving system is the same as the usual electronics.

There are three basic circuit designs for the front ends configurations: **low impedance (LZ), high impedance (HZ),** and **transimpedance (TZ)** [28]. The circuit topologies for LZ and HZ preamplifiers are the same (Figure 5.21a). The difference is the *impedance level.* The photodetectors connected to the preamplifiers may be PINs or APDs. For simplicity, we will consider PIN diodes. A reverse biased PIN diode may be represented by an ideal current source i_s, a capacitance C_d (on the order of a few pF), a shunt resistance R_d (a few hundred MΩ), and a series resistance R_s (a few Ω). In actual circuits, there may also be

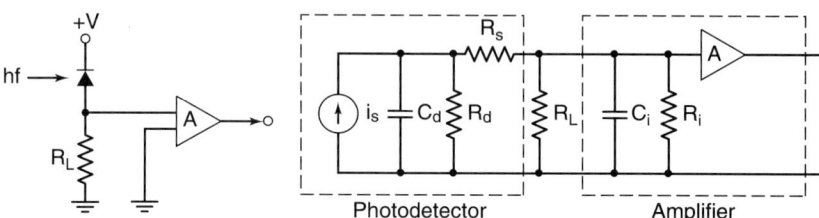

(a) A photodetector in a low-impedance or high-impedance amplifier circuit

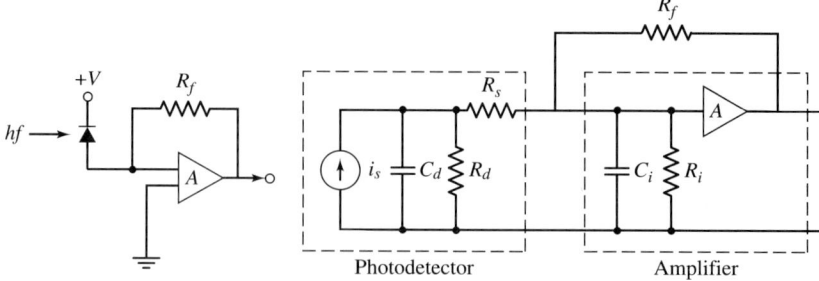

(b) A photodetector in a transimpedance amplifier circuit

Figure 5.21 Circuit topology for LZ, HZ, and TX preamplifiers

a current-limiting resistor between the diode and the biasing voltage source, and a bypass capacitor (0.01 to 0.1 μF), which is used to short the ac signals to ground. The amplifier characteristics may be represented by an input resistance R_i, an input capacitance C_i, and an ideal amplifier with a voltage amplification $A(\omega_m)$ and an infinite input impedance. From the equivalent circuit shown in Figure 5.21a, we have the input voltage to the ideal amplifier:

$$e_{in}(\omega_m) = \frac{R_{eq}}{1 + j\omega_m R_{eq} C_{eq}} i_s(\omega_m)$$

where R_{eq} and C_{eq} are the equivalent resistance and capacitance of R_d, R_s, R_L, R_i, C_d, and C_i. R_s is small and can be ignored. Thus,

$$R_{eq} \sim \frac{R_d R_L R_i}{(R_d R_L + R_d R_i + R_L R_i)}$$

and $C_{eq} = C_d + C_i$. The output voltage $e_{out}(\omega_m)$ is simply $A(\omega_m) e_{in}(\omega_m)$. Thus, the **current-to-voltage transfer function** for the LZ or HZ preamplifiers is

$$H_{lz\,hz}(\omega_m) = \frac{e_{out}(\omega_m)}{i_s(\omega_m)} = \frac{A(\omega_m) R_{eq}}{1 + j\omega_m R_{eq} C_{eq}} \quad (5.61)$$

From (5.61), we see that the frequency response of LZ and HZ preamplifiers is determined by the amplifier gain $A(\omega_m)$ and the time constant $R_{eq} C_{eq}$ of the input circuit. Broadband amplifiers with bandwidths in the 9 to 10 GHz range are available commercially [29]. This means that the bandwidths of LZ and HZ front ends are essentially limited by the RC time constant of the input circuit.

The impedance level of preamplifiers used for LZ front ends is nominally 50 Ω. As an example, assume $C_{eq} \sim 2$ pF and $R_{eq} \sim R_i = 50$ Ω. The RC time constant is about 100 ps, which corresponds to a bandwidth of a few GHz. In other words, the bandwidth of LZ preamplifiers is quite broad. However, since the impedance level is low, the noise level is relatively high compared to other circuit topologies. To reduce the noise level, HZ preamplifiers, which have a high input impedance, may be used.

An HZ front end is also known as an **integrating front end,** because charges generated by the photodetector within the RC time constant are accumulated in the capacitance. For HZ preamplifiers, the input resistance is on the order of 1 MΩ. Thus, the equivalent resistance is approximately

$$R_{eq} \sim \frac{R_L R_i}{R_L + R_i}$$

For $R_{eq} \sim 1$ MΩ and $C_{eq} \sim 2$ pF, the RC time constant is on the order of 2 μs, which corresponds to a bandwidth of less than 100 kHz, a rather narrow bandwidth. In addition, the dynamic range is limited, due to the gain at the low-frequency range.

The presence of a feedback resistor R_f in the TZ topology (Figure 5.21b) makes it quite distinct from the HZ and LZ configurations. Let the open-loop amplification factor of the ideal amplifier be $A(\omega_m)$. Then, the input voltage to the amplifier is

$$e_{in}(\omega_m) = \frac{i_s(\omega_m)}{\dfrac{1}{R_{eq}} + \dfrac{1 + A(\omega_m)}{R_f} + j\omega_m C_{eq}}$$

The current–to–voltage transfer function of a TZ amplifier is

$$H_{tz}(\omega_m) = \frac{e_{out}(\omega_m)}{i_s(\omega_m)} = -\frac{A(\omega_m)}{\dfrac{1}{R_{eq}} + \dfrac{1 + A(\omega_m)}{R_f} + j\omega_m C_{eq}} \tag{5.62}$$

When the open-loop gain $A(\omega_m)$ is large, the transfer function is approximately

$$H_{tz}(\omega_m) \approx -\frac{R_f}{1 + \dfrac{j\omega_m R_f C_{eq}}{A(\omega_m)}} \tag{5.63}$$

Although the signal gain of a TZ front end is nominal, its bandwidth is quite wide. The RC time constant is reduced to $R_f C_{eq}/A(\omega_m)$. In the limit of very high amplification, the transfer function H_{tz} is simply $-R_f$ and is independent of frequency. In addition to a broad bandwidth, the dynamic range is improved by the negative feedback. Thus, the TZ configuration is preferred in many applications.

Our analysis of preamplifiers is incomplete without consideration of their noise characteristics. Depending on the circuit topology, the noise introduced by a preamplifier may be represented by a shunt current source and a series voltage source. These are too complex for this text, and we refer the reader to the two articles by Personick [29].

REFERENCES

1. Krause, P. W.; L. D. McGlauchlin; and R. B. McQuistan. *Elements of Infrared Technology: generation, transmission and detection.* New York, NY: John Wiley & Sons, Inc., 1962.
2. Anderson, L. K.; M. DiDomenico, Jr.; and M. B. Fisher. "High-speed photodetectors for microwave demodulation of light." In *Advances in Microwaves* 5, ed. L. Young. New York, NY: Academic Press, 1970.
3. Seib, D. H.; and L. W. Aukerman. "Photodetectors for the 0.1 to 1.0 μm spectral region." In *Advances in Electronics and Electron Physics* 34, ed. L. Marton. New York, NY: Academic Press, 1973, pp. 95–221.
4. Keyes, R. J. *Optical and Infrared Detectors.* New York, NY: Springer-Verlag, 1980.
5. Smith, R. G. "Photodetectors for fiber transmission systems." *Proc. IEEE* 68, 1980, pp. 1247–1253.
6. Poehler, T. O. "Detectors." *Physical Optics and Light Measurements.* Chapter 6, ed. D. Malacara. New York, NY: Academic Press, 1988.

7. DeLange, O. E. "Optical heterodyne detection." *IEEE Spectrum,* 1968, pp. 77–85.
8. Jacobs, S. F. "Optical heterodyne (coherent) detection." *Am. J. Phys.* 56, 1988, pp. 235–245.
9. Golay, M. J. E. "A pneumatic infra-red detector." *Rev. Sci. Ins.* 18, 1947, pp. 357–362.
10. Hecht, E.; and A. Zajac. *Optics.* 2nd ed. Reading, MA: Addison Wesley Publishing Co., 1987.
11. Melchior, H. "Demodulation and photodetection techniques." In *Laser Handbook,* ed. F. T. Arecchi; and E. O. Schulz-Dubois. Amsterdam, The Netherlands: Elsevier-North Holland, 1972.
12. Piotrowski, J.; W. Galus; and M. Grudzien. "Near room-temperature IR photodetectors." *Infrared Phys.* 31, 1991, pp. 1–48.
13. Product Manual, EG&G Judson Inc. Montgomeryville, PA, 1988.
14. Emmons, R. B.; S. R. Hawkins; and K. F. Cuff. "Infrared detectors: an overview." *Opt. Eng.* 14, 1975, pp. 21–30.
15. Forrest, S. R. "Optical detectors: three contenders." *IEEE Spectrum,* 1986, pp. 76–84.
16. Forrest, S. R. "Optical detectors for lightwave communication." In *Optical, Fiber Telecommunications II,* ed. S. E. Miller; and I. P. Kaminow. New York, NY: Academic Press, 1988, pp. 569–599.
17. Olson, H. M. "p-i-n diodes." *Microwave Semiconductor Devices and their Circuit Applications.* Chapter 9. ed. H. A. Watson. New York, NY: McGraw-Hill Book Company, 1969.
18. Stillman, G. E. "Detectors for optical-waveguide communications." *Optical-Fiber Transmission.* Chapter 11. ed. E. E. Basch. New York, NY: H. W. Sams & Co., 1987, pp. 335–374.
19. Melchior, H.; M. B. Fisher; and F. R. Arams. "Photodetectors for optical communication systems." *Proc. IEEE* 58, 1970, pp. 1466–1486.
20. Gartner, W. W. "Depletion-layer photoeffects in semiconductors."*Phys. Rev.* 115, 1959, pp. 84–86.
21. Anderson, L. K.; and B. J. McMurtry. "High-speed photodetectors." *Proc. IEEE* 54, 1966, pp. 1335–1349.
22. Campbell, J. C.; W. T. Tsang; G. L. Qua; and J. E. Bowers. "InP/InGaAsP/InGaAs avalanche photodiodes with 70 GHz gain-bandwidth product." *Appl. Phys. Lettr.* 51, 1987, pp. 1454–1456.
23. Kasper, B. L. "Optical detectors and receiver design." *IEEE/OSA 1988 Optical Fiber Communications* January 25–28, 1988, New Orleans, LA.
24. Pierce, J. R. "Physical sources of noise." *Proc. IRE* 44, 1956, pp. 601–608.
25. van der Ziel, A. *Noise in Solid State Devices and Circuits.* Chapter 1, New York, NY: John Wiley & Sons, Inc., 1986.
26. Smith, R. G.; and S. D. Personick. "Receiver design for optical fiber communication systems." *Semiconductor Devices for Optical Communication.* Chapter 4, ed. H. Kressel. New York, NY: Springer-Verlag, 1980, pp. 89–160.
27. Kazovsky, L. G. "Optical Heterodyning versus optical homodyning." *J. Opt. Comm.* 6, 1985, pp. 18–24.

28. Muoi, T. V. "Optical receivers." *Optoelectronic Technology and Lightwave Communications Systems.* Chapter 16, ed. C. Lin. New York, NY: Van Norstrand Reinhold, 1989.
29. Personick, S. D. "Receiver design for digital fiber optical communication systems, parts I and II." *Bell Syst. Tech. J.* 52, 1973, pp. 843–886.
30. Muoi, T. V., private communication.

ADDITIONAL READING

1. Jenkins, F. A.; and H. E. White. *Fundamentals of Optics.* 4th ed. New York, NY: McGraw-Hill Book Company, 1976.
2. Karim, M. A. *Electro-optical Devices and Systems.* Boston, MA: PWS-Kent Publishing Co., 1990.
3. Nussbaum, A.; and R. A. Phillips. *Contemporary Optics for Scientists and Engineers.* Englewood Cliffs, NJ: Prentice-Hall Inc., 1976.
4. Pedrotti, F. L.; and L. S. Pedrotti. *Introduction to Optics.* Englewood Cliffs, NJ: Prentice-Hall Inc., 1991.
5. Saleh, B. E. A.; and M. C. Teich. *Fundamentals of Photonics.* New York, NY: John Wiley & Sons, Inc., 1991.
6. Senior, J. M. *Optical Fiber Communications, Principles and Practice.* 2nd ed. Englewood Cliffs, NJ: Prentice-Hall Inc., 1992.
7. Yariv, A. *Introduction to Optical Electronics.* 3rd ed. New York, NY: Holt, Rinehart & Winston, 1984.

Problems

1. Show that (5.37), (5.38), and (5.39) are correct for random numbers *n* characterized by a Poisson distribution function (5.36).

    Hint: Use the series expansion for e^x to establish two identities:

    $$\sum_{n=0}^{\infty} n \frac{x^n}{n!} = x e^x \quad \text{and} \quad \sum_{n=0}^{\infty} n^2 \frac{x^n}{n!} = (x^2 + x)e^x$$

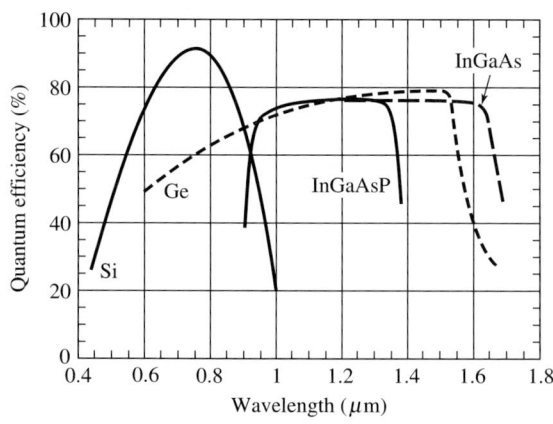

2. The spectral responses of Si, Ge, InGaAsP, and InGaAs PIN diodes are shown in the figure on page 302 (after [30]). They are in the form of quantum efficiency versus wavelength plots. Convert these curves to current responsivity versus wavelength plots. Be very careful in labeling the axes.

3. Consider a high-speed photodetector with a square cross-section that is used in a coherent detection scheme to monitor incoming radiation ($\lambda = 1$ μm). Suppose that the signal beam and the local oscillator beam are not aligned perfectly, and that the angle between the two beams is ζ (see Figure 5.20). Then, the electrical signal output is a function of ζ. Plot the heterodyne detection response $I(\zeta)/I(0)$ as a function of angular misalignment ζ if the detector area is 50 μm², and estimate the angle ζ when $I(\zeta)/I(0)$ is reduced to 0.5. Repeat for square detectors with an area of 2000 μm² and 8000 μm². Would it pay to use a detector with a large area?

4. Consider an incoherent detection system consisting of a PIN diode and a 50 Ω load. The system is designed to operate at 300 °K and has a bandwidth Δf_m of 6 MHz. The PIN diode has a current responsivity of 0.52 A/W at 0.9 μm and a dark current of 2 nA. The input optical power is of the form given by equation (5.45). The background radiation is negligible.
   a. Estimate the power level P_0 required such that the mean square current due to shot noise is twice as strong as that due to the dark current.
   b. Estimate the power level P_0 required such that the mean square noise current due to shot noise is twice as strong as that due to the thermal noise in the load.
   c. Plot the signal-to-noise ratio as a function of the input optical power, with P_0 ranging from 1nW to 1μW. Assume $P_0 = P_m$.
   d. Suppose that the operating temperature is reduced to 4.2 °K, but other parameters remain the same. Replot the signal-to-noise ratio as a function of the input optical power.

5. A detector with a current responsivity of 0.8 A/W at 1.3 μm is used in a direct detection system that has a bandwidth of 3 MHz. The input signal is of the form given by equation (5.45) with $P_0 = P_m$.
   a. What is the noise equivalent power if the dark current is the dominate noise term and the dark current is 2 nA?
   b. What is the noise equivalent power if the thermal noise due to a load of 300 Ω is the dominate noise term and the operating temperature is 300 °K?
   c. What is the noise equivalent power if the shot noise is the dominate noise term?

MODULATION AND DEFLECTION OF OPTICAL BEAMS

CHAPTER 6

CHAPTER OUTLINE

6.1 Introduction
6.2 State of Polarization
    6.2.1 Linearly Polarized Waves
    6.2.2 Circularly Polarized Waves
    6.2.3 Elliptically Polarized Waves
    6.2.4 Wave Superposition
    6.2.5 Polarizers and Analyzers
6.3 Simplified Acoustooptic, Electrooptic, and Magnetooptic Effects
    6.3.1 Electric Polarizability and the Clausius–Mosotti Relation
    6.3.2 Simple Acoustooptic Effects
    6.3.3 Simple Electrooptic Effects
    6.3.4 Simple Magnetooptic Effects
    6.3.5 Comments
6.4 Faraday Rotation, and Magnetooptic Modulators and Isolators
    6.4.1 Faraday Rotation
    6.4.2 MO Amplitude Modulators
    6.4.3 MO Isolators
6.5 Index Ellipsoids
    6.5.1 Isotropic Media
    6.5.2 Uniaxial Media
    6.5.3 Biaxial Media
    6.5.4 Wave Propagation along the Principal Axes
6.6 Linear Electrooptic Effects
    6.6.1 Linear EO Coefficients
    6.6.2 ADP/KDP Crystals
6.7 Electrooptic Modulators
    6.7.1 Longitudinal EO Modulators
    6.7.2 Transverse EO Modulators

   6.7.3 Longitudinal vs Transverse EO Modulators, and MO vs EO Modulators
   6.7.4 EO Material Figure of Merit
 6.8 Elasticity and Acoustooptic Effects
   6.8.1 One-Dimensional Elasticity
   6.8.2 Longitudinal Acoustic Waves
   6.8.3 Acoustooptic Effects
 6.9 Acoustooptic Modulators and Deflectors
   6.9.1 Raman–Nath diffraction
   6.9.2 Bragg Diffraction
   6.9.3 Parameter Q and Transition Region Diffraction
   6.9.4 Resolvable Spots
   6.9.5 AO Material Figure of Merit
Problems
References
Additional Reading

6.1 INTRODUCTION

Information retrieval, material processing, and other applications often require that optical beams be deflected quickly and accurately to specific directions, using some device. While mechanical devices, such as the galvanometers, taut bands, tuning forks, torsion rods, rotating disks, etc., shown in Figure 6.1, are useful as light scanners or choppers [1,2], their frequency response is limited by their mechanical resonant frequencies and their harmonics. For example, the upper frequency limit of torsion rod choppers is about 20 kHz. For operation at the megahertz range or higher, electromagnetic devices must be used.

  Information transmission requires that optical beams be modulated by varying one or more of the wave parameters: *frequency, amplitude, phase, polarization,* and *propagation direction.* Amplitude, frequency, and phase modulations are relatively easy to understand and are well known in radio and microwave technologies. Polarization modulation is less known, although it also plays an important role in microwave and optical systems.

  In order to fully utilize the huge bandwidth potential of optical beams, modulation in the gigahertz range is required. For semiconductor injection lasers, superluminescent diodes, or light emitting diodes, direct amplitude modulation can be achieved by varying the electric current injected into the semiconductor junctions, as discussed in Chapter 4. However, the emission frequency also changes, introducing "frequency chirping" and "mode hopping," which may be detrimental to many systems. For other lasers, direct modulation is very difficult, if not impossible. One solution is to use external modulators, in which modulation is achieved by passing light through a medium for which the refractive index or attenuation coefficient is varied by applied electric, magnetic, or

6.1 Introduction

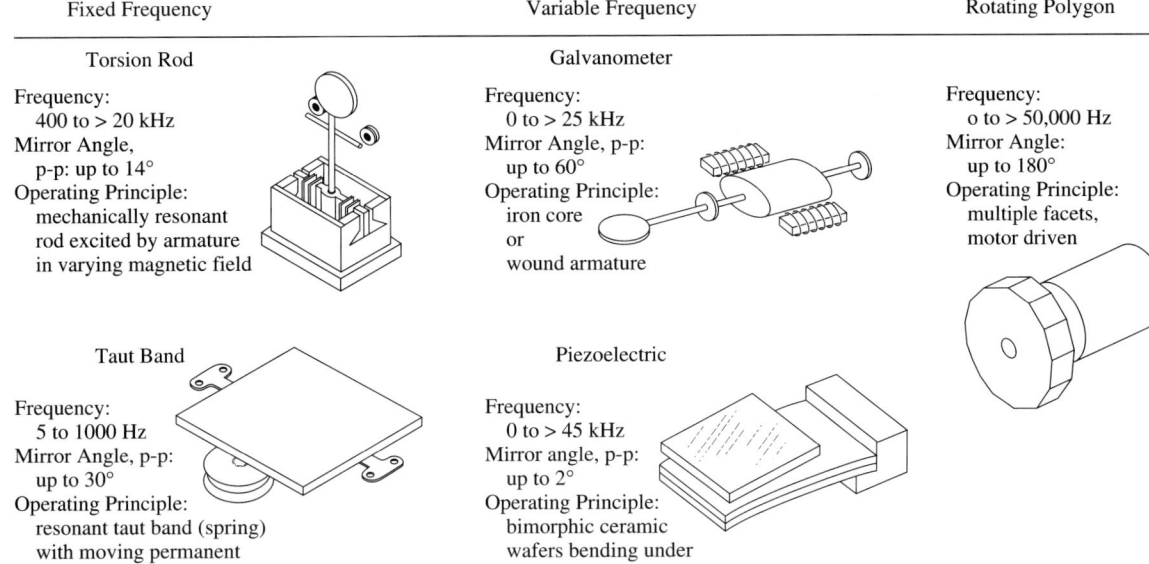

Figure 6.1 Various mechanical scanners and choppers [2]

acoustic fields. Alternatively, **free-charge carrier absorption** or the **Franz–Keldysh effect** may be used to accomplish *direct* amplitude modulation.

Phase modulation may be achieved by **electrooptic (EO)** or **magnetooptic (MO) effects.** Since the phase changes of different field components can be different, the state of polarization (SOP) can change as the waves propagate, implying that the SOP may also be altered by EO and MO effects. Phase or SOP modulation must be converted to amplitude variations. In other words, the amplitude modulation in EO and MO devices is achieved indirectly. When optical beams interact with acoustic waves, the optical beams are redirected, the optical frequency is shifted, and the amplitude is modulated.

The Franz–Keldysh effect concerns the change in the semiconductor bandgap when electric fields are applied. Varying the applied electric fields allows the semiconductor bandgap to be tuned, which in turn affects the optical absorption. The resulting amplitude modulation is rapid and direct.

Six MO effects are known [3]. They are: the Zeeman effect, inverse Zeeman effect, Voigt effect, Cotton–Mouton effect, Faraday effect, and Kerr magnetooptic effect. All are different because they are based on different physical processes. Rather than examine the subtle differences between these various effects, we will simply note that the **Faraday effect** is the one most often used. The Faraday effect, also known as **Faraday rotation,** causes the electric field vectors of the optical fields to rotate as the waves propagate.

In EO effects, the refractive index of a medium is changed by the modulating electric fields. There are five EO effects known, again each involving a different physical process. They are: the Stark effect, inverse Stark effect, electric double refraction, Kerr effect, and Pockels effect. Qualitative descriptions of these can be found in reference [3]. The **Pockels effect** is probably the most useful in modulation applications.

As noted earlier, the evolution of the SOP must be converted into an amplitude variation. In MO and EO amplitude modulators, this is accomplished by placing an MO or EO material between a polarizer and an analyzer. The polarizer preselects a particular SOP, and modulating magnetic or electric fields are applied to change the SOP. The analyzer then allows a specific, predetermined field component to pass, and it rejects the orthogonal field component. These concepts are discussed in more detail in section 6.2.

The physical processes involved are explained through simple physical pictures of various effects, in section 6.3. Details of MO modulators and related devices are discussed in section 6.4. Section 6.5 introduces the concept of the **index ellipsoid,** in terms of which linear EO effects are then described in detail in section 6.6. This leads to a discussion of the basic setup of EO amplitude modulators. Longitudinal and transverse EO modulators are discussed in section 6.7, along with the figure of merit of EO materials. Although the basic scheme of EO amplitude modulators is similar to that of MO amplitude modulators, there are essential differences, which are enumerated in section 6.7.

The **acoustooptic effect (AO)** may be understood in terms of the periodic spatial variation of the refractive index, induced by the acoustic waves. As the acoustic waves propagate, so does the spatial variation in the refractive index. Because of the motion and periodic changes of the refractive index, the optical beams are deflected spatially and shifted spectrally. The last section of this chapter is devoted to AO modulators and deflectors, the figure of merit of AO material, and resolvable spots.

6.2 STATE OF POLARIZATION

Sinusoidal voltages can be expressed as $V(t) = V_o \cos(\omega t + \phi)$ in the time domain representation, or as $V(\omega) = V_o e^{j\phi}$ in the frequency domain representation. In both representations, V_o is the amplitude and ϕ the phase relative to a given reference frame. The two representations are related through the following:

$$V(t) = Re[V(\omega)e^{j\omega t}]$$

where Re signifies the real part of a complex quantity.

Time-harmonic vectors can be expressed in the same manner. Each vector component is expressed as a sinusoidal function with an amplitude and a phase term. For example, the x component of an electric field is given by $\mathscr{E}_x(t) = E_{xo}\cos(\omega t + \phi_x)$ in the time domain, or as $E_x = E_{xo}e^{j\phi_x}$ in the frequency domain. In this chapter, as well as elsewhere in this book, script letters, such as $\mathscr{E}$ and $\mathscr{D}$, refer to the actual physical quantities, which are real functions of time. The corresponding phasor quantities, such as E and D, are complex functions of ω. Since the dependence of $\mathscr{E}$ and $\mathscr{D}$ on t, and of E and D on ω, is obvious, the arguments t and ω are often omitted for brevity.

Of particular interest are waves that have field components transverse to the direction of propagation. Consider the x and y field components of waves propagating in the $+z$ direction. To introduce the concept of state of polarization, focus on an arbitrary point in space and consider the field components at that point. In the time domain representation, the electric field vector at that point is

$$\mathcal{E} = \mathbf{a}_x \mathcal{E}_x + \mathbf{a}_y \mathcal{E}_y = \mathbf{a}_x E_{xo} \cos(\omega t + \phi_x) + \mathbf{a}_y E_{yo} \cos(\omega t + \phi_y) \quad (6.1)$$

where E_{xo} and E_{yo} are real and positive. $\mathbf{a}_x$ and $\mathbf{a}_y$ are unit vectors in the x- and y-directions. In the frequency domain representation, the electric field vector corresponding to (6.1) is

$$\mathbf{E} = \mathbf{a}_x E_x + \mathbf{a}_y E_y = \mathbf{a}_x E_{xo} e^{j\phi_x} + \mathbf{a}_y E_{yo} e^{j\phi_y} \quad (6.2)$$

Since ϕ_x and ϕ_y may be different, the two field components may vary differently as functions of time. As a result, the instantaneous magnitude and direction of $\mathcal{E}$ may change. The **state of polarization** describes the evolution of the electric field vectors as functions of time at a given location. Depending on the ratio E_{yo}/E_{xo} and the phase difference $\phi_y - \phi_x$, the "tip" of $\mathcal{E}$ may trace a *linear, circular,* or *elliptical trajectory,* in a *left-hand* or *right-hand sense.* Since the physical meanings of linear or circular trajectories are particularly significant, and are easy to understand, we will consider these special cases first before discussing the general case.

6.2.1 Linearly Polarized Waves

Suppose the two components of the electric fields are in time phase, that is, $\phi_x = \phi_y$. We can then simplify (6.1) to be

$$\mathcal{E} = (E_{xo} \mathbf{a}_x + E_{yo} \mathbf{a}_y) \cos(\omega t + \omega_x)$$

While the length of the electric field vector changes as a cosine function, the field vector itself always points in the same direction, $E_{xo} \mathbf{a}_x + E_{yo} \mathbf{a}_y$. Since the "tip" of $\mathcal{E}$ traces a straight line as time changes, we refer to the fields with $\phi_x = \phi_y$ as **linearly polarized (LP) fields** and the waves as **linearly polarized waves.**

6.2.2 Circularly Polarized Waves

If the two field components have the same amplitude, that is, $E_{xo} = E_{yo}$ and they are out of phase with each other by $\phi_y - \phi_x = -\pi/2$, then (6.2) can be written as

$$\mathbf{E} = E_{xo} e^{j\phi_x}(\mathbf{a}_x + e^{-j\pi/2} \mathbf{a}_y) = E_{xo} e^{j\phi_x}(\mathbf{a}_x - j\mathbf{a}_y) \quad (6.3)$$

In the time domain, the electric field vector is

$$\mathcal{E} = E_{xo}[\mathbf{a}_x \cos(\omega t + \phi_x) + \mathbf{a}_y \sin(\omega t + \phi_x)]$$

As time changes, the "tip" of the electric field vector traces a circular path. Now, recall that we are considering waves moving in the $+z$ direction. With our right thumb pointing in the $+z$ direction, our right-hand fingers curl in the same direction as the motion of the "tip" of the electric fields, that is, counterclockwise for observers looking toward the light source. By definition, then, waves having field components specified by (6.3), that is, $E_{xo} = E_{yo}$ and $\phi_y - \phi_x = -\pi/2$, are **right-hand circularly polarized (RHCP)** waves.

Similarly, for fields with $E_{xo} = E_{yo}$ and $\phi_y - \phi_x = +\pi/2$, the frequency and time domain representations are

$$\mathbf{E} = E_{xo}e^{j\phi_x}(\mathbf{a_x} + e^{j\pi/2}\mathbf{a_y}) = E_{xo}e^{j\phi_x}(\mathbf{a_x} + j\mathbf{a_y}) \qquad (6.4)$$

$$\mathcal{E} = E_{xo}[\mathbf{a_x}\cos(\omega t + \phi_x) - \mathbf{a_y}\sin(\omega t + \phi_x)]$$

In this case, for waves propagating in the $+z$ direction, with field components given by (6.4), the "tip" of the field vector traces a circle in the left-hand sense, and the fields are **left-hand circularly polarized (LHCP)** waves.[1] To observers facing the approaching waves, the field vector rotates in the clockwise direction.

6.2.3 Elliptically Polarized Waves

Except for the two special cases just discussed, the "tip" of an electric field vector generally traces an elliptical trajectory. Thus, (6.1) and (6.2) describe **elliptically polarized** fields. The shape and orientation of the elliptical trajectory depend on the amplitude ratio and the phase difference. The sense of rotation, that is, the right-handedness or left-handedness, is determined by the phase difference alone. To elaborate, from (6.1) we obtain

$$\left(\frac{\mathcal{E}_x}{E_{xo}}\right)\sin\Delta\phi = \cos(\omega t + \phi_x)\sin\Delta\phi$$

$$\left(\frac{\mathcal{E}_x}{E_{xo}}\right)\cos\Delta\phi - \left(\frac{\mathcal{E}_y}{E_{yo}}\right) = \sin(\omega t + \phi_x)\sin\Delta\phi$$

These equations can be combined to give

$$\left(\frac{\mathcal{E}_x}{E_{xo}}\right)^2 + \left(\frac{\mathcal{E}_y}{E_{yo}}\right)^2 - 2\left(\frac{\mathcal{E}_x}{E_{xo}}\right)\left(\frac{\mathcal{E}_y}{E_{yo}}\right)\cos\Delta\phi = \sin^2\Delta\phi \qquad (6.5)$$

Equation (6.5) represents an ellipse inscribed in a $2E_{xo} \times 2E_{yo}$ rectangle, as shown in Figure 6.2. The ellipse is known as the **polarization ellipse.** Note that the major and minor axes do not necessarily coincide with the x and y axes. However, if the coordinates are rotated, (6.5) can be reduced to the canonical form for ellipses [5], in which the lengths of the major and minor axes $2E_{\text{mj}}$ and $2E_{\text{mn}}$ are readily identified. Here, we assume that E_{mj} and E_{mn} are positive, that $E_{\text{mj}} \geq E_{\text{mn}}$, and that the angle θ of the major axis relative to the x axis is known. The angle θ is known as the **azimuth** of the ellipse. The shape of the ellipse may be expressed in terms of the **ellipticity**

---

[1]In this book, we adopt the IEEE definition for the right-handedness or left-handedness of waves. In much physics and optics literature, the terminology used is just the opposite; waves having field components given by (6.3) and (6.4) are identified, respectively, as the **left-** and **right-hand circularly polarized waves** [4].

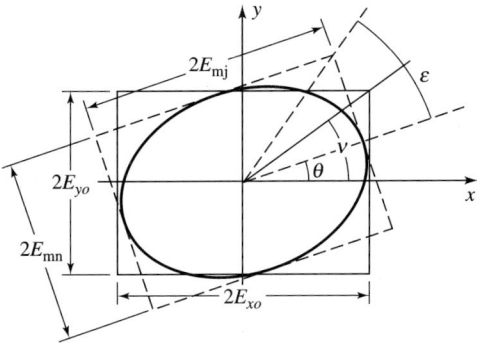

Figure 6.2 Parameters of elliptically polarized waves

$$\text{Ellipticity} = \frac{E_{mn}}{E_{mj}} \tag{6.6}$$

or the **visibility (VS)**

$$VS = \left|\frac{E_{mj}^2 - E_{mn}^2}{E_{mj}^2 + E_{mn}^2}\right| = \frac{\left|1 - \left(\frac{E_{mn}^2}{E_{mj}^2}\right)\right|}{\left|1 + \left(\frac{E_{mn}^2}{E_{mj}^2}\right)\right|} \tag{6.7}$$

both of which are functions of E_{mn}/E_{mj}.

As before, the sense of rotation can be deduced by tracing the "tip" of $\mathcal{E}$ as a function of time. For $0 < \Delta\phi < \pi$, waves rotate in the left-hand sense. That is, the "tip" of $\mathcal{E}$ rotates in the same way as our left hand fingers curl with the left thumb pointing in the direction of propagation. Waves with $\pi < \Delta\phi < 2\pi$, which is the same as $-\pi < \Delta\phi < 0$, have a right-hand sense of rotation. The limiting case of $\Delta\phi = 0$ or π corresponds to linearly polarized waves.

The SOP, which is specified by E_{yo}/E_{xo} and $\Delta\phi$, can also be expressed in terms of ellipticity, azimuth, and the sense of rotation. Thus there are two equivalent means of specifying the SOP. The transformation from E_{yo}/E_{xo} and $\Delta\phi$ to E_{mn}/E_{mj}, θ, and the sense of rotation is facilitated by the following relationships:

$$\varepsilon_{lr}\frac{E_{mn}}{E_{mj}} = \tan\varepsilon, \quad -\pi/4 \leq \varepsilon \leq \pi/2 \tag{6.8}$$

$$\sin 2\varepsilon = (\sin 2\nu)\sin\Delta\phi \tag{6.9}$$

$$\tan 2\theta = (\tan 2\nu)\cos\Delta\phi, \quad 0 \leq \theta < \pi \tag{6.10}$$

$$\frac{E_{yo}}{E_{xo}} = \tan\nu, \quad 0 \leq \nu \leq \pi/2 \tag{6.11}$$

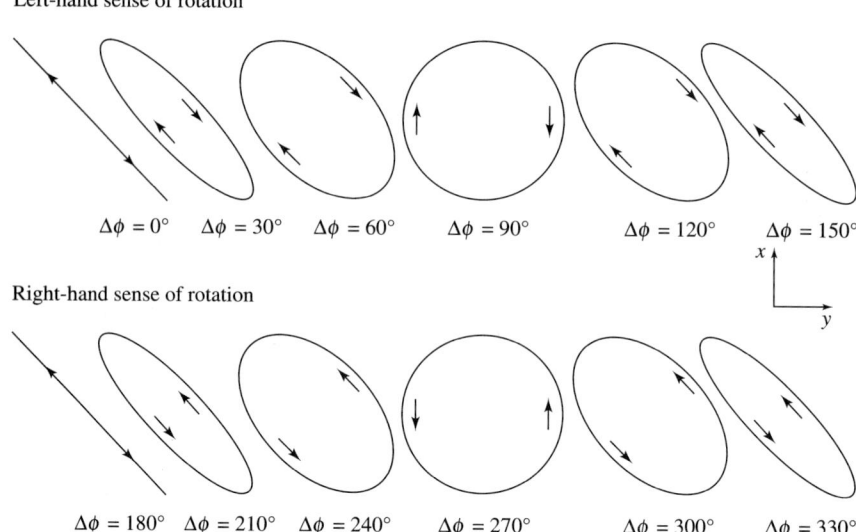

Figure 6.3 Elliptically polarized waves propagating in the $+z$ direction with $E_{xo} = E_{yo}$

In (6.8), ε_{lr} is $+1$ for the left-handed rotation and -1 for the right-handed rotation. It should be emphasized that E_{mj} and E_{mn} are positive and that $E_{mn} \leq E_{mj}$. Detailed derivations for these relationships can be found in [6]. The physical meanings of E_{yo}/E_{xo}, E_{mn}/E_{mj}, $\Delta\phi$, and θ are easily understood. Although ε and ν also have their own geometric meaning, as shown in Figure 6.2, we simply view ε and ν as auxiliary variables introduced as a convenient means of representing the ratios E_{mn}/E_{mj} and E_{yo}/E_{xo}.

Example

Consider fields for which $E_{xo} = E_{yo}$ and $\Delta\phi$ is unspecified. From (6.11), we have $\nu = +\pi/4$. Since $\tan 2\nu = \tan \pi/2 \rightarrow \infty$, 2θ is $+\pi/2$ or $-\pi/2$, depending on the sign of $\cos\Delta\phi$. Thus, the major axis of the polarization ellipse is $+45$ degrees from the x axis if $\cos\Delta\phi$ is positive, and -45 degrees if $\cos\Delta\phi$ is negative. For $\Delta\phi = 0$, or π, we obtain from (6.9) that $\varepsilon = 0$ and the ellipticity is 0, which corresponds to LP fields. The field vector is $+45$ degrees relative to the x axis if $\cos\Delta\phi > 0$, or -45 degrees if $\cos\Delta\phi < 0$. For $\Delta\phi = \pm\pi/2$, the fields are circularly polarized and the ellipticity is 1. If $\Delta\phi \neq 0, \pm\pi/2$, or $\pm\pi$, we deduce from (6.9) that $\varepsilon = \Delta\phi/2$. The ellipticity is then related directly to $\Delta\phi$, that is, $E_{mn}/E_{mj} = |\tan\varepsilon| = |\tan(\Delta\phi/2)|$. These cases are shown in Figure 6.3.

6.2.4 Wave Superposition

Waves are considered to be **plane waves** if the constant-phase surfaces of the waves are planes. Further, if the wave amplitude is the same everywhere on the constant-phase planes, the waves are **uniform plane waves.** In addition, uniform plane waves in isotropic media are transverse electromagnetic (TEM) waves, since the electric and magnetic field vectors are normal to the direction of propagation. This means that the electric fields of uniform plane waves propagating in the $+z$ direction are confined to the xy plane, which is the type of wave considered in the last three subsections.

Equation (6.2) implies that elliptically polarized fields can be considered to be the superposition of two orthogonal LP fields. In *isotropic media,* such as free space, all field components propagating in all directions experience the same refractive index. Consider plane waves with a free-space wave vector **k** propagating in the z direction in isotropic media with an index n. Treat the fields as the superposition of x-polarized and y-polarized fields and let the field components at $z = 0$ be $E_x(0)$ and $E_y(0)$. At z, the fields are

$$\mathbf{E}(z) = \mathbf{a}_x E_x(0) e^{-jknz} + \mathbf{a}_y E_y(0) e^{-jknz} \tag{6.12}$$

Since the refractive index n is the same for both field components, the two field components given in (6.12) have the same complex exponential term, e^{-jknz}. Thus, the phase difference between the two field components at z is exactly the same as the phase difference at $z = 0$. In addition, there is no amplitude change. Therefore, there is no change in the state of polarization as waves propagate in an isotropic medium.

Alternatively, we can rewrite (6.2) to show that elliptically polarized fields can also be considered the superposition of two *counter-rotating CP* fields.

$$\mathbf{E} = \left(\frac{E_x + jE_y}{2} + \frac{E_x - jE_y}{2}\right)\mathbf{a}_x - j\left(\frac{E_x + jE_y}{2} - \frac{E_x - jE_y}{2}\right)\mathbf{a}_y$$

$$= \left(\frac{E_x + jE_y}{\sqrt{2}}\right)\mathbf{a}_R + \left(\frac{E_x - jE_y}{\sqrt{2}}\right)\mathbf{a}_L$$

In these equations,

$$\mathbf{a}_R = \frac{\mathbf{a}_x - j\mathbf{a}_y}{\sqrt{2}} \quad \text{and} \quad \mathbf{a}_L = \frac{\mathbf{a}_x + j\mathbf{a}_y}{\sqrt{2}} \tag{6.13}$$

are the **basis vectors** for RHCP and LHCP fields. Unit vectors $\mathbf{a}_x$ and $\mathbf{a}_y$ can also be expressed in terms of $\mathbf{a}_R$ and $\mathbf{a}_L$,

$$\mathbf{a}_x = \frac{\mathbf{a}_R + \mathbf{a}_L}{\sqrt{2}} \quad \text{and} \quad \mathbf{a}_y = j\frac{\mathbf{a}_R - \mathbf{a}_L}{\sqrt{2}} \tag{6.14}$$

The magnitudes of the RHCP and LHCP fields are $|(E_x + jE_y)|/\sqrt{2}$ and $|(E_x - jE_y)|/\sqrt{2}$, respectively. If we write the amplitudes of the two circularly polarized waves as $E_R(0)$ and $E_L(0)$ at $z = 0$, then the fields at z are

$$\mathbf{E}(z) = \mathbf{a}_R E_R(0) e^{-jknz} + \mathbf{a}_L E_L(0) e^{-jknz} \tag{6.15}$$

Again, the refractive index n is the same for the two counter-rotating CP fields, and we conclude from (6.15) that there is no change in the SOP as these waves propagate in an isotropic media. Therefore, waves propagating in an isotropic medium can be expressed as the superposition of either two linearly polarized fields, as given by (6.12), or two circularly polarized fields, as represented by (6.15). More importantly, there is no change in the SOP as waves propagate.

In *anisotropic media,* different field components may propagate with different velocities, and the phase velocities, and therefore the refractive indices, are functions of the direction of propagation. Although the decomposition of elliptically polarized waves as two orthogonal LP waves or two counter-rotating CP waves is still valid, the difference in the phase velocities and refractive indices must be taken into account. Because of the phase velocity difference, the SOP *evolves* as the waves propagate. The anisotropic medium is **linearly birefringent** if two orthogonal LP waves propagate with different velocities, and is **circularly birefringent** if two counter-rotating CP waves propagate with different velocities. Detailed discussions on this complex subject are contained in Chapter 10. For the present, we simply note that plane waves propagating in linear birefringent media are represented by

$$\vec{E}(z) = \mathbf{a}_x E_x(0)e^{-jkn_x z} + \mathbf{a}_y E_y(0)e^{-jkn_y z} \qquad (6.16)$$

For circular birefringent media, the waves are given by

$$\mathbf{E}(z) = \mathbf{a}_R E_R(0)e^{-jkn_R z} + \mathbf{a}_L E_L(0)e^{-jkn_L z} \qquad (6.17)$$

Note in particular the indices n_x and n_y in the exponents of (6.16), as well as n_R and n_L in the exponents of (6.17). These imply that the x component of **E** propagates with a phase velocity of c/n_x and the y component of **E** propagates with a phase velocity of c/n_y. Similarly, the phase velocities of the RHCP and LHCP waves in (6.17) are c/n_R and c/n_L, respectively.

6.2.5 Polarizers and Analyzers

Analyzers and **polarizers** are polarizing optical elements used to select a particular field component and block the orthogonal field component. Light emerging from a *perfect* linear polarizer or analyzer is perfectly linearly polarized. The position of the polarizing element in the optical system determines whether it is called a polarizer or an analyzer. The direction of polarization of the emergent light is referred to as the **polarizer's axis** or the **analyzer's axis.** For polarizers and analyzers operating in the transmission mode, the direction is also referred to as the **transmission axis.** In schematic diagrams, these directions are indicated by heavy lines or arrows. Mathematically, the function of a polarizer or analyzer can be represented by the scalar product of two vectors. Let the transmission axis of a linear polarizer or analyzer be given by a unit vector $\mathbf{a}_t$. The field component passing through the linear analyzer or polarizer without attenuation is $\mathbf{a}_t \cdot \mathbf{E}$, and field components normal to $\mathbf{a}_t$ are blocked completely. The electric field emerging from the polarizer is therefore $(\mathbf{a}_t \cdot \mathbf{E}) \mathbf{a}_t$. In many optical systems, the transmission axis of the polarizer and the analyzer are perpendicular to each other, and they are known as **crossed polarizers.**

6.3 SIMPLIFIED ACOUSTOOPTIC, ELECTROOPTIC, AND MAGNETOOPTIC EFFECTS

As indicated in Section 6.1, the refractive index may be changed by applying strong magnetic, electric, or acoustic fields to the medium. These effects are referred to as the acoustooptic, electrooptic, and magnetooptic effects, respectively. To understand how this change occurs, recall that the definition of the electric displacement $\mathcal{D}$ is

$$\mathcal{D} = \varepsilon_0 \mathcal{E} + \mathcal{P} = \varepsilon_0 \varepsilon_r \mathcal{E} = \varepsilon_0 n^2 \mathcal{E} \tag{6.18}$$

where $\mathcal{P}$ is the polarization, $\mathcal{E}$ the electric field intensity, and ε_0 the vacuum permittivity [4]. The SI units for $\mathcal{D}$ (and $\mathcal{P}$) and $\mathcal{E}$ are C/m² and V/m, respectively. The relative dielectric constant ε_r and the refractive index n are dimensionless quantities.

6.3.1 Electric Polarizability and the Clausius–Mosotti Relation

A material medium may be viewed as the collection of a large number of molecules. The polarization $\mathcal{P}$ is the net electric dipole moment per unit volume. If there are N identical molecules per unit volume, and the effective electric dipole moment of a single module is p (C-m), the polarization, that is, the vector sum of all electric dipole moments in a volume of 1 m³, is

$$\mathcal{P} = \sum_1^N p = N\,p \tag{6.19}$$

To the first order of approximation, both the induced dipole moment and the effective dipole moment per molecule due to the reorientation of molecules are linearly proportional to the local field $\mathcal{E}_{loc}$. In isotropic media, p is parallel with $\mathcal{E}_{loc}$

$$p = \varepsilon_0 \alpha_e \mathcal{E}_{loc} \tag{6.20}$$

where α_e is the **electric polarizability**, which has a dimension of m³. The local field intensity $\mathcal{E}_{loc}$ is the total field experienced by a molecule at a given location, including the fields due to the external sources, dipole moments in the immediate neighborhood of the point in question, and all dipole moments elsewhere in the medium. For isotropic media, it can be shown that [7]

$$\mathcal{E}_{loc} = \mathcal{E} + \frac{\mathcal{P}}{3\varepsilon_0} \tag{6.21}$$

By substituting (6.20) and (6.21) into (6.19), we obtain an explicit expression for $\mathcal{P}$

$$\mathcal{P} = \frac{N\alpha_e \varepsilon_0}{1 - \frac{N\alpha_e}{3}} \mathcal{E}$$

which, when combined with (6.18), yields

$$\mathcal{D} = \varepsilon_0 \varepsilon_r \mathcal{E} = \varepsilon_0 \left(1 + \frac{N\alpha_e}{1 - \frac{N\alpha_e}{3}}\right) \mathcal{E}$$

In terms of N and α_e, the terms ε_r and n can be written as

$$n^2 = \varepsilon_r = \frac{1 + \frac{2N\alpha_e}{3}}{1 - \frac{N\alpha_e}{3}}$$

The relation can also be expressed as

$$\frac{n^2 - 1}{n^2 + 2} = \frac{\varepsilon_r - 1}{\varepsilon_r + 2} = \frac{1}{3} N\alpha_e \qquad (6.22)$$

which is the **Clausius–Mosotti relation.** If the medium contains several molecular species and the density of species i is N_i and the electric polarizability is α_{ei}, (6.22) may be generalized to

$$\frac{n^2 - 1}{n^2 + 2} = \frac{1}{3} \sum_i N_i \alpha_{ei} \qquad (6.23)$$

Although (6.22) and (6.23) are written specifically for linear isotropic media, they can also help in understanding various effects in anisotropic media, as well. Specifically, the principal effect of strong electric and magnetic fields is a change in the electric polarizability α_e. The main effect of acoustic fields is a change in the molecular density N.

6.3.2 Simple Acoustooptic Effects

The basic mechanism responsible for acoustooptic effects can be understood in terms of (6.22) or (6.23). If the density N is changed by ΔN for any reason, we have

$$\frac{6n\Delta n}{(n^2 + 2)^2} = \frac{1}{3} \alpha_e \Delta N$$

Combining this equation with (6.22) and eliminating α_e, we obtain

$$\frac{6n\Delta n}{(n^2 + 2)^2} = \frac{n^2 - 1}{n^2 + 2} \frac{\Delta N}{N}$$

To relate ΔN to acoustic waves, consider the forces applied to simple one-dimensional objects and ignore the possible changes in cross sectional areas. By definition, **tensile strain** S is the fractional change of length of an object. With the change of the cross sectional area ignored, the fractional volume change

$\Delta V/V$ is related linearly to the fractional change in length, which is simply S. Also, since there is no change in the total mass of the object, the fractional change in mass density is related directly to the fractional volumetric change, $\Delta N/N = -\Delta V/V = -S$. Therefore, from the last equation we obtain

$$\Delta n = -\frac{1}{2} n^3 \left[\frac{(n^2 - 1)(n^2 + 2)}{3n^4} \right] S \qquad (6.24)$$

Note that Δn varies linearly with S. The term in the square bracket is the dimensionless **photoelastic constant** p

$$p = \frac{(n^2 - 1)(n^2 + 2)}{3n^4} \qquad (6.25)$$

As an example consider water. In the visible spectrum, the index of water is about 1.33, and we obtain $p \approx 0.31$ from (6.25), which agrees surprisingly well with the measured value of 0.31. In general, Δn is much smaller than n. For example, $|\Delta n|$ is about 3.6×10^{-6} for strain on the order of 10^{-5}. Since Δn is small, we have as an approximation,

$$-\frac{2\Delta n}{n^3} = \frac{1}{n^2}(1 - \frac{2\Delta n}{n} - 1) \approx \frac{1}{n^2(1 + \frac{\Delta n}{n})^2} - \frac{1}{n^2} = \frac{1}{(n + \Delta n)^2} - \frac{1}{n^2}$$

Thus, (6.24) can be written as

$$\frac{1}{(n + \Delta n)^2} \approx \frac{1}{n^2} + pS + \ldots \qquad (6.26)$$

Further discussion on the AO effect is contained in section 6.8.

6.3.3 Simple Electrooptic Effects

Several physical processes are involved in electrooptic effects. In the presence of applied electric fields, the "electron cloud" may either be displaced from its original position, or distorted from its original shape. The **electronic polarizability** α_{el} is the effect associated with this displacement or deformation, as shown schematically in Figure 6.4a. Simple calculations based on Coulomb's law show that, for atoms of radius r_a, $\alpha_{el} = 4\pi r_a^3$ [7,8]. In nonpolar materials, the net electric dipole moment per molecule in the absence of electric fields is zero. However, in the presence of strong electric fields, the electric charge distribution may change, and the effective center of the positive charge may shift away from the effective center of the negative charge. As a result, there is a net *change* in the dipole moment per molecule. Figure 6.4b schematically illustrates the effects of electric fields on nonpolar CO_2 molecules. The net change of dipole moment per molecule is represented by the **ionic polarizability** α_{ion}. In polar materials, such as NaCl, the net dipole moment per molecule, in the absence of applied

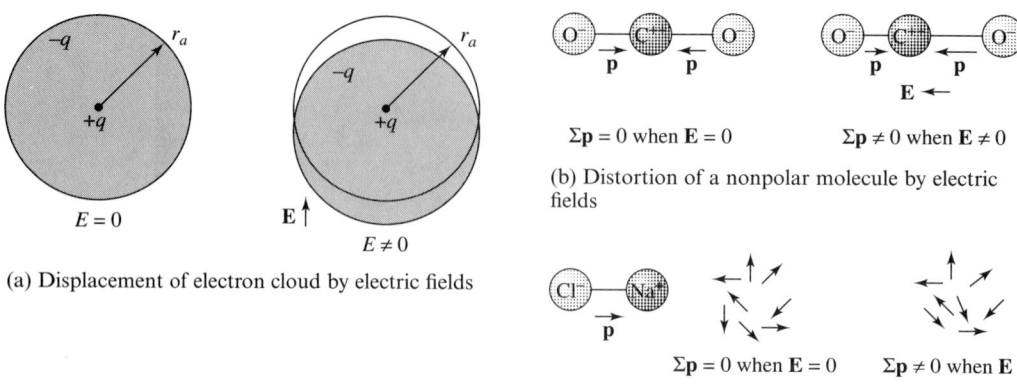

Figure 6.4 Change and reorientation of molecular dipole moments by electric fields

electric fields, is *not* zero. However, the polar molecules are oriented randomly, due to random thermal motion, and the net dipole moment per unit volume vanishes. In the presence of electric fields, the net change of the dipole moment may be quite small; however, the reorientation may be considerable. This is illustrated in Figure 6.4c. The reorientation of molecules leads to an increase in the dipole moment per unit volume. The polarizability α_{reor}, due to the molecular reorientation, is $p^2/(3kT)$, where T is the temperature in Kelvin. Therefore, total electric polarizability is

$$\alpha_e = \alpha_{\text{el}} + \alpha_{\text{ion}} + \alpha_{\text{reor}} \tag{6.27}$$

The contribution due to each process depends not only on the molecular and crystalline structure of the material, but also on the frequency of the optical beam. As depicted schematically in Figure 6.5, the main contribution to α_e in the infrared and visible spectra is the electronic polarizability. In the microwave or millimeter frequency ranges, the ionic polarizability is the dominating factor. At lower frequencies, the effect of molecular reorientation is important.

To express the EO effects quantitatively, suppose that the external electric field has one component $\mathscr{E}_m$, and that the index in the absence and in the presence of applied fields is n and $n + \Delta n$, respectively. The subscript m is used to distinguish the modulating field from the optical field to be modulated. The EO effects are then expressed as a power series of $\mathscr{E}_m$,

$$\frac{1}{(n + \Delta n)^2} - \frac{1}{n^2} = r\mathscr{E}_m + R\mathscr{E}_m^2 + \cdots \tag{6.28}$$

The first term on the right side of (6.28) indicates the linear dependence of $(n + \Delta n)^{-2} - n^{-2}$ on $\mathscr{E}_m$. This linear EO effect is known as the **Pockels effect,** and the constant r is the **linear EO coefficient.** The quadratic term on the right side of (6.28) is the **Kerr effect** and the constant R is the **Kerr coefficient.** Kerr effects are present in all materials, including gases, liquids, and crystalline solids.

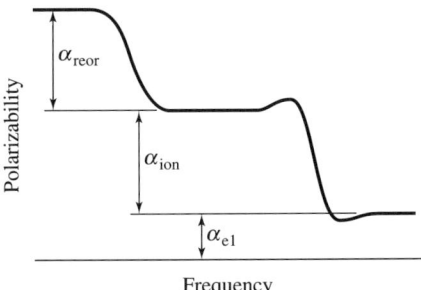

Figure 6.5 Frequency dependence of the electronic, ionic, and reorientational polarizability

However, the Pockels effect exists only in crystals which have no center of inversion. In our discussions, we will only examine direct effects, and will ignore all secondary effects, including the change of index by strain induced indirectly through the piezoelectric or electrostrictive effects from applied electric fields [8]. In materials where both EO effects are present, the Kerr effect is so much weaker than the Pockels effect that it can be ignored. The linear EO coefficient r has a unit of m/V and is typically on the order of 10^{-12} to 10^{-11} m/V. Since $\mathscr{E}_m$ is less than 10^7 V/m prior to material breakdown, $r\mathscr{E}_m$ is on the order of 10^{-4}, and, as noted previously, $R\mathscr{E}_m^2$ is even smaller. We can therefore ignore Kerr effect terms in the remainder of our discussion.

Since Δn is very small in comparison to n, we can approximate

$$\frac{1}{(n + \Delta n)^2} \approx \frac{1}{n^2} - \frac{2\Delta n}{n^3} + \ldots$$

and obtain from (6.28)

$$\Delta n \approx -\frac{1}{2} n^3 \, r\mathscr{E}_m + \ldots \tag{6.29}$$

Additional discussions on EO effects can be found in Section 6.6, in which (6.29) is generalized to include the effects of three components of the modulating electric fields.

6.3.4 Simple Magnetooptic Effects

Of the six magnetooptic effects mentioned in the introduction, the Faraday effect is the one most often used in engineering applications. Therefore, our discussion is confined to the Faraday effect and devices based on that effect. To understand the origin of Faraday effects, consider electronic motion in strong magnetic fields. Electrons in isolated atoms are bound to the nucleus by the Coulomb force. In crystalline solids, there is also the force due to the crystal lattice. In the absence of external fields, an electron is therefore bound in an equilibrium position. If, for any reason, an electron moves away from the equilibrium position, it is pushed back to that position by a restoring force in the radial direction.

In the absence of other forces, then, the equation of motion is simply

$$m\frac{d^2\mathcal{R}}{dt^2} = -K\mathcal{R}$$

where $\mathcal{R}$ is the electron position relative to the equilibrium position and K is a constant representing the restoring force. In the presence of external fields, the equation must be amended by a **Lorentz force** term representing the force due to the electric and magnetic fields. To consider MO effects, assume that a strong magnetic field $\mathbf{B}_{dc}$ exists, in addition to the fields $\mathcal{E}$ and $\mathcal{B}$ of the optical beam. The equation of motion becomes

$$m\frac{d^2\mathcal{R}}{dt^2} = -K\mathcal{R} - e[\mathcal{E} + \frac{d\mathcal{R}}{dt} \times \mathcal{B} + \frac{d\mathcal{R}}{dt} \times \mathbf{B}_{dc}]$$

where $\frac{d\mathcal{R}}{dt}$ is the electron velocity. For optical fields, $|\mathcal{B}|/|\mathcal{E}|$ is on the order of n/c, where n is the index of the medium. This means that the force due to $\mathcal{B}$ is smaller than that of $\mathcal{E}$ by a factor of $|\frac{n}{c}\frac{d\mathcal{R}}{dt}|$. Neglecting the force due to the magnetic fields $\mathcal{B}$ of the optical beam, we have the simplified equation of motion

$$m\frac{d^2\mathcal{R}}{\mathbf{dt}^2} = -K\mathcal{R} - e[\mathcal{E} + \frac{d\mathcal{R}}{dt} \times \mathbf{B}_{dc}]$$

If there is no dc magnetic field, we can show that $\mathcal{R}$ is in the same direction as $\mathcal{E}$. In the presence of $\mathbf{B}_{dc}$, however, the situation is much more complicated. Depending on the direction of $\mathbf{B}_{dc}$ relative to $\mathcal{E}$, the electron may move in a circular, helical, or other trajectory. To simplify the calculations, we consider sinusoidally time-varying optical fields with an angular frequency ω. The electrons also move sinusoidally with the same frequency. Therefore, considerable simplification is realized by casting all terms in the phasor notation. In the *frequency domain representation* (i.e., $e^{+j\omega t}$), the simplified equation of motion can be written as

$$[K - \omega^2 m - j\omega e\mathbf{B}_{dc} \times]\mathbf{R} = -e\mathbf{E} \quad (6.30)$$

where

$$\mathbf{R} = \mathbf{a}_x R_x + \mathbf{a}_y R_y \quad \text{and} \quad \mathbf{E} = \mathbf{a}_x E_x + \mathbf{a}_y E_y$$

are the frequency domain representations of $\mathcal{R} = \mathbf{a}_x \mathcal{R}_x + \mathbf{a}_y \mathcal{R}_y$ and $\mathcal{E} = \mathbf{a}_x \mathcal{E}_x + \mathbf{a}_y \mathcal{E}_y$, respectively. To be specific, we assume $\mathbf{B}_{dc}$ is along the direction of wave propagation, that is, the z direction, and that $\mathbf{E}$ is in the xy plane. Then (6.30) can be written in component form,

$$(K - \omega^2 m)R_x + j\omega B_{dc} R_y = -eE_x$$
$$-j\omega B_{dc} R_x + (K - \omega^2 m)R_y = -eE_y$$

From these equations we can solve for R_x and R_y in terms of E_x and E_y and obtain

$$R_x + jR_y = \frac{-e(E_x + jE_y)}{(K - \omega^2 m) + \omega e B_{dc}} \quad (6.31)$$

$$R_x - jR_y = \frac{-e(E_x - jE_y)}{(K - \omega^2 m) - \omega e B_{dc}} \quad (6.32)$$

For RHCP waves, $\mathbf{E} = E_R(\mathbf{a}_x - j\mathbf{a}_y)/\sqrt{2}$. By substituting $E_x = E_R/\sqrt{2}$ and $E_y = -jE_R/\sqrt{2}$ into (6.31) and (6.32), we obtain

$$R_x = jR_y = -\frac{eE_R}{\sqrt{2}} \frac{1}{(K - \omega_m^2) + \omega e B_{dc}}$$

The electrons are pushed away from their equilibrium positions by RHCP waves, and the displacement $|\mathbf{R}|$ is

$$|\mathbf{R}| = \sqrt{R_x^2 + R_y^2} = \frac{eE_R}{(K - \omega^2 m) + \omega e B_{dc}} \quad (6.33)$$

If the disturbing fields are LHCP waves, for which $\mathbf{E} = E_L(\mathbf{a}_x + j\mathbf{a}_y)/\sqrt{2}$, we obtain, from (6.31) and (6.32),

$$R_x = -jR_y = -\frac{eE_L}{\sqrt{2}} \frac{1}{(K - \omega^2 m) - \omega e B_{dc}}$$

Thus, the electron position relative to the equilibrium position is

$$|\mathbf{R}| = \sqrt{R_x^2 + R_y^2} = \frac{eE_L}{(K - \omega^2 m) - \omega e B_{dc}} \quad (6.34)$$

Note that $\mathbf{R}$, and therefore $|\mathbf{R}|$, under the influence of RHCP waves is different from that due to the excitation from LHCP waves. In other words, the dipole moments induced by right-hand and left-hand circularly polarized waves are different. Also, the net dipole moment $\mathcal{P}$ per unit volume due to RHCP waves, and the index of refraction as seen by such waves, differ from those of LHCP waves. Because of the difference in $\mathcal{P}$, the refractive index n_R of RHCP waves is different from n_L of LHCP waves.

6.3.5 Comments

It is important to emphasize that AO, EO, and MO effects are very complicated, and the discussions presented in this section are greatly simplified in order to explain the basic physical mechanisms involved. In general, the refractive index depends on the direction of wave propagation and on the state of polarization of the waves. For anisotropic media, the index is a **second order tensor,** as is Δn. The modulating electric fields $\mathbf{E}_m$ are vector quantities. We would need a third order linear EO tensor $\overset{\leftrightarrow}{\mathbf{r}}$, with 16 tensor components, to describe the Pockels effect completely.

The AO effects are even more complicated. As will be noted in section 6.8 and Appendix B, strain and stress are second order tensor quantities and each has six independent components. The **photoelastic tensor** $\overset{\leftrightarrow}{\mathbf{p}}$ is a fourth order tensor and has 36 tensor components. This means that 36 photoelastic coefficients would be needed to describe the AO effects in crystalline solids.

6.4 FARADAY ROTATION, AND MAGNETOOPTIC MODULATORS AND ISOLATORS

The Faraday effect was discovered in 1845 in Faraday's attempt to establish a link between magnetic fields and optical beams. In his experiments, LP waves were directed into glass blocks which were subjected to large dc magnetic fields along the direction of wave propagation. Faraday found that the polarization of LP fields rotated as the beams propagated in the glass. The rotation of polarization, commonly known as the Faraday rotation, has been found in all materials, including liquids and gases. In the following, we will show that Faraday rotation is a direct consequence of the fact that $n_R \neq n_L$. We will also relate Faraday rotation to material properties and strong magnetic fields. The basic configuration of MO modulators and isolators will then be discussed.

6.4.1 Faraday Rotation

In the terminology discussed in Section 6.2, isotropic and unmagnetized media are nonbirefringent, and RHCP and LHCP waves propagate with the same velocities. However, when a medium is subjected to a strong dc magnetic field, the dipole moment p and the polarization seen by the RHCP and LHCP waves are different, as indicated in (6.33) and (6.34). As a result, the two counter-rotating CP waves propagating along the direction of the strong magnetic field have different phase velocities, c/n_R and c/n_L. Suppose that at $z = 0$, the waves are linearly polarized in the x direction; that is,

$$\mathbf{E}(0) = E_{xo} \mathbf{a}_x = \frac{E_{xo}}{\sqrt{2}} (\mathbf{a}_R + \mathbf{a}_L)$$

For $z > 0$, the waves may be written as

$$\mathbf{E}(z) = \frac{E_{xo}}{\sqrt{2}} (\mathbf{a}_R \, e^{-jkn_R z} + \mathbf{a}_L \, e^{-jkn_L z})$$

The expression can be rearranged in terms of $\mathbf{a}_x$ and $\mathbf{a}_y$. When this is done, the equation becomes

$$\mathbf{E}(z) = E_{xo}[\mathbf{a}_x\cos\phi_F(z) + \mathbf{a}_y\sin\phi_F(z)]e^{-jk(n_R+n_L)z/2} \qquad (6.35)$$

where

$$\phi_F(z) = \frac{k(n_L - n_R)z}{2} \qquad (6.36)$$

In the time domain, the electric fields are

$$\mathcal{E}(z;\,t) = Re[\mathbf{E}(z)\,e^{j\omega t}] = E_{xo}[\mathbf{a}_x\cos\phi_F(z) + \mathbf{a}_y\sin\phi_F(z)]\cos(\omega t - k\frac{n_R + n_L}{2}z)$$

The electric field vector $\mathcal{E}(z;\,t)$ is at an angle $\phi_F(z)$ with respect to the x axis, and $\phi_F(z)$ increases linearly with z. This means that the field vectors rotate as the waves propagate. This is the polarization rotation effect discovered by Faraday.

In paramagnetic and diamagnetic materials, the angle of rotation ϕ_F is linearly proportional to the component of the magnetic field intensity H_m in the direction of propagation $\mathbf{a}_k$,

$$\phi_F(z) = V\,(\mathbf{H}_m \cdot \mathbf{a}_k)\,z = (V\,H_m\cos\theta_{hk})\,z \qquad (6.37)$$

where θ_{hk} is the angle between $\mathbf{H}_m$ and $\mathbf{a}_k$. The proportionality constant V is the **Verdet constant,** and is the angle of rotation per magnetic field intensity projected onto the direction $\mathbf{a}_k$ for unit path length. The constant V has a dimension of rad/A. Figure 6.6 shows the Verdet constant of several glass materials and garnets [9].

From (6.36) and (6.37), we can relate the Verdet constant to n_R and n_L,

$$V = \frac{k(n_L - n_R)}{2\,\mathbf{H}_m \cdot \mathbf{a}_k} = \frac{\pi\,(n_L - n_R)}{\lambda\,\mathbf{H}_m \cdot \mathbf{a}_k} \qquad (6.38)$$

In ferromagnetic materials, the rotation can be expressed in terms of the magnetization $\mathbf{M}_m$:

$$\phi_F(z) = K(\mathbf{M}_m \cdot \mathbf{a}_k)\,z = (K\,M_m\cos\theta_{mk})z \qquad (6.39)$$

where K is the **Kundt constant** and θ_{mk} is the angle between the magnetization $\mathbf{M}_m$ and $\mathbf{a}_k$.

6.4.2 MO Amplitude Modulators

Finally, we turn our attention to MO amplitude modulators. Figure 6.7a depicts a basic MO modulator setup, which consists of an MO material of length ℓ placed between a polarizer and an analyzer. A modulating magnetic field $\mathbf{H}_m$[2] is applied along $\mathbf{a}_k$. The optical fields entering the MO material are linearly polar-

---

[2] In the SI system, the unit for the magnetic field intensity H is ampere–turns per meter (A/m). The magnetic field intensity can also be expressed in oersted (Oe) and 1.0 A/m $= 4\pi \times 10^{-3}$ Oe. The unit for the magnetic flux density B is tesla (T), which is labeled as weber per meter square (Wb/m²) in older literature. Numerically, 1.0 T $= 1.0$ Wb/m² $= 1.0 \times 10^4$ Gauss.

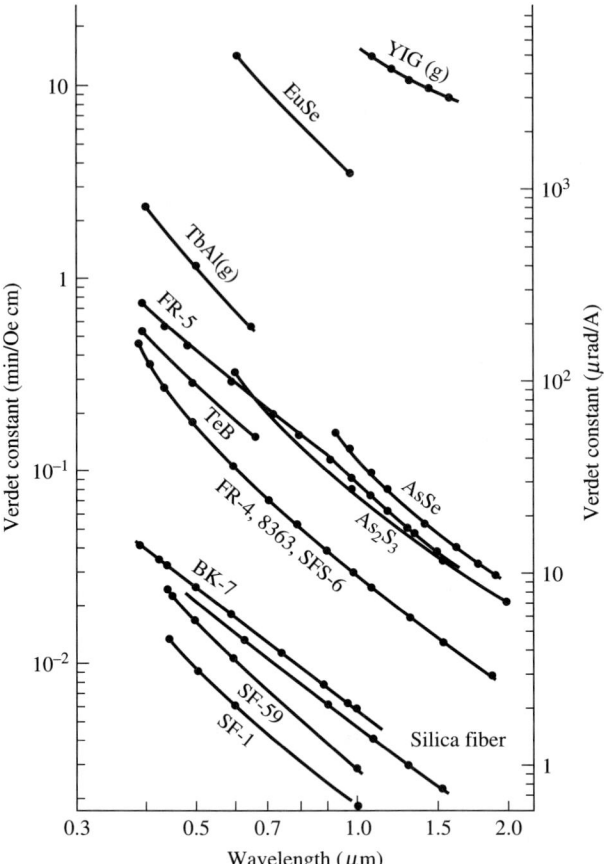

Figure 6.6 Verdet constant of some glasses and garnets [9]

ized in the x direction by the polarizer, that is, $\mathbf{E}(0) = E_{in}\mathbf{a}_x$. Under the influence of $\mathbf{H}_m$, the waves remain linearly polarized throughout the MO material, however, the electric field vector rotates as the wave propagates, as described by (6.35). The rotation of the field vectors is depicted schematically in the vector diagram immediately below the modulator in Figure 6.7a. Upon exiting the MO material, the waves are linearly polarized at an angle $\phi_F(\ell)$ from the x axis. There is no further change in the SOP. The field component selected by the analyzer is $E_{out} = \mathbf{a}_y \cdot \mathbf{E}(\ell)$ Thus, $|E_{out}| = |E_{in}\sin\phi_F(\ell)|$. The input beam intensity I_{in} is proportional to $|E_{in}|^2$, and the output beam intensity I_{out} is proportional to $|E_{out}|^2$, which means that

$$I_{out} = I_{in}\sin^2\phi_F(\ell) \tag{6.40}$$

Note that I_{out} is not linear with respect to ϕ_F. In fact, for small $\phi_F(\ell)$, the output intensity is a quadratic function of ϕ_F, as follows:

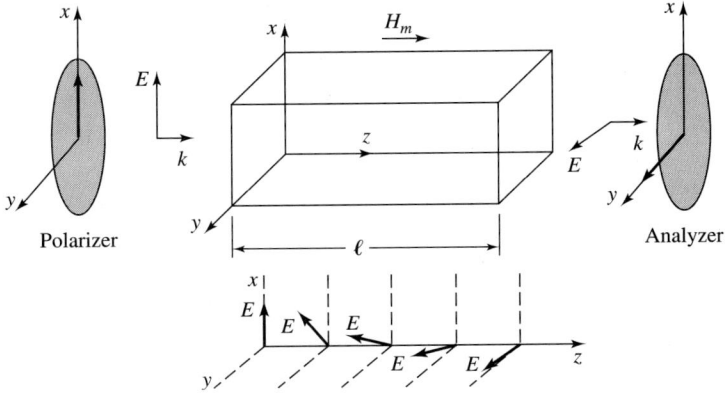

(a) MO modulator with crossed polarizer and analyzer

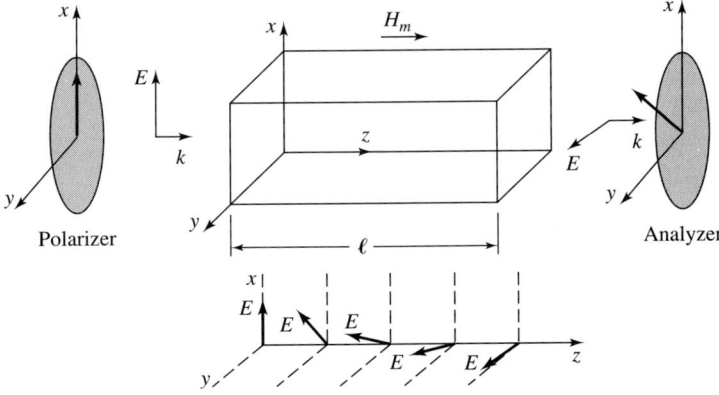

(b) MO modulator with polarizer and analyzer axes at 45 degrees

Figure 6.7 Schematic diagram of simple MO amplitude modulators

$$I_{out} \approx I_{in}\, \phi_F^2(\ell)$$

If the magnetic field $\mathbf{H}_m$ is produced by a current in a magnetic coil, $\mathbf{H}_m$ changes linearly with the electric current, but the output intensity I_{out} does not.

Next, consider a variation in the arrangement of Figure 6.7a. If the transmission axis of the polarizer is at 45 degrees to the axis of the analyzer, as shown in Figure 6.7b, the fields emerging from the analyzer are

$$E'_{out} = \frac{\mathbf{a}_x + \mathbf{a}_y}{\sqrt{2}} \cdot \mathbf{E}(\ell)$$

Therefore,

$$|E'_{out}| = \left| \frac{E_{in}}{\sqrt{2}} [\cos\phi_F(\ell) + \sin\phi_F(\ell)] \right|$$

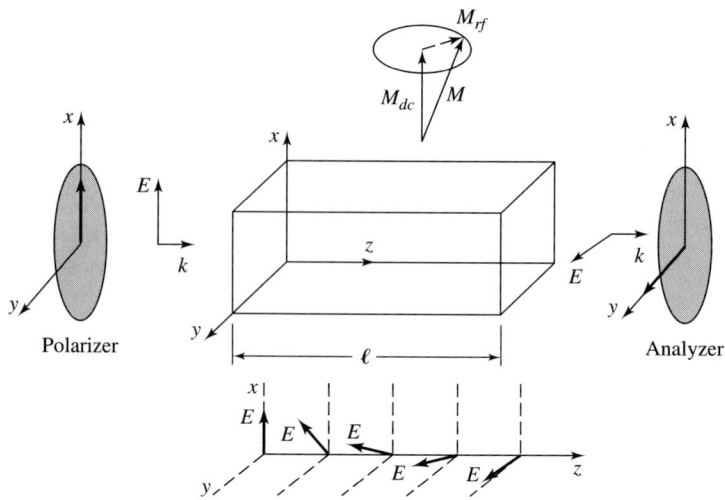

Figure 6.8 Schematic diagram of simple MO amplitude modulator with the dc magnetization normal to, and the rf magnetization parallel to, the wave vector.

and the output intensity becomes

$$I'_{out} = I_{in} \frac{1 + \sin 2\phi_F(\ell)}{2} \quad (6.41)$$

For $\phi_F(\ell) \ll 1$, I'_{out} may be approximated as

$$I'_{out} \approx I_{in} \left[\frac{1}{2} + \phi_F(\ell)\right]$$

which has a component that varies linearly with $\phi_F(\ell)$. As previously stated, the term ϕ_F itself varies linearly with the current in the magnetic coil.

While the MO modulation schemes shown in Figures 6.7a and b are simple conceptually, the electric current needed for modulating magnetic fields can be quite large. Also, the bandwidth is limited by the coil inductance. Hence, the configurations shown in Figures 6.7 a and b are not very useful in high-frequency modulation applications. On the other hand, the setup shown in Figure 6.7b may be used for optical isolators, which are discussed in the next subsection.

An alternative and more useful arrangement is shown in Figure 6.8. A suitable MO material, such as a paramagnetic crystal doped with rare earth ions or a ferromagnetic garnet [yttrium iron garnet (YIG)], is placed under the influence of radio frequency (rf) or microwave magnetic fields, in addition to a strong dc magnetic field. The MO material is biased into saturation by the dc field normal to $\mathbf{a}_k$. Because of the rf magnetic field in parallel with $\mathbf{a}_k$, the magnetization precesses about the dc magnetic field, as shown in Figure 6.8. Near the spin resonance in paramagnetic crystals, or the ferromagnetic resonance in ferromagnetic materials, the rf magnetization has a significant component in the direction of $\mathbf{a}_k$. In essence, the rf magnetization component causes the SOP to rotate. With the scheme shown in Figure 6.8, amplitude modulation in the gigahertz range can be achieved.

6.4.3 MO Isolators

The amplitude, phase, and frequency stability of all sources, including lasers, are affected by reflection. Yet, if we use attenuators to reduce the strength of the reflected signals, we also reduce the transmitted signals. Instead, we use **isolators,** which are devices that transmit the signals in one direction, with minimal attenuation, while imposing considerable loss on signals traveling in the opposite direction. The setup shown in Figure 6.7b is most useful as an optical isolator.

In Faraday rotation, the SOP rotates relative to the magnetic field $\mathbf{H}_m$, rather than the direction of wave propagation. This means that for waves propagating in either direction, the SOP rotates in the direction of increasing angle relative to the x axis. Consider the setup shown in Figure 6.7b, where the polarizer axis is 45 degrees relative to the analyzer axis. For waves propagating in the $+z$ direction, the SOP rotates from the x axis to 45 degrees relative to the x axis. If the incoming waves being selected by the polarizer are linearly polarized along the x axis, the SOP at $z = \ell$ coincides perfectly with the analyzer axis. Therefore, waves propagating in the $+z$ direction and linearly polarized along the x axis at the input are allowed to pass freely without attenuation. For waves propagating in the $-z$ direction, the situation is entirely different. Upon entering the analyzer from the right, which now serves as a polarizer, the waves are linearly polarized at an angle of 45 degrees relative to the x axis. As the wave proceeds toward the left, the SOP rotates continuously in the direction of increasing angle with respect to the x axis. At $z = 0$, $\mathbf{E}$ is now linearly polarized at an angle of $45° + 45° = 90°$ relative the x axis, that is, the y direction. The electric field $\mathbf{E}$ has become orthogonal to the polarizer axis, which now serves as an analyzer. Therefore, waves going in the $-z$ direction are blocked completely by the polarizer. This is the basic principle of optical **Faraday isolators.**

It should be emphasized that not all MO effects are nonreciprocal. While the Faraday effect is *nonreciprocal,* other magnetooptic effects, such as the Voigt effect, may be *reciprocal.* Devices based on reciprocal effects cannot function as isolators.

6.5 INDEX ELLIPSOIDS

For the remainder of this chapter, we will mainly be concerned with EO and AO materials that are nonmagnetic insulators. For nonmagnetic media, $\mathbf{B} = \mu_o \mu_r \mathbf{H}$ where $\mu_r \approx 1$. EO and AO materials are usually anisotropic, and the SOP of waves propagating in anisotropic media evolves in a complicated manner.

To begin our discussion on this subject, it is helpful to review wave propagation in isotropic media. The propagation of plane waves in nonconductive, electrically isotropic media is as follows:

- The motion of constant-phase wave fronts is taken as the direction of propagation and is represented by a wave vector **k**. The transport of power, as represented by the **Poynting vector P** $= \frac{1}{2}Re(\mathbf{E}\times\mathbf{H}^*)$, is in the same direction as **k**. In the expression, * signifies the complex conjugate of the complex quantity.
- The electric field vector **E** is always in parallel with **D**, and **B** is with **H**, and all field vectors are perpendicular to **k**. In addition **E**, **D**, and **k** are coplanar vectors. Also, **B**, **H**, and **k** are coplanar vectors, in a different plane from **E** and **D**. This is shown in Figure 6.9a.
- Unit vectors $\mathbf{a}_E$, $\mathbf{a}_H$, and $\mathbf{a}_k$ follow the right-hand rule and form the orthogonal vector triplet, $\mathbf{a}_E \times \mathbf{a}_H = \mathbf{a}_k$.
- The SOP remains unchanged as the waves propagate.
- The phase velocity of the plane waves is independent of $\mathbf{a}_k$ and the SOP.

The dielectric properties of electrically anisotropic media are much more complicated. The relationship of **E** to **D** is not characterized by a simple scalar dielectric constant ε_r; (6.18) is replaced by a tensor relationship. In an arbitrary coordinate system (x', y', z'), the electric constitutive relation of **E** and **D**, in matrix form, is:

$$\begin{bmatrix} D_{x'} \\ D_{y'} \\ D_{z'} \end{bmatrix} = \varepsilon_o \begin{bmatrix} \varepsilon_{rx'x'} & \varepsilon_{rx'y'} & \varepsilon_{rx'z'} \\ \varepsilon_{ry'x'} & \varepsilon_{ry'y'} & \varepsilon_{ry'z'} \\ \varepsilon_{rz'x'} & \varepsilon_{rz'y'} & \varepsilon_{rz'z'} \end{bmatrix} \begin{bmatrix} E_{x'} \\ E_{y'} \\ E_{z'} \end{bmatrix} \quad (6.42)$$

where ε_o is the free-space permittivity and the 3×3 matrix is the relative dielectric constant tensor. The matrix relationship in (6.42) can be greatly simplified by taking advantage of the symmetry of the matrix elements. Since $\varepsilon_{rij} = \varepsilon_{rji}^*$, it is always possible to choose a new coordinate system (x, y, z) such that the relative dielectric constant tensor is reduced to a *diagonal matrix*

$$\begin{bmatrix} D_x \\ D_y \\ D_z \end{bmatrix} = \varepsilon_o \begin{bmatrix} \varepsilon_{rx} & 0 & 0 \\ 0 & \varepsilon_{ry} & 0 \\ 0 & 0 & \varepsilon_{rz} \end{bmatrix} \begin{bmatrix} E_x \\ E_y \\ E_z \end{bmatrix} = \varepsilon_o \begin{bmatrix} n_1^2 & 0 & 0 \\ 0 & n_2^2 & 0 \\ 0 & 0 & n_3^2 \end{bmatrix} \begin{bmatrix} E_x \\ E_y \\ E_z \end{bmatrix} \quad (6.43)$$

The new coordinates $x, y,$ and z are the **principal axes** of the medium; ε_{rx}, ε_{ry}, and ε_{rz} are the **principal values** of the relative dielectric constant tensor; and n_1, n_2, and n_3 are the **principal indices** of the medium.

Because of the complications introduced by the electric constitutive relation (6.42) or (6.43), wave propagation in anisotropic media is also complicated. We could solve Maxwell's equations in conjunction with the appropriate constitutive relations. From these tedious studies, the basic properties of wave propagation in anisotropic media emerge. The first two are:

1. Although **E**, **D**, and **k** are coplanar vectors, **E** and **D** are not necessarily in parallel. While **D** is always perpendicular to **k**, **E** may have one component in parallel with, and another component normal to, **k**. As noted earlier,

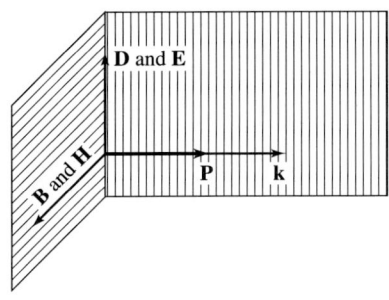 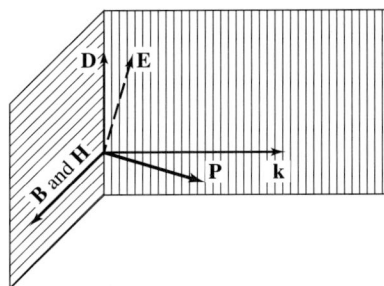

(a) Isotropic media   (b) Anisotropic media

Figure 6.9 **E, D, B, H, k** and **P** vectors of plane waves in isotropic and anisotropic media

H and **B** are in parallel and they are perpendicular to **k**. This is shown in Figure 6.9b. Also note that **k** and **P** are not necessarily in the same direction.

2. In general, the state of polarization of **D** evolves as the wave propagates. However, on the plane normal to a given **k**, there exist two unique and mutually perpendicular directions $\mathbf{a}_{p1}$ and $\mathbf{a}_{p2}$. Waves propagating along **k** while **D** is along $\mathbf{a}_{p1}$ or $\mathbf{a}_{p2}$ remain linearly polarized. We refer to $\mathbf{a}_{p1}$ or $\mathbf{a}_{p2}$ as the **principal directions,** and waves with **D** polarized along the principal directions are the **normal modes** propagating along $\mathbf{a}_k$. We identify the phase velocities of the normal modes as v_{p1} and v_{p2}, and the refractive indices associated with v_{p1} and v_{p2} are $n_{p1} = c/v_{p1}$ and $n_{p2} = c/v_{p2}$.

From Maxwell's equations, the dependence of n_{pj} ($j = 1$ or 2) on $\mathbf{a}_k$ and **D** can be determined. Once $\mathbf{a}_{pj}$ and n_{pj} are determined as functions of $\mathbf{a}_k$ and **D**, a geometrical representation for the index can be constructed. For a given $\mathbf{a}_k$, two vectors with vector lengths proportional to n_{pj} are drawn from the origin along $\mathbf{a}_{pj}$. The surface traced by these vectors as $\mathbf{a}_k$ varies is the **index ellipsoid,** which is also known as the **optical indicatrix** [10,11].

In an arbitrary coordinate system (x', y', z'), the equation for an ellipsoid is of the form

$$A'x'^2 + B'y'^2 + C'z'^2 + 2D'y'z' + 2E'z'x' + 2F'x'y' = 1$$

where A', B', C', etc., are constants. The equation can also be written as

$$\frac{x'^2}{n'^2_1} + \frac{y'^2}{n'^2_2} + \frac{z'^2}{n'^2_3} + \frac{2y'z'}{n'^2_4} + \frac{2z'x'}{n'^2_5} + \frac{2x'y'}{n'^2_6} = 1 \tag{6.44}$$

where ($1/n_i'^2$) are the elements of the **relative dielectric impermeability tensor.** By rotating the coordinates, we can reduce equation (6.44) to a canonical form [4]

$$Ax^2 + By^2 + Cz^2 = 1$$

In this form, the cross-product terms xy, yz, and zx are absent, and an ellipsoid is uniquely specified by the three constants A, B, and C. The rotated axes x, y, and z are the **principal axes of the ellipsoid.** These axes of the ellipsoid are also the principal axes of the medium introduced previously, in (6.43), and the constants A, B, and C are related to the principal indices of the medium. In terms of the principal axes and the principal indices of the medium, the index ellipsoid in the canonical form becomes

$$\frac{x^2}{n_1^2} + \frac{y^2}{n_2^2} + \frac{z^2}{n_3^2} = 1 \tag{6.45}$$

The tensor elements of the relative dielectric impermeability tensor in the rotated coordinate system are usually different from those in the original coordinate system, that is, $n_i \neq n'_i$. Depending on the values of the principal indices, materials may be classified as isotropic, uniaxial, or biaxial. For **isotropic media,** *three* principal indices are identical. For **uniaxial** and **biaxial media,** *two* and *none* of the principal indices are the same, respectively. Once the index ellipsoid for a medium is known, considerable information about waves propagating in the medium can be deduced.

The properties of wave propagation in anisotropic media are:

3. For a given $\mathbf{a}_k$, a plane perpendicular to $\mathbf{a}_k$ and through the origin is drawn. The intersection of this plane with the index ellipsoid is, in general, an ellipse. The major and minor axes of the ellipse are the principal directions $\mathbf{a}_{pj}$ mentioned in 2. Waves with $\mathbf{D}$ in the principal directions are the normal modes.
4. Once the wave vector and the principal directions are known, a vector can be drawn from the origin in the principal direction $\mathbf{a}_{pj}$. The intersection of that vector with the index ellipsoid gives the refractive index n_{pj} for the normal mode j propagating in the direction of $\mathbf{k}$ with $\mathbf{D}$ linearly polarized in $\mathbf{a}_{pj}$.
5. Once $\mathbf{D}$ is known, we can use the constitutive relation (6.43) to calculate $\mathbf{E}$ associated with $\mathbf{D}$.
6. There is an alternate way of determining $\mathbf{E}$. It can be shown that the $\mathbf{E}$ associated with a given $\mathbf{D}$ is normal to the *ellipsoidal surface* at that point. Thus, $\mathbf{E}$ can be obtained by a simple geometric construction. A plane tangent to the ellipsoid at the point corresponding to $\mathbf{D}$ is determined. A vector perpendicular to the tangent plane is drawn from the origin. This is the direction of $\mathbf{E}$.

It is worthwhile repeating that $\mathbf{D}$ is always perpendicular to the wave vector $\mathbf{k}$, but $\mathbf{E}$ is not necessarily so, as shown in Figure 6.9b. Even though $\mathbf{E}$ is usually the quantity of interest, it may be simpler to work with $\mathbf{D}$ rather than $\mathbf{E}$ in considering waves propagating in anisotropic media. An arbitrary $\mathbf{D}$ in a plane normal to $\mathbf{k}$ can always be expressed as the superposition of the two normal modes. For example, consider waves propagating in the z direction. Let $\mathbf{D}$ at $z = 0$ be

$$\mathbf{D}(0) = \mathbf{a}_{p1} D_{p1} + \mathbf{a}_{p2} D_{p2}$$

If we assume the normal modes $\mathbf{a}_{pj} D_{pj}$ propagate with indices n_{pj}, then we have, at $z > 0$,

$$\mathbf{D}(z) = \mathbf{a}_{p1} D_{p1} e^{-jkn_{p1}z} + \mathbf{a}_{p2} e^{-jkn_{p2}z}$$

Using these equations, we can describe wave propagation in various media.

6.5.1 Isotropic Media

Gases, liquids, and many solid materials are optically isotropic. In isotropic materials, all directions are equivalent (Figure 6.11a). For example, in crystals with cubic structures, such as Si, Ge, GaAs, and CdS, the three crystal axes are equivalent [6, 7], and the materials are optically isotropic. Also, the three principal indices of an isotropic medium are identical, that is, $n_1 = n_2 = n_3$, and it is only necessary to use a single index to describe the index ellipsoid. If we let the single index be n, then the index ellipsoid is

$$\frac{x^2}{n^2} + \frac{y^2}{n^2} + \frac{z^2}{n^2} = 1 \qquad (6.46)$$

which is simply a sphere (Figure 6.10a). All lines through the center of the sphere are normal to the spherical surface. Thus **D** and **E** are always in parallel and the index n is the same for all **k** and **D**. These properties of wave propagation in isotropic media were mentioned at the beginning of section 6.5.

6.5.2 Uniaxial Media

Crystals with **trigonal, tetragonal,** or **hexagonal** symmetry have a unique crystallographic direction or axis. When these crystals are rotated 120°, 90°, or 60° with respect to this particular axis, the rotated crystal lattices are indistinguishable from those prior to rotation (Figure 6.11b). In the terminology of crystallography, the particular axis is the c axis. These materials are optically uniaxial and the c axis is known as the **optic axis** of the uniaxial media [6, 7].

The index ellipsoids of uniaxial media are ellipsoids of revolution, that is spheroids (Figure 6.10b). They can be generated by rotating an ellipse with respect to one of its semiaxes, which is also the c axis. If we choose the c axis as the z axis, then the ellipsoid is given by

$$\frac{x^2}{n_o^2} + \frac{y^2}{n_o^2} + \frac{z^2}{n_e^2} = 1 \qquad (6.47)$$

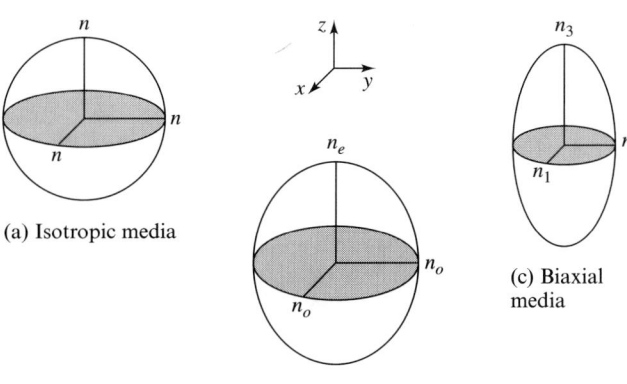

Figure 6.10 Index ellipsoids

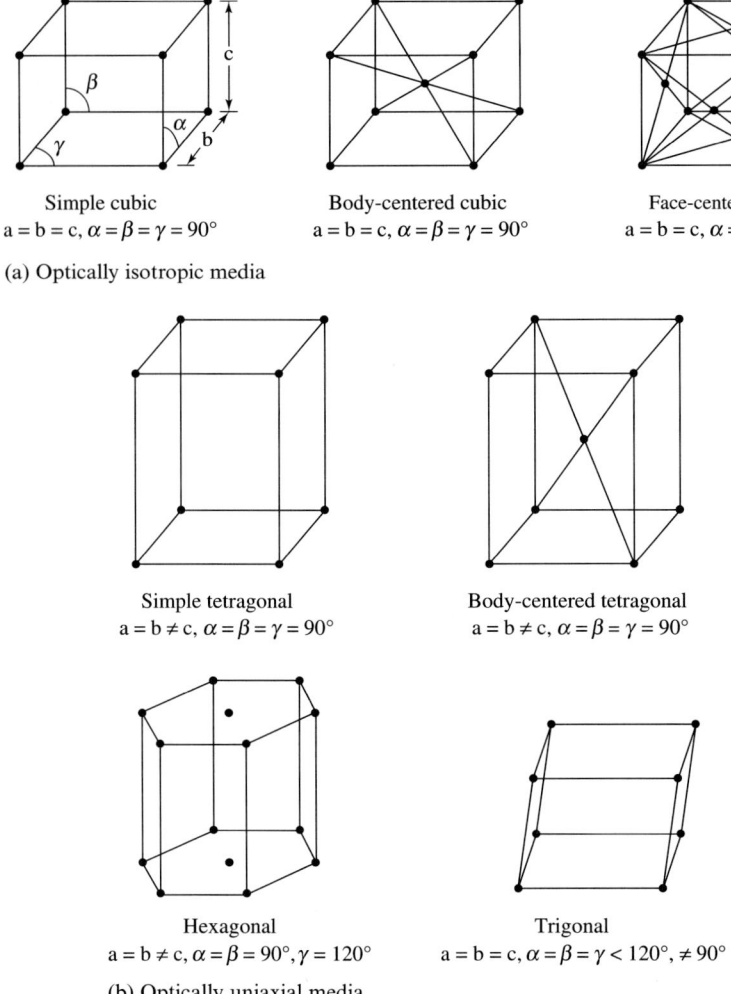

Figure 6.11 Crystal lattices in three-dimensional space

Two of three principal indices are the same. They are the **ordinary index**, n_o, of the medium. The third index is the **extraordinary index**, n_e. Materials with $n_e > n_o$ are known as **positive uniaxial** materials; those with $n_e < n_o$ are **negative uniaxial** materials.

Waves propagating along the optic axis are easy to describe. With **k** along the optic axis, all directions normal to $\mathbf{a}_z$ are the principal directions, all **D** vectors normal to $\mathbf{a}_z$ are the normal modes, and $v_{p1} = v_{p2}$. This is shown in Figure 6.12a. Plane waves propagating along the optic axis have a polarization-independent index, and **D** is in parallel with **E**. These waves are the ordinary waves, and the index of refraction is the ordinary index of refraction n_o.

6.5 Index Ellipsoids

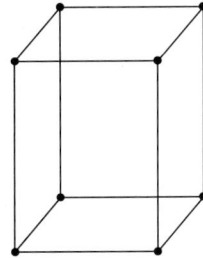

Simple orthorhombic
$a \neq b \neq c, \alpha = \beta = \gamma = 90°$

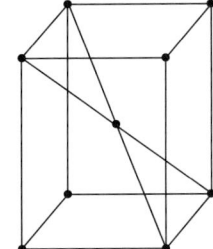

Body-centered Orthorhombic
$a \neq b \neq c, \alpha = \beta = \gamma = 90°$

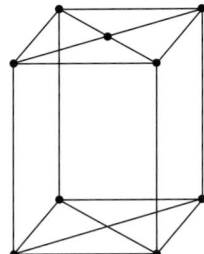

Base-centered orthorhombic
$a \neq b \neq c, \alpha = \beta = \gamma = 90°$

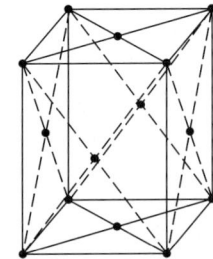

Face-centered orthorhombic
$a \neq b \neq c, \alpha = \beta = \gamma = 90°$

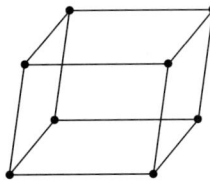

Simple monoclinic
$a \neq b \neq c, \alpha = \gamma = 90° \neq \beta$

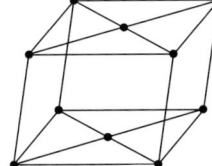

Base-centered monoclinic
$a \neq b \neq c, \alpha = \gamma = 90° \neq \beta$

Simplle triclinic
$a \neq b \neq c, \alpha \neq \beta \neq \gamma$

(c) Optically biaxial media

Figure 6.11 (*Continued*)

For waves propagating in directions other than the optic axis, the situation is more complicated. Consider waves traveling normal to the optic axis, in the y direction. For waves with **D** polarized in the x direction, the index of refraction is the ordinary index of refraction n_o, and the **E** accompanying **D** is in parallel with **D**. These vectors are shown as $\mathbf{D}_{or}$ and $\mathbf{E}_{or}$ in Figure 6.12b. Waves with the displacement vector along the optic axis are the extraordinary waves, and propagate with the extraordinary index of refraction n_e. This is shown as $\mathbf{D}_{ex}$ in Figure 6.12b. Note that $\mathbf{E}_{ex}$ is also in parallel with $\mathbf{D}_{ex}$. Also note that unit vectors $\mathbf{a}_x$ and $\mathbf{a}_z$ are the principal directions for waves propagating in the y direction. An arbitrary $\mathbf{D}_3$ in the xz plane can be viewed as the linear superposition of $\mathbf{D}_{or}$ and $\mathbf{D}_{ex}$. The electric field vectors associated with $\mathbf{D}_{or}$ and $\mathbf{D}_{ex}$ can

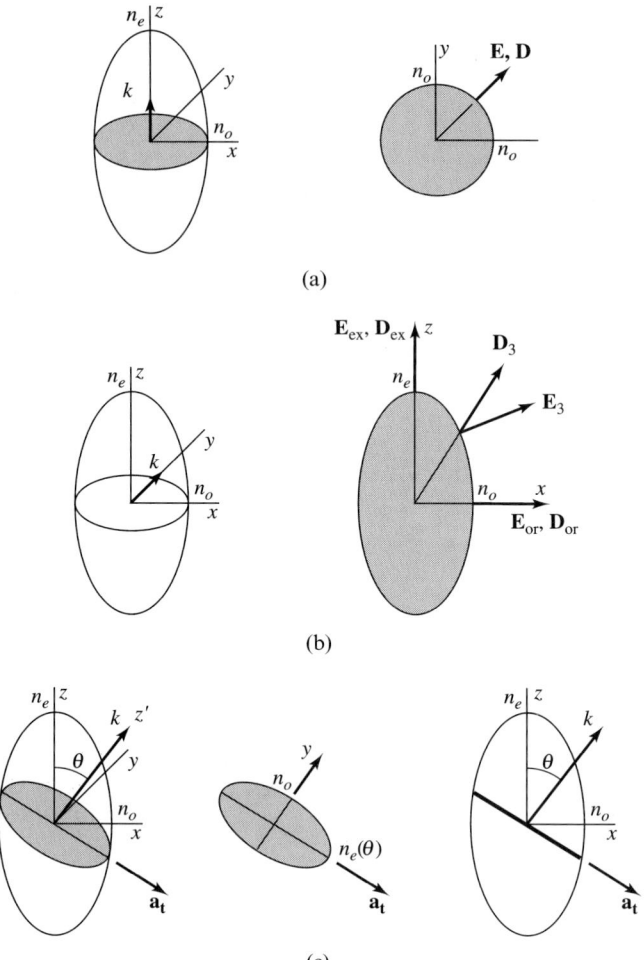

Figure 6.12 Determination of $n_e(\theta)$ of uniaxial materials

also be determined, and can be combined to give the electric field vector $\mathbf{E}_3$ associated with $\mathbf{D}_3$. However, $\mathbf{D}_3$ and $\mathbf{E}_3$ are not in parallel.

Finally, consider waves propagating at an arbitrary angle θ with respect to the optic axis (Figure 6.12c). Waves with $\mathbf{D}$ perpendicular to the plane defined by $\mathbf{a}_z$ and $\mathbf{k}$ (i.e., perpendicular to the paper) are the ordinary waves, for which $\mathbf{D}$ and $\mathbf{E}$ are in parallel and the index n_o is independent of θ. Waves with $\mathbf{D}$ in the plane defined by $\mathbf{a}_z$ and $\mathbf{k}$ are the extraordinary waves, for which the index $n_e(\theta)$ is a function of θ, and $\mathbf{D}$ and $\mathbf{E}$ are not in parallel. To determine $n_e(\theta)$, we note that the index ellipsoid is rotationally symmetric with respect to the z axis, and all vertical planes are equivalent. Without loss of generality, we choose $\mathbf{k}$ to be in the xz plane and $\mathbf{D}$ in the direction $\mathbf{a}_t$, as shown in Figure 6.12c. The coordinates of a point on the index ellipsoid are then given by

6.5 Index Ellipsoids

$$x^2 = n_e^2(\theta)\cos^2\theta, \qquad y^2 = 0, \qquad z^2 = n_e^2(\theta)\sin^2\theta$$

Using these relations in (6.47), we obtain

$$n_e^2(\theta)\left[\frac{\cos^2\theta}{n_o^2} + \frac{\sin^2\theta}{n_e^2}\right] = 1$$

Therefore,

$$\frac{1}{n_e^2(\theta)} = \frac{\cos^2\theta}{n_o^2} + \frac{\sin^2\theta}{n_e^2} \tag{6.48}$$

In two special cases, $n_e(0°) = n_o$ and $n_e(90°) = n_e$. For other values of θ, we use (6.48) to evaluate $n_e(\theta)$.

Example

Consider a uniaxial medium with $n_o = 1.500$ and $n_e = 1.700$. Suppose that the wave vector **k** is 30 degrees from the optic axis (z axis). Label the direction of **k** as the z' direction. At $z' = 0$,

$$\mathbf{D}(0) = D_t(0)\mathbf{a}_t + D_y(0)\mathbf{a}_y$$

where $D_t(0)$ and $D_y(0)$ are amplitude constants and $\mathbf{a}_t$ is shown in Figure 6.12c. Expressions for **E** and **D** as functions of z' are desired.

Obviously, waves with **D** linearly polarized in the directions of $\mathbf{a}_t$ and $\mathbf{a}_y$ are normal modes. Waves with **D** in the $\mathbf{a}_y$ direction are the ordinary waves and have an index of n_o. To evaluate the index for waves with **D** in the $\mathbf{a}_t$ direction, we obtain from (6.48)

$$n_e(30°) = \left[\frac{\cos^2 30°}{2.25} + \frac{\sin^2 30°}{2.89}\right]^{-1/2} = 1.543$$

Thus, for $z' > 0$

$$\mathbf{D}(z') = \mathbf{a}_t D_t(0)\, e^{-j1.543 k z'} + \mathbf{a}_y D_y(0)\, e^{-j1.500 k z'}$$

To determine **E**, we convert $\mathbf{a}_t$ back to $\mathbf{a}_x$ and $\mathbf{a}_z$

$$\mathbf{a}_t = \mathbf{a}_x \cos 30° - \mathbf{a}_z \sin 30°$$

In terms of x, y, and z components, **D** is

$$\mathbf{D}(z') = \mathbf{a}_x\, 0.8660 D_t(0)\, e^{-j1.543 k z'} + \mathbf{a}_y D_y(0)\, e^{-j1.500 k z'} - \mathbf{a}_z\, 0.5000 D_t(0)\, e^{-j1.543 k z'}$$

Then **E** is obtained from (6.43)

$$\mathbf{E}(z') = \frac{1}{\varepsilon_o}[\mathbf{a}_x \frac{0.8660 D_t(0)}{2.25} e^{-j1.543 k z'}$$
$$+ \mathbf{a}_y \frac{D_y(0)}{2.25} e^{-j1.500 k z'} - \mathbf{a}_z \frac{0.5000 D_t(0)}{2.89} e^{-j1.543 k z'}]$$

6.5.3 Biaxial Media

For crystals with **orthorhombic, monoclinic,** or **triclinic** symmetry, all crystallographic directions are distinct, and none are equivalent (Figure 6.11c). Nevertheless, two unique directions exist. [It is left as an exercise for the readers to show that these axes exist (see Problem 3).] For waves propagating along these directions, the index of refraction is polarization independent. These directions are the optic axes of the crystals, and they do not necessarily coincide with the crystallographic axes. Since there are two optic axes, these materials are **biaxial media** [6, 7].

The index ellipsoid of biaxial media is given by (6.45) with x, y, and z as the principal axes and n_1, n_2, and n_3 as the principal indices. For orthorhombic materials, three principal axes, though distinct, coincide exactly with the crystals axes. For materials with monoclinic symmetry, one of the principal axes is oriented along one of the crystal axes. For triclinic crystals, none of the principal axes is pointed along the crystal axes. In other words, two (monoclinic) or all (triclinic) of the principal axes of the index ellipsoid are oriented differently from the crystal axes.

Waves propagating in biaxial media are too complicated to be considered further in this text. We note, however, that uniaxial EO materials under the influence of electric fields usually become optically biaxial. Thus, the general expression (6.44) is useful in describing the index ellipsoid of EO materials under the influence of modulating electric fields.

6.5.4 Propagation along the Principal Axes

For waves propagating along the principal axes, there is an interesting and special property that can be used to treat wave propagation problems more effectively. Recall that for waves propagating along an arbitrary direction **k**, there are two principal directions $\mathbf{a}_{p1}$ and $\mathbf{a}_{p2}$. For waves propagating along one of the principal axes of the medium, the other principal axes are precisely the principal directions. Since the principal axes are normal to the ellipsoidal surface, the **E** and **D** vectors along these directions are parallel. To be specific, let x, y, and z be the principal axes of a medium with principal indices n_1, n_2 and n_3. For waves propagating in the x direction, the principal directions are exactly the y and z directions. Waves with **D** along the y and z directions are the normal modes. The **E** associated with these normal modes are also along the y and z directions. In other words, **E** and **D** of the normal modes are parallel if **k** is along one of the principal axes. It should be emphasized that this special feature holds *only* for waves propagating along the principal axes; it is not necessarily valid for waves propagating along other directions.

6.6 LINEAR ELECTROOPTIC EFFECTS

6.6.1 Linear EO Coefficients

In this section, we will consider the effects of modulating electric fields on the optical properties of media. Since the modulating electric fields are also time-harmonic fields, it is advantageous to use the frequency domain representation $\mathbf{E_m}$ in lieu of the time domain representation $\mathcal{E}_m$ to describe those fields. Note that the frequencies of the modulating fields are much lower than the optical frequencies. As far as the optical beams are concerned, the modulating fields are low-frequency or even dc fields. The effects of the modulating fields are to distort and/or rotate the index ellipsoid. Equation (6.29) is the simplest equation for describing the EO effect and is useful for simple situations. The general expression describing EO effects is much more complicated, because different modulating field components can lead to different effects in the various optical fields. These effects can best be expressed in terms of changes in the elements of the relative dielectric impermeability tensor

$$\Delta\left[\frac{1}{n_i^2}\right] = \frac{1}{n_{iE}^2} - \frac{1}{n_{i0}^2}, \quad i = 1, 2, \ldots 6 \qquad (6.49)$$

where $(1/n_{iE}^2)$ and $(1/n_{i0}^2)$ are the matrix elements of the relative dielectric impermeability tensor with and without the modulating electric fields, respectively. In the presence of n_4, n_5, and n_6, the general expression (6.44) must be used to describe the index ellipsoid. Let the modulating fields be represented by

$$\mathbf{E_m} = E_{mx}\mathbf{a_x} + E_{my}\mathbf{a_y} + E_{mz}\mathbf{a_z}$$

The linear equation describing the change in $1/n_i^2$ can be written in matrix form.

$$\begin{bmatrix} \Delta\frac{1}{n_1^2} \\ \Delta\frac{1}{n_2^2} \\ \Delta\frac{1}{n_3^2} \\ \Delta\frac{1}{n_4^2} \\ \Delta\frac{1}{n_5^2} \\ \Delta\frac{1}{n_6^2} \end{bmatrix} = \begin{bmatrix} \frac{1}{n_{1E}^2} - \frac{1}{n_{10}^2} \\ \frac{1}{n_{2E}^2} - \frac{1}{n_{20}^2} \\ \frac{1}{n_{3E}^2} - \frac{1}{n_{30}^2} \\ \frac{1}{n_{4E}^2} - \frac{1}{n_{40}^2} \\ \frac{1}{n_{5E}^2} - \frac{1}{n_{50}^2} \\ \frac{1}{n_{6E}^2} - \frac{1}{n_{60}^2} \end{bmatrix} = \begin{bmatrix} r_{11} & r_{12} & r_{13} \\ r_{21} & r_{22} & r_{23} \\ r_{31} & r_{32} & r_{33} \\ r_{41} & r_{42} & r_{43} \\ r_{51} & r_{52} & r_{53} \\ r_{61} & r_{62} & r_{63} \end{bmatrix} \begin{bmatrix} E_{mx} \\ E_{my} \\ E_{mz} \end{bmatrix}$$

The unit of the linear EO coefficients r_{ij} is m/V. Although there are potentially 18 independent EO coefficients, many of them are zero [10,11]. Three groups of EO materials often used in EO modulators and their EO coefficients are described in the following paragraphs.

GaAs, GaP, ZnTe and CdTe are cubic crystals with three nonzero, numerically identical EO coefficients: $r_{41} = r_{52} = r_{63}$. For these materials, the EO tensor is of the form

$$\begin{bmatrix} 0 & 0 & 0 \\ 0 & 0 & 0 \\ 0 & 0 & 0 \\ r_{41} & 0 & 0 \\ 0 & r_{41} & 0 \\ 0 & 0 & r_{41} \end{bmatrix} \tag{6.51}$$

The crystal symmetry of potassium dihydrogen phosphate (KH_2PO_4, or KDP) is identical to that of ammonium dihydrogen phosphate ($NH_4H_2PO_4$, or ADP). If these crystals are grown in heavy water (D_2O), some hydrogen atoms are replaced by deuterium atoms and the EO coefficients become larger. Such materials are potassium dideuterium phosphate (KD_2PO_4) and ammonium dideuterium phosphate ($NH_4D_2PO_4$), also known as D-KDP and D-ADP, or as KD*P and AD*P. Since these materials have the same EO properties, they are grouped together and referred to as ADP/KDP crystals. ADP/KDP crystals have a tetragonal symmetry with three nonzero EO coefficients (r_{41}, r_{52}, and r_{63}), two of which are identical, ($r_{41} = r_{52}$). Thus, the EO matrix of ADP/KDP crystals is

$$\begin{bmatrix} 0 & 0 & 0 \\ 0 & 0 & 0 \\ 0 & 0 & 0 \\ r_{41} & 0 & 0 \\ 0 & r_{41} & 0 \\ 0 & 0 & r_{63} \end{bmatrix} \tag{6.52}$$

Lithium niobate ($LiNbO_3$) and lithium tantalate ($LiTaO_3$) crystals have trigonal symmetry, and have a relatively complicated EO matrix of the form

$$\begin{bmatrix} 0 & -r_{22} & r_{13} \\ 0 & r_{22} & r_{13} \\ 0 & 0 & r_{33} \\ 0 & r_{51} & 0 \\ r_{51} & 0 & 0 \\ -r_{22} & 0 & 0 \end{bmatrix} \tag{6.53}$$

Table 6.1 lists the EO coefficients and refractive indices of selected EO materials at 0.633 μm [11].

Table 6.1 EO coefficients of selected materials (excerpted from [11])

Material	EO coefficient $\times 10^{-12}$ m/V	Refractive index	Dielectric constant
GaP	$r_{41}=-.097$	$n=3.32$	$\varepsilon_r=10$
ZnTe	$r_{41}=4.3$	$n=2.99$	$\varepsilon_r=10.1$
ADP	$r_{41}=23.41$ $r_{63}=7.83$	$n_o=1.5220$ $n_e=1.4773$	$\varepsilon_{rx}=\varepsilon_{ry}=56$ $\varepsilon_{rz}=15$
KDP	$r_{41}=8$ $r_{63}=11$	$n_o=1.5074$ $n_e=1.4667$	$\varepsilon_{rx}=\varepsilon_{ry}=42$ $\varepsilon_{rz}=21$
LiNbO$_3$	$r_{13}=8.6, r_{22}=3.4$ $r_{33}=30.8, r_{51}=28$	$n_o=2.286$ $n_e=2.200$	$\varepsilon_{rx}=\varepsilon_{ry}=43$ $\varepsilon_{rz}=28$
LiTaO$_3$	$r_{13}=7.5, r_{22}=1$ $r_{33}=33, r_{51}=20$	$n_o=2.176$ $n_e=2.180$	$\varepsilon_{rx}=\varepsilon_{ry}=41$ $\varepsilon_{rz}=43$

1. $\lambda=0.633$ μm
2. The EO coefficients and the relative dielectric constants for ADP and KDP are for low frequency modulations, (audio frequencies or lower). The data for other materials are for high frequency applications (radio frequency or higher).

6.6.2 ADP/KDP Crystals

In preparation for the discussion of EO modulators in the next section, we consider the EO effects in ADP/KDP crystals. Choose the optic axis of the uniaxial material as the z axis. In the absence of modulating electric fields, $n_1 = n_2 = n_o$, $n_3 = n_e$, and $(1/n_i^2) = 0$ for $i = 4, 5,$ and 6. Assuming that the modulating electric fields are along the z axis, $\mathbf{E}_m = \mathbf{a}_z E_{mz}$, and we obtain $\Delta(1/n_6^2) = r_{63}E_{mz}$ from (6.50). There is no change in any other matrix elements of the relative dielectric impermeability tensor. Thus, (6.44) becomes

$$\frac{1}{n_o^2}x^2 + \frac{1}{n_o^2}y^2 + \frac{1}{n_e^2}z^2 + 2r_{63}E_{mz}\,xy = 1 \qquad (6.54)$$

The z axis remains a principal direction for ADP/KDP, even in the presence of the modulating fields $\mathbf{a}_z E_{mz}$. However, the material is not uniaxial, and the x and y directions are not principal directions in the presence of $E_{mz}\mathbf{a}_z$. If a plane is drawn through the origin and perpendicular to the z axis, the intercept of the plane with the ellipsoid (6.54) is an ellipse and the major and minor axes of the ellipse are not along the x and y axes. To find the major and minor axes of the ellipse, we rotate the x and y coordinates with respect to the z axis, as shown in Figure 6.13. Let the rotated coordinates be (x'', y'', z); then,

$$x = x''\cos\phi - y''\sin\phi, \qquad y = x''\sin\phi + y''\cos\phi$$

Substituting the above equation into (6.54), we obtain

$$(\frac{1}{n_o^2} + r_{63}E_{mz}\sin2\phi)x''^2 + (\frac{1}{n_o^2} - r_{63}E_{mz}\sin2\phi)y''^2 + \frac{1}{n_e^2}z^2 + 2r_{63}E_{mz}\cos2\phi x''y'' = 1$$

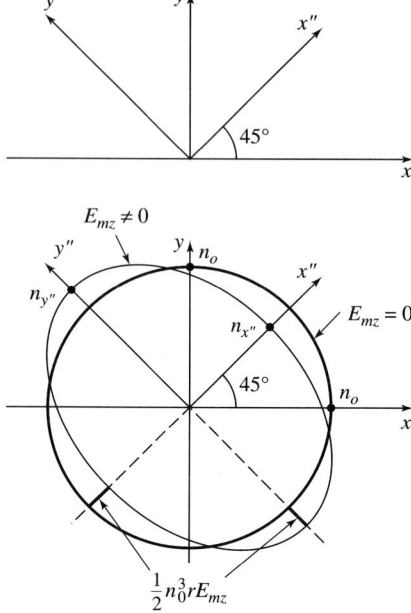

Figure 6.13 Effect of a modulating field $\mathbf{a}_z E_{mz}$ on the index ellipsoid of KDP/ADP materials

For $\phi = \pm 45$ degrees, the cross product $x''y''$ term vanishes, and the x'' and y'' axes are the minor and major axes, respectively. Let $\phi = +45$ degrees[3], and let $n_{x''}$ and $n_{y''}$ be the indices of refraction for waves propagating along the z axis and polarized along the x'' and y'' axes. Then

$$\frac{1}{n_{x''}^2} = \frac{1}{n_o^2} + r_{63} E_{mz}$$

$$\frac{1}{n_{y''}^2} = \frac{1}{n_o^2} - r_{63} E_{mz}$$

By taking into account that $|r_{63} E_{mz}| \ll 1/n_o^2$, we obtain explicit expressions for $n_{x''}$ and $n_{y''}$

$$n_{x''} \approx n_o - \frac{1}{2} n_o^3 r_{63} E_{mz} \qquad (6.55)$$

$$n_{y''} \approx n_o + \frac{1}{2} n_o^3 r_{63} E_{mz} \qquad (6.56)$$

In terms of the rotated coordinates x'', y'', and z, and the indices $n_{x''}$, $n_{y''}$, and n_e, the index ellipsoid becomes

---

[3]If $\phi = -45$ degrees, the resulting axes are 90 degrees from the x'' and y'' axes.

$$\frac{1}{n_{x''}^2} x''^2 + \frac{1}{n_{y''}^2} y''^2 + \frac{1}{n_e^2} z^2 = 1 \qquad (6.57)$$

In summary, ADP/KDP materials, which are uniaxial materials without the modulating fields, become biaxial in the presence of $\mathbf{a}_z E_{mz}$. The electric field $\mathbf{a}_z E_{mz}$ has the effect of respectively increasing and decreasing the indices of refraction for optical beams polarized along the x'' and y'' directions. However, the electric field has no effect on beams polarized along the z axis. Also, the rotated coordinates x'', y'', and z are the principal axes for ADP/KDP in the presence of $\mathbf{a}_z E_{mz}$. As discussed previously, waves propagating along a principal axis are particularly simple in that $\mathbf{E}$ and $\mathbf{D}$ are parallel if $\mathbf{D}$ is polarized along the other principle axes. These features are used to analyze EO modulators.

6.7 ELECTROOPTIC MODULATORS

As discussed in the last section, the relative dielectric impermeability of EO materials can be varied by electric fields. The presence of modulating electric fields can change the phase delay of waves propagating in the EO media. In addition, the phase changes for different components of the optical field are different. As a result, the SOP of the wave evolves as the wave propagates. By converting the evolution of the SOP to amplitude variations, we produce amplitude modulation. This is the basic mechanism in EO amplitude modulators [10,11]. To convert the polarization changes to amplitude variations, an EO crystal is placed between a polarizer and an analyzer, a setup that is similar to that of MO modulators. The polarizer serves as a polarization selector and the analyzer serves as a polarization discriminator. There are two basic EO modulator configurations. The difference between them is in the direction of the modulating electric field relative to the propagation of the optical beam. If the modulating electric field is along the direction of optical beam propagation, the setup is called a **longitudinal EO modulator configuration.** If the modulating field is perpendicular to the optical beam, the arrangement is known as the **transverse modulator configuration.** A subscript L is used to designate terms associated with longitudinal modulators, and T is used for transverse modulator terms. Also, we will use ADP/KDP crystals in the discussions illustrating the operating of EO modulators. For simplicity, assume that the front and back facets of the EO material are coated with antireflecting layers to minimize reflections from the interface between air and the ADP/KDP material.

6.7.1 Longitudinal EO Modulators

Figure 6.14 depicts the basic setup for longitudinal EO modulators. An ADP/KDP crystal of length ℓ is subjected to an applied voltage V_m. In the presence of a modulating electric field $E_{mz} = V_m/\ell$ in the z direction, the principal axes of ADP/KDP are the x'', y'', and z axes, as discussed in the last section. The polarizer axis (y direction) is 45 degrees relative to the x'' axis. For waves propagating in the z direction with $\mathbf{D}$ polarized along the x'' or y'' axis, $\mathbf{E}$ is also in

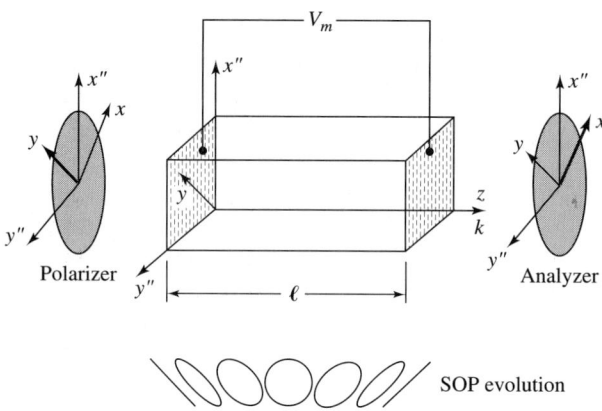

Figure 6.14 Schematic diagram of a longitudinal EO modulator with a KDP/ADP crystal

the x'' and y'' directions, and the indices are n''_x or n''_y, as given in (6.55) or (6.56), respectively. Let the electric field of the optical beam at $z = 0$ be

$$\mathbf{E}(0) = \mathbf{a}_{x''}E_{0x''} + \mathbf{a}_{y''}E_{0y''} \tag{6.58}$$

Also assume that the two field components are in time phase at the input. The field at $z > 0$ is

$$\mathbf{E}(z) = \mathbf{a}_{x''} E_{0x''} e^{-jk(n_o - \frac{1}{2}n_o^3 r_{63} E_{mz})z} + \mathbf{a}_{y''} E_{0y''} e^{-jk(n_o + \frac{1}{2}n_o^3 r_{63} E_{mz})z} \tag{6.59}$$

Although the field components are in time phase at $z = 0$, they are out of phase for $z > 0$, and the phase difference is

$$\Delta\Phi_L(z) = -k(\frac{1}{2} n_o^3 r_{63} E_{mz} z + \frac{1}{2} n_o^3 r_{63} E_{mz} z) = -k\, n_o^3 r_{63} E_{mz} z \tag{6.60}$$

Due to the phase difference $\Delta\Phi_L(z)$, one field component is either retarded or advanced relative to the other component, and the phase difference is known as the **retardation**. Depending on the value of $\Delta\Phi_L(z)$, $\mathbf{E}(z)$ given by (6.59) may be linearly, circularly, or elliptically polarized. Consider the special case where $E_{0x''} = E_{0y''}$. When $\Delta\Phi_L(z) = 0, 2\pi, 4\pi, \ldots$, $\mathbf{E}(z)$ is linearly polarized along -45 degrees from the x'' axis. When $\Delta\Phi_L(z) = \pi, 3\pi, 5\pi, \ldots$, $\mathbf{E}(z)$ is also linearly polarized; however, the field vector is $+45$ degrees from the x'' axis. When $\Delta\Phi_L(z)$ is odd integer multiples of $\pi/2$, $\mathbf{E}(z)$ becomes circularly polarized. Except for these two retardation values, the waves are elliptically polarized.

The retardation at $z = \ell$ changes linearly with the applied voltage V_m, since

$$\Delta\Phi_L(\ell) = -kn_o^3 r_{63} E_{mz}\ell = -kn_o^3 r_{63} V_m \tag{6.61}$$

We define the **half-wave voltage**, $V_{L\pi}$, as the voltage required to change the retardation by $\pm\pi$, as follows:

$$V_{L\pi} = \frac{\pi}{k\, n_o^3 r_{63}} = \frac{\lambda}{2 n_o^3 r_{63}} \qquad (6.62)$$

In terms of the half-wave voltage, the retardation at $z = \ell$ becomes

$$\Delta\Phi_L(\ell) = -\pi V_m / V_{L\pi} \qquad (6.63)$$

In Figure 6.14, the polarizer axis is along the y direction. Waves emerging from the polarizer and incident upon the EO material are therefore polarized in the y direction. Let the amplitude of y polarized fields entering the EO material be E_{in}. The corresponding light intensity I_{in} is proportional to $|E_{in}|^2$. By resolving the input fields into x'' and y'' components and noting that $\hat{a}_y \cdot \hat{a}_{x''} = \hat{a}_y \cdot \hat{a}_{y''} = 1/\sqrt{2}$, we obtain, for $z = 0$,

$$\mathbf{E}(0) = E_{in}\mathbf{a}_y = \frac{E_{in}}{\sqrt{2}}(\mathbf{a}_{x''} + \mathbf{a}_{y''})$$

Corresponding to (6.58), we have $E_{0x''} = E_{0y''} = E_{in}/\sqrt{2}$. At $z = \ell$, the electric fields are, from (6.59) and (6.60),

$$\mathbf{E}(\ell) = \frac{E_{in}}{\sqrt{2}} e^{-jkn_o\ell}[\mathbf{a}_{x''}e^{-j\Delta\Phi_L(\ell)/2} + \mathbf{a}_{y''}e^{j\Delta\Phi_L(\ell)/2}]$$

There is no further change in the SOP for waves outside the EO crystal. The polarization of waves arriving at the analyzer is the same as that emerging from the ADP/KDP crystal.

The analyzer plays an important role in determining the modulator output. To emphasize this point, consider three special cases. First, remove the analyzer completely. In the absence of an analyzer, the output intensity is independent of the modulating fields, and the arrangement shown in Figure 6.14 without an analyzer does not function as an amplitude modulator.

Second, keep the analyzer, and assume that its axis is normal to the polarizer axis, as shown in Figure 6.14. Noting that $\mathbf{a}_x \cdot \mathbf{a}_{x''} = -\mathbf{a}_x \cdot \mathbf{a}_{y''} = 1/\sqrt{2}$, the field component selected by the analyzer is

$$\mathbf{a}_x \cdot \mathbf{E}(\ell) = \frac{E_{in}}{\sqrt{2}} e^{-jkn_o\ell}\left[\frac{1}{\sqrt{2}} e^{-j\Delta\Phi_L(\ell)/2} - \frac{1}{\sqrt{2}} e^{j\Delta\Phi_L(\ell)/2}\right]$$
$$= -j E\, e^{-jkn_o\ell} \sin(\Delta\Phi_L(\ell)/2)$$

The light intensity I_{out} emerging from the analyzer is proportional to $|\mathbf{a}_x \cdot \mathbf{E}(\ell)|^2$. Thus, the ratio of the output to the input intensity is

$$\frac{I_{out}}{I_{in}} = \sin^2 \frac{\Delta\Phi_L(\ell)}{2} = \sin^2 \frac{\pi V_m}{2 V_{L\pi}} \qquad (6.64)$$

Finally, consider the case in which the analyzer axis is aligned with the polarizer axis. In Figure 6.14, the analyzer axis would be in the y direction. The field component selected by the analyzer would then be

344 Chapter 6 Modulation and Deflection of Optical Beams

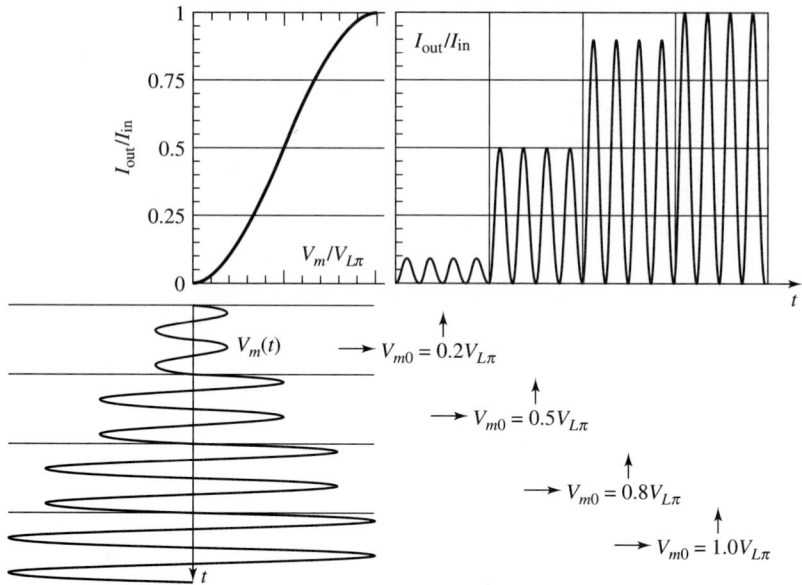

Figure 6.15 Response of an EO modulator with crossed polarizer and analyzer

$$\mathbf{a}_y \cdot \mathbf{E}(\ell) = \frac{E_{in}}{\sqrt{2}} e^{-jkn_o\ell} \left[\frac{1}{\sqrt{2}} e^{-j\Delta\Phi_L(\ell)/2} + \frac{1}{\sqrt{2}} e^{j\Delta\Phi_L(\ell)/2} \right]$$

$$= E_{in} e^{-jkn_o\ell} \cos(\Delta\Phi_L(\ell)/2)$$

The output intensity I'_{out} is now proportional to $|\mathbf{a}_y \cdot \mathbf{E}(\ell)|^2$, and the ratio of the output to the input intensity is

$$\frac{I'_{out}}{I_{in}} = \cos^2 \frac{\pi V_m}{2 V_{L\pi}} \tag{6.65}$$

which is complementary to (6.64)

If the modulating voltage $V_m(t)$ is a function of time, the outputs $I_{out}(t)$ and $I'_{out}(t)$ are, also. However, $I_{out}(t)$ and $I'_{out}(t)$ are not true replicas of $V_m(t)$. To illustrate the input–output relationship, assume that the modulating voltage is a simple sine function, $V_m(t) = V_{m0} \sin\omega_m t$, and plot the modulating voltages with $V_{m0} = 0.2V_{L\pi}$, $0.5V_{L\pi}$, $0.8V_{L\pi}$, and $1.0V_{L\pi}$ as shown in the lower left-hand corner of Figure 6.15. Also, plot each modulating signal as a function of time over two periods. The corresponding output $I_{out}(t)/I_{in}$ is shown in the upper right-hand corner of Figure 6.15. While $V_m(t)$ can be positive or negative, the output $I_{out}(t)$ is always positive. Therefore, $I_{out}(t)$ resembles a "rectified" version of $V_m(t)$. Actually, the output has sharp minima, and relatively smooth maxima. As V_{m0} increases, the output waveform becomes distorted. For V_{m0} greater than $V_{L\pi}$, the distortion becomes quite obvious. If we plot $I'_{out}(t)/I_{in}$ in the same manner, the waveform is even more distorted.

Obviously, the input–output relationship is not linear. To linearize this relationship, return to the crossed polarizers arrangement shown in Figure 6.14. Also, superimpose an rf component onto the dc component. Assume the modulating voltage is given by $V_m(t) = V_{dc} + V_{rf}\sin\omega_m t$, and that $V_{dc} = V_{L\pi}/2$. Then,

$$V_m(t) = \frac{1}{2} V_{L\pi} + V_{rf}\sin\omega_m t \tag{6.66}$$

and (6.64) becomes

$$\frac{I_{out}(t)}{I_{in}} = \frac{1}{2}\left[1 + \sin\left(\frac{\pi V_{rf}}{V_{L\pi}}\sin\omega_m t\right)\right] \tag{6.67}$$

The input–output relationship is plotted in Figure 6.16 for $V_{rf} = 0.2V_{L\pi}$, $0.5V_{L\pi}$, $0.8V_{L\pi}$, and $1.0V_{L\pi}$. We see from this figure that the rf component of the output resembles the rf component of the input if V_{rf} is less than $0.5V_{L\pi}$. In particular, if V_{rf} is much smaller than $V_{L\pi}$, then

$$\frac{I_{out}(t)}{I_{in}} \approx \frac{1}{2}[1 + \frac{\pi V_{rf}}{V_{L\pi}}\sin\omega_m t] \tag{6.68}$$

This shows that the time-varying portion of $I_{out}(t)/I_{in}$ varies linearly with respect to the rf component of V_m. If the electric bias V_{dc} is not equal to $V_{L\pi}/2$, the input–output relationship can still be linearized, but the range of linear operation is smaller.

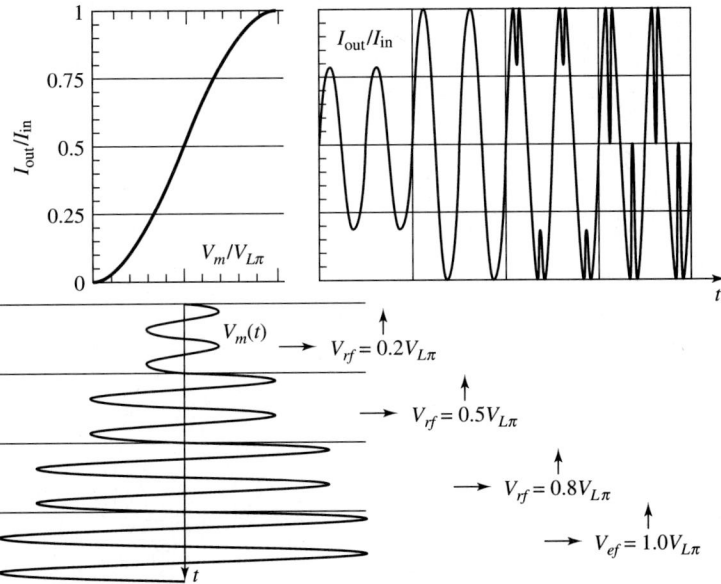

Figure 6.16 Response of an EO modulator with crossed polarizer and analyzer, and an electric bias for linear operation

6.7.2 Transverse EO Modulators

In transverse EO modulators, the modulating electric fields are transverse to the optical path. ADP/KDP crystals can be used in this arrangement. The modulating electric field is in the z direction (Figure 6.17), $E_{mz} = V_m/d$. The optical beams propagate in the y'' direction and **D** and **E** are in the $x''-z$ plane. The light emerging from the polarizer is again linearly polarized. By resolving the electric fields into x'' and z components, we obtain

$$\mathbf{E}(y'') = \mathbf{a}_z \frac{E_{in}}{\sqrt{2}} e^{-jkn_e y''} + \mathbf{a}_{x''} \frac{E_{in}}{\sqrt{2}} e^{-jk(n_o - \frac{1}{2} n_o^3 r_{63} E_{mz}) y''} \tag{6.69}$$

Note that the applied electric fields have no effect on the z polarized waves. The retardation at $y'' = \ell$ is

$$\Delta \Phi_T(\ell) = k(n_e - n_o + \frac{1}{2} n_o^3 r_{63} \frac{V_m}{d}) \ell \tag{6.70}$$

which has a voltage-independent part

$$[\Delta \Phi_T(\ell)]_{op} = k(n_e - n_o) \ell$$

and a voltage-dependent part

$$[\Delta \Phi_T(\ell)]_v = \frac{1}{2} k n_o^3 r_{63} \frac{\ell}{d} V_m$$

The voltage-independent part is the retardation due to the index difference $n_e - n_o$ of the ADP/KDP materials, and is purely optical in origin. Its effect may be compensated for by using a retarder plate, which provides an optical bias, or by the combination use of an optical bias and a dc voltage.

The voltage-dependent part $[\Delta \Phi_T(\ell)]_v$ changes linearly with the crystal length ℓ. Therefore, the applied voltage V_m can be reduced by using longer crystals. A

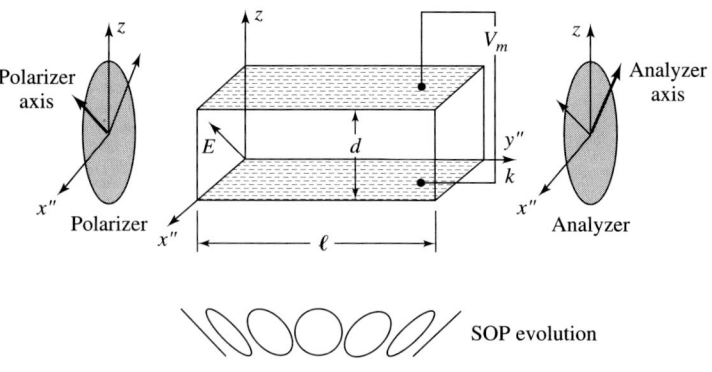

Figure 6.17 Schematic diagram of a transverse EO modulator with a KDP/ADP crystal

half-wave voltage for the transverse EO modulator is defined as the voltage required to change the voltage-dependent part of the retardation by $\pm \pi$. Thus,

$$V_{T\pi} = \frac{\lambda}{n_o^3 r_{63}} \frac{d}{\ell} \qquad (6.71)$$

As $V_{T\pi}$ varies with the aspect ratio d/ℓ of the modulator, a transverse EO modulator with a small aperture and a long crystal requires less modulating voltage and power. A typical transverse EO modulator designed for visible and near IR wavelength has an aperture on the order of a few square millimeters and a length of 5 to 10 cm. For operations with a 3-dB bandwidth on the order of 100 MHz, an rf power on the order of a few mW/MHz is required.

If the EO crystal is placed between a polarizer and an analyzer, changes in $\Delta \Phi_T$ can be converted to amplitude modulations. The analysis follows in exactly the same manner as that for longitudinal EO modulators.

6.7.3 Longitudinal vs transverse EO modulators, and MO vs EO modulators

It is instructive to compare the features of longitudinal EO modulators with their transverse counterparts. For longitudinal EO modulators, the features are as follows:

1. $V_{L\pi}$ is independent of the crystal length.
2. The electrode capacitance associated with a large aperture can be reduced by using long crystals.
3. The electrodes must be optically transparent. Electrically conducting and optically transparent materials, such CdO, SnO, or InO, may be used in modulators designed for visible or near IR spectra. Thin, conductive wire meshes or grids can be used for longer wavelengths.

The features of transverse EO modulators are:

1. $V_{T\pi}$ can be reduced by increasing the crystal length and decreasing the aperture.
2. As a result, the electrode capacitance increases and the RF bandwidth decreases.
3. Since the electrodes are not in the optical path, many opaque conductors may be used as electrodes.

The main drawback of EO modulators is the large voltage needed for the modulation. Even for good EO materials, $V_{L\pi}$ and $V_{T\pi}$ are in the kilovolt range. Since $V_{L\pi}$ is independent of the crystal length ℓ, as indicated in (6.62), the half-wave voltage cannot be varied by using shorter or longer crystals. However, it is possible to reduce the operating voltage by arranging a few EO crystals in series and driving the electrodes in parallel. Figure 6.18 shows the cascading of four transverse EO modulators [12].

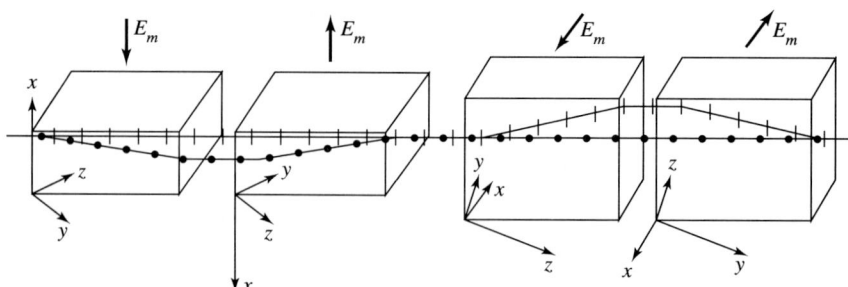

Figure 6.18 Cascade operation of four transverse EO modulators [12]

Although the basic setup of EO and MO modulators is similar, their actual operations are significantly different.

1. Waves remain LP throughout MO materials and the SOP rotates as the waves propagate; that is, the ellipticity is 0. In EO materials, however, both the ellipticity and azimuth of the polarization ellipse change as the waves travel. Thus, the SOP evolves as waves travel in the modulator.

2. The MO modulators based on the Faraday rotation effect are *nonreciprocal* devices, and may be used as optical isolators. In contrast, EO modulators are always *reciprocal,* and reciprocal devices cannot be used as optical isolators.

6.7.4 EO Material Figure of Merit

From the electric circuit point of view, the electrodes on an EO crystal are the electrodes of a capacitor. At radio and microwave frequencies, it is necessary to include the capacitance of the electrodes and the inductance of the lead wires in the circuit design. For simplicity, assume that the electrode area is the same as the cross-sectional area of the EO crystal, as shown in Figure 6.19a. The capacitance C is $\varepsilon_o \varepsilon_r dh/\ell$, where dh is the cross section area, ℓ the length, and ε_r the relative dielectric constant of the EO material at the modulating frequency f_m. Since f_m is much lower than the optical frequencies, the refractive index n at the optical frequencies can be quite different from $\sqrt{\varepsilon_r}$ at the modulating frequencies.

Consider a shunt RLC circuit shown in Figure 6.19a. Let Z be the input impedance of the RLC circuit. A plot of $|Z|$ as a function of f_m, given in Figure 6.19b, shows that $|Z|$ peaks at the center frequency $1/(2\pi\sqrt{LC})$ with a peak value of R, and decreases to a value of $R/\sqrt{2}$ at approximately $\dfrac{1}{2\pi}\left[\dfrac{1}{\sqrt{LC}} \pm \dfrac{1}{2RC}\right]$. Therefore, the 3-dB bandwidth of the shunt RLC circuit is $\Delta f_m \approx 1/(2\pi RC)$. Also note that Z is precisely R at the center frequency.

6.7 Electrooptic Modulators

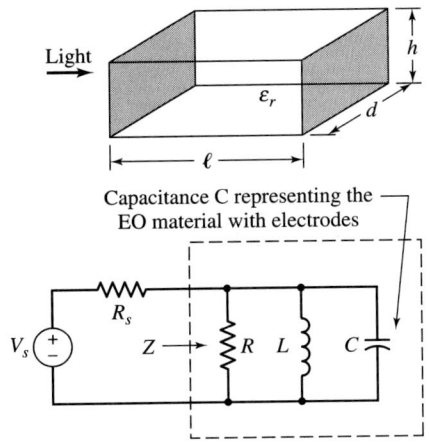

(a) Equivalent shunt RLC circuit

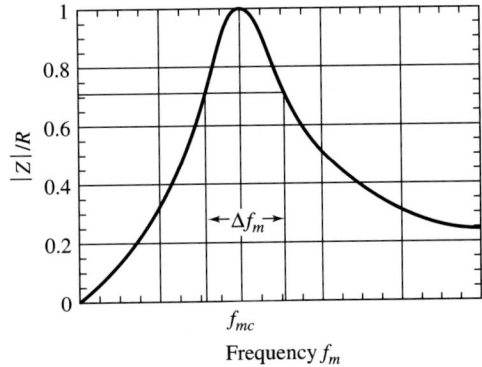

(b) Frequency response of an shunt RLC circuit

Figure 6.19 Equivalent circuit of an EO modulator

Suppose that the modulating source at the center frequency is represented by a Thevenin equivalent circuit with an open-circuit voltage source V_s and a source resistance R_s. To extract the maximum power from the voltage source, choose $R_s = R$. To reduce reflections and ringing caused by the impedance mismatch, it is necessary to match the source and load impedances. They should also be matched to the characteristic impedance of the transmission lines connecting the source to the EO crystal.

The time-average rf power P_{rf} from the modulating source can be related to the retardation of the EO material and the bandwidth of the electric circuit. To do so, we will use the longitudinal EO modulator (Figure 6.14) as an example. As noted previously, the shunt RLC circuit is purely resistive at the center frequency. The electrical power absorbed by the circuit is

$$P_{rf} = \frac{V_m^2}{2R} = \frac{\Delta \Phi_L^2(\ell) \lambda^2}{4\pi^2 n_o^6 r_{63}^2} \frac{1}{2R} = \Delta \Phi_L^2(\ell) \frac{\lambda^2 \varepsilon_o}{4\pi} \Delta f_m \frac{\varepsilon_r}{n_o^6 r_{63}^2} \frac{dh}{\ell} \quad (6.72)$$

Therefore,

$$\frac{P_{rf}}{\Delta \Phi_L^2(\ell) \Delta f_m} = \frac{\lambda^2 \varepsilon_o}{4\pi} \frac{\varepsilon_r}{n_o^6 r_{63}^2} \frac{dh}{\ell} \quad (6.73)$$

The rf power required for a given retardation and modulating bandwidth depends on the factor $n_o^6 r_{63}^2 / \varepsilon_r$, which involves only the optical properties of the EO material, and the **geometrical factor** dh/ℓ, which is related to the shape of the EO crystal. Although the geometrical factor may be reduced by using a longer EO material with a smaller cross-sectional area, these terms cannot be entirely arbitrary. There is a limiting value of dh/ℓ, dictated by the shape of the diffraction-limited Gaussian beams. For a Gaussian beam propagating in free space with a beam radius w_o at its beam waist, the Rayleigh range is $z_R = \pi w_o^2/\lambda$ [see (2.61)], and the beam radius at the Rayleigh range is $\sqrt{2}w_o$. With a cross-sectional area of $dh = \pi(\sqrt{2}w_o)^2$ and a crystal length $\ell = 2z_R = 2\pi w_o^2/\lambda$, the minimum ratio is

$$\frac{dh}{\ell} = \frac{\pi(\sqrt{2}\,w_o)^2}{2\pi w_o^2/\lambda} = \lambda$$

If the index of the medium in n_o, the minimum ratio is λ/n_o. Thus, for diffraction-limited Gaussian beams, we have

$$\frac{P_{rf}}{\Delta\Phi_L^2 \Delta f_m} = \frac{\lambda^3 \varepsilon_o}{4\pi} \frac{\varepsilon_r}{n_o^7 r_{63}^2}$$

The term $n_o^7 r_{63}^2/\varepsilon_r$ depends entirely on the physical properties of the ADP/KDP material used in the longitudinal EO modulator configuration, and is the material figure of merit for the ADP/KDP material in that configuration.

Similar results may be obtained for other EO modulators. Although different EO coefficients and refractive index values would be involved, the final expression for $P_{rf}/(\Delta\Phi_L^2 \Delta f_m)$ will have the same form. Henceforth, the subscripts of the refractive index and the linear EO coefficient are dropped, and we introduce the following **material figure of merit** for EO materials:

$$F = n^7 r^2/\varepsilon_r \tag{6.74}$$

For different EO modulators with different configurations, the pertinent index of refraction n, EO coefficient r, and relative dielectric constant ε_r should be used. In terms of the material figure of merit F, the performance of a modulator is

$$\frac{P_{rf}}{\Delta\Phi^2 \Delta f_m} = \frac{\lambda^3 \varepsilon_o}{4\pi} \frac{1}{F} \tag{6.75}$$

Example

The pertinent material properties of KDP at 0.633 μm are $r_{63} = 11 \times 10^{-12}$ m/V, $n_o = 1.5074$, and $\varepsilon_{rz} = 21$.[4] Then, the half-wave voltage and the material figure of merit are

$$V_{L\pi} = \frac{\lambda}{2n_o^3 r_{63}} = \frac{0.633 \times 10^{-6}}{2 \times 1.5074^3 \times (11 \times 10^{-12})} = 8400 \text{ V}$$

$$F = \frac{n_o^7 r_{63}^2}{\varepsilon_{rz}} = \frac{1.5074^7 \times (11 \times 10^{-12})^2}{21} = 1.017 \times 10^{-22} \text{ (m/V)}^2$$

---

[4]Since KDP crystals are uniaxial materials, $\varepsilon_{rx} = \varepsilon_{ry} \neq \varepsilon_{rz}$. For the EO modulator under consideration, the modulating electric field is along the z axis. Thus the relevant relative dielectric constant is ε_{rz}.

In any practical engineering design, it is prudent to include a **safety factor** in the design. If a safety factor S is introduced in each transverse dimension, then the cross-sectional area becomes dhS^2. Thus, for the diffraction-limited Gaussian beam case, we have

$$\frac{P_{rf}}{\Delta\Phi^2\Delta f_m} = \frac{\lambda^3\varepsilon_o}{4\pi}\frac{S^2}{F} \qquad (6.76)$$

In practice, a safety factor in the range of $3 < S < 6$ may be chosen. If a smaller value of S is used, the alignment could be difficult and the mechanical stability of the optical elements could be jeopardized.

It is possible and highly desirable to incorporate an EO modulator into integrated optic systems, which are discussed in Chapter 7. For integrated optic modulators, the alignment problem does not exist and a safety factor close to 1 may be used. This means that a reduction of $P_{rf}/(\Delta\Phi^2\Delta f_m)$ by a factor of 10 is achievable. In addition, optical beams can be confined by the waveguide structure in one or two transverse directions. Therefore, $d/\sqrt{\ell}$ and/or $h/\sqrt{\ell}$ are not restricted by the diffraction-limited value. However, the distribution of the optical fields in the waveguides is not uniform in the transverse plane, nor is the distribution of the modulating electric field. To account for these nonuniformities, an **overlap factor** ξ must be introduced, and (6.73) becomes

$$\frac{P_{rf}}{\Delta\Phi^2\Delta f_m} = \frac{\lambda^2\varepsilon_o}{4\pi}\frac{\varepsilon_r}{n^6r^2}\frac{d}{\sqrt{\ell}}\frac{h}{\sqrt{\ell}}\frac{S^2}{\xi} \qquad (6.77)$$

For integrated optic modulators based on *planar thin-film waveguides,* one of the geometrical factors, such as $d/\sqrt{\ell}$, is bounded by the diffraction-limited value $\sqrt{\lambda/n}$, and (6.77) becomes

$$\left.\frac{P_{rf}}{\Delta\Phi^2\Delta f_m}\right|_{pl\ wg} = \frac{\varepsilon_o\lambda^2}{4\pi}\left(\frac{\varepsilon_r}{n^6r^2}\right)\left(\frac{\lambda}{n}\right)^{1/2}\frac{h}{\sqrt{\ell}}\frac{S}{\xi} \qquad (6.78)$$

If the modulator is built on *strip* or *channel waveguides,* the optical beams are confined in two spatial directions, the interaction length ℓ may increase considerably, and the value of dh/ℓ is governed only by the crystal size, the fabrication precision, and the waveguide loss. By setting S to 1, we obtain

$$\left.\frac{P_{rf}}{\Phi^2\Delta f_m}\right|_{2d\ wg} = \frac{\varepsilon_o\lambda^2}{4\pi}\left(\frac{\varepsilon_r}{n^6r^2}\right)\frac{dh}{\ell}\frac{1}{\xi} \qquad (6.79)$$

6.8 ELASTICITY AND ACOUSTOOPTIC EFFECTS

While AO effects exist in both isotropic and anisotropic media, AO effects in isotropic materials are generally weak. Unfortunately, AO effects in anisotropic media are too complicated to be considered in this book. Here, we will consider only isotropic elastic materials, despite their shortcomings, with an emphasis on

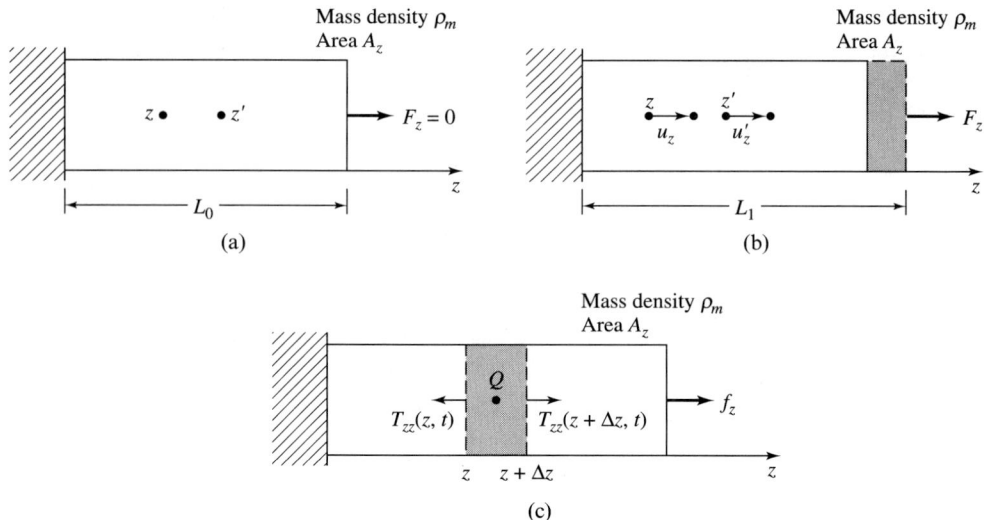

Figure 6.20 One-dimensional strain and stress

the physical processes involved in AO interactions. In this section, the concepts of strain and stress in simple one-dimensional systems are introduced. Longitudinal acoustic waves in isotropic media are also discussed. Topics not immediately needed for the discussions on AO modulators and deflectors are presented in Appendix B.

6.8.1 One-Dimensional Elasticity

For our discussions, we adopt a macroscopic viewpoint and consider materials as a matter continuum. Consider a bar of length L_0 and cross-sectional area A_z. The bar is fixed at one end, and a force is applied to the other end, as shown in Figure 6.20a. For simplicity, we will study the longitudinal deformation only, and will ignore effects associated with lateral dimensions. Consider two arbitrary points in the bar. Let the coordinates of these points be z and z' in the absence of an applied force. In the presence of an applied force F_z, the two points move to $z + u_z$ and $z' + u'_z$, respectively (Figure 6.20b). Thus, in response to the applied force, the distance between the two points changes from $z' - z$ to $(z' + u'_z) - (z + u_z)$. The fractional change in length is

$$\frac{[(z' + u'_z) - (z + u_z)] - [z' - z]}{z' - z} = \frac{u'_z - u_z}{z' - z}$$

The *limiting value* of the fractional change in length in the z direction as z approaches z' is the **tensile strain component** S_{zz},

$$S_{zz} = \lim_{z' \to z} \frac{u'_z - u_z}{z' - z} = \frac{\partial u_z}{\partial z} \qquad (6.80)$$

In other words, **strain** is defined as the *percentage deformation* of the material. If S_{zz} is positive, it represents a percentage elongation; if S_{zz} is negative, it shows a fractional compression.

Stress is the *force per unit cross-sectional area*. The tensile stress component T_{zz} is the force in the z direction per unit area of a plane or surface normal to the z axis,

$$T_{zz} = \lim_{A_z \to 0} \frac{F_z}{A_z} \tag{6.81}$$

As the stress increases, the strain in the material also increases. As long as the strain is within the **elastic limit** of the material, or the onset of material fatigue has not occurred, the strain varies linearly with the stress. When the stress is removed, the strain vanishes and the material returns to its original length. The linear relationship between strain and stress is known as **Hooke's law**

$$T_{zz} = Y S_{zz} \tag{6.82}$$

where the proportionality constant Y is **Young's modulus.** Strain S_{zz} is a dimensionless quantity, and Y and T_{zz} have the dimension of N/m². For most materials, Hooke's law is satisfied if the strain components are on the order of 10^{-4} to 10^{-3} or less.

The work W applied to stretch or compress an elastic bar is converted to an acoustic potential energy. To calculate the resulting energy stored in the bar, assume that the force f_z increases gradually from 0 to F_z, and that the bar length ℓ changes from the initial length L_0 to a final value L_1 (Figure 6.2b). At a given instant, ℓ and f_z are related by

$$\ell \approx L_0 (1 + S_{zz}) = L_0 (1 + \frac{f_z}{YA_z})$$

When the force increases by df_z, the strain changes by dS_{zz} and the bar length by

$$d\ell \approx \frac{L_0}{Y} \frac{df_z}{A_z}$$

The change of work done to the bar is

$$dW = f_z\, d\ell \approx \frac{L_0}{YA_z} f_z\, df_z \tag{6.83}$$

The total work done as the force increases from 0 to F_z is

$$W = \int_0^{F_z} \frac{L_0}{YA_z} f_z df_z = \frac{1}{2} \frac{L_0}{YA_z} F_z^2 = \frac{1}{2}(L_0 A_z) \frac{T_{zz}^2}{Y}$$

This is the **total acoustic energy** stored in a bar of volume $L_0 A_z$. The acoustic energy density stored in the material is

$$U_a = \frac{W}{L_0 A_z} = \frac{1}{2} \frac{T_{zz}^2}{Y} = \frac{1}{2} Y S_{zz}^2 = \frac{1}{2} T_{zz} S_{zz} \tag{6.84}$$

6.8.2 Longitudinal Acoustic Waves

If the applied force is a function of *time*, the strain, stress, and stored energy are, also. Consider an infinitesimal section of bar Δz surrounding a material point Q located at z (Figure 6.20c). Let the displacement of Q in the z direction be $u_z(z; t)$. The velocity and acceleration, respectively, of Q are

$$\frac{\partial u_z(z; t)}{\partial t} \quad \text{and} \quad \frac{\partial^2 u_z(z; t)}{\partial t^2}$$

Let the mass density of the material be ρ_m. The mass of the infinitesimal section is then $\rho_m A_z \Delta z$. The forces acting on the left and right faces of the section are $-\mathcal{T}_{zz}(z; t)A_z$ and $\mathcal{T}_{zz}(z + \Delta z; t)A_z$, respectively. By Newton's law, the net force acting on the infinitesimal section is related to the acceleration of the material,

$$(\rho_m A_z \Delta z) \frac{\partial^2}{\partial t^2} u_z(z;t) = [\mathcal{T}_{zz}(z + \Delta z;t) - \mathcal{T}_{zz}(z;t)] A_z \approx \frac{\partial}{\partial z} \mathcal{T}_{zz}(z;t) \Delta z A_z$$

In the limit $\Delta z \to 0$, we have an **equation of motion**

$$\rho_m \frac{\partial^2}{\partial t^2} u_z(z;t) = \frac{\partial}{\partial z} \mathcal{T}_{zz}(z;t) \tag{6.85}$$

Substituting (6.80) and (6.82) into (6.85), we obtain a second-order partial differential equation for $u_z(z;t)$:

$$\rho_m \frac{\partial^2}{\partial t^2} u_z(z;t) = Y \frac{\partial^2}{\partial z^2} u_z(z;t) \tag{6.86}$$

This is the one-dimensional **acoustic wave equation** describing the displacement of a point at an arbitrary location in the material. Since the material points move in the same direction as the wave propagation, the waves are **longitudinal acoustic waves** (Figure 6.21a).

The differential work done to a bar by an external force $f_z(z;t)$ is shown in (6.83). The energy transferred into a short section of bar of cross section A_z in the time interval Δt is $\dfrac{(u_z' - u_z)f_z}{\Delta t}$. In the limit of $\Delta t \to 0$ and $A_z \to 0$, the power transferred to the bar per unit cross section is $\dfrac{f_z(z;t)}{A_z} \dfrac{\partial}{\partial t} u_z(z;t)$, which may be written as $\mathcal{T}_{zz}(z;t) \dfrac{\partial}{\partial t} u_z(z;t)$. The power transferred from the bar per unit cross-sectional area is

$$\mathcal{P}_z(z;t) = -\mathcal{T}_{zz}(z;t) \frac{\partial}{\partial t} u_z(z;t) \tag{6.87}$$

Note the minus sign in the above equation.

When the applied force changes sinusoidally with a frequency f_a, so does the

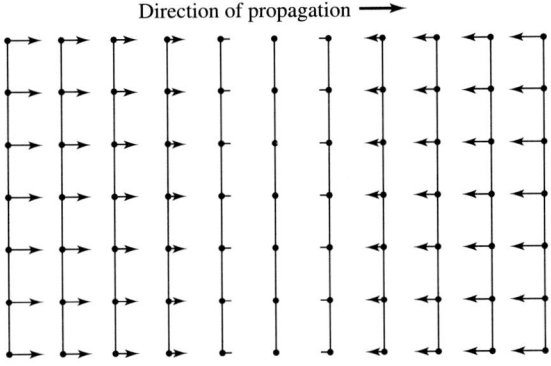

(a) Longitudinal acoustic waves

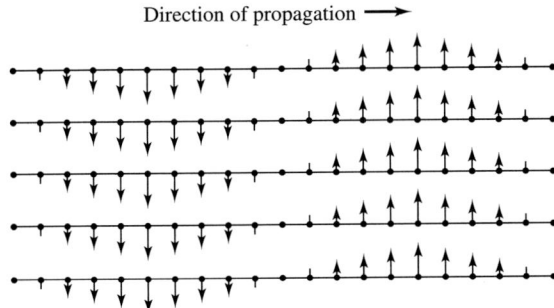

(b) Transverse acoustic waves

Figure 6.21 Longitudinal and transverse acoustic waves Small dots represent material particles, and small arrows denote the direction and magnitude of particle motion.

displacement $u_z(z;t)$. In terms of the frequency-domain functions $u_z(z)$ and $T_{zz}(z)$, the time-domain functions $u_z(z;t)$ and $\mathfrak{T}_{zz}(z;t)$ may be written as

$$u_z(z;t) = Re[u_z(z)e^{j\Omega t}] \qquad \mathfrak{T}_{zz}(z;t) = Re[T_{zz}(z)e^{j\Omega t}]$$

where $\Omega = 2\pi f_a$ is the acoustic angular frequency. Similar phasor quantities can be defined for other terms of interest. Then the one-dimensional acoustic wave equation (6.86) becomes

$$\frac{d^2}{dz^2}u_z(z) + \Omega^2 \frac{\rho_m}{Y} u_z(z) = 0 \qquad (6.88)$$

The solution is of the form

$$u_z(z) = c_1 e^{-jK_e z} + c_2 e^{jK_e z}$$

where c_1 and c_2 are constants yet to be determined. The first and second terms, respectively, represent the acoustic waves propagating in the $+z$ and $-z$ directions. The propagation constant or acoustic wave vector K_ℓ and the velocity of propagation V_ℓ of longitudinal acoustic waves in isotropic media are, respectively,

$$K_\ell = \Omega \sqrt{\frac{\rho_m}{Y}} \qquad (6.89)$$

$$V_\ell = \frac{\Omega}{K_\ell} = \sqrt{\frac{Y}{\rho_m}} \qquad (6.90)$$

Table 6.2 lists the longitudinal acoustic wave velocity of several materials [12]. Note that the acoustic wave velocity is typically five orders of magnitude slower than the speed of light. The time-average acoustic power transfer per unit cross-sectional area is, from (6.87),

$$P_{z\,av}(z) = \frac{1}{2} Re\left[-T_{zz}(z)[j\Omega\, u_z(z)]^*\right] = \frac{1}{2} Re\left[j\Omega T_{zz}(z) u_z^*(z)\right] \qquad (6.91)$$

where * indicates the complex conjugate of the quantity.

Consider longitudinal waves propagating in the +z direction:

$$u_z(z) = c_1 e^{-jK_\ell z}$$

From (6.80) and (6.82), we have the strain and stress associated with the longitudinal acoustic waves:

$$S_{zz}(z) = -jK_\ell c_1 e^{-jK_\ell z} \qquad (6.92)$$

$$T_{zz}(z) = -jK_\ell Y c_1 e^{-jK_\ell z} \qquad (6.93)$$

The time-average power density transferred by the longitudinal waves moving in the +z direction is

Table 6.2 Comparison of AO materials [13]

Material	ρ_m mg/m³	V_l km/s	n	p	Loss dB	M/M_{water} dB
Water	1.0	1.5	1.33	0.31	0	1.0
Dense flint glass	4.8	3.80	1.72	0.30	−12	0.06
Extra dense flint glass	6.3	3.10	1.92	0.25	−9	0.12
Fused quartz (SiO_2)	2.2	5.97	1.46	0.20	−22	0.006
Polystyrene	1.06	2.35	1.59	0.31	−1	0.8
KRS-5	7.4	2.11	2.60	0.21	+2	1.6
Lithium niobate ($LiNbO_3$)	4.7	7.40	2.25	0.15	−19	0.012
Lithium fluoride (LiF)	2.6	6.00	1.39	0.13	−29	0.001
Rutile (TiO_3)	4.26	10.30	2.60	0.05	−29	0.001
Sapphire (Al_2O_3)	4.0	11.00	1.76	0.17	−29	0.001

$$P_{z\,av}(z) = \frac{1}{2}\Omega K_\ell Y |c_1|^2 = \frac{1}{2}\rho_m V_\ell^3 |S_{zz}|^2 = \frac{1}{2}\frac{1}{\rho_m V_\ell}|T_{zz}|^2 \qquad (6.94)$$

Thus far, our discussion has been restricted to the tensile stress applied to, and the strain induced in, one-dimensional objects. We have ignored all effects associated with the lateral dimensions. For objects with a finite cross section, elongation is accompanied by a reduction of the cross-sectional area. Thus, tensile stress simultaneously produces longitudinal elongation and lateral compression in solids. When the concepts of elongation and compression are generalized to distortion, it is necessary to use a 3×3 strain tensor $\overleftrightarrow{S}$ to describe the strain. $\overleftrightarrow{S}$ has nine elements, six of which are independent. Detailed discussions on this topic are in Appendix B.

If the material points move in directions perpendicular to the direction of acoustic wave propagation (Figure 6.21b), the waves are known as **transverse acoustic** (or **shear**) **waves**. A different wave equation or set of wave equations is needed to describe such waves. In the same medium and at the same frequency f_a, transverse acoustic waves propagate with a different velocity, and have a different acoustic wavelength and propagation constant. We identify these terms as V_t, Λ_t, and K_t, respectively. Further details are included in Appendix B.

6.8.3 Acoustooptic Effects

In elastic media, the relative dielectric impermeability is perturbed by the strain. If the medium is anisotropic, the effects are different for different strain components. In other words, the strain affects not only the numerical value of the refractive index, but also the shape and orientation of the index ellipsoid. The acoustooptic effects, also known as **elastooptic** or **photoelastic effects,** refer to the change induced in the relative dielectric impermeability tensor by the strain. For a weak strain, the change in the tensor is a linear function of strain, and a photoelastic tensor $\overleftrightarrow{p}$ may be used to describe the linear relationship.[5] Since the relative dielectric impermeability tensor and the strain tensor each have nine tensor components, six of which are independent, $\overleftrightarrow{p}$ has 81 tensor components, 36 of which are independent. Clearly, problems involving AO effects are quite complicated. Elastic problems involving liquids, as discussed in section 6.3, are exceptions, because there is no shearing strain in liquids.

6.9 ACOUSTOOPTIC MODULATORS AND DEFLECTORS

In AO media, moving index gratings are produced by the strain accompanying acoustic waves. Because of the motion of periodic index variations, the optical beams are spatially deflected, frequency shifted, and amplitude modulated. Depending on the interaction length and the strength of the AO interaction, the effects on the optical waves may be classified as **Raman–Nath diffraction** (also

---

[5]For simplicity, the effect due to antisymmetric rotation tensor has been ignored [10].

known as **Debye–Sears diffraction**), diffraction in the transition region, or **Bragg diffraction**.

In discussing AO effects, it is necessary to refer to the frequency, wavelength, and other properties of optical and acoustic waves. To avoid confusion, we will adopt the following convention. The frequency and angular frequency of optical beams are designated f and ω, and the speed of light, wavelength, and propagation constant of optical beams in vacuum are c, λ, and k, respectively. These quantities are related as follows:

$$\omega = 2\pi f, \quad \lambda = c/f, \quad k = 2\pi/\lambda = \omega/c$$

The corresponding quantities for acoustic waves in a material medium are f_a, Ω, V, Λ, and K, respectively. The relationships between the various acoustic wave quantities are

$$\Omega = 2\pi f_a, \quad \Lambda = V/f_a, \quad K = 2\pi/\Lambda = \Omega/V$$

where V stands for V_ℓ or V_t, Λ for Λ_ℓ or Λ_t, and K for K_ℓ or K_t.

6.9.1 Raman–Nath Diffraction

The discussion on AO interaction proceeds in stages of increasing complexity [13]. The first consideration is the effects of *dielectric slab index changes* on optical beams. The second study concerns the effects of *time-varying index changes*, and the third study examines the effects of *moving index gratings*. To depict wave motion, we will initially use the time-domain representation, and will convert various terms back to the frequency-domain representation toward the end of the discussion.

Stationary Time-Independent Index Gratings Consider optical beams propagating in the x direction in an isotropic dielectric slab with an index n, as depicted in Figure 6.22a. In the time-domain representation, the optical beams are given by

$$\mathcal{E}(x;t) = \mathbf{E}_{in}\cos(\omega t - knx)$$

In the presence of a static pressure, the index n changes to $n + \Delta n$, the phase term is changed by ($k\, \Delta n\, x$), and the optical fields are represented by

$$\mathcal{E}(x;t) = \mathbf{E}_{in}\cos[\omega t - k(n + \Delta n)x]$$

While the index change is very small, the optical path length in the acoustic medium may be very long in terms of the wavelength. As a result, the phase change due to a small index variation can be substantial. As a numerical example, consider a water layer subjected to a strain of 10^{-5}. From (6.26), we have $\Delta n = -3.65 \times 10^{-6}$. After wave propagation in a water layer 1.0 cm in thickness, the optical phase change is about 20.7 degrees, for a wavelength of 0.633 μm.

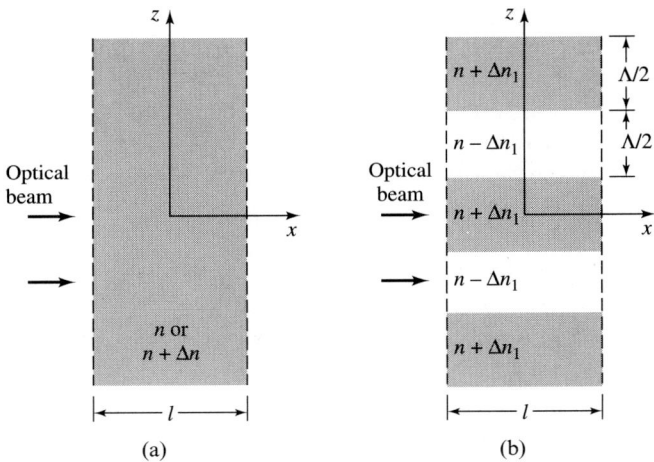

Figure 6.22 Interaction of optical waves with acoustic waves (after [13], © 1967 IEEE)

Stationary Time-Varying Index Gratings Next, suppose the index variations are produced by standing acoustic waves. The index change has a time-varying component,

$$n(t) = n + \Delta n_1 \cos\Omega t$$

The optical fields therefore have a time-varying phase term

$$\mathcal{E}(x;t) = \mathbf{E}_{in}\cos[\omega t - k(n + \Delta n_1\cos\Omega t)x]$$

To examine the meaning of this term, rewrite the equation as

$$\mathcal{E}(x;t) = Re\left[\mathbf{E}_{in}\, e^{j\omega t}\, e^{-jkx(n + \Delta n_1 \cos\Omega t)}\right]$$

and make use of the identity

$$e^{-jx\cos\phi} = \sum_{m=-\infty}^{\infty} (-j)^m\, e^{jm\phi} J_m(x) \tag{6.95}$$

where $J_m(x)$ is the Bessel function of order m. In terms of Bessel functions, $\mathcal{E}(x;t)$ may be written as

$$\mathcal{E}(x;t) = Re\left\{\mathbf{E}_{in}\sum_{m=-\infty}^{\infty}\left[(-j)^m\, e^{j(\omega+m\Omega)t} J_m(k\Delta n_1 x)\, e^{-jknx}\right]\right\}$$

The equation can now be cast in terms of cosine functions, as follows:

$$\mathcal{E}(x;t) = \mathbf{E}_{in}\sum_{m=-\infty}^{\infty}\left\{J_m(k\Delta n_1 x)\cos[(\omega+m\Omega)t - knx - m\frac{\pi}{2}]\right\}$$

The mth component of $\mathcal{E}(x;t)$ is frequency shifted to $f + mf_a$ and is amplitude modulated according to $J_m(k\Delta n_1 x)$.

Traveling Index Gratings Finally, consider the effects of moving index gratings caused by traveling acoustic waves of frequency f_a moving in the z direction with a velocity V and an acoustic wavelength Λ. Note that acoustic wave propagation is normal to the direction of optical wave propagation. Suppose that the AO interaction is confined to a region of width ℓ, which is short in a sense to be discussed later. In the region $0 < x < \ell$, the index is given by

$$n(z;t) = n + \Delta n_1 \cos(\Omega t - Kz)$$

Again, making use of the identity (6.95), we have

$$\mathcal{E}(x,z;t) = Re\left\{\mathbf{E}_{in} e^{j\omega t} e^{-jknx} [\sum_{-\infty}^{\infty} (-j)^m e^{jm(\Omega t - Kz)} J_m(k\Delta n_1 x)]\right\}$$

$$= Re\left\{\sum_{-\infty}^{\infty} [\mathbf{E}_{in} (-j)^m e^{j(\omega + m\Omega)t} e^{-j(knx + mKz)} J_m(k\Delta n_1 x)]\right\}$$

Each term of $\mathcal{E}(x,z;t)$ may be interpreted in the terminology used for describing frequency-modulated signals. The m = 0 term is the "carrier" term

$$\mathcal{E}_0(x,z;t) = Re\left\{\mathbf{E}_{in} e^{j\omega t} e^{-jknx} J_0(k\Delta n_1 x)\right\} = \mathbf{E}_{in} J_0(k\Delta n_1 x)\cos(\omega t - knx)$$

This has the same frequency and propagates in the same direction as the incident beam. However, the amplitude changes as $J_0(k\Delta n_1 x)$. The carrier, therefore, is the *undiffracted beam* whose amplitude is changed by the acoustic waves. The sideband terms (m = ±1, ±2, . . .)

$$\mathcal{E}_m(x,z;t) = \mathbf{E}_{in} J_m(k\Delta n_1 x) \cos[(\omega + m\Omega)t - (knx + mKz)]$$

are frequency shifted to $f + mf_a$ and amplitude modulated according to $|J_m(k\Delta n_1 x)|$. In addition, each sideband propagates in a different direction. Thus, each sideband term corresponds to a diffracted beam. Take the mth diffracted beam as an example. The constant-phase plane is

$$knx + mKz = 2\pi\left(\frac{nx}{\lambda} + \frac{mz}{\Lambda}\right) = 2\pi C$$

where C is a constant. The constant-phase plane intercepts the x and z axes at $x = C\lambda/n$ and $z = C\Lambda/m$, respectively. In other words, the plane is inclined at an angle

$$\theta = \tan^{-1}\frac{m\lambda}{n\Lambda}$$

with respect to the z axis, as shown in Figure 6.23. Since λ is usually much smaller than Λ, the arctangent function may be approximated by its argument

$$\theta \approx m\lambda/(n\Lambda)$$

Thus, the mth diffracted beam is deflected by an angle $m\lambda/(n\Lambda)$, as depicted schematically in Figure 6.24.

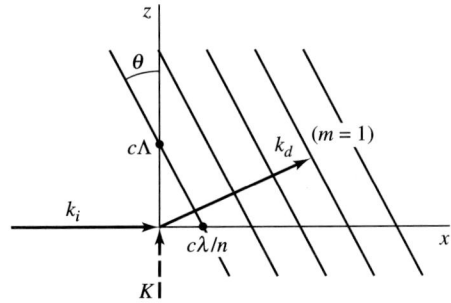

Figure 6.23 Diffraction of optical beams by acoustic waves

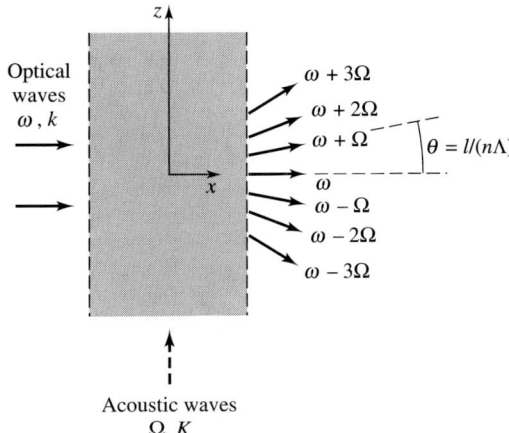

Figure 6.24 Spatial deflection and spectral shift of optical waves by acoustic waves

Recall that the intensity I is proportional to $|\mathbf{E}|^2$. The intensity of the mth order diffraction beam is,[6]

$$I_m(x) = I_{in} [J_m(|k\Delta n_1 x|)]^2 \qquad (6.99)$$

As Δn_1 increases linearly with the strain, a particular diffraction order may grow at the expense of others. When $|k\Delta n_1 x|$ is small, the zeroth order beam is the dominant term. As $|k\Delta n_1 x|$ increases, $I_{\pm 1}(x)$ and $I_{\pm 2}(x)$ increase, while $I_0(x)$ decreases. The variations of $I_m(x)/I_{in}$ as functions of $|k\Delta n_1 x|$ are plotted in Figure 6.25. For example, at $|k\Delta n_1 x| = 2.4048$, I_0 vanishes, while $I_{\pm 1} \approx 0.27\, I_{in}$ and $I_{\pm 1} \approx 0.19\, I_{in}$. Higher-order diffraction orders, while present, are generally weak, but the total optical power is conserved for all values of $|k\Delta n_1 x|$. This can be demonstrated using the identity

$$J_0^2(x) + 2\sum_{m=1}^{\infty} J_m^2(x) = 1$$

In the above discussion, we assumed that the optical waves propagate in a direction normal to that of the acoustic waves, that is, $\mathbf{k} \cdot \mathbf{K} = 0$. In general, when the optical wave vector and the acoustic wavefronts intersect at an angle θ', that is, $\mathbf{a}_k \cdot \mathbf{a}_K = -\sin\theta'$, the intensity of the mth diffraction order is [14–16]

$$I_m(x) = I_{in} J_m^2 \left(\frac{k\Delta n_1 x}{\cos\theta'}\right) \qquad (6.100)$$

---

[6]For an integer m, $J_{-m}(x) = (-1)^m J_m(x)$ and $J_m(-x) = (-1)^m J_m(x)$.

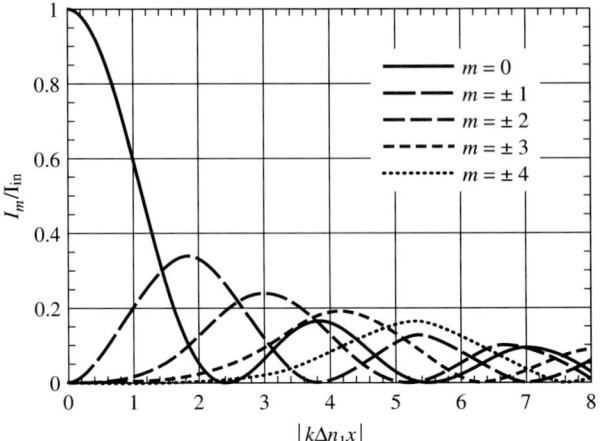

Figure 6.25 Intensity variation of first five diffraction orders

6.9.2 Bragg Diffraction

Bragg Condition When the interaction length ℓ is long, all diffraction orders are present and are weak, unless the optical and acoustic waves intersect at a particular angle. This angle is determined by the *constructive interference* of the diffractions from (a) all points on an arbitrary acoustic wavefront and (b) adjacent wavefronts separated by an acoustic wavelength Λ. When the condition of constructive interference is met, one and only one diffraction order ($m = +1$ or -1) becomes very strong, and all other diffraction orders are suppressed. This is the preferred AO interaction in many applications. In discussing interference of waves, we are referring to interference in the far zone. Let the wave vectors of the incident and diffracted optical beams and of the acoustic waves be $\mathbf{k}_i$, $\mathbf{k}_d$, and $\mathbf{K}$, respectively. Since $\mathbf{k}_i$ is nearly, not exactly, normal to $\mathbf{K}$, $\mathbf{k}_i$ may have a component opposing or following $\mathbf{K}$ (see Figures 6.26a and b, and 6.26d and e, respectively.)

To establish the condition of constructive interference, consider optical beams intercepting an arbitrary acoustic wavefront at two points separated by a distance x, as shown in Figure 6.26a. The diffracted waves interfere constructively in the far field if the path length difference $\overline{OQ} - \overline{O'Q'}$ is an integer multiple of λ/n

$$\overline{OQ} - \overline{O'Q'} = x(\cos\theta_d - \cos\theta_i) = m'(\lambda/n)$$

where m' is an integer, and θ_i and θ_d are the angles between k_i and k_d with respect to the acoustic wavefronts. Note that x is an arbitrary distance and can assume any value. The condition for constructive interference is met for all values of x if and only if $m' = 0$ and $\theta_i = \theta_d$.

Next, consider waves diffracted by two adjacent wavefronts separated by Λ (Figure 6.26b). The condition for constructive interference requires

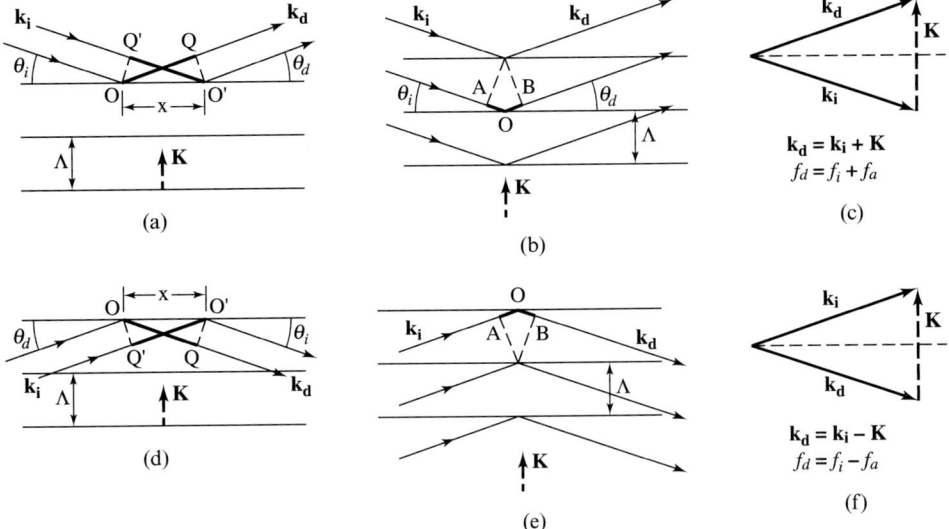

Figure 6.26 Bragg diffraction

$$\overline{OA} + \overline{OB} = 2\Lambda\sin\theta_i = m''(\lambda/n)$$

where m'' is again an integer. This condition is met only if m'' is 1. In summary, the condition of constructive interference in the far zone is met if $\cos\theta_d = \cos\theta_i$ and $2\Lambda\sin\theta_i = \lambda/n$.

Two requirements are satisfied simultaneously if

$$\sin\theta_i = \sin\theta_d = \frac{\lambda}{2n\Lambda}$$

This incident angle θ_i is known as the **Bragg angle** θ_B

$$\sin\theta_B = \frac{\lambda}{2n\Lambda} \quad (6.101)$$

Usually, $\lambda \ll \Lambda$; then θ_B is small and may be approximated by

$$\theta_B \approx \frac{\lambda}{2n\Lambda} \quad (6.102)$$

It is informative to construct a simple vector diagram showing the geometrical relationship between the wave vectors such that

$$\mathbf{k}_d = \mathbf{k}_i + \mathbf{K} \quad (6.103)$$

as depicted in figure 6.26c. This shows that, in the Bragg diffraction regime, the wave vector of the diffracted optical beam is the *vector sum* of the incident wave vector and the acoustic wave vector. Also, the frequency of the diffracted beam is *up-shifted* to

$$f_d = f_i + f_a \quad (6.104)$$

If the incident optical waves and the acoustic waves move in the same direction, as shown in Figures 6.26d and e, then we have a different vector relation in lieu of (6.103), as follows:

$$\mathbf{k}_d = \mathbf{k}_i - \mathbf{K} \tag{6.105}$$

and the diffracted beam is frequency *down-shifted* (Figure 6.26e) to

$$f_d = f_i - f_a \tag{6.106}$$

As a result of Bragg diffraction, power in the incident optical beam is split into two beams: an undiffracted beam, and a diffracted beam. The undiffracted beam propagates in the same direction as the incident beam. The diffracted beam is at an angle of $2\theta_B = \lambda/(n\Lambda)$ relative to the undiffracted beam. This is also the angle between the undiffracted and first-order diffracted beam in the Raman–Nath diffraction regime given in (6.98).

Thus far, the undiffracted and diffracted optical beams and the acoustic waves discussed have been *inside* the acoustic medium with index n. For many applications and experiments, the incident optical beam propagates in air before impinging on the acoustic medium, where the interaction occurs. After interacting with the acoustic waves, the undiffracted and diffracted beams return to air. All measurements are done outside the interacting medium. Since the index of air is very close to 1, we treat air as a vacuum, as an approximation. For simplicity, we also assume that the acoustic waves propagate along the air/dielectric boundary, with the acoustic wavefronts normal to the boundary, as shown in Figure 6.27. Corresponding to the angles θ_i and θ_d *inside* the acoustic medium, we have the angles θ_i' and θ_d' *outside* the acoustic medium. The relationship between these angles can be found with Snell's law, as follows:

$$1\sin\theta_i' = n\sin\theta_i \qquad 1\sin\theta_d' = n\sin\theta_d$$

Since $\theta_i = \theta_d = \theta_B$, and all the angles are small, then $\theta_i' = \theta_d' \approx n\theta_B$. For beams outside the interacting medium, the angle between the undiffracted and diffracted beams is $2\theta_B'$, and

$$2\theta_B' \approx 2n\theta_B \approx \frac{\lambda}{\Lambda} \tag{6.107}$$

Since n is greater than 1, the angle θ_B in the acoustic medium with an index n is smaller than the angle θ_B' in the air.

Diffracted Beam Amplitude For long interaction lengths, if the diffracted waves generated in the interaction region add constructively, the diffracted beam may be very strong. Therefore, we are interested in the intensity of the diffracted beam as a function of the interaction length ℓ and the index perturbation Δn_1. Consider the evolution of the undiffracted beam $\mathcal{E}_0(x,z;t)$ and the first-order diffracted beam $\mathcal{E}_1(x,z;t)$ in the presence of the moving index gratings (6.96). Let the amplitudes of the undiffracted and first-order diffracted beams be e_0 and e_{+1}. Then,

6.9 Acoustooptic Modulators and Deflectors

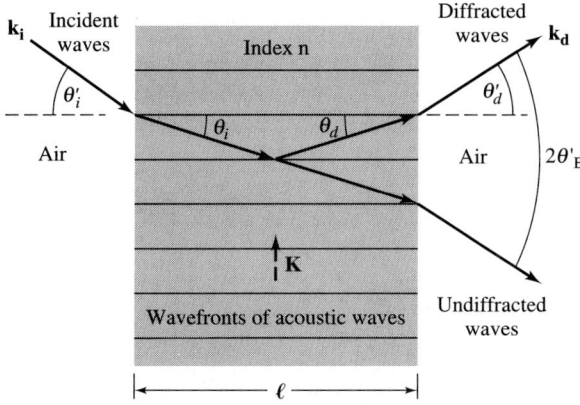

Figure 6.27 Diffracted and undiffracted beams outside the interacting medium

$$\mathcal{E}_0(x,z;t) = Re[e_0(x)\, e^{j\omega t}\, e^{-jknx}]$$
$$\mathcal{E}_{+1}(x,z;t) = Re[e_{+1}(x)\, e^{j(\omega+\Omega)t} e^{-j(knx+Kz)}]$$

Assume that the Bragg condition (6.103) is satisfied, and that the angle between the undiffracted and the first-order frequency *up-shifted* diffracted beams is θ_B. In an infinitesimal region Δx, the change in $e_{+1}(x)$ must be proportional to $e_0(x)$ and $k\Delta n_1 \Delta x$. Similarly, the change in $e_0(x)$ must be proportional to $e_{+1}(x)$ and $k\Delta n_1 \Delta x$. Through detailed analysis, it can be shown that the proportionality constant is $\dfrac{j}{2\cos\theta_B}$ [14–18]. Therefore, changes in $e_0(x)$ and $e_{+1}(x)$ are

$$\Delta e_{+1}(x) = \frac{j}{2}\frac{k\Delta n_1 \Delta x}{\cos\theta_B} e_0(x)$$
$$\Delta e_0(x) = \frac{j}{2}\frac{k\Delta n_1 \Delta x}{\cos\theta_B} e_{+1}(x)$$

In the limit of $\Delta x \to 0$, the equations become a set of coupled differential equations

$$\frac{de_{+1}(x)}{dx} = \frac{j}{2}\frac{k\Delta n_1}{\cos\theta_B} e_0(x) \qquad (6.108)$$

$$\frac{de_0(x)}{dx} = \frac{j}{2}\frac{k\Delta n_1}{\cos\theta_B} e_{+1}(x) \qquad (6.109)$$

The solutions to the coupled differential equations are the linear combinations of the sine and cosine functions of $\dfrac{k\Delta n_1 x}{2\cos\theta_B}$. If $e_0 = E_{in}$ and $e_{+1} = 0$ at $x = 0$, then,

$$e_0(x) = E_{in}\cos(\frac{k\Delta n_1}{2\cos\theta_B}x)$$
$$e_{+1}(x) = jE_{in}\sin(\frac{k\Delta n_1}{2\cos\theta_B}x)$$

The intensities of the undiffracted and first-order frequency up-shifted diffracted beams are proportional to $|e_0(x)|^2$ and $|e_1(x)|^2$, respectively, as follows:

$$I_0(x) = I_{in}\cos^2(\frac{k\Delta n_1}{2\cos\theta_B}x) \tag{6.110}$$

$$I_{+1}(x) = I_{in}\sin^2(\frac{k\Delta n_1}{2\cos\theta_B}x) \tag{6.111}$$

If x is small, $I_0(x) \approx I_{in}$ and $I_{+1}(x)$ increases as $\frac{(k\Delta n_1 x)^2}{4\cos^2\theta_B}$. As x increases, the diffracted beam may be larger than the undiffracted beam. In the Bragg region, it is possible for all incident power to be converted to the diffracted beam. This occurs if $|k\Delta n_1 x|$ is odd multiples of π. In all cases, however, $I_0(x) + I_{+1}(x) = I_{in}$; in other words, the total optical power is conserved.

If the condition of (6.105) is satisfied, the power exchange is between the undiffracted beam and the first-order, frequency *down-shifted,* diffracted beam, that is, the $m = -1$ term. Then, in lieu of (6.111), we have

$$I_{-1}(x) = I_{in}\sin^2(\frac{k\Delta n_1}{2\cos\theta_B}x) \tag{6.112}$$

6.9.3 Parameter Q and Transition Region Diffraction

In discussing Raman–Nath diffraction, we stipulated that the interaction length was short. For Bragg diffraction, we assumed that the interaction region was long. In either case, we did not discuss these stipulations further. To appreciate the rationale behind these stipulations, we return to the index gratings with length ℓ and periodicity Λ [13]. We also replace the index gratings with a stack of dielectric slabs of length ℓ and width $\Lambda/2$. The refractive indices of the slabs alternate between $n + \Delta n_1$ and $n - \Delta n_1$ (Figure 6.22b). From elementary physical optics, we know that an optical beam with a vacuum wavelength λ in a medium with an index n is diffracted by a slit, and the diffracted beam diverges with a half-angle $\lambda/(nd)$, where d is the slit width. For our problem, we have an optical beam impinging on a slab with an index $n \pm \Delta n_1$ and thickness $\Lambda/2$. The diffracted beam spreads with a half-angle $2\lambda/(n\Lambda)$. After traveling a distance ℓ, the diffracted beam is displaced laterally in the z direction by a distance $2\lambda\ell/(n\Lambda)$. If the lateral displacement is greater than $\Lambda/2$, the diffracted beam is located at the center of the neighboring slab, which has an index $n \mp \Delta n_1$. Therefore, the effects due to an index perturbation ($\pm\Delta n_1$) of one slab are negated by the index change ($\mp \Delta n_1$) in the neighboring slab. To prevent this cancellation from occurring, and to guarantee that the effects of the index change on the optical beam accumulate, the interaction length must be limited by the condition $\ell \dfrac{2\lambda}{n\Lambda} \leq \dfrac{\Lambda}{2}$, which is equivalent to $\dfrac{4\lambda \ell}{n\Lambda^2} \leq 1$.

The interactions of acoustic waves with optical beams in acoustic media has been analyzed numerically by Klein and Cook [17]. To summarize their results, they introduce two dimensionless parameters. One is a parameter Q

$$Q = \frac{2\pi \lambda \ell}{n \Lambda^2} \tag{6.113}$$

which is used to quantify the interaction length. To represent the strength of the interaction, they define a parameter v

$$v = \frac{k \Delta n_1 \ell}{\cos \theta_B} \tag{6.114}$$

Figure 6.28 is one of the many plots based on their numerical study. The figure depicts the values of I_0 and I_{+1} as functions of Q, for $v = \pi$. Their study confirms that the results based on Raman–Nath diffraction are valid approximations for $Q \leq 0.3$. For $Q \geq 7$, more than 90 percent of the incident power is converted to I_{+1}. Consequently, it is customary to use the parameter Q to distinguish three regions of diffraction. The region in which $Q \ll 1$ corresponds to **Raman–Nath diffraction**; the region in which $Q \gg 1$ corresponds to **Bragg diffraction;** and the **transition region** is the region in which Q is in the intermediate range. In practice, the transition region corresponds to Q in the range $0.3 < Q < 10$. However, it should also be emphasized that there is really no sharp demarcation between the various regions. The inequality given is mainly for convenience.

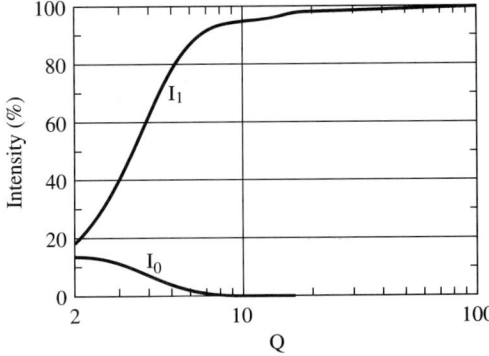

Figure 6.28 Undiffracted and first-order diffracted light intensities as a function of Q at Bragg incidence with $v = \pi$ ([17], © 1967 IEEE)

6.9.4 Resolvable Spots

Acoustooptic interactions may be used to deflect optical beams at high speed. In such applications (Figure 6.29), the incident optical beams and the acoustic waves intersect at a fixed angle. However, the acoustic frequency f_a may be scanned. With the angle fixed, either condition (6.103) or (6.105) is satisfied at exactly one frequency for a given isotropic medium. At other frequencies, the condition is only approximately met. A change in f_a will cause $\mathbf{k_d}$ to change by a small angle. Let the change in the acoustic wave frequency be Δf_a and the corresponding angular change in $\mathbf{k_d}$ be $\Delta \theta_d$ (Figure 6.30). Then,

$$\Delta \theta_d = \frac{\Delta K}{k_d} = \frac{\lambda}{nV} \Delta f_a$$

The frequency change Δf_a is usually limited by the bandwidth of the transducers used to convert the electrical signals to acoustic waves. Let $(\Delta f_a)_{max}$ be the transducer bandwidth and, hence, the maximum value of Δf_a. The optical beams can then be scanned over the range

$$(\Delta \theta_d)_{max} = \frac{\lambda}{nV} (\Delta f_a)_{max}$$

If we recall that a Gaussian beam with a waist radius w_o diverges with a half-angle $\lambda/(nw_o\pi)$, then the number of resolvable spots of an AO deflector is

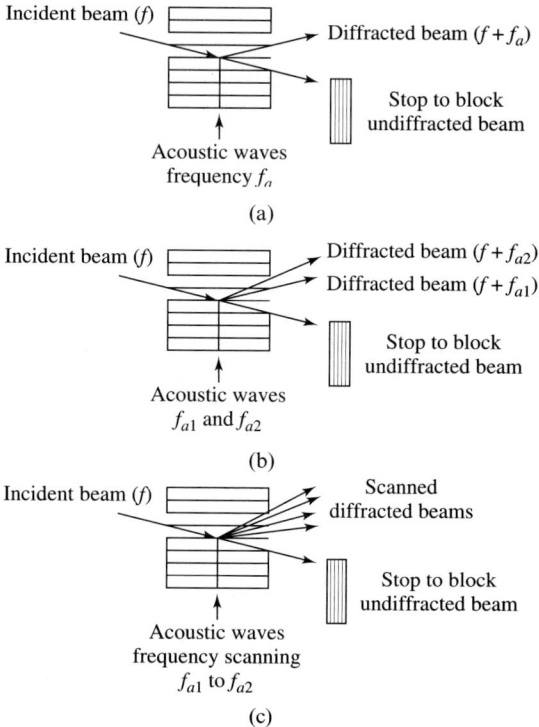

Figure 6.29 Three ways of utilizing acoustooptic interaction

6.9 Acoustooptic Modulators and Deflectors

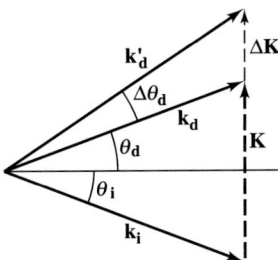

Figure 6.30 Vector diagram showing scanning of optical beams by acoustic beams by changing the acoustic wave frequency

$$N = \frac{(\Delta\theta_d)_{max}}{2\lambda/(nw_o\pi)} = \frac{\pi}{4}\frac{2w_o}{V}(\Delta f_a)_{max} \qquad (6.115)$$

The term $(2w_o/V)$ is the transit time of the acoustic waves traversing the full width of the optical beam. In view of (6.115), we see that AO materials with a low acoustic velocity V are preferred in AO deflector applications.

6.9.5 AO Material Figure of Merit

Since AO effects exist in all materials, it is useful to establish a criterion for comparing such materials. Consider optical beams of free-space wavelength λ interacting with an acoustic beam of width ℓ and height h. Let the pertinent strain component be S and the velocity of the acoustic waves be V. The time-average acoustic wave power P_{av}, from (6.94), is

$$P_{av} = \frac{1}{2}(h\ell)\rho_m V^3 |S|^2 \qquad (6.116)$$

The optical phase change due to the interaction with acoustic beams is, from (6.26),

$$\Delta\phi = \frac{2\pi\ell\Delta n}{\lambda} = -\frac{\pi\ell}{\lambda}n^3 pS$$

Combining this equation with (6.116) we obtain,

$$\frac{P_{av}}{\Delta\phi^2} = \frac{\lambda^2}{2\pi^2}\frac{h}{\ell}\frac{1}{\frac{n^6 p^2}{\rho_m V^3}} \qquad (6.117)$$

In (6.117), $P_{av}/\Delta\phi^2$ is the product of a *geometric factor* h/ℓ and a factor that depends solely on material properties. The *material factor*

$$M_2 = \frac{n^6 p^2}{\rho_m V^3} \qquad (6.118)$$

is known as the **material figure of merit** of AO materials for beam deflection applications. For other applications, such as AO modulators, different figures of merit are required [14,15]. As indicated in (6.115) and (6.118), the acoustic wave velocity is an important material property in determining AO interactions. Because acoustic waves move very slowly in liquids compared to acoustic waves in solids, M_2 for liquids is very large. For example, at visible wavelengths, the M_2 of lithium niobate, an excellent AO material, is much smaller than that of water:

$$\frac{(M_2)_{\text{LiNbO}_3}}{(M_2)_{\text{H}_2\text{O}}} \approx 0.012$$

Acoustic properties, including $M_2/(M_2)_{\text{H}_2\text{O}}$, of many acoustic media are listed in Table 6.2 [13]. Note that the attenuation of acoustic waves in liquids is usually high at high frequencies, which means that liquids are useful mainly in low-frequency applications.

PROBLEMS

1. With reference to Figure 6.2 and starting from (6.5), show that the ellipticity E_{mn}/E_{mj} and the azimuth θ are given by (6.8)–(6.11) in terms of E_{yo}/E_{xo} and $\Delta\phi$.
2. Starting from (6.5), show that E_{yo}/E_{xo} and $\Delta\phi$ can be obtained from (6.8)–(6.11), if the ellipticity E_{mn}/E_{mj} and the azimuth θ are given.
3. Consider a biaxial medium with $n_3 > n_2 > n_1$. Suppose the x and z axes are rotated by an angle θ to x' and z', and that

$$\theta = \tan^{-1}\left[\frac{n_3^2(n_2^2 - n_1^2)}{n_1^2(n_3^2 - n_2^2)}\right]^{1/2}$$

   a. Show that, in terms of the rotated axes x' and z', the index ellipsoid becomes

$$\frac{x'^2}{n_2^2} + \frac{y^2}{n_2^2} + z'^2\left(\frac{\sin^2\theta}{n_1^2} + \frac{\cos^2\theta}{n_3^2}\right) - 2x'z'\sin\theta\cos\theta\left(\frac{1}{n_1^2} - \frac{1}{n_3^2}\right) = 1$$

   b. Consider waves propagating along the z' direction. Show that the index is n_2 and that it is independent of the SOP.
4. Consider plane waves propagating in the y direction in a uniaxial material (Figure 6.12) that has an ordinary and an extraordinary index of 1.5 and 1.7, respectively. At $y = 0$, the fields are $\mathbf{E} = 3.0\mathbf{a}_x + 2.0\mathbf{a}_z$ V/m. What are the ellipticity and azimuth (relative to the x axis) of the fields at $y = 0.625, 1.250, 1.875,$ and 2.500λ?
5. Consider plane waves propagating in the z direction in a uniaxial material (Figure 6.12) that has an ordinary and an extraordinary index of 1.5 and 1.7 respectively. At $z = 0$, the fields are $\mathbf{E} = 3.0\mathbf{a}_x + 2.0\mathbf{a}_y$ V/m. What are the ellipticity and azimuth (relative to the x axis) of the fields at $z = 0.625, 1.250, 1.875,$ and 2.500λ?
6. For a certain uniaxial material, $n_o = 1.500$ and $n_e = 1.700$. Consider waves propagating along an angle $\theta = 30°$ with respect to the z axis.
   a. What is the index of refraction for waves polarized with $\mathbf{D}$ in the y direction?
   b. What is the index of refraction for waves polarized with $\mathbf{D}$ in the xz plane?

c. Given $\mathbf{D} = D_1(\mathbf{a}_y + \mathbf{a}_t)$, where $\mathbf{a}_t$ is a unit vector in xz plane, as shown in Figure 6.12c, and D_1 is a constant. Express $\mathbf{E}$ in terms of D_1, n_o, n_e, and ε_o.

7. Uniform plane waves propagate in a uniaxial medium with $n_o = 1.7$ and $n_e = 1.5$. At a certain point, Q_a, the flux density is $\mathbf{D} = 3\,\mathbf{a}_x - j3\,\mathbf{a}_t$ C/m². As shown in the figure, the unit vector $\mathbf{a}_t$ and the direction of propagation $\mathbf{a}_k$ are in the yz plane. After traveling a distance of 2.5λ in the $\mathbf{a}_k$ direction, to point Q_b, the waves become linearly polarized.
   a. Write the expression for $\mathbf{E}$ at point Q_a.
   b. Determine the direction of propagation $\mathbf{a}_k$ relative to the z axis (i.e., determine θ)

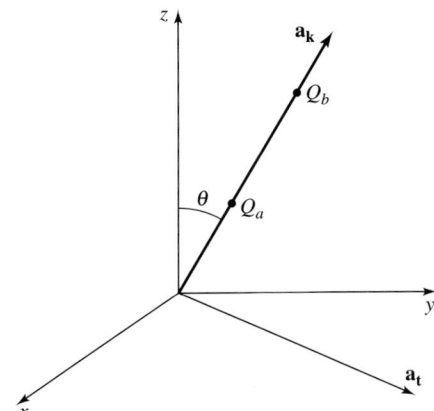

   c. Write the expression for $\mathbf{D}$ at point Q_b.

8. For LiNbO₃ and LiTaO₃ in the absence of applied electric fields,
$$n_1 = n_2 = n_o, \quad n_3 = n_e, \quad \frac{1}{n_4^2} = \frac{1}{n_5^2} = \frac{1}{n_6^2} = 0$$
The electrooptic matrix is given in (6.53).
   a. Show that if $\mathbf{E}_m = \mathbf{a}_z E_{mz}$, the index ellipsoid is given by
$$\left(\frac{1}{n_o^2} + r_{13} E_{mz}\right) x^2 + \left(\frac{1}{n_o^2} + r_{13} E_{mz}\right) y^2 + \left(\frac{1}{n_e^2} + r_{33} E_{mz}\right) z^2 = 1$$
   b. Given that $|r_{13} E_{mz}|$ and $|r_{33} E_{mz}|$ are much smaller than $1/n_e^2$ and $1/n_o^2$, show that the principal indices are
$$n_x = n_y \approx n_o - \frac{1}{2} n_o^3 r_{13} E_{mz} \qquad n_z \approx n_e - \frac{1}{2} n_e^3 r_{33} E_{mz}$$

9. Using the results obtained in problem 8, show that the retardation has voltage-independent and voltage-dependent terms. Show that for the voltage-dependent part only, the half-wave voltage of a transverse EO modulator made of LiNbO₃ or LiTaO₃ is
$$V_{T\pi} = \frac{\lambda}{|n_e^3 r_{33} - n_o^3 r_{13}|} \frac{d}{\ell}$$

10. In the absence of an applied electric field, the index ellipsoid of a certain material is

$$\frac{x^2}{n_o^2} + \frac{y^2}{n_o^2} + \frac{z^2}{n_o^2} = 1$$

and the electrooptic matrix is given by (6.51).

a. Show that if $\mathbf{E_m} = \mathbf{a}_z E_{mz}$, the index ellipsoid becomes

$$\frac{x^2}{n_o^2} + \frac{y^2}{n_o^2} + \frac{z^2}{n_o^2} + 2r_{41}E_{mz}xy = 1$$

b. Show that the equation in (a) may be written as

$$(\frac{1}{n_o^2} + r_{41}E_{mz})x'^2 + (\frac{1}{n_o^2} - r_{41}E_{mz})y'^2 + (\frac{1}{n_o^2})z^2 = 1$$

where x' and y' are axes rotated by $\pi/4$ with respect to the x and y axes.

c. Given that $|r_{41}E_{mz}| \ll 1$, show that the indices of refraction along the new principal axes, x', y', and z, are

$$n_{x'} \approx n_o - \frac{1}{2}n_o^3 r_{41} E_{mz}, \qquad n_{y'} \approx n_o + \frac{1}{2}n_o^3 r_{41} E_{mz}, \qquad n_z = n_o$$

d. Consider an optical beam of wavelength λ propagating in the y' direction with $\mathbf{E} = E_o(\mathbf{a}_{x'} + \mathbf{a}_z)$ at $y' = 0$. Find the expression for $\mathbf{E}$ for $y' > 0$.

11. Materials with hexagonal symmetry of the $\bar{6}m2$ class are uniaxial media with ordinary and extraordinary indices n_o and n_e. The electrooptic matrix of the material is of the form

$$\begin{bmatrix} 0 & -r_{22} & 0 \\ 0 & r_{22} & 0 \\ 0 & 0 & 0 \\ 0 & 0 & 0 \\ 0 & 0 & 0 \\ -r_{22} & 0 & 0 \end{bmatrix}$$

a. What is the effect of electric fields $\mathbf{E_m} = \mathbf{a}_z E_{mz}$ on the indices?

b. Show that if $\mathbf{E_m} = \mathbf{a}_y E_{my}$, the index ellipsoid is given by

$$(\frac{1}{n_o^2} - r_{22} E_{my})x^2 + (\frac{1}{n_o^2} + r_{22} E_{my})y^2 + (\frac{1}{n_e^2})z^2 = 1$$

c. Given that $|r_{22} E_{my}| \ll 1/n_o^2$, show that, in the presence of $E_{my}\mathbf{a}_y$, the indices along the principal axes are

$$n_x \approx n_o + \frac{1}{2}n_o^3 r_{22} E_{my}, \qquad n_y \approx n_o - \frac{1}{2}n_o^3 r_{22} E_{my}, \qquad n_z = n_e$$

d. Using the material in a transverse EO modulator configuration with the applied electric field in the y direction, show that the half-wave voltage is

$$V_{T\pi} = \frac{\lambda d}{n_o^3 r_{22} \ell}$$

12. A schematic diagram for a longitudinal EO modulator is shown in Figure 6.14. An ADP crystal is used in lieu of a KDP crystal. If $\lambda \approx 0.546\mu m$, then $n_o \approx 1.5266$,

$n_e \approx 1.4808$, $r_{41} \approx 23.76 \times 10^{-12}$ m/V, and $r_{63} \approx 8.56 \times 10^{-12}$ m/V. At radio frequencies, $\varepsilon_{rx} = \varepsilon_{ry} \approx 56$ and $\varepsilon_{rz} \approx 15$.

a. Find the modulating voltage V_m which maximizes the fraction of optical power passing through the modulator. Identify this voltage as V_{mM}.

b. Ignore the reflection at the ADP/air interfaces and calculate the fraction of power transmitted through the modulator for $V_m = V_{mM}$.

c. Let V_m vary from 0 to V_{mM}, and plot the fraction of power transmitted through the modulator as a function of V_m. Again ignore the reflection at the ADP/air interfaces.

d. Repeat (c) taking into account the reflections at the ADP/air interfaces. Note that there are *two* ADP/air interfaces.

13. Using the longitudinal EO modulator with an ADP crystal specified in problem 12, consider the state of polarization of the waves emerging from the right face of the ADP crystal. What is the ellipticity and azimuth of the waves when $V_m = 0$? Repeat for $0.25V_{mM}$, $0.5V_{mM}$, $0.75V_{mM}$, and V_{mM}.

14. Design a transverse EO modulator (Figure 6.17) with a KDP crystal. The relevant material parameters of KDP crystals at 0.546 μm are:

$$n_o \approx 1.5115, \quad n_e \approx 1.4698$$
$$r_{41} \approx 8.77 \times 10^{-12} \text{ m/V}, \quad r_{63} \approx 10.3 \times 10^{-12} \text{ m/V}$$
$$\varepsilon_{rx} = \varepsilon_{ry} \approx 42, \quad \varepsilon_{yz} \approx 21$$

a. What is the peak-to-peak amplitude of the sinusoidal voltage and the dc bias voltage which would give a modulation factor [7] of 10 percent at 100 MHz?

b. Suppose that the incident optical beam has a beam waist radius of 0.15 cm and the rf bandwidth of the modulator is 30 MHz (FWHP) centered around 100 MHz. The rf voltage source has a source impedance of 50 Ω. Specify the length and radius of the KDP crystal. Justify your choice.

15. Consider the diffraction of light ($\lambda = 0.633$ μm) by acoustic waves in water. Suppose the interaction length ℓ is 1.5 cm.

a. What is the upper limit of the acoustic wave frequency f_a if the interaction is to be considered Raman–Nath diffraction?

b. What is the lower limit of f_a if the interaction is to be considered Bragg diffraction?

16. Again, consider the diffraction of light in water. The acoustic waves are such that the strain is approximately 3.0×10^{-5}. Assume that the condition of Raman–Nath diffraction is satisfied. Estimate the intensity of the undiffracted beam (i.e., the zeroth order beam), and the first- and second-order diffracted beams if:

a. $\ell = 1.0$ cm.

b. $\ell = 2.0$ cm.

c. $\ell = 1.0$ cm and S is changed to 1×10^{-4}.

17. An optical beam with $\lambda = 0.9$ μm interacts with acoustic waves in an acoustic material with $n = 1.46$, and $V = 6.0 \times 10^3$ m/s. The optical beam impinges on the acoustic medium at a right angle (90°). The index perturbation due to acoustic waves is

---

[7] According to the IEEE Standard Dictionary, *modulation factor* is "the ratio (usually expressed in percent) of the peak variation of the envelop from the reference value. The reference value is usually taken to be the amplitude of unmodulated wave."

$$\Delta n(z;t) = 2\times 10^{-4} \cos(\Omega t - Kz)$$

where $\Omega = 2\pi f_a$, $K = \Omega/V$, and $f_a = 15$ MHz.

a. What is the maximum interaction length ℓ if the interaction is to be classified as Raman–Nath diffraction?

b. What is the angle, in degrees and measured in air, between the undiffracted and first-order diffracted beams?

c. Choose the interaction length ℓ such that power in the first-order diffracted beam is maximized. Assume that the diffraction is still in the Raman–Nath region.

d. Choose the interaction length ℓ such that power in the zeroth order beam is completely depleted. Again, assume that the diffraction is in the Raman–Nath region.

18. By interacting with acoustic waves of 130 MHz frequency in a certain medium, light waves of $\lambda = 1.2$ μm are diffracted into many diffraction orders. However, only two beams, b and c, have been found. No other beams, diffracted or undiffracted, have been found. Nor can the orders of diffraction of these beams be determined. When measured outside the acoustic medium, $|\theta'_b - \theta'_c| = 6°$. When measured inside the acoustic medium, $|\theta_b - \theta_c| = 4°$. Experiments also show that $|f_b - f_c| = 260$ MHz.

a. Calculate the acoustic wave velocity.

b. Calculate the index of refraction.

c. If the interaction length ℓ is 0.6 mm, can the diffraction be classified as Raman–Nath diffraction?

REFERENCES

1. Beisa, L. *Laser Applications.* New York, NY: Academic Press, 1974.
2. Reich, S. "The use of electro-mechanical mirror scanning devices." SPIE Proceedings, *Laser Scanning Components and Techniques* 84, (Aug. 24–25, 1976), Bellingham, WA.
3. Jenkins, F. A.; and H. E. White. *Fundamentals of Optics.* Chapter 32. 4th ed. New York, NY: McGraw-Hill Book Co., 1976.
4. Kraus, J. D. *Electromagnetics.* 4th ed. New York, NY: McGraw-Hill Book Co., 1991.
5. Lass, H. *Vector and Tensor Analysis.* New York, NY: McGraw-Hill Book Co., 1950.
6. Born, M.; and E. Wolf. *Principles of Optics.* Chapter 1. 6th ed. Pergamon Press, 1980.
7. Kittle, C. *Introduction to Solid State Physics.* Chapter 6. New York, NY: John Wiley & Sons, Inc., 1976.
8. Nelson, D. F. *Electric, Optic, and Acoustic Interactions in Dielectrics.* Chapter 14. New York, NY: Wiley Interscience, 1979.
9. Donati, S.; V. Annovazzi-Lodi; and T. Tambosso. "Magneto-optical fibre sensors for electrical industry: analysis of performances." *IEE Proceedings* 135, pt. J., (1988), pp. 372–382.
10. Kaminow, I. P. *An Introduction to Electrooptic Devices.* New York, NY: Academic Press, 1974.

11. Yariv, A.; and P. Yeh. *Optical Waves in Crystals.* New York, NY: John Wiley & Sons, Inc., 1984.
12. Librecht, Francois; and F. M. Librecht. "ADP 45° X-cut four-crystal light modulator." *Appl. Opt.* 11, (1972), pp. 472–473.
13. Adler, R. "Interaction between light and sound." *IEEE Spectrum* 4, (May 1967), p. 42.
14. Chang, I. C. "Acousto-optic devices and applications." *IEEE Trans. of Sonics and Ultrasonics* SU-23, (January 1976), pp. 2–22.
15. Korpel, A. "Acousto-optic." In *Applied Solid State Science, Advances in Materials and Device Research* 3, ed. R. Wolfe. New York, NY: Academic Press, 1972. Also, Korpel, A. "Acousto-optics—a review of fundamentals." *Proc. IEEE* 69, (January 1981), pp. 48–53.
16. Kino, G. S. *Acoustic waves: devices, imaging, and analog signal processing.* Englewood Cliffs, NJ: Prentice-Hall, 1987.
17. Klein, W. R.; and B. D. Cook. "Unified approach to ultrasonic light diffraction." *IEEE Trans. on Sonics and Ultrasonics* SU-14, (1967), pp. 123–134.
18. Uchida, N.; and N. Niizeki. "Acoustooptic deflection materials and techniques." *Proc. IEEE* 61, (1973), pp. 1073–1092.

ADDITIONAL READING

1. Banerjee, P. P.; and T. C. Poon. *Principles of Applied Optics.* Homewood, IL: Aksen Associates, Inc., 1991.
2. Guenther, B. D. *Modern Optics.* John Wiley & Sons, Inc., 1990.
3. Jenkins, F. A.; and H. E. White. *Fundamentals of Optics.* 4th ed. New York, NY: McGraw-Hill Book Co., 1976.
4. Karim, M. A. *Electro-optical Devices and Systems.* Boston, MA: PWS-Kent Publishing Co., 1990.
5. Nussbaum, A.; and R. A. Phillips. *Contemporary Optics for Scientists and Engineers.* Englewood Cliffs, NJ: Prentice Hall, Inc., 1976.
6. Pedrotti, F. L.; and L. S. Pedrotti. *Introduction to Optics.* Englewood Cliffs, NJ: Prentice Hall, Inc., 1992.
7. Saleh, B. E. A.; and M. C. Teich. *Fundamentals of Photonics.* New York, NY: John Wiley & Sons, Inc., 1991.
8. Wilson, J.; and J. F. B. Hawkes. *Optoelectronics, An Introduction.* 2nd ed. Englewood Cliffs, NJ: Prentice Hall, Inc., 1989.
9. Yariv, A. *Introduction to Optical Electronics.* 4th ed. New York, NY: Saunder College Pub., 1991.

ELASTICITY FOR THREE-DIMENSIONAL OBJECTS

APPENDIX B

APPENDIX OUTLINE

B.1 Strain Tensor
B.2 Stress Tensor
B.3 Hooke's Law and Elastic Constants
B.4 Elastic Equations of Motion
B.5 Elastic Waves in Isotropic Media
    B.5.1 Longitudinal Acoustic Waves
    B.5.2 Transverse Acoustic Waves
B.6 Abbreviated Indices
References

B.1 STRAIN TENSOR

This appendix discusses strain, stress, and acoustic waves in three-dimensional objects. Tensile strain is the fractional elongation or compression of an object. For three-dimensional objects, other types of deformations also exist. Consider a rectangle in the xy plane, as shown in Figure B1a. Three types of deformation are possible: **elongation** or **compression** (Figure B1b), **rotation** (Figure B1c), and **shearing distortion** (Figure B1d). Elongation or compression is deformation parallel to a length. For elongation in the x direction, the tensile strain component is

$$S_{xx} = \frac{\partial u_x}{\partial x} \tag{B1}$$

Also shown in Figure B1b is an elongation in the y direction. The tensile strain component accompanying the elongation in the y direction is

$$S_{yy} = \frac{\partial u_y}{\partial y} \tag{B2}$$

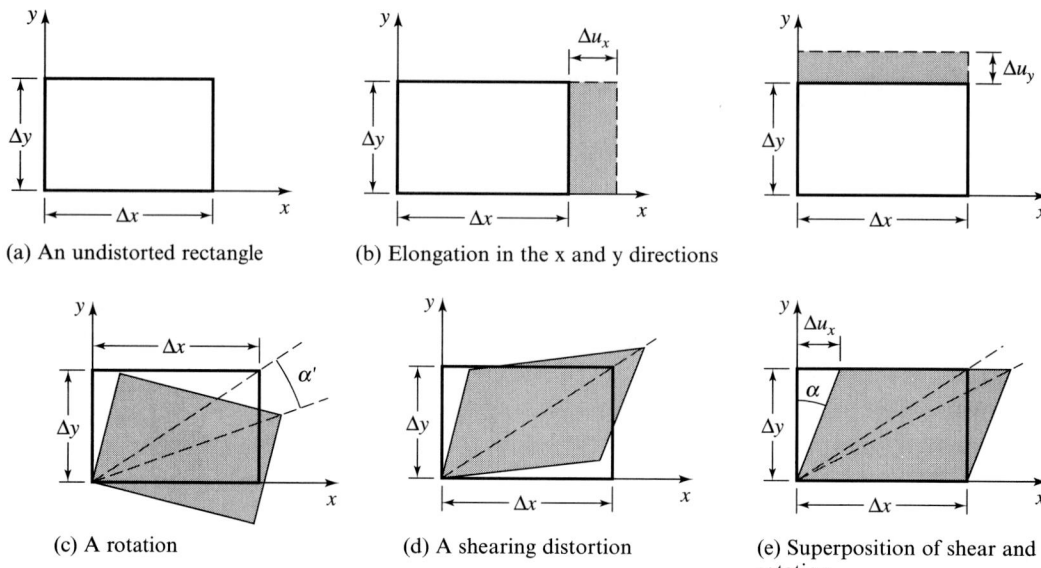

Figure B.1 Deformation of a rectangle

In **rotation** deformation (Figure B1c), the entire rectangle is rotated by an angle, and there is no change in the shape of the rectangle. Consequently, the distance between two arbitrary points in the rectangle remains unchanged. In **shearing distortion,** the rectangle becomes a rhomboid (Figure B1d). The diagonal of the original rectangle, although stretched, remains in its original position. In addition, the distance between two arbitrary points is changed.

An arbitrary deformation in the direction normal to a length may be viewed as the superposition of a rotation and a shearing distortion. Consider a distortion such as that shown in Figure B1e. The angle α is used as a measure for quantifying the distortion. The angle α is given by

$$\tan\alpha \approx \frac{\Delta u_x}{\Delta y} = \frac{1}{2}\left[\frac{\Delta u_x}{\Delta y} + \frac{\Delta u_y}{\Delta x}\right] + \frac{1}{2}\left[\frac{\Delta u_x}{\Delta y} - \frac{\Delta u_y}{\Delta x}\right] \tag{B3}$$

The terms in the first bracket represent the **shearing strain,** and those in the second bracket represent the **rotation.** In the limit of $\Delta x \to 0$ and $\Delta y \to 0$, $\frac{\Delta u_x}{\Delta y}, \frac{\Delta u_y}{\Delta x}$, etc. approach partial derivatives. Thus, the shearing strain component of the distortion shown in Figure B1e is

$$S_{xy} = S_{yx} = \frac{1}{2}\left(\frac{\partial u_y}{\partial x} + \frac{\partial u_x}{\partial y}\right) \tag{B4}$$

Similarly, by considering rectangles in yz and zx planes, we have additional strain components, as follows:

$$S_{zz} = \frac{\partial u_z}{\partial z} \tag{B5}$$

$$S_{zy} = S_{yz} = \frac{1}{2}(\frac{\partial u_z}{\partial y} + \frac{\partial u_y}{\partial z}) \tag{B6}$$

$$S_{xz} = S_{zx} = \frac{1}{2}(\frac{\partial u_x}{\partial z} + \frac{\partial u_z}{\partial x}) \tag{B7}$$

It is convenient to assemble all strain components into a compact matrix form. The **tensile strain** components, S_{xx}, S_{yy}, and S_{zz}, are the diagonal elements, and the shearing strain components, S_{xy}, S_{yz}, and S_{zx}, etc., are the off-diagonal matrix elements. In the matrix form, then, a strain tensor becomes

$$\overleftrightarrow{S} = \begin{bmatrix} S_{xx} & S_{xy} & S_{xz} \\ S_{yx} & S_{yy} & S_{yz} \\ S_{zx} & S_{zy} & S_{zz} \end{bmatrix} \tag{B8}$$

A pictorial representation of a three-dimensional object's deformation associated with each strain component is given in Figure B2.

B.2 STRESS TENSOR

Stress is defined as the *force per unit area*. For three-dimensional objects, it is necessary to specify the orientation of the elementary area and the force component acting on it. Figure B3a shows an elementary area A_x normal to the x axis. If the force acting on the elementary area A_x is

$$\mathbf{F} = \mathbf{a}_x F_x + \mathbf{a}_y F_y + \mathbf{a}_z F_z$$

the stress has three components, which are:

$$\begin{aligned} T_{xx} &= \lim_{A_x \to 0} \frac{F_x}{A_x} \\ T_{yx} &= \lim_{A_x \to 0} \frac{F_y}{A_x} \\ T_{zx} &= \lim_{A_x \to 0} \frac{F_z}{A_x} \end{aligned} \tag{B9}$$

The first subscript designates the force component, and the second subscript specifies the orientation of the elementary area. Similarly, we define the stress components for elementary areas A_y and A_z normal to the y axis and z axis, respectively (Figures B3b and B3c). The stress at an arbitrary point in the material is uniquely specified by the nine tensor components. The unit of a stress component is N/m².

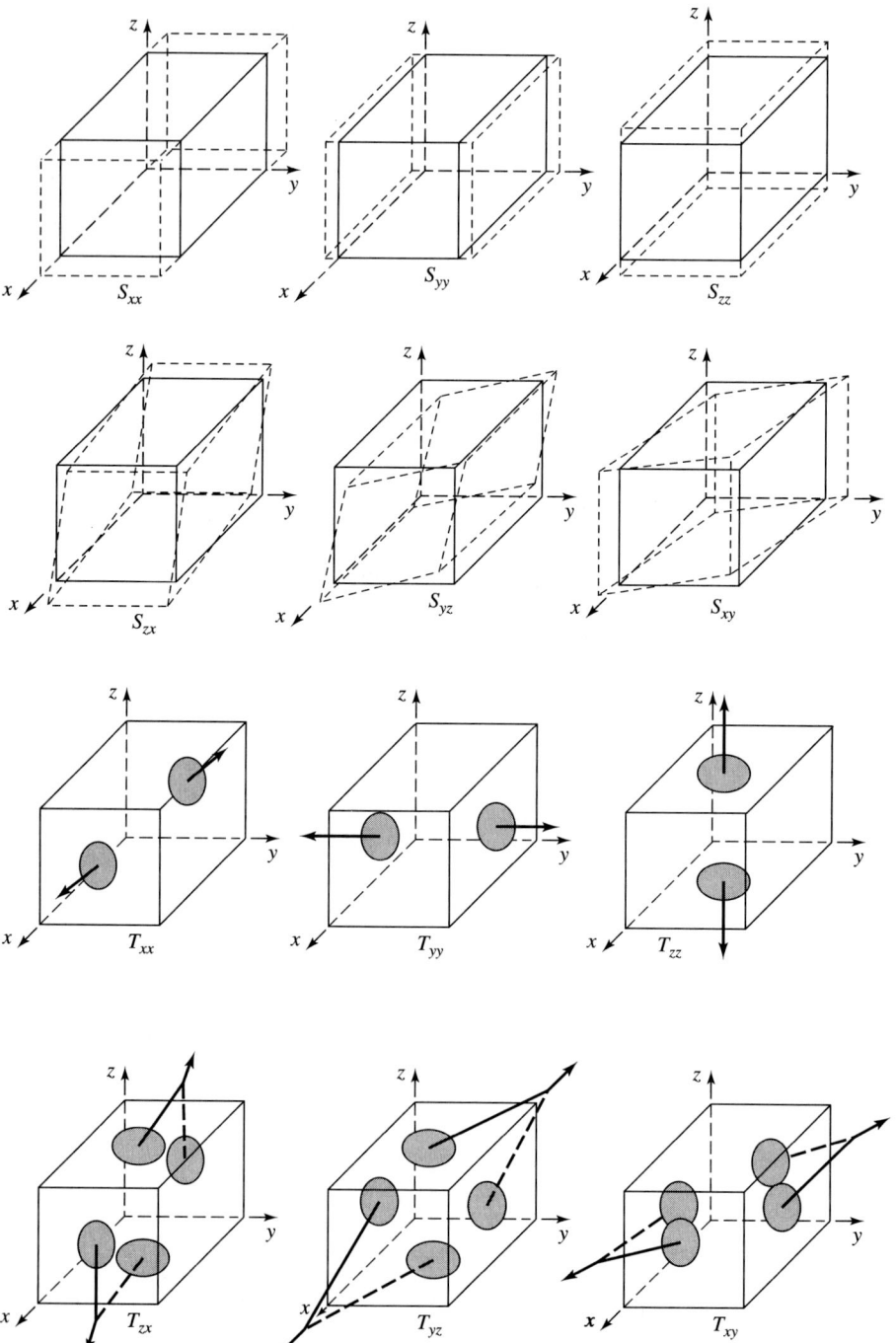

Figure B.2 Graphical representation of strain and stress tensor components (after [1])

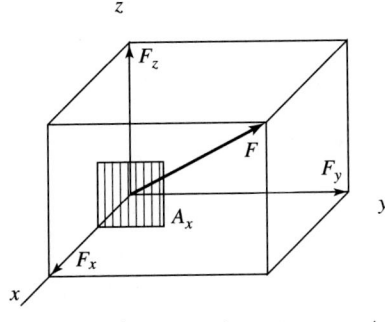

$$T_{xx} = \lim_{A_x \to 0} \frac{F_x}{A_x}$$

$$T_{yx} = \lim_{A_x \to 0} \frac{F_y}{A_x}$$

$$T_{zx} = \lim_{A_x \to 0} \frac{F_z}{A_x}$$

(a) Force acting on an elementary area A_x

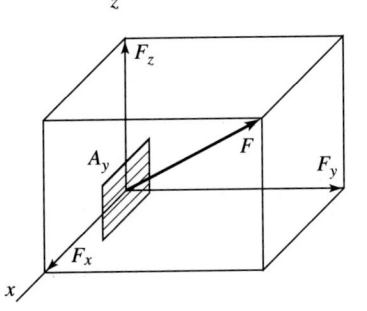

$$T_{xy} = \lim_{A_y \to 0} \frac{F_x}{A_y}$$

$$T_{yy} = \lim_{A_y \to 0} \frac{F_y}{A_y}$$

$$T_{zy} = \lim_{A_y \to 0} \frac{F_z}{A_y}$$

(b) Force acting on an elementary area A_y

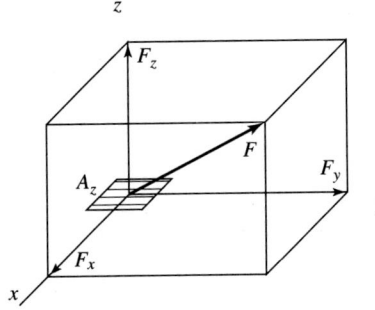

$$T_{xz} = \lim_{A_z \to 0} \frac{F_x}{A_z}$$

$$T_{yz} = \lim_{A_z \to 0} \frac{F_y}{A_z}$$

$$T_{zz} = \lim_{A_z \to 0} \frac{F_z}{A_z}$$

(c) Force acting on an elementary area A_z

Figure B.3 Definition of components of a stress tensor

Alternatively, the components of a stress tensor may be interpreted as follows. A box is drawn around the point of interest, and T_{ij} designates the i component of force per unit area acting on a surface *normal to* and *facing* the $+j$ direction, where i and j stand for x, y, or z (Figure B4). If the material is in rotational equilibrium, the total torque acting on any elemental volume vanishes which means that

$$T_{xy} = T_{yx}, \qquad T_{xz} = T_{zx}, \qquad T_{yz} = T_{zy}$$

In other words, six of the nine stress components are independent. A pictorial representation of these six stress components is also depicted in Figure B2. From the definition of the stress components, we see that $T_{x(-z)} = -T_{xz}$.

Again, it is convenient to cast a stress tensor in matrix form, as follows:

$$\overset{\leftrightarrow}{T} = \begin{bmatrix} T_{xx} & T_{xy} & T_{xz} \\ T_{yx} & T_{yy} & T_{yz} \\ T_{zx} & T_{zy} & T_{zz} \end{bmatrix} \tag{B10}$$

Once the stress tensor $\overset{\leftrightarrow}{T}$ is specified, we can calculate the force from $\overset{\leftrightarrow}{T}$ on an area A oriented in an arbitrary direction. Let the normal to the surface be (Figure B5)

$$\mathbf{n} = n_x \mathbf{a}_x + n_y \mathbf{a}_y + n_z \mathbf{a}_z \tag{B11}$$

and we consider the projection of area A on the xy plane. The projected surface (shaded horizontally) has an area An_z and is facing the $-\mathbf{a}_z$ direction. The force acting on the projected area is

$$\mathbf{F}_c = An_z[\mathbf{a}_x T_{x(-z)} + \mathbf{a}_y T_{y(-z)} + \mathbf{a}_z T_{z(-z)}] = -An_z[\mathbf{a}_x T_{xz} + \mathbf{a}_y T_{yz} + \mathbf{a}_z T_{zz}] \tag{B12}$$

Similarly, the forces acting on the projected areas on the zx plane (shaded vertically) and yz plane are

$$\mathbf{F}_b = -An_y[\mathbf{a}_x T_{xy} + \mathbf{a}_y T_{yy} + \mathbf{a}_z T_{zy}] \tag{B13}$$

$$\mathbf{F}_a = -An_x[\mathbf{a}_x T_{xx} + \mathbf{a}_y T_{yx} + \mathbf{a}_z T_{zx}] \tag{B14}$$

The total force acting on the area A is the vector sum of $\mathbf{F}_a$, $\mathbf{F}_b$, and $\mathbf{F}_c$,

$$\mathbf{F} = A \{\mathbf{a}_x[n_x T_{xx} + n_y T_{xy} + n_z T_{xz}] \tag{B15}$$
$$+ \mathbf{a}_y[n_x T_{yx} + n_y T_{yy} + n_z T_{yz}] + \mathbf{a}_z[n_x T_{zx} + n_y T_{zy} + n_z T_{zz}]\}$$

Therefore, the force per unit area acting on a surface A by $\overset{\leftrightarrow}{T}$ is, in the matrix notation,

$$\begin{bmatrix} T_{xn} \\ T_{yn} \\ T_{zn} \end{bmatrix} = \begin{bmatrix} T_{xx} & T_{xy} & T_{xz} \\ T_{yx} & T_{yy} & T_{yz} \\ T_{zx} & T_{zy} & T_{zz} \end{bmatrix} \begin{bmatrix} n_x \\ n_y \\ n_z \end{bmatrix} \tag{B16}$$

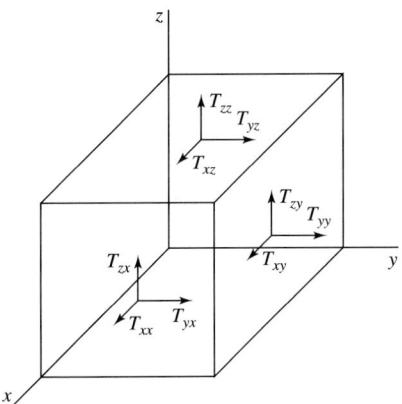

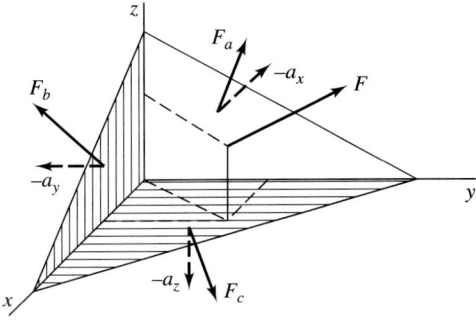

Figure B.5 Force on an arbitrarily oriented surface

Figure B.4 Components of a stress tensor

B.3 HOOKE'S LAW AND ELASTIC CONSTANTS

As in the one-dimensional problem, the stress–strain relation is linear if the material is not stressed beyond the **elastic limit.** When a three-dimensional object is stretched in an arbitrary direction, the lateral dimensions contract simultaneously. Therefore, the stress component T_{xx} produces S_{yy} and S_{zz} in addition to S_{xx}, as shown schematically in Figure B6. The linear relationship between T_{xx} and S_{xx} is extended to S_{yy} and S_{zz}. For isotropic elastic materials,

$$S_{yy} = S_{zz} = -\frac{\sigma}{Y} T_{xx} \tag{B17}$$

where the **Poisson ratio** σ is the ratio of the longitudinal elongation to the lateral contraction produced by the same tensile stress; that is,

$$\sigma = -\frac{S_{yy}}{S_{xx}} = -\frac{S_{zz}}{S_{xx}} \tag{B18}$$

Since elongation is accompanied by contraction in the lateral directions, S_{xx} and accompanying S_{yy} and S_{zz} have opposite signs. Thus, the need for a minus sign in (B18) is quite obvious. For most materials, σ is about ⅓. For isotropic materials, Young's modulus and Poisson's ratio are *independent* of stress direction and magnitude. When all tensile strain components T_{xx}, T_{yy}, and T_{zz} are present, the total strain is the *superposition* of contributions from each stress component,

$$\begin{aligned}
S_{xx} &= \frac{1}{Y} T_{xx} - \frac{\sigma}{Y} T_{yy} - \frac{\sigma}{Y} T_{zz} \\
S_{yy} &= -\frac{\sigma}{Y} T_{xx} + \frac{1}{Y} T_{yy} - \frac{\sigma}{Y} T_{zz} \\
S_{zz} &= -\frac{\sigma}{Y} T_{xx} - \frac{\sigma}{Y} T_{yy} + \frac{1}{Y} T_{zz}
\end{aligned} \tag{B19}$$

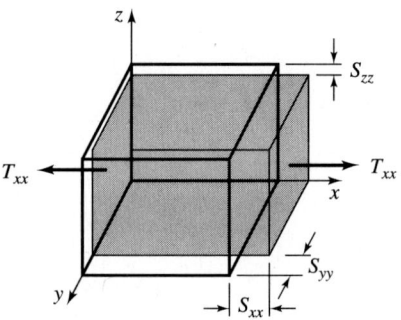

Figure B.6 Elongation and contraction due to tensile stress

No *shearing stress* or *shear strain* components are involved in (B19). The shearing stress components are related to the shearing strain components as follows:

$$T_{ij} = 2\mu S_{ij}, \quad (i \neq j, \quad i,j = x, y, \text{ or } z) \quad \text{(B20)}$$

where μ is the **shear modulus** or **modulus of rigidity**.

For computational purposes, it may be desirable to express the stress components in terms of strain components. Equation (B19) may be used to solve for T_{xx}, T_{yy}, and T_{zz}, resulting in the following, as an example:

$$T_{xx} = \frac{Y\sigma}{(1 - 2\sigma)(1 + \sigma)}(S_{yy} + S_{zz}) + \frac{Y(1 - \sigma)}{(1 + \sigma)(1 - 2\sigma)} S_{xx} \quad \text{(B21)}$$

There are three elastic constants, Y, σ, and μ. For isotropic materials, two of these three are independent. For example, Y can be expressed in terms of μ and σ.

$$Y = 2\mu(1 + \sigma) \quad \text{(B22)}$$

These constants can be combined in several ways. For example, **Lame's coefficients** μ and λ are as follows,[1]

$$\lambda = \frac{Y\sigma}{(1 + \sigma)(1 - 2\sigma)} \quad \text{(B23)}$$

$$\mu = \frac{Y}{2(1 + \sigma)} \quad \text{(B24)}$$

In terms of Lame's coefficients, (B21) becomes

$$T_{xx} = \lambda(S_{xx} + S_{yy} + S_{zz}) + 2\mu S_{xx} \quad \text{(B25)}$$

---

[1] Here, we follow the generally accepted notations for Lame's coefficients, with the understanding that Lame's coefficient λ should not be confused with the usual symbol for wavelength.

Similarly,

$$T_{yy} = \lambda(S_{xx} + S_{yy} + S_{zz}) + 2\mu S_{yy} \tag{B26}$$

$$T_{zz} = \lambda(S_{xx} + S_{yy} + S_{zz}) + 2\mu S_{zz} \tag{B27}$$

In addition, the fractional change of volume is

$$\frac{\Delta V}{V} = S_{xx} + S_{yy} + S_{zz} \tag{B28}$$

This implies that it is quite meaningful to group together the tensile strain components in (B25), (B26), and (B27).

B.4 ELASTIC EQUATIONS OF MOTION

The equation of motion for elastic media is simply a mathematical expression for **Newton's second law.** Consider an arbitrary point (x, y, z) in the medium, surrounded by a volume element of dimensions $2\Delta x$, $2\Delta y$, and $2\Delta z$, with (x, y, z) as its center (Figure B7). For convenience, we refer to the point as a "particle," and consider its displacement, velocity, and acceleration etc. In our discussions, stress, strain, and other variables are functions of time and position. We also assume that the elementary volume under consideration is sufficiently small that the variations of stress and strain in the elementary volume are gradual.

Let the mass density of the material be ρ_m and the particle displacement be $\vec{u}(x,y,z;t)$. The total mass in the volume element is then $\rho_m \cdot 8\Delta x \Delta y \Delta z$, and the particle acceleration is $\dfrac{\partial^2 \vec{u}(x,y,z;t)}{\partial t^2}$. Consider the front and back surfaces, $(2\Delta y)(2\Delta z)$, of the volume element. The force from the stress on the *front surface* is

$$4\Delta y \Delta z [\mathbf{a}_x \mathcal{T}_{xx}(x + \Delta x, y, z; t) + \mathbf{a}_y \mathcal{T}_{yx}(x + \Delta x, y, z; t) + \mathbf{a}_z \mathcal{T}_{zx}(x + \Delta x, y, z; t)]$$

For the *back surface*, the normal is pointed in the $-\mathbf{a}_x$ direction, and the force is

$$4\Delta y \Delta z [\mathbf{a}_x \mathcal{T}_{x(-x)}(x - \Delta x, y, z; t) + \mathbf{a}_y \mathcal{T}_{y(-x)}(x - \Delta x, y, z; t)$$
$$+ \mathbf{a}_z \mathcal{T}_{z(-x)}(x - \Delta x, y, z; t)]$$
$$= -4\Delta y \Delta z [\mathbf{a}_x \mathcal{T}_{xx}(x - \Delta x, y, z; t) + \mathbf{a}_y \mathcal{T}_{yx}(x - \Delta x, y, z; t)$$
$$+ \mathbf{a}_z \mathcal{T}_{zx}(x - \Delta x, y, z; t)]$$

Combining these expressions, we have the force acting on the volume element via the *front and back surfaces*

$$\mathcal{F}_{fb} \approx 8\Delta x \Delta y \Delta z \frac{\partial}{\partial x}[\mathbf{a}_x \mathcal{T}_{xx}(x,y,z;t) + \mathbf{a}_y \mathcal{T}_{yx}(x,y,z;t) + \mathbf{a}_z \mathcal{T}_{zx}(x,y,z;t)]$$

Similarly, the force acting on the *top and bottom* and the *right and left* surfaces, respectively, are

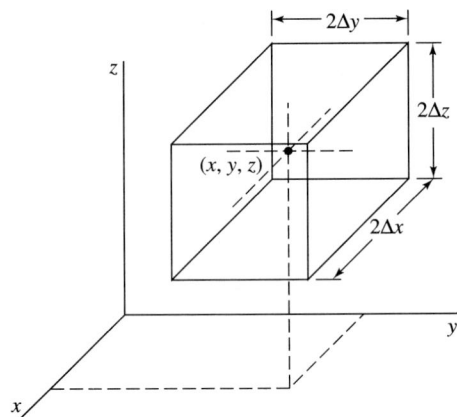

Figure B.7 Volume element used in considering the equation of motion

$$\mathscr{F}_{tb} \approx 8\Delta x \Delta y \Delta z \frac{\partial}{\partial z}[\mathbf{a}_x \mathscr{T}_{xz}(x,y,z;t) + \mathbf{a}_y \mathscr{T}_{yz}(x,y,z;t) + \mathbf{a}_z \mathscr{T}_{zz}(x,y,z;t)]$$

$$\mathscr{F}_{rl} \approx 8\Delta x \Delta y \Delta z \frac{\partial}{\partial y}[\mathbf{a}_x \mathscr{T}_{xy}(x,y,z;t) + \mathbf{a}_y \mathscr{T}_{yy}(x,y,z;t) + \mathbf{a}_z \mathscr{T}_{zy}(x,y,z;t)]$$

Therefore, from Newton's second law, we have

$$\rho_m(8\Delta x \Delta y \Delta z)\frac{\partial^2 \vec{u}}{\partial t^2} \approx \left[\frac{\partial}{\partial x}(\mathbf{a}_x \mathscr{T}_{xx} + \mathbf{a}_y \mathscr{T}_{yx} + \mathbf{a}_z \mathscr{T}_{zx})\right.$$

$$+ \frac{\partial}{\partial y}(\mathbf{a}_x \mathscr{T}_{xy} + \mathbf{a}_y \mathscr{T}_{yy} + \mathbf{a}_z \mathscr{T}_{zy})$$

$$\left. + \frac{\partial}{\partial z}(\mathbf{a}_x \mathscr{T}_{xz} + \mathbf{a}_y \mathscr{T}_{yz} + \mathbf{a}_z \mathscr{T}_{zz})\right]8\Delta x \Delta y \Delta z$$

After factoring out the term $8\Delta x \Delta y \Delta z$ and taking the limit $\Delta x \Delta y \Delta z \to 0$, we obtain an equation for each component of acceleration:

$$\rho_m \frac{\partial^2 u_x}{\partial t^2} = \frac{\partial \mathscr{T}_{xx}}{\partial x} + \frac{\partial \mathscr{T}_{xy}}{\partial y} + \frac{\partial \mathscr{T}_{xz}}{\partial z}$$

$$\rho_m \frac{\partial^2 u_y}{\partial t^2} = \frac{\partial \mathscr{T}_{yx}}{\partial x} + \frac{\partial \mathscr{T}_{yy}}{\partial y} + \frac{\partial \mathscr{T}_{yz}}{\partial z} \qquad (B29)$$

$$\rho_m \frac{\partial^2 u_z}{\partial t^2} = \frac{\partial \mathscr{T}_{zx}}{\partial x} + \frac{\partial \mathscr{T}_{zy}}{\partial y} + \frac{\partial \mathscr{T}_{zz}}{\partial z}$$

These are the time domain **equations of motion** for elastic media. In the derivation, no assumption is made relative to the nature of the media. Therefore, (B29) is valid for isotropic and anisotropic, elastic and piezoelectric, media.

B.5 ELASTIC WAVES IN ISOTROPIC MEDIA

For isotropic elastic media, the stress components are related to the strain components through Hooke's law, as given by (B20) and (B25)–(B27). By expressing the strain components in terms of the particle displacement, (B1), (B2), and (B4)–(B7), (B29) becomes

$$\rho_m \frac{\partial^2 u_x}{\partial t^2} = (\lambda + \mu) \frac{\partial}{\partial x} \left(\frac{\partial u_x}{\partial x} + \frac{\partial u_y}{\partial y} + \frac{\partial u_z}{\partial z} \right) + \mu \left(\frac{\partial^2 u_x}{\partial x^2} + \frac{\partial^2 u_x}{\partial y^2} + \frac{\partial^2 u_x}{\partial z^2} \right)$$

$$\rho_m \frac{\partial^2 u_y}{\partial t^2} = (\lambda + \mu) \frac{\partial}{\partial y} \left(\frac{\partial u_x}{\partial x} + \frac{\partial u_y}{\partial y} + \frac{\partial u_z}{\partial z} \right) + \mu \left(\frac{\partial^2 u_y}{\partial x^2} + \frac{\partial^2 u_y}{\partial y^2} + \frac{\partial^2 u_y}{\partial z^2} \right)$$

$$\rho_m \frac{\partial^2 u_z}{\partial t^2} = (\lambda + \mu) \frac{\partial}{\partial z} \left(\frac{\partial u_x}{\partial x} + \frac{\partial u_y}{\partial y} + \frac{\partial u_z}{\partial z} \right) + \mu \left(\frac{\partial^2 u_z}{\partial x^2} + \frac{\partial^2 u_z}{\partial y^2} + \frac{\partial^2 u_z}{\partial z^2} \right)$$
(B30)

These equations can be expressed in a compact manner through the use of the vector and operator notations, although the physical meanings of the terms are somewhat obscured. In the vector and operator notations [2], (B29) and (B30) become, respectively,

$$\rho_m \frac{\partial^2 \vec{u}}{\partial t^2} = \nabla \cdot \overleftrightarrow{\mathcal{T}} \tag{B31}$$

$$(\lambda + \mu)\nabla(\nabla \cdot \vec{u}) + \mu(\nabla \cdot \nabla)\vec{u} = \rho_m \frac{\partial^2 \vec{u}}{\partial t^2} \tag{B32}$$

where $\vec{u}$ in the matrix form is

$$\vec{u} \equiv \begin{bmatrix} u_x \\ u_y \\ u_z \end{bmatrix} \tag{B33}$$

and the operator $\nabla \cdot$ is

$$\nabla \cdot \equiv \begin{bmatrix} \frac{\partial}{\partial x} & 0 & 0 & 0 & \frac{\partial}{\partial z} & \frac{\partial}{\partial y} \\ 0 & \frac{\partial}{\partial y} & 0 & \frac{\partial}{\partial z} & 0 & \frac{\partial}{\partial x} \\ 0 & 0 & \frac{\partial}{\partial z} & \frac{\partial}{\partial y} & \frac{\partial}{\partial x} & 0 \end{bmatrix} \tag{B34}$$

For time-harmonic ($e^{j\Omega t}$) strain and stress, the equations can be expressed in terms of the frequency-domain representation **u** for the particle displacement, in lieu of the time-domain representation $\vec{u}$. Then, (B30) becomes

$$(\lambda + \mu)\frac{\partial}{\partial x}\left(\frac{\partial u_x}{\partial x} + \frac{\partial u_y}{\partial y} + \frac{\partial u_z}{\partial z}\right) + \mu\left(\frac{\partial^2 u_x}{\partial x^2} + \frac{\partial^2 u_x}{\partial y^2} + \frac{\partial^2 u_x}{\partial z^2}\right) + \Omega^2 \rho_m u_x = 0$$

$$(\lambda + \mu)\frac{\partial}{\partial y}\left(\frac{\partial u_x}{\partial x} + \frac{\partial u_y}{\partial y} + \frac{\partial u_z}{\partial z}\right) + \mu\left(\frac{\partial^2 u_y}{\partial x^2} + \frac{\partial^2 u_y}{\partial y^2} + \frac{\partial^2 u_y}{\partial z^2}\right) + \Omega^2 \rho_m u_y = 0$$

$$(\lambda + \mu)\frac{\partial}{\partial z}\left(\frac{\partial u_x}{\partial x} + \frac{\partial u_y}{\partial y} + \frac{\partial u_z}{\partial z}\right) + \mu\left(\frac{\partial^2 u_z}{\partial x^2} + \frac{\partial^2 u_z}{\partial y^2} + \frac{\partial^2 u_z}{\partial z^2}\right) + \Omega^2 \rho_m u_z = 0$$

(B35)

In the operator form, the equation is

$$(\lambda + \mu)\nabla(\nabla\cdot\mathbf{u}) + \mu(\nabla\cdot\nabla)\mathbf{u} + \Omega^2 \rho_m \mathbf{u} = 0 \tag{B36}$$

B.5.1 Longitudinal Acoustic Waves

Consider plane elastic waves propagating along the z direction, with the particle motion also in the z direction. Then, $\mathbf{u} = \mathbf{a}_z u_z$. Since the particle motion is in the same direction as the wave propagation, the acoustic waves are classified as **longitudinal acoustic waves.** Noting that $\partial/\partial x = \partial/\partial y = 0$, (B35) becomes

$$(\lambda + 2\mu)\frac{\partial^2 u_z}{\partial z^2} + \Omega^2 \rho_m u_z = 0$$

which leads to a propagation velocity of

$$V_\ell = \sqrt{\frac{\lambda + 2\mu}{\rho_m}} = \sqrt{\frac{Y(1 - \sigma)}{\rho_m(1 + \sigma)(1 - 2\sigma)}} \tag{B37}$$

Note that the longitudinal waves discussed in section 6.8 of Chapter 6 are really special cases with $\sigma = 0$.

B.5.2 Transverse Acoustic Waves

Acoustic waves with particles moving in directions perpendicular to the direction of wave propagation are **transverse acoustic waves.** Let the particle motion be in the x direction and the wave propagation be in the z direction. Then, $\mathbf{u} = \mathbf{a}_x u_x$, and (B35) becomes

$$\mu\frac{\partial^2 u_x}{\partial z^2} + \Omega^2 \rho_m u_x = 0$$

Consequently, the velocity of transverse acoustic waves is

$$V_t = \sqrt{\frac{\mu}{\rho_m}} = \sqrt{\frac{Y}{\rho_m 2(1 + \sigma)}} \tag{B38}$$

Note that V_ℓ and V_t are different, and that the ratio of V_ℓ/V_t depends on the Poisson ratio

$$\frac{V_\ell}{V_t} = \sqrt{\frac{2(1-\sigma)}{1-2\sigma}} \tag{B39}$$

As σ varies from 0 to 0.4, V_ℓ/V_t varies from $\sqrt{2}$ to $\sqrt{6}$. In general, $\sqrt{2}V_t < V_\ell$.

B.6 ABBREVIATED INDICES

Taking advantage of the symmetry of the strain and stress components, $S_{ij} = S_{ji}$ and $T_{ij} = T_{ji}$, and adopting the following abbreviated or contracted indices,

$$11 \text{ or } xx \to 1 \qquad 23 \text{ and } 32, \text{ or } yz \text{ and } zy \to 4$$
$$22 \text{ or } yy \to 2 \qquad 13 \text{ and } 31, \text{ or } xz \text{ and } zx \to 5$$
$$33 \text{ or } zz \to 3 \qquad 12 \text{ and } 21, \text{ or } xy \text{ and } yx \to 6$$

we can express the strain and stress tensors in (B6) and (B10) as 1×6 column matrices,

$$\overset{\leftrightarrow}{S} = \begin{bmatrix} S_1 \\ S_2 \\ S_3 \\ S_4 \\ S_5 \\ S_6 \end{bmatrix} = \begin{bmatrix} S_{xx} \\ S_{yy} \\ S_{zz} \\ 2S_{yz} \\ 2S_{zx} \\ 2S_{xy} \end{bmatrix} = \begin{bmatrix} S_{xx} \\ S_{yy} \\ S_{zz} \\ 2S_{zy} \\ 2S_{xz} \\ 2S_{yx} \end{bmatrix} \tag{B40}$$

$$\overset{\leftrightarrow}{T} = \begin{bmatrix} T_1 \\ T_2 \\ T_3 \\ T_4 \\ T_5 \\ T_6 \end{bmatrix} = \begin{bmatrix} T_{xx} \\ T_{yy} \\ T_{zz} \\ T_{yz} \\ T_{zx} \\ T_{xy} \end{bmatrix} = \begin{bmatrix} T_{xx} \\ T_{yy} \\ T_{zz} \\ T_{zy} \\ T_{xz} \\ T_{yx} \end{bmatrix} \tag{B41}$$

Hooke's law can also be compressed in a compact form

$$\begin{bmatrix} T_1 \\ T_2 \\ T_3 \\ T_4 \\ T_5 \\ T_6 \end{bmatrix} = \begin{bmatrix} c_{11} & c_{12} & c_{13} & c_{14} & c_{15} & c_{16} \\ c_{21} & c_{22} & c_{23} & c_{24} & c_{25} & c_{26} \\ c_{31} & c_{32} & c_{33} & c_{34} & c_{35} & c_{36} \\ c_{41} & c_{42} & c_{43} & c_{44} & c_{45} & c_{46} \\ c_{51} & c_{52} & c_{53} & c_{54} & c_{55} & c_{56} \\ c_{61} & c_{62} & c_{63} & c_{64} & c_{65} & c_{66} \end{bmatrix} \begin{bmatrix} S_1 \\ S_2 \\ S_3 \\ S_4 \\ S_5 \\ S_6 \end{bmatrix} \tag{B42}$$

In the operator form, Hooke's law is

$$\overset{\leftrightarrow}{T} = \overset{\leftrightarrow\leftrightarrow}{c}\overset{\leftrightarrow}{S} \tag{B43}$$

Many matrix elements of $\overset{\leftrightarrow\leftrightarrow}{c}$ are either zero or can be expressed in terms of other matrix elements. For example, various matrix elements of **isotropic solids** are

$$c_{11} = c_{22} = c_{33} = \lambda + 2\mu$$
$$c_{12} = c_{21} = c_{13} = c_{31} = c_{23} = c_{32} = \lambda$$
$$c_{44} = c_{55} = c_{66} = \mu$$

Other components of c_{ij} are zero. Therefore, $\overleftrightarrow{c}$ of isotropic solids is of the form

$$\overleftrightarrow{c} = \begin{bmatrix} c_{11} & c_{12} & c_{12} & 0 & 0 & 0 \\ c_{12} & c_{11} & c_{12} & 0 & 0 & 0 \\ c_{12} & c_{12} & c_{11} & 0 & 0 & 0 \\ 0 & 0 & 0 & c_{44} & 0 & 0 \\ 0 & 0 & 0 & 0 & c_{44} & 0 \\ 0 & 0 & 0 & 0 & 0 & c_{44} \end{bmatrix} \tag{B44}$$

As discussed in section 6.6 of Chapter 6, the relative dielectric impermeability tensor, $1/n_i^2$, and the strain tensor $\overleftrightarrow{S}$, have six independent components each. Therefore, the photoelastic tensor $\overleftrightarrow{p}$ has 36 components. In the abbreviated indices, and in matrix form, the AO effect can be expressed as

$$\begin{bmatrix} \Delta(1/n_1^2) \\ \Delta(1/n_2^2) \\ \Delta(1/n_3^2) \\ \Delta(1/n_4^2) \\ \Delta(1/n_5^2) \\ \Delta(1/n_6^2) \end{bmatrix} = \begin{bmatrix} p_{11} & p_{12} & p_{13} & p_{14} & p_{15} & p_{16} \\ p_{21} & p_{22} & p_{23} & p_{24} & p_{25} & p_{26} \\ p_{31} & p_{32} & p_{33} & p_{34} & p_{35} & p_{36} \\ p_{41} & p_{42} & p_{43} & p_{44} & p_{45} & p_{46} \\ p_{51} & p_{52} & p_{53} & p_{54} & p_{55} & p_{56} \\ p_{61} & p_{62} & p_{63} & p_{64} & p_{65} & p_{66} \end{bmatrix} \begin{bmatrix} S_1 \\ S_2 \\ S_3 \\ S_4 \\ S_5 \\ S_6 \end{bmatrix} \tag{B45}$$

Many matrix elements of $\overleftrightarrow{p}$ are also zero. For *isotropic solids*, $\overleftrightarrow{p}$ is of the form

$$\overleftrightarrow{p} = \begin{bmatrix} p_{11} & p_{12} & p_{12} & 0 & 0 & 0 \\ p_{12} & p_{11} & p_{12} & 0 & 0 & 0 \\ p_{12} & p_{12} & p_{11} & 0 & 0 & 0 \\ 0 & 0 & 0 & p_{66} & 0 & 0 \\ 0 & 0 & 0 & 0 & p_{66} & 0 \\ 0 & 0 & 0 & 0 & 0 & p_{66} \end{bmatrix} \tag{B46}$$

and $p_{66} = (p_{11} - p_{12})/2$. For **liquids**, $p_{11} = p_{12}$, and therefore p_{66} is also zero.

REFERENCES

B.1. Beam, W. R. *Electronics of Solids.* New York, NY: McGraw-Hill Book Company, 1965.

B.2. Auld, B. A. *Acoustic Fields and Waves in Solids.* New York, NY: John Wiley & Sons, Inc., 1973.

INTEGRATED OPTICS

CHAPTER 7

CHAPTER OUTLINE

7.1 Introduction
7.2 Guided Waves: A Physical Picture
    7.2.1 Reflection by Conducting Boundaries
    7.2.2 Parallel-Plate Waveguides
7.3 Guided Wave Phase and Group Velocities
    7.3.1 Phase Velocity
    7.3.2 Group Velocity
    7.3.3 Phase and Group Velocities in Terms of ω and β
7.4 Reflection by Planar Dielectric Boundaries
7.5 Step Index Thin-Film Waveguides
    7.5.1 Generalized Parameters
    7.5.2 Characteristic Equations and Fields
    7.5.3 Modes Guided by Step Index Thin-film Waveguides
7.6 Graded Index Thin-Film Waveguides
7.7 Prism Couplers
7.8 Film Index and Thickness Measurement: An Application
    7.8.1 Two Measured Coupling Angles
    7.8.2 Three or More Measured Coupling Angles
7.9 Optical Directional Couplers
    7.9.1 Coupled Mode Theory
    7.9.2 Multilayer Directional Coupler Example
    7.9.3 Waveguide Directional Coupler Types
References
Additional Reading
Problems

7.1 INTRODUCTION

In metallic waveguides, electromagnetic waves in the radio and microwave frequency ranges are confined, and are bounced in a zigzag fashion, by conducting boundaries. The same concept is also useful at optical frequencies, if effective means of reflecting the optical waves are available. The effectiveness of a waveguide material can be inferred by examining the reflection coefficient at the operating frequency. At audio, radio, microwave, or millimeter-wave frequencies, electrical conductors are very good reflectors. For example, for uniform plane waves at 10 GHz incident perpendicular to an air–gold boundary, 99.97 percent of the incoming power is reflected. In contrast, for waves at 474 THz ($\lambda = 0.633$ μm), only 94.83 percent of the incident power is reflected by the air–gold boundary. Metals, such as Al, Ag, or Au, are good reflectors at millimeter-wave or lower frequencies, but they are quite lossy at optical frequencies. Since *all* electrical conductors are lossy at optical frequencies, other materials must be used as optical waveguide materials. This leads us to dielectric thin-film waveguides and optical fibers.

The basic principle behind dielectric thin-film waveguides is the same as that of optical fibers, and the mathematics is much simpler. Therefore, we will use thin-film waveguides as a vehicle to introduce the concepts of waveguiding by dielectric boundaries. Optical planar waveguides and other integrated optic waveguides are the basic building blocks of many **optoelectronic integrated circuits** (OEIC), which find application as the transmitting and receiving components of communication systems and consumer products, such as CD players. Therefore **integrated optics** is an important subject in its own right. However, the objective of this chapter is to introduce the concepts of modes, phase and group velocities of guided modes, and generalized parameters used extensively in characterizing optical fibers. A brief discussion on optical directional couplers is also included, at the end of the chapter.

7.2 GUIDED WAVES: A PHYSICAL PICTURE

Consider uniform plane waves of frequency f and vacuum wavelength λ propagating in isotropic, lossless, nonmagnetic media. For uniform plane waves propagating in the direction $\mathbf{a}_k$, the vacuum wave vector is

$$\mathbf{k} = (2\pi/\lambda)\mathbf{a}_k$$

In a medium with a refractive index n, the wave vector is $n\mathbf{k}$. The constant-phase wavefronts are perpendicular to $\mathbf{a}_k$. In Figure 7.1, lines are drawn to depict wavefronts at any given instant in time. Lines of $+$ and $-$ signs represent the peaks and valleys of the electric fields, respectively. These wavefronts are referred to as the $+$ and $-$ wavefronts, respectively. The separation between two consecutive wavefronts of the same polarity is λ/n, which means that the phase difference between two consecutive $+$, or two consecutive $-$, wavefronts is 2π.

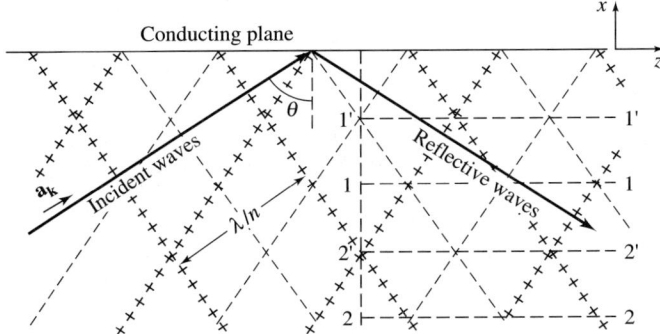

Figure 7.1 Reflection of uniform plane waves by a conducting plane

7.2.1 Reflection by Conducting Boundaries

We will now consider plane waves impinging upon a conducting boundary. Let the angle between the incident wave vector and the normal to the conducting boundary be θ. For convenience, assume that the incident electric fields lie in a plane perpendicular to the paper, and that the reflected electric fields are also normal to the paper. The total field in the region is the superposition of the incident and reflected waves, and is transverse to the direction of propagation. At the perfectly conducting surface, the tangential component of the total electric field must be zero. Fields with these features are known as **transverse electric (TE) waves**. Thus the incident and reflected fields must have opposite signs for all points on the conductor. That is, the incident + wavefronts become − wavefronts upon reflection. The direction of propagation is changed, and there is a sudden phase change of π at the perfectly conducting boundary.

7.2.2 Parallel-Plate Waveguides

In Figure 7.1, we see that the + and − wavefronts of the incident waves and the − and + wavefronts of the reflected waves intersect at plane 1. If another conducting plane is placed at plane 1, the condition of vanishing tangential electric field is automatically satisfied along plane 1. Because of the presence of the second conducting plane, waves are bounced back and forth between the two conducting planes. Note that the + and − wavefronts of the incident fields also intersect the − and + wavefronts of the reflected waves at planes 2, 3, etc. The second conducting plane can be placed at any of these locations, and the waves would be bounced back and forth with no loss.

With a conducting plane at $x = 0$ and another at plane 1 or plane 2, a **parallel-plate waveguide** is formed. Suppose the second conducting plane is placed at plane 2. Since two + wavefronts intersect at plane 2′ and two − wavefronts at 1′, the total field will have a peak at 2′ and a valley at plane 1′. On the other hand, + and − wavefronts intersect at plane 1. Thus, the total electric field changes from a null at plane 2 to a positive peak at plane 2′, a null at plane 1, and then a valley at plane 1′, before returning to a null again at the top conducting plane.

Characteristic Equations of Modes Guided by Parallel-Plate Waveguides In the previous discussion, we assume that the angle of incidence θ is fixed, and that the second conducting plane is positioned to fit existing incident and reflected fields. Usually, however, the situation is reversed: the spacing between conducting planes is fixed, and the waves traveling in the waveguide without incurring excessive loss are those with angles θ satisfying a certain condition. This condition is known as the **characteristic equation.** The equation admits a finite number of discrete solutions for θ. Each acceptable angle corresponds to a specific field distribution, and each field distribution is referred to as a mode.

To derive the characteristic equation, consider a waveguide with two perfectly conducting planes separated by a distance h. Let the index of the medium between the conducting planes be n. Consider a wavefront $\overline{C'A'}$, shown as a dashed line in Figure 7.2a, and trace rays at points A' and C' to establish the condition for θ. The ray originating from point A' follows a straight path to point A and then to point B. The total phase delay is

$$2\pi \frac{\overline{A'A} + \overline{AB}}{\lambda/n} = kn\,(\overline{A'A} + \overline{AB})$$

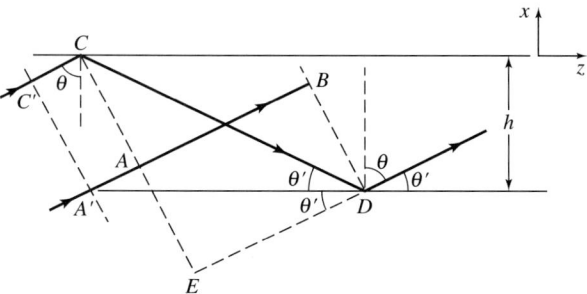

(a) $\theta > \pi/4$

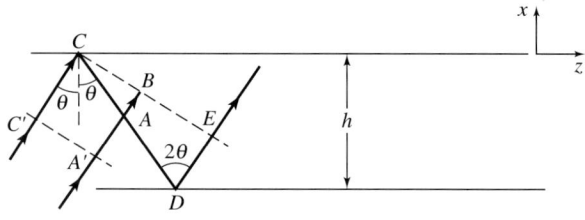

(b) $\theta < \pi/4$

Figure 7.2 Zigzag paths of guided waves in a parallel-plate waveguide

The ray originating from C' is reflected at point C by the top conducting plane. In addition to a change in the ray direction, there is also a phase shift of π. At the lower conducting plane there is a second reflection and an additional phase shift of π. Thus, a ray originating from C' follows a zigzag path and is reflected twice, and the total phase change produced by this path is

$$kn\,(\overline{C'C} + \overline{CD}) - \pi - \pi$$

Begin with two points A' and C' on a given wavefront and arrive at points B and D. The phase difference between these two rays is

$$(kn\,\overline{CD} - \pi - \pi) - kn\,\overline{AB}$$

Points B and D are on the same wavefront again if the phase difference between these two rays is either 0 or integer multiples of 2π:

$$(kn\,\overline{CD} - \pi - \pi) - kn\,\overline{AB} = 2m\pi \tag{7.1}$$

where m is an integer. Simple geometrical considerations reveal that, for the case of $\theta > \pi/4$ (Figure 7.2a),

$$\overline{CD} = \frac{h}{\cos\theta}, \qquad \overline{AB} = \overline{ED} = \overline{CD}\cos2\theta'$$

where $\theta' = \pi/2 - \theta$, as shown in Figure 7.2a. Therefore,

$$\overline{CD} - \overline{AB} = \overline{CD}\,(1 - \cos2\theta') = \frac{h}{\cos\theta}(2\sin^2\theta') = \frac{h}{\cos\theta}(2\cos^2\theta) = 2h\cos\theta$$

and (7.1) can be written as

$$2knh\cos\theta - \pi - \pi = 2m\pi \tag{7.2}$$

For the case of $\theta < \pi/4$ (Figure 7.2b), the details are different but the final result is the same. Equation (7.2) is the characteristic equation, also known as the **dispersion relation,** of **TE modes** guided by parallel-plate waveguides with two conducting planes.

If the incident and reflected *magnetic fields* are perpendicular to the paper, the fields are known as **transverse magnetic (TM) modes.** TM modes can also be viewed as waves bounced repeatedly by conducting boundaries. However, there is a crucial difference between them and TE modes. For TM modes, there is no phase shift due to reflection by the conducting boundaries. Therefore, the characteristic equation for TM modes guided by parallel-plate waveguides is

$$2knh\cos\theta - 0 - 0 = 2m\pi \tag{7.3}$$

In (7.2) and (7.3), $kn\cos\theta$ is the wave vector projected onto the x direction, and π or 0 are the phase shifts caused by the reflection at the conducting boundary. Thus, the terms on the left-hand side of (7.2) and (7.3) are the total round-trip phase changes of waves zigzagging between conducting boundaries. Usually, a phase difference of 2π or 0 on the left-hand side of (7.2) or (7.3) is of no importance. To relate the present discussion to that to be presented in section

7.5, we purposely and explicitly keep the phase terms at 2π or 0 in (7.2) and (7.3).

Field Distributions of Modes Guided by Parallel-Plate Waveguides As indicated previously, the frequency of waves and the spacing between conducting planes are fixed, and only waves with an angle satisfying the characteristic equations (7.2) and (7.3) can propagate in the waveguide structure without attenuation. For each value of m, there is a solution for θ and a corresponding field distribution. Figures 7.3a, b, and c show the zigzagged paths and corresponding field distributions of the first three TE modes guided by a parallel-plate waveguide. These modes are identified as TE_0, TE_1, and TE_2, respectively, where the subscript signifies the value of m. To understand the field distribution of each mode, trace along the vertical dashed line. In Figure 7.3a, the + and − wavefronts cross at only two conducting planes. Therefore, the electric fields of TE_0 vanish only at the conducting planes. In Figure 7.3b, the + and − wavefronts cross at the midpoint, in addition to the points at the two conducting planes. Thus, the

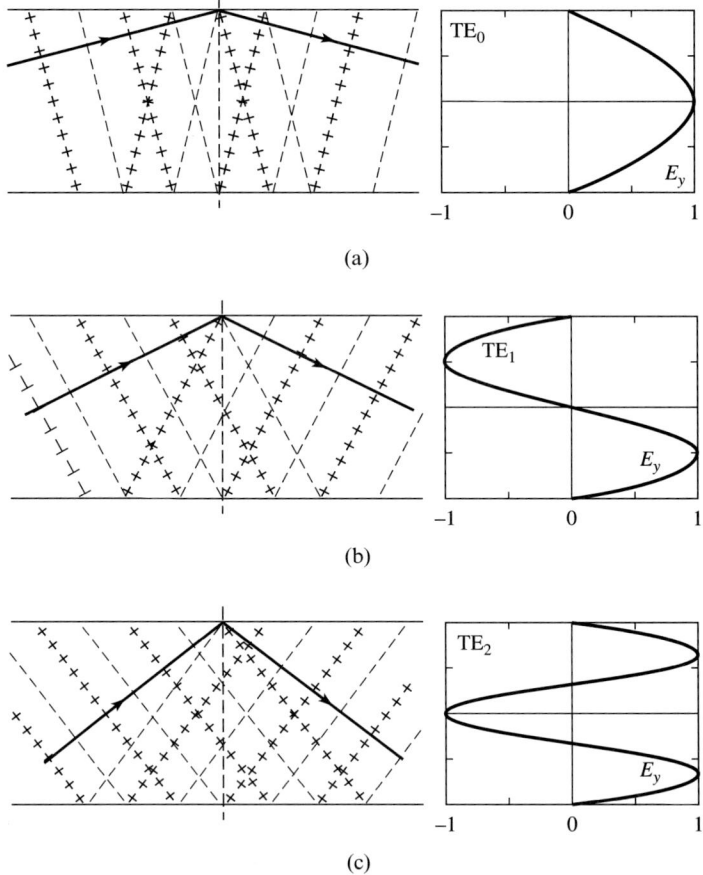

Figure 7.3 Field distribution of TE_0, TE_1, and TE_2 modes for a parallel-plate waveguide with $h = 2.5\lambda/n$.

field distribution of TE_1 has a null at the midpoint. For the TE_2 modes, shown in Figure 7.3c, there are two nulls, in addition to the nulls at the conducting boundaries.

As a numerical example, consider a parallel-plate waveguide with $h = 3\lambda$ and $n = 1$. For such a waveguide, (7.2) becomes

$$\cos\theta = \frac{m+1}{6}$$

Hence, there are five propagating modes, for which θ is 80.4°, 70.5°, 60.0°, 48.2°, and 33.6°, and m is 0, 1, 2, 3, and 4, respectively. The cutoff mode, if exists, corresponds to a solution with $\theta = 0°$.

In general, an arbitrary waveguide supports a finite number of modes. The lower-order modes correspond to small values of m and large values of θ. Since θ is the angle between the incident wave vector and the normal to the waveguide boundary, a large θ means the wave vector is mainly in parallel with the waveguide axis. This corresponds to rays impinging upon the waveguide boundaries at a shallow angle.

7.3 GUIDED WAVE PHASE AND GROUP VELOCITIES

7.3.1 Phase Velocity

In the last section, we viewed guided modes as waves following zigzag paths with an angle θ satisfying (7.2) or (7.3). Now focus on a specific mode corresponding to a specific θ. To an observer moving a distance of λ/n in the direction $\mathbf{a}_k$ from one $-$ wavefront to the next $-$ wavefront, the phase changes by 2π (Figure 7.4a). However, to an observer moving in the $+z$ direction, the phase changes by 2π each time the observer moves a distance of $\overline{AA'}$. We define the **guide wavelength** λ_{gu} as the repeat distance experienced by an observer moving in the $+z$ direction. From Figure 7.4a, we have a simple geometrical relationship:

$$\lambda_{gu}\sin\theta = \frac{\lambda}{n}$$

Therefore,

$$\lambda_{gu} = \frac{\lambda}{n\sin\theta} \tag{7.4}$$

A **phase velocity** associated with λ_{gu} is

$$v_{ph} = f\lambda_{gu} = \frac{c}{n\sin\theta} \tag{7.5}$$

where f is the frequency of the wave. To the observer, the constant-phase wavefront of the guided waves "appears" to move with the phase velocity in the waveguide direction. We introduce the **propagation constant** β

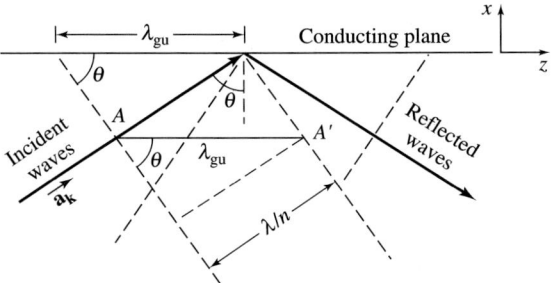
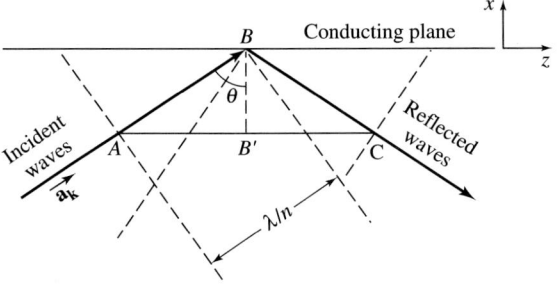

Figure 7.4 Phase and group velocities

$$\beta = \frac{\omega}{v_{ph}} = \frac{2\pi}{\lambda_{gu}} = \frac{2\pi}{\lambda} n\sin\theta = kn\sin\theta \tag{7.6}$$

and the **effective index of refraction** N

$$N = \frac{c}{v_{ph}} = \frac{\beta}{k} = n\sin\theta \tag{7.7}$$

for the guided mode.

7.3.2 Group Velocity

If a narrow electromagnetic pulse with a center frequency f is emitted in the $\mathbf{a_k}$ direction by a source at A (Figure 7.4b), the pulse will be reflected once by the conducting boundary before reaching point C. The total path length is $\overline{AB} + \overline{BC}$, and the time it takes to move from A to C is $(\overline{AB} + \overline{BC})n/c$. Thus, in traveling from A to C, the pulse effectively moves with a velocity

$$\frac{\overline{AC}}{(\overline{AB} + \overline{BC})n/c} = \frac{c}{n}\frac{2\overline{AB'}}{2\overline{AB}} = \frac{c}{n}\sin\theta$$

This is the **group velocity** of the guided waves

$$v_{gr} = \frac{c}{n}\sin\theta \tag{7.8}$$

From (7.5) and (7.8), it follows that

$$v_{ph}v_{gr} = c^2/n^2 \tag{7.9}$$

Equation 7.9 is important because it relates the phase velocity to the group velocity. Also, from (7.5) and (7.8), we see that $v_{ph} \geq c/n \geq v_{gr}$. This is acceptable because the phase velocity is not associated with the motion of a physical quantity, and it can therefore be faster than the speed of light. However, the group velocity is related to the motion of the energy and the envelop of an electromagnetic pulse, and must be slower than the speed of light.

7.3.3 Phase and Group Velocities in Terms of ω and β

The phase and group velocities can also be expressed in terms of β and the angular frequency ω. In fact, the expression for the phase velocity has already been given in (7.6):

$$v_{ph} = \frac{\omega}{\beta}$$

To derive an expression for the group velocity, consider sinusoidal waves propagating in the waveguides. Suppose first that simple sinusoidal waves of the form $\cos\omega t$ are used, and let the propagation constant of a mode be β. The wave motion in the waveguide is described by $\cos(\omega t - \beta z)$. However, simple sine or cosine waves carry no information. Therefore, consider next amplitude-modulated sinusoidal waves, $\cos\omega_m t \cos\omega t$, where the angular frequency ω_m of the modulation is much smaller than the angular frequency ω of the carrier. An amplitude-modulated cosine function can be viewed as the superposition of two spectral components:

$$\mathcal{V}(0;t) = \cos\omega_m t \cos\omega t = \frac{1}{2}[\cos(\omega + \omega_m)t + \cos(\omega - \omega_m)t]$$

Since β is a function of ω, each spectral component travels in the waveguide with a different propagation constant. For convenience, we introduce

$$\omega_1 = \omega + \omega_m \quad \text{and} \quad \omega_2 = \omega - \omega_m$$

and write

$$\beta_1 = \beta(\omega + \omega_m) \quad \text{and} \quad \beta_2 = \beta(\omega - \omega_m)$$

The fields at z are the superposition of two spectral components

$$\mathcal{V}(z;t) = \frac{1}{2}[\cos(\omega_1 t - \beta_1 z) + \cos(\omega_2 t - \beta_2 z)] \tag{7.10}$$

Since $\omega_m \ll \omega$, we can approximate β_1 and β_2 as

$$\beta_1 = \beta(\omega + \omega_m) \approx \beta(\omega) + \omega_m \frac{d\beta}{d\omega}$$

$$\beta_2 = \beta(\omega - \omega_m) \approx \beta(\omega) - \omega_m \frac{d\beta}{d\omega}$$

Therefore,

$$\cos(\omega_1 t - \beta_1 z) \approx \cos[(\omega t - \beta(\omega)z) + \omega_m (t - \frac{d\beta}{d\omega} z)]$$

$$\cos(\omega_2 t - \beta_2 z) \approx \cos[(\omega t - \beta(\omega)z) - \omega_m (t - \frac{d\beta}{d\omega} z)]$$

Using these expressions, (7.10) becomes

$$\mathcal{V}(z;t) \approx \cos[\omega t - \beta z] \cos[\omega_m (t - \frac{d\beta}{d\omega} z)]$$

The first cosine function represents the carrier signal of angular frequency ω moving with a phase velocity ω/β. The second cosine function describes the motion of the amplitude modulation moving with a velocity

$$v_{gr} = \frac{d\omega}{d\beta} \qquad (7.11)$$

which is an alternative way of expressing the group velocity given in (7.8).

7.4 REFLECTION BY PLANAR DIELECTRIC BOUNDARIES

As a precursor to the detailed analysis of dielectric thin-film waveguides, we consider the reflection of time-harmonic ($e^{+j\omega t}$) uniform plane waves by a planar boundary separating two dielectric media. Only the essential results are given here. Details on the derivations can be found in many textbooks on electromagnetics, including references [1], [2], and [3]. Consider a planar boundary separating two lossless, nonmagnetic, isotropic dielectric media with refractive indices n_1 and n_2, respectively (Figure 7.5a). A **plane of incidence** is defined by the normal to the interface and the wave vector of the incident waves. Any incident plane waves can be viewed as the superposition of TE waves and TM waves. For TE waves, the electric fields of incident, reflected, and transmitted waves are normal to the plane of incidence. For TM waves, the incident, reflected, and transmitted magnetic fields are perpendicular to the plane of incidence. The directions of propagation of the incident and transmitted waves are related through Snell's law,

$$n_1 \sin\theta_1 = n_2 \sin\theta_2$$

where θ_1 and θ_2, respectively, are the angles between the incident and transmitted wave vectors and the normal to the boundary. In particular, θ_1 is referred to as the **angle of incidence**.

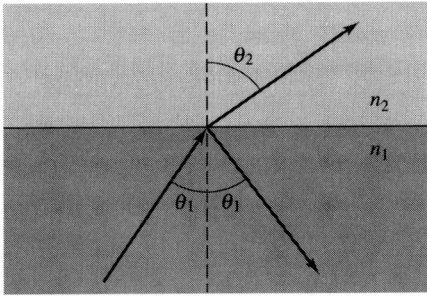

(a) Reflection by a dielectric boundary

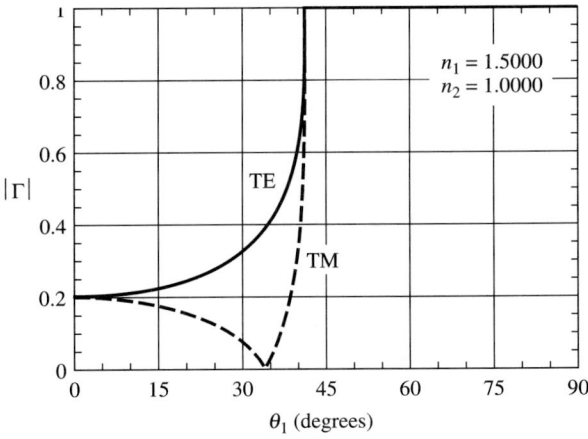

(b) Amplitude of reflection coefficient

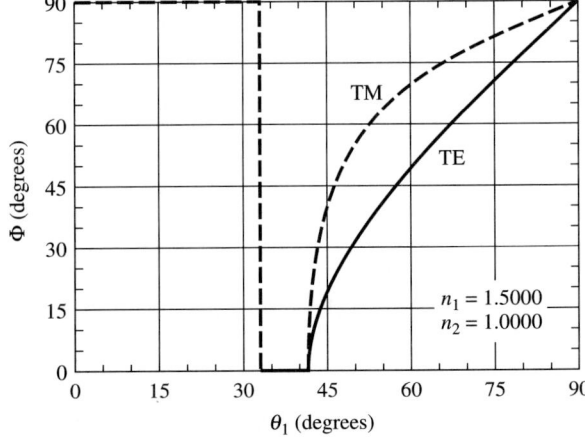

(c) Phase of reflection coefficient

Figure 7.5 Reflection coefficients of a boundary between two dielectric media with $n_1 = 1.5000$ and $n_2 = 1.0000$

In the case of TE waves, let the electric field intensity of the incident waves be E_i, and that of the reflected fields be E_r. Then, the **reflection coefficient**, Γ_{TE}, for TE polarization is the ratio E_r/E_i, and the coefficient depends on n_1, n_2, and θ_1, through the relationship

$$\Gamma_{TE} = |\Gamma_{TE}| e^{j2\phi_{TE}} = \frac{E_r}{E_i} = \begin{cases} \dfrac{n_1\cos\theta_1 - \sqrt{n_2^2 - n_1^2\sin^2\theta_1}}{n_1\cos\theta_1 + \sqrt{n_2^2 - n_1^2\sin^2\theta_1}}, & \text{for } n_2 > n_1\sin\theta_1 \\[2ex] \dfrac{n_1\cos\theta_1 + j\sqrt{n_1^2\sin^2\theta_1 - n_2^2}}{n_1\cos\theta_1 - j\sqrt{n_1^2\sin^2\theta_1 - n_2^2}}, & \text{for } n_2 < n_1\sin\theta_1 \end{cases} \quad (7.12)$$

where ϕ_{TE} is

$$\phi_{TE} = \tan^{-1}\left[\frac{\sqrt{n_1^2\sin^2\theta_1 - n_2^2}}{n_1\cos\theta_1}\right] \quad (7.13)$$

For TM waves, let the magnetic field intensity of the incident and reflected waves be H_i and H_r. Then, the reflection coefficient, Γ_{TM}, is defined as H_r/H_i. In terms of n_1, n_2, and θ, we have

$$\Gamma_{TM} = |\Gamma_{TM}| e^{j2\phi_{TM}} = \frac{H_r}{H_i} = \begin{cases} \dfrac{n_2^2\cos\theta_1 - n_1\sqrt{n_2^2 - n_1^2\sin^2\theta_1}}{n_2^2\cos\theta_1 + n_1\sqrt{n_2^2 - n_1^2\sin^2\theta_1}}, & \text{for } n_2 > n_1\sin\theta_1 \\[2ex] \dfrac{n_2^2\cos\theta_1 + jn_1\sqrt{n_1^2\sin^2\theta_1 - n_2^2}}{n_2^2\cos\theta_1 - jn_1\sqrt{n_1^2\sin^2\theta_1 - n_2^2}}, & \text{for } n_2 < n_1\sin\theta_1 \end{cases} \quad (7.14)$$

where ϕ_{TM} is

$$\phi_{TM} = \tan^{-1}\left[\frac{n_1\sqrt{n_1^2\sin^2\theta_1 - n_2^2}}{n_2^2\cos\theta_1}\right] \quad (7.15)$$

In (7.12) and (7.14), we have chosen

$$\sqrt{n_2^2 - n_1^2\sin^2\theta_1} = -j\sqrt{n_1^2\sin^2\theta_1 - n_2^2}$$

for $n_1\sin\theta_1 > n_2$. This choice is dictated by the requirement that, for $n_1\sin\theta_1$ greater than n_2, fields in the low-index region must decay exponentially from the boundary.

Figures 7.5b and c depict $|\Gamma_{TE}|$, $|\Gamma_{TM}|$, ϕ_{TE}, and ϕ_{TM} as functions of θ_1 for $n_1 = 1.5000$ and $n_2 = 1.0000$. If $n_1\sin\theta_1 < n_2$, then Γ_{TE} and Γ_{TM} are real quantities. While Γ_{TE} is positive, Γ_{TM} can be positive or negative. Therefore, ϕ_{TE} and ϕ_{TM} are either 0° or 90°. If $n_1\sin\theta_2 > n_2$, then Γ_{TE} and Γ_{TM} are complex quantities of the form of $(a + jb)/(a - jb)$. Therefore, $|\Gamma_{TE}| = |\Gamma_{TM}| = 1$. In other words, the waves are completely reflected by the dielectric boundary when $\sin\theta_1$ is greater than n_2/n_1. This is known as **total internal reflection**, and the angle $\theta_1 = \sin^{-1}(n_2/n_1)$ is the **critical angle**.

It is important to note that no power is lost in the total internal reflection process. If there are two dielectric interfaces and the conditions of total internal reflection are satisfied at both boundaries, then waves may be trapped and guided by the two dielectric boundaries. This is the basic waveguiding mechanism of dielectric waveguides and optical fibers.

7.5 STEP INDEX THIN-FILM WAVEGUIDES

A typical thin-film waveguide structure consists of three lossless dielectric regions, including the **cover, film,** and **substrate** (Figure 7.6a). Let the indices of these regions be n_c, n_f, and n_s, respectively. The cover and substrate regions are infinitely thick, while the film thickness h is finite. The lateral dimension in the y direction is very large in terms of wavelengths and film thickness h. This waveguide is also known as a **dielectric slab waveguide** or a **planar waveguide**. To trap waves in the film region, n_f must be larger than n_s and n_c. Usually, although not always, the cover region is simply air or vacuum. We refer the region with the smallest index as the cover region, and we assume throughout our discussion that $n_f > n_s > n_c$.

As shown in Figure 7.6a, the index in each region is the same everywhere in that region, and the index changes abruptly at the interfaces between the cover, film, and substrate regions. This type of index distribution is known as the **step index** profile. In other dielectric waveguides, the index changes gradually, particularly near the film/substrate "boundary." In fact, the film and substrate regions often merge into one, and the index varies continuously as a function of position. Waveguides with a continuously varying index $n_f(x)$ (Figure 7.6b) are referred to as **graded index** waveguides. In most of our discussions, we will restrict ourselves to step index waveguides.

Except in two important respects, the physics of thin-film waveguides is similar to that of parallel-plate waveguides (section 7.2). First, $|\Gamma_{TE}|$ and $|\Gamma_{TM}|$ due to perfectly conducting boundaries is 1 for all angles of incidence. At the dielectric boundaries, total internal reflection occurs only if the angle of incidence is greater than the critical angle. Second, the phase shift due to reflection at the

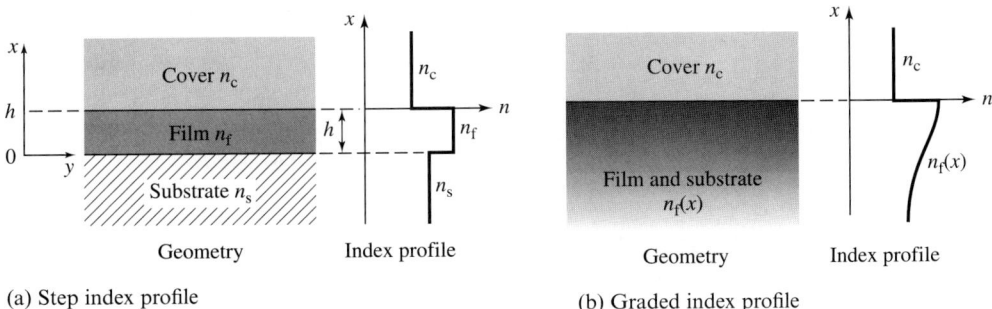

Figure 7.6 Structures and index profiles of thin-film waveguides.

conducting boundaries is π or 0. At the dielectric boundaries of a step index waveguide and under the condition of total internal reflection, the phase shift is $2\phi_{TE}$ or $2\phi_{TM}$, depending on the polarization of the incident fields. Let the phase shifts at the film/substrate and film/cover interfaces be $2\phi_{fs}$ and $2\phi_{fc}$, respectively. Explicit expressions for ϕ_{fs} and ϕ_{fc} can be found by substituting n_f for n_1 and n_s and n_c for n_2 in (7.13) or (7.15). Using $2\phi_{fs}$ and $2\phi_{fc}$ in lieu of the phase term π in (7.2), we have

$$2n_f kh\cos\theta - (2\phi_{fc} + 2\phi_{fs}) = 2m\pi \tag{7.16}$$

This equation is simply a statement that the round-trip phase change in the direction perpendicular to the direction of propagation is an integer multiple of 2π [4]. Equation (7.16) can be written as

$$\tan(n_f kh\cos\theta) = \tan(m\pi + \phi_{fc} + \phi_{fs}) = \frac{\tan\phi_{fc} + \tan\phi_{fs}}{1 - \tan\phi_{fc}\tan\phi_{fs}} \tag{7.17}$$

When ϕ_{fc} and ϕ_{fs} are expressed in terms of n_c, n_f, n_s, and θ, (7.17) becomes

$$\tan(n_f kh\cos\theta) = \frac{n_f\cos\theta\left[\sqrt{n_f^2\sin^2\theta - n_c^2} + \sqrt{n_f^2\sin^2\theta - n_s^2}\right]}{n_f^2\cos^2\theta - \sqrt{n_f^2\sin^2\theta - n_c^2}\sqrt{n_f^2\sin^2\theta - n_s^2}} \tag{7.18}$$

Equation (7.18) is the characteristic equation of TE modes guided by thin-film waveguides. Similarly, the characteristic equation for TM modes is

$$\tan(n_f kh\cos\theta) = \frac{n_f\cos\theta\left[n_s^2\sqrt{n_f^2\sin^2\theta - n_c^2} + n_c^2\sqrt{n_f^2\sin^2\theta - n_s^2}\right]}{n_c^2 n_s^2\cos^2\theta - n_f^2\sqrt{n_f^2\sin^2\theta - n_c^2}\sqrt{n_f^2\sin^2\theta - n_s^2}} \tag{7.19}$$

Example

No analytical solutions for (7.18) and (7.19) are known. Usually, we resort to numerical methods to solve for θ. As an example, consider TE modes guided by a dielectric waveguide with a thin film of index 1.6000 and thickness 2.4 μm sandwiched between cover and substrate regions with $n_c = n_s = 1.3000$[1]. Let the operating wavelength be 0.8 μm. Since $n_s = n_c$, the waveguide structure is symmetric with respect to the film region. Equation (7.18) can be simplified to

$$\tan(n_f kh\cos\theta) = \frac{2n_f\cos\theta\sqrt{n_f^2\sin^2\theta - n_s^2}}{n_f^2\cos^2\theta - n_f^2\sin^2\theta + n_s^2} \tag{7.20}$$

If $\theta < 54.3°$, then $n_f^2\sin^2\theta - n_s^2 < 0$, and the right-hand side of (7.20) is imaginary, while the left-hand side is real. Thus, no real solution for θ exists for $\theta < 54.3°$. This means that there is no guided mode with $\theta < 54.3°$. The physical

---

[1] For usual integrated optic waveguides, the index difference $n_f - n_s$ is on the order of a few percent of n_f or less. A large index difference is chosen here to illustrate the nature of the solution.

explanation is as follows. At the interface between the film and substrate media, the critical angle is 54.3°. For $\theta < 54.3°$, waves are not total internally reflected by the dielectric boundaries. Instead, the waves are partially transmitted into the cover and substrate regions. Each time the waves are reflected, a fraction of the incident power is transmitted to the cover or substrate regions. As a result, the waves decay as they zigzag in the film region.

Acceptable solutions exist only in the range of $\theta > 54.3°$. To find the acceptable values for θ, plot both sides of (7.20) as functions of θ, as shown by the solid and dashed lines in Figure 7.7a. Only angles $\theta > 54.3°$ are shown. The six intersections found are: 56.98°, 62.58°, 68.22°, 73.77°, 79.23°, and 84.63°. Each intersection corresponds to a guided mode. Using these values of θ in (7.16), we obtain the corresponding values of m: $m = 5, 4, 3, 2, 1$, and 0, respectively. For

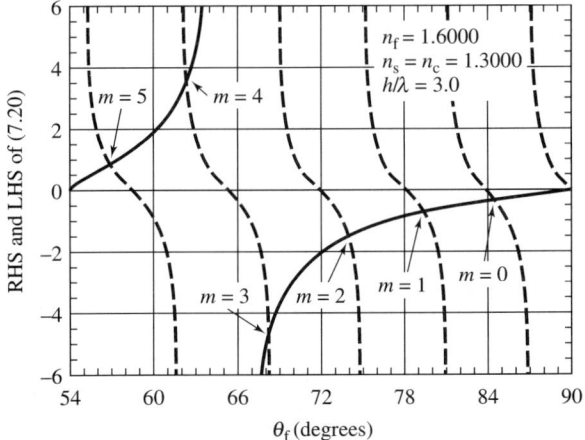

(a) Graphical solution of characteristic equation (7.20)

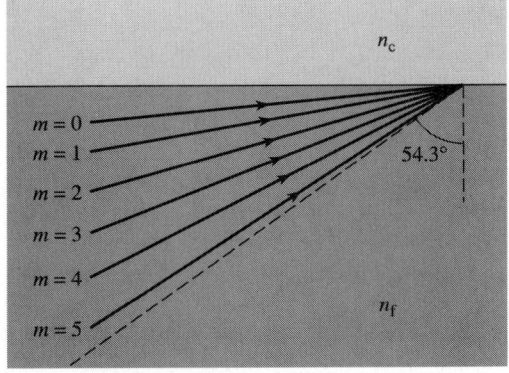

(b) Ray angles for different modes

Figure 7.7 Numerical example of thin-film waveguide with $n_f = 1.6000$, $n_s = n_c = 1.3000$, and $h = 3\lambda$.

modes with a smaller m, the angle θ is closer to 90° than for those with a larger m (Figure 7.7b). This means that the lower-order modes propagate mostly along the waveguide axis, while the higher-order modes follow mostly zigzag paths and deviate considerably from the waveguide axis.

7.5.1 Generalized Parameters

The characteristic equations (7.18) and (7.19) are very complicated, and no analytical solution is known. In addition, for most practical waveguides of interest, the indices n_c, n_f, and n_s are numerically close. Therefore, it is also no trivial matter to solve the equations numerically. A set of universal curves valid for all cases of interest is presented. Generalized parameters have been introduced for this purpose [5–9]. Three generalized parameters are sufficient to cover all TE modes guided by dielectric thin-film waveguides with a step index profile. The generalized parameters are: the **generalized frequency** or **generalized film thickness**,

$$V = kh\sqrt{n_f^2 - n_s^2} \tag{7.21}$$

the **asymmetry** of the waveguide structure,

$$a = \frac{n_s^2 - n_c^2}{n_f^2 - n_s^2} \tag{7.22}$$

and the **normalized guide index**

$$b = \frac{N^2 - n_s^2}{n_f^2 - n_s^2} \tag{7.23}$$

All generalized parameters are expressed in terms of the index *differences*, rather than the indices themselves. In terms of these generalized parameters, we cast $khn_f\cos\theta$ as

$$khn_f\cos\theta = kh\sqrt{n_f^2 - N^2}$$
$$= kh\left[n_f^2 - n_s^2\right]^{1/2}\left[\frac{n_f^2 - n_s^2 + n_s^2 - N^2}{n_f^2 - n_s^2}\right]^{1/2}$$
$$= V\sqrt{1 - b}$$

Similarly, we can express ϕ_{fc} and ϕ_{sc} in terms of a and b. From (7.14) we have

$$\phi_{fc} = \tan^{-1}\left[\frac{(n_f^2\sin^2\theta - n_c^2)^{1/2}}{n_f\cos\theta}\right] = \tan^{-1}\left[\frac{N^2 - n_c^2}{n_f^2 - N^2}\right]^{1/2} = \tan^{-1}\sqrt{\frac{a + b}{1 - b}}$$

Consequently, for *TE modes*, the characteristic equation becomes

$$V\sqrt{1 - b} = m\pi + \tan^{-1}\sqrt{\frac{a + b}{1 - b}} + \tan^{-1}\sqrt{\frac{b}{1 - b}}, \quad m = 0, 1, 2\cdots \tag{7.24}$$

For *TM modes*, one additional parameter, $c = n_s^2/n_f^2$, is needed, and the characteristic equation becomes [9]

$$V\sqrt{1-b} = m\pi + \tan^{-1}\frac{1}{d}\sqrt{\frac{a+b}{1-b}} + \tan^{-1}\frac{1}{c}\sqrt{\frac{b}{1-b}}, \quad m = 0, 1, 2, \cdots \quad (7.25)$$

where

$$d = n_c^2/n_f^2 = c - a(1-c)$$

For a given waveguide geometry and set of refractive indices, the terms a, c, and V can be readily calculated. Then, b can be solved from (7.24) or (7.25) for a specific value of m. Further discussion on the characteristic equations is included later in this chapter. However, without actually solving (7.24) or (7.25), we can see that b has a value between 0 and 1. From (7.23), we obtain immediately $N = n_s$ when $b = 0$, and $N = n_f$ when $b = 1$. In general, the effective refractive index of a guided mode is in the range

$$n_c \leq n_s < N < n_f$$

The corresponding propagation constant β is in the range

$$kn_c \leq kn_s < \beta < kn_f$$

7.5.2 Characteristic Equations and Fields

In this section, we will present a rigorous derivation for the characteristic equations (7.24) and (7.25). It is based on a systematic approach to the study of waves guided by thin-film waveguides. Initially, we will demonstrate that waves guided by thin-film waveguides can be classified as TE or TM modes. Then, we will derive a wave equation for each mode. By solving the wave equation and using the boundary conditions, we rigorously derive the characteristic equation for each mode, as well as expressions for the field components. The boundary conditions are: the continuation of the tangential components of the electric and magnetic field intensity; and, the normal component of the electric and magnetic flux density.

Classification of Fields To evaluate time-harmonic waves guided by a step index thin-film waveguide, begin with time-harmonic ($e^{j\omega t}$) Maxwell's equations for isotropic, nonmagnetic media with $\mu = \mu_o$ and $\varepsilon = \varepsilon_o n^2$, as follows:

$$\nabla \times \mathbf{E}(x,y,z) = -j\omega\mu_o \mathbf{H}(x,y,z) \quad (7.26)$$

$$\nabla \times \mathbf{H}(x,y,z) = j\omega\varepsilon_o n^2 \mathbf{E}(x,y,z) \quad (7.27)$$

$$\nabla \cdot \mu_o \mathbf{H}(x,y,z) = 0 \quad (7.28)$$

$$\nabla \cdot \varepsilon_o n^2 \mathbf{E}(x,y,z) = 0 \quad (7.29)$$

where n is the refractive index of the medium.

The waveguide geometry is shown in Figure 7.6a. Since the waveguide geometry and the refractive indices n_c, n_f, and n_s are independent of z, and since waves propagating in the z direction are of interest, all field components must vary as $e^{-j\beta z}$, and the propagation constant β must be determined. It is convenient to express the fields in terms of their transverse and longitudinal parts, as follows:

$$\mathbf{E}(x,y,z) = [\mathbf{e}_t(x,y) + \mathbf{a}_z e_z(x,y)]e^{-j\beta z} \tag{7.30}$$

$$\mathbf{H}(x,y,z) = [\mathbf{h}_t(x,y) + \mathbf{a}_z h_z(x,y)]e^{-j\beta z} \tag{7.31}$$

where $\mathbf{a}_z$ is a unit vector in the $+z$ direction. A subscript t is used to denote components *transverse* to the z direction. Therefore, $\mathbf{e}_t$ and $\mathbf{h}_t$ are field components normal to the propagation direction.

For $\mathbf{E}$ and $\mathbf{H}$ given in (7.30) and (7.31), the partial differentiation of a field quantity with respect to z amounts to multiplication of the field quantity by $-j\beta$. Since the waveguide structure is very wide in the y direction, as shown in Figure 7.6a, all field components are independent of y. Therefore, we set $\dfrac{\partial}{\partial y}$ to zero, as well. In other words, $\mathbf{e}_t$, e_z, $\mathbf{h}_t$ and h_z are all functions of x only. With these observations, we can rewrite Maxwell's equations in component form as:

$$j\beta e_y = -j\omega\mu_o h_x \tag{7.32}$$

$$-j\beta e_x - \frac{de_z}{dx} = -j\omega\mu_o h_y \tag{7.33}$$

$$\frac{de_y}{dx} = -j\omega\mu_o h_z \tag{7.34}$$

$$j\beta h_y = j\omega n^2 \varepsilon_o e_x \tag{7.35}$$

$$-j\beta h_x - \frac{dh_z}{dx} = j\omega n^2 \varepsilon_o e_y \tag{7.36}$$

$$\frac{dh_y}{dx} = j\omega n^2 \varepsilon_o e_z \tag{7.37}$$

These equations separate naturally into two groups: (7.32), (7.34), and (7.36) involve only e_y, h_x, and h_z; and, (7.33), (7.35), and (7.37) contain only e_x, e_z, and h_y. For the first group, the electric field is in the y direction, which is transverse to the direction of propagation. These fields are TE modes. In the second group, the magnetic field is transverse to the z direction, making these fields the TM modes. Note in particular that one group is completely uncoupled from the other; therefore, each group can be considered individually.

TE Modes From (7.32), and (7.34), we can express h_x and h_z in terms of e_y:

$$h_x = -\frac{\beta}{\omega\mu_o} e_y \tag{7.38}$$

$$h_z = \frac{j}{\omega\mu_o} \frac{de_y}{dx} \tag{7.39}$$

Substituting these expressions in (7.36), we obtain

7.5 Step Index Thin-Film Waveguides

$$\frac{d^2e_y}{dx^2} + (k^2n^2 - \beta^2)e_y = 0 \qquad (7.40)$$

where $k^2 = \omega^2\mu_o\varepsilon_o$.

Recall that n_c, n_f, and n_s are indices of the cover, film, and substrate regions, respectively, and the propagation constant β for a guided wave is between kn_f and kn_s. This means that $(k^2n_f^2 - \beta^2)$ is positive, while $(k^2n_s^2 - \beta^2)$ and $(k^2n_c^2 - \beta^2)$ are negative.

The final task is to solve the differential equation, subject to the boundary conditions at $x = 0$ and $x = h$. First, however, it would be helpful to anticipate the form of the solution for each region. Recall that the key physical process contributing to the losslessness of guided modes is total internal reflection at the film/substrate and film/cover boundaries. Total internal reflection means there is no reduction in wave amplitude when the waves are reflected. In the film region, the waves are bounced back and forth by the two dielectric boundaries. When these waves are superimposed, there is constructive and destructive interference. Because of this interference, fields in the film region oscillate as a function of x. Therefore, in the film region, we expect e_y to be the linear combination of sine and cosine functions. Again, because of total internal reflection at the dielectric boundaries, fields in the lower index regions, that is, the cover and substrate regions, decrease exponentially from the dielectric boundaries. Accordingly, e_y in the cover and substrate regions decreases as exponential functions of x.

In the substrate region where $n = n_s$, $k^2n_s^2 - \beta^2$ is negative. Therefore, (7.40) can be written as

$$\frac{d^2e_y}{dx^2} - \gamma_s^2 e_y = 0$$

where $\gamma_s = \sqrt{\beta^2 - k^2n_s^2}$.

The two linearly independent solutions to the differential equation are $e^{\pm\gamma_s x}$. Since we expect the field of a guided mode to decrease from the film/substrate boundary, we write, for $x \leq 0$

$$e_y = C_1 e^{\gamma_s x} \qquad (7.41)$$

where C_1 is a constant.

In the thin-film region where $n = n_f$, $k^2n_f^2 - \beta^2$ is positive. The two linearly independent solutions to the differential equation (7.40) are $\sin k_f x$ and $\cos k_f x$ where $k_f = \sqrt{k^2n_f^2 - \beta^2}$. We can then combine the two functions as $\cos(k_f x + \phi)$. Depending on the value of ϕ, $\cos(k_f x + \phi)$ may contain a sine or cosine function, or the linear combination of two functions. Since we expect the field of a guided mode to oscillate as a function of x, we have, for the film region of $0 \leq x \leq h$,

$$e_y = C_2 \cos(k_f x + \phi) \qquad (7.42)$$

where C_2 and ϕ are unknown constants.

In the cover region where $n = n_c$, the two linearly independent solutions to the wave equation (7.40) are $e^{\pm \gamma_c x}$, where $\gamma_c = \sqrt{\beta^2 - k^2 n_c^2}$. Since the fields are expected to decay from the film/substrate boundary, we write, for $x \geq h$,

$$e_y = C_3'' e^{-\gamma_c x}$$

where C_3'' is an unknown constant. For convenience, C_3'' is written as $C_3 e^{\gamma_c h}$, and the above expression becomes

$$e_y = C_3 e^{-\gamma_c(x-h)} \tag{7.43}$$

Thus far, we have four unknown constants: C_1, C_2, C_3, and ϕ. We choose these constants such that the boundary conditions are satisfied. By doing so, we can reduce the four constants to one constant representing the wave amplitude, and a characteristic equation.

The relevant boundary conditions at $x = 0$ and $x = h$ are the continuation of e_y and $\dfrac{de_y}{dx}$ corresponding to the continuation of the tangential component of the electric field, e_y, and the tangential component of the magnetic field h_z. In addition, since all regions are nonmagnetic materials, the permeability is the same everywhere. From (7.38), we see that the continuation of e_y guarantees the continuation of the normal component of the magnetic flux density as well. In short, all boundary conditions are satisfied when the conditions on e_y and $\dfrac{de_y}{dx}$ are met.

The boundary conditions at $x = 0$ require that

$$C_1 = C_2 \cos\phi$$
$$\gamma_s C_1 = -k_f C_2 \sin\phi$$

Eliminating C_1 and C_2 from these two equations, we obtain

$$\tan\phi = -\frac{\gamma_s}{k_f}$$

Therefore,

$$-\phi = \tan^{-1}\frac{\gamma_s}{k_f} + m'\pi \tag{7.44}$$

where $m' = 0, 1, 2, \ldots$.

The boundary conditions at $x = h$ are

$$C_3 = C_2 \cos(k_f h + \phi)$$
$$\gamma_c C_3 = k_f C_2 \sin(k_f h + \phi)$$

Thus,

$$k_f h + \phi = \tan^{-1}\frac{\gamma_c}{k_f} + m''\pi \tag{7.45}$$

where m'' is also an integer. By combining (7.44) with (7.45), we obtain,

$$k_f h = \tan^{-1}\frac{\gamma_s}{k_f} + \tan^{-1}\frac{\gamma_c}{k_f} + m'\pi + m''\pi \qquad (7.46)$$

From the definition of the generalized parameters, (7.21)–(7.23), we can show that

$$\gamma_c h = V\sqrt{a+b}$$
$$k_f h = V\sqrt{1-b}$$
$$\gamma_s h = V\sqrt{b}$$

We can also combine $m' + m''$ as a single integer m. Then, (7.46) becomes

$$V\sqrt{1-b} = m\pi + \tan^{-1}\sqrt{\frac{b}{1-b}} + \tan^{-1}\sqrt{\frac{a+b}{1-b}}, \qquad m = 0, 1, 2, \ldots$$

This is the characteristic equation (7.24) in the last section.

The constants C_2, ϕ, and C_3 can also be cast in terms of C_1. As a result, the field expressions can be expressed in terms of C_1 and the generalized parameters, as follows:

$$e_y = C_1 e^{V\sqrt{b}(x/h)}, \qquad \text{for } x \leq 0 \qquad (7.47)$$

$$e_y = C_1\left[\cos(V\sqrt{1-b}\,\frac{x}{h}) + \sqrt{\frac{b}{1-b}}\sin(V\sqrt{1-b}\,\frac{x}{h})\right], \qquad \text{for } 0 \leq x \leq h \qquad (7.48)$$

$$e_y = C_1\left[\cos(V\sqrt{1-b}) + \sqrt{\frac{b}{1-b}}\sin(V\sqrt{1-b})\right]e^{-V\sqrt{a+b}(x-h)/h}, \qquad \text{for } x \geq h \qquad (7.49)$$

The constant C_1 can be viewed as the amplitude constant. Expressions for other field components can be obtained by substituting (7.47)–(7.49) into (7.38) and (7.39).

TM Modes We can derive the characteristic equation for TM modes in the same manner. From (7.33) and (7.35), we can express e_x and e_z in terms of h_y:

$$e_x = \frac{\beta}{\omega n^2 \varepsilon_o} h_y \qquad (7.50)$$

$$e_z = -\frac{j}{\omega n^2 \varepsilon_o}\frac{dh_y}{dx} \qquad (7.51)$$

Substituting the resulting expressions in (7.37), we obtain a differential equation for h_y

$$\frac{d^2 h_y}{dx^2} + (k^2 n^2 - \beta^2) h_y = 0 \qquad (7.52)$$

The boundary conditions are the continuation of e_z and h_y at $x = 0$ and $x = h$. These conditions are the same as requiring h_y and $\frac{1}{n^2}\frac{dh_y}{dx}$ to be continuous at these boundaries. By matching h_y and $\frac{1}{n^2}\frac{dh_y}{dx}$, we obtain the characteristic equation for the TM modes,

$$V\sqrt{1-b} = m\pi + \tan^{-1}\frac{1}{c}\sqrt{\frac{b}{1-b}} + \tan^{-1}\frac{1}{d}\sqrt{\frac{a+b}{1-b}}, \quad m = 0, 1, ,2, , \ldots$$

where c and d were defined in connection with (7.25). Similarly, we also obtain expressions for h_y

$$h_y = C_1' e^{V\sqrt{b}(x/h)}, \quad \text{for } x \leq 0 \tag{7.53}$$

$$h_y = C_1'\left[\cos(V\sqrt{1-b}\,\frac{x}{h}) + \frac{1}{c}\sqrt{\frac{b}{1-b}}\sin(V\sqrt{1-b}\,\frac{x}{h})\right], \quad \text{for } 0 \leq x \leq h \tag{7.54}$$

$$h_y = C_1'\left[\cos(V\sqrt{1-b}) + \frac{1}{c}\sqrt{\frac{b}{1-b}}\sin(V\sqrt{1-b})\right]e^{-V\sqrt{a+b}(x-h)/h}, \quad \text{for } x \geq h \tag{7.55}$$

where C_1' is the amplitude constant. Equations describing other field components can be obtained by substituting (7.53)–(7.55) into (7.50) and (7.51). Detailed derivations for the characteristic equation for TM modes and (7.53)–(7.55) are left as an exercise for the reader.

7.5.3 Modes Guided by Step Index Thin-Film Waveguides

bV Diagram We have now formally derived the characteristic equations, and can examine the solutions to these equations. For given values of a, c, and V, b may be solved numerically from (7.24) and (7.25). Plots of b versus V with a as a parameter are given in Figure 7.8 for TE modes. The plots for TM modes are not given here since they are very similar to those depicted in Figure 7.8. Knowing the parameter b, the effective index N of a guided mode can be determined from (7.23)

$$N = [n_s^2 + b(n_f^2 - n_s^2)]^{1/2} \tag{7.56}$$

Equations (7.5)–(7.8) can then be used to calculate β, θ, v_{ph}, and v_{gr}.

To illustrate the use of the bV diagram, we return to the numerical example studied previously. With $n_f = 1.6000$, $n_s = n_c = 1.3000$, and $h/\lambda = 3.000$, we have $V = 17.58$. From Figure 7.8, we obtain (with a little imagination?) $b \approx 0.98$ for mode $m = 0$. Using this value in conjunction with (7.23), we obtain $N \approx 1.594$. Thus,

$$\theta \approx \sin^{-1}(1.594/1.600) = 84.6°$$

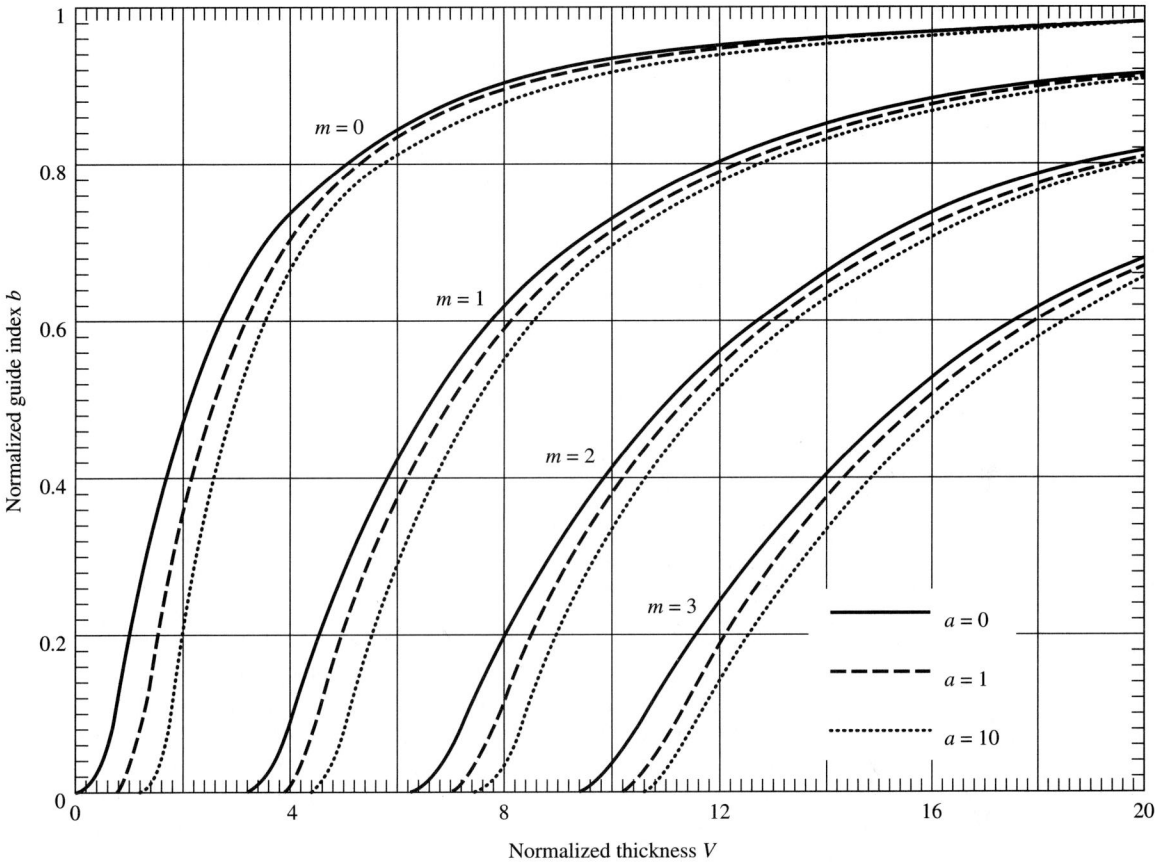

Figure 7.8 The bV diagram of TE modes guided by a step index thin-film waveguide [8]

which is very close to the numerical value obtained graphically from Figure 7.7a. For other modes, N and θ can be obtained in the same manner.

We can visualize the shapes of bV curves by considering the case of parallel-plate waveguides with conducting plates. By defining $V = knh$ and $b = N^2/n^2$ for a metallic waveguide, we can write (7.2) as

$$V\sqrt{1-b} = (m+1)\pi \qquad (7.57)$$

Thus,

$$b = 1 - \left[\frac{(m+1)\pi}{V}\right]^2 \qquad (7.58)$$

The dependence of b on V, as given in (7.58), is plotted as dots in Figure 7.9. For comparison, the bV plot for TE modes guided by a symmetric thin-film waveguide, that is, $a = 0$, is also depicted in Figure 7.9, as solid lines. The two sets of curves clearly have the same general shape.

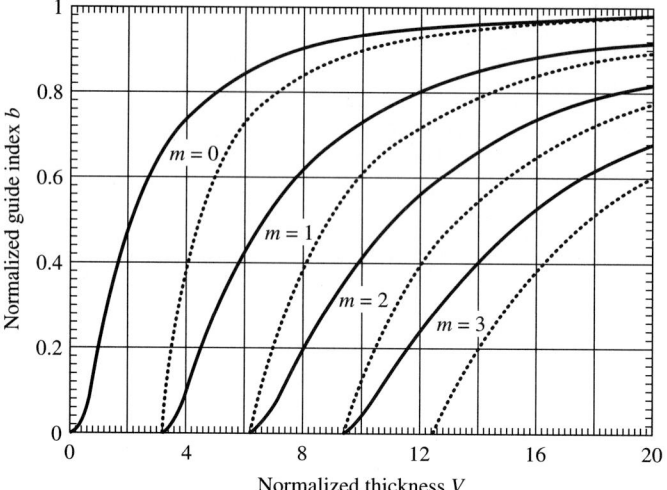

Figure 7.9 Comparison of the bV diagram of TE modes guided by a symmetric dielectric waveguide (solid lines) to that of a parallel-plate waveguide with conducting plates (dotted lines).

Cutoff Thicknesses and Cutoff Frequencies For conducting parallel-plate waveguides, as well as thin-film dielectric waveguides, there are **cutoff thicknesses** or **cutoff frequencies** for *higher-order modes*. The *exception* is the lowest-order TE mode ($m = 0$) of the symmetric thin-film waveguide, which has no cutoff.[2]

Number of Guided Modes For thin-film waveguides with V less than 10, we can use Figure 7.8 to calculate the number of guided modes. Also note from Figure 7.8 that, for symmetric waveguides, that is, $n_c = n_s$, the cutoff V value for mode m is approximately $m\pi$. In other words, each time V increases by π, there is one additional TE mode. The same is true for TM modes. Therefore, for symmetric thin-film waveguides with V greater than 10, the number of TE and TM modes supported by the waveguide is

$$\tilde{n} \approx 2(\frac{V}{\pi} + 1) = \frac{4h\sqrt{n_f^2 - n_s^2}}{\lambda} + 2 \qquad (7.59)$$

The number of TE and TM modes guided by **asymmetric waveguides** is approximately the same as that of **symmetric waveguides**. This means that (7.59) is also useful in estimating the number of modes guided by asymmetric thin-film waveguides.

Guided Mode Fields Typical distributions for $|E_y|$ of the TE_0, TE_1, and TE_2 modes of *asymmetric* waveguides are presented in Figures 7.10a, b, and c. In

---

[2]For parallel-plate waveguides with conducting plates, the TEM mode, which is the limiting case of the TM mode, has no cutoff.

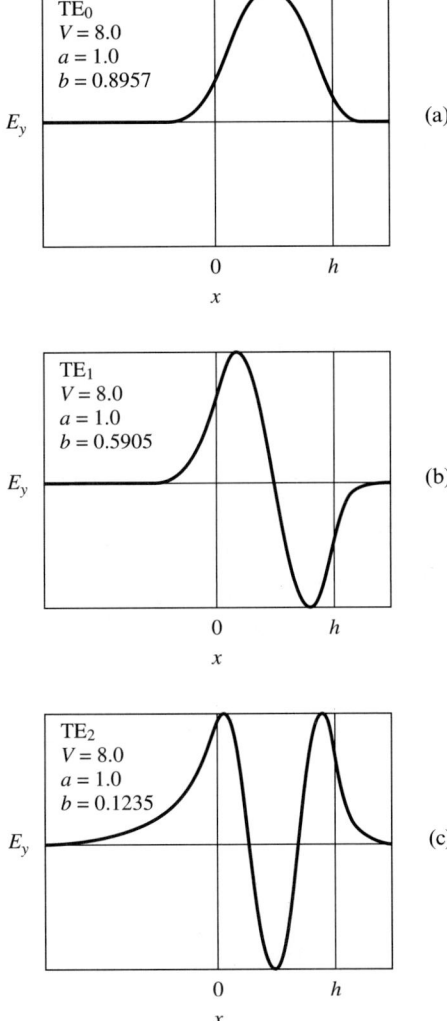

Figure 7.10 Field distributions of the first three TE modes guided by a step index thin-film waveguide

these figures, the film/substrate and film/cover boundaries are marked by vertical lines. In the film region, the field distributions of a guided mode are described by oscillatory functions, as given in (7.48). In the cover and substrate regions, the fields decay exponentially from the film/cover and film/substrate boundaries, and are expressed in terms of the exponential functions given by (7.47) and (7.49). For all modes, fields in the cover and substrate regions are confined in the thin layers near the film/cover or film/substrate boundaries. Beyond these thin layers, the fields are very weak, unless the mode is near the

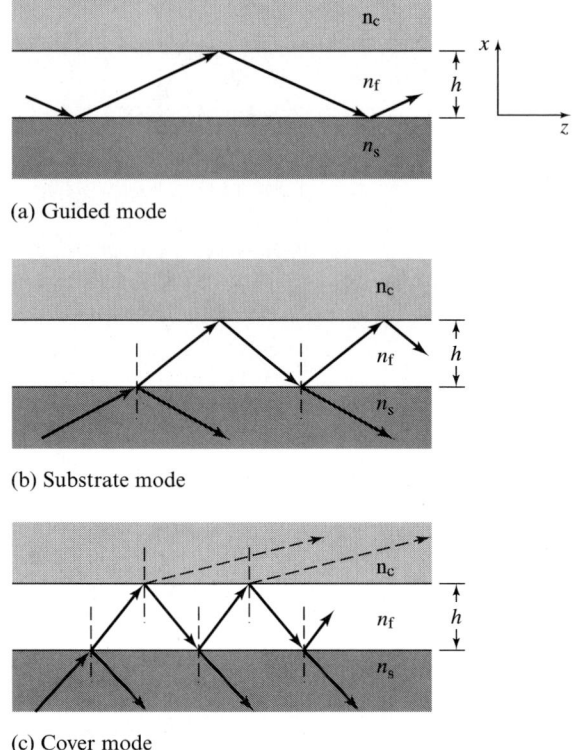

Figure 7.11 Various modes of dielectric slab waveguides ($n_f > n_s > n_c$)

cutoff. A careful examination of these plots also shows that the fields in the substrate region are stronger than those in the cover region. This is expected, since the waveguide is asymmetric and $n_s > n_c$.

Cover and Substrate Modes A thin-film waveguide supports a continuum of radiation modes in the cover and substrate regions, in addition to a finite number of guided modes. These radiation modes, which are referred to as the **cover (or air)** and the **substrate modes,** are not confined in the film region, nor guided by the dielectric boundaries. For *substrate modes,* if $n_s > n_c$, total internal reflection occurs at the cover/film interface, but not at the film/substrate boundary. Therefore, the field in the cover region decreases exponentially from the film/cover boundary. However, the field in the substrate region does not decay exponentially from the film/substrate boundary. Depending on the excitation, the field of a substrate mode in the substrate region may remain constant independent of x, or may vary as an oscillatory function of x, due to the interference between the incident and reflected waves. For the *cover mode,* there is no total internal reflection at either boundaries. Therefore, the field has a constant amplitude, or has crests and troughs in one or the other region. In other words, the fields of a cover mode penetrate into the substrate region, as well as the cover region. This is shown schematically in Figure 7.11.

Numerical Aperture of Symmetric Step Index Waveguides Numerical aperture (NA) is an important parameter that quantifies the ability of an optical element or component to receive and radiate fields. Here, we consider the ability of a symmetric step index waveguide to capture incident rays from a surrounding air medium. Consider incident rays impinging on the waveguide end at an angle ϕ_{air} relative to the normal to the waveguide end. At the film/cover boundary, the angle between the refracted rays and the normal to the boundary is θ_f. In the cover region, the angle between the transmitted rays and the normal is θ_c (Figure 7.12). If ϕ_{air} is sufficiently small, θ_f will be sufficiently large, and the rays are internally reflected at the film/cover boundary. Rays, such as Ray 1 in Figure 7.12, that enter at the film region are trapped there by the two dielectric boundaries. As ϕ_{air} increases, θ_f decreases. Beyond a certain value, rays are no longer totally internally reflected by the film/cover boundary, and a fraction of incident power leaks into, and is subsequently lost in, the cover region. Ray 2 in Figure 7.12 is an example. For rays that remain trapped in the film region, ϕ_{air} must be smaller than the critical value. The sine function of the critical angle is defined as the **numerical aperture** (NA) of the waveguide. In other words, NA is the sine of the largest incident angle for which rays, once launched into the film region, will remain trapped in the region by total internal reflection at the film/cover boundary. Since the waveguide we are evaluating has a symmetric structure, similar results are found for rays impinging on the film/substrate boundary.

To find the critical value for ϕ_{air}, assume that the medium surrounding the waveguide is air or vacuum, with an index of 1. Then, ϕ_f and ϕ_{air} are related by Snell's law,

$$1 \sin\phi_{air} = n_f \sin\phi_f \qquad (7.60)$$

Also, $n_f \sin\theta_f = n_c \sin\theta_c$. For total internal reflection, $(n_f/n_c)\sin\theta_f$ must be greater than 1. Since $\theta_f = \dfrac{\pi}{2} - \phi_f$, we obtain

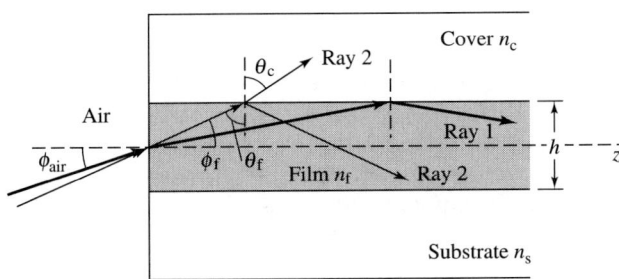

Figure 7.12 Numerical aperture of symmetric waveguide with a step index profile

$$\sin\theta_f = \cos\phi_f = \sqrt{1 - \sin^2\phi_f}$$

Consequently, the condition for total internal reflection at the film/cover boundary becomes

$$(n_f/n_c)\sqrt{1 - \sin^2\phi_f} \geq 1$$

which can be written as

$$1 - \sin^2\phi_f \geq (n_c/n_f)^2$$

By making use of (7.60), we obtain

$$\sin\phi_{air} \leq \sqrt{n_f^2 - n_c^2}$$

The numerical aperture is therefore

$$NA = \sin\phi_{air\,max} = \sqrt{n_f^2 - n_c^2} \qquad (7.61)$$

For waveguides with a constant film index, the numerical aperture is independent of the location of the incident ray. This is a special feature of step index waveguides.

7.6 GRADED INDEX THIN-FILM WAVEGUIDES

In thin-film waveguides with step index profiles, rays follow straight-line trajectories until they reach the boundaries, at which the rays bend abruptly. The numerical aperture is independent of the position of the incident rays. In contrast, for **graded index waveguides,** n_f is a function of x, and the ray trajectories bend continuously. In addition, the numerical aperture is a function of the position of the entrance rays. These features are also present in graded index fibers.

To derive an equation describing curved trajectories in graded index waveguides, divide the film region into many thin layers, each with a constant index. Eventually, the layer thickness will reduce to an infinitesimal value. Since the indices are different for different layers (Figure 7.13a), rays are refracted at, or even totally internally reflected by, the boundaries. Let the index of the jth layer be n_{fj}, and let a ray have an angle θ_j with respect to the normal to the boundaries between layers. Applying Snell's law, we have

$$n_{fj}\sin\theta_j = n_{fj+1}\sin\theta_{j+1}$$

By repeatedly applying Snell's law, we can trace the rays from one boundary to another. For a given trajectory, the product $n_{fj}\sin\theta_j$ is a constant for all layers and can be denoted by $\mathfrak{N}$. This relationship remains valid as the layer thickness vanishes. In the limit of an infinitesimal layer thickness, we replace n_{fj} and θ_j by $n_f(x)$ and $\theta(x)$. Then, the equation becomes

$$n_f(x)\sin\theta(x) = \mathfrak{N} \qquad (7.62)$$

For a curved trajectory (Figure 7.13b), an infinitesimal path length ds is given by

7.6 Graded Index Thin-Film Waveguides

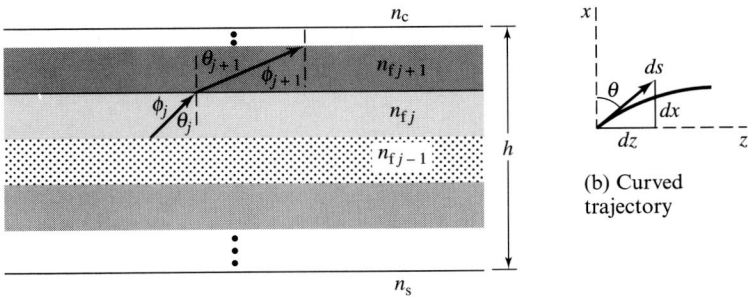

(a) Approximation using layered structure

(b) Curved trajectory

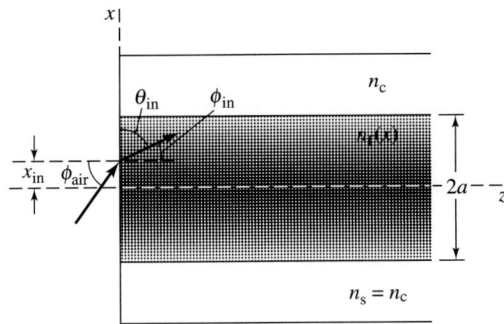

(c) Symmetrically graded index profile

Figure 7.13 Rays in, and the local numerical aperture of, a thin-film waveguide with a symmetrically graded index profile.

$$ds^2 = dx^2 + dz^2$$

and

$$\frac{dx}{ds} = \cos\theta(x), \qquad \frac{dz}{ds} = \sin\theta(x)$$

Combining these equations with (7.62), we obtain

$$\left(\frac{dx}{dz}\right)^2 = \cot^2\theta(x) = \frac{1}{\sin^2\theta(x)} - 1 = \left[\frac{n_f(x)}{\mathfrak{N}}\right]^2 - 1$$

Differentiating both sides of the equation with respect to s, we obtain

$$\frac{d}{ds}\left(\frac{dx}{dz}\right)^2 = \frac{1}{\mathfrak{N}^2}\frac{d}{ds}[n_f(x)]^2$$

That is,

$$2\frac{dx}{dz}\frac{d^2x}{dz^2}\frac{dz}{ds} = \frac{1}{\mathfrak{N}^2}\frac{d}{dx}[n_f(x)]^2 \frac{dx}{ds}$$

Recognizing that $\dfrac{dx}{dz}\dfrac{dz}{ds} = \dfrac{dx}{ds}$, we obtain a differential equation that describes the ray trajectories in graded index thin-film waveguides [10],

$$\frac{d^2x}{dz^2} = \frac{1}{2\mathcal{N}^2}\frac{d}{dx}(n_f(x))^2 \tag{7.63}$$

This equation is known as the **ray equation**.

Once $n_f(x)$ is specified, we can solve (7.63) to obtain equations which describe the trajectories. As an example, consider rays in a symmetric thin-film waveguide with a parabolic index profile. For convenience, we also shift the origin to the film center, as shown in Figure 7.13c, and let the film thickness be $2a$. The distribution of the refractive index in the film region is then given by

$$n_f^2(x) = n_{f0}^2\left(1 - \frac{\Delta}{a^2}x^2\right), \qquad -a \le x \le a \tag{7.64}$$

where Δ is a constant and n_{f0} is the index at the waveguide center. Substituting (7.64) in (7.63), we have a simple differential equation,

$$\frac{d^2x}{dz^2} = -\frac{n_{f0}^2 \Delta}{\mathcal{N}^2 a^2}x \tag{7.65}$$

Consider a ray entering the waveguide at $(x_{in}, 0)$ with an angle ϕ_{in} (Figure 7.13c). The constant $\mathcal{N}$ is

$$\mathcal{N} = n_f(x_{in})\sin\theta_{in} = n_f(x_{in})\cos\phi_{in}$$

Solving the differential equation (7.65), we obtain

$$x(z) = A_1\cos\left(\frac{n_{f0}\sqrt{\Delta}}{\mathcal{N}a}z\right) + A_2\sin\left(\frac{n_{f0}\sqrt{\Delta}}{\mathcal{N}a}z\right) \tag{7.66}$$

where the constants A_1 and A_2 are determined by the initial conditions of the incoming ray. Explicitly,

$$A_1 = x_{in}$$
$$A_2 = \frac{n_f(x_{in})a\,\sin\phi_{in}}{n_{f0}\sqrt{\Delta}}$$

Combining the sine and cosine functions, we obtain, for $-a \le x \le a$,

$$x(z) = A\sin\left(\frac{n_{f0}\sqrt{\Delta}}{\mathcal{N}a}z + \psi\right) \tag{7.67}$$

where

$$A = \sqrt{x_{in}^2 + \frac{[n_f(x_{in})a\,\sin\phi_{in}]^2}{n_{f0}^2\Delta}}$$

$$\psi = \sin^{-1}\frac{x_{in}}{A}$$

Thus, rays in a parabolic index waveguide follow sinusoidal trajectories with a spatial periodicity of $2\pi \mathcal{H} a/(n_{f0}\sqrt{\Delta})$. Rays are confined in the film region if the amplitude constant A is less than a. From this condition, we obtain the condition for rays trapped in the film region:

$$\sin^2\phi_{in} n_{f0}^2 (1 - \frac{\Delta}{a^2} x_{in}^2) \leq n_{f0}^2 (1 - \frac{x_{in}^2}{a^2})\Delta$$

To consider the numerical aperture, let us suppose that the cover index precisely matches $n_f(a)$; that is, $n_c = n_{f0}(1 - \Delta)$. Then

$$n_f^2(x) - n_c^2 = n_{f0}^2 \Delta [1 - \frac{x^2}{a^2}]$$

Applying Snell's law to the end face, we have

$$\sin\phi_{air} = n_f(x_{in}) \sin\phi_{in} \leq \sqrt{n_f^2(x_{in}) - n_c^2}$$

Note that the maximum value of $\sin\phi_{air}$ is a function of x_{in}. We define the **local numerical aperture** for the graded index waveguide as

$$NA(x) = \sqrt{n_f^2(x) - n_c^2} \qquad (7.68)$$

This expression can be compared to (7.61) for the numerical aperture of step index waveguides.

7.7 PRISM COUPLERS

There are three basic methods of coupling light into or out of thin-film waveguides; **end fire coupling, grating coupling,** and **prism coupling** (Figure 7.14). The prism coupling method is also useful in identifying the propagating modes.

A prism coupler is a high-index prism, with an index n_p and a base angle θ_p, placed in close proximity to, and separated by a thin air gap from, the thin-film waveguide region (Figure 7.14c). The prism index must be greater than the film index, if the prism is to serve as a coupler. Consider an optical beam directed toward the film/prism boundary at an angle θ_2. If the projection of the phase velocity of plane waves in the prism exactly matches c/N_m, where N_m is the effective index of the guided mode m, then the guided mode m may be excited. This is the basic principle of prism coupling. Referring to Figure 7.14c and using Snell's law, we obtain the following relationships,

$$n_c \sin\theta = n_p \sin\theta_1$$
$$n_p \sin\theta_2 = n_f \sin\theta_f$$
$$\theta_p + (\frac{\pi}{2} + \theta_1) + (\frac{\pi}{2} - \theta_2) = \pi$$

Once the coupling angle θ is known, we can deduce θ_1, θ_2, and θ_f

$$n_f \sin\theta_f = n_p \sin\theta_2 = n_p \sin(\theta_p + \theta_1) \qquad (7.69)$$

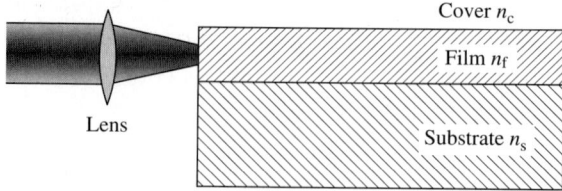

(a) End-fire coupling with a lens

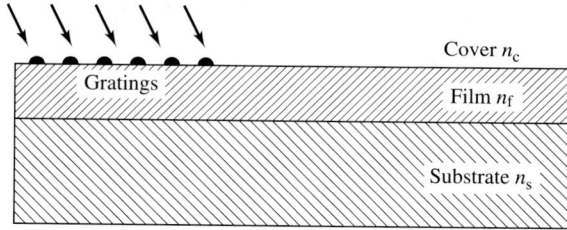

(b) Gratings coupling

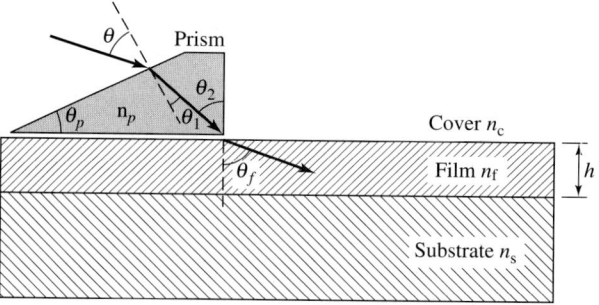

(c) Prism coupling

Figure 7.14 Three methods of coupling

The effective index of refraction N_m is

$$N_m = n_f \sin\theta_f = n_p \sin(\theta_p + \theta_1)$$
$$= \sin\theta_p \sqrt{n_p^2 - n_c^2 \sin^2\theta} + n_c \sin\theta \cos\theta_p \qquad (7.70)$$

The coupling angle θ is different from different modes. Of particular interest is the case in which the cover region is simply air (i.e., $n_c = 1$). The effective index of refraction of the guided mode is then (Figures 7.14c and 7.15)

$$N_m = \sin\theta_p \sqrt{n_p^2 - \sin^2\theta} + \cos\theta_p \sin\theta \qquad (7.71)$$

or, in terms of θ' depicted in Figure 7.15,

$$N_m = \sin\theta_p \sqrt{n_p^2 - \sin^2\theta'} - \cos\theta_p \sin\theta' \qquad (7.72)$$

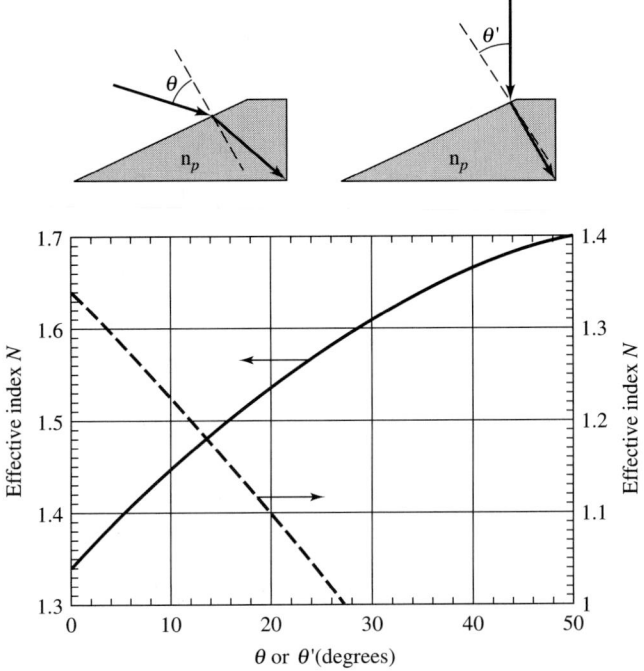

Figure 7.15 Effective refractive index measured with a glass prism with $n_p = 1.75$ and $\theta_p = 50°$

The key components of a prism coupler is the prism itself. The following factors should be taken into account in selecting prisms for prism couplers:

1. The prism index n_p should be larger, not just *slightly* larger, than the film index.
2. To avoid spurious responses due to multiple and repeated reflections from the prism surfaces, prisms with angles of 30°, 45°, 60°, 90°, or other rational fractions of 360° should be avoided [11].
3. For the arrangements depicted in Figures 7.14c and 7.15, the prism edge from which the guided modes are launched should be sharp and well defined. This edge is crucial to the operation of the prism coupler.
4. The prism should have a flat top, rather than a sharp apex, so that mechanical pressure can be applied uniformly to the prism.

As an example, consider a planar waveguide consisting of a Corning 7059 glass film ($n_f = 1.555$) of thickness $h = 1.0$ μm, and a thick, pyrex glass plate with $n_s = 1.450$. Assume air is the cover region. As far as the guided modes are concerned, the pyrex glass plate is considered to be infinitely thick. Solving (7.18) or (7.24) numerically, or estimating from Figure 7.8, we conclude that two TE modes are supported by the air/Corning 7059/pyrex structure, with $N_0 = 1.536$, and $N_1 = 1.478$. Now, suppose a glass prism with an index of

$n_p = 1.750$ and a base angle of $\theta_p = 50°$ is used for coupling. For $\theta' > 27°$, the air mode is excited. Although the beam is partially reflected from the pyrex/air boundary on the back of the pyrex glass plate, the reflection coefficient is small ($|\Gamma|^2 \approx 0.034$). The light beam, if not blocked by an object, would emerge from behind the pyrex glass. As the incident angle is changed to $\theta' < 27°$ or $\theta < 10.4°$, a substrate mode is excited. The optical beam is totally internally reflected at the air/pyrex boundary, and the beam is trapped in the pyrex glass plate. Each time the trapped beam is incident upon and reflected by the boundary, a dot appears at the air/pyrex boundaries. Thus a series of dots is visible at the air/pyrex glass plate boundary. The spacing between the dots depends on the thickness of the pyrex glass plate and the angle θ or θ'. For a pyrex glass plate of thickness of 0.1 cm, the separation between dots may vary from 0.10 cm to 1.27 cm as the angle changes from $\theta' = 25.0°$ to $\theta = 10.0°$. As θ approaches $13.6°$, a continuous streak appears near the Corning 7059 glass film, corresponding to the TE_1 mode with $N_1 = 1.478$. For $\theta = 20.0°$, the TE_0 mode with $N_0 = 1.536$ is excited.

From this example, we can discern three points useful for the proper interpretation and successful operation of prism coupling:

1. If possible, there should be no reflective objects behind the substrate, so that light emerging from behind the substrate is not reflected back into the substrate and the film. If this is not possible, then the space behind the substrate should be made accessible. By inserting a probe into the space behind the substrate, we can differentiate between light reflected by the object behind the substrate and light of the substrate mode trapped by the air/substrate boundary.

2. As the coupling angle changes, the spacing between the dots on the substrate boundaries increases. The limiting case of widely separated dots should be very close to the highest-order guided mode.

3. From the coupling angle θ or θ' where the separation between dots is quite large, we can use (7.71) or (7.72) to estimate n_s.

7.8 FILM INDEX AND THICKNESS MEASUREMENT: AN APPLICATION

One of the most useful features of prism coupling is the relationship between the coupling angle θ (or θ') and the effective index. When θ or θ' is measured experimentally, the effective index can be computed from (7.71) and (7.72), provided n_p and θ_p are known. When several coupling angles are determined, they can be used in conjunction with the characteristic equation, (7.18) or (7.19), to estimate n_f and h [11]. In most cases, n_s, n_c, and λ are known, while the film index and thickness are unknown. If only one of the coupling angles is measured, and the effective index N_m is computed from (7.71) or (7.72), we have one equation, (7.18) or (7.19), and two unknowns. There is insufficient information to solve for n_f and h. If two coupling angles are determined experimentally, n_f and h can then be evaluated, in principle. In cases where three or more coupling angles are measured, iterative numerical calculations may be used to evaluate n_f and h from the experimentally measured data. A computer program may be written to carry out the iterative computations. The theoretical foundation and the algorithm for the computer program are presented in this section [11].

To facilitate this discussion, we rearrange (7.18) and (7.19) as

$$\frac{2\pi h}{\lambda} \sqrt{n_f^2 - N_m^2} = \Psi_m(n_f, n_s, n_c, N_m) \tag{7.73}$$

where

$$\Psi_m(n_f, n_s, n_c, N_m) = m\pi + \phi_{fs}(n_f, n_s, N_m) + \phi_{fc}(n_f, n_c, N_m) \tag{7.74}$$

$$\phi_{fs}(n_f, n_s, N_m) = \tan^{-1}\left\{\left(\frac{n_f}{n_s}\right)^{2\rho}\left(\frac{N_m^2 - n_s^2}{n_f^2 - N_m^2}\right)\right\}^{1/2} \tag{7.75}$$

$$\phi_{fc}(n_f, n_c, N_m) = \tan^{-1}\left\{\left(\frac{n_f}{n_c}\right)^{2\rho}\left(\frac{N_m^2 - n_c^2}{n_f^2 - N_m^2}\right)\right\}^{1/2} \tag{7.76}$$

where $\rho = 0$ for TE modes and $\rho = 1$ for TM modes.

7.8.1 Two Measured Coupling Angles

Suppose that two coupling angles are measured, and that N_i and N_j are computed from (7.71) or (7.72). Substituting N_i and N_j into (7.73) and rearranging the resultant equations, we obtain

$$n_f^2 = \frac{N_i^2 \Psi_j^2 - N_j^2 \Psi_i^2}{\Psi_j^2 - \Psi_i^2} \tag{7.77}$$

This is a nonlinear equation for n_f^2, since Ψ_i and Ψ_j are also functions of n_f^2. A trial-and-error method can be used to solve for n_f^2. Specifically, a trial value for n_f^2 is assumed and (7.77) is used to compute a new n_f^2. If the new n_f^2 is very different from the old n_f^2, the new n_f^2 is used to compute a newer n_f^2. This process is repeated until the difference between the "new" and the "newest" n_f^2 is negligible, or satisfactory accuracy is achieved. Once n_f^2 is known, it is a simple matter to determine h/λ from (7.73).

7.8.2 Three or More Measured Coupling Angles

The cases in which three or more coupling angles are measured, assuming the waveguide supports three or more modes, are much more complicated. Suppose that ℓ coupling angles are measured and $\ell \geq 3$. Each measured coupling angle corresponds to a value for N_m, which in turn leads to an equation with a specific N_m. There are ℓ equations, and yet there are only two unknowns. One cannot uniquely solve for two unknowns, n_f and h, from these equations unless $\ell - 2$

equations are redundant. In the ideal situation where the experimental results are precise and error free, $\ell - 2$ equations are redundant. In real measurements, however, there are experimental errors or inaccuracies. It would not be possible to determine n_f and h uniquely from the measured coupling angles. Therefore, instead of "solving" for n_f and h from measured data, the **best estimates** for n_f and h are sought. Reasonable values of n_f and h are assumed, and are used in conjunction with known values of n_s and n_c to compute the effective refractive index from the characteristic equation (7.73). We label the effective refractive index thus obtained as $N_m(n_f, h)$. The values of n_f and h are searched such that N_m matches $N_m(n_f, h)$. Following Ulrich and Torge's suggestion [11], we define an error sum as

$$\sigma(n_f, h) = \sum_m [N_m - N_m(n_f, h)]^2 \tag{7.78}$$

If the experimental results are perfectly accurate, we search for n_f and h such that σ vanishes. In the presence of experimental errors and inaccuracies, we look for the values of n_f and h that minimize σ. The minimum is found by requiring that

$$\frac{\partial \sigma}{\partial n_f} = 0 \tag{7.79}$$

$$\frac{\partial \sigma}{\partial h} = 0 \tag{7.80}$$

Any numerical method may be used to solve for n_f and h from the simultaneous equations (7.79) and (7.80). For example, a gradient method may be used. The basic idea of the gradient method is as follows. Suppose a relief or a surface of σ is plotted as a function of (n_f, h). Then, σ at (n_f, h) is evaluated. If σ vanishes at a certain point, this point corresponds to the exact solution. If $\sigma > 0$, we search for a point (n_f, h) with a smaller value of σ. If we "stand" on the relief plot, we "look" for and "move" in the direction of steepest descent, that is, in the direction of the negative gradient. The process is repeated until the point (n_f, h) with the smallest σ is reached. This point corresponds to the best estimate of n_f and h.

7.9 OPTICAL DIRECTIONAL COUPLERS

Dielectric waveguides, in various forms, including thin-film waveguides, are useful in connecting one optical component to another. They are also the building blocks of many optical devices, such as optical directional couplers. An optical directional coupler consists of two parallel waveguides in close proximity, as shown in Figure 7.16. Because of the interaction between the two waveguides, power fed into one waveguide can be transferred to the adjacent waveguide. The percentage of power transfer can be varied by controlling the coupling constant, the interaction length, or the phase mismatch between the two waveguides. If

7.9 Optical Directional Couplers

electrooptic materials are used as part of the waveguiding structure, these parameters can be controlled electrically. Therefore, directional couplers based on electrooptic materials can also be used as **optical switches** or **intensity modulators**. Since the power transfer characteristics of directional couplers are wavelength dependent, directional couplers can also be, and often are, used as **optical filters**. A directional coupler can be used as a power splitter or a power combiner.

The coupling between waveguides is usually weak, unless the waveguides are very close. However, if the waveguides have different propagation constants, no significant power exchange between them is possible, even if they are very close. Even if the two waveguides are identical or similar and having identical or similar propagation constants, substantial power exchange between waveguides is possible only if additional conditions are met. In short, for a *complete power transfer,* two necessary conditions must be satisfied. In the following discussions, we use a **coupled mode theory** to describe the operation of optical directional couplers, and to establish the necessary conditions for the complete power transfer between waveguides. We begin by considering two isolated, uncoupled waveguides. Then we study the power exchange between the two coupled waveguides. Although we frequently refer to thin-film waveguides as examples, the treatment and conclusions are independent of waveguide type, which means that the theory is valid for directional couplers based on various types of dielectric waveguides, as well as for transmission line couplers, microwave waveguide couplers, surface acoustic wave couplers, and, more importantly, optical fiber couplers, to be discussed in Chapter 10.

7.9.1 Coupled Mode Theory

Recall that waves guided by isolated thin-film waveguides are given by (7.30) and (7.31). The propagation of the waves is represented by $e^{-j\beta z}$, and there is no change in the wave amplitude. Let the function $A(z)$ represent the wave propagation and amplitude in an isolated waveguide of *any type.* Then, $A(z)$ must be a product of an amplitude constant, a function independent of z describing the field distribution in the transverse plane and $e^{-j\beta z}$. The differential equation for $A(z)$ is

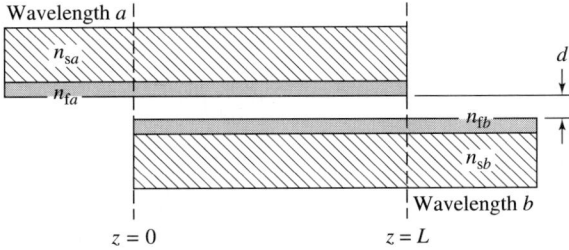

Figure 7.16 Two coupled thin-film waveguides

$$\frac{dA}{dz} + j\beta A = 0$$

For convenience, choose the amplitude constant and the field distribution such that power carried by the waveguide is $|A(z)|^2$. In the presence of a second waveguide, the propagation constant may change only slightly; however, the perturbation in the wave amplitude may be considerable.

Let the two parallel waveguides be a and b, as depicted in Figure 7.16. Suppose that the two waveguides are separated by a distance d, and that they interact in the region $0 \leq z \leq L$. First, consider two waveguides that are far apart ($d \to \infty$); therefore, there is no interaction between them. Although each isolated waveguide may support many modes, we will consider only one mode each, guided by an isolated waveguide. Let the propagation constants of these guided modes β_a and β_b, respectively. The differential equations governing the propagation of the two modes are:

$$\frac{dA_a}{dz} + j\beta_a A_a = 0 \tag{7.81}$$

$$\frac{dA_b}{dz} + j\beta_b A_b = 0 \tag{7.82}$$

When the separation d is reduced, the interaction increases. Nevertheless, the interaction is still sufficiently weak that the propagation of waves in one waveguide is perturbed only slightly by the presence of the other waveguide. Therefore, it is still meaningful to use β_a and β_b to describe wave propagation in these waveguides. To account for the presence of the other waveguide and the interaction between the waveguides, it is necessary to modify (7.81) and (7.82). For this purpose, we introduce a **coupling constant** κ, and $A_a(z)$ and $A_b(z)$ are then described by

$$\frac{dA_a}{dz} + j\beta_a A_a = j\kappa A_b \tag{7.83}$$

$$\frac{dA_b}{dz} + j\beta_b A_b = j\kappa A_a \tag{7.84}$$

Note that the two equations are coupled by the terms on the right-hand sides of the equations. (7.83) and (7.84) are the **coupled mode equations** [12]. The next task is to solve for A_a and A_b, subject to the boundary conditions at $z = 0$. The solution is greatly facilitated by converting the two first-order differential equations with two unknowns to a second-order differential equation with one unknown. This is done by differentiating one equation with respect to z and using the other equation. Recall that β_a and β_b are constants independent of z. Since two thin-film waveguides are in parallel, κ is also a constant independent of z. Therefore, only A_a and A_b are differentiated. The result for A_a is

$$\frac{d^2A_a}{dz^2} + j(\beta_a + \beta_b)\frac{dA_a}{dz} + (\kappa^2 - \beta_a\beta_b)A_a = 0 \quad (7.85)$$

A similar equation is obtained for A_b. Solutions to a second-order differential equation with constant coefficients are well known. Solving these equations for A_a and A_b, we obtain

$$A_a(z) = \left[A_a(0)\left(\cos\sigma z - j\frac{\delta\beta}{2\sigma}\sin\sigma z\right) + A_b(0)\left(j\frac{\kappa}{\sigma}\sin\sigma z\right)\right]e^{-j\beta_{av}z} \quad (7.86)$$

$$A_b(z) = \left[A_a(0)\left(j\frac{\kappa}{\sigma}\sin\sigma z\right) + A_b(0)\left(\cos\sigma z + j\frac{\delta\beta}{2\sigma}\sin\sigma z\right)\right]e^{-j\beta_{av}z} \quad (7.87)$$

where $\beta_{av} = (\beta_a + \beta_b)/2$ is the average propagation constant, $\delta\beta = \beta_a - \beta_b$ is the difference of the propagation constants and $\sigma = \sqrt{\kappa^2 + (\delta\beta/2)^2}$. Here, $A_a(0)$ and $A_b(0)$ are the boundary conditions at $z = 0$.

Suppose that power fed into waveguide a at $z = 0$ is 1, and that no power is fed to waveguide b. Then, $A_a(0) = 1$, $A_b(0) = 0$, and (7.86) and (7.87) are reduced to

$$A_a(z) = \left[\cos\sigma z - j\frac{\delta\beta}{2\sigma}\sin\sigma z\right]e^{-j\beta_{av}z} \quad (7.88)$$

$$A_b(z) = \left[j\frac{\kappa}{\sigma}\sin\sigma z\right]e^{-j\beta_{av}z} \quad (7.89)$$

The power at z carried by each waveguide is

$$P_a(z) = |A_a(z)|^2 = 1 - \frac{\kappa^2}{\sigma^2}\sin^2\sigma z \quad (7.90)$$

$$P_b(z) = |A_b(z)|^2 = \frac{\kappa^2}{\sigma^2}\sin^2\sigma z \quad (7.91)$$

From these equations, we can see that $\delta\beta$ and κ are the key parameters affecting the coupler characteristics. To further illustrate the effects of $\delta\beta$ and κ, plot $P_a(z)$ and $P_b(z)$ as functions of κz for three values of $\delta\beta/\kappa$ (Figure 7.17). We now have five important observations:

1. Although power exchanges continuously between two waveguides, the total power $P_a(z) + P_b(z)$ remains unchanged for all values of z. In other words, the *total power is conserved.*

2. A complete power transfer from one waveguide to the other is not possible unless $\delta\beta = 0$. Therefore, *one necessary condition* for a complete power transfer is: the two waveguides must have identical propagation constants. This condition is referred to as the **phase match condition.** Note that the condition is on the propagation constants, not the waveguides themselves. The two waveguides can be different, that is, they can have different geome-

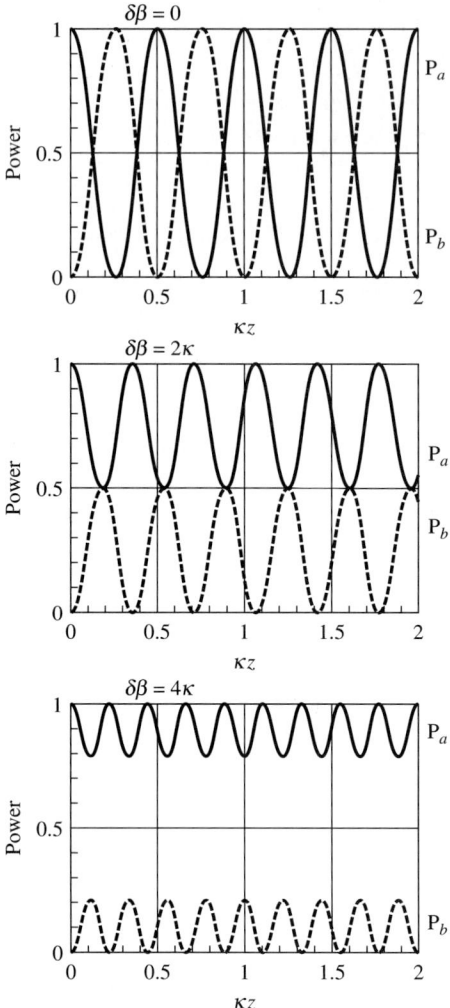

Figure 7.17 Power exchange in coupled waveguides

tries, index profiles, or index difference, and can still have identical propagation constants at discrete wavelengths.

3. Under the phase match condition, the power transfer is 100 percent if κL is an odd multiple of $\pi/2$, where L is the interaction length. This is the second necessary condition for a complete power transfer. The shortest interaction length needed for a complete power transfer is $\pi/(2\kappa)$, which is generally known as the **coupling length.** A small coupling constant corresponds to a long coupling length, which requires a long interaction length for a 100 percent power transfer. Because a long interaction length is diffi-

cult to maintain, κ can be increased by keeping the waveguiding regions close, or by "pushing" the fields out from high-index regions. For optical directional couplers, the spacing between film regions is typically on the order of a few micrometers or less.

4. As $|\delta\beta|$ increases, the maximum power transfer to the other waveguide decreases as $\kappa/\sqrt{\kappa^2 + (\delta\beta/2)^2}$.

5. Regardless of the condition of phase matching, it is always possible to have the total power returned to the original waveguide. All that is required is that σL be an integer multiple of π.

As noted earlier, we have only considered one mode for each isolated waveguide. Although a waveguide may support many modes, the propagation constants of different modes are different, except for degenerated modes which may exist in some waveguides. Ignoring the degenerate modes, at most only one propagation constant of waveguide a equals, exactly or approximately, a propagation constant of waveguide b. The modes treated in the coupled mode equation (7.83) and (7.84) are precisely the modes with equal or approximately equal propagation constants.

7.9.2 Multilayer Directional Coupler Example

As an example, consider the experimental directional coupler reported by Ihaya, et al. [13]. Their directional coupler consists of three glass layers on a pyrex glass substrate (Figure 7.18). Layers 1 and 3 are Corning 7059 glass films with an index 1.551. Situated between the two Corning 7059 glass layers is a low-index vycor glass layer. Vycor glass is chosen because its index (1.470) closely matches that of pyrex glass substrate (1.472). The layer thicknesses are given in Figure 7.18. In the experiments, light is coupled into guide 3 by a prism on the left. Due to the interaction between the two guides, light is transferred to guide 1. Output is then taken from the prism on the right, which is in contact with the vycor glass film above guide 1. Consequently, light from the output prism is proportional to the light transferred from guide 3 to guide 1. Note that layer 3 is thicker than layer 1. A thick layer 3 is required to ensure synchronization of the phase velocities of the two waveguides. To demonstrate that the design is

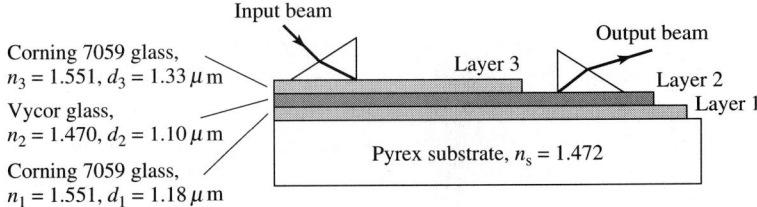

Figure 7.18 A directional coupler consisting of three glass layers and a glass substrate (after [13])

correct, let us estimate the propagation constants of two "isolated" waveguides, referred to as waveguides a and b, as in the last subsection. As an approximation, treat the structure consisting of the pyrex glass substrate, the lower Corning 7059 glass layer, layer 1, and the vycor glass layer, layer 2 as waveguide a. In other words, consider that the vycor glass layer as infinitely thick. With $n_s = 1.472$, $n_f = 1.551$, and $n_c = 1.470$, and a film thickness of 1.18 µm, we obtain $a = 0.024$ and $V = 5.724$ at $\lambda = 0.633$ µm. Since the asymmetry a is very small, assume it is 0. Therefore, for the TE_0 mode, we have, from Figure 7.8, $b = 0.839$. Thus, $N_a = 1.538$. The upper Corning 7059 glass layer, layer 3, bound by the vycor glass layer on one side and air on the other side, is treated as waveguide b. Again, we take the vycor glass layer as infinitely thick. Thus, for waveguide b, we have $a = 4.74$ and $V = 6.53$. From the curves in figure 7.8 corresponding to $a = 1$ and 10, we can interpolate the values corresponding to $a = 4.74$ and $V = 6.53$. For the TE_0 mode, we obtain $b = 0.849$ and $N_b = 1.539$. Clearly, with the indices and layer thicknesses given in Figure 7.18, the propagation constants, β_a and β_b, of the "isolated waveguides" are very close.

7.9.3 Waveguide Directional Coupler Types

In thin-film waveguides, waves bounce back and forth between the film/cover and film/substrate boundaries as they propagate in the z direction. In Figure 7.6a, we note that there is no boundary of any kind in the y direction. Therefore, the fields are independent of y and are not confined in any way in the y direction. Since the waveguide structure is independent of one lateral direction, that is, the y direction in Figure 7.6a, these waveguides are referred to as **two-dimensional waveguides.**

It is often desirable to make the most economical use of the available "real estate." For example, we might wish to reduce the component size. Also, in systems requiring a large number of components, the spacing between components must be kept to a minimum. However, for systems with densely packed components, crosstalk or unwanted interactions between components could be serious problems. To minimize these effects, the fields must be confined in *two* lateral directions. For this purpose, geometrical or dielectric discontinuities are introduced in the second lateral direction. This leads to various types of **three-dimensional waveguides.**

There are many types of three-dimensional waveguides, three of which are shown in Figure 7.19. The thin-film waveguide in Figure 7.19a is a two-dimensional waveguide. If the film layer on both sides is removed, with an etching process, while the film region in the central portion remains intact, the resulting structure (Figure 7.19b) is known as a **raised stripe waveguide,** which is three-dimensional. If the index of a region near the substrate surface is increased, by an ion implantation technique or a diffusion process, the high-index region would be embedded in the substrate. The resulting three-dimensional structure is referred to as a **buried** or **embedded strip waveguide.** A ridge wave-

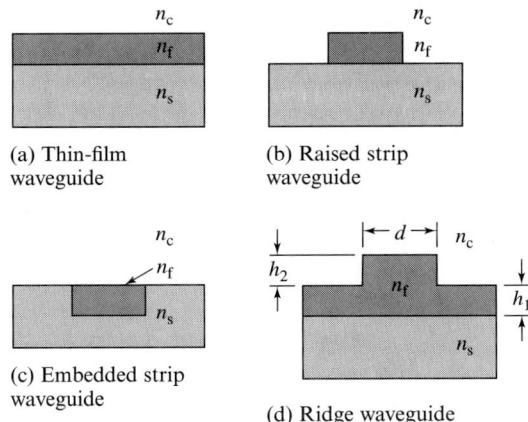

(a) Thin-film waveguide
(b) Raised strip waveguide
(c) Embedded strip waveguide
(d) Ridge waveguide

Figure 7.19 Four types of dielectric waveguides

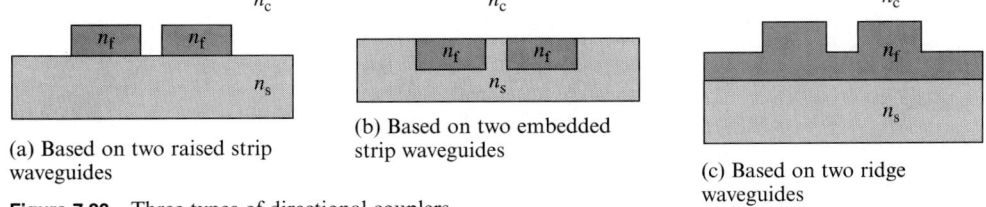

(a) Based on two raised strip waveguides
(b) Based on two embedded strip waveguides
(c) Based on two ridge waveguides

Figure 7.20 Three types of directional couplers

guide (Figure 7.19d) is very similar to a raised strip waveguide, except that the film regions on the two sides are not removed completely. In all three-dimensional waveguides, including those shown in Figures 7.19b, c, and d, there are geometrical or dielectric boundaries in the y as well as the x direction. As a result, waves are confined in both lateral dimensions. In essence, the waves are bounced back and forth between four boundaries.

When two strip waveguides are placed side by side, the result is a directional coupler based on strip waveguides. Three examples are shown in Figure 7.20. Most, if not all, optical directional couplers are based on three-dimensional waveguides. The operation principle for all directional couplers is the same, whether they are based on two-dimensional or three-dimensional waveguides. The difference is in the values of β_a, β_b, and κ.

REFERENCES

1. Paris, D. T.; and F. K. Hurd. *Basic Electromagnetic Theory.* New York, NY: McGraw-Hill Book Company, 1969.
2. Ramo, S.; J. R. Whinnery; and T. Van Duzer. *Fields and Waves in Communication Electronics.* 3rd ed. New York, NY: John Wiley & Sons, Inc., 1993.

3. Haus, H. A. *Waves and Fields in Optoelectronics.* Englewood Cliffs, NJ: Prentice Hall, Inc., 1984.
4. Tien, P. K. "Light waves in thin-films and integrated optics." *Appl. Opt.* 10, (1971), pp. 2395–2413.
5. Kogelnik, H. "Theory of dielectric waveguides." In *Integrated Optics.* Chapter 2. ed. T. Tamir, Berlin, New York, NY: Springer-Verlag, 1982.
6. Hunsperger, R. G. *Integrated Optics: Theory and Technology.* 2nd ed. Berlin, New York, NY: Springer-Verlag, 1984.
7. Nishihara, N.; M. Harana; and T. Suhara. *Optical Integrated Circuits.* New York, NY: McGraw-Hill Book Company, 1989.
8. Kogelnik, H.; and V. Ramaswamy. "Scaling rules for thin film optical waveguides." *Appl. Opt.* 13, (1974), pp. 1857–1862.
9. Bennett, G. A.; and C. L. Chen. "Wavelength dispersion of optical waveguides." *Appl. Opt.* 19, (1980), pp. 1990–1995.
10. Ghatak, A. *Optics.* New York, NY: McGraw-Hill Book Company, 1977.
11. Ulrich, R.; and R. Torge. "Measurement of thin-film parameters with a prism coupler." *Appl. Opt.* 12, (1973), pp. 2901–2908.
12. Yariv, A. *Introduction to Optical Electronics.* 4th ed. Philadelphia, PA: Saunders College Pub., 1991.
13. Ihaya, A.; H. Furuta; and H. Noda. "Directional coupling between thin film optical guides." *Fujitsu Scien. and Tech. J.* 9, no. 2, (1973), pp. 101–119.

ADDITIONAL READING

1. Barnoski, M. K. *Fundamentals of Optical Fiber Communications.* 2nd ed. New York, NY: Academic Press, 1981.
2. Cheo, P. K. *Fiber Optics & Optoelectronics.* 2nd ed. Englewood Cliffs, NJ: Prentice Hall, Inc., 1990.
3. Hayt, W. H., Jr. *Engineering Electromagnetics.* 5th ed. New York, NY: McGraw-Hill Book Company, 1988.
4. Jones, W. B., Jr. *Introduction to Optical Fiber Communication Systems.* New York, NY: Holt, Rinehart and Winston, 1988.
5. Palais, J. C. *Fiber Optic Communications.* 2nd ed. Englewood Cliffs, NJ: Prentice Hall, Inc., 1988.
6. Saleh, B. E. A.; and M. C. Teich. *Fundamentals of Photonics.* New York, NY: John Wiley & Sons, Inc., 1991.
7. Suematsu, Y.; and K. I. Iga. *Introduction to Optical Fiber Communications.* New York, NY: John Wiley & Sons, 1982.
8. Wilson, J.; and J. F. B. Hawkes. *Optoelectronics, An Introduction.* 2nd ed. Englewood Cliffs, NJ: Prentice Hall, Inc., 1989.

PROBLEMS

1. Show that equation (7.2) is also valid for rays with $\theta < \pi/4$, which are shown in Figure 7.2b.

2. How many TE modes are supported by a thin-film optical waveguide with $n_c = 1.000$, $n_f = 1.540$, $n_s = 1.500$, and $h = 2.5$ µm, at $\lambda = 0.633$ µm? Estimate the effective refractive index, and the phase and group velocities of the TE_0 and TE_1 modes.

3. Consider a dielectric thin-film waveguide with $n_f = 1.500$, $n_s = 1.450$, and $n_c = 1.330$. The film thickness and the operational wavelength are h and λ, respectively.
   a. Given $h/\lambda = 2.5$, estimate the effective index of refraction, phase velocity, and group velocity of the second TE mode guided by the structure.
   b. What is the maximum h/λ if there is to be one and only one mode?

4. Consider a symmetric waveguide with a film index n_f, a cover and substrate index n_c, and a film thickness h. Calculate the numerical aperture of the waveguide if it is immersed in a medium with an index of n_1.

5. Consider a thin-film waveguide with an unspecified thickness h. The indices of various regions are $n_f = 1.60000$, and $n_s = n_c = 1.59000$.
   a. If the group velocity of TE_0 mode is $0.623047c$, where c is the speed of light in vacuum, estimate h/λ.
   b. Estimate the effective index of the TE_0 mode guided by the waveguide with h/λ given in (a).

6. Consider a thin-film waveguide with film, substrate, and cover indices n_f, n_s, and n_c, and a film thickness h. Assume that $n_s = n_c$ and that the index difference $n_f - n_s$ is much smaller than n_f and n_s.
   a. If the effective refractive index of the TE_0 mode is $(n_f + n_s)/2$, estimate h/λ.
   b. Estimate the group velocity of the TE_0 mode.
   c. Is the waveguide with h/λ given in (a) single-moded?

7. A glass prism with $n_p = 1.750$ and $\theta_p = 50°$ is used to couple light ($\lambda = 0.633$ µm) into a thin-film waveguide. The waveguide is formed by depositing a 4-µm film of index 1.500 on a substrate of index 1.483. Calculate the incident angles required to excite all TE modes supported by the waveguide.

8. A dielectric film of unknown thickness and index is deposited on a glass substrate. At 0.633 µm, the glass index is 1.4830. Measurements show that the effective indices of TE_0, and TE_1 modes are 1.4984 and 1.4936, respectively. Estimate the film index and thickness.

9. Derive (7.25).
10. Derive (7.47)–(7.49).
11. Starting from Maxwell's equations, derive the characteristic equation for TM modes guided by thin-film waveguides.
12. Derive (7.53)–(7.55).
13. Starting from (7.83) and (7.84), derive (7.86) and (7.87).

OPTICAL FIBERS

CHAPTER 8

CHAPTER OUTLINE

8.1 Introduction
8.2 Simple Characteristics of Step Index Fibers
    8.2.1 Rays and Numerical Aperture
    8.2.2 Traditional Mode Designation
    8.2.3 Generalized Parameters
8.3 Linearly Polarized Modes Guided by Weakly Guided Step Index Fibers
    8.3.1 Basic Field Properties
    8.3.2 Boundary Conditions
    8.3.3 Characteristic Equation
    8.3.4 Fields of Linearly Polarized Modes
    8.3.5 Mode Designation
    8.3.6 Single-Mode Operation
    8.3.7 Principal Mode Number
    8.3.8 Total Number of Guided Modes
8.4 Information Capacity
    8.4.1 Intermodal Dispersion
    8.4.2 Intramodal Dispersion
    8.4.3 Zero-Dispersion Wavelength
8.5 Fiber Fabrication Processes
    8.5.1 Outside Vapor Deposition
    8.5.2 Vapor Phase Axial Deposition
    8.5.3 Modified Chemical Vapor Deposition
    8.5.4 Plasma Modified Chemical Vapor Deposition
8.6 Fiber Losses
    8.6.1 Absorption
    8.6.2 Intrinsic Scattering
    8.6.3 Waveguide Imperfections
References
Additional Reading
Problems

8.1 INTRODUCTION

Waveguiding structures are needed at optical frequencies because of intrinsic atmospheric absorption and the effects of atmospheric disturbance on the propagation of visible and near IR radiation (Figure 8.1). Since all electrical conductors are lossy at optical frequencies, low-loss dielectric materials are used instead for optical waveguiding applications. Many material combinations can be, and have been, used to form optical fibers (Figure 8.2). For example, silicate-based fibers are used extensively in communication applications in the visible and near IR wavelength regions. On the other hand, plastic fibers based on polymethyl methacrylate (PMMA) are useful only in a limited spectral range. Fluoride and chloride glasses are transparent in the infrared regions, but they are toxic and/or hygroscopic and must be handled very carefully. Chalcogenide glasses are potentially useful for longer wavelengths, but they are still in the research and development stage. In this chapter, we will concentrate on silicate-based glass fibers.

Basically, an optical fiber consists of a **core region** and a **cladding region.** Although there may be one or more layers of protective coating outside the cladding, their effects on the optical properties of the fiber are minimal. The

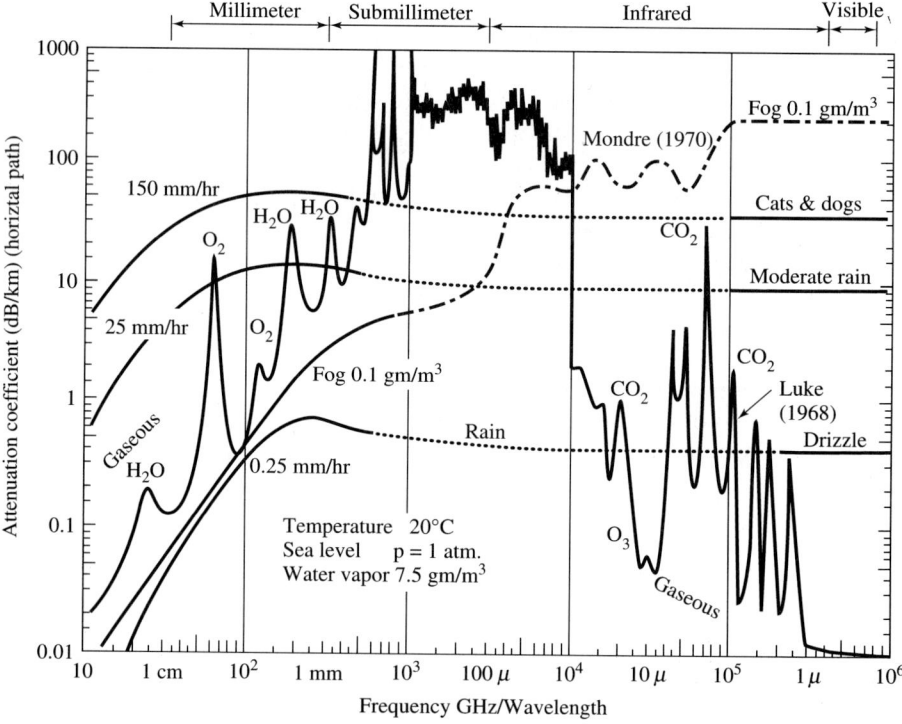

Figure 8.1 Effects of gaseous constituents and precipitation on the transmission of electromagnetic waves through the atmosphere [1]

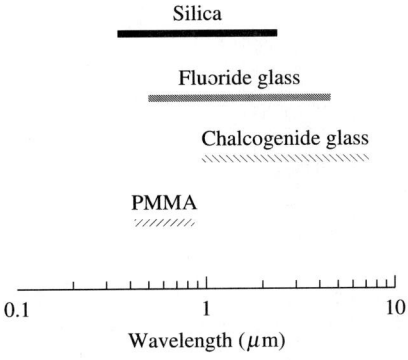

Figure 8.2 Transparent regions of four transparent materials with loss less than 1000 dB/km [2]

main functions of the coating material are: (1) to protect the core and cladding against mechanical damage; (2) to provide mechanical strength; and (3) to seal out moisture or other contaminations. The transmission characteristics of optical fibers depend on the size, shape, and the index profile of the core and cladding. Although a fiber core can be of any shape, most fibers of current interest have a nominally circular core, as shown in Figure 8.3a. Special-purpose fibers may have an elliptical core (Figure 8.3b), cladding, or both. Figure 8.3c shows several polarization-maintaining fibers with various noncircular cross sections. There are also fibers with two or more cores. Only fibers with a circular core are discussed in this chapter.

In all fibers, fields are concentrated mainly in the core region, and in the cladding region near the core/cladding boundary. Since the fields are extremely weak near the outer boundary of the cladding region, we treat the cladding as infinitely large, for simplicity and as an approximation. We also assume that the fibers are infinitely long, and we consider only time-harmonic waves ($e^{j\omega t}$) that propagate in the axial z direction.

Depending on the variation of the core index, a fiber may have a **step index** or a **graded index profile**. In step index fibers (Figure 8.4a), the core and cladding indices n_{co} and n_{cl} are constants, and the **index difference** Δ is

$$\Delta = \frac{n_{co} - n_{cl}}{n_{co}} \tag{8.1}$$

For graded index fibers, the core index profile (Figure 8.4b) is specified by the index difference Δ and the **index profile parameter** g. The index in the core region is then

$$n_{co}(r) = n_{co}\left[1 - 2\Delta\left(\frac{r}{a}\right)^g\right]^{1/2} \tag{8.2}$$

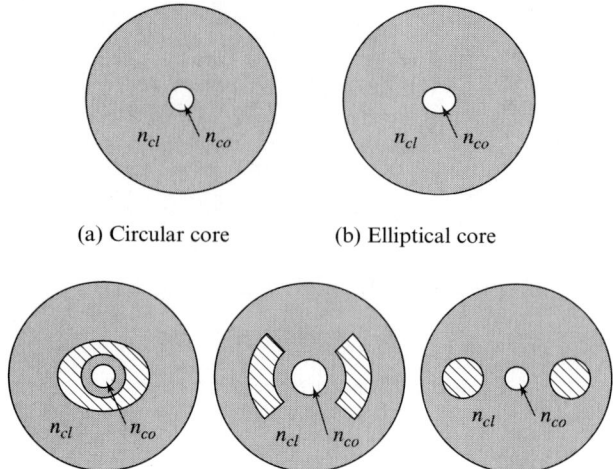

(a) Circular core   (b) Elliptical core

(c) Polarization maintaining fibers with various stress applying regions

Figure 8.3 Fibers with various cross sections

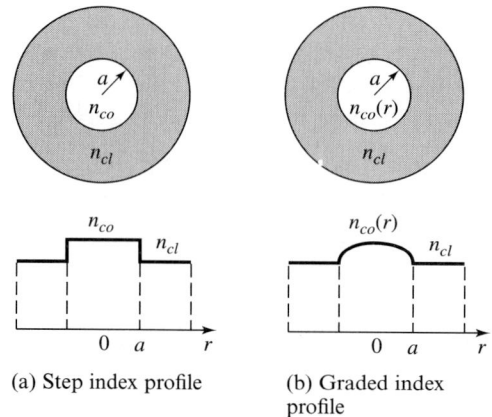

(a) Step index profile   (b) Graded index profile

Figure 8.4 Fibers with various index profiles

where $r \leq a$ and a is the core radius. In (8.2), r is the radial distance from the fiber axis. The cladding index is roughly uniform throughout the cladding region and is given by $n_{cl} = n_{co}[1 - 2\Delta]^{1/2} \approx 1 - \Delta$. We may view the step index profile as a special case of the graded index profile with $g \to \infty$. As noted in section 8.4 and Appendix D, the fiber bandwidth is particularly broad if the fiber has a *parabolic* or approximately parabolic index profile, that is, if $g = 2$.

In general, the optical characteristics of fibers are very complicated, unless the index difference is very small. Fortunately, for most fibers of practical interest, Δ is typically only a few percent or less. Fibers having a small Δ and a core

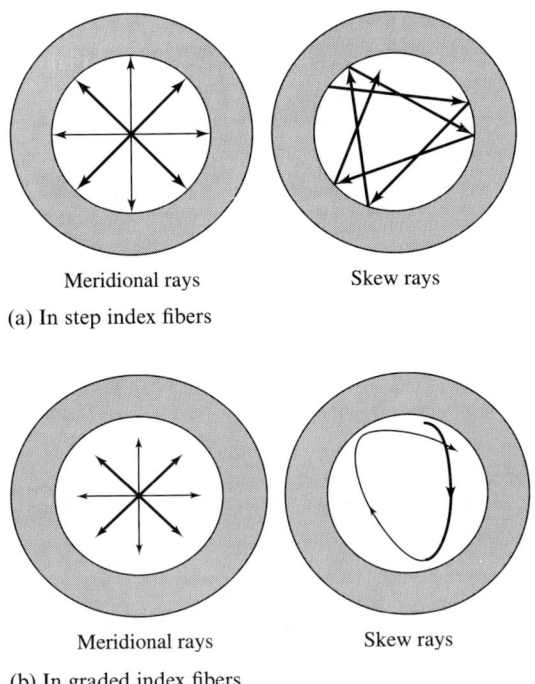

Figure 8.5 Meridional and skew rays in step index and graded index fibers

much larger than the wavelength are commonly referred to as **weakly guided fibers.**

8.2 SIMPLE CHARACTERISTICS OF STEP INDEX FIBERS

8.2.1 Rays and Numerical Aperture

The concept of rays, introduced in Chapters 2 and 7, is useful in describing waves in fibers that support multiple modes. The waves bouncing back and forth in *thin-film waveguides* were meridional rays. In optical fibers, both meridional and skew rays are involved. The rays that cross the optical axis are the **meridional rays** and those that wander in a fiber but never actually cross the optical axis are the **skew rays.** In step index fibers, rays bend abruptly at the core/cladding boundary (Figure 8.5a). In graded index fibers, rays bend smoothly and continuously and never reach the core/cladding boundaries (Figure 8.5b). This is analogous to the curved ray trajectories in graded index thin-film waveguides.

As mentioned in Chapter 7, the numerical aperture (NA) is a useful parameter in quantifying the ability of optical components, including optical fibers, to collect incoming light and radiate outgoing beams. The numerical aperture of

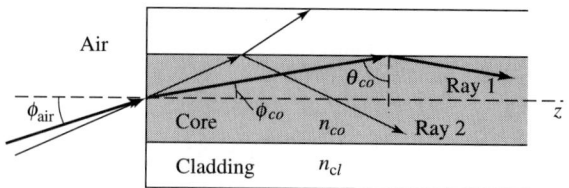

Figure 8.6 Numerical aperture of step index fibers

step index fibers can be studied in the same manner as that of the *NA* of thin-film waveguides. The numerical aperture related to the capture of meridional rays (Figure 8.6) is

$$NA = \sin\phi_{air\,max} = \sqrt{n_{co}^2 - n_{cl}^2} \qquad (8.3)$$

This equation is precisely the same as (7.61), with n_f and n_s replaced by n_{co} and n_{cl}. For weakly guided fibers, the expression can be further simplified to

$$NA = \sqrt{n_{co}^2 - n_{cl}^2} \approx n_{co}\sqrt{2\Delta} \qquad (8.4)$$

This derivation assumes that the medium outside the fiber is vacuum. If the outside medium has an index n_{os}, we have, in lieu of (8.4),

$$NA = \frac{1}{n_{os}}\sqrt{n_{co}^2 - n_{cl}^2} \approx \frac{n_{co}}{n_{os}}\sqrt{2\Delta} \qquad (8.5)$$

8.2.2 Traditional Mode Designation

In metallic transmission lines and waveguides, the fields are classified as **transverse electromagnetic (TEM), transverse electric (TE),** or **transverse magnetic (TM)** modes [3,4]. For TE modes, E_z vanishes; for TM modes, H_z is zero; and for TEM modes, E_z and H_z are both zero. Waves guided by two dimensional thin-film waveguides can also be grouped as TE and TM modes. This is because in a two-dimensional waveguide, the waveguide geometry extends *indefinitely* in one of two transverse directions. In contrast, for optical fibers, the two transverse dimensions are *finite,* and the question of field classification becomes quite complicated. In addition to TE and TM modes, modes exist for which neither E_z nor H_z vanishes. These modes are labeled **hybrid modes.** Since the designation for hybrid modes is somewhat arbitrary, we will follow the practice commonly used in microwave technology. If H_z and E_z have the *same* signs far from the cutoff, the hybrid mode is labeled an **HE mode.** If H_z and E_z have *opposite* signs far from cutoff, we have an **EH mode** [5,6].

Each mode has a specific field distribution and propagation constant, which are dependent on the operating wavelength λ, the core radius, and the index profile. These characteristics are determined by Maxwell equations. It is often convenient to express the propagation constant β in terms of an effective refractive index N. For example, if we write $\beta = kN$, then the N of guided modes in fibers is in the range of $n_{cl} < N < n_{co}$. To describe the field distributions, we use two integers, ℓ and m. The first integer, the **azimuthal mode number** ℓ, signifies the azimuthal variation of the fields. The second integer, the **radial mode number,** m, specifies the field variations in the radial direction. Thus, for step index fibers with circular cross sections, the field distributions can be identified as the $TE_{\ell m}$, $TM_{\ell m}$, $EH_{\ell m}$, and $HE_{\ell m}$ modes, or the linear combination of these modes.

8.2.3 Generalized Parameters

The generalized parameters V and b introduced in Chapter 7 are also useful in characterizing modes propagating in fibers. For step index fibers, the **generalized frequency** V and **generalized guide index** b are

$$V = \frac{2\pi a}{\lambda} \sqrt{n_{co}^2 - n_{cl}^2} \tag{8.6}$$

$$b = \frac{N^2 - n_{cl}^2}{n_{co}^2 - n_{cl}^2} \tag{8.7}$$

Using the definition of NA (8.4), we see that the generalized frequency V is the product of a **geometric factor,** $2\pi a/\lambda$, and an **optical factor** (NA).

In terms of b, V, and Δ, the exact characteristic equation for fibers with a **perfectly circular core** and a **step index profile** may be written as [5–12]

$$\left[\frac{1}{V\sqrt{1-b}} \frac{J_\ell'(V\sqrt{1-b})}{J_\ell(V\sqrt{1-b})} + \frac{(1-\Delta)^2}{V\sqrt{b}} \frac{K_\ell'(V\sqrt{b})}{K_\ell(V\sqrt{b})} \right] \tag{8.8}$$

$$\left[\frac{1}{V\sqrt{1-b}} \frac{J_\ell'(V\sqrt{1-b})}{J_\ell(V\sqrt{1-b})} + \frac{1}{V\sqrt{b}} \frac{K_\ell'(V\sqrt{b})}{K_\ell(V\sqrt{b})} \right] = \ell^2 \frac{b + (1-\Delta)^2(1-b)}{(V^2 b(1-b))^2}$$

where $J_\ell(z)$ and $K_\ell(z)$ are Bessel and modified Bessel functions of order ℓ, and ' indicates the differentiation with respect to the argument of the Bessel and modified Bessel functions. A brief introduction to the Bessel and modified Bessel functions can be found in Appendix C. Detailed derivations for (8.8) are omitted here.

Figure 8.7 depicts bV plots for the seven lowest-order modes guided by fibers with $\Delta = 0.2$. However, it should be noted that, for most fibers, Δ is rather small and many curves merge together. In plotting Figure 8.7, we chose a large Δ, primarily to keep the curves apart.

The HE_{11} mode is the **dominant** or **fundamental mode** for fibers with a circular cross section. For $V < 2.4048$, all modes, except the HE_{11} mode are cutoff modes. Therefore, fibers having circular cross sections are **single-mode fibers** if V is less than 2.4048. As depicted in one of the inserts of Figure 8.7, the transverse electric fields associated with the HE_{11} mode are approximately linearly polarized. There are two possible **linearly independent polarizations:** The transverse fields of HE_{11} modes may be nearly polarized in the x direction, or the y direction.

For step index fibers, the next three modes are the TE_{01}, TM_{01}, and HE_{21} modes. The cut-offs for TE_{01} and TM_{01} are $V = 2.4048$. The cutoff value of HE_{21}, while near 2.4048, is dependent on Δ. As shown in Figure 8.7, the bV

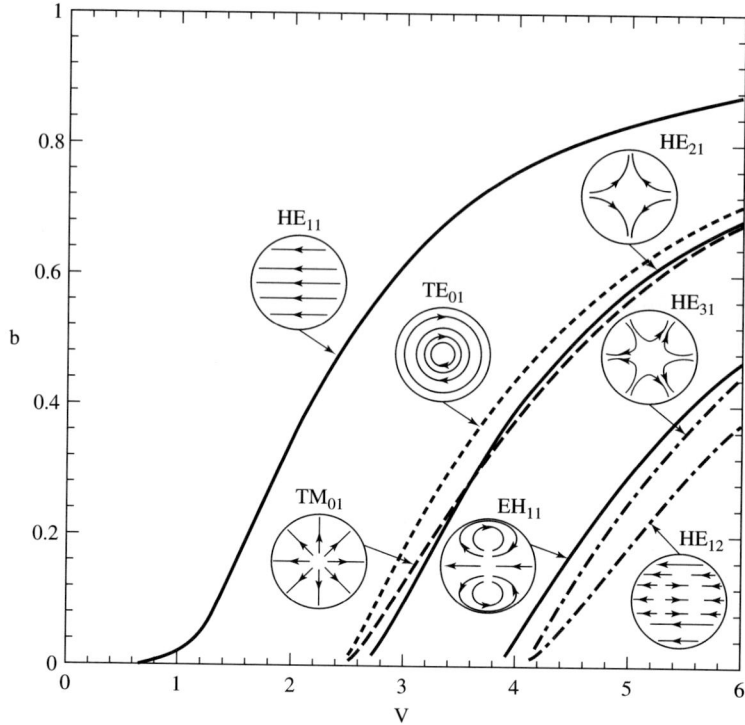

Figure 8.7 Dispersion of a fiber with $\Delta = 0.2$ and electric field lines of various modes

curves of the TE_{01}, TM_{01}, and HE_{21} modes are nearly the same, and they become indistinguishable if Δ is small. Individually, these modes are not linearly polarized; however, they can be *combined* to give *nearly* linearly polarized fields. Figure 8.8 depicts linearly polarized fields formed by superimposing the fields of the TM_{01} and HE_{21} modes, or the TE_{01} and HE_{21} modes.

For $3.8317 < V\ 5.1356$, three additional modes come into being: HE_{12}, EH_{11}, and HE_{31}. The cutoffs of the HE_{12} and HE_{31} modes are dependent on Δ, while that of the EH_{11} mode is independent of Δ. In weakly guided fibers, these modes become degenerate, and when properly combined, the superimposed fields also become linearly polarized. In summary, either the modes of weakly guided fibers are approximately **linearly polarized,** or several nearly degenerate modes may be combined to form approximately linearly polarized fields.

8.3 LINEARLY POLARIZED MODES GUIDED BY WEAKLY GUIDED STEP INDEX FIBERS

In this section, we will develop the characteristic equation and the fields of linearly polarized (LP) modes guided by weakly guided fibers. The development closely follows that originally presented by Gloge [13]. For linearly polarized

8.3 Linearly Polarized Modes Guided by Weakly Guided Step Index Fibers

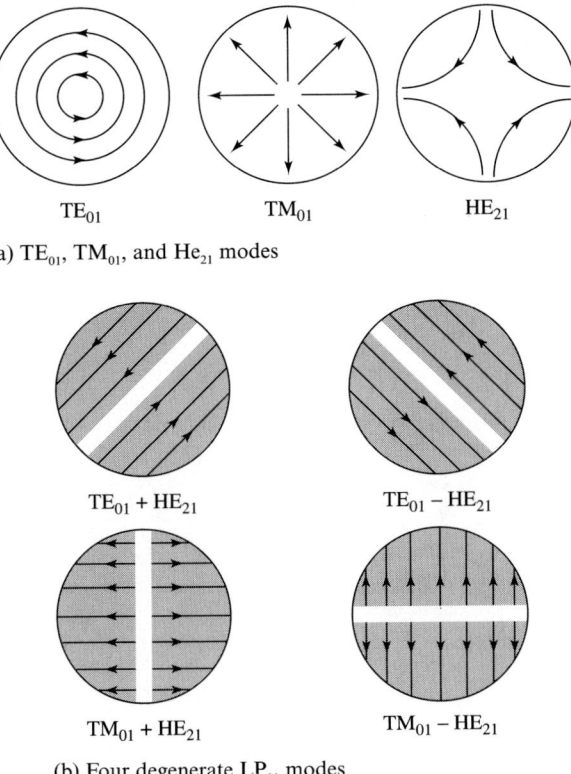

(a) TE_{01}, TM_{01}, and He_{21} modes

(b) Four degenerate LP_{11} modes

Figure 8.8 Superposition of TE_{01}, TM_{01}, and HE_{21} modes to form LP_{11} modes [5]

modes, one of the transverse field components is much stronger than the orthogonal transverse field component and the longitudinal field component. When the weak transverse field component is ignored, a complicated *vector* field problem is reduce to a *scalar* field problem. Also, the magnetic fields and the longitudinal electric field component can be expressed in terms of the strong transverse electric field. When the transverse electric field component is determined, all other field components of LP modes are determined.

8.3.1 Basic Field Properties

The formal study of the fields and characteristic equation of modes guided by weakly guided fibers begins with the time-harmonic ($e^{j\omega t}$) Maxwell equations. Since the fiber geometry and the refractive index profile are independent of z, and since we are interested in waves propagating in the z direction, it follows that all field components will vary as $e^{-j\beta z}$, and where β is the propagation constant. Following the procedure used in the last chapter to study fields guided by thin-film waveguides, we express the fields and the ∇ operator in terms of their transverse and longitudinal parts,

$$E(x,y,z) = [\mathbf{e}_t(x,y) + \mathbf{a}_z e_z(x,y)]e^{-j\beta z}$$
$$H(x,y,z) = [\mathbf{h}_t(x,y) + \mathbf{a}_z h_z(x,y)]e^{-j\beta z}$$
$$\nabla = \nabla_t - j\beta \mathbf{a}_z = [\mathbf{a}_x \frac{\partial}{\partial x} + \mathbf{a}_y \frac{\partial}{\partial y}] - j\beta \mathbf{a}_z$$

where the subscript t denotes components transverse to the z direction. Then, Maxwell equations for isotropic nonmagnetic media, given in (7.26)–(7.29), become

$$\nabla_t \times \mathbf{e}_t = -j\omega\mu_o h_z \mathbf{a}_z \tag{8.9}$$
$$\mathbf{a}_z \times (j\beta \mathbf{e}_t + \nabla_t e_z) = j\omega\mu_o \mathbf{h}_t \tag{8.10}$$
$$\nabla_t \times \mathbf{h}_t = j\omega\varepsilon_o n^2 e_z \mathbf{a}_z \tag{8.11}$$
$$\mathbf{a}_z \times (j\beta \mathbf{h}_t + \nabla_t h_z) = -j\omega\varepsilon_o n^2 \mathbf{e}_t \tag{8.12}$$
$$j\beta e_z = \nabla_t \cdot \mathbf{e}_t \tag{8.13}$$
$$j\beta h_z = \nabla_t \cdot \mathbf{h}_t \tag{8.14}$$

By manipulating (8.9), (8.12), and (8.13), we obtain the *wave equation* for $\mathbf{e}_t$,

$$\nabla_t^2 \mathbf{e}_t + (k^2 n^2 - \beta^2)\mathbf{e}_t = 0 \tag{8.15}$$

where $k^2 = \omega\mu_o\varepsilon_o$ and n stands for the index n_{co} or n_{cl} of the *core* or *cladding* regions, respectively. A similar equation for $\mathbf{h}_t$ can be obtained from (8.10), (8.11), and (8.14).

Fibers are considered to be "weakly guiding" if $\Delta \to 0$ while the generalized frequency V remains finite. The condition is met if $n_{cl} \approx n_{co}$ and the core radius is large compared to λ. Under these circumstances, the transverse field components should be much stronger than the longitudinal components, and the variation in the transverse direction should be much smaller than the change in the longitudinal direction. Thus,

$$|\nabla_t e_z| \ll |\beta \mathbf{e}_t| \quad \text{and} \quad |\nabla_t h_z| \ll |\beta \mathbf{h}_t|$$

Then, from (8.10) and (8.12), we have

$$\mathbf{h}_t \approx \frac{\beta}{\omega\mu_o} \mathbf{a}_z \times \mathbf{e}_t \tag{8.16}$$

$$\mathbf{h}_t \approx \frac{\omega\varepsilon_o n^2}{\beta} \mathbf{a}_z \times \mathbf{e}_t \tag{8.17}$$

We see that (8.16) has a factor $\beta/(\omega\mu_o)$, and (8.17) has a factor $\omega\varepsilon_o n^2/\beta$. However, if these equations are to hold simultaneously, the two factors must be the same. Therefore, $\beta \approx kn$, and (8.16) and (8.17) become

$$\mathbf{h}_t \approx \frac{n}{\eta_o} \mathbf{a}_z \times \mathbf{e}_t \tag{8.18}$$

where $\eta_o = \sqrt{\mu_o/\varepsilon_o}$ is the **intrinsic impedance of free space**.

In view of (8.18), we note that the transverse field components $\mathbf{e}_t$ and $\mathbf{h}_t$ of weakly guided fibers are related in the same manner as the electric and magnetic fields of uniform plane waves in isotropic nonmagnetic media [12]. As we know, uniform plane waves are *transverse electromagnetic waves*. Further analysis shows that (8.18) is accurate to the order of Δ. This is a basic property of fields of weakly guided fibers.

8.3.2 Boundary Conditions

It is clear from (8.13), (8.14), and (8.16) that e_z, $\mathbf{h}_t$, and h_z can be expressed in terms of $\mathbf{e}_t$. The next task is to determine $\mathbf{e}_t$ from (8.15) and the boundary conditions. There are two linearly independent and mutually orthogonal solutions each of which corresponds to a polarized field. The two polarized fields are independent of and orthogonal to each other. For convenience, the two orthogonal polarizations are labeled the *x*-polarized and *y*-polarized electric fields. In the following, we will focus on *x*-polarized electric fields only.

At the core/cladding interface, the tangential components of the electric and magnetic field intensity and the normal components of the electric and magnetic flux density are continuous. When the continuity for the tangential components of the field intensity vectors are satisfied, the continuity for the normal components of the flux density vectors are *automatically* satisfied [3, 4].

Without loss of generality, we express $\mathbf{e}_t$ in terms of a *Fourier cosine series*. Since $\sin\ell\phi$ and $\cos\ell\phi$ are orthogonal functions in the range $(0, 2\pi)$, the continuation of the total tangential field components amounts to the continuation of the individual Fourier components of the fields. Therefore, all boundary conditions can be satisfied with one Fourier component each for each region. In short, we need only consider each Fourier component individually.

Consider a single cosine term of the Fourier series. For the core region and the cladding region, respectively, we write,

$$\mathbf{e}_t(r,\phi) = e_{co}(r) \cos\ell\phi \, \mathbf{a}_x \qquad (8.19)$$

and

$$\mathbf{e}_t(r,\phi) = e_{cl}(r) \cos\ell\phi \, \mathbf{a}_x \qquad (8.20)$$

where $e_{co}(r)$ and $e_{cl}(r)$ are the fields in the core and cladding regions respectively. They are yet to be determined.

In terms of $e_{co}(r)$, the other field components in the core region are

$$e_z(r,\phi) = -\frac{j}{\beta}\left[\frac{de_{co}}{dr}\cos\ell\phi\cos\phi + \frac{e_{co}}{r}\ell\sin\ell\phi\sin\phi\right] \qquad (8.21)$$

$$\mathbf{h}_t(r,\phi) = \frac{\beta}{\omega\mu_o} e_{co} \cos\ell\phi \, \mathbf{a}_y \qquad (8.22)$$

$$h_z(r,\phi) = -\frac{j}{\omega\mu_o}\left[\frac{de_{co}}{dr}\cos\ell\phi\sin\phi - \frac{e_{co}}{r}\ell\sin\ell\phi\cos\phi\right] \qquad (8.23)$$

Corresponding expressions can be written for the field components in the cladding region.

The continuation of e_ϕ and e_z at $r = a$ requires that

$$e_{co}\Big|_{r=a} = e_{cl}\Big|_{r=a} \tag{8.24}$$

and

$$\frac{de_{co}}{dr}\Big|_{r=a} = \frac{de_{cl}}{dr}\Big|_{r=a} \tag{8.25}$$

When these conditions are satisfied, h_ϕ and h_z are also continuous. Thus, all boundary conditions are satisfied.

8.3.3 Characteristic Equation

To determine $e_{co}(r)$ and $e_{cl}(r)$, substitute (8.19) and (8.20) into the wave equation (8.15). For the core region, a differential equation for $e_{co}(r)$ is obtained:

$$\frac{1}{r}\frac{d}{dr}\left(r\frac{de_{co}}{dr}\right) + \left(k^2 n_{co}^2 - \beta^2 - \frac{\ell^2}{r^2}\right) e_{co} = 0 \tag{8.26}$$

It is convenient to cast $(k^2 n_{co}^2 - \beta^2)$ in terms of the normalized parameters defined in (8.6) and (8.7), as follows:

$$k^2 n_{co}^2 - \beta^2 = k^2(n_{co}^2 - N^2) = k^2(n_{co}^2 - n_{cl}^2)\left(1 - \frac{N^2 - n_{cl}^2}{n_{co}^2 - n_{cl}^2}\right) = \frac{V^2(1 - b)}{a^2}$$

Then, (8.26) becomes

$$\frac{1}{r}\frac{d}{dr}\left(r\frac{de_{co}}{dr}\right) + \left(\frac{V^2(1-b)}{a^2} - \frac{\ell^2}{r^2}\right) e_{co} = 0$$

This is a *Bessel differential equation*. As noted in Appendix C, a Bessel differential equation has two linearly independent solutions: the *Bessel functions* of the first and second kind of order ℓ, J_ℓ and Y_ℓ. However, $Y_\ell(V\sqrt{1-b}\,\frac{r}{a})$ is not acceptable for the present problem, since it is singular at $r = 0$. Therefore, $e_{co}(r)$ can only be expressed in terms of the Bessel function of the first kind,

$$e_{co}(r) = C \frac{J_\ell(V\sqrt{1-b}\,\frac{r}{a})}{J_\ell(V\sqrt{1-b})} \tag{8.27}$$

where C is a constant yet to be determined.

For the cladding region, there is a differential equation identical to (8.26), with $e_{co}(r)$ substituted by $e_{cl}(r)$, and $(k^2 n_{co}^2 - \beta^2)$ becoming

$$k^2 n_{cl}^2 - \beta^2 = k^2(n_{cl}^2 - N^2) = -k^2(n_{co}^2 - n_{cl}^2)\left(\frac{N^2 - n_{cl}^2}{n_{co}^2 - n_{cl}^2}\right) = -V^2 \frac{b}{a^2}$$

Note in particular that, for guided modes, $(k^2 n_{cl}^2 - \beta^2)$ is negative. The differential equation for $e_{cl}(r)$ is a *modified Bessel differential equation,* and $e_{cl}(r)$ decays as r increases. A modified Bessel differential equation admits two linearly independent solutions: I_ℓ and K_ℓ. As the function argument increases, I_ℓ increases and K_ℓ decreases exponentially. Thus, $e_{cl}(r)$ may be expressed in terms of the modified Bessel function K_ℓ

$$e_{cl}(r) = C' \frac{K_\ell(V\sqrt{b}\frac{r}{a})}{K_\ell(V\sqrt{b})} \tag{8.28}$$

and C' is also an unknown constant. From the boundary conditions (8.24) and (8.25), we have

$$C = C'$$

and

$$CV\sqrt{1-b}\,\frac{J'_\ell(V\sqrt{1-b})}{J_\ell(V\sqrt{1-b})} = C'V\sqrt{b}\,\frac{K'_\ell(V\sqrt{b})}{K_\ell(V\sqrt{b})}$$

Combining these equations, we have the characteristic equation

$$V\sqrt{1-b}\,\frac{J'_\ell(V\sqrt{1-b})}{J_\ell(V\sqrt{1-b})} - V\sqrt{b}\,\frac{K'_\ell(V\sqrt{b})}{K_\ell(V\sqrt{b})} = 0$$

Using the recurrence relation of Bessel and modified Bessel functions [(C.3)–(C.6)], we can rewrite the previous equation as

$$V\sqrt{1-b}\,\frac{J_{\ell-1}(V\sqrt{1-b})}{J_\ell(V\sqrt{1-b})} + V\sqrt{b}\,\frac{K_{\ell-1}(V\sqrt{b})}{K_\ell(V\sqrt{b})} = 0 \tag{8.29}$$

This is the *characteristic equation for weakly guided fibers.* Equation (8.29) can also be obtained directly from (8.9) by letting $\Delta \to 0$ while keeping V finite.

To determine the propagation constant and the fields, it is necessary to solve for b for a given V and ℓ. It would not be surprising to note that no analytical expression for b is known and we have to resort to numerical techniques to evaluate b. To illustrate the methods involved, we plot, as an example, the left-hand side of (8.29), with $V = 6.0$ and $\ell = 0$, as a function of b (Figure 8.9). We can see that there are two roots. The root with the largest b is labeled the first root; that is, for $V = 6.0$ and $\ell = 0$, the first root is $b = 0.883$, and the second root is $b = 0.403$. For different values of V and ℓ, the roots are different.

Each root corresponds to a linearly polarized mode. These linearly polarized modes are labeled as the *LP modes* by Gloge [13], to distinguish them from the *traditional mode designation* discussed in the last section. The bV relationships of various LP modes of weakly guided fibers are shown in Figure 8.10 [13]. They are qualitatively similar to the dispersion curves shown in Figure 8.7. However, a crucial difference exists. Under the *"weakly guiding"* approximation, bV curves are rather insensitive to changes in Δ. On the other hand, the curves depicted in Figure 8.7 are for a specific value of Δ, that is, $\Delta = 0.2$.

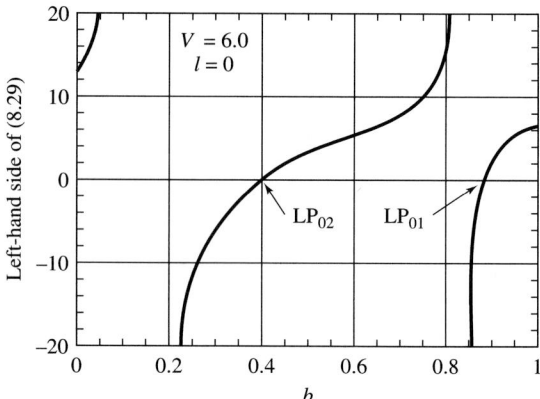

Figure 8.9 Plot of the left hand side of (8.29) as a function of b, with $V = 6.0$ and $\ell = 0$.

8.3.4 Fields of Linearly Polarized Modes

Once b is determined, we have an expression for $\mathbf{e}_t$ from (8.19), (8.20), (8.27), and (8.28). When $\mathbf{e}_t$ is known, other field components are readily deduced, from (8.14)–(8.16). For example, the fields of LP modes with $\ell = 0$ are

$$E_x(r,\phi,z) = \begin{cases} \dfrac{\eta_o}{n_{co}} H_y(r,\phi,z) \\[6pt] \dfrac{\eta_o}{n_{cl}} H_y(r,\phi,z) \end{cases} = E_0 e^{-jkNz} \begin{cases} \dfrac{J_0(V\sqrt{1-b}\, r/a)}{J_0(V\sqrt{1-b})} & 0 \le r \le a \\[8pt] \dfrac{K_0(V\sqrt{b}\, r/a)}{K_0(V\sqrt{b})} & r \ge a \end{cases} \quad (8.30)$$

$$E_z(r,\phi,z) = \dfrac{jE_0}{kaN} e^{-jkNz} \cos\phi \begin{cases} \dfrac{V\sqrt{1-b}\, J_1(V\sqrt{1-b}\, r/a)}{J_0(V\sqrt{1-b})} & 0 \le r \le a \\[8pt] \dfrac{V\sqrt{b}\, K_1(V\sqrt{b}\, r/a)}{K_0(V\sqrt{b})} & r \ge a \end{cases} \quad (8.31)$$

$$H_z(r,\phi,z) = \dfrac{jE_0}{\eta_o ka} e^{-jkNz} \sin\phi \begin{cases} \dfrac{V\sqrt{1-b}\, J_1(V\sqrt{1-b}\, r/a)}{J_0(V\sqrt{1-b})} & 0 \le r \le a \\[8pt] \dfrac{V\sqrt{b}\, K_1(V\sqrt{b}\, r/a)}{K_0(V\sqrt{b})} & r \ge a \end{cases} \quad (8.32)$$

Following the same procedure, we can also derive expressions for fields of LP modes having $\ell \ne 0$ (see Problem 7). In these expressions, E_0 is an amplitude constant. All other terms are cast in terms of V and b. Recall that N can be determined from b, and $\beta = kN$.

8.3 Linearly Polarized Modes Guided by Weakly Guided Step Index Fibers

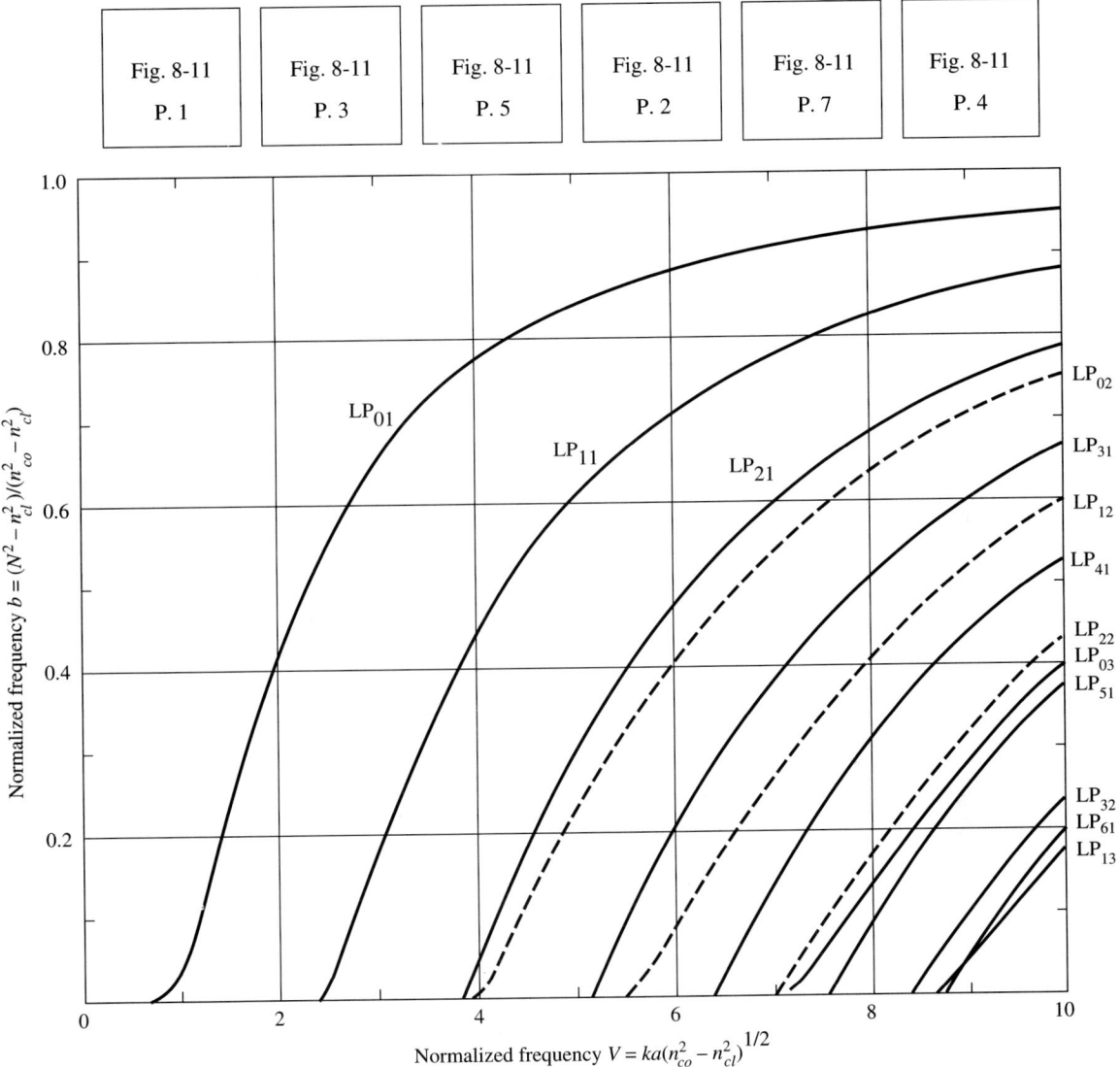

Figure 8.10 bV characteristics of weakly guided fibers [13]

Up to this point, we have only considered modes with transverse electric fields in the x direction. Modes with the transverse electric fields in the y direction can be examined in the same manner. The two states of polarization are linearly independent and mutually orthogonal. If the fiber is perfectly circular in all respects, the two states of polarization are degenerate. For fibers with *nominally* circular cross sections, the two states of polarization are *nearly* degenerate.

8.3.5 Mode Designation

Two mode numbers, the azimuthal mode number ℓ and the radial mode number m, are used to designate LP modes. Thus, linearly polarized modes are identified as $LP_{\ell m}$ modes. The subscript ℓ refers to the angular variation of the fields. Specifically, fields of $LP_{\ell m}$ modes vary as $\sin\ell\phi$ or $\cos\ell\phi$. Also, if ℓ is zero, the fields are independent of ϕ. Thus, the transverse electric fields of LP_{0m}, as given by (8.30), are independent of ϕ.

The subscript m, refers to the mth root of (8.29) for a given V and ℓ. As noted previously, the root with the largest b is the first root. In general, the *field distribution* $\mathbf{e}_t(r,\phi)$ of the $LP_{\ell m}$ mode has ℓ maxima and ℓ minima as a function of ϕ. The *intensity distribution* depicts $|\mathbf{e}_t(r,\phi)|^2$ as a function of r and ϕ, and has 2ℓ bright spots on a circle of constant r, as shown in Figure 8.11. The radial mode number m specifies the number of field maxima in the radial direction, including, in particular, the maximum at the origin for modes with $\ell = 0$.

In summary, there are two possible ways to designate modes in weakly guided fibers. They are: the *traditional mode designation,* and the *LP mode designation.* The two are equivalent, as established by Gloge [13]:

$$LP_{\ell m} \leftrightarrow HE_{\ell+1\,m},\, EH_{\ell-1\,m} \qquad \text{for } \ell \neq 1$$
$$LP_{1\,m} \leftrightarrow HE_{2\,m},\, TE_{0\,m},\, TM_{0\,m} \qquad \text{for } \ell = 1$$

In the LP mode designation, the dominant mode is the LP_{01} mode. In the traditional mode designation, the dominant mode is the HE_{11} mode. The first 20 modes are listed in Table 8.1.

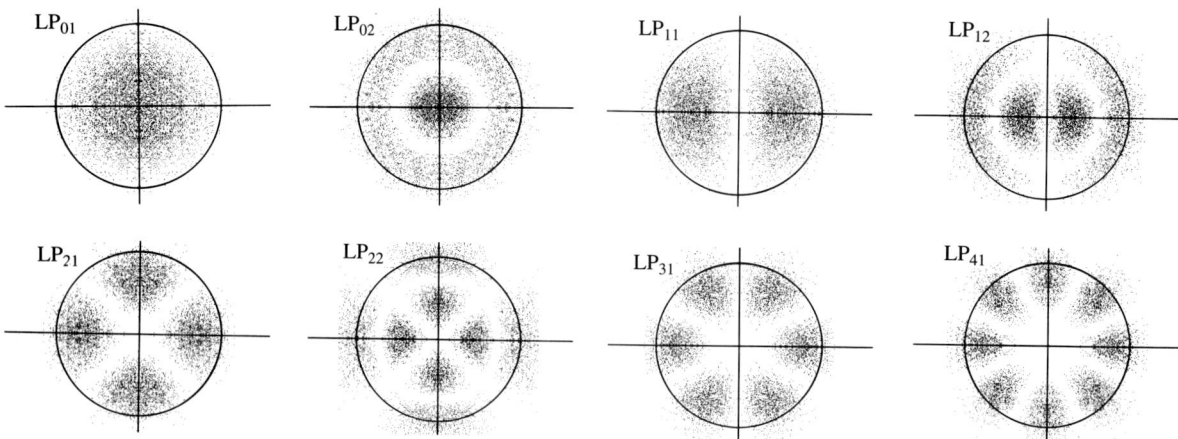

Figure 8.11 Calculated intensity distribution of LP modes guided by a step index fiber with $V = 7.1$. The density of dots is proportional to $|\mathbf{e}_t|^2$

Table 8.1 Traditional versus LP mode designation

Normalized frequency (V)	Traditional mode designation	LP mode designation	Additional no. of modes	Total no. of modes
0–2.4048	HE_{11}	LP_{01}	2	2
2.4048–3.8317	$TE_{01}, TM_{01}, HE_{21}$	LP_{11}	4	6
3.8317–5.1356	EH_{11}, HE_{31}	LP_{21}	4	10
	HE_{12}	LP_{02}	2	12
5.1356–5.5201	EH_{21}, HE_{41}	LP_{31}	4	16
5.5201–6.3802	$TE_{02}, TM_{02}, HE_{22}$	LP_{12}	4	20

[1]All hybrid modes are doubly-degenerate modes. The TE_{0m} and TM_{0m} modes are nondegenerate.
[2]The following relationships apply:
$LP_{\ell m} \leftrightarrow HE_{\ell+1m}, EH_{\ell-1m}$ for $\ell \neq 1$
$LP_{1m} \leftrightarrow HE_{2m}, TE_{0m}, TM_{0m}$ for $\ell = 1$

8.3.6 Single-Mode Operation

The information capacity of step index multimode fibers is rather limited. However, the information capacity can be greatly increased if the core radius is reduced such that one and only one mode is supported by the fiber. Fibers supporting one and only one mode are known as **single-mode** or **mononode** fibers.

As shown in Figure 8.10, the lowest-order mode, LP_{01}, has no cutoff. The second-lowest-order mode is LP_{11}. If LP_{11} mode is cut off, all higher-order modes are also cut off. To find the cutoffs of LP_{11} and the higher-order modes, we note that b tends to 0 as the cutoff condition is approached. Using the power series expansion of $K_\ell(z)$ [(C.9) and (C.10)], we can show that

$$\lim_{b \to 0} \left[V\sqrt{b} \, \frac{K_{\ell-1}(V\sqrt{b})}{K_\ell(V\sqrt{b})} \right] \to 0$$

for all values of ℓ. Then, (8.29) is simplified as

$$J_{\ell-1}(V) = 0 \qquad \text{for } \ell \geq 1 \tag{8.33}$$

when b approaches 0. Therefore, the cutoff condition for the LP_{11} mode is that V is a root of $J_0(V) = 0$. Examination of any mathematics table [14] will show that the first root of $J_0(V)$ is $V = 2.4048$. Therefore, for single mode-operation, V should be less than 2.4048. The same conclusion is also reached by examining Figure 8.10. From the definition of V, we have the following condition for the single-mode operation:

$$\frac{a}{\lambda} < \frac{2.4048}{2\pi n_{co}\sqrt{2\Delta}} = \frac{0.2706}{n_{co}\sqrt{\Delta}} \tag{8.34}$$

As a simple example, consider fibers with $n_{co} \approx 1.48$ operating at $\lambda \approx 1.3\mu m$. From (8.34) we can see that the fiber is single-moded if the core radius is less than 2.38 μm for $\Delta = 0.010$, or 3.36 μm for $\Delta = 0.005$, or 4.34 μm for $\Delta = 0.003$. This explains that most single-mode fibers have radii in the range of 3 to 5 μm (diameters between 6 to 10 μm).

As the second example, suppose the glass cladding of the fiber considered in the last example is removed, leaving the glass fiber core with an air cladding. The *bare fiber* has core and cladding indices of 1.48 and 1.0, respectively, and is *not* a weakly guided fiber, since Δ is as large as 0.32. The inequality $V < 2.4048$ is still a valid condition for single-mode operation, but it is necessary to use the *exact expression* for V to estimate the core radius for single-mode operation,

$$\frac{2\pi}{\lambda} a \sqrt{n_{co}^2 - n_{cl}^2} < 2.4048$$

At $\lambda = 1.3\mu m$, the radius must be smaller than 0.456 μm if a bare fiber is to be single-moded.

As noted earlier, a fiber is single-moded if V is less than 2.4048. However, if V is *too* small, a significant portion of power will reside in the cladding region. Most single-mode fibers of practical interest are designed to operate with a V between 1.5 and 2.4. Numerical calculations show that, for a V between 1.5 and 2.5, the b of the LP_{01} mode may be approximated by [15]

$$b \approx \left(1.1428 - \frac{0.996}{V}\right)^2 \tag{8.35}$$

and the error is less than 0.1 percent.

8.3.7 Principal Mode Number

As indicated previously, two mode numbers, ℓ and m, are needed to specify a mode. Under certain conditions, they can be combined into a single index. Consider LP modes *far from cutoff,* that is, $b \to 1$. When $V\sqrt{b}$ is large, we use the asymptotic expression (C.12) for the modified Bessel function to show that $V\sqrt{b}\, K_{\ell-1}(V\sqrt{b})/K_\ell(V\sqrt{b})$ tends to V as b approaches 1. Thus, (8.29) is satisfied *only* if

$$J_\ell(V\sqrt{1-b}) \approx 0, \quad \ell = 0,1,2,\ldots$$

Using the asymptotic expression (C.11) for Bessel functions with a large argument, we can show that this condition becomes

$$V\sqrt{1-b} - \frac{\ell\pi}{2} - \frac{\pi}{4} \approx \left(m - \frac{1}{2}\right)\pi, \quad \ell = 0,1,2,3,\ldots, \quad m = 1,2,3,\ldots.$$

Therefore, for modes with large ℓ and m and far from the cutoff condition, b may be approximated as

$$b \approx 1 - (\frac{\pi}{2V})^2 (\ell + 2m - \frac{1}{2})^2$$

Note that b depends on $\ell + 2m$, rather than ℓ or m individually. In other words, modes far from cutoff have the same effective index of refraction and propagation as a group, if they have the same $\ell + 2m$. Other characteristics are also similar. For example, when a fiber is truncated, fields radiate into the air in a cone that has a half-angle [10, 13]

$$\theta_h \approx \frac{(\ell + 2m)\lambda}{4a}$$

Note the appearance of $\ell + 2m$. We can now introduce the **principal mode number** for step index fibers:

$$m_p = \ell + 2m \tag{8.36}$$

It should be emphasized that that the principal mode number is mainly useful for describing modes *far from the cutoff condition*.

8.3.8 Total Number of Guided Modes

For V less than 10, we can use Figure 8.10 to identify the number of modes guided by a weakly guided fiber. However, typical multimode fibers have a V much larger than 10, and Figure 8.10 is useless. To estimate the number of guided modes for fibers with a large V, observe that, for a given ℓ and m, there are *two possible states of polarization*, linear polarization in the x or y directions, and that for each state of polarization, there are *two possible angular variations*, $\sin\ell\phi$ and $\cos\ell\phi$. This means that each set of ℓ and m corresponds to *four* possible modes, including the states of polarization. It is useful to visualize a set of ℓ and m as a point on the ℓm plane (Figure 8.12). Each array point (l, m) corresponds to four LP modes. Therefore, the total number of modes is four

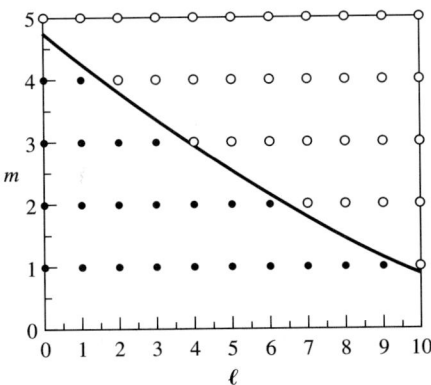

Figure 8.12 Array points on the ℓm plane

times the total number of array points. Now, instead of considering individual array points, we relate those points to a surrounding area on the ℓm plane. To establish the relationship, we observe that ℓ and m are integers, and that the separation between two neighboring integers ℓ or two neighboring integers m is precisely 1. Thus, we may view each array point as occupying one *unit area* in the ℓm plane.

Next, we estimate the area on the ℓm plane. For this purpose, we return to the cutoff condition (8.33). To allow for the possibility that ℓ may be as large as, though not larger than, V, we approximate $J_\ell(V)$ (8.33) by the asymptotic expression for Bessel functions with a large order (C.13):

$$\cos\left[\sqrt{V^2 - (\ell - 1)^2} - (\ell - 1)\cos^{-1}\frac{\ell - 1}{V} - \frac{\pi}{4}\right] \approx 0$$

From this equation, we can see that no root exists if $(\ell - 1) > V$. One or more roots may exist if $(\ell - 1) < V$. For each root, an integer m may be found such that

$$\sqrt{V^2 - (\ell - 1)^2} - (\ell - 1)\cos^{-1}\frac{\ell - 1}{V} - \frac{\pi}{4} \approx (m - \frac{1}{2})\pi \qquad (8.37)$$

In searching for the upper limits of ℓ and m, we ignore the constants 1 and 1/2 in (8.37), which then becomes

$$m \approx \frac{1}{\pi}[\sqrt{V^2 - \ell^2} - \ell \cos^{-1}\frac{\ell}{V}]$$

Since ℓ and m are large integers, we approximate them as continuous variables and plot m as a function of ℓ. The plot is shown as the solid curve in Figure 8.12. The array points are the solid dots below the solid curve. The area below the solid curve is

$$\int_0^V m\, d\ell$$

The *total number of guided modes* is

$$M_{tot} = 4\int_0^V m\, d\ell \approx \frac{4}{\pi}\int_0^V [\sqrt{V^2 - \ell^2} - \ell\cos^{-1}\frac{\ell}{V}]d\ell$$

A factor of 4 has been included in this equation to account for the four-fold degeneracy for circular fibers with a step index profile. The integral can be evaluated in a closed form, as follows:[1]

---

[1]
$$\int x\cos^{-1}x\, dx = \frac{(2x^2 - 1)\cos^{-1}x - x\sqrt{1 - x^2}}{4}$$
$$\int \sqrt{a^2 - x^2}\, dx = \frac{x\sqrt{a^2 - x^2} + a^2\sin^{-1}(x/a)}{2}$$

$$M_{tot}\bigg|_{step\ index} = \frac{V^2}{2} \tag{8.38}$$

As an numerical example, consider a step index fiber with a core diameter of 56 μm and a numerical aperture of 0.14. If $\lambda = 0.647$ μm, the value of V is

$$V = \frac{2\pi}{\lambda} a \sqrt{n_{co}^2 - n_{cl}^2} = \frac{2\pi}{0.647 \times 10^{-6}} \times \frac{56 \times 10^{-6}}{2} \times 0.14 = 38.1$$

and there are $38.1^2/2 \approx 726$ guided modes.

8.4 INFORMATION CAPACITY

Communication systems with a wider bandwidth can carry more signal channels and can transmit signals faster. There are a number of factors limiting the overall bandwidth of communications systems, including the bandwidth of the transmitter and receiver electronics, and the bandwidth of the transmitting medium. In this context, we are interested in the fiber bandwidth.

It is well known that when pulses are applied to an electric circuit, the output is distorted because of damping or overshooting. In particular, the rise and fall times increase from a smaller value at the input to a larger value at the output. As a result, the output pulses are broadened relative to the input pulses. This is true for all pulses, electrical and optical. For fibers, the distortion and broadening occur because group velocity depends on the mode and the wavelength.

As discussed in Chapter 7, pulse envelops travel at the speed of the group velocity v_{gr}. The transit time for a pulse traversing a fiber of length L is, from (7.11),

$$\tau = \frac{L}{v_{gr}} = L\frac{d\beta}{d\omega} = \frac{L}{c}\frac{d\beta}{dk} = -\frac{L}{2\pi c}\lambda^2 \frac{d\beta}{d\lambda} \tag{8.39}$$

If several modes are excited, pulse broadening occurs, because different modes move with different v_{gr}, even though they have the same frequency. This is known as **intermodal** or **multimode dispersion,** and it is the dominant pulse broadening mechanism in multimode fibers. To illustrate these effects, consider a fiber that supports four modes. Figures 8.13a and b depict *three narrow pulses* fed into one end of the four-mode fiber, and *three groups of pulses* emerging from the other end. For each pulse fed into the four-mode fiber, a pulse *group* consisting of four closely-spaced pulses appears at the output. The separation δt between the pulses within a pulse group varies linearly with the fiber length. If the four-mode fiber is short, the separation δt is very narrow, and it is not possible to distinguish the four separate pulses. For typical multimode fibers, which support a multitude of modes, the separation δt between pulses *is* extremely small, even for long fiber sections. Consequently, instead of a large number of narrow pulses, a single broadened pulse corresponding to the envelope of these narrow pulses appears at the output. This is shown in Figure 8.13c. Although the intermodal broadening also depends on the finite spectral width

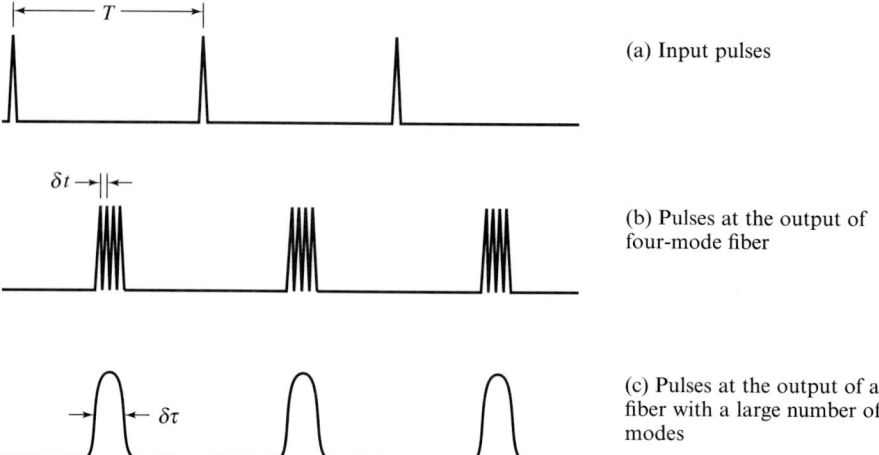

Figure 8.13 Distortion and broadening of pulses

of the signal, the effects are negligibly small compared to those attributable to the presence of several modes in multimode fibers.

If there is one and only one guided mode, as is the case of single-mode fibers, or if there is single-mode and single-polarization, as in certain special fibers, the pulses are still broadened. This occurs because all signals of a finite duration have a finite spectral width, and different spectral components travel with a different v_{gr}, even for the same mode. This effect is the **intramodal** or **chromatic dispersion.** For weakly guided fibers, the intramodal dispersion can be divided further, as an approximation, into **material dispersion** and **waveguide dispersion.**

If the input pulses are temporally too closely spaced, it is not possible to separate the broadened pulses at the output. Therefore, the maximum information that can be transmitted by a system is limited by the pulse broadening of the system. Let T be the time separation between pulses (Figure 8.13a). The **pulse repetition rate** is $1/T$. For simplicity, assume that the input pulse width is negligibly narrow in comparison with T. If the output pulse width is increased to $\delta\tau$, and if T is less than $2\delta\tau$, two neighboring pulses will merge and become indistinguishable. This means that the maximum pulse repetition rate is limited to $1/(2\delta\tau)$.

8.4.1 Intermodal Dispersion

Step Index Fibers As noted previously, in multimode fibers, many modes are excited, and each mode propagates with a different group velocity. Since different modes arrive at the receiving end at different times, the pulses are distorted and broadened. Because the lower-order modes propagate mainly along the waveguide axis, while the higher-order modes follow a more zigzag path, the lowest-order mode arrives first, and the highest-order mode arrives last. The time separation between the arrivals of the fastest and the slowest modes is used as an estimate of the pulse width at the output.

The lowest-order mode, that is, the fundamental mode, has a group velocity of [from (7.8)]

$$v_{gr\,fs} = \frac{c}{n_{co}}\sin\theta \bigg|_{\theta=\pi/2} = \frac{c}{n_{co}}$$

It is the fastest mode so far as the group velocity is concerned. The highest-order mode corresponds to a wave impinging on the core/cladding boundary with an angle of incidence close to the critical angle. Therefore, the group velocity of the highest-order mode is rather slow and is given by

$$v_{gr\,sl} = \frac{c}{n_{co}}\sin\theta \bigg|_{\theta=\sin^{-1}(n_{cl}/n_{co})} = \frac{cn_{cl}}{n_{co}^2}$$

The arrival of the highest-order mode lags behind that of the lowest-order mode by

$$\delta\tau_{mm} = \frac{L}{v_{gr\,sl}} - \frac{L}{v_{gr\,fs}} = \frac{L}{c}\left(\frac{n_{co}^2}{n_{cl}} - n_{co}\right) = \frac{L}{c}n_{co}\left(\frac{\Delta}{1-\Delta}\right) \qquad (8.40)$$

where the subscript mm refers to a multimode fiber. When Δ is small, we can express $\delta\tau_{mm}$ directly in terms of NA, as follows:

$$\delta\tau_{mm} \approx \frac{L}{c}n_{co}\Delta = \frac{L}{c}\frac{(NA)^2}{2n_{co}} \qquad \text{(step index profile)} \qquad (8.41)$$

Equation (8.40) or (8.41) can be used as an estimate of the pulse broadening due to **intermodal dispersion** in step index fibers. It is clear that $\delta\tau_{mm}$ may be reduced by reducing Δ or, equivalently, the numerical aperture. However, the bending loss also increases when Δ decreases, making the fibers very difficult to handle. Alternatively, then, either multimode fibers with a graded index profile or single-mode fibers could be used, to reduce intermodal dispersion.

Graded Index Fibers To appreciate the advantages of graded index fibers, return to the discussion on rays in section 8.2. Rays corresponding to lower-order modes move mainly along the fiber axis, while those of higher-order modes follow zigzag paths. In step index fibers, rays corresponding to the higher-order modes journey all the way to the core/cladding boundary, and then turn abruptly, as indicated in Figure 8.5a. In graded index fibers, rays corresponding to the higher-order modes bend *gradually* toward the fiber axis, and they do so in a shorter period of time. Thus, the path lengths for higher-order modes in graded index fibers are physically shorter than those in step index fibers, assuming the index differences and the core radii of the two fibers are identical. In addition, the index in the region near the core/cladding boundary of a graded index fiber is smaller than the index near the core/cladding boundary of a step index fiber, and the waves propagate faster than waves near the fiber axis. Consequently, there is a *decrease in the path length* and an *increase in the propagation velocity* for the higher-order modes of graded index fibers, and the higher-order modes arrive sooner. This is particularly true for the highest-order mode in graded index fibers. As a result, $\delta\tau_{mm}$ of graded index fibers is smaller than those in step index fibers with the same index difference and core radius. In short, the intermodal dispersion of graded index fibers is substantially reduced.

Details of the intermodal pulse broadening of fibers with a *parabolic* index profile are presented in Appendix D. For our purposes, it is sufficient to note here that, corresponding to (8.40), we have

$$\delta\tau_{mm} \approx \frac{L}{c} n_{co} \frac{\Delta^2}{2} = \frac{L}{c} \frac{(NA)^4}{8n_{co}^3} \qquad \text{(parabolic index profile)} \qquad (8.42)$$

Note that $\delta\tau_{mm}$ for graded index fibers is proportional to Δ^2. It can be shown that the communications capacity is maximized if the core index has an *approximate parabolic index profile,* that is, $g \approx 2$, where g is the index profile parameter from (8.2) [13].

Example

Consider a fiber with $n_{co} = 1.458$ and $\Delta = 0.013$. If the fiber has a step index profile, then from (8.41) we obtain

$$\frac{\delta\tau_{mm}}{L} \approx \frac{n_{co}\Delta}{c} = \frac{1.458 \times 0.013}{3 \times 10^8} = 63 \text{ ns/km}$$

For a 1-km fiber link, the minimum pulse separation is $T \approx 2\delta\tau_{mm} \approx 126$ ns, and the maximum pulse repetition rate is $1/(126 \times 10^{-9}) \approx 8$ MHz. If the link length is 2 km, then $\delta\tau_{mm}$ and T are doubled and the maximum pulse repetition rate is 4 MHz. Therefore, for the step index fiber considered, the **bandwidth–length product** is 8 MHz–km.

If the fiber has a parabolic index profile, then from (8.42) we obtain

$$\frac{\delta\tau_{mm}}{L} \approx \frac{n_{co}\Delta^2}{2c} = \frac{1.458 \times 0.013^2}{2 \times 3 \times 10^8} = 0.41 \text{ ns/km}$$

The corresponding maximum pulse repetition rate is 1.21 GHz for a 1-km link. The bandwidth–length product is 1.21 GHz–km. This is an increase of a factor of 152 compared to the bandwidth–length product of step index fibers with the same index difference.

8.4.2 Intramodal Dispersion

As mentioned previously, all pulses of a finite temporal duration have a finite spectral width. Since v_{gr} is a function of λ, *different spectral components* arrive at the far end at different times, and the pulses are distorted and broadened. This is intramodal dispersion in single-mode fibers. Single-mode fibers are dispersive on two accounts. First, the normalized guide index b is a function of V, which is a function of λ, n_{co}, and n_{cl}. Second, the indices themselves are functions of λ.

Consider signals with a finite spectral width $\Delta\lambda$ centered at λ_0. As a simple estimate, consider the arrival time of two spectral components $\lambda_0 \pm \dfrac{\Delta\lambda}{2}$,

$$\tau(\lambda_0 \pm \frac{\Delta\lambda}{2}) \approx \tau(\lambda_0) \pm \frac{\Delta\lambda}{2}\frac{d\tau}{d\lambda} + \ldots$$

The arrivals of each of these spectral components are separated by a time interval

$$\delta\tau = \tau(\lambda_0 + \frac{\Delta\lambda}{2}) - \tau(\lambda_0 - \frac{\Delta\lambda}{2}) \approx \Delta\lambda\frac{d\tau}{d\lambda} \tag{8.43}$$

which shows that $\delta\tau$ increases linearly with $\Delta\lambda$ and length L. Therefore, we define the **dispersion** $\mathcal{D}$ as the *broadening per unit spectral width for a fiber of unit length*

$$\mathcal{D} \equiv \frac{1}{L}\frac{\delta\tau}{\Delta\lambda} \approx \frac{1}{L}\frac{d\tau}{d\lambda} = -\frac{1}{2\pi c}\frac{d}{d\lambda}(\lambda^2\frac{d\beta}{d\lambda}) \tag{8.44}$$

Since $k = 2\pi/\lambda$, $\mathcal{D}$ can be expressed directly in terms of k

$$\mathcal{D} = -\frac{k^2}{2\pi c}\frac{d^2\beta}{dk^2} \tag{8.45}$$

To determine an expression for β for single-mode fibers, we refer to (8.7) and note that, for weakly guided fibers,

$$\beta = kN = k[n_{cl}^2 + (n_{co}^2 - n_{cl}^2)b]^{1/2} \approx kn_{co}(1 - \Delta + \Delta b)$$

Since core and cladding materials are basically similar, the dispersions of these materials are also similar. In other words, Δ is roughly independent of λ. By ignoring $\dfrac{d\Delta}{dk}$, we have

$$\frac{d\beta}{dk} \approx (kn_{co})'(1 - \Delta + \Delta b) + kn_{co}\Delta\frac{db}{dk} \tag{8.46}$$

where

$$(kn_{co})' \equiv \frac{d}{dk}(kn_{co}) = n_{co} + k\frac{dn_{co}}{dk} = n_{co} - \lambda\frac{dn_{co}}{d\lambda}$$

For most glass materials in the visible and near IR regions, $\left|\lambda\dfrac{dn}{d\lambda}\right|$ is on the order of 0.01, as shown in Figure 8.14 [16]. In other words, $\left|\dfrac{\lambda}{n}\dfrac{dn}{d\lambda}\right|$ is on the

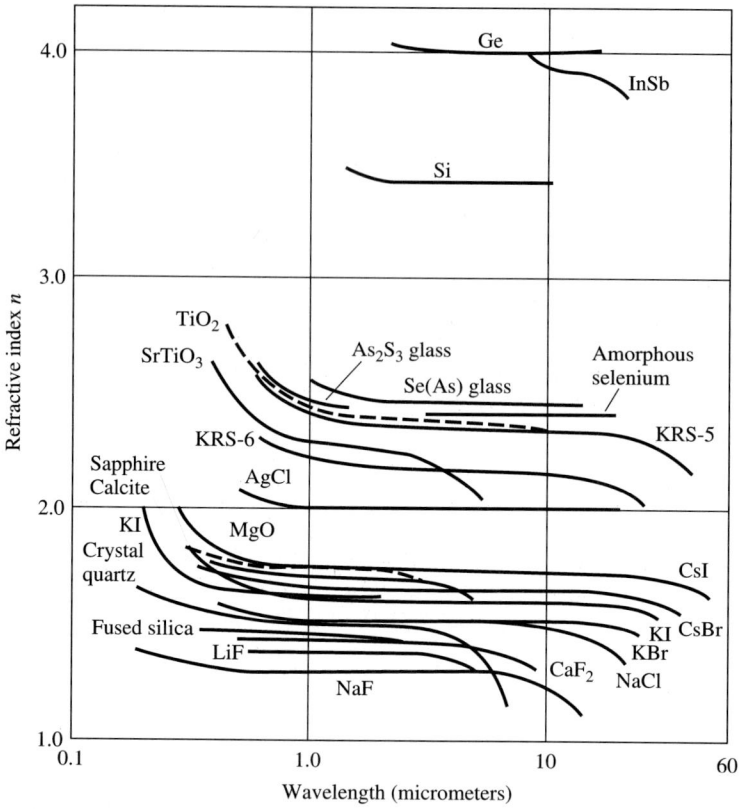

(a) Refractive index n

Figure 8.14 Refractive indices and dispersion | $dn/d\lambda$ | as functions of wavelength, for several crystals and glasses [16], reproduced with permission.

order of Δ. Therefore, $(kn_{co})'$ may be approximated by n_{co}, and $(kn_{cl})'$ may be approximated by n_{cl}. Using these approximations, we obtain a simple approximation for $\dfrac{dV}{dk}$, as follows:

$$\frac{dV}{dk} = \frac{a[kn_{co}(kn_{co})' - kn_{cl}(kn_{cl})']}{[(kn_{co})^2 - (kn_{cl})^2]^{1/2}} \approx \frac{a[kn_{co}^2 - kn_{cl}^2]}{[(kn_{co})^2 - (kn_{cl})^2]^{1/2}} = \frac{V}{k}$$

It then follows that

$$\frac{db}{dk} = \frac{db}{dV}\frac{dV}{dk} \approx \frac{V}{k}\frac{db}{dV}$$

Based on these approximations, we obtain

$$\frac{d\beta}{dk} \approx (kn_{cl})' + n_{co}\Delta\left[\frac{d(Vb)}{dV} - 1\right] \tag{8.47}$$

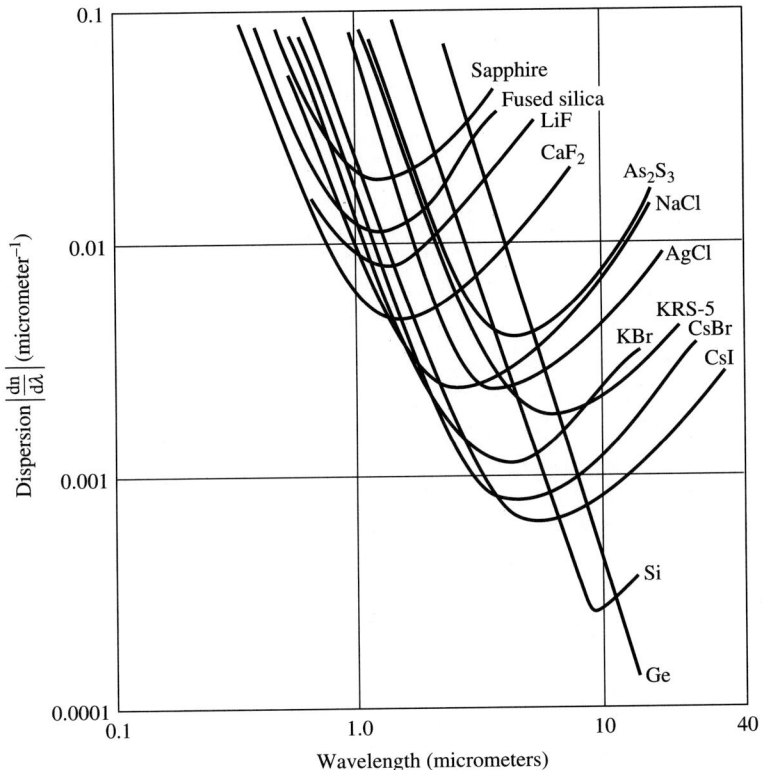

(b) Dispersion $|dn/d\lambda|$

Figure 8.14 (*Continued*)

Further differentiation leads to

$$\mathcal{D} = -\frac{k^2}{2\pi c}\left[(kn_{co})'' + \left(\Delta\frac{dn_{co}}{dk}\left[\frac{d(Vb)}{dV} - 1\right]\right) + \left(\frac{n_{co}\Delta}{k}V\frac{d^2(Vb)}{dV^2}\right)\right] \quad (8.48)$$

where

$$(kn_{co})'' \equiv \frac{d^2}{dk^2}(kn_{co}) = \frac{\lambda^3}{2\pi}\frac{d^2n_{co}}{d\lambda^2}$$

Plots of $\dfrac{d(Vb)}{dV}$ and $V\dfrac{d^2(Vb)}{dV^2}$ are given in Figure 8.15 [17, 18]. Note that $\dfrac{d(Vb)}{dV}$ is on the order of 1. Therefore, the second term in (8.48) is negligible in comparison with the other terms. With the second term ignored, we have

$$\mathcal{D} \approx -\frac{k^2}{2\pi c}\left(\frac{\lambda^3}{2\pi}\frac{d^2n_{co}}{d\lambda^2} + \frac{n_{co}\Delta}{k}V\frac{d^2(Vb)}{dV^2}\right) \quad (8.49)$$

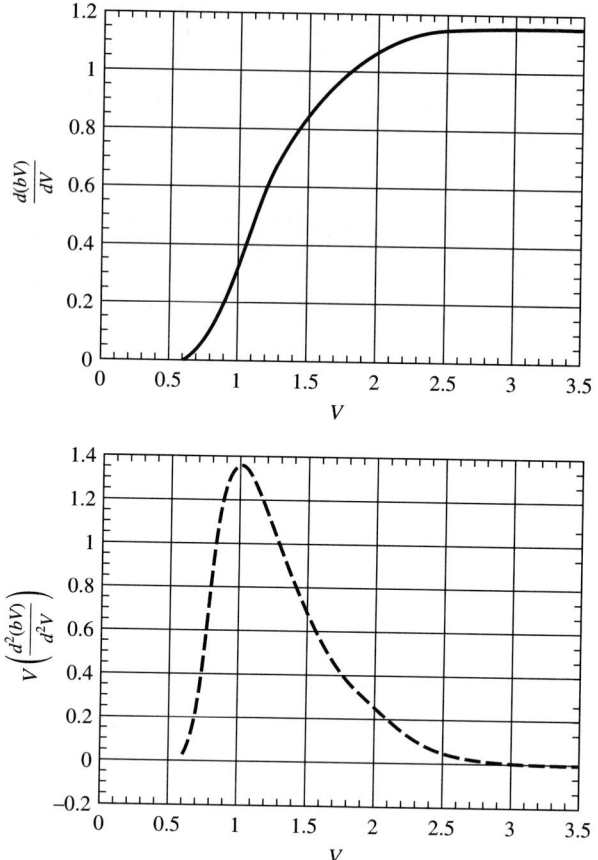

Figure 8.15 $\dfrac{d(Vb)}{dV}$ and $V\dfrac{d^2(Vb)}{dV^2}$ as functions of the normalized parameter V [17, 18]

Equation (8.49) is the total intramodal dispersion of weakly guided single-mode fibers with a step index profile.

To appreciate the physical meanings of the two terms in (8.49), consider two idealized cases. First, consider the case of $V > 2$, which means that $V\dfrac{d^2(Vb)}{dV^2}$ is less than 0.2, as shown in Figure 8.15. The dominant term of (8.49) then is the first term, which is

$$\mathcal{D}_{mt} \approx -\frac{k^2}{2\pi c}\frac{\lambda^3}{2\pi}\frac{d^2 n_{co}}{d\lambda^2} = -\frac{\lambda}{c}\frac{d^2 n_{co}}{d\lambda^2} \qquad (8.50)$$

Equation 8.50 is exactly the dispersion of uniform plane waves propagating in a material with an index n_{co}. Recall that for single-mode fibers operating with a V greater than 2.0, the fields are mainly in the core region. Therefore, the fiber dispersion is essentially the material dispersion of the core and the dispersion effect due to the waveguiding is minimal. We therefore identify the *first term* of (8.49) as the **material dispersion.** Figure 8.14 can be used to estimate the material dispersion of various crystals and glasses. Figure 8.16 depicts $\mathcal{D}_{mt}$ as a function of wavelength for several silicate-based glass materials [19]. In particular, $\mathcal{D}_{mt}$ is about 60 ps/(km–nm) at wavelengths near 0.9 μm. Figure 8.16 also shows that $\mathcal{D}_{mt}$ decreases as the wavelength increases, and vanishes for λ near 1.25 μm.

To study the physical meaning of the second term of (8.49), assume that the core and cladding materials are nondispersive. When the second derivative of n_{co} with respect to k is ignored, (8.49) reduces to

$$\mathcal{D}_{wg} = -\frac{n_{co}\Delta}{\lambda c} V \frac{d^2(Vb)}{dV^2} \qquad (8.51)$$

Equation (8.51) is the **waveguide dispersion,** and it represents the dispersion of a fictitious fiber with wavelength-independent materials.

In summary, when the dispersion of Δ is ignored, the intramodal dispersion may be viewed as the superposition of the material and waveguide dispersions, and the two dispersion effects are additive. We should caution, however, that this simple interpretation, though intuitively pleasing, is not valid for fibers with a small core or for those operating with a small V. For such fibers, material and

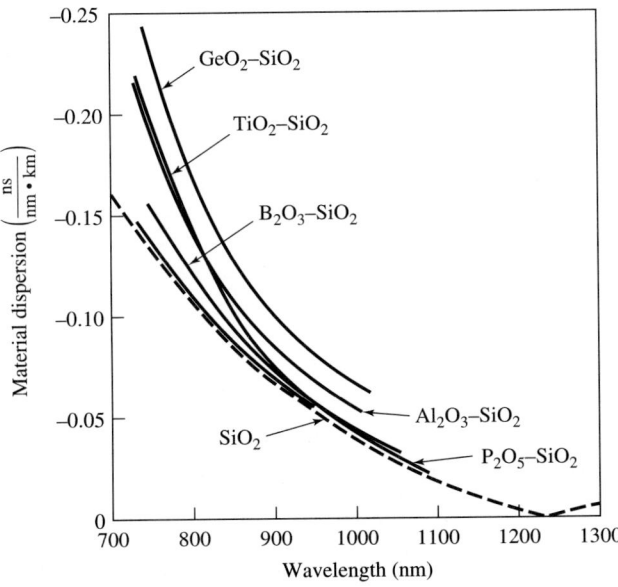

Figure 8.16 Material dispersion of silicate based glasses (modified from [19]). To be consistent with the definition of $\mathcal{D}_{mt}$, a negative sign has been added to the ordinate.

Table 8.2 Comparison of various dispersion effects at 0.9 μm in typical fibers

Fibers	Dispersion	Index profile	Δλ	δτ
Multimode	Intermodal	Step index	—	$\delta\tau_{mm} \approx 60$ ns/km
Multimode	Intermodal	$g = 2$	—	$\delta\tau_{mm} \approx 600$ ps/km
Single mode	Intramodal	Step index	1 nm	$\delta\tau_{mt} \approx 60$ ps/km
Single mode	Intramodal	Step index	1 nm	$\delta\tau_{wg} \approx 6$ ps/km

waveguide dispersions are not additive [20]. Table 8.2 lists typical values of various pulse broadening effects at 0.9 μm.

8.4.3 Zero-Dispersion Wavelength

It is clear from Table 8.2 that the material dispersion is the dominant pulse broadening mechanism in single mode fibers. If single mode fibers are designed to operate in regimes where $d^2n/d\lambda^2$ is small or even zero, the fiber information capacity can be greatly increased. For many materials $|dn/d\lambda|$ has a minimum in the near IR region, as shown in Figure 8.14. At this wavelength, $d^2n/d\lambda^2$ vanishes and the material is truly dispersion free. For fused silica fibers, the zero-dispersion wavelength is about 1.25 μm. Also, the material dispersion $\mathcal{D}_{mt}$ is positive for $\lambda > 1.25$ μm and negative for shorter wavelengths, while the waveguide dispersion $\mathcal{D}_{wg}$ is always negative. Therefore, it is possible to choose a value for V such that the two dispersion effects cancel at a desired wavelength. Thus, the *total* fiber dispersion vanishes at a desired wavelength [21,22]. This is the **zero-dispersion wavelength** of the fiber.

8.5 FIBER FABRICATION PROCESSES

A fiber consists of a core, a cladding layer, and one or more layers of protective coating. Either glass or polymers may be used as the core and/or cladding material. For example, there are fibers with a *glass core* and *plastic cladding,* or a *plastic core* and *plastic cladding.* Most low-loss fibers have a *glass core* and a *glass cladding,* and are called **all-glass fibers.** In this section, we will discuss the fabrication of all-glass fibers.

Glass is an amorphous solid solution. The basic ingredient of oxide glasses used for visible and near IR spectral applications is fused silica (SiO_2).[2] Such glass may also contain GeO_2, B_2O_3, As_2O_3, P_2O_5, V_2O_5, and other oxides. Small amounts of F or the oxides GeO_2, B_2O_3, and P_2O_5 may be added as dopants to adjust the refractive index and increase the mechanical strength, optical transparency, chemical durability, and stability of the glass. For example, the refrac-

---

[2]Quartz is SiO_2 in the crystalline form. Typical glass jars are made of soda–lime glass, which is silica with Na_2O and CaO added to reduce the melting point.

8.5 Fiber Fabrication Processes

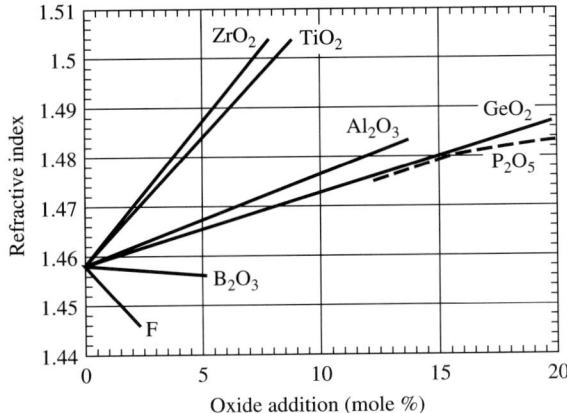

Figure 8.17 Refractive index of doped silica glasses at 0.5893 μm [23]

tive index of silica-based glasses can be increased by the addition of ZrO_2, TiO_2, Al_2O_3, GeO_2, or P_2O_5, and can be decreased by B_2O_3 or F (Figure 8.17) [23]. Components such as CaO, BaO, K_2O, Na_2O, etc., may be introduced to modify the melting point and other physical and chemical properties of the glass. Depending on the SiO_2 content, glasses may be classified as **multicomponent oxide glasses** or **high-silica glasses**. Multicomponent oxide glasses contain considerable amounts of Na_2O, B_2O_3, and CaO, in addition to SiO_2. In contrast, in **high-silica glass,** SiO_2 is the main material.

Most low-glass fibers are made by the double crucible method or one of the three-step fabrication processes. In the **double crucible method** [24], the core and cladding materials are melted simultaneously in two concentric crucibles, and the fiber is drawn from the bottom orifice (Figure 8.18). In a **three-step process,** there are three basic steps: preform forming, drawing, and coating. Of the three, preform forming is the most crucial step.

There are several techniques for making preforms. A **preform** is an enlarged version of the fiber, with the same geometric shape, core-radius-to-cladding-radius ratio, and index profile. Details of the **preform forming** techniques are presented in the following subsections. In the **drawing process,** a preform is heated and drawn into the fiber, and the geometric relation and index profile are retained. For example, a fiber[3] with a core diameter of 50 μm and a cladding diameter of 125 μm may be drawn from a preform with a core diameter of 0.8 cm and a cladding diameter of 2.0 cm. Finally, the composite structure consisting of a core and a cladding is coated with protective materials.

In the early days, preforms were made by placing a high-index glass rod,

---

[3] In some literature, fibers with a core diameter of 50 μm and a cladding diameter of 125 μm are referred to as 50/125 fibers. If there is a 400 μm diameter jacket, they are referred to as 50/125/400 fibers.

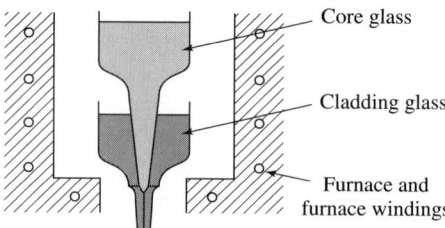

Figure 8.18 Fiber drawing by the double crucible method [24]

corresponding to the fiber core, inside a lower-index glass tube, corresponding to the cladding. The glass rods and tubes were formed by the conventional glass-making process of *melting in a crucible*. Most modern preforms are manufactured by some form of **vapor deposition technology,** four of which are described here. In vapor deposition processes, submicron glass particles (0.01 to 0.1 μm) are formed by oxidation reactions of chlorides with oxygen at an elevated temperature (1300 to 1700 °C). The chemical reactions involved include

$$SiCl_4 + O_2 \rightarrow SiO_2 + 2Cl_2$$
$$GeCl_4 + O_2 \rightarrow GeO_2 + 2Cl_2$$
$$4POCl_3 + 3O_2 \rightarrow 2P_2O_5 + 6Cl_2$$
$$4BCl_3 + 3O_2 \rightarrow 2B_2O_5 + 6Cl_2$$

8.5.1 Outside Vapor Deposition

In the **outside vapor deposition (OVD) process** [24]–[26], a thin ceramic target rod or a mandrel is held in a lathe, as shown in Figure 8.19. A hydrogen–oxygen burner is placed near the target rod, and is moved laterally along the lathe. At the same time, the target rod is rotated by the lathe while it is heated by the traveling burner. Silicon tetrachloride ($SiCl_4$), germanium tetrachloride ($GeCl_4$), phosphorous oxychloride ($POCl_3$), BCl_3, oxygen (O_2), and other reactants are fed into the burner, along with the fuel gas (CH_4 or H_2). The thermally-activated oxidation reactions occurring in the flame cause fine glass particles, known as **soot,** to be formed. A stream of these glass particles is ejected from the flame onto the rotating target rod, where the glass particles stick. The burner is moved back and forth along the target rod, depositing successive layers, forming a porous soot preform. A typical preform may require 50 to 100 passes of the flame along the entire length. Adjusting the composition of the reactants and the flow rate effectively controls the refractive index of each layer. The soot preform, with the ceramic rod removed, is then sintered and collapsed into a glass preform. Since glass particles are deposited on the outside surface of a target rod, this process is known as the outside vapor deposition process [24]–[26].

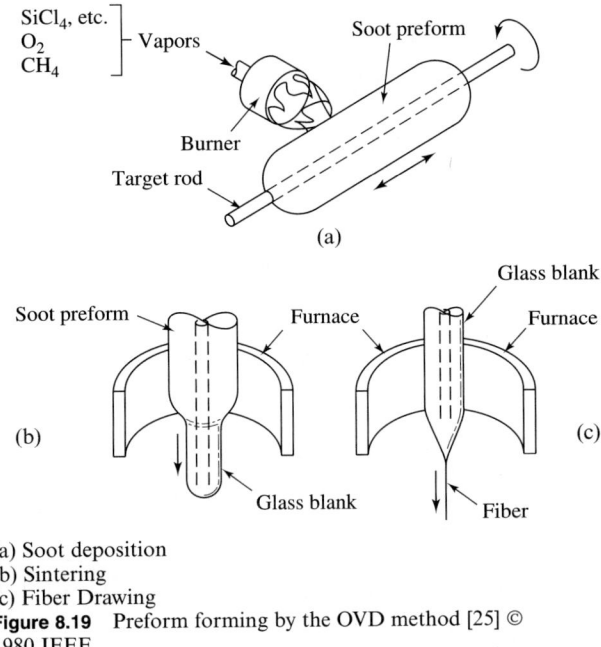

(a) Soot deposition
(b) Sintering
(c) Fiber Drawing
Figure 8.19 Preform forming by the OVD method [25] © 1980 IEEE

8.5.2 Vapor Phase Axial Deposition

The **vapor phase axial deposition (VAD) process** (Figure 8.20) can be viewed as a combination of an OVD facility and a crystal pulling machine. Like the OVD process, various reactants are fed through hydrogen–oxygen burners. However, the soot layers are grown *axially* from the end of the seed rod or porous preform, rather than cylindrically around the surface of the target rod, and several burners may be used. Soot from the central burner forms the core, while other burners provide the glass particles for the cladding region. The compositions and flow rates of reactants destined for different regions may be varied individually to accommodate the different index profiles required for these regions. Also, the deposition rate can be increased by using multiflame torches and substituting $SiHCl_3$ for $SiCl_4$.

When a sufficient number of glass particles is accumulated at the end of the rod, the porous preform is pulled to a heating zone, where the porous preform is heated to a higher temperature, melted, and then consolidated into a glass preform. The seed rod is cut off after the preform is fabricated. In principle, since no starting target rod is needed, there is no limit to the preform length [27]–[28].

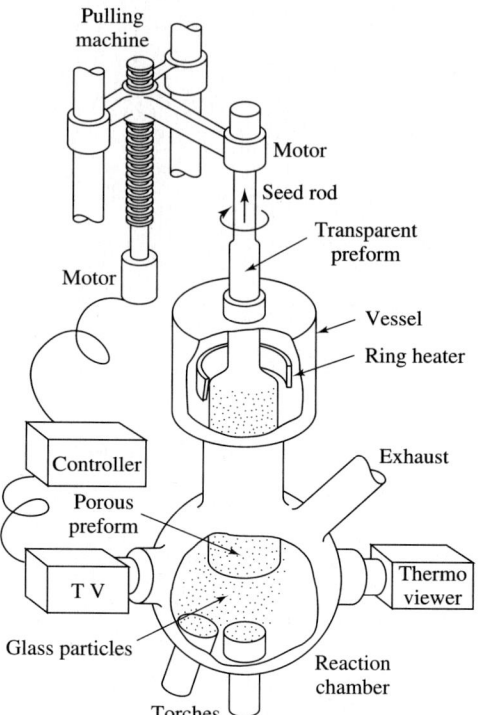

Figure 8.20 Preform forming by the VAD method [27] © 1980 IEEE

8.5.3 Modified Chemical Vapor Deposition

In the **modified chemical vapor deposition (MCVD) process,**[4] a pure silica tube is placed on a lathe. An oxygen stream saturated with the chlorides of silicon, germanium, boron, and other reactant is fed into the tube, which is heated from the outside by a hydrogen–oxygen flame (Figure 8.21). The torch traverses the entire length of the lathe, oxidation reactions take place inside the tube, and soot particles attach to the inner surface of the tube downstream from the heat zone generated by the burner. The deposited layers are subsequently fused onto the tube as the burner moves. Gaseous byproducts are discharged from the other end of the tube.

When the deposition process is complete, the flow of reactants is shut off and the motion of the burner is slowed. The tube with the deposited soot layers is heated to a higher temperature, and the composite structure is sintered and collapsed to form a glass preform. The original glass tube corresponds to the cladding region and the newly deposited soot layers correspond to the core re-

---

[4]In the original chemical vapor deposition (CVD) method, borosilicate (B_2O_3) and pure silica (SiO_2) layers are formed by the reactions of silane (SiH_4), germane (GeH_4), and diborane (B_2H_6) with CO_2 and O_2. These reactions produce water as a byproduct.

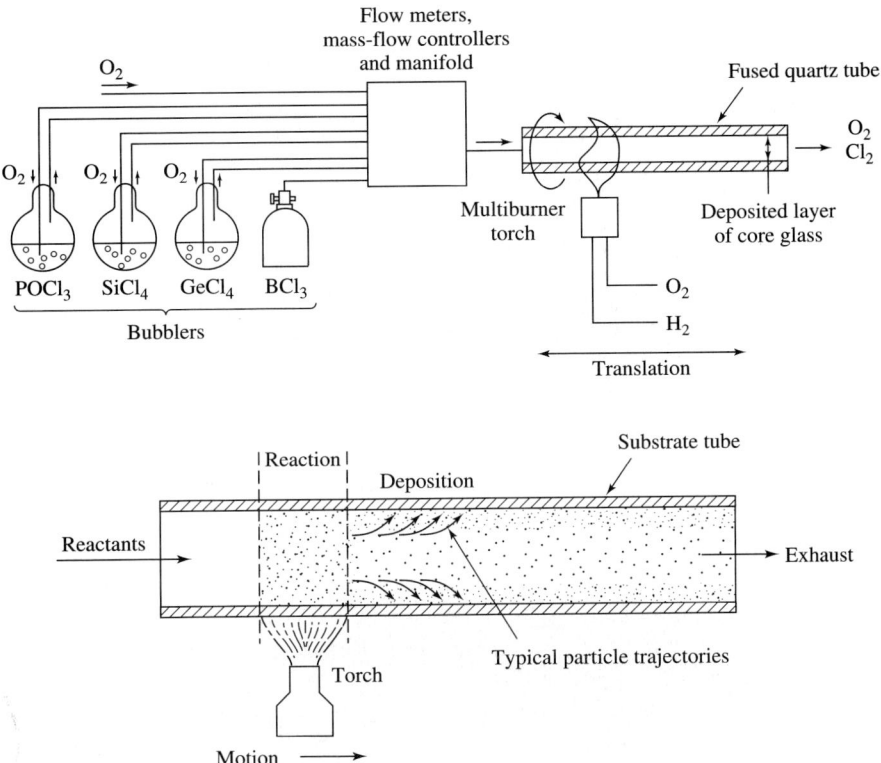

Figure 8.21 Preform forming by the MCVD method [30, 31] © 1980 IEEE

gion. Also, since the glass layers are built radially inward from the inner surface of the starting glass tube, the MCVD process is also known as the **inside vapor deposition (IVD) process** [29]–[31]. In the MCVD process, reactions take place in an enclosed environment completely separated from the flame. This is an important feature of the MCVD process.

8.5.4 Plasma Modified Chemical Vapor Deposition

The **plasma modified chemical vapor deposition (PCVD) process** is very similar to the MCVD process, except that the traveling burner is replaced by a moving microwave cavity (Figure 8.22). Oxidation reactions are stimulated by the localized microwave-generated (2.45 GHz) plasma inside the microwave cavity.

Many variations of PCVD have been proposed. In one version, the moving microwave cavity is replaced by a stationary cavity extending the full length of the glass tube, and a pulsed microwave source is used to generate the plasma. This is known as the **plasma-impulse chemical vapor deposition (PICVD) process.** If, however, a surface wave launcher is used to excite the plasma, the process is

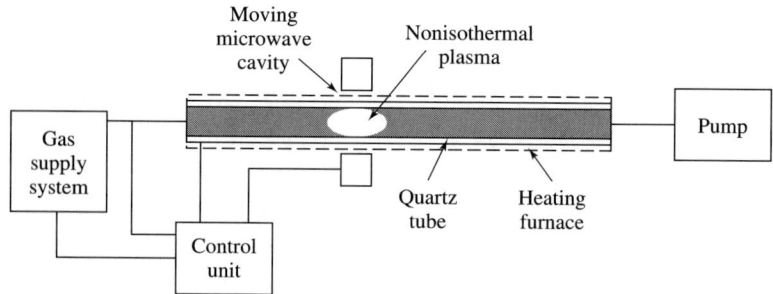

Figure 8.22 Preform forming by the PCVD method [33] © 1986 IEEE

known as the **surface plasma chemical vapor deposition (SPCVD) process** [32]–[33].

8.6 FIBER LOSSES

All fibers have losses, which may be intrinsic to and distributed throughout the fibers. Loss is also introduced when fibers are joined with other fiber sections, optical sources, detectors, or other optical components. Figure 8.23 schematically depicts various mechanisms contributing to fiber loss [34]. In this section, we are mainly interested in those losses attributable to the fibers themselves.

There are three basic loss mechanisms: absorption, scattering, and handling or cabling losses. **Absorption loss** is due to the interaction of light with electronic states or molecular vibration modes. The resulting power loss is converted into heat. In the **scattering process,** optical power in the incoming guided mode is converted to other guided modes, or to unguided waves in the forward and backward directions. Before fibers are shipped to the field for installation, it is necessary to wrap the fibers in many layers of packing materials and to place it in plastic tubings for protection. Fibers are also bundled with other fibers or electrical transmission lines to form fiber cables. Additional metallic shields may be added to provide further protection in a process known as cabling. After cabling, the fiber cables are coiled onto large circular drums for shipping. In the field, the fiber cables are uncoiled. Clearly, fibers are being handled and stressed during the cabling and installation process. Additional losses are induced when fibers are stressed.

The total power loss in fibers is the sum of all the losses mentioned here. These losses are discussed in detail in the following paragraphs.

8.6.1 Absorption

By the Glass Forming Materials and Dopants   As discussed in the last section, SiO_2, P_2O_5, B_2O_3, and GeO_3, etc., are either the glass-forming materials or the dopants needed to adjust the physical and chemical properties of the glass.

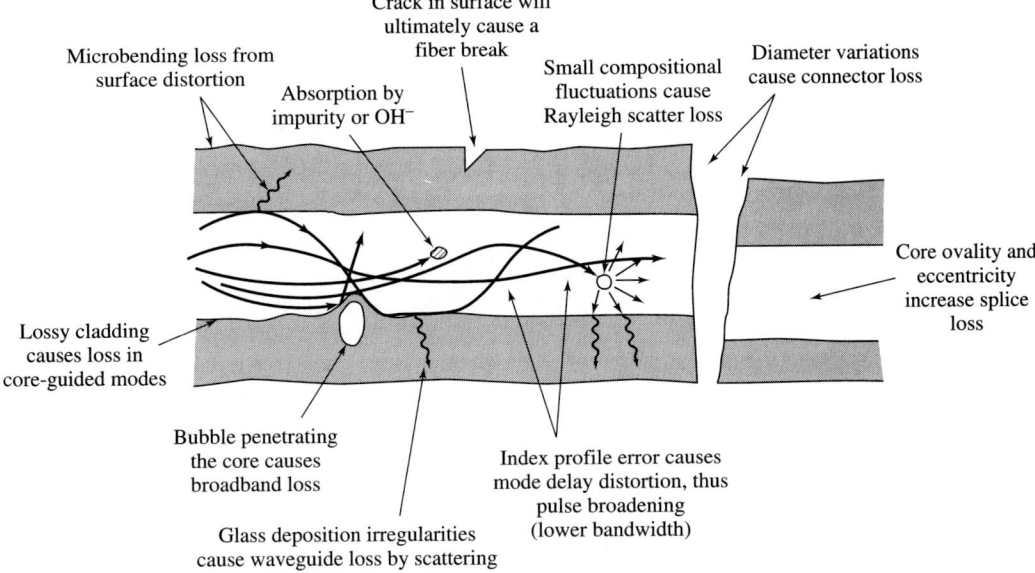

Figure 8.23 Various mechanisms causing fiber loss [34]

These materials have *electronic transitions* in the ultraviolet (UV) regions and *molecular vibrational modes* in the infrared (IR) regions. Photons with sufficient energy to elevate electrons from lower energy levels to higher levels are absorbed by the material. In addition, light with frequencies corresponding to molecular vibrational modes is absorbed by the material. These losses are due to the *intrinsic material properties,* and they are present even if no impurity of any kind is present.

By Impurities Unwanted impurities are inevitably introduced into fibers during the manufacturing process, and these impurities absorb energy. Most transition metal ions have absorption bands in the visible and near IR regions. The main culprit of fiber energy loss is the hydroxyl ion (OH^-). Although the *fundamental* stretching vibrational mode of OH^- (2.72 μm) is outside the visible and near IR regions, the second through the fifth harmonic overtones of the vibrational mode (1.38, 0.95, 0.72 and 0.60 μm) do contribute to losses in these regions. In addition, these harmonics combine with the fundamental vibrational mode of tetrahedral SiO_4 molecules to contribute to fiber loss in the near IR and visible regions.

Figures 8.24a and b show the total attenuation of a fiber with GeO_2 and Al_2O_3-doped cores operating in the visible and near IR regions [35, 36]. Many of the sharp peaks in these curves can be attributed to OH^-. Table 8.3 lists the attenuation due to a given impurity concentration level.

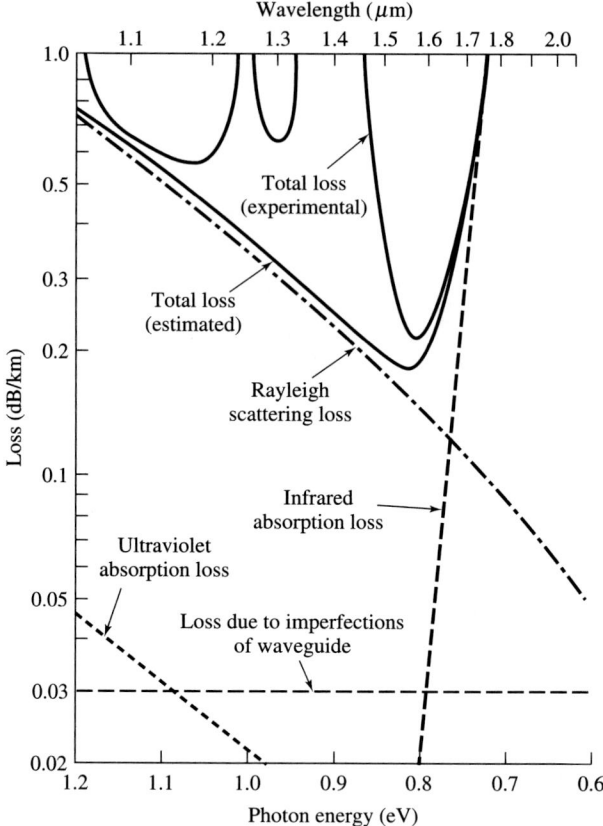

(a) GeO_2 doped silica core [35]

Figure 8.24 Transmission loss in silica based fibers

Table 8.3 Loss due to hydroxyl ions and other impurities [34]

Impurity	Concentration	Add loss at wavelength
OH^-	1 ppm	50 dB/km at 1.38 μm 2.4 dB/km at 1.13 μm 1 dB/km at 0.95 μm
FE^{+2}	1 ppb	0.7 dB/km at 1.10 μm
Cu^{+2}	1 ppb	0.4 dB/km at 0.85 μm
Cr^{+3}	1 ppb	1.0 dB/km at 0.65 μm

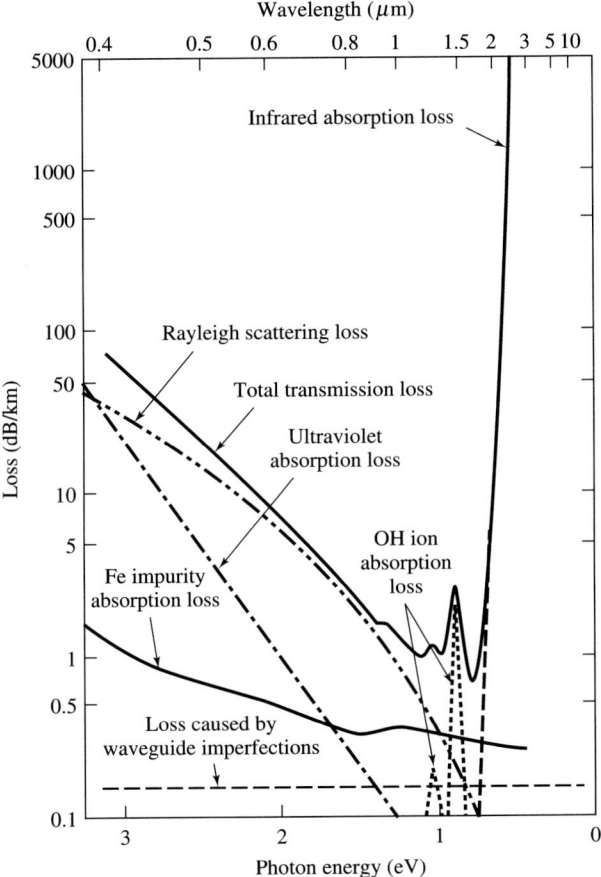

(b) Al_2O_3 doped silica core [36] © 1983 IEEE

Figure 8.24 (*Continued*)

8.6.2 Intrinsic Scattering

Since glasses are amorphous materials, density fluctuations in them are inevitable, and such fluctuations are present in all glass media. In multicomponent oxide glasses, there are compositional fluctuations, as well. This means that submicron-sized grains are distributed throughout glass materials. These density variations and compositional changes lead to microscopic variations in the refractive index, which in turn causes scattering. The process of absorption and immediate re-emission of waves by scatterers smaller than the wavelength is known as **Rayleigh scattering.** It is well known that the power loss due to Rayleigh scattering varies as λ^{-4}.

To appreciate this λ^{-4} dependence, we consider a tiny scatterer, when illuminated by incoming waves, as an electric dipole antenna with a dipole moment

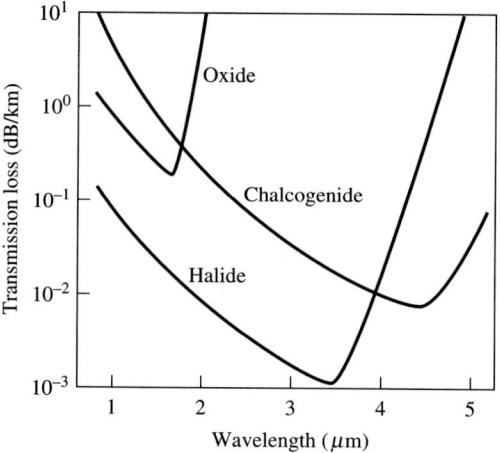

Figure 8.25 Projected loss minimum of IR glasses [39] © 1984 IEEE

p. The radiation by short dipole antennas is discussed in many elementary texts on electromagnetics [37, 38]. In particular, the time-average power radiated by an elementary dipole antenna of length h and current I is

$$P_{av} = \left(\frac{Ih}{\lambda}\right)^2 \eta_o \frac{\pi}{3}$$

The antenna current I is related to the electric charge q through the relationship $I = j\omega q$. By rewriting the expression in terms of the dipole moment $p = qh$, we obtain

$$P_{av} = \frac{4\pi^3 c}{3\varepsilon_o} \frac{p^2}{\lambda^4} \tag{8.52}$$

Note the dependence on λ^{-4}. Because of this dependence, Rayleigh scattering is the dominant loss term at UV and visible wavelengths, as shown in Figures 8.24a and b.

With the advent of modern fabrication techniques, the concentration of OH^- and transition metal ions in fibers is reduced to a minute level, resulting in a greatly reduced impurity absorption. Nevertheless, the absorption peaks due to these impurities are still noticeable as shown in Figures 8.24a and b. Except for the isolated absorption lines, the total fiber loss is bounded by the Rayleigh scattering in the UV and visible regions, and by the absorption due to molecular vibrations in the IR regions.

For silica-based fibers, a minimum loss occurs near 1.5 to 1.6 μm. If other host materials are used in lieu of SiO_2 to form the glass, the loss due to molecular vibrations can be pushed to longer wavelengths. At longer wavelengths, the Rayleigh scattering is also reduced. Thus, fibers based on halides or chalcogenides may have loss minima in the 3 to 4 μm region, as depicted in Figure 8.25 [39].

8.6.3 Waveguide Imperfections

Waveguide imperfections, such as changes in the radius, eccentricity, ellipticity, and core index profile, as well as impurities or air bubbles in the core and cladding regions, also cause scattering (Figure 8.23). Since scattering loss due to waveguide imperfections is roughly independent of the wavelength, this loss component is represented by a horizontal line in Figures 8.24a and b.

REFERENCES

1. Wood, L. E. "Optical communication systems for short haul applications." *Opt. Eng.* 13, (1974), pp. 409–415.
2. Izawa, T.; and S. Sudo. *Optical Fibers: Materials and Fabrication.* KTK Scientific Publications, Tokyo, Japan: Kluwer Academic Publications, 1986.
3. Ramo, S.; J. R. Whinnery; and T. Van Duzer. *Fields and Waves in Communication Electronics.* 3rd ed. New York, NY: John Wiley & Sons, Inc., 1994.
4. Paris, D. T.; and F. K. Hurd. *Basic Electromagnetic Theory.* New York, NY: McGraw-Hill Book Company, 1969.
5. Snitzer, E. "Cylindrical dielectric waveguide modes." *J. of Opt. Soc. of Am.* 51, (1961), pp. 491–498; and Snitzer, E.; and H. Osterberg. "Observed dielectric waveguide modes in the visible spectrum." *J. of Opt. Soc. of Am.* 51, (1961), pp. 499–505.
6. Yeh, C. "Guided waves modes in cylindrical optical fibers." *IEEE Transactions on Education* E-30, (1987), pp. 43–51.
7. Marcuse, D. *Light Transmission Optics.* Princeton, NJ: Van Nostrand Reinhold Co., 1972.
8. Adams, M. J. *An Introduction to Optical Waveguides.* New York, NY: John Wiley & Sons, Inc., 1981.
9. Okoshi, T. *Optical Fibers.* New York, NY: Academic Press, 1982.
10. Cherin, A. H. *An Introduction to Optical Fibers.* New York, NY: McGraw-Hill Book Company, 1983.
11. Snyder, A. W.; and J. D. Love. *Optical Waveguide Theory.* London, England: Chapman and Hall, Ltd., 1983.
12. Keiser, G. *Optical Fiber Communications.* 2nd ed. New York, NY: McGraw-Hill Book Company, 1991.
13. Gloge, D. "Weakly guided fibers." *Appl. Opt.* 10, (1971), pp. 2252–2258.
14. Abramowitz, M.; and I. A. Stegun, (ed.) *Handbook of Mathematical Functions, with Formulas, Graphs, and Mathematical Tables,* New York, NY: Dover Publications, Inc., (1965).
15. Rudolph, H. D.; and E. G. Neumann. "Approximations for the eigenvalues of the fundamental mode of a step index glass fiber waveguide." *Nachrichtentechn. Z.* 29, (1976), pp. 328–329.
16. Ballard, S. S.; J. S. Browder; and J. F. Ebersole. "Refractive index of special crystals and certain glasses." Chapter 6. *American Institute of Physics Handbook.* 3rd ed. ed. G. E. Gray. New York, NY: McGraw-Hill Book Company, 1972.
17. Gloge, D. "Dispersion in weakly guiding fibers." *Appl. Opt.* 10, (1971), pp. 2442–2445.

18. Chang, C. T. "Minimum dispersion in a single-mode step-index optical fiber." *Appl. Opt.* 18, (1979), pp. 2516–2522.
19. Love, R.; and D. B. Keck. "Characteristics of optical fiber waveguides." *Opt. Eng.* 17, (1978), pp. 114–119.
20. Marcuse, D. "Interdependence of waveguide and material dispersion." *Appl. Opt.* 18, (1979), pp. 2930–2932.
21. Jeunhomme, J. "Dispersion minimization in single-mode fibres between 1.3 μm and 1.7 μm." *Electron. Lett.* 15, (1979), pp. 478–479.
22. White, K. I.; and B. P. Nelson. "Zero total dispersion in step-index monomode fibres at 1.30 and 1.55 μm." *Electron. Lett.* 15, (1979), pp. 396–397.
23. Bagley, B. G.; C. R. Kurkjian; J. W. Mitchell; G. E. Peterson; and A. R. Tynes. "Materials, Properties and Choices." Chapter 7. *Optical Fiber Telecommunications*, ed. S. E. Miller; and A. G. Chynoweth. New York, NY: Academic Press, 1979.
24. Schultz, P. C. "Progress in optical waveguides process and materials." *Appl. Opt.* 18, (1979), pp. 3684–3693.
25. Schultz, P. C. "Fabrication of optical waveguides by the outside vapor deposition process." *Proc. IEEE* 68, (1980), pp. 1187–1190.
26. VanDewoestine, R. V.; and A. J. Morrow. "Developments in optical waveguide fabrication by the outside vapor deposition process." *IEEE J. Lightwave Technol.* LT-4, (1986), pp. 1020–1025.
27. Izawa, T.; and N. Inagaki. "Materials and processes for preform fabrication–vapor-phase axial deposition." *Proc. IEEE* 68, (1980), pp. 1184–1187.
28. Murata, H. "Recent developments in vapor phase axial deposition." *IEEE J. Lightwave Technol.* LT-4, (1986), pp. 1026–1033.
29. MacChesney, J. B.; P. B. O'Connor; and H. M. Presby. "A new technique for the preparation of low-loss and graded-index optical fibers." *Proc. IEEE* 62, (1974), pp. 1280–1281.
30. MacChesney, J. B. "Materials and processes for preform fabrication–modified chemical vapor deposition and plasma chemical vapor deposition." *Proc. IEEE.* 68, (1980), pp. 1181–1184.
31. Partus, F. P.; and M. A. Saifi. "Lightguide preform manufacture." *The Western Electric Engineer* 26, (Winter, 1980), pp. 39–47.
32. Kuppers, D.; H. Lydtin; and F. Meijer. "Preparation methods for optical fibers applied in Philips Research." *1977 International Conference on Integrated Optics and Optical Fiber Communication, Tokyo, Japan*, (July 18–20, 1977), pp. 319–322.
33. Lydtin, H. "PCVD: a technique suitable for large-scale fabrication of optical fibers." *IEEE J. Lightwave Technol.* LT-4, (1986), pp. 1034–1038.
34. Jefferies, J. A.; and R. J. Klaiber. "Lightguide theory and its implications in manufacturing." *The Western Electric Engineer* 26, (1980), pp. 13–23.
35. Miya, T.; Y. Terunma; T. Hosaka; and T. Miyashita. "Ultimate low-loss single-mode fiber at 1.55 μm." *Electron. Lett.* 15, (1979), pp. 106–108.
36. Ohmori, Y.; T. Miya; and M. Horiguchi. "Transmission-loss characteristics of Al_2O_3-doped silica fibers." *IEEE J. Lightwave Technol.* LT-1, (1983), pp. 50–55.
37. Hayt, W. H., Jr. *Engineering Electromagnetics.* New York, NY: McGraw-Hill Book Company, 1989.

38. Iskander, M. F. *Electromagnetic Fields and Waves.* New York, NY: Prentice Hall, Inc. 1992.
39. Tran, D. C.; G. H. Siegel, Jr.; and B. Bendow. "Heavy metal fluoride glasses as fibers, a review." *IEEE J. Lightwave Technol.* LT-2, (1984), pp. 566–586.

ADDITIONAL READING

1. Barnoski, M. K. *Fundamentals of Optical Fiber Communications.* 2nd ed. New York, NY: Academic Press, 1981.
2. Basch, E. E. *Optical Fiber Transmission.* Indianapolis, IN: Howard W. Sams & Co. Inc., (1988).
3. Cheo, P. K. *Fiber Optics & Optoelectronics.* 2nd ed. New York, NY: Prentice Hall, Inc., 1990.
4. Jones, K. A. *Introduction to Optical Electronics.* New York, NY: Harper & Row Publishers, 1987.
5. Jones, W. B., Jr. *Introduction to Optical Fiber Communication Systems.* Holt, Reinhart and Winston, 1988.
6. Karim, M. A. *Electro-optical Devices and Systems.* Boston, MA: PWS-Kent Publishing Co., 1990.
7. J. D. Kraus, *Electromagnetics.* 4th ed. New York, NY: McGraw-Hill Book Co., 1991.
8. Midwinter, J. E.; and Y. L. Guo. *Optoelectronics and Lightwave Technology.* New York, NY: John Wiley & Sons, Inc., 1992.
9. Palais, J. C. *Fiber Optic Communications.* 2nd ed. New York, NY: Prentice Hall, Inc., 1988.
10. Senior, J. M. *Optical Fiber Communications, Principles, and Practice.* 2nd ed. New York, NY: Prentice Hall, Inc., 1992.
11. Suematsu, Y.; and K. I. Iga. *Introduction to Optical Fiber Communications.* New York, NY: John Wiley & Sons, Inc., 1982.
12. Wilson, J.; and J. F. B. Hawkes. *Optoelectronics, An Introduction.* 2nd ed. New York, NY: Prentice Hall, Inc., 1989.
13. Yariv, A. *Introduction to Optical Electronics.* 4th ed. New York, NY: Sauners College Pub., 1991.

PROBLEMS

1. Consider an optical fiber with $n_{co} = 1.540$, $n_{cl} = 1.500$, and $a = 4.0$ µm. How many LP modes are supported by the fiber operating at $\lambda = 1.55$ µm? At what wavelength would the fiber become single-moded? If the indices are roughly independent of wavelength, in what wavelength range would the fiber support the LP_{01} and LP_{11} modes?
2. Estimate the intermodal dispersion of a step index fiber with $n_{co} = 1.540$, $n_{cl} = 1.500$, and $a = 25.0$ µm, operating at 1.55 µm. What is the multimode dispersion if the fiber has a parabolic index profile?

3. A fiber with $n_{co} = 1.480$ and $n_{cl} = 1.470$ is intended for single-mode operation at $\lambda = 1.22$ μm. However, the core is too large; the core radius is actually 3.4 μm and the fiber supports more than one mode.
   a. How many LP modes are guided by the fiber?
   b. As accurately as possible, estimate the bandwidth–length product of the fiber. Ignore the intramodal dispersion effects.
      Remark: The equation for the bandwidth–product derived in section 8.4 is not applicable for the fiber in question, since it supports a few, but not very many, modes.

4. Consider a bare fiber ($n_{cl} = 1.000$) made with polycrystalline thallium bromoiodide (trade name KRS-5) with a core index of 2.37 at 10.6 μm and a core radius of 250 μm.
   a. Is the condition for "weakly guided" satisfied?
   b. Starting from the definition for numerical aperture, calculate the numerical aperture of the bare KRS-5 fiber.
   c. Suppose the fiber is cladded with a material that has an index of 2.20. What should the radius be if the cladded KRS-5 fiber is to be single-moded at 10.6 μm?

5. A weakly guided single-mode fiber is designed for operation at 1.5 μm. Let $n_{co} = 1.458$ at 1.5 μm. As an approximation, assume n_{co} and Δ (yet to calculated) are roughly constant over the wavelength range of interest.
   a. Choose a reasonable value for Δ and the core radius such that $V = 2.20$ at $\lambda = 1.50$ μm.
   b. At what wavelength would the fiber designed in (a) cease to be single-moded?
   c. How many LP modes does the fiber designed in (a) support at $\lambda = 0.80$ μm?

6. Consider a weakly guided fiber with a numerical aperture of 0.15, a core index of 1.465, and a core radius of 6.5 μm.
   a. How many LP modes does the fiber support at $\lambda = 1.3$ μm?
   b. Suppose a narrow pulse with a center wavelength of 1.3 μm is fed into the fiber. How many pulses emerge from the far end, if the fiber is 1-km long?
   c. Estimate the pulse separation between neighboring pulses. Ignore the effects due to intramodal dispersion.
   d. What is the time separation between the first and last pulse?
   e. Suppose the core radius is changed to 7.4 μm and repeat (d).

7. Derive expressions for E_x and E_z of $LP_{\ell m}$ modes for $\ell \neq 0$.

8. Derive (8.38).

BESSEL AND MODIFIED BESSEL FUNCTIONS

APPENDIX C

APPENDIX OUTLINE

C.1 Differential Equations
C.2 Power Series Expansions And Asymptotic Expansions
References

C.1 DIFFERENTIAL EQUATIONS

A Bessel differential equation is a second-order ordinary differential equation of the form of

$$\frac{1}{z}\frac{d}{dz}\left(z\frac{df}{dz}\right) + \left(1 - \frac{\ell^2}{z^2}\right)f = 0 \tag{C1}$$

The equation has two linearly independent solutions. They are the **Bessel functions** of the **first kind** $J_\ell(z)$, and the **second kind**, $Y_\ell(z)$, if ℓ is a real, positive number. Two functions are oscillatory functions of z. $J_\ell(0)$ is either 1 ($\ell = 0$) or 0 ($\ell \neq 0$), while $Y_\ell(z)$ is singular at $z = 0$. The solutions can also be expressed as the superposition of the **Hankel functions** of the first and second kinds, $H_\ell^{(1)}(z)$ and $H_\ell^{(1)}(z)$. Again, for a real, positive ℓ,

$$H_\ell^{(1)}(z) = J_\ell(z) + j\,Y_\ell(z)$$
$$H_\ell^{(2)}(z) = J_\ell(z) - j\,Y_\ell(z)$$

A differential equation of the form of

$$\frac{1}{z}\frac{d}{dz}\left(z\frac{df}{dz}\right) - \left(1 + \frac{\ell^2}{z^2}\right)f = 0 \tag{C2}$$

is known as the modified Bessel differential equation. Note in particular the sign change in one of the terms. The two linearly independent solutions are the **modified Bessel functions** of the **first kind,** $I_\ell(z)$, and the **second kind,** $K_\ell(z)$. $I_\ell(z)$ is an exponentially increasing function of z, while $K_\ell(z)$ is an exponentially decreasing function. Also note that $I_\ell(0)$ is either 1 ($\ell = 0$) or 0 ($\ell \neq 0$), and $K_\ell(z)$ is singular at $z = 0$.

The Bessel and modified Bessel functions are complicated functions. Therefore, only the relations and identities needed in our discussion of fibers are listed here; that is, we will only be concerned with $J_\ell(z)$ and $K_\ell(z)$.

The **recurrence relations** for Bessel and modified Bessel functions of integer order ℓ are,

$$J'_\ell(z) = J_{\ell-1}(z) - \frac{\ell}{z}J_\ell(z) = -J_{\ell+1}(z) + \frac{\ell}{z}J_\ell(z) \tag{C3}$$

$$J_{\ell-1}(z) + J_{\ell+1}(z) = \frac{2\ell}{z}J_\ell(z) \tag{C4}$$

$$K'_\ell(z) = -K_{\ell-1}(z) - \frac{\ell}{z}K_\ell(z) = -K_{\ell+1}(z) + \frac{\ell}{z}K_\ell(z) \tag{C5}$$

$$K_{\ell-1}(z) - K_{\ell+1}(z) = -\frac{2\ell}{z}K_\ell(z) \tag{C6}$$

C.2 POWER SERIES EXPANSIONS AND ASYMPTOTIC EXPANSIONS

The leading term, or terms, of the **power series expansions** for $J_\ell(z)$ and $K_\ell(z)$ are:

$$J_0(z) \approx 1 - \frac{z^2}{4} \tag{C7}$$

$$J_\ell(z) \approx (\frac{z}{2})^\ell \frac{1}{\Gamma(\ell + 1)}, \qquad \ell \neq -1, -2, -3, \ldots, \tag{C8}$$

$$K_0(z) \approx -\ln z \tag{C9}$$

$$K_\ell(z) \approx \frac{\Gamma(\ell)}{2} (\frac{2}{z})^\ell, \qquad \ell \geq 1 \tag{C10}$$

where $\Gamma(\ell)$ is Gamma function and for integer ℓ, $\Gamma(\ell) = (\ell-1)!$.

The power series expansions are useful in evaluating those functions with small arguments. If z is large, it is necessary to use the **asymptotic expressions**. For $z \gg \ell$, the asymptotic expressions are

$$J_\ell(z) \approx \sqrt{\frac{2}{\pi z}} \left[\cos(z - \frac{\ell\pi}{2} - \frac{\pi}{4}) + O(|z|^{-1}) \right] \tag{C11}$$

$$K_\ell(z) \approx \sqrt{\frac{2}{\pi z}} e^{-z} \left[1 + O(|z|^{-1}) \right] \tag{C12}$$

If ℓ is also large and $z > \ell$, then

$$J_\ell(z) \approx \sqrt{\frac{2}{\pi(z^2 - \ell^2)^{1/2}}} \left[\cos(\sqrt{z^2 - \ell^2} - \ell \cos^{-1}\frac{\ell}{z} - \frac{\pi}{4}) + O(\ell^{-1}) \right] \tag{C13}$$

If $\ell > z,$ yet another asymptotic expression should be used. However, it is not needed for our discussion and is therefore not presented here. Details can be found in [C.1] and [C.2].

References:

C.1 Watson, G. N. *A Treatise on the Theory of Bessel Functions,* 2d ed., London, England: Cambridge University Press, 1966

C.2 Abramowitz, M.; and I. A. Stegun, (ed.) *Handbook of Mathematical Functions with Formulas, Graphs, and Mathematical Tables,* New York, NY: Dover Publications, Inc., 1965.

WEAKLY GUIDED FIBERS: PARABOLIC INDEX PROFILE

APPENDIX D

APPENDIX OUTLINE

D.1 Approximate Characteristic Equation
D.2 Parabolic Index Profile
D.3 Intermodal Dispersion
References

D.1 APPROXIMATE CHARACTERISTIC EQUATION

For graded index fibers, the core index is a function of r, as given in (8.2). Since no exact characteristic equation for graded index fibers with an arbitrary g is known, an approximate characteristic equation may be established. To do so, consider uniform plane waves propagating in a medium with an index n. The wave vector is $\mathbf{k_1} = n\mathbf{k}$, where $\mathbf{k}$ is the vacuum wave vector. In cylindrical coordinates, the wave vector can be written as

$$\mathbf{k_1} = k_{1r}\mathbf{a_r} + k_{1\phi}\mathbf{a_\phi} + \beta\mathbf{a_z}$$

where β is the propagation constant. Clearly,

$$k_{1r}^2 + k_{1\phi}^2 + \beta^2 = k^2 n^2$$

For structures with a cylindrical symmetry, the fields vary in the ϕ direction as $\sin\ell\phi$, $\cos\ell\phi$, or a combination of the two functions. At a distance r from the axis, field variation in the ϕ direction has a periodicity $2\pi r/\ell$. Therefore, $k_{1\phi}$ is

$$k_{1\phi} = \frac{2\pi}{2\pi r/\ell} = \frac{\ell}{r}$$

In terms of ℓ and β, k_{1r}^2 is

$$k_{1r}^2 = k^2 n^2 - \beta^2 - (\ell^2/r^2) \tag{D1}$$

Also, k_{1r}^2 can be positive or negative, depending on n, k, β, ℓ, and r.

In graded index fibers, the core index n is a function of r. Thus, k_{1r}^2 can be positive in one region and negative elsewhere. If k_{1r}^2 is positive, then k_{1r} is a *real* quantity, meaning that the fields in the region vary in an oscillatory fashion in the radial direction. If k_{1r}^2 is negative, then k_{1r} is *imaginary,* and the fields decay exponentially in the radial direction. The points where k_{1r}^2 vanishes are *demarcations* between the two regions and are referred to as the **turning points** or **caustics.** In most graded index fibers, there are two turning points. We identify them as R_1 and R_2. In the following discussions, assume that k_{1r}^2 is negative for $r < R_1$ and for $r > R_2$, and that k_{1r}^2 is positive for the region between R_1 and R_2.

The discussion on thin-film waveguides (Chapter 7) notes that fields in the film region vary in an oscillatory manner, and the characteristic equation (7.16) is equivalent to a condition that requires the total round-trip phase delay in the transverse direction to be integer multiples of 2π. For the graded index fibers in question, the fields vary in r in an oscillatory manner in the region between the two turning points. Therefore, the approximate characteristic equation is of the form

$$2\int_{R_1}^{R_2} \left(k^2 n^2(r) - \beta^2 - \frac{\ell^2}{r^2}\right)^{1/2} dr - \phi_1 - \phi_2 = 2m\pi \tag{D2}$$

where $m = 1, 2, 3, \ldots$ is the **radial mode number,** and ϕ_1 and ϕ_2 are the phase shifts at the turning points R_1 and R_2, respectively. Rigorous derivations of [D1, D2] show that ℓ^2 in (D2) should be replaced by $(\ell^2 - 1/4)$. However, the phase terms, ϕ_1 and ϕ_2, and the constant $1/4$ are not very important for higher-order modes. Since we are mainly interested in multimode graded index fibers with a large number of modes, these terms are ignored, as an approximation.

D.2 PARABOLIC INDEX PROFILE

Since the general case with an arbitrary g is very complicated, we will only consider the special case of fibers with a parabolic index profile ($g = 2$). For this special case, the turning points, R_1 and R_2, and, more importantly, the radial mode number m, can be readily obtained. Substituting (8.2) into (D2) and using $g = 2$ and $\beta = kN$, we obtain

$$m = \frac{k}{\pi} \int_{R_1}^{R_2} \left[n_{co}^2\left(1 - 2\Delta\left(\frac{r}{a}\right)^2\right) - N^2 - \frac{\ell^2}{k^2 r^2}\right]^{1/2} dr \tag{D3}$$

By setting the integrand to zero, we have explicit expressions for the turning points,

$$\left.\begin{matrix}R_1^2\\R_2^2\end{matrix}\right\} = \frac{a^2}{4n_{co}^2\Delta}\left[(n_{co}^2 - N^2) \mp \left((n_{co}^2 - N^2)^2 - 8n_{co}^2\Delta\frac{\ell^2}{k^2 a^2}\right)^{1/2}\right]$$

which are functions of ℓ and ka.

The integral in (D3) can be evaluated in a closed form, with the following result:

$$m = \frac{ka(n_{co}^2 - N^2)}{4n_{co}\sqrt{2\Delta}} - \frac{\ell}{2}$$

The effective index of refraction N of modes guided in fibers with a parabolic index profile is then given by

$$N^2 = n_{co}^2 - \frac{2n_{co}\sqrt{2\Delta}}{ka}(\ell + 2m) \tag{D4}$$

Again, note that modes with the same $(\ell + 2m)$ have the same effective index of refraction. Therefore, the **principal mode number** ($m_p = \ell + 2m$) defined in (8.36) for step index fibers is also useful for fibers with a parabolic index profile.

For guided modes, N must be greater than n_{cl}. From (D4), we have

$$m_p \leq kan_{co}\sqrt{\frac{\Delta}{2}} = \frac{V}{2}$$

The largest principal mode number, $m_{p\,max}$, is $V/2$. The largest ℓ and m are $V/2$ and $V/4$, respectively.

To estimate the total number of modes, observe that ℓ can assume any value between 0 and $m_{p\,max}$, and that for a given ℓ, m can be any integer between 1 and $(m_{p\,max} - \ell)/2$. Therefore, the *total number of modes* for fibers with a *parabolic index profile* is

$$M_{tot}\bigg|_{parabolic} = 4 \sum_{\ell=0}^{m_{p\,max}} \left(\frac{m_{p\,max} - \ell}{2}\right) \approx m_{p\,max}^2 = \frac{V^2}{4} \tag{D5}$$

This equation should be compared with (8.38) for step index fibers.

D.3 INTERMODAL DISPERSION

To estimate the pulse broadening in multimode fibers with a parabolic index profile, we use (D4) to calculate $\delta\tau_{mm}$. Since $\beta = kN$, we have, from (D4),

$$\beta = \left[\frac{kn_{co}}{a}\left(kn_{co}a - 2\sqrt{2\Delta}(\ell + 2m)\right)\right]^{1/2}$$

To find the group velocity, differentiate the above expression with respect to k. Since core and cladding materials are basically similar, the dispersions of these materials are also similar. In other words, Δ is roughly independent of λ. By ignoring $\frac{d\Delta}{dk}$, we have

$$\frac{d\beta}{dk} \approx \frac{(kn_{co})'}{\sqrt{kn_{co}a}} \frac{kn_{co}a - \sqrt{2\Delta}(\ell + 2m)}{[kn_{co}a - 2\sqrt{2\Delta}(\ell + 2m)]^{1/2}} \tag{D6}$$

where $(kn_{co})'$ was given in Chapter 8.

For low-order modes, including the fundamental mode, we can ignore $[\sqrt{\Delta}\,(\ell + 2m)]$ in (D6) and obtain $\dfrac{d\beta}{dk} \approx (kn_{co})'$. For the highest-order mode,

$$\ell + 2m \approx kan_{co}\sqrt{\Delta/2}$$

and

$$\frac{d\beta}{dk} \approx (kn_{co})'\,\frac{1-\Delta}{\sqrt{1-2\Delta}}$$

Recall that

$$\frac{1}{v_{gr}} = \frac{1}{c}\frac{d\beta}{dk} \quad \text{and} \quad (kn_{co})' \approx n_{co}$$

Hence, we have

$$\delta\tau_{mm} = \frac{L}{v_{gr\,sl}} - \frac{L}{v_{gr\,fs}} \approx \frac{L}{c}(kn_{co})'\left(\frac{1-\Delta}{\sqrt{1-2\Delta}} - 1\right) \tag{D7}$$

which corresponds to (8.40). Expressing (D7) in terms of a power series in Δ and keeping the leading term, we have the expression for the intermodal pulse broadening of multimode fibers with a parabolic index profile:

$$\delta\tau_{mm} \approx \frac{L}{c}\,n_{co}\,\frac{\Delta^2}{2} \quad \text{(parabolic index profile)} \tag{D8}$$

This was given as (8.42) in Chapter 8.

References

1. Gloge, D.; and E. A. J. Marcatili. "Multimode theory of graded fibers." *Bell Syst. Tech. J.* 52, (1973), pp. 1563–1578.
2. Gloge, D. "Propagation effects in optical fibers." *IEEE Trans. Microwave Theory Tech.*, MTT-23, (1975), pp. 106–120.

MULTIMODE FIBER COMPONENTS AND SYSTEMS

CHAPTER 9

CHAPTER OUTLINE

9.1 Introduction
9.2 Taps and Star Couplers
    9.2.1 Passive Components Based on Thin-Film Beam Splitters
    9.2.2 Passive Components Based on GRIN Lenses
    9.2.3 Passive Components Based on Spherical Reflectors
    9.2.4 Fused Biconical Taper Fiber Directional Couplers
    9.2.5 Star Couplers with a Mixer Rod or Plate
9.3 Multimode Fiber Switches
    9.3.1 Mechanical Switches
    9.3.2 Electrowetting Switches
    9.3.3 Liquid Crystal Switches
9.4 Power Coupled into Fibers
    9.4.1 Coupling Efficiency
    9.4.2 Lens Effects
9.5 Connector Loss
    9.5.1 Intrinsic Loss
    9.5.2 Extrinsic Loss
    9.5.3 Normalized Mismatch Parameters
9.6 Risetime, Pulse Width, and Bandwidth
    9.6.1 Empirical Bandwidth–Risetime Relationship
    9.6.2 Bandwidth–Pulse Width Relationship
9.7 Power Budget and Risetime Budget Example
References
Problems

9.1 INTRODUCTION

In the preceding chapters, we discussed the optical sources, detectors, modulators and fibers. In this chapter, we will study optical fiber communication systems based on these components. Since multimode fibers have limited bandwidth-length product, their use is restricted to short haul systems, such as local networks (LANs). However, the core diameter of multimode fibers is much larger than that of single mode fibers, the alignment of multimode fiber components is much less critical than that required for single mode components. In addition, the 0.8 μm LEDs and detectors compatible with multimode fibers are reliable and inexpensive. Therefore, for simple communication systems with link lengths of a few kilometers or shorter, multimode graded index fibers may be the preferred choice. In this chapter, our discussions are confined entirely to multimode fiber systems.

In communications networks, terminals may either be linked directly to each other, or may be connected to a main bus or a central node. Each terminal consists of a transmitter and a receiver. Depending on the system architecture, a LAN can have one or a combination of *three basic topologies,* as shown in Figure 9.1 [1]. In the **bus configuration** (Figure 9.1a), each terminal is connected to a common bus, also known as the trunk. Communication between terminals is through the bus. The key components of the bus topology are the **taps.** This arrangement is similar to the architecture commonly used in LANs based on coaxial transmission lines. For a given system length and number of terminals, the total fiber length needed for a bus configuration is shorter than for other configurations. When new terminals are needed, taps are simply added, with minimum disturbance to existing connections and terminals.

If the ends of the bus are joined, as depicted in Figure 9.1b, a **ring configuration** is formed. In the special case where all taps merge to a common central junction, the setup is a **star configuration** (Figure 9.1c). The central junction, which has a large number of input and output ports, is referred to as the **star coupler,** which is the key component in a star configuration. In a star configuration, the signal-path redundancy is greatly increased at the expense of total fiber length. Taps, star couplers, and other passive components are discussed in

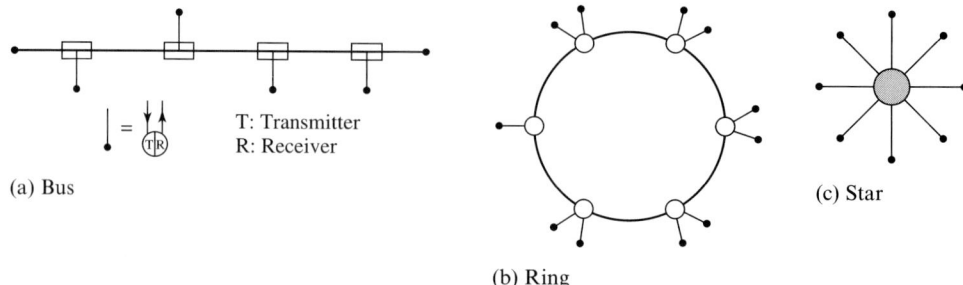

Figure 9.1 Basic topologies of local networks ([1], © 1985 IEEE)

section 9.2. When a network must be rearranged, or the paths to the terminals must be changed, optical switches are used. Optical switches are discussed in section 9.3.

In any communication system, it is crucial to insure that the signals arriving at a terminal must not be too weak to be extracted from noise, nor distorted beyond recognition. **Loss** is one of the factors limiting the maximum system length.

From a power consideration standpoint, we can determine: (a) the maximum system length for a given power level available from the transmitter; and (b) the minimum power level required by a receiver in a specific detection scheme. The minimum signal level required by a receiver depends on the type of detector and signal processing electronics used, the detection scheme adapted, and the signal-to-noise ratio (SNR) or the bit error rate (BER) dictated by the system requirements [2, 3]. These topics are too complicated to be discussed in this text. Therefore, for the present, we will simply assume that the minimum power level required by the receiver is known, and we select optical sources and components to meet those requirements. Factors included in the consideration are: source-to-fiber coupling, fiber attenuation, insertion loss introduced by the connectors and various components, and fiber-to-detector coupling. These topics are discussed in sections 9.4 and 9.5. **Pulse broadening** also affects the system length, as well as the pulse repetition rate. The limitation imposed by the risetime or bandwidth requirement is discussed in section 9.6. The chapter concludes with a simple example.

9.2 TAPS AND STAR COUPLERS

9.2.1 Passive Components Based on Thin-Film Beam Splitters

A passive component that distributes power from an input port to two output ports is known as a **1×2 coupler** or a **tee coupler**. Such a passive component is also referred to as a **tap** or a **power divider**, depending on the power splitting ratio in the output ports. If the input power is divided approximately evenly in the two output ports, the 1×2 coupler is referred to as a *power divider*. On the other hand, if only a small fraction of the input power is diverted to one port, while a large fraction goes to the other port, the 1×2 coupler is referred to as a *tap*. Since it is a passive, reciprocal device, a 1×2 coupler can also function as a *2×1 coupler*, in which signals fed into the two input ports are *combined* in a single output port. In such an application, this passive component is referred to as a **power combiner**.

In the conventional optic or bulk optic form, a 1×2 coupler can simply be a beam splitter, which directs beams from an input port to two output ports, as shown in Figure 9.2a. Since a finite fiber separation is required for placement of the beam splitter, and since the beam emerging from the input fiber diverges quickly, the exiting beam must be collimated to reduce power loss. Conventional

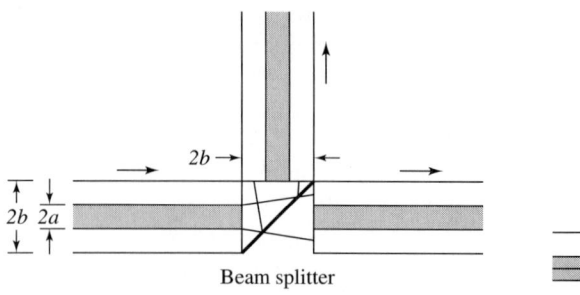

(a) A tee coupler with a thin-film beam splitter

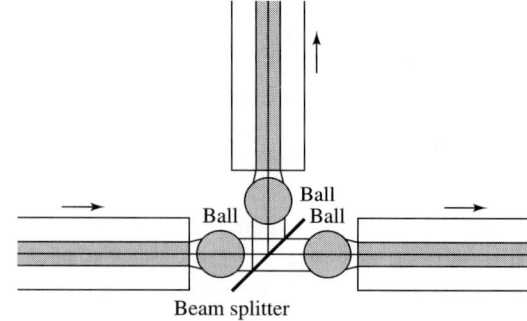

(b) A tee coupler with three glass balls and a thin-film beam splitter

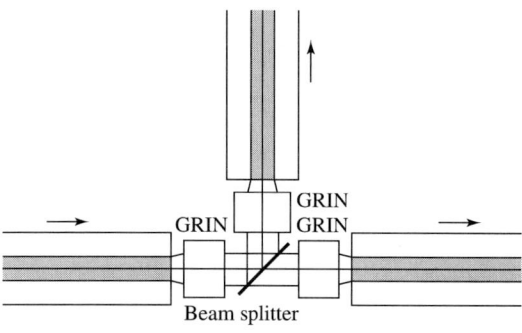

(c) A tee coupler based on a thin-film beam splitter and three quarter-pitch GRIN lenses

Figure 9.2 Simple tee couplers

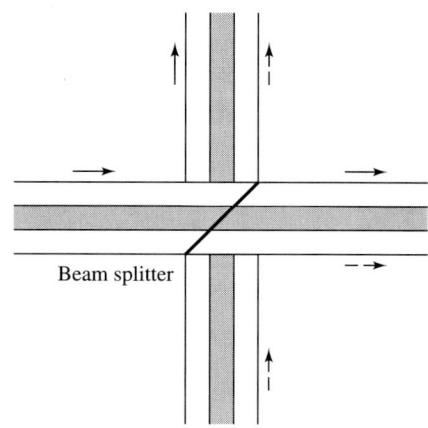

(d) A tee coupler with a thin-film splitter deposited on the fiber

lenses, spherical balls, or **graded refractive index (GRIN) lenses** are used for this purpose, as well as to focus the beams to the output fibers, as shown in Figures 9.2b and c. In addition, by making the beam radius of the collimated beam large, the alignment of the optical components is also made easier. An alternate arrangement involves cleaving the fiber at 45 degrees relative to the fiber axis, and depositing dielectric films directly on the cleaved fiber endface to serve as the beam splitter. This arrangement is shown in Figure 9.2d.

The use of lenses to improve fiber coupling is studied quantitatively in section 9.4. In this section, we will restrict ourselves to qualitative descriptions.

9.2.2 Passive Components Based on GRIN Lenses

Conventional lenses and GRIN lenses are used in optical components, including optical fiber components, to focus light or transform the beam size. The function of such lenses is derived from the fact that the *optical path lengths* in

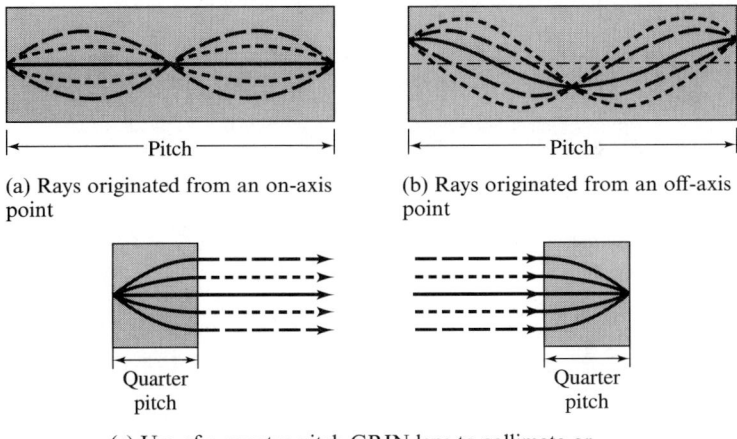

Figure 9.3 Rays in GRIN lenses, and the use of quarter-pitch GRIN lenses

different parts of the lens are different. In conventional lenses, the refractive index is the same everywhere in the lens, and the lens shape is contoured such that individual rays follow separate trajectories with different physical lengths. In contrast, a GRIN lens has two flat ends, different rays have only slightly different path lengths, and the required optical path length difference is realized by varying the refractive index as a function of position. This is the basic difference between conventional lenses and GRIN lenses.

Since the emerging optical beams diverge rather quickly, the beams must be collimated before further processing or transformation. Lenses with a large aperture, short focal length, or preferably both, are used for this purpose. Because GRIN lenses are inherently small in size and usually have short focal lengths, they are particularly suited to fiber applications. In addition, because GRIN lenses have a long cylindrical surface and flat ends, they are easy to mount.

A GRIN lens is a transparent cylinder with an index that varies as a function of the radius r [6–8]. In many GRIN lenses, $n(r)$ can be approximated by a quadratic equation

$$n(r) = n_a \left[1 - \left(\frac{2\pi^2 r^2}{p^2}\right)\right] \qquad (9.1)$$

where n_a is the refractive index on the lens axis.

In Chapter 7, we showed that meridional rays in a medium with a parabolic index profile follow sinusoidal trajectories with a periodicity p. Figures 9.3a and b depict rays that originate from on-axis and off-axis points. These curves clearly show that the rays return to their original radial position and slope relative to the z axis after traveling a distance p. The periodicity p is of special significance in GRIN lenses; it is commonly referred to as the **pitch**. The length of a GRIN lens is often specified as a fraction of the pitch. For example, a quarter-pitch

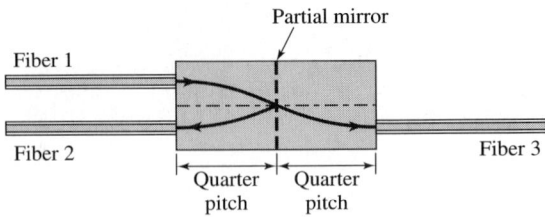

(a) A 1×2 coupler

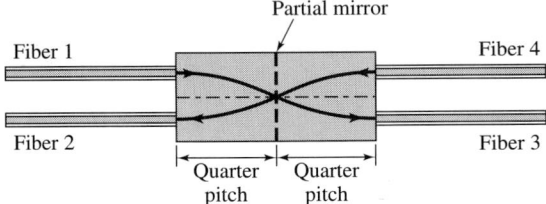

(b) A 2×2 coupler

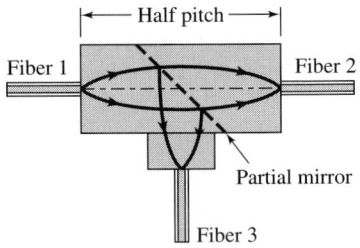

(c) A tee coupler

Figure 9.4 1×2 and 2×2 taps based on half-pitch GRIN lenses

GRIN lens converts a beam diverging from the center of an endface to a collimated beam at the other endface, or *vice versa,* as shown in Figure 9.3c. The GRIN lenses shown in Figure 9.2c are quarter-pitch rods.

Taps or power dividers can be designed to take advantage of the periodic nature of off-axis rays. As shown in Figure 9.3b, off-axis rays go to a point on the opposite side of the axis, after traveling a distance of $p/2$. In the arrangement shown in Figure 9.4a, a partial mirror is placed between two quarter-pitch GRIN lenses, allowing a portion of the input power from fiber 1 to be transmitted to fiber 3, with the rest reflected by the partial mirror to fiber 2. The arrangement shown in Figure 9.4a therefore works as a 1×2 coupler. If a fourth fiber is added, the arrangement can function as a multiplexer or demultiplexer, because the input from fiber 1 is *directed* to fibers 2 and 3, while the input from fiber 4 is *distributed* to fibers 2 and 3 (Figure 9.4b). The partial mirror can also be at 45 degrees relative to the lens axis, as shown in Figure 9.4c. Thus a tee coupler is formed.

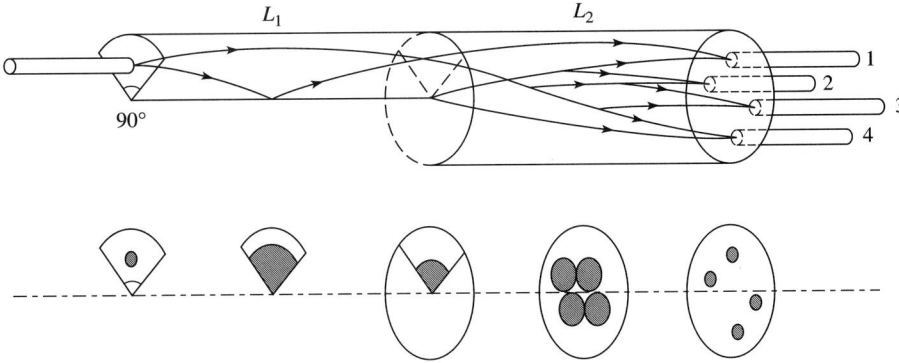

Figure 9.5 A 1×4 tap based on a 90-degree GRIN wedge and a GRIN rod [9]

The most ingenious use of GRIN rods is depicted in Figure 9.5. A quarter-pitch 90 degree GRIN wedge and a quarter-pitch GRIN rod are cascaded. In the 90 degree wedge, one-fourth of the input light goes directly into the GRIN rod. The rest is total internally reflected by the wedge surface before reaching the GRIN rod. Because of the periodic nature of the off-axis rays and the reflection by the wedge boundary, the rays are split into three parts in the GRIN rod. Thus four outputs emerge from the GRIN rod. In other words, the combination of a quarter-pitch 90 degree GRIN wedge and a quarter-pitch GRIN rod functions as a *1×4 star coupler.*

If the 90-degree GRIN wedge is replaced by a 45-degree quarter-pitch GRIN wedge, seven-eighths of the input is reflected by the 45 degree wedge surfaces before reaching at the GRIN rod. Because of the reflection at the wedge boundaries and the periodic nature of the off-axis rays, this part of input splits into seven ray bundles. One-eighth of the input reaches the GRIN rod directly and without encountering the wedge boundary. This portion of input forms the eighth ray bundle. Thus, the input splits into eight ray bundles at the output. In other words, by joining a 45-degree quarter-pitch GRIN wedge with a quarter-pitch GRIN rod, a 1×8 star coupler is formed [9].

9.2.3 Passive Components Based on Spherical Reflectors

It is well known that light emitted from the *center* of a spherical surface is reflected back to its original position, i.e., the center. On the other hand, rays emitted by an off-center emitter (point A in Figure 9.6a) are reflected to an image point at the opposite side of the center (point A′ in Figure 9.6a). If paraxial rays are involved, the emitter and its image are symmetric with respect to the center. A family of fiber components has been designed to take advantage of this approximately symmetric relationship. The simplest one is a power divider with *two tilted spherical surfaces,* as shown in Figure 9.6b [10]. Point O_2 is the center of the spherical surface on the top, and O_3 is the center of the spheri-

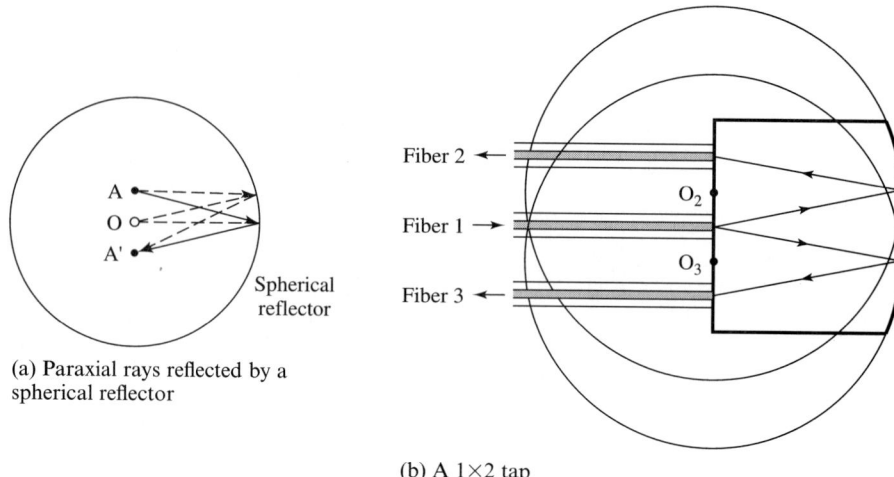

Figure 9.6 A 1×2 tap based on spherical reflectors [10]

cal surface below. Fibers 1 and 2 are symmetrical with respect to O_2, and fibers 1 and 3 are symmetrical with respect to O_3. Power emerging from fiber 1 is evenly split to fibers 2 and 3.

9.2.4 Fused Biconical Taper Fiber Directional Couplers

There are a number of passive components based on multimode fiber couplers. A multimode fiber directional coupler is simply two multimode fibers positioned very close to each other. The operation of such couplers is similar to that of thin-film directional couplers, discussed in Chapter 7. If two fibers are positioned close to each other in an extended region, power can be coupled from one fiber to the other. The two fibers interact when their cores are kept close. Interaction also increases when the core radii decrease, because a substantial fraction of the power resides *outside* the core region if the fiber core is small. To reduce the spacing between fiber cores and to expose the core, fiber jackets are completely stripped. Portions of the cladding are also removed by mechanical lapping or chemical etching. In addition, the fibers are twisted together, stretched under tension, and heated in the fabrication process to soften the stressed sections. The resulting stretched, twisted fiber sections are tapered and fused, and have a biconical shape. These components are generally known as **fused biconical taper directional couplers** [10–13].

Although the operating principle of multimode fiber directional couplers is similar to that of thin-film directional couplers, it is difficult to apply the coupled mode theory developed in Chapter 7, because there are so many modes in a typical multimode fiber. It is also difficult to account for the tapering in the coupling region. On the other hand, the large number of modes makes it pos-

9.2 Taps and Star Couplers

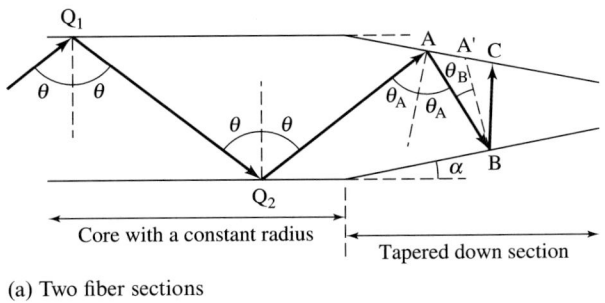

(a) Two fiber sections

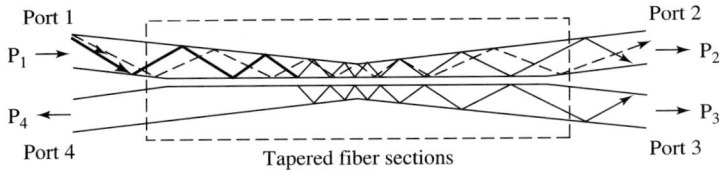

(b) Operating principle

Figure 9.7 Tapered multimode fiber directional couplers

sible to understand the operation of multimode tapered fiber directional couplers in terms of rays.

To understand the operation of such couplers, follow the meridional rays that zigzag in various fiber sections, including in particular the tapered sections. Figure 9.7a depicts the cross sections of two fiber sections. The first has a constant core radius, followed by a tapered section in which the core radius decreases gradually. The core/cladding boundaries are shown as thin solid lines and normals to the boundary are depicted as dashed lines. Consider a ray (heavy solid lines) that impinges on the core/cladding boundary at point Q_1 with an incident angle θ relative to the normal to the core-cladding boundary. In the section with a constant core radius, the two core/cladding boundaries are parallel. There is no change in the incident angle as the ray bounces back and forth in the core region. The incident angles at points Q_1 and Q_2 are the same. If θ is *greater* than the critical angle $\sin^{-1}(n_{cl}/n_{co})$, the ray is totally internally reflected by the core/cladding boundary. A ray with θ *greater* than the critical angle remains guided as it propagates in the untapered section.

The situation changes when the ray enters the tapered section, where the two boundaries incline at an angle relative to each other and to the fiber axis. Let the angle between the core/cladding boundary and the fiber axis be $\pm\alpha$. Also, assume that the ray enters the tapered down region and impinges on the top boundary at point A (Figure 9.7a). When reflected, the ray turns toward and intersects the lower boundary at point B. It then intersects the top boundary at point C, and so on. Let the incident angles at points A, B, C, etc., be θ_A, θ_B, θ_C, etc. Since the boundaries of the tapered section are at an angle relative to the fiber axis, $\theta_A \neq \theta_B \neq \theta_C \neq \cdots$. To establish a relationship between these angles, consider the triangle AA′B shown in Figure 9.7a. Since $\angle A'AB = \frac{\pi}{2} - \theta_A$, $\angle ABA' = \theta_B$, and $\angle AA'B = \frac{\pi}{2} + 2\alpha$. Therefore,

$$\angle A'AB + \angle ABA' + \angle AA'B = \left(\frac{\pi}{2} - \theta_A\right) + \theta_B + \left(\frac{\pi}{2} + 2\alpha\right) = \pi$$

We obtain immediately

$$\theta_B = \theta_A - 2\alpha$$

In the same manner, we can also show

$$\theta_C = \theta_B - 2\alpha = \theta_A - 4\alpha$$

In other words, each time a ray hits a boundary of a tapered down section, the incident angle *decreases* by 2α. After a ray is reflected n times by the boundaries in the tapered down section, the incident angle reduces to $(\theta_A - 2n\alpha)$. Although the angle α is usually small, a ray can be reflected so many times in the tapered section that $(\theta_A - 2n\alpha)$ becomes smaller than the critical angle. The ray in question is no longer totally internally reflected by the core/cladding boundary. In the absence of total internal reflection, fields can reach into the cladding region. If there is another fiber close by, fields in the cladding region will be shared by two fibers; that is, the two fibers will become coupled through the tapered sections. This accounts for the interaction in tapered fiber couplers.

In a tapered fiber directional coupler, there are two fibers. Each has a tapered *down* section, followed by a tapered *up* section where the core radius gradually increases. We can show that each time a ray intersects a boundary of a *tapered up section*, the angle of incidence increases by 2α. Consequently, upon leaving a tapered down section and entering a tapered up section, rays will remain guided by the tapered up section, and by the following fiber section with constant core radius.

To reiterate, consider rays in the top fiber of Figure 9.7b. First, consider a ray with a large angle of incidence, shown schematically as dashed lines in the figure. This ray generally follows the fiber axis. Since the angle of incidence θ is sufficiently large, $(\theta - 2n\alpha)$ remains greater than the critical angle, even after many reflections by the core/cladding boundary in the tapered down section. The ray is guided throughout the top fiber and emerges from port 2.

Second, consider a ray with a smaller angle of incidence, represented by heavy lines in the top fiber and then split into thin lines in the top and neighboring fibers signifying rays in both fibers. After having been bounced several times in the tapered down section, the ray has an angle of incidence that is smaller than the critical angle. The ray is then partially transmitted into the cladding region and the core of the neighboring fiber. The power originally confined in the top fiber is now shared by the bottom fiber. As a result, power emerges from ports 2 and 3.

The operation of a multimode tapered fiber directional coupler can also be understood in terms of the fields outside the core regions. In the tapered section, the core radii are small, and a large portion of the fields is outside the cores. This is particularly true for higher-order modes. The fields outside the cores result in a substantial power exchange between neighboring fibers.

In an ideal directional coupler, all power fed into one port is transmitted to the port or ports on the other side of the coupling region, which in our discussions is also the tapered region, and no power is reflected to fibers on the same side of the coupling region. In other words, power is transmitted to ports in the forward direction, and none is transmitted in the backward direction. This is what it means by the word "directional." In reality, however, power transfer to the ports in the forward direction is less than 100 percent, and a small portion of the input power is reflected to the ports in the backward direction. In addition, a small amount of power is lost due to scattering and radiation in the tapered section. To quantify the performance of directional couplers, we refer to the 2×2 directional coupler shown in Figure 9.7b. Suppose ports 1 and 3 are the input and output ports, and the powers in these ports are P_1 and P_3, respectively. The **insertion loss** (in dB) is defined as

$$IL_{13} = -10 \log\left(\frac{P_3}{P_1}\right) \quad (9.2)$$

If port 2 is the output port, the insertion loss between ports 1 and 2 is

$$IL_{12} = -10 \log\left(\frac{P_2}{P_1}\right) \quad (9.3)$$

The distribution of power among the two output ports is described by the **coupling ratio,** in percentages, as

$$CR_2 = \frac{P_2}{(P_2 + P_3)} \quad (9.4)$$

and

$$CR_3 = \frac{P_3}{(P_2 + P_3)} \quad (9.5)$$

In a real directional coupler, $(P_2 + P_3)$ is less than P_1. The **excess loss,** which quantifies the power loss, is defined (in dB) as

$$EL = -10 \log\left[\frac{P_2 + P_3}{P_1}\right] \quad (9.6)$$

The directionality of a coupler is quantified by the **directional isolation** (in dB) as

$$IS = -10 \log\left[\frac{P_4}{P_1}\right] \quad (9.7)$$

The coupling ratio, the insertion loss, etc. can be controlled by varying the length and angle of the tapered sections.

Directional couplers can have several input and output ports. Figure 9.8 depicts five examples of fused biconical taper fiber directional couplers. The 2×2 fiber coupler shown in Figure 9.8a is probably the simplest, most basic fiber coupler.

In the 8×8 **transmission star coupler** (Figure 9.8b), input from any one of the fibers on one side of the coupling region is distributed to eight fibers on the other side. However, power is not necessarily uniformly distributed among the outgoing ports. To improve the uniformity of the transmitted power, two 8×8 transmission star couplers can be connected in series, as shown in Figure 9.8c. For an 8×8 **reflection star coupler** (Figure 9.8d), power from any fiber is looped back and redistributed to all fibers on the *same* side of the coupling region, including the input fiber.

The 2×8 **transmission–reflection star coupler** shown in Figure 9.8e is a hybrid of the other two star couplers. Inputs to fibers 10 and 11 are transmitted to fibers 1–8 on the other side of the coupling region. In addition, inputs to fibers 1–8 are transmitted to fibers 10 and 11 on the other side of the tapered region, as well as to fibers 1–8 on the same side of that region.

9.2.5 Star Couplers with a Mixer Rod or Plate

Many components were originally designed as taps and later modified to function as star couplers. For those taps and star couplers sharing the same basic design, the difference is the number of ports. We generally refer to junctions with only a few ports as **taps;** and those with a multitude of ports as **star couplers.** There are also passive components that had their genesis as star couplers, including star couplers with a mixer rod or mixer plate.

Figure 9.9a is a star coupler with a mixer plate [14]. The polished ends of two groups of fibers are joined to the polished ends of a thin slab, which is the **mixer plate.** The mixer plate is a high-index glass plate with a thickness equal to the core diameter. It is sandwiched between two low-index sheets, which serve as the cladding regions. The high-index mixer plate and the two low-index plates together form a planar waveguide structure. The mixer plate can thus be viewed as the limiting case of fused cores, and the low-index plates are the cladding regions surrounding the cores.

The operation of a star coupler with a mixer plate is as follows. In the y direction, beams emitted by an input fiber are multiply reflected by the high-index and low-index boundaries. In the x direction, the beams diverge until they are reflected several times by the end surfaces of the mixer plate. Because of the beam divergence in the x direction and the multiple reflections in the x and y directions, the input power is distributed to all output fibers approximately uniformly. Ray tracing studies show that an approximately uniform distribution can be achieved if the length-to-width ratio is greater than 20 [15].

Polymers can be used to form the mixer plate, as well as a portion of the input and output waveguides [16]. Such an arrangement, shown in Figure 9.10, is an *integrated optic star coupler.*

9.2 Taps and Star Couplers

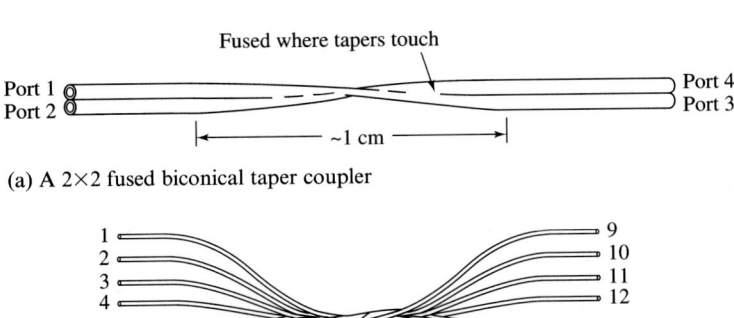

(a) A 2×2 fused biconical taper coupler

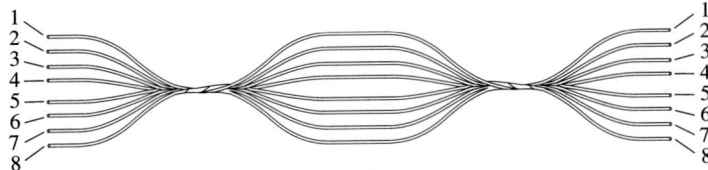

(b) An 8×8 transmission star coupler

(c) Two 8×8 transmission star couplers connected in series

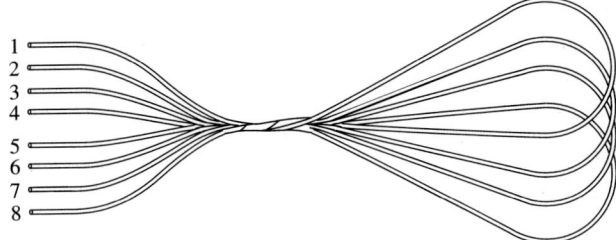

(d) An 8×8 reflection star coupler

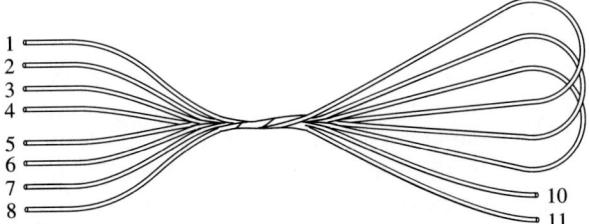

(e) A 2×8 hybrid transmission–reflection star coupler

Figure 9.8 Five couplers based on fused biconical taper couplers [7]

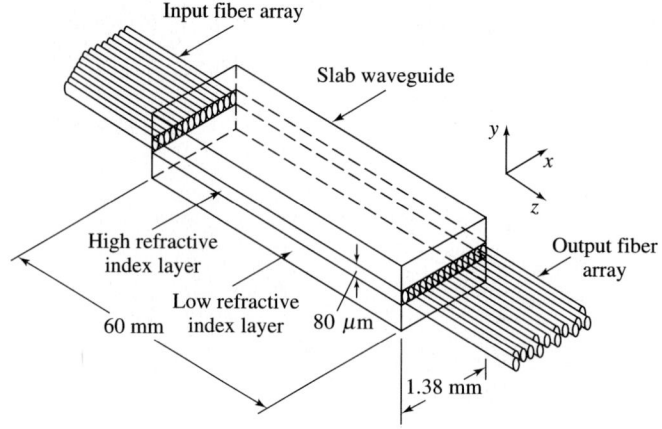

(a) A star coupler with a mixer plate

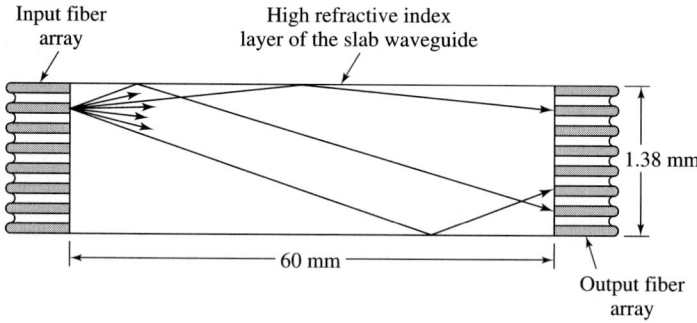

(b) Rays in the mixer plate

Figure 9.9 A multimode star coupler with a glass sheet as the mixer plate [14]

9.3 MULTIMODE FIBER SWITCHES

When a terminal must be added to or removed from a network, optical switches are used for the operation. In the connection and disconnection, switching need not be very fast. A switching speed on the order of a few milliseconds or even 10–20 ms is adequate. In this section, we will describe the operation of three types of multimode fiber switches: **mechanical switches, electrowetting switches,** and **liquid crystal switches.**

In a mechanical switch, a fiber is physically moved by some mechanism. In an electrowetting switch, a mercury slug moves in an electrolyte in response to the applied voltage. In a liquid crystal switch, the liquid crystal molecules are rotated by the applied electric field. Since the motion of a physical mass is involved, the switching speed of these switches is not very fast.

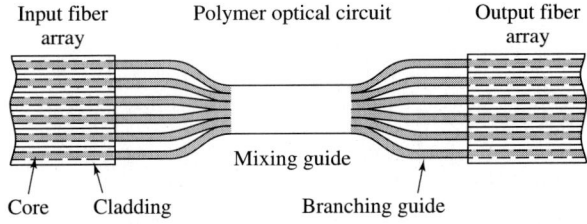

Figure 9.10 A star coupler with a polymer mixer plate and branching waveguides [16]

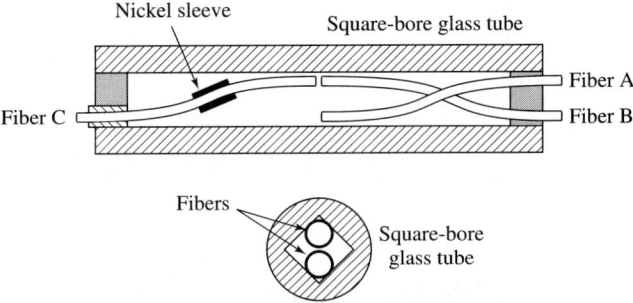

Figure 9.11 A mechanical single-pole double-throw fiber switch [17]

9.3.1 Mechanical Switches

Several mechanisms have been proposed for mechanical switches, and many designs are possible. Figure 9.11 shows a mechanical optical fiber switch as reported by Hale and Kompfner [11]. Two fibers, A and B, are fixed at opposite corners of a square tube. A third, movable fiber, C, is located at the other end of the square tube. Attached to the movable fiber is a nickel sleeve, which responds to an externally applied magnetic field, thereby moving the fiber. Motion can also be achieved with an electromagnetic actuator [18].

When fiber C is in the position shown, power from that fiber is fed to fiber B. When fiber C is pushed to the opposite corner, power is coupled to fiber A. Thus, the switch shown in Figure 9.11 is a single-pole double-throw switch. Although the mechanical motions are generally slow, the distance fiber C must travel is rather small. For a typical multimode fiber with a cladding diameter of 125 µm, a transverse displacement of $125\sqrt{2}$ µm is sufficient for optical switching.

9.3.2 Electrowetting Switches

Figure 9.12a shows the essential components of an electrowetting switch. A reflecting slug is completely surrounded by a transparent electrolyte, which is contained in a thin capillary tube that has an indentation in each end. The electro-

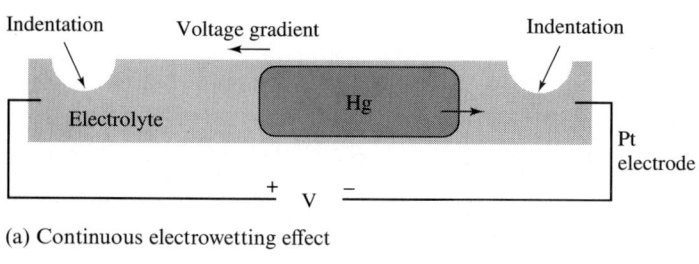

Figure 9.12 An electrowetting switch

lyte is chosen such that it does not react with the reflecting slug and no electric charge will be transferred between the slug and the electrolyte. Many material combinations are possible. In the switches reported by Jackel et al. [19], a mercury slug is immersed in 0.1M–H_2SO_4. Since there is no charge transfer across the Hg/electrolyte interface, the interface is viewed as a thin capacitor with positive charges on the Hg side and negative charges on the electrolyte side. An electric voltage can be applied through platinum electrodes imbedded in the electrolyte. Because of the applied voltage, a voltage gradient is established at the Hg/electrolyte interface, which in turn sets up a surface tension gradient along the interface. The Hg slug is pushed toward the region that has a lower surface tension. With the voltage polarity shown in Figure 9.12a, the Hg slug moves to the right in a 0.1M–H_2SO_4 solution. Eventually, the motion of the slug is stopped by the indentation built into the capillary tube. When the voltage polarity is reversed, the motion of the slug is also reversed.

To use the electrowetting effect as a switch, the thin capillary tube containing the Hg slug and electrolyte is placed between two quarter-pitch GRIN lenses, as shown in Figure 9.12b and c. As a voltage is applied to the electrodes, the Hg slug is moved into or away from the optical path. When the optical beam is *not* blocked by the Hg slug, light from fiber 1 is transmitted to fiber 4. This is the

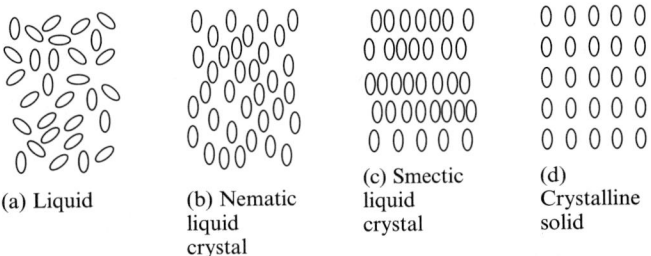

Figure 9.13 Long-range ordering or disordering of liquids, liquid crystals, and crystalline solids

transmission state shown in Figure 9.12b. When the light beam is intercepted and reflected by the Hg slug, light is diverted to fiber 2. This is the reflection state depicted in Figure 9.12c.

9.3.3 Liquid Crystal Switches

Liquid crystals are fluids with large, nonspherical organic molecules. Depending on the orientational and positional orders of the molecules, these liquids are classified as being in the **nematic, smectic,** or **cholesteric phase.** To understand the distinction between these phases, begin with isotropic liquids and crystalline solids, since various liquid crystal phases are actually the intermediate states between amorphous liquids and crystalline solids. In the usual isotropic liquids, molecules are oriented and positioned randomly, as shown schematically in Figure 9.13a. In crystalline solids, there is long-range orientational and positional ordering, as shown in Figure 9.13d.

In the nematic phase (Figure 9.13b), the molecules, depicted as long ellipses, show a **long-range orientational order;** however, there is no long-range correlation in their positions. On average, the molecules are oriented parallel to each other, although the centers of mass of the molecules are located randomly relative to each other.

In the smectic phase, there is a **short-range correlation in their positions,** in addition to the long-range orientational order. Because of the short-range order in their positions, layers are formed. Figure 9.13c depicts *one of the 12 possible subgroups* of smectic liquid crystals.

In the cholesteric phase (not shown), molecular layers are also formed, and molecules in each layer can also be aligned. However, the orientation changes gradually and periodically from layer to layer.

We will concentrate mainly on the use of nematic liquid crystals in optical applications. For convenience, we will refer to the average orientation of the molecular axis as the **nematic axis.** Consider the electrical properties of nematic liquid crystals; that is, consider their interaction with dc or audiofrequency electric fields. Low-frequency electric fields along the nematic axis "see" the relative dielectric constant $\varepsilon_{\parallel}$, while the electric fields normal to the nematic axis "experience" the dielectric constant $\varepsilon_\perp$. Because of this dielectric anisotropy and the fact that the medium is a fluid, the molecules are free to align with, or be normal to, the applied electric field when a low-frequency electric field above a certain threshold value is applied. If $(\varepsilon_\parallel - \varepsilon_\perp)$ is positive, the nematic axis aligns with the applied electric field; if $(\varepsilon_\parallel - \varepsilon_\perp)$ is negative, the nematic axis becomes normal to the electric field.

Optically, nematic liquid crystals are uniaxial materials with the ordinary and extraordinary refractive indices n_o and n_e, respectively. For all nematic liquid crystals, n_e is greater than n_o. Optical waves polarized with their electric fields *along* the nematic axis propagate with the extraordinary index of refraction n_e. Optical beams polarized *normal* to the nematic axis experience the ordinary index of refraction n_o. If the optical beam is polarized at an angle θ relative to the nematic axis, equation (6.48) is used to calculate the index of refraction. In particular, for the optical electric field at an angle θ relative to the nematic axis, the index is given by

$$\frac{1}{n_e^2(\theta)} = \frac{\cos_2\theta}{n_e^2} + \frac{\sin_2\theta}{n_o^2} \tag{9.8}$$

As noted earlier, liquid crystal molecules reorient in response to an applied low-frequency electric field. Because of the reorientation of the nematic axis, the optical properties also change. Since the molecules are large, their response time is on order of a few milliseconds or slower.

Next, consider an optical beam incident upon the interface between a glass with an index n_g, and a nematic liquid crystal with the nematic axis normal to the paper (Figure 9.14a). For simplicity, choose $n_g = n_e$. The angle of incidence is such that $n_g \sin\theta > n_o$. For the waves with electric fields *normal* to the plane of incidence (the paper), the electric field $\mathbf{E}_\perp$ "sees" the extraordinary index of refraction and is transmitted across the interface. On the other hand, for waves with electric fields *in* the plane of incidence, the electric field $\mathbf{E}_\parallel$ is normal to the nematic axis and "sees" n_o. Since $n_g \sin\theta > n_o$, the waves are total internally reflected by the boundary.

Assume that a low-frequency electric field is applied, and the nematic liquid crystal molecules are aligned normal to the glass surface, as shown in Figure 9.14b. With $\mathbf{E}_\perp$ normal to the nematic axis, the waves "see" the ordinary index of refraction n_o, instead of n_e, and are totally internally reflected by the boundary. The fields in the plane of incidence "see" an index $n_e(\frac{\pi}{2} - \theta)$. If $n_e(\frac{\pi}{2} - \theta)$ is greater than $n_g \sin\theta$, then $\mathbf{E}_\parallel$ is transmitted across the boundary, as is the case shown in Figure 9.14b. Not shown in Figure 9.14 is the case in which $\mathbf{E}_\parallel$ is reflected by the boundary when $n_e(\frac{\pi}{2} - \theta)$ is less than $n_g \sin\theta$.

With the previous discussions in mind, consider an optical switch based on nematic liquid crystals [20]. The switch consists of two glass prisms and a thin layer of nematic liquid crystal, as shown in Figures 9.14c and d. The glass index

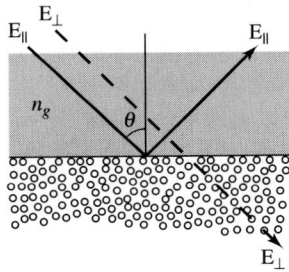

(a) Molecules aligned in parallel with the glass surface

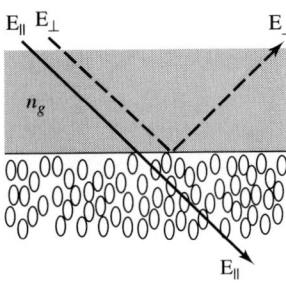

(b) Molecules aligned normal to the glass surface

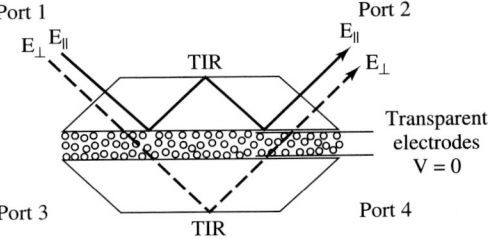

(c) A liquid crystal switch with no voltage applied to the electrodes

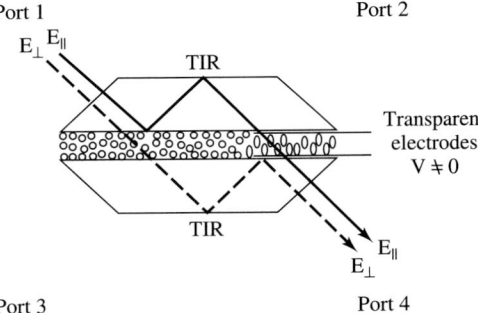

(d) A liquid crystal switch with a voltage applied to the electrodes

Figure 9.14 A liquid crystal switch [20]

n_g is again chosen to match the extraordinary index of the liquid crystal. A portion of the prism base is covered by transparent electrodes. Four GRIN lenses are used to collimate the light from the fibers into and out of the glass prisms. (For clarity, these GRIN lenses are not shown in the figures.) Because of the surface treatment of the prism bases, the molecules are aligned with the glass surface in the absence of an applied voltage. In the presence of an applied voltage, the molecules rotate to the normal orientation relative to the prism base.

To understand the switch operation, suppose that incoming beams impinge upon the glass prism from port 1. For input beams with arbitrary polarization, the electric fields may be decomposed into two polarizations $\mathbf{E}_\parallel$ and $\mathbf{E}_\perp$. At the first glass/liquid crystal boundary, the nematic axis is aligned with the glass surface, in an arrangement identical to that shown in Figure 9.14a. The field $\mathbf{E}_\parallel$ is totally internally reflected, while $\mathbf{E}_\perp$ is transmitted to the second glass prism. The reflected beam goes toward the glass/air boundary at the top of the first prism, and the transmitted beam travels to the glass/air boundary at the bottom of the second prism. Both polarizations are totally internally reflected by the glass/air boundaries. If no voltage is applied to the electrodes, as shown in Figure 9.14c, $\mathbf{E}_\parallel$ is again reflected by the glass/liquid crystal boundary, and $\mathbf{E}_\perp$ is transmitted across that boundary. As a result, two polarizations are combined in port 2. If a voltage is applied to the electrodes, the nematic axis becomes normal to the glass surface, and the situation is similar to Figure 9.14b. Field $\mathbf{E}_\perp$ is now reflected by the the glass/liquid crystal interface, and $\mathbf{E}_\parallel$ is transmitted across that interface. Two polarizations are combined in port 4. Thus, the input beam can be directed to port 2 or port 4, depending on the absence or presence of an applied voltage.

9.4 POWER COUPLED INTO FIBERS

The coupling between various optical components, including fibers, is an important factor affecting the link length of a communications system. For systems with incoherent sources, it is possible to treat the coupling problem in terms of fields, and then to express the final results in terms of power, as was done by Marcuse [21]. A simpler approach is to treat the problem directly in terms of the radiance. In Chapter 4, we defined **radiance** $N(\theta,\phi;x_s,y_s,z_s)$ as the power $dP(\theta,\phi;x_s,y_s,z_s)$ contained in a solid angle $d\Omega$ radiated by an elementary source area dS_s, as follows:

$$N(\theta,\phi;x_s,y_s,z_s) = \frac{dP(\theta,\phi;x_s,y_s,z_s)}{dS_s d\Omega}$$

The solid angle $d\Omega$ defines a cone centered in the direction (θ, ϕ), where θ and ϕ are angles relative to the normal $\mathbf{n}_s$ of the source surface element dS_s located at (x_s, y_s, z_s) (Figure 9.15a). In terms of θ and ϕ, $d\Omega = \sin\theta d\theta d\phi$. It can also be expressed in terms of variables associated with the receiver surface element. With reference to Figure 9.15b, we note that

$$d\Omega = \frac{\cos\theta_r}{r_{rs}^2} dS_r$$

where r_{rs} is the distance between two elemental surface areas and θ_r is the angle relative to the normal $\mathbf{n}_r$ to dS_r.

The total power radiated by the source is [22, 23]

$$\int_{S_s}\int_{\Omega} N(\theta,\phi;x_s,y_s,z_s)dS_s d\Omega \tag{9.9}$$

which can be written in the equivalent form

$$\int_{S_s}\int_{S_r} N(x_r,y_r,z_r;x_s,y_s,z_s)\frac{\cos\theta_r}{r_{sr}^2} dS_s dS_r \tag{9.10}$$

Either (9.9) or (9.10) is the starting point for many coupling or junction problems involving incoherent radiation. Depending on the coordinate system chosen, the source surface element dS_s can be cast in various form. For example, if the source surface is normal to the z_s axis, then dS_s is $dx_s dy_s$ in cartesian coordinates, or $r_s dr_s d\phi_s$ in cylindrical coordinates. The term dS_r can be written in a similar manner.

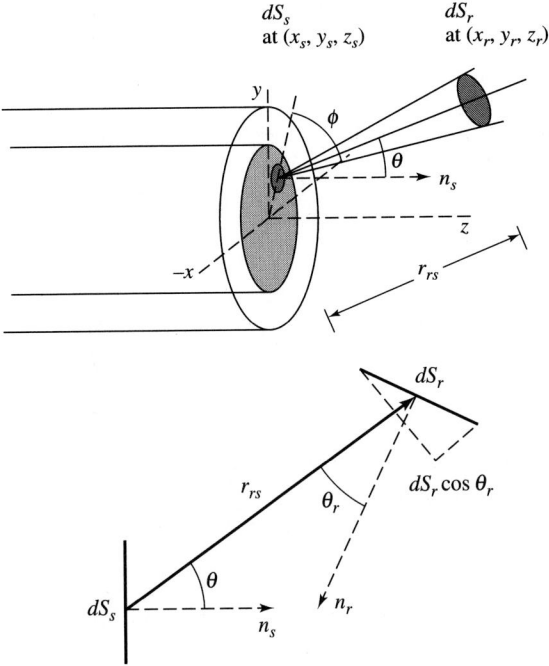

Figure 9.15 Geometry of an elementary receiving area dS_r and a source area dS_s

As an application of (9.9), consider an **incoherent source** with a total radiated power of P_{sr}. Suppose the source has a circular emitting area with a radius a_s, and that all points on the radiating surface emit with equal strength in all directions. According to **Lambert's cosine law,** an incoherent source has a radiance pattern of the form

$$N(\theta,\phi;x_s,y_s,z_s) = N_0\cos\theta \qquad 0\leq\theta\leq\pi/2, \qquad 0\leq\phi\leq2\pi \qquad (9.11)$$

The cosine function is needed to account for the projection of dS_s in the direction θ, and N_0 is a constant to be determined. To relate N_0 to the total power radiated by the source, we note that

$$P_{sr} = \int_0^{a_s}\left\{\int_0^{2\pi}\left[\int_0^{2\pi}\left(\int_0^{\pi/2} N_0\cos\theta\,\sin\theta\,d\theta\right)d\phi\right]d\phi_s\right\}r_s dr_s = a_s^2\pi^2 N_0$$

That is,

$$N_0 = \frac{P_{sr}}{a_s^2\pi^2} \qquad (9.12)$$

As a second example, suppose that a lens of radius a_L is placed at a distance l from the circular source discussed in the last section, and consider the power

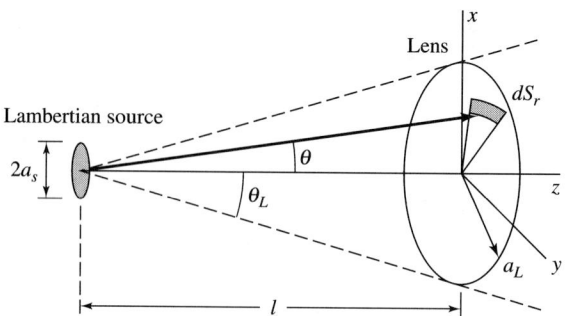

Figure 9.16 A large lens facing a small light source

intercepted by the lens. For simplicity, assume that the source and the lens are parallel, as shown in Figure 9.16, and that a_L is much larger than a_s. Under this condition, the angle θ_L subtended by the lens is approximately $\tan^{-1} a_L/l$. The power intercepted by the lens is, from (9.9) and (9.12),

$$P_{\text{lens}} = \int_0^{a_s} \left\{ \int_0^{2\pi} \left[\int_0^{2\pi} \left(\int_0^{\theta_L} N_0 \cos\theta \, \sin\theta \, d\theta \right) d\phi \right] d\phi_s \right\} r_s dr_s \quad (9.13)$$

$$= a_s^2 \pi^2 N_0 \sin^2\theta_L \approx P_{sr} \frac{a_L^2}{z_L^2 + l^2}$$

As discussed in Chapter 4, not all power emitted by a source is *captured* by a fiber, even under perfect conditions. The power *coupled* to a fiber is even less, if the conditions are not perfect. There may be a transverse offset δ, a longitudinal separation l, an angular misalignment θ, or an area mismatch between the emitting surface and the fiber section, as shown in Figure 9.17a. In the following, we will use (9.9) or (9.10) to examine the effects of area mismatch on the coupling.

To reduce the complexity of the problem, we restrict our considerations to step index fibers only. The final results for graded index multimode fibers are given without derivation. Although power in a fiber is not confined entirely to the core, under normal operating conditions, the percentage of power contained in the cladding is very small. We ignore that portion of the power in the cladding region as an approximation.

9.4.1 Coupling Efficiency

Suppose an incoherent source of radius a_s is aligned perfectly and is *in contact* with the fiber core, as shown in Figure 9.17b. Assume that the incoherent source emits equally from all points on the emitting surface and uniformly in all directions. In other words, the radiance of each point is given by (9.11). The total radiated power is P_{sr}.

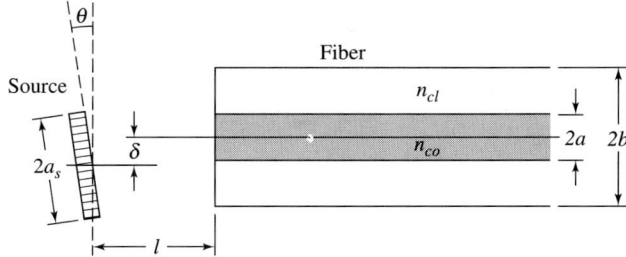

(a) Transverse offset, longitudinal separation, and angular misalignment

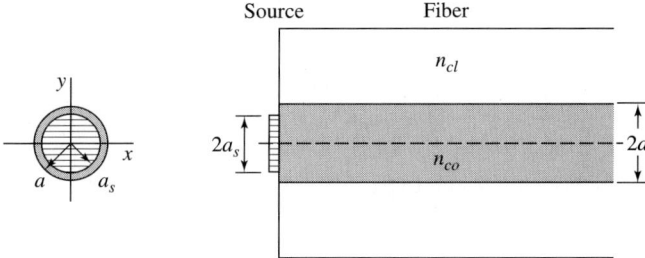

(b) A source in contact with the fiber

Figure 9.17 Coupling from a source to a step index fiber

Consider the case in which the emitting surface is smaller than or equal to the fiber core area (Figure 9.17b). Since $a_s \leq a$, a point on the emitting surface is also a point on the endface of the core. Therefore, the power captured by the fiber is

$$P_{fb} = \int_0^{a_s} \left\{ \int_0^{2\pi} \left[\int_0^{2\pi} \int_0^{\theta_{c\,fb}} N_0 \cos\theta \, \sin\theta \, d\theta d\phi \right] d\phi_s \right\} r_s dr_s = \pi^2 a_s^2 N_0 \sin^2\theta_{c\,fb}$$

where the angle $\theta_{c\,fb}$ is determined by the numerical aperture of the fiber, $\sin\theta_{c\,fb} = NA$. Using (9.12), we obtain

$$P_{fb} = NA^2 P_{sr} \qquad a_s \leq a$$

The **coupling** or **launching efficiency** is defined as the percentage of power radiated by a source coupled to a fiber. For *step index fibers* with $a_s \leq a$, the coupling efficiency is

$$\eta_{c\,step} \equiv \frac{P_{fb}}{P_{sr}} = NA^2 \qquad a_s \leq a \qquad (9.14)$$

Note that $\eta_{c\,step}$ is independent of a_s as long as a_s is smaller than a.

For sources with $a_s \geq a$, none of the light emitted from regions outside the fiber core is captured by the fiber. Thus, for $a_s \geq a$, we have

$$P_{fb} = \int_0^a \left\{ \int_0^{2\pi} \left[\int_0^{2\pi} \int_0^{\theta_{cfb}} N_0 \cos\theta \sin\theta \, d\theta d\phi \right] d\phi_s \right\} r_s dr_s = NA^2 \frac{a^2}{a_s^2} P_{sr}$$

Note that the upper limit in the first integral has been changed from a_s to a. The coupling efficiency is now given by

$$\eta_{c\,step} = NA^2 \frac{a^2}{a_s^2} \qquad a_s \geq a \qquad (9.15)$$

The coupling efficiency for *graded index* fibers can be obtained in the same manner. Recall that the index profile of graded index fibers is specified by the index difference Δ and the index profile parameter g, as noted in section 8.1. Since the core index is a function of position, the numerical aperture is also. We use the local numerical aperture $NA(r)$ in lieu of NA. The mathematical manipulations involved in calculating the coupling efficiency for graded index fibers are more complicated. However, the final results are quite simple [23], as follows:

$$\eta_{c\,gr} = \begin{cases} [NA(0)]^2 \left[1 - \frac{2}{g+2}\left(\frac{a_s}{a}\right)^g\right], & \text{for } a_s \leq a \\ [NA(0)]^2 \frac{g}{g+2} \frac{a^2}{a_s^2}, & \text{for } a_s > a \end{cases} \qquad (9.16)$$

In general, the coupling efficiency of step index fibers is greater than that of graded index fibers with the same core radius and index difference. This occurs because the NA of step index fibers is the same as the $NA(0)$ of graded index fibers with the same Δ, and $NA(0) > NA(r)$ for $r \neq 0$. For example, from (9.14)–(9.16), we can determine that twice as much power is *captured* by step index fibers as by fibers with a parabolic index profile with the same a and Δ. Also, for step index fibers, η_c is independent of a_s/a as long as a_s is smaller than a. For a specific or given graded index fiber, however, the fraction of power *coupled* to the fiber increases as the emitting area decreases. If the emitting area is larger than the fiber core, the power emitted from the area outside the core is wasted, and the percentage of power captured by the fibers is greatly reduced.

If the coupling efficiency is expressed in terms of dB, it is also referred to as the **insertion loss**. From (9.14) and (9.15), we obtain

$$IL_{c\,step} = 10 \log(\eta_{c\,step}) = \begin{cases} 20 \log(NA) & a_s \leq a \\ 20 \log(NA) + 20 \log(a/s_s) & a_s \geq a \end{cases}$$

$$(9.17)$$

For graded index fibers, the insertion loss is

$$IL_{c\,gr} = 10\log(\eta_{c\,gr})$$

$$= \begin{cases} 20\log[NA(0)] + 10\log\left[1 - \dfrac{2}{g+2}\left(\dfrac{a_s}{a}\right)^g\right] & a_s \leq a \\ 20\log[NA(0)] + 10\log\left[\dfrac{g}{g+2}\dfrac{a^2}{a_s^2}\right] & a_s \geq a \end{cases} \quad (9.18)$$

An examination of (9.17) and (9.18) shows that the coupling efficiency is limited by two factors. First, light is unguided by the fiber unless the incident angle is within the critical angle. Second, light cannot be captured by the fiber unless light impinges on the core area. The first terms in (9.17) and (9.18) represent the incident angle restriction imposed by the numerical aperture. The second terms of (9.17) and (9.18) relate to the area where rays impinge on the fiber. Also recall that (9.14)–(9.18) are derived for incoherent sources with a radiant pattern of the type given by (9.11). For more directed sources, (9.14)–(9.18) must be modified.

Up to this point, only perfect alignment cases have been considered. In the presence of angular misalignments, gaps, or offsets, the rays must be traced from the emitting surface to the fiber ends [24, 25]. Figures 9.18a and b show the effects of offsets and gaps on the coupling of light into multimode fibers [23]. In particular, the power P_{fb} coupled to the fiber as a fraction of the total available power ΔP_{sr} is plotted as a function of the longitudinal separation and the transverse offset normalized with respect to the core radius. In general, the coupling efficiency remains constant if l and δ are less than the core radius, and η_c decreases precipitously if l or δ is greater than a.

9.4.2 Lens Effects

Since only a small fraction of power can be coupled directly from an incoherent source to a fiber, it is natural to inquire if lenses or other focusing elements may be used to improve the coupling efficiency. Figure 9.19 shows the use of conventional lenses, glass balls, GRIN lenses, or spherical tips to improve the coupling efficiency. The effects of lenses are twofold. First, the emitting surface is imaged by the lens to the fiber endface. Second, the angle at which the ray impinges on the fiber is also altered by the lens. To quantify these effects, we refer to the arrangement depicted in Figure 9.20a, where a thin lens with an aperture radius a_L and a focal length f is used to image an object at z_{in} to an image at z_{out}. In other words, z_{out} and z_{in} are the conjugate points of a thin lens, and they are related via the **lens formula**

$$\frac{1}{z_{out}} + \frac{1}{z_{in}} = \frac{1}{f} \quad (9.19)$$

The **lateral** or **transverse magnification** M is given by

$$|M| \equiv \left|\frac{h_{out}}{h_{in}}\right| = \left|\frac{z_{out}}{z_{in}}\right| = \left|1 - \frac{z_{out}}{f}\right| = \left|(1 - \frac{z_{in}}{f})^{-1}\right| \quad (9.20)$$

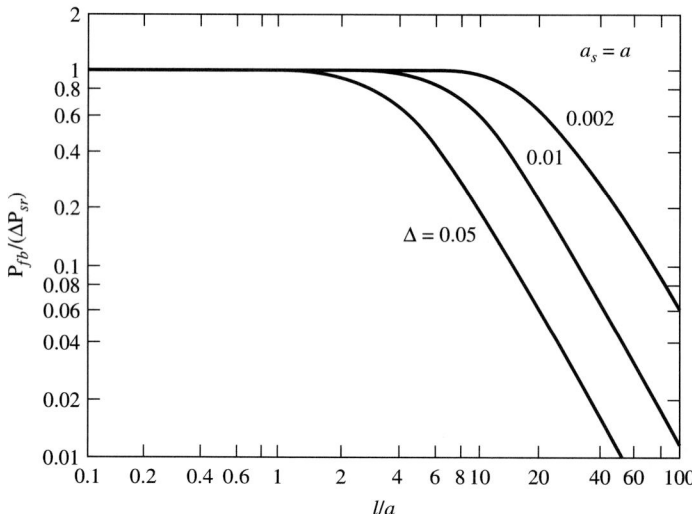

(a) End separation

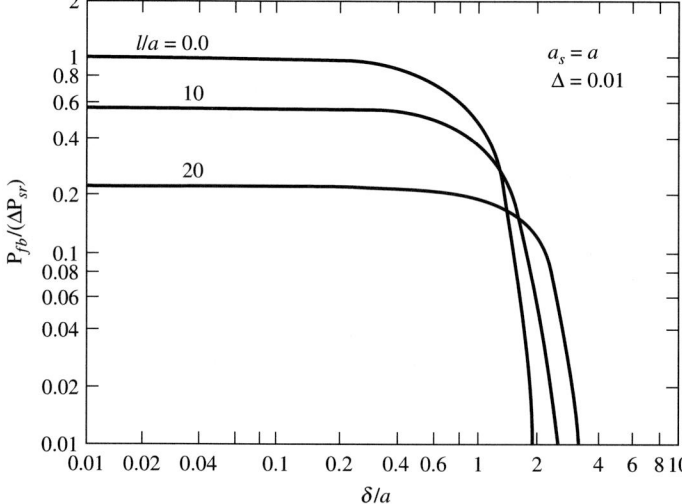

(b) Transverse offset

Figure 9.18 Effects of end separation and transverse offset on the power coupled to a graded index fiber with g = 2 [23]

where $|h_{in}|$ and $|h_{out}|$ are the object and image heights. Since we are mainly interested in the image of an area, the sign of M is immaterial. Equations (9.19) and (9.20) were given previously as (2.105) and (2.106). They can also be found in several texts on optics [26–28].

We can use the ABCD matrix of a thin lens (see equation 2.27) to examine the effects of the thin lens on the ray angles. Consider a ray leaving the edge of

9.4 Power Coupled into Fibers

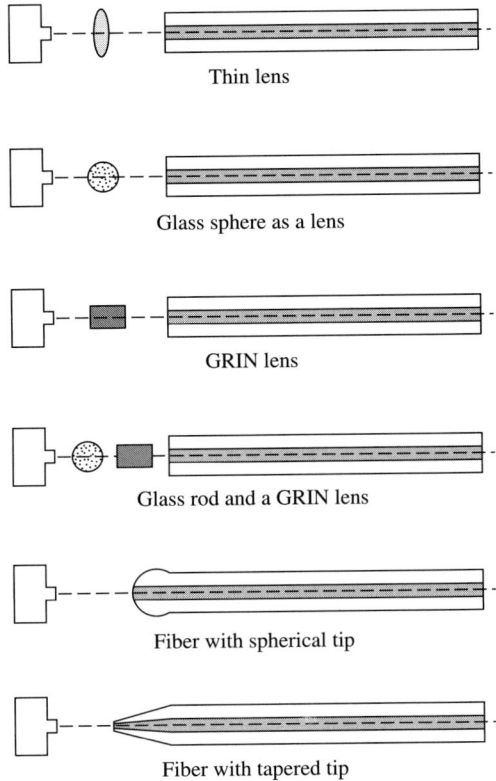

Figure 9.19 Various means of improving the coupling to a fiber

an object with a ray angle of θ_{in}. The ray intersects the thin lens at height $h_{in} + z_{in}\tan\theta_{in}$, which is approximately $h_{in} + z_{in}\theta_{in}$ under the paraxial approximation. Using (2.27), we obtain the ray angle leaving the lens

$$\theta'_{in} = -\frac{h_{in} + z_{in}\theta_{in}}{f}$$

The maximum ray angle impinging on the lens is of particular interest. The steepest ray angle $\theta_{in\,mx}$ impinging on the lens from the object corresponds to rays emanating from the tip of the object to the opposite edge of the lens. With reference to Figure 9.20b, we can show that

$$\tan\theta_{in\,mx} = \frac{a_L + |h_{in}|}{z_{in}} = \frac{|M|[1 + (|h_{in}|/a_L)]}{2(1 + |M|)f/(2a_L)} \qquad (9.21)$$

This derivation is left to the reader (see problem 1.) For rays leaving the lens, the steepest angle $\theta_{out\,mx}$ corresponds to rays going from the bottom of the lens edge to the opposite side of the image plane, as follows:

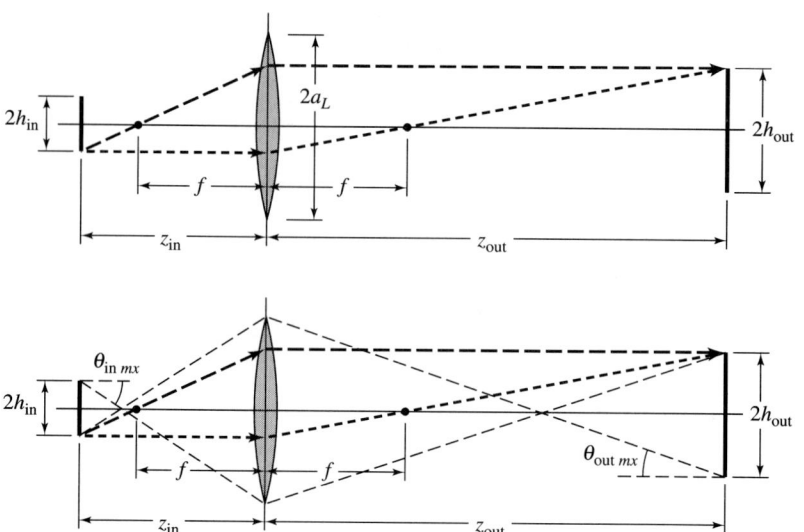

Figure 9.20 Effects of lenses

$$\tan\theta_{out\ mx} = \frac{a_L + |h_{out\ mx}|}{z_{out}} = \frac{1 + (|h_{out}|/a_L)}{2(1 + |M|)f/(2a_L)} \quad (9.22)$$

When $a_L \gg |h_{in}|$ and $|h_{out}|$, (9.21) and (9.22) may be approximated by

$$\tan\theta_{in\ mx} \approx \frac{|M|}{2(1 + |M|)\,f\#} \quad (9.23)$$

$$\tan\theta_{out\ mx} \approx \frac{1}{2(1 + |M|)\,f\#} \quad (9.24)$$

where $f\# \equiv f/(2a_L)$ is the f-number of the lens. Under the paraxial approximation, $\theta_{out\ mx}$ and $\theta_{in\ mx}$ are small, and we obtain $\theta_{in\ mx} \approx |M|\,\theta_{out\ mx}$. This means that when an object is magnified by a factor $|M|$, the ray angles are reduced by a factor $1/|M|$.

Next, consider the use of lenses to improve the coupling efficiency. As shown in Figure 9.21, an ideal thin lens with a focal length f and aperture radius a_L is placed between a source and a step index fiber with a numerical aperture of NA. The lens is positioned to image a source of radius a_s located at z_{in} to the fiber endface at z_{out}. If the image radius matches perfectly with the fiber core, the magnification is $|M| = a/a_s$. At the fiber endface, rays within a cone of angle $\theta_{c\ fb} = \sin^{-1} NA$ are captured by the fiber. This angle corresponds to an angle of $|M|\,\theta_{c\ fb}$ at the emitting surface. Assuming that $\theta_L = \tan^{-1}(a_L/z_{in})$ is greater than $|M|\,\theta_{c\ fb}$, then the power coupled into the fiber through the lens is, from (9.9), (9.11), and (9.12),

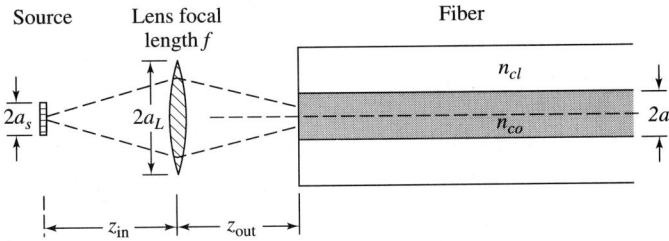

Figure 9.21 Coupling to a fiber through a simple lens

$$P_{fb\ lens} = \int_0^{a_s} \left\{ \int_0^{2\pi} \left[\int_0^{2\pi|M|} \int_0^{\theta_c\ fb} N_0 \cos\theta\ \sin\theta\ d\theta d\phi \right] d\phi_s \right\} r_s dr_s$$

$$= \pi^2 a_s^2 N_0 \sin^2(|M|\ \theta_c\ fb) \approx P_{sr}\ NA^2 M^2$$

In other words, power coupled into step index fibers with the help of a lens is

$$P_{fb\ lens} \approx \left(\frac{a}{a_s}\right)^2 NA^2 P_{sr} \qquad (9.25)$$

Comparing (9.18) with (9.25), we can see that lenses can be used to improve power coupling to a fiber if a_s is *smaller* than a. An improvement by a factor of a^2/a_s^2 is possible if a_s is smaller than a, and if the *f*-number of the lens is small enough that θ_L is greater than $|M|\ \theta_c\ fb$. However, if a_s is *larger* than a, derivations similar to those given for (9.19) and (9.25) show that no improvement in power coupling is possible.

Our discussions have only considered the effects of ideal thin lenses. For real lenses, the lens aberration must be taken into account in determining the coupling efficiency. This has been studied by Nice [29], along with the use of ball lenses and graded index rods for coupling light into fibers.

9.5 CONNECTOR LOSS

Often, fibers of various lengths are connected. The fibers may be identical or may have different geometries, numerical apertures, or index profiles. The loss caused by different fiber geometries, numerical apertures, or index profiles is known as the **intrinsic loss,** which is analogous to the loss introduced in connecting transmission lines with different characteristic impedances. If *identical fibers* are joined, loss also arises because of transverse offset, longitudinal gaps, or angular misalignment. The loss associated with the nonperfect joining of identical fibers is identified as **extrinsic loss.**

In this section, we will study the loss incurred in joining fiber 1, with a core radius a_1 and a numerical aperture NA_1, to fiber 2, with a core radius a_2 and a numerical aperture NA_2. Let P_1 be the power *carried* by fiber 1 and P_2

be the power *transmitted* to fiber 2. The insertion loss (in dB) in joining the two fibers is

$$IL = 10 \log (P_2/P_1)$$

In most cases of practical interest, the insertion loss due to various individual imperfections is small. When several imperfections are present, the total insertion loss is approximately the sum of the individual insertion losses. The effects of multiple reflections on the insertion loss are small and can be ignored.

9.5.1 Intrinsic Loss

To evaluate intrinsic loss, assume that step index fibers are joined perfectly in all respects. However, the core radii, numerical apertures, or both may be different. Also assume that the incoming fiber is sufficiently long that the power P_1 is uniformly distributed in the fiber. Since every point on the endface of fiber 1 acts as a source with a radiance of

$$N(\theta,\phi;x_s,y_s,z_s) = \frac{P_1}{a_1^2 \pi^2 NA_1^2} \cos\theta$$

the power launched into fiber 2 is obtained by substituting the above equation into (9.9). In the integration, the upper limit of r_s is the smaller a_1 and a_2, and the limit of θ is the smaller of $\sin^{-1} NA_1$ and $\sin^{-1} NA_2$. In the case where $a_1 \geq a_2$ and $NA_1 \geq NA_2$, we have

$$P_2 = \frac{P_1}{a_1^2 \pi^2 NA^2} \int_0^{a_2} \left\{ \int_0^{2\pi} \left[\int_0^{2\pi} \left(\int_0^{\sin^{-1} NA_2} \cos\theta \sin\theta d\theta \right) d\phi \right] d\phi_s \right\} r_s dr_s \quad (9.26)$$

$$= P_1 \left(\frac{a_2}{a_1}\right)^2 \left(\frac{NA_2}{NA_1}\right)^2$$

Therefore, in the case where $a_1 \geq a_2$ and $NA_1 \leq NA_2$, we obtain

$$\frac{P_2}{P_1} = \left(\frac{a_2}{a_1}\right)^2 \quad (9.27)$$

Corresponding expressions can be obtained for the case in which $a_1 \leq a_2$ (see problem 2). The insertion loss due to the joining of two fibers with different core radii and numerical apertures is therefore

$$IL = \begin{cases} 20 \log (a_2/a_1) + 20 \log(NA_2/NA_1), & a_1 \geq a_2 \text{ and } NA_1 \geq NA_2 \\ 20 \log(a_2/a_1), & a_1 \geq a_2 \text{ and } NA_1 \leq NA_2 \\ 20 \log(NA_2/NA_1), & a_1 \leq a_2 \text{ and } NA_1 \geq NA_2 \\ 0, & a_1 \leq a_2 \text{ and } NA_1 \leq NA_2 \end{cases} \quad (9.28)$$

The results can be interpreted as follows. First, suppose that $NA_1 \leq NA_2$. All rays that impinge on the core area of fiber 2 from fiber 1 are retained by fiber 2. If a_1 is also smaller than a_2, all incoming power is received and retained by fiber 2 and there is no loss in the joining fiber 1. On the other hand, if $a_1 > a_2$, a portion of the incoming power is not intercepted by fiber 2. Since P_1 is uniformly distributed in fiber 1, the power density in fiber 1 is $P_1/(\pi a_1^2)$. The power intercepted by the core of fiber 2 and therefore captured by that fiber is

$$\frac{P_1}{\pi a_1^2} \pi a_2^2 = P_1 \frac{a_2^2}{a_1^2}$$

This explains the presence of the factor $(a_2/a_1)^2$ in (9.26) and (9.27). Therefore, the insertion loss due to **core area mismatch** alone is

$$IL_a = \begin{cases} 20 \log (a_2/a_1), & \text{if } a_1 \geq a_2 \\ 0, & \text{if } a_1 \leq a_2 \end{cases} \qquad (9.29)$$

The consequence of different numerical apertures may be studied in the same manner. The result leads to the insertion loss due **numerical aperture mismatch**

$$IL_{NA} = \begin{cases} 20 \log (NA_2/NA_1), & \text{if } NA_1 \geq NA_2 \\ 0, & \text{if } NA_1 \leq NA_2 \end{cases} \qquad (9.30)$$

In our discussions, only fibers with step index profiles are considered. While (9.28)–(9.30) are not truly valid for graded index fibers, they may be used as an estimate of the upper bound for each loss term, because the power density in a graded index fiber is weaker in regions near the cladding. Also, the index profile parameter g of different graded index fibers could be different, which would lead to an additional intrinsic loss term [30].

9.5.2 Extrinsic Loss

If fibers are not properly aligned, loss is incurred even if the fibers are identical [30, 31]. The imperfection may be a gap, a transverse offset, or an angular misalignment, as shown in Figure 9.22a, b, and c. Assume the fibers are identical step index fibers. Then,

$$a_1 = a_2 = a \quad \text{and} \quad NA_1 = NA_2 = NA$$

The medium outside the step index fibers may be air or a material with an index n_m.

End Separation Suppose the fiber endfaces are separated by a distance l, and the index of the medium outside the fibers is n_m. When a fiber with a numerical aperture NA is immersed in a medium with an index of n_m, rays originating from the fiber are confined to a cone of angle

$$\theta_m = (\sin^{-1} NA)/n_m$$

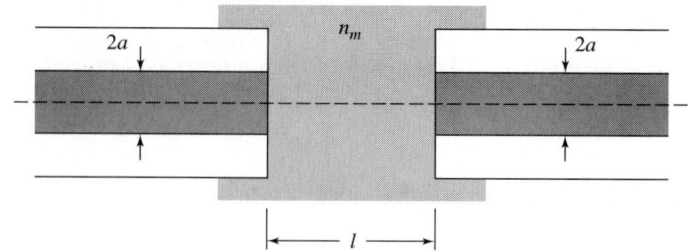

(a) End separation

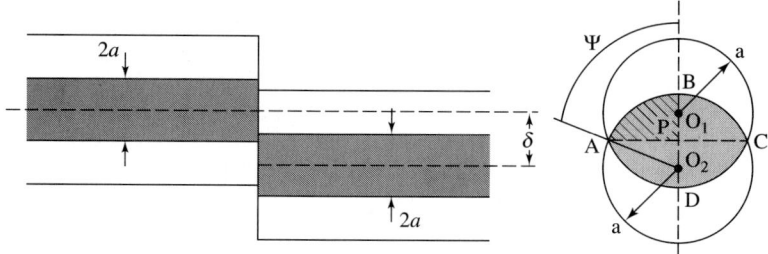

(b) Transverse offset

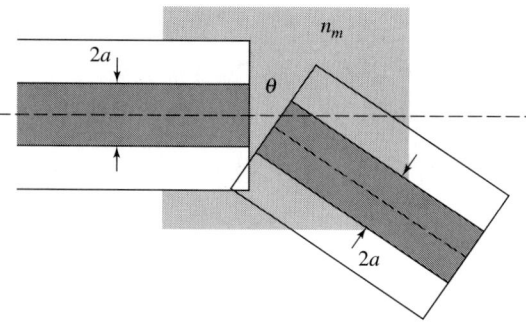

(c) Angular misalignment

Figure 9.22 Longitudinal, transverse, and angular misalignments in joining two identical fibers ($a_1 = a_2 = a$ and $NA_1 = NA_2 = NA$)

as indicated in Chapter 8. We could again evaluate the integral (9.9) and include in the calculation only those rays arriving at fiber 2 with an angle less than θ_m [24, 25]. A simpler approach, however, is to observe that rays leaving fiber 1 diverge as a cone with an angle θ_m. At the entrance of fiber 2, the cross-sectional area of the cone is $\pi(a + l\tan\theta_m)^2$. Again, assuming that power is uniformly distributed across the area, the power accepted by fiber 2 is

$$P_2 = P_1 \frac{\pi a^2}{\pi(a + l\tan\theta_m)^2}$$

The insertion loss due to an **end separation** l is therefore

$$IL_{es} = 10 \log \frac{a^2}{(a + l\tan\theta_m)^2} = -20 \log \left[1 + \frac{l}{a}\tan\frac{\sin^{-1}NA}{n_m}\right] \quad (9.31)$$

Transverse Offset To consider the loss due to a transverse offset or displacement δ (Figure 9.22b), again assume that P_1 is uniformly distributed in the core area of fiber 1. Also, it is only necessary to include the power contained in the area where the two fibers overlap. The overlapping area ABCD is shown in Figure 9.22b as the shaded and dotted areas. Note that area ABCD is four times that of the shaded area ABP. By observing that

$$\text{Area(ABP)} = \text{Area(ABO}_2) - \text{Area(APO}_2)$$

and

$$\text{Area(ABO}_2) = \frac{1}{2}\psi a^2 = \frac{1}{2}a^2 \cos^{-1}\frac{\delta}{2a}$$

$$\text{Area(APO}_2) = \frac{1}{2}\overline{PO_2}\ \overline{AP} = \frac{1}{2}\frac{\delta}{2}\sqrt{a^2 - \left(\frac{\delta}{2}\right)^2}$$

we can show that the insertion loss due to a **transverse offset** δ is

$$IL_{td} = 10 \log \left[\frac{1}{\pi}\left(2\cos^{-1}\frac{\delta}{2a} - \frac{\delta}{a}\sqrt{1 - \left(\frac{\delta}{2a}\right)^2}\right)\right] \quad (9.32)$$

Fresnel Reflection In weakly guided fibers, rays propagate mostly along the fiber axis. Therefore, rays impinging upon the endface are approximately normal to the endface. As an approximation, then, use the reflection coefficient of plane waves incident normally upon a boundary,

$$\Gamma = \frac{n_m - n_{co}}{n_m + n_{co}}$$

The percentage of power transmitted through the boundary is

$$1 - |\Gamma|^2 = \frac{4n_m n_{co}}{(n_m + n_{co})^2}$$

Therefore, the insertion loss due to **Fresnel reflection** at the air/fiber core boundary is

$$IL_{Fr} = 10 \log \left[\frac{4n_m n_{co}}{(n_m + n_{co})^2}\right] \quad (9.33)$$

For silica-based fibers, the core index is about 1.46, which results in an insertion loss of about -0.155 dB due to Fresnel reflection at the air/fiber boundary. Usually, there are two boundaries, one from fiber 1 to the outside medium and the other from the outside medium to fiber 2. Therefore, the *total* insertion loss is about -0.31 dB. If an index-matching liquid is used, or if the fiber ends are coated with antireflection layers, the Fresnel loss is reduced considerably.

Angular Misalignment The effect of an angular misalignment, or a tilt (Figure 9.22c), is quite complicated and has been studied by Thiel and Hawk [30]. Their study shows that the insertion loss due to an **angular misalignment** θ is

$$IL_{ang} = 10 \log \left\{ \cos\theta \left[\left(\frac{1}{2} - \frac{p\sqrt{1-p^2}}{\pi} - \frac{\sin^{-1}p}{\pi} \right) \right. \right.$$
$$\left. \left. + q \left(\frac{1}{2} + \frac{r\sqrt{1-r^2}}{\pi} + \frac{\sin^{-1}r}{\pi} \right) \right] \right\} \quad (9.34)$$

where

$$p = \frac{\cos\theta_m(1-\cos\theta)}{\sin\theta_m \sin\theta} \qquad q = \frac{\cos^3\theta_m}{(\cos^2\theta_m - \sin^2\theta)^{3/2}}$$

$$r = \frac{\cos^2\theta_m(1-\cos\theta) - \sin^2\theta}{\sin\theta_m \cos\theta_m \sin\theta} \qquad \sin\theta_m = \frac{NA}{n_m}$$

It should be noted that in deriving (9.31)–(9.34), we assumed that the fibers to be joined are step index fibers. The corresponding results for graded index fibers are too complicated to be presented here and they can be found in [32–34].

9.5.3 Normalized Mismatch Parameter

To aid in the comparison of various loss terms, we introduce a normalized mismatch parameter for each effect. The **normalized mismatch parameters** for core radii and numerical aperture mismatches, end separation, and transverse offset are as follows:

$$\delta_a = \frac{a_1 - a_2}{a_1}$$

$$\delta_{NA} = \frac{NA_1 - NA_2}{NA_1}$$

$$\delta_{es} = \frac{l}{a} \tan \frac{\sin^{-1} NA}{n_m}$$

$$\delta_{td} = \frac{\delta}{a}$$

In terms of these normalized mismatch parameters, the insertion losses given in (9.29)–(9.32) become:

$$IL_a = \begin{cases} 20 \log(1-\delta_a) & \text{if } a_1 \geq a_2,\ NA_1 = NA_2 \\ 0 & \text{if } a_1 \leq a_2,\ NA_1 = NA_2 \end{cases} \quad (9.35)$$

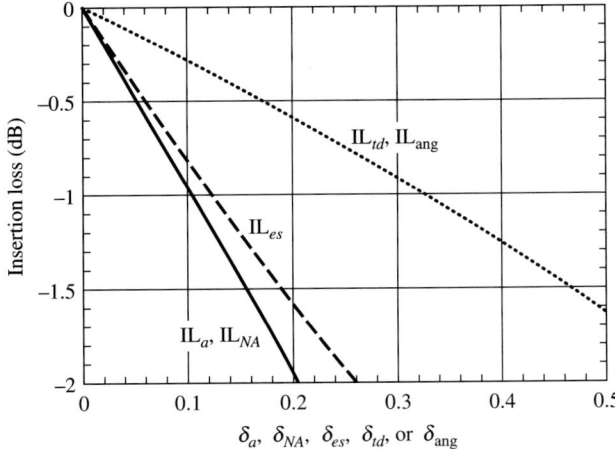

Figure 9.23 Insertion loss as function of normalized mismatch parameters

$$IL_{NA} = \begin{cases} 20 \log (1 - \delta_{NA}) & \text{if } NA_1 \geq NA_2, a_1 = a_2 \\ 0 & \text{if } NA_1 \leq NA_2, a_1 = a_2 \end{cases} \quad (9.36)$$

$$IL_{es} = -20 \log (1 + \delta_{es}) \quad (9.37)$$

$$IL_{td} = 10 \log \left[\frac{1}{\pi}\left(2\cos^{-1}\frac{\delta_{td}}{2} - \delta_{td}\sqrt{1 - \left(\frac{\delta_{td}}{2}\right)^2}\right)\right] \quad (9.38)$$

In Figure 9.23, IL_a, IL_{NA}, and IL_{es} are plotted as functions of δ_a, δ_{NA}, and δ_{es}, respectively. The insertion loss due to an angular misalignment depends on NA and n_m, in a very complicated manner. However, when IL_{ang} is plotted as a function of $\delta_{ang} = \theta/\theta_m$ for several values of NA and n_m, the curves are very similar to the curve for the transverse offset. Therefore, in Figure 9.23, the plots for IL_{ang} and IL_{td} are shown as a single curve, as an approximation.

As a numerical example, consider the loss in connecting two fibers with a core radius of 50 μm, a numerical aperture of 0.3, and a core index n_{co} of 1.458. Let the transverse offset and the end separation be 5 μm and 15 μm, respectively. Also assume that there is no angular misalignment. Table 9.1 lists the results with and without an index-matching liquid.

Note that by using an index-matching material, the loss due to end separation is reduced, as well as the expected reduction of Fresnel loss.

9.6 RISETIME, PULSE WIDTH, AND BANDWIDTH

As mentioned in section 9.1, pulse broadening is an important factor limiting the information capacity and link length of a communications system. Pulse broadening and risetime are parameters frequently used to quantify the time-domain response of a system. The corresponding term in the frequency domain

Table 9.1 Insertion loss of a typical connector

	No index matching liquid $n_m = 1$	Index matching liquid $n_m = 1.46$
IL_{td}	−0.285 dB	−0.285 dB
IL_{es}	−0.783 dB	−0.531 dB
IL_{Fr}	−0.306 dB	0 dB
IL_{ang}	0 dB	0 dB
IL_{total}	−1.374 dB	−0.816 dB

is bandwidth. Although the frequency-domain and time-domain responses are related through the Laplace transform, no simple, exact relationship between the bandwidth and risetime exists. However, an *approximate, empirical* relationship has been noted for simple circuits under certain conditions [35–37]. This empirical relationship is useful in an order-of-magnitude estimate of the information capacity and link length. The purpose of this section is to demonstrate that empirical relationship between bandwidth and risetime.

The frequency response of a system is the response under sinusoidal steady-state excitations and is expressed in terms of the transfer function. Consider a first order, low pass system. Suppose the frequency response peaks at the *peak frequency* f_{pk} and decreases to $1/\sqrt{2}$ of the peak value at the 3-dB or **half-power frequency** f_{3dB} (Figure 9.24a). The **bandwidth** of the system is then defined as

$$BW = |f_{3dB} - f_{pk}| \qquad (9.39)$$

The time-domain response of a system is the response under the condition of a step function excitation. The response changes from an initial value at $t = 0^+$ to a final value as t approaches infinity (Figure 9.24b). If the initial value is 0, then the risetime is the time needed for the transient response to increase from 10% to 90% of the final value. This is usually referred to as the 10%-to-90% risetime, or simply the **risetime**.

Also, when several systems are cascaded and each system has a risetime RT_i, then the risetime of the cascaded system is

$$RT_t = \left[\sum_1^n RT_i^2\right]^{1/2} \qquad (9.40)$$

9.6.1 Empirical Bandwidth–Risetime Relationship

To demonstrate the empirical relationship between bandwidth and risetime, consider a simple series RC circuit. Simple analysis will show that the transfer function of a series RC circuit is

$$\frac{|V_c(f)|}{|V_s(f)|} = \frac{1}{\sqrt{1 + (2\pi f\, RC)^2}}$$

which has a peak of 1 at dc and drops to $1/\sqrt{2}$ at $f_{3dB} = 1/(2\pi\, RC)$. Therefore, the bandwidth of a series RC circuit is

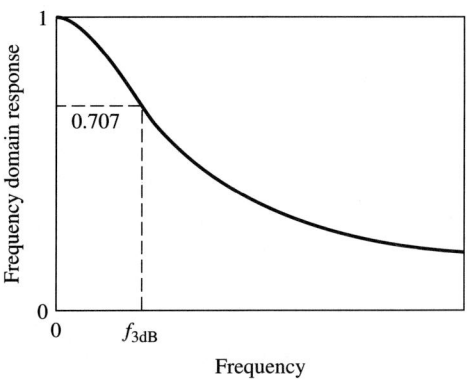
(a) Normalized frequency domain response

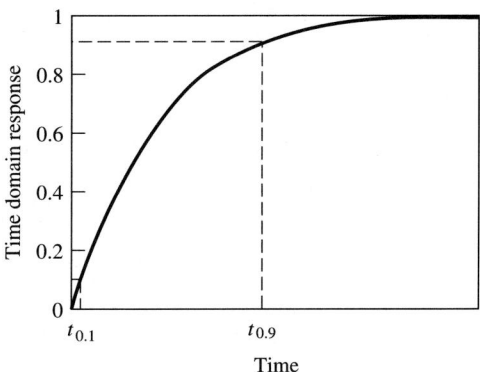
(b) Normalized time domain response

Figure 9.24 Bandwidth and risetime of a first-order system

$$BW = |f_{3dB} - f_{pk}| = \frac{1}{2\pi RC} \tag{9.41}$$

To evaluate the transient response of the RC circuit, suppose that the capacitor is initially uncharged. With a step-function voltage $V_o U(t)$ applied to the RC circuit, the voltage across the capacitor is

$$v_c(t) = V_0(1 - e^{-t/RC})U(t)$$

The initial and final values are 0 and V_o, respectively. The times required for $v_c(t)$ to reach 10% and 90% of the final value are $t_{0.1} = -RC \, ln0.9$ and $t_{0.9} = -RC \, ln0.1$, respectively. Therefore, the 10%-to-90% risetime is

$$RT = t_{0.9} - t_{0.1} = RC \, ln9 = 2.19RC \tag{9.42}$$

Combining (9.41) with (9.42), we have the **bandwidth–risetime product** of a series RC circuit:

$$BW \times RT = \frac{ln \, 9}{2\pi} \approx 0.35 \tag{9.43}$$

For second and higher-order systems, the transient response is not necessarily monotonic. Overshoots appear frequently. For these circuits, the bandwidth–risetime products are dependent on the overshoot of the response. Calculations show that if the overshoot is less than 5 percent, the bandwidth–risetime product is approximately a constant and the constant is 0.35. For an overshot larger than 5 percent, the constant is about 0.45 [36, 38].

9.6.2 Bandwidth–Pulse Width Relationship

In Chapter 8, we discussed the effects of fiber dispersion on optical signals including the pulse broadening due to various dispersion effects. A direct relationship can be established between the pulse width and the bandwidth. For this purpose, consider Gaussian pulses, which have a special property: the Fourier transform of a Gaussian function is also a Gaussian function [39]. Let the Gaussian function be

$$g(t) = e^{-\sigma^2 t^2}$$

where σ is a constant. Since the function $g(t)$ has a peak value of 1 at $t = 0$ and a value of 1/2 at $t = \pm\sqrt{ln2}/\sigma$, the full width between half-power points (FWHP) is $PW = 2\sqrt{ln2}/\sigma$. In other words, σ is inversely proportional to the pulse width.

The Fourier transform of $g(t)$ is

$$G(f) = \int_{-\infty}^{\infty} g(t)e^{-i2\pi ft}dt = \frac{\sqrt{\pi}}{\sigma} e^{-\pi^2 f^2/\sigma^2}$$

The function $G(f)$ has a peak at dc and a 3-dB frequency of $\sqrt{ln2}\sigma/\pi$. The bandwidth of the Gaussian pulses is $\sqrt{ln2}\sigma/\pi$. Thus, the bandwidth–pulse width product of a Gaussian pulse is

$$BW \times PW = \frac{2ln2}{\pi} \approx 0.44 \qquad (9.44)$$

9.7 POWER BUDGET AND RISETIME BUDGET EXAMPLE

In the design of a communications system, the losses in various parts of the system and the risetime or bandwidth of each component must be taken into consideration. This is accomplished through the **power budget** and the **risetime budget,** considerations which are described here with the use of a specific example that pulls together many of the concepts discussed in this text.

Assume a system is designed to transmit 8 MHz signals. An LED with an emitting area 200 μm in diameter is selected as the optical source. The emission from the LED may be approximated by

$$J(\theta) = \begin{cases} J_2\cos^2\theta, & 0 \leq \theta \leq \frac{\pi}{2} \\ 0 & \frac{\pi}{2} \leq \theta \leq \pi \end{cases}$$

with a total power output of 3 mW, as measured in air. A PIN diode with an active area 200 μm in diameter is used as the detector. A minimum power of 2 μW, as measured in air, is required by the receiver. The transmission medium is a step index fiber with an attenuation of 10 dB/km, a numerical aperture of 0.3, a core index of 1.458, and a core radius of 50 μm. Two connectors are needed, and the connector insertion loss is -1.374 dB per connector. A safety margin of 5 dB is specified to ensure reliability of operation. Also assume that the bandwidth of the transmitter, including the LED, and the receiver, including the PIN diode, is very broad in comparison with that of the fiber. Based on all of this information, estimate the maximum fiber length of the system.

9.7 Power Budget and Risetime Budget Example

First, study the system length from the *power viewpoint*. With reference to 1 mW a power of 3 mW is +4.771 dBm.

For a fiber with an *NA* of 0.3, the critical angle $\theta_{c\ fb}$ is 17.46 degrees. The fraction of power contained in a cone of angle 17.46 degrees is

$$\frac{2\pi \int_0^{\theta_{c\ fb}} J_2 \cos^2\theta \sin\theta d\theta}{2\pi \int_0^{\pi/2} J_2 \cos^2\theta \sin\theta d\theta} = 0.132$$

which gives an insertion loss of -8.794 dB. The difference between the LED emitting area and the fiber core cross-sectional area leads, from (9.29), to an insertion loss of -6.020 dB. The Fresnel loss is, from (9.33), -0.153 dB. Therefore, the total LED-to-fiber coupling loss is

$$-8.794 \text{ dB} - 6.020 \text{ dB} - 0.153 \text{ dB} = -14.967 \text{ dB}$$

If the fiber length in km is L, then the loss due to fiber attenuation is $-10L$ dB. Two connectors will result in an insertion loss of -2.748 dB. Also included is a safety margin of 5 dB. Since the detector area is larger than the fiber core cross-sectional area, there is no loss due to area mismatch. However, it is necessary to account for the Fresnel loss of -0.153 dB due to the fiber/air boundary. Taking these factors into account, we obtain the signal level, in dBm, at the detector

$$4.771 - 14.967 - 10L - 2.748 - 5.00 - 0.153 = -18.097 - 10L$$

Since the minimum detectable signal is 2 μW, which is -26.989 dBm, the fiber length is limited by the requirement

$$-18.097 - 10L \geq -26.989$$

In other words, from power considerations, the system length is limited to 0.889 km.

Next, examine the system length from the *risetime viewpoint*. We do so by studying the bandwidth. Since the bandwidth of the transmitter and the receiver is very broad, the risetime of these components is very short. In view of (9.40) we can ignore the risetime of these components and consider only the bandwidth of the fiber. For step index fibers with a numerical aperture of 0.3 and a core index of 1.458, we have, from (8.41), a length–bandwidth product of 4.86 km–MHz. For 8 MHz signals, the system length must be less than 0.607 km, which is less than the limit already established by the power considerations. Therefore, for the step index fiber system being examined, the link length is limited to 0.607 km by the risetime (i.e., bandwidth), rather than the power losses in the system.

If a parabolic index fiber is used in lieu of the step index fiber, and if all other parameters and requirements remain the same, the situation is reversed. From

(8.42), we have a length–bandwidth product of 459 MHz-km for the parabolic index fiber, which shows that the system length is limited by the power consideration. In other words, if parabolic index fibers are used, the link length may be increased to 0.889 km.

REFERENCES

1. Matsushita, S.; K. Kawai; and H. Uchida. "Fiber-optics devices for local network applications." *IEEE J. Lightwave Technol.* LT-3, (1985), pp. 544–555.
2. Ziemmer, R. E.; and W. H. Tranter. *Principles of Communications, Systems, Modulations and Noise.* Chapters 6 and 7. 2nd ed. Boston, MA: Houghton Mifflin Company, 1985.
3. Agrawal, G. P. *Fiber-optic Communication Systems.* Chapters 4, 5, 6, and 7. New York, NY: Wiley Interscience, 1992.
4. Morra, P.; and E. Vezzoni. "Fiber-optic splices, connectors, and couplers." Chapter 3, In *Fiber Optics Handbook for Engineers and Scientists,* ed. F. A. Allard. New York, NY: McGraw-Hill Book Company, 1990.
5. Morra, P.; and E. Vezzoni. "Splices, connectors, and passive devices." Chapter 9, In *Fiber Optics Communication Handbook,* ed. F. Tosco. Blue Ridge Summit, PA: TAB Professional and Reference Books, 1990.
6. Kobayashi, K.; R. Ishikawa; K. Minemura; and S. Sugimoto. "Micro-optic devices for fiber optic communications." *Fiber and Integrated Optics* 1, (1979), pp. 1–17.
7. Tomlinson, W. J. "Applications of GRIN-rod lenses in optical fiber communication systems." *Appl. Opt.* 19, (1980), pp. 1127–1138.
8. Iga, K.; Y. Kokubun; and M. Oikawa. *Fundamentals of Microoptics. Distributed-index, Microlens, and Stacked Planar Optics.* Tokyo, Japan: Academic Press, Inc., 1984.
9. Ishikawa, R.; K. Kaede; H. Nishimoto; K. Minemura; and S. Matsushita. "Kaleidoscope micro-optic star coupler." *Electron. Lett.* 16, (1980), pp. 248–250.
10. Agrawal, A. K. "Review of optical fiber couplers." *Fiber and Integrated Optics* 6, (1985), pp. 27–53.
11. Kawasaki, B. S.; D. C. Johnson; and K. O. Hill. "Configurations, performance and applications of biconical taper optical fibre coupling structures." *Can. J. Phys.* 61, (1983), pp. 352–360.
12. Szarka, P.; A. Lightstone; J. W. Y. Lit; and R. Hughes. "A review of biconical taper couplers." *Fiber and Integrated Optics* 3, (1980), pp. 285–298.
13. Li, Y. F.; and J. W. Y. Lit. "Coupling efficiency of a multimode biconical taper coupler." *J. Opt. Soc. Am. A* 2, (1985), pp. 1301–1306.
14. Nosu, K.; and R. Watanabe. "Slab waveguide star coupler for multimode optical fibers." *Electron. Lett.* 16, (1980), pp. 608–609.
15. Stockmann, M.; and H. H. Witte. "Planar star coupler for multimode fibers." *Appl. Opt.* 19, (1980), pp. 2584–2588.
16. Takato, N.; and T. Kurokawa. "Polymer waveguide star coupler." *Appl. Opt.* 21, (1982), pp. 1940–1942.

17. Hale, P. G.; and R. Kompfner. "Mechanical optical-fiber switch." *Electron. Lett.* 12, (1976), p. 388.
18. Rawson, E. G.; and M. D. Bailey. "A fiber optical relay for bypassing computer network repeaters." *Opt. Eng.* 19, (1980), pp. 628–629.
19. Jackel, J. L.; S. Hackwood; J. J. Veselka; and G. Beni. "Electrowetting switch for multimode optical fibers." *Appl. Opt.* 22, (1983), pp. 1765–1770.
20. Soref, R. A. "Low-loss fiber optic switches based on liquid crystals." In "Physics of Fiber Optics" *Advances in Ceramics* 2, ed. B. Bendow; and S. S. Mitra. Westerville, OH: The American Ceramics Society, Inc., 1980.
21. Marcuse, D. "Excitation of parabolic-index fibers with incoherent sources." *Bell Syst. Tech. J.* 54, (1975), pp. 1507–1531.
22. Longhurst, R. S. *Geometrical and Physical Optics.* Chapter 18. 2nd ed. New York, NY: John Wiley & Sons, Inc., 1967.
23. Marcuse, D.; D. Gloge; and E. A. J. Marcatili. "Guiding properties of fibers." Chapter 3, In *Optical Fiber Telecommunications,* ed. S. E. Miller; and A. G. Chynoweth. New York, NY: Academic Press, 1979.
24. Di Vita, P.; and R. Vannucci. "Geometrical theory of coupling errors in dielectric optical waveguides." *Opt. Comm.* 14, (1975), pp. 139–144.
25. Di Vita, P.; and U. Rossi. "Theory of power coupling between multimode optical fibers." *Opt. and Quantum Electron.* 10, (1978), pp. 107–117.
26. Hecht, E. *Optics.* Chapter 5. 2nd ed. Reading, MA: Addison-Wesley Publishing Co., 1987.
27. Pedrotti, F. L.; and L. S. Pedrotti. *Introduction to Optics.* Chapter 4. 2nd ed. New York, NY: Prentice Hall, Inc., 1993.
28. Jenkins, F. A.; and H. E. White. *Fundamentals of Optics.* 4th ed. New York, NY: McGraw-Hill Book Company, 1976.
29. Nice, A. "Lens coupling in fiber-optic devices: efficiency limits." *Appl. Opt.* 20, (1981), pp. 3136–3145.
30. Thiel, F. L.; and R. M. Hawk. "Optical waveguide cable connection." *Appl. Opt.* 15, (1976), pp. 2785–2791.
31. Tsuchiya, H.; H. Nakagome; N. Shimizu; and S. Ohara. "Double eccentric connectors for optical fibers." *Appl. Opt.* 16, (1977), pp. 1323–1331.
32. Gloge, D. "Offset and tilt loss in optical fiber splices." *Bell Syst. Tech. J.* 55, (1976), pp. 905–916.
33. Miller, C. M. "Transmission vs transverse offset for parabolic-profile fiber splices with unequal core diameters." *Bell Syst. Tech. J.* 55, (1976), pp. 917–927.
34. van Etten, W.; W. Lambo; and P. Simmons. "Loss in multimode fiber connections with a gap." *Appl. Opt.* 24, (1985), pp. 970–976.
35. Terman, F. E. *Electronic and Radio Engineering.* 4th ed. New York, NY: McGraw-Hill Book Company, 1955, p. 289.
36. Valley, G. E., Jr.; and H. Wallman. (ed.) "*Vacuum-tube Amplifiers.*" Editor-in-chief, L. N. Ridenour *MIT Radiation Laboratory Series* 18, New York, NY: McGraw-Hill Book Company, 1948, p. 80.
37. Terman, F. E.; and J. M. Petit. *Electronic Measurement.* New York, NY: McGraw-Hill Book Company, 1952, pp. 259–260, and 327–332.

38. Ghausi, M. S. *Principles and Design of Linear Active Circuits*. Chapter 16. New York, NY: McGraw Hill Book Company, 1965.
39. McGillem, C. D.; and G. R. Cooper. *Continuous and Discrete Signal and System Analysis*. Chapter 5. New York, NY: Holt, Rinehart and Winston, 1974.

PROBLEMS

1. Derive (9.21) and (9.22).
2. Consider the intrinsic loss in joining two fibers with different core radii and numerical apertures. Show that if $a_1 \leq a_2$, then

$$\frac{P_2}{P_1} = \begin{cases} (NA_2/NA_1)^2, & \text{if } NA_1 \geq NA_2 \\ 1, & \text{if } NA_1 \leq NA_2 \end{cases}$$

3. Consider the coupling from a multimode fiber with a circular core to a square waveguide. Show that the insertion loss is

$$IL = 10 \log [4p^2/\pi] \quad \text{for } p < \frac{1}{\sqrt{2}}$$

$$IL = 10 \log [1 - \frac{4}{\pi}(\sin^{-1}\sqrt{1-p^2} - p\sqrt{1-p^2})] \quad \text{for } \frac{1}{\sqrt{2}} \leq p < 1$$

$$IL = 0 \quad \text{for } 1 \leq p$$

where p is the ratio of the width of the square waveguide to the diameter of the fiber core.

4. Consider the coupling from the square waveguide to the multimode fiber. Show that the insertion loss is

$$IL = 0 \quad \text{for } p < \frac{1}{\sqrt{2}}$$

$$IL = 10 \log[\sqrt{p^{-2}-1} - \frac{1}{p^2}(\sin^{-1}\sqrt{1-p^2} - \frac{\pi}{4})] \quad \text{for } \frac{1}{\sqrt{2}} \leq p < 1$$

$$IL = 10 \log[\frac{\pi}{4p^2}] \quad \text{for } 1 \leq p$$

where p is the ratio of the width of the square waveguide to the diameter of the fiber core.

5. Consider an optical fiber communications system. The electronic components are designed such that the risetimes of the transmitter and the receiver are negligibly short. The relevant parameters of the optical components are as listed. Estimate the maximum length of the fiber system if 5 MHz signals are to be transmitted by the system.

   LED
   Output power: 3 mW
   Active area: 200 μm in diameter.
   Radiation pattern: $J(\theta) = J_o$ for $\theta \leq 45°$, $J(\theta) = 0$ for $\theta > 45°$
   Fiber
   $\alpha = 7$ dB/km, $a = 50$ μm, $n_{co} = 1.4600$ and $n_{cl} = 1.4380$

Detector
    Minimum signal required: 2 μW
    Active area: 200 μm in diameter.
Alignment
    Perfect alignment, with zero transverse offset and zero spacing between fiber
        and LED and detector.
Safety Margin
    3 dB
Connectors
    Four connectors required.
    No index matching liquid used.
    Angular alignment: ±3°,
    Lateral misalignment: 10 μm,
    Spacing between connector ends: 10 μm,

6. Consider the same communication system given in problem 5. What is the maximum system length if the signal bandwidth is 20 MHz?

CHAPTER 10

POLARIZATION EFFECTS: SINGLE-MODE FIBERS AND COMPONENTS

CHAPTER OUTLINE

10.1 Introduction
10.2 Birefringent Media and Jones Calculus
    10.2.1 Isotropic Media
    10.2.2 Linearly Birefringent Media
    10.2.3 Circularly Birefringent Media
    10.2.4 Linearly Birefringent Media in Magnetic Field
    10.2.5 Polarizers
    10.2.6 Axes Rotation
    10.2.7 Continuous Axis Rotation
    10.2.8 Lossless Media
    10.2.9 SOP Experimental Determination
    10.2.10 Jones Matrices Element Measurements
10.3 Single-Mode Fiber Birefringence
    10.3.1 Geometrical or Form Birefringence
    10.3.2 Internal or Built-In Stress Birefringence
    10.3.3 External Applied Stress or Fields Birefringence
    10.3.4 Polarization Beat Length
    10.3.5 Fiber Axis Rotation
    10.3.6 Jones Matrices: Birefringent Fibers
10.4 Single-Mode Fiber Polarization Components
    10.4.1 Single-mode Fiber Polarizers
    10.4.2 Fiber Retarders
    10.4.3 Fiber Polarization Controllers
10.5 Single-Mode Fiber Directional Couplers
    10.5.1 Twin Core Couplers
    10.5.2 Biconical Tapered Structure Couplers
    10.5.3 Polished Fiber Couplers

10.6 Erbium-Doped Fiber Amplifiers
    10.6.1 Erbium-Doped Fibers
    10.6.2 Rate Equations
    10.6.3 Normalized Variables
    10.6.4 Simple Amplification Characteristics
Problems
References

10.1 INTRODUCTION

In this chapter we will concentrate on single-mode fibers and single-mode fiber components. Since most single-mode fibers support two nearly degenerate polarization modes, they are actually single-mode double-polarization fibers. Single-mode single-polarization fibers do exist. But, most single-mode fibers in use, or to be used in the near future, remain single-mode double polarization fibers. We follow the customary practice of referring to single-mode double-polarization fibers simply as single-mode fibers. Because the propagation constants of the two polarization modes are not identical, single-mode fibers are birefringent. Jones calculus is introduced in section 10.2, for use in characterizing the polarization properties of birefringent components in general and birefringent fibers in particular. The physical origin of the fiber birefringence is discussed in section 10.3.

As noted in Chapter 8, the main advantage of single-mode fibers over multimode fibers is the bandwidth–length product. To take advantage of the large bandwidth available in single-mode fibers, a coherent detection scheme is usually employed in single-mode fiber communication systems. In all coherent detection systems, the state of polarization (SOP) of the input signals must match that of the local oscillator, as noted in Chapter 5. If the SOPs of the two signals do not match, the sensitivity of the coherent detection system can be seriously degraded. Polarization components at the front end of coherent receivers are used to select or control the SOP.

There are two ways to select or maintain the SOP. One is to attenuate the unwanted polarization mode using polarizers. The other is to use polarization components to convert the input fields of an arbitrary SOP to the preferred SOP, without reducing the signal intensity. While classical or conventional optical components, also referred to as *bulk optical components,* may be used in these applications, the mechanical stability and alignment problems may be considerable. These difficulties may be lessened if *all-fiber components* are used. In section 10.4, we discuss fiber polarizers and polarization transformers. Jones matrices are used to describe the operation of these components quantitatively.

Section 10.5 is devoted to single-mode fiber directional couplers, which are useful in switching and routing applications. The operations of fiber directional couplers are examined in terms of the coupled mode theory discussed in Chapter 7. In the last section, we discuss erbium-doped fiber amplifiers, which are ideally suited for compensating for the loss in the fibers.

10.2 BIREFRINGENT MEDIA AND JONES CALCULUS

As stated in Chapter 6, any plane waves may be viewed as the superposition of two *orthogonal linearly polarized fields*. For plane waves propagating in the z direction, the two orthogonal field components are the x and y components. In the time domain, the electric fields are

$$\mathcal{E}(t) = \mathbf{a}_x E_{xo}\cos(\omega t + \phi_x) + \mathbf{a}_y E_{yo}\cos(\omega_x + \phi_y)$$

where E_{xo} and E_{yo} are real quantities representing the amplitudes of the two components. The corresponding expression in phasor notation is

$$\mathbf{E} = E_x \mathbf{a}_x + E_y \mathbf{a}_y \qquad (10.1)$$

where $E_x = E_{xo}e^{j\phi_x}$ and $E_y = E_{yo}e^{j\phi_y}$ are complex quantities. These field components propagate with phase velocities v_x and v_y, or have the propagation constants $\beta_x = kn_x = \omega/v_x$ and $\beta_y = kn_y = \omega/v_y$.

Fields can also be considered as the superposition of two *counter rotating circularly polarized fields,* the right-hand field and the left-hand field, as follows:

$$\mathbf{E} = \frac{E_x + jE_y}{\sqrt{2}} \mathbf{a_R} + \frac{E_x - jE_y}{\sqrt{2}} \mathbf{a_L}$$

and

$$\mathbf{a_R} = \frac{\mathbf{a}_x - j\mathbf{a}_y}{\sqrt{2}}$$

$$\mathbf{a_L} = \frac{\mathbf{a}_x + j\mathbf{a}_y}{\sqrt{2}}$$

where $\mathbf{a_R}$ and $\mathbf{a_L}$ are basis vectors for right-hand and left-hand circularly polarized fields, respectively. The phase velocities of the two fields are v_r and v_l, and the propagation constants are $\beta_r = kn_R = \omega/v_r$ and $\beta_l = kn_L = \omega/v_l$, respectively.

Now we are ready to introduce the concept of birefringence. A medium is **linearly birefringent** if $\beta_x \neq \beta_y$, and **circularly birefringent** if $\beta_r \neq \beta_l$.

For waves propagating in isotropic media, such as free-space, the phase velocity is independent of the SOP; and the SOP remains unchanged as the waves propagate. In birefringent media, the phase velocities of *orthogonal* field components are different. As a result, the SOP evolves as the waves propagate. **Jones calculus** is an elegant method of relating the SOP at a given point in the medium to the SOP at another point. The heart of the Jones calculus is the 2 × 2 matrix, used to characterize the birefringent media or devices, and the 1 × 2 column matrix describing the electric fields. The 2 × 2 matrices are known as the **Jones matrices,** and the 1 × 2 column matrices are the **Jones vectors** [1].

To introduce the Jones calculus, we cast the fields given in (10.1) in a column matrix form, as follows:

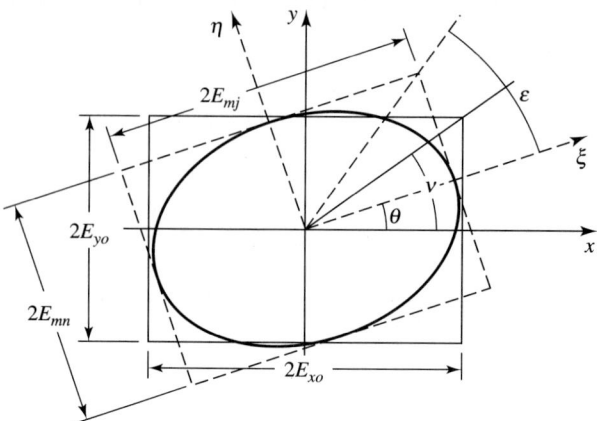

Figure 10.1 Parameters of elliptically polarized waves

$$\mathbf{E} = \begin{bmatrix} E_x \\ E_y \end{bmatrix} = \begin{bmatrix} E_{xo}e^{j\phi_x} \\ E_{yo}e^{j\phi_y} \end{bmatrix} \quad (10.2)$$

This is the Jones vector describing the fields. Depending on the relative amplitude of the orthogonal field components and the phase difference between them, a field may be a **linearly, circularly,** or **elliptically polarized field,** as described in Chapter 6. For example, the Jones vectors of the horizontally and vertically linearly polarized fields, and the right-hand and left-hand circularly polarized fields, are

$$\begin{bmatrix} E_x \\ 0 \end{bmatrix}, \quad \begin{bmatrix} 0 \\ E_y \end{bmatrix}, \quad \begin{bmatrix} E_x \\ -jE_x \end{bmatrix}, \quad \text{and} \quad \begin{bmatrix} E_x \\ jE_x \end{bmatrix} \quad (10.3)$$

respectively. If $\Delta\phi = \phi_y - \phi_x$ and v is defined such that $\tan v = E_{yo}/E_{xo}$ (Figure 10.1), the Jones vector (10.2) can be written as

$$\mathbf{E} = \sqrt{E_{xo}^2 + E_{yo}^2}\, e^{j\phi_x} \begin{bmatrix} \cos v \\ e^{j\Delta\phi}\sin v \end{bmatrix} \quad (10.4)$$

This is the expression for fields in an arbitrary coordinate system. The expression in (10.4) can be simplified by selecting a new coordinate system and time reference. If the coordinate system is rotated from (x,y) to (ξ, η) and the time reference is shifted from t to t', then $\mathcal{E}(t)$ given in (10.1) can be written as (see Problem 1)

$$\mathcal{E}(t') = \mathbf{a}_\xi E_{mj}\cos\omega t' - \mathbf{a}_\eta E_{mn}\sin\omega t' = \mathbf{a}_\xi E_{mj}\cos\omega t' + \mathbf{a}_\eta E_{mn}\cos(\omega t' + \frac{\pi}{2})$$

where $\mathbf{a}_\xi$ and $\mathbf{a}_\eta$ are unit vectors along the major and minor axes of the polarization ellipse, as shown in Figure 10.1, and E_{mj} and E_{mn} are the real amplitudes of the field components along the major and minor axes. It is left as an exercise for the reader (problem 2) to show that

$$E_{mj}^2 + E_{mn}^2 = E_{xo}^2 + E_{yo}^2$$

Physically, this expression means that the diagonal of the solid rectangle shown in Figure 10.1 and that of the dashed rectangle in the same figure are of equal length. In terms of the ξ and η coordinates, the Jones vector assumes the simple form

$$\mathbf{E} = e^{j(\phi_x - \delta)} \begin{bmatrix} E_{mj} \\ jE_{mn} \end{bmatrix} \qquad (10.5)$$

where the phase angle δ is given by the relation $\omega t + \phi_x = \omega t' + \delta$.

To summarize, the Jones vector of any elliptically polarized field, when referenced to an *arbitrary* coordinate system x and y, is given by (10.4). However, if the coordinates *coincide* with the major and minor axes of the polarization ellipse, the Jones vector reduces to (10.5).

In many applications where the SOP is important, the detailed descriptions of the fields in terms of the full Jones vectors (10.2), (10.4), or (10.5) are not necessary. Since the SOP depends on the ratio of the amplitudes of orthogonal field components, and the relative phase between them, it is convenient to reduce the Jones vectors to the simplest, most easily recognizable form. This is done by eliminating the factor common to the two field components. Then, (10.2) becomes

$$\mathbf{E}_n = \begin{bmatrix} E_{nx} \\ E_{ny} \end{bmatrix}$$

and $\mathbf{E}_n$ is defined as the **normalized Jones vector.** The matrix elements E_{nx} and E_{ny} of the normalized Jones vector are chosen such that $|E_{nx}|^2 + |E_{ny}|^2 = 1$. For example, the normalized Jones vector for linearly polarized fields pointed in the direction θ relative to the x axis is

$$\begin{bmatrix} \cos\theta \\ \sin\theta \end{bmatrix} \qquad (10.6)$$

Corresponding to the full Jones vectors listed in (10.3), the normalized Jones vectors are

$$\begin{bmatrix} 1 \\ 0 \end{bmatrix}, \quad \begin{bmatrix} 0 \\ 1 \end{bmatrix}, \quad \frac{1}{\sqrt{2}} \begin{bmatrix} 1 \\ -j \end{bmatrix}, \quad \text{and} \quad \frac{1}{\sqrt{2}} \begin{bmatrix} 1 \\ j \end{bmatrix} \qquad (10.7)$$

These and other normalized Jones vectors are listed in Table 10.1.

A *Jones matrix* is a matrix expression describing the birefringent characteristics of a medium or device. Suppose the input and output Jones vectors are

$$\mathbf{E}_{in} = \begin{bmatrix} E_{xin} \\ E_{yin} \end{bmatrix}$$

$$\mathbf{E}_{out} = \begin{bmatrix} E_{xout} \\ E_{yout} \end{bmatrix}$$

Table 10.1 Normalized Jones vectors ($e^{j\omega t}$)

Polarization	Orientation	Jones vector
Linear	Horizontal	$\begin{bmatrix} 1 \\ 0 \end{bmatrix}$
Linear	Vertical	$\begin{bmatrix} 0 \\ 1 \end{bmatrix}$
Linear	θ	$\begin{bmatrix} \cos\theta \\ \sin\theta \end{bmatrix}$
Circular	Right hand	$\frac{1}{\sqrt{2}}\begin{bmatrix} 1 \\ -j \end{bmatrix}$
Circular	Left hand	$\frac{1}{\sqrt{2}}\begin{bmatrix} 1 \\ j \end{bmatrix}$
Elliptical	—	$\begin{bmatrix} \cos\nu \\ e^{j\Delta\phi}\sin\nu \end{bmatrix}$

For a birefringent medium, $\mathbf{E}_{in}$ and $\mathbf{E}_{out}$ are linearly related. The relationship can be expressed in terms of a 2 × 2 matrix, as follows:

$$\begin{bmatrix} E_{xout} \\ E_{yout} \end{bmatrix} = \begin{bmatrix} J_{11} & J_{12} \\ J_{21} & J_{22} \end{bmatrix} \begin{bmatrix} E_{xin} \\ E_{yin} \end{bmatrix}$$

This can be written in an abbreviated notation,

$$\mathbf{E}_{out} = \mathbf{J}\mathbf{E}_{in}$$

where the Jones matrix for the birefringent medium or device is

$$\mathbf{J} = \begin{bmatrix} J_{11} & J_{12} \\ J_{21} & J_{22} \end{bmatrix}$$

Since Jones matrices are used to characterize the birefringent properties of a medium or component, common, multiplicative phase terms are often ignored since they have no effect on the birefringence of the medium or component.

The Jones matrix of a complicated or composite optical device is the **ordered product** of the Jones matrices of the constituent elements. This property of Jones matrices is analogous to the properties of ABCD matrices discussed in Chapter 2; that is, the ABCD matrix of a complete system can be constructed from the ABCD matrices of elementary building blocks. In the following subsections, we will derive Jones matrices for *simple* polarization components. We will then use these Jones matrices as vehicles to study the polarization characteristics of complicated polarization components or systems.

10.2.1 Isotropic Media

In isotropic media, for which $\beta_x = \beta_y$, the output field components at $z = L$ are related to the input field components at $z = 0$ in the following manner:

$$E_{xout} = E_{xin} e^{-j\beta_x L}$$
$$E_{yout} = E_{yin} e^{-j\beta_y L}$$

We recognize that the two propagation constants are identical, and we write β in lieu of β_x and β_y. The relationship may be written in a matrix form

$$\begin{bmatrix} E_{xout} \\ E_{yout} \end{bmatrix} = e^{-j\beta L} \begin{bmatrix} 1 & 0 \\ 0 & 1 \end{bmatrix} \begin{bmatrix} E_{xin} \\ E_{yin} \end{bmatrix} \quad (10.9)$$

Therefore, the Jones matrix for an isotropic medium is simply an identity matrix with a multiplicative constant $e^{-j\beta L}$. If we express the fields in terms of normalized Jones vectors, the exponential term is ignored, and the relationship is further simplified to

$$\begin{bmatrix} E_{n\,xout} \\ E_{n\,yout} \end{bmatrix} = \begin{bmatrix} 1 & 0 \\ 0 & 1 \end{bmatrix} \begin{bmatrix} E_{n\,xin} \\ E_{n\,yin} \end{bmatrix}$$

In other words, the Jones matrix for an *isotropic lossless medium* is simply an identity matrix

$$\begin{bmatrix} 1 & 0 \\ 0 & 1 \end{bmatrix}$$

If the isotropic medium is lossy, with an attenuation constant α and a propagation constant β, the exponential term in (10.9) becomes $e^{-(\alpha+j\beta)L}$, and the Jones matrix for a *lossy isotropic medium* is

$$e^{-(\alpha+j\beta)L} \begin{bmatrix} 1 & 0 \\ 0 & 1 \end{bmatrix} \quad (10.10)$$

Again, we ignore the phase term. Then the Jones matrix (10.10) becomes simply

$$\begin{bmatrix} e^{-\alpha L} & 0 \\ 0 & e^{-\alpha L} \end{bmatrix}$$

10.2.2 Linearly Birefringent Media

For *anisotropic lossless media,* the propagation constants for different field components are different. Therefore, such media are birefringent. Let the propagation constants for the x and y field components be β_x and β_y respectively. Then the input–output relationships are:

$$E_{xout} = E_{xin} e^{-j\beta_x L} = E_{xin} e^{-j(\beta_x+\beta_y)L/2} e^{-j(\beta_x-\beta_y)L/2}$$
$$E_{yout} = E_{yin} e^{-j\beta_y L} = E_{yin} e^{-j(\beta_x+\beta_y)L/2} e^{j(\beta_x-\beta_y)L/2}$$

Since $\beta_x \ne \beta_y$, the phase of E_{yout} is delayed by $(\beta_y - \beta_x)L$ relative to that of E_{xout}. The total phase delay $\delta_l = (\beta_y - \beta_x)L$ incurred as the wave propagates in the medium is known as the **linear retardation.** In matrix form, the input–output relationship is

$$\begin{bmatrix} E_{xout} \\ E_{yout} \end{bmatrix} = \begin{bmatrix} e^{-j\beta_x L} & 0 \\ 0 & e^{-j\beta_y L} \end{bmatrix} \begin{bmatrix} E_{xin} \\ E_{yin} \end{bmatrix} = e^{-j(\beta_x + \beta_y)L/2} \begin{bmatrix} e^{j\delta_l/2} & 0 \\ 0 & e^{-j\delta_l/2} \end{bmatrix} \begin{bmatrix} E_{xin} \\ E_{yin} \end{bmatrix}$$

Thus, the Jones matrix for a birefringent medium with a linear retardation δ_l is

$$\begin{bmatrix} e^{j\delta_l/2} & 0 \\ 0 & e^{-j\delta_l/2} \end{bmatrix}$$

The linear retardation δ_l is positive if $\beta_y > \beta_x$, that is, if $v_x - v_y$. In other words, δ_l is positive if the phase of the x component "moves" faster than that of the y component. If δ_l is positive, the x axis is the **fast axis** of the birefringent medium. If δ_l is negative, then the x axis is the **slow axis.**

A **linear retarder** is a linearly birefringent medium that retards the field component in the slow axis by δ_l relative to the field component along the fast axis. Linear retarders with a retardation of $\pi/2$ and π are commonly referred to as **quarterwave** and **halfwave plates,** respectively.

To characterize a linear retarder, we specify the linear retardation and the orientation of the fast axis relative to a reference direction of our choosing. Let the Jones matrix for a linear retarder be $\mathbf{R}_l(\theta, \delta_l)$, where θ refers to the direction of the fast axis relative to the x axis and δ_l is the linear retardation in radians. For example, assume the Jones matrix of a quarterwave plate with the fast axis along the x axis is

$$\mathbf{R}_l\left(0, \frac{\pi}{2}\right) = \begin{bmatrix} e^{j\pi/4} & 0 \\ 0 & e^{-j\pi/4} \end{bmatrix} = \frac{1+j}{\sqrt{2}} \begin{bmatrix} 1 & 0 \\ 0 & -j \end{bmatrix}$$

Similarly, a halfwave plate with the fast axis along the x axis has a Jones matrix

$$\mathbf{R}_l(0, \pi) = \begin{bmatrix} e^{j\pi/2} & 0 \\ 0 & e^{-j\pi/2} \end{bmatrix} = j \begin{bmatrix} 1 & 0 \\ 0 & -1 \end{bmatrix}$$

In general, the Jones matrix of a linear retarder with its fast axis along the x axis is

$$\mathbf{R}_l(0, \delta_l) = \begin{bmatrix} e^{j\delta_l/2} & 0 \\ 0 & e^{-j\delta_l/2} \end{bmatrix} \tag{10.11}$$

10.2.3 Circularly Birefringent Media

When an isotropic medium is subjected to a strong magnetic field, the medium becomes circularly birefringent because of the Faraday rotation (see Chapter 6). The evolution of the SOP in such a medium is evaluated by expressing the field as the superposition of two circularly polarized fields. The incoming fields are

$$\mathbf{E}_{in} = \mathbf{E}(0) = \frac{E_{xin} + jE_{yin}}{\sqrt{2}} \mathbf{a}_R + \frac{E_{xin} - jE_{yin}}{\sqrt{2}} \mathbf{a}_L$$

and the fields elsewhere in the medium are

$$\mathbf{E}(z) = \frac{E_{xin} + jE_{yin}}{\sqrt{2}} e^{-j\beta_r z} \mathbf{a}_R + \frac{E_{xin} - jE_{yin}}{\sqrt{2}} e^{-j\beta_l z} \mathbf{a}_L$$

At $z = L$ the output fields are

$$\mathbf{E}_{out} = \mathbf{E}_{(L)} = \frac{E_{xin} + jE_{yin}}{\sqrt{2}} e^{-j\beta_r L} \mathbf{a}_R + \frac{E_{xin} - jE_{yin}}{\sqrt{2}} e^{-j\beta_l L} \mathbf{a}_L$$

$$= e^{-j(\beta_r + \beta_l)L/2} \left[\frac{E_{xin} + jE_{yin}}{\sqrt{2}} e^{-j\delta_c/2} \mathbf{a}_R + \frac{E_{xin} - jE_{yin}}{\sqrt{2}} e^{+j\delta_c/2} \mathbf{a}_L \right]$$

where $\delta_c = (\beta_r - \beta_l)L$ is the **circular retardation.** In terms of cartesian coordinates, the output fields are

$$\mathbf{E}_{out} = e^{-j(\beta_r + \beta_l)L/2} \left[[E_{xin}\cos\frac{\delta_c}{2} + E_{yin}\sin\frac{\delta_c}{2}] \mathbf{a}_x + [-E_{xin}\sin\frac{\delta_c}{2} + E_{yin}\cos\frac{\delta_c}{2}] \mathbf{a}_y \right]$$

In the column matrix form, the output electric fields are

$$\begin{bmatrix} E_{xout} \\ E_{yout} \end{bmatrix} = e^{-j(\beta_r + \beta_l)L/2} \begin{bmatrix} \cos\delta_c/2 & \sin\delta_c/2 \\ -\sin\delta_c/2 & \cos\delta_c/2 \end{bmatrix} \begin{bmatrix} E_{xin} \\ E_{yin} \end{bmatrix}$$

A **circular retarder** is a circularly birefringent medium that introduces a circular retardation between two circularly polarized fields. The Jones matrix for a circular retarder is therefore

$$\mathbf{R}_c(\delta_c) = \begin{bmatrix} \cos\delta_c/2 & \sin\delta_c/2 \\ -\sin\delta_c/2 & \cos\delta_c/2 \end{bmatrix} \tag{10.12}$$

10.2.4 Linearly Birefringent Media in Magnetic Field

For linearly birefringent media under the influence of a strong magnetic field in the direction of propagation, the Jones matrix is

$$\mathbf{J} = \begin{bmatrix} \cos\frac{\delta}{2} + j\frac{1-\sigma^2}{1+\sigma^2}\sin\frac{\delta}{2} & \frac{2\sigma}{1+\sigma^2}\sin\frac{\delta}{2} \\ -\frac{2\sigma}{1+\sigma^2}\sin\frac{\delta}{2} & \cos\frac{\delta}{2} - j\frac{1-\sigma^2}{1+\sigma^2}\sin\frac{\delta}{2} \end{bmatrix} \tag{10.13}$$

where δ and σ are functions of the linear birefringence and the Faraday rotation [2]. Precise definitions of δ and σ and detailed derivations for (10.13) are given in Appendix E.

10.2.5 Polarizers

A **perfect polarizer** transmits fields polarized along a preferred direction, with no attenuation, while rejecting fields in the orthogonal direction. If the fields polarized in the x direction are transmitted without loss and the fields in the y direction are blocked completely, the input and output fields are related by

$$E_{xout} = E_{xin}e^{-j\beta_x L}$$

$$E_{yout} = 0$$

Ignoring the exponential factor, the Jones matrix is,

$$\mathbf{P}(0) = \begin{bmatrix} 1 & 0 \\ 0 & 0 \end{bmatrix} \tag{10.14}$$

The argument of **P** identifies the direction of the preferred axis. For example, a perfect polarizer passing fields polarized along the y axis is represented by $\mathbf{P}\left(\frac{\pi}{2}\right)$, and the Jones matrix is

$$\mathbf{P}\left(\frac{\pi}{2}\right) = \begin{bmatrix} 0 & 0 \\ 0 & 1 \end{bmatrix} \tag{10.15}$$

In general, a perfect polarizer transmitting linearly polarized fields at an arbitrary angle θ with respect to the x axis has a Jones matrix of the form

$$\mathbf{P}(\theta) = \begin{bmatrix} \cos^2\theta & \cos\theta\sin\theta \\ \cos\theta\sin\theta & \sin^2\theta \end{bmatrix} \tag{10.16}$$

10.2.6 Axes Rotation

For most media and devices thus far discussed, the fast axis is assumed to be along the x axis. For media with the fast axis in an arbitrary direction, a matrix relation is needed to describe the rotation of the coordinate axes. Figure 10.2 depicts two coordinate systems (x,y) and (x',y'). An arbitrary vector **E** can be expressed in terms of the (x,y) coordinates as

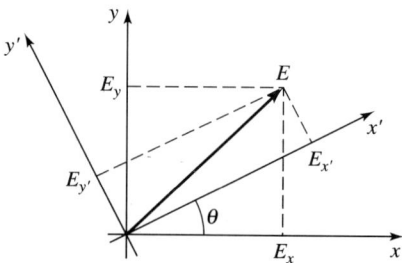

Figure 10.2 Rotation of coordinates

$$\mathbf{E} = E_x \mathbf{a}_x + E_y \mathbf{a}_y$$

or in terms of the (x',y') coordinates as

$$\mathbf{E} = (E_x \cos\theta + E_y \sin\theta)\mathbf{a}_{x'} + (-E_x \sin\theta + E_y \cos\theta)\mathbf{a}_{y'}$$

The Jones vector of $\mathbf{E}$ in the xy coordinate system is

$$\begin{bmatrix} E_x \\ E_y \end{bmatrix}$$

In the $x'\,y'$ coordinates, this becomes

$$\begin{bmatrix} E_x \cos\theta + E_y \sin\theta \\ -E_x \sin\theta + E_y \cos\theta \end{bmatrix}$$

These two column matrices are related through the simple matrix relationship

$$\begin{bmatrix} E_x \cos\theta + E_y \sin\theta \\ -E_x \sin\theta + E_y \cos\theta \end{bmatrix} = \begin{bmatrix} \cos\theta & \sin\theta \\ -\sin\theta & \cos\theta \end{bmatrix} \begin{bmatrix} E_x \\ E_y \end{bmatrix}$$

Therefore, the **rotation matrix** describing the coordinate transformation is:

$$\mathbf{S}(\theta) = \begin{bmatrix} \cos\theta & \sin\theta \\ -\sin\theta & \cos\theta \end{bmatrix} \tag{10.17}$$

Equation (10.17) is used to study the effects of axis rotation. Let $\mathbf{J}(0)$ be the Jones matrix of a polarization device with its fast axis in the x axis. If the fast axis is rotated by θ, then $\mathbf{J}(\theta)$ is obtained by *premultiplying* $\mathbf{J}(0)$ by $\mathbf{S}(-\theta)$ and *postmultiplying* it by a $\mathbf{S}(\theta)$ (see problem 5). That is,

$$\mathbf{J}(\theta) = \mathbf{S}(-\theta)\mathbf{J}(0)\mathbf{S}(\theta) \tag{10.18}$$

The term $\mathbf{S}(-\theta)$ is also known as the **counter rotation matrix.** The expression (10.18) is a **similarity transformation.** Simple algebra will confirm that the $\mathbf{P}(\theta)$ given by (10.16) is indeed $\mathbf{S}(-\theta)\mathbf{P}(0)\mathbf{S}(\theta)$, where $\mathbf{P}(0)$ is given by (10.14). The Jones matrices of linear and circular polarizers and retarders are listed in Table 10.2. Also included in the table is the Jones matrix for rotation operation.

In the following paragraphs, four examples are given to demonstrate the use of Jones calculus. Three of the examples illustrate the use of polarization components to vary the SOP without affecting the light intensity. The last example sets the stage for understanding the experimental setup and manipulations involved in measuring the SOP and the operation of fiber polarization controllers.

Example 1

Consider the effect of a *circular retarder* on linearly polarized fields. Let the input electric fields be oriented at an angle θ with respect to the x axis. The Jones vector is given by (10.6) and the output fields are

$$\begin{bmatrix} \cos\frac{\delta_c}{2} & \sin\frac{\delta_c}{2} \\ -\sin\frac{\delta_c}{2} & \cos\frac{\delta_c}{2} \end{bmatrix} \begin{bmatrix} \cos\theta \\ \sin\theta \end{bmatrix} = \begin{bmatrix} \cos(\theta - \frac{\delta_c}{2}) \\ \sin(\theta - \frac{\delta_c}{2}) \end{bmatrix}$$

From this, we see that the output fields are at an angle of $(\theta - \delta_c/2)$ relative to the x axis. This means that the effect of a circular retarder is to rotate linearly polarized fields by an angle $-\delta_c/2$, and the intensity is not changed.

Example 2

Next, consider the effect of a *halfwave* plate on a linearly polarized input given by (10.6). The fast axis of the wave plate is at θ_h with respect to the x axis. The output is

$$\begin{bmatrix} \cos\theta_h & -\sin\theta_h \\ \sin\theta_h & \cos\theta_h \end{bmatrix} \begin{bmatrix} e^{j\pi/2} & 0 \\ 0 & e^{-j\pi/2} \end{bmatrix} \begin{bmatrix} \cos\theta_h & \sin\theta_h \\ -\sin\theta_h & \cos\theta_h \end{bmatrix} \begin{bmatrix} \cos\theta \\ \sin\theta \end{bmatrix} = j\begin{bmatrix} \cos(2\theta_h - \theta) \\ \sin(2\theta_h - \theta) \end{bmatrix}$$

The output remains linearly polarized, but the fields are rotated from θ to $2\theta_h - \theta$. To elaborate, consider a halfwave plate for which $\theta_h = 0$. Because of

Table 10.2 Jones matrices ($e^{j\omega t}$)

Component	Axis	Symbol	Jones matrix
Linear polarizer	0°	**P**(0)	$\begin{bmatrix} 1 & 0 \\ 0 & 0 \end{bmatrix}$
Linear polarizer	90°	$\mathbf{P}(\frac{\pi}{2})$	$\begin{bmatrix} 0 & 0 \\ 0 & 1 \end{bmatrix}$
Circular polarizer, right hand			$\frac{1}{\sqrt{2}}\begin{bmatrix} 1 & j \\ -j & 1 \end{bmatrix}$
Circular polarizer, left hand			$\frac{1}{\sqrt{2}}\begin{bmatrix} 1 & -j \\ j & 1 \end{bmatrix}$
Linear retarder	0°	$\mathbf{R}_l(0, \delta_l)$	$\begin{bmatrix} e^{j\delta_l/2} & 0 \\ 0 & e^{-j\delta_l/2} \end{bmatrix}$
Circular retarder		$\mathbf{R}_c(\delta_c)$	$\begin{bmatrix} \cos\delta_c/2 & \sin\delta_c/2 \\ -\sin\delta_c/2 & \cos\delta_c/2 \end{bmatrix}$
Rotation operation	θ	$\mathbf{S}(\theta)$	$\begin{bmatrix} \cos\theta & \sin\theta \\ -\sin\theta & \cos\theta \end{bmatrix}$

the linear retardation of π, an incoming field pointing in an arbitrary direction θ is directed to $-\theta$. When the fast axis of the halfwave plate is rotated from 0 to θ_h, the output field rotates from $-\theta$ to $2\theta_h - \theta$. In other words, the rotation of the field is twice the angle θ_h. Also, as in example 1, the output intensity is not changed.

Example 3

Now consider the effect of a *quarterwave* plate on waves linearly polarized in the x direction. Suppose that the fast axis of the quarterwave plate is at 45 degrees from the x axis. The output is then

$$\frac{1}{\sqrt{2}}\begin{bmatrix} 1 & -1 \\ 1 & 1 \end{bmatrix}\begin{bmatrix} e^{j\pi/4} & 0 \\ 0 & e^{-j\pi/4} \end{bmatrix}\frac{1}{\sqrt{2}}\begin{bmatrix} 1 & 1 \\ -1 & 1 \end{bmatrix}\begin{bmatrix} 1 \\ 0 \end{bmatrix} = \frac{1}{\sqrt{2}}\begin{bmatrix} 1 \\ j \end{bmatrix}$$

This means that linearly polarized waves are transformed to left-hand circularly polarized waves by the quarterwave plate. Again, there is no change in the light intensity.

Similarly, we can show that circularly polarized waves are converted to linearly polarized waves by a quarterwave plate, without reducing their intensity (problem 3).

Example 4

Finally, consider the effect of a quarterwave plate on an arbitrary input. For convenience, choose the major axis of the input polarization ellipse as the x axis. Therefore, the Jones matrix of the input is given by (10.5), and the exponential term in front of the Jones vector is dropped for simplicity.

Let the fast axis of the quarterwave plate be at an angle θ_q relative to the x axis. The output of the quarterwave plate is then

$$\begin{bmatrix} \cos\theta_q & -\sin\theta_q \\ \sin\theta_q & \cos\theta_q \end{bmatrix}\begin{bmatrix} e^{j\pi/4} & 0 \\ 0 & e^{-j\pi/4} \end{bmatrix}\begin{bmatrix} \cos\theta_q & \sin\theta_q \\ -\sin\theta_q & \cos\theta_q \end{bmatrix}\begin{bmatrix} E_{mj} \\ jE_{mn} \end{bmatrix} \quad (10.19)$$

$$= e^{j\pi/4}\begin{bmatrix} \cos\theta_q(E_{mj}\cos\theta_q - E_{mn}\sin\theta_q) + j\sin\theta_q(E_{mn}\cos\theta_q - E_{mj}\sin\theta_q) \\ \cos\theta_q(E_{mj}\sin\theta_q + E_{mn}\cos\theta_q) + j\sin\theta_q(E_{mj}\cos\theta_q + E_{mn}\sin\theta_q) \end{bmatrix}$$

In general, the output is an elliptically polarized field. The output is linearly polarized *only* if the two field components are in phase. The two components are *in time phase* if

$$\frac{\sin\theta_q(E_{mn}\cos\theta_q - E_{mj}\sin\theta_q)}{\cos\theta_q(E_{mj}\cos\theta_q - E_{mn}\sin\theta_q)} = \frac{\sin\theta_q(E_{mj}\cos\theta_q + E_{mn}\sin\theta_q)}{\cos\theta_q(E_{mj}\sin\theta_q + E_{mn}\cos\theta_q)}$$

which may be reduced to

$$(E_{mj}^2 - E_{mn}^2)\sin 2\theta_q = 0 \quad (10.20)$$

Equation (10.20) is satisfied if one of two conditions is met. First, if the input is circularly polarized, that is, if $E_{mj}^2 = E_{mn}^2$, then the output is linearly polarized for all values of θ_q. Second, if the fast axis of the quarterwave plate coincides with the major or minor axis of the input polarization ellipse, that is, if $\theta_q = 0°$ or 90° then the output is linearly polarized. As a check, set θ_q to zero. Then, from (10.19), we obtain

$$e^{j\pi/4} \begin{bmatrix} E_{mj} \\ E_{mn} \end{bmatrix}$$

which is a field linearly polarized in the direction of $\tan^{-1} E_{mn}/E_{mj}$. Similarly, by setting θ_q to 90°, we have an output

$$e^{-j\pi/4} \begin{bmatrix} E_{mj} \\ -E_{mn} \end{bmatrix}$$

which corresponds to a linearly polarized electric field at an angle $-\tan^{-1} E_{mn}/E_{mj}$.

In conclusion, for any plane waves incident on a quarterwave plate, it is always possible to choose two orientations, that is, two values of θ_q, such that the output is a linearly polarized field. These orientations correspond to the cases in which the fast axis of the quarterwave plate is aligned with either the major or minor axis of the input polarization ellipse. Under these conditions, the output is linearly polarized at an angle $\pm\tan^{-1}(E_{mn}/E_{mj})$ with respect to the major axis of the input polarization ellipse. For the special case in which the input is circularly polarized, the output is linearly polarized for any value of θ_q.

10.2.7 Continuous Axis Rotation

Suppose a linearly birefringent medium is twisted such that the fast axis varies continuously. Physically, a continuous rotation of the axes is viewed as the sum of a large number of infinitesimal rotations. Therefore, the effects of continuous axis rotation can be studied by using (10.17) repeatedly and letting the rotation angle reduce to zero in the limit. However, such an approach is too difficult and cumbersome to be included here. Appendix F shows how coupled mode equations are developed to describe the continuous axis rotation, and a Jones matrix is derived based on the coupled mode equations. The results are as follows.

Let the linear retardation of the medium be $\delta_l = (\beta_y - \beta_x)L$, for no continuous axis rotation. If the axis rotates at a rate of ξ per unit length, we define $\delta_r = 2\xi L$. Then, the Jones matrix for a linearly birefringent medium with a continuous axis rotation is [3, 4]

$$\mathbf{J} = \begin{bmatrix} \cos\frac{\delta_t}{2} + j\frac{\delta_l}{\delta_t}\sin\frac{\delta_t}{2} & \frac{\delta_r}{\delta_t}\sin\frac{\delta_t}{2} \\ -\frac{\delta_r}{\delta_t}\sin\frac{\delta_t}{2} & \cos\frac{\delta_t}{2} - j\frac{\delta_l}{\delta_t}\sin\frac{\delta_t}{2} \end{bmatrix} \quad (10.21)$$

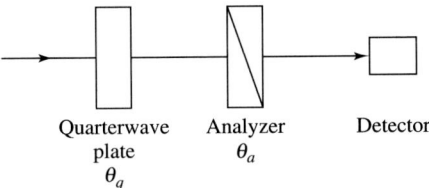

Figure 10.3 Experimental setup for the SOP measurement

where $\delta_t = \sqrt{\delta_i^2 + \delta_r^2}$.

10.2.8 Lossless Media

A matrix is *unitary* if the inverse matrix equals the transposed complex conjugate matrix.

$$\mathbf{J}^{-1} = (\mathbf{J}^*)^t \tag{10.22}$$

where the superscripts *, t and -1 signify the complex conjugation, transposition, and inversion of the matrices, respectively. For a 2×2 unitary matrix, three of the four matrix elements are independent. The most general form of a 2×2 unitary matrix is

$$\begin{bmatrix} \cos\theta\, e^{j\phi} & -\sin\theta\, e^{-j\psi} \\ \sin\theta\, e^{j\psi} & \cos\theta\, e^{-j\phi} \end{bmatrix} \tag{10.23}$$

which has three independent, real variables θ, ϕ, and ψ.

It has been shown by Jones that the Jones matrices of lossless media or devices are unitary matrices [1]. Thus the matrix of given in (10.23) with three independent variables θ, ϕ and ψ is the Jones matrix for a lossless medium in of the most general form. The Jones matrices for isotropic lossless media, linearly birefringent media, linear retarders, and circular retarders are special cases of (10.23). It is simple to show that matrix given in (10.13) is also a unitary matrix.

10.2.9 SOP Experimental Determination

Figure 10.3 depicts the setup for determining the SOP. The experiment consists of varying the angle θ_q of the quarterwave plate and angle θ_a of the analyzer and monitoring the variation of the detector output I. In the experiment, the quarterwave plate is rotated until the emerging waves are linearly polarized. As discussed in Example 4, two possible angles θ_q exist. These θ_q give the directions of the major and minor axes of the polarization ellipse of the field under test. To see if the emerging field is indeed linearly polarized, we use the analyzer and the detector. Specifically, the variation of the detector output is monitored as the analyzer is rotated. Let the maximum and minimum detector reading be I_{max}

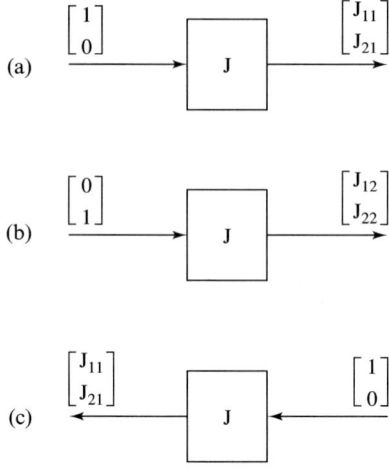

Figure 10.4 Experimental determination of the Jones matrix

and I_{min} respectively. Under the ideal conditions, the I_{min} vanishes completely when the field under test is linearly polarized. More realistically, the ratio I_{min}/I_{max} becomes very small when the field is linearly polarized or approximately linearly polarized.

Suppose the emerging field is linearly polarized when the quarterwave plate angle is θ_q. Furthermore, suppose the emerging field is linearly polarized in the direction θ_a. Then the detector output reaches to its maximum and minimum when the analyzer angle is θ_a and $\theta_a + 90°$ respectively. From the angle difference $\theta_a - \theta_q$, we obtain $\pm \tan^{-1}(E_{mn}/E_{mj})$. The **visibility** of the input field is then

$$VS = \left| \frac{E_{mj}^2 - E_{mn}^2}{E_{mj}^2 + E_{mn}^2} \right| = \left| \frac{1 - \tan^2(\theta_a - \theta_q)}{1 + \tan^2(\theta_a - \theta_q)} \right| = |\cos 2(\theta_a - \theta_q)| \quad (10.24)$$

To summarize, the experimental determination of SOP consists of adjusting the quarterwave plate angle, θ_q, until the field emerging from it is linearly polarized. The analyzer is used to determine if the emerging field is indeed linearly polarized and the direction of the linearly polarized field. From θ_q, we deduce the direction of the major and minor axes of the polarization ellipse. The visibility is calculated from the angle difference $\theta_a - \theta_q$.

10.2.10 Jones Matrices Element Measurements

The four matrix elements of the optical component can be determined experimentally, except for a multiplicative constant, by measuring the response of the component under three input conditions. The experiments are conducted in three steps. First, as the input, we use fields linearly polarized along an arbitrary direction, which for convenience is labeled as the x direction (Figure 10.4a). The output is then

$$\begin{bmatrix} J_{11} & J_{12} \\ J_{21} & J_{22} \end{bmatrix} \begin{bmatrix} 1 \\ 0 \end{bmatrix} = \begin{bmatrix} J_{11} \\ J_{21} \end{bmatrix}$$

By measuring the output SOP, we determine the complex ratio J_{11}/J_{21}. For future reference, we refer to the complex ratio as k_1.

Second, we choose as the input those fields linearly polarized in the orthogonal direction, that is, the y direction (Figure 10.4b), and determine the complex ratio J_{12}/J_{22}, identifying it as k_2.

Third, we let an excitation linearly polarized in the x direction be incident from the opposite side of the component and determine the resulting output SOP (Figure 10.4c). We thus obtain J_{11}/J_{12}, which is labeled k_3.

In other words, we experimentally determine

$$k_1 = \frac{J_{11}}{J_{21}}$$

$$k_2 = \frac{J_{12}}{J_{22}}$$

$$k_3 = \frac{J_{11}}{J_{12}}$$

In the following, we show that all matrix elements, except for a multiplicative constant, can be expressed in terms of k_1, k_2, and k_3. To begin, note that

$$\frac{J_{11}}{J_{22}} = \frac{J_{12}}{J_{22}} \frac{J_{11}}{J_{12}} = k_2 k_3$$

$$\frac{J_{21}}{J_{22}} = \frac{J_{12}}{J_{22}} \frac{J_{11}}{J_{12}} \frac{J_{21}}{J_{11}} = \frac{k_2 k_3}{k_1}$$

The Jones matrix is written as

$$\mathbf{J} = \begin{bmatrix} J_{11} & J_{12} \\ J_{21} & J_{22} \end{bmatrix} = J_{22} \begin{bmatrix} J_{11}/J_{22} & J_{12}/J_{22} \\ J_{21}/J_{22} & 1 \end{bmatrix}$$

Therefore, the matrix elements of $\mathbf{J}$ can be expressed in terms of k_1, k_2, and k_3, as follows:

$$\mathbf{J} = J_{22} \begin{bmatrix} k_2 k_3 & k_2 \\ k_2 \frac{k_3}{k_1} & 1 \end{bmatrix} \qquad (10.25)$$

The multiplicative constant J_{22} is not determined. Although it does affect the amplitude and phase of the transmitted waves, this term has no effect on the polarization characteristics of the optical component being tested.

10.3 SINGLE-MODE FIBER BIREFRINGENCE

If a fiber has a perfectly circular core and cladding, a rotationally symmetric index profile, and freedom from mechanical, electrical, and magnetic disturbances, then it is truly nonbirefringent. However, imperfections are introduced, intentionally or otherwise, in the fabrication process, or stress may be introduced in the cabling or handling processes following fabrication. Bending and twisting are inevitable during installation, or in experiments. In the presence of external disturbances, nonbirefringent fibers become birefringent. In short, nonbirefringent fibers rarely exist.

In general, fiber birefringence can be traced to three possible origins or causes: *geometry, internal stress,* and *external stress.* These stresses or causes lead to linear or circular birefringence, as well as rotation of the fast axis. [5, 6].

10.3.1 Geometrical or Form Birefringence

Birefringence due to noncircular geometries is known as **geometrical** or **form birefringence.** All real fibers have nominally circular cores and claddings. Typically, the deviation from the perfectly circular shape is on the order of 0.1 percent, due to the variations in the melting, collapse, and drawing processes. As a first-order approximation, these noncircular shapes can be represented by ellipses. For some fibers, the core and/or cladding are made elliptical on purpose. In any case, the linear birefringence due to an elliptical core is a function of

$$\frac{(a_{mj}^2 - a_{mn}^2)}{(a_{mj}^2 + a_{mn}^2)}$$

where a_{mj} and a_{mn} are the semimajor and semiminor axes of the elliptical core, and the *fast axis* is along the direction of the *minor axis* [6].

10.3.2 Internal or Built-In Stress Birefringence

The thermal expansion coefficients of the core and cladding materials can be quite different. While there is no stress when a composite glass structure is softened at high temperature, considerable stress may be developed when the structure is cooled to room temperature. For fibers with an asymmetric cross section, the thermally induced lateral stress is anisotropic. Because of this anisotropic mechanical stress, and as a result of elastooptic effects, fibers become optically anisotropic. Many polarization-maintaining fibers are designed to take advantage of the frozen-in optical anisotropy. Figure 10.5 shows the cross sections of three polarization-maintaining fibers [6]. In these fibers, the *fast axis* is along the direction with the *larger transverse compressive stress.*

Physical twisting can be purposely introduced to, and frozen into, fibers during fabrication. For example, a fiber preform may be spun during the drawing process. Geometrical twisting is frozen into the fibers when cooled. Fibers thus formed are known as **spun fibers** [3]. If fibers are spun sufficiently, the effects of linear birefringence due to the noncircular cross section or the anisotropic ther-

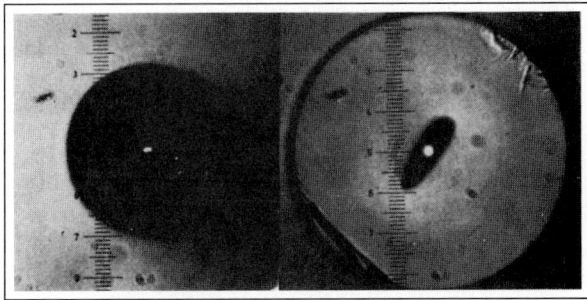

(a) Elliptical core and cladding
(b) Circular core and elliptical cladding

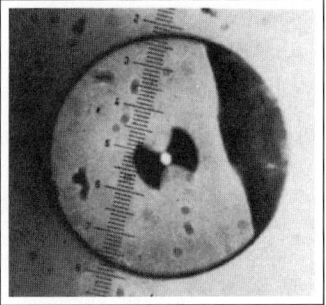

(c) Circular core and stress lobes

Figure 10.5 Cross section of three highly birefringent fibers ([6] ©1983 IEEE)

mal stress can be overwhelmed by the circular birefringence introduced by spinning.

Most fibers are *linearly birefringent,* because of geometry and internal stress. Spun fibers are the exceptions. Spun fibers may be used effectively as *circularly birefringent* fibers.

10.3.3 External Applied Stress or Fields Birefringence

When nonbirefringent fibers are subjected to lateral stress, free bending, bending and stretching against a solid cylinder, or to a strong electric field, they become linearly birefringent [3, 5]. Suppose a force F in the y direction is applied uniformly on a fiber of length L (Figure 10.6a). The *lateral force* per length acting on the fiber is F/L. The linear birefringence due to the **lateral force** is [6]

$$\beta_x - \beta_y = -\frac{2\pi}{\lambda} \frac{2n_{co}^3 (p_{11} - p_{12})(1 + \sigma)}{\pi Y} \frac{1}{b} \frac{F}{L} \qquad (10.26)$$

where b is the cladding radius, p_{11} and p_{12} are the photoelastic constants, Y is Young's modulus, and σ is the Poisson ratio. Values for these constants for silica

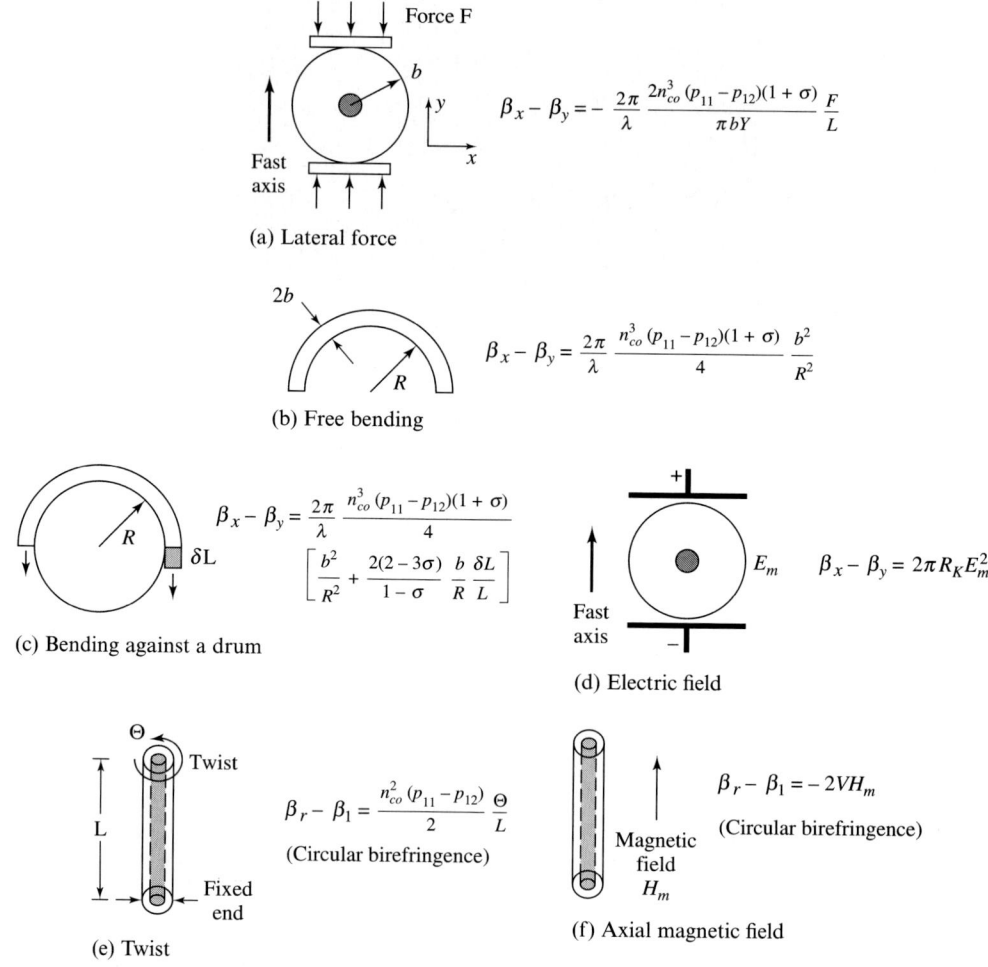

Figure 10.6 Birefringence due to external stress or fields

glass fibers are listed in Table 10.3. When these numerical values are used, we have,

$$\beta_x - \beta_y \approx 4.54 \times 10^{-5} \frac{1}{b} \frac{F}{L} \text{ rad/m} \qquad (10.27)$$

where b and L are in meters and F is in Newtons.

Note that the linear retardation $\delta_l = (\beta_y - \beta_x)L$, depends only on F/b. For a single-mode fiber with a cladding diameter of 125 μm, a linear retardation of π is realized when a lateral force of 4.32 N is applied to the fiber. Therefore, the stressed fiber section functions as a halfwave plate.

When a fiber with a cladding radius of b is bent into a circular coil of radius R without stretching (Figure 10.6b), the linear birefringence induced by **free bending** is

$$\beta_x - \beta_y = \frac{2\pi}{\lambda} \frac{2n_{co}^3 (p_{11} - p_{12})(1 + \sigma)}{4} \frac{b^2}{R^2} \qquad (10.28)$$

Numerically, the linear birefringence at 0.633 μm is

$$\beta_x - \beta_y \approx -1.36 \times 10^6 \frac{b^2}{R^2} \text{ rad/m} \qquad (10.29)$$

In contrast, when a fiber is stretched and bent against a drum or a solid cylinder of radius R, the linear birefringence due to *bending under tension* is

$$\beta_x - \beta_y = \frac{2\pi}{\lambda} \frac{2n_{co}^3 (p_{11} - p_{12})(1 + \sigma)}{4} \frac{b}{R} \left[\frac{b}{R} + \frac{2(2 - 3\sigma)}{1 - \sigma} \frac{\delta L}{L} \right] \qquad (10.30)$$

where δL is the elongation produced by stretching (Figure 10.6c). Again, making use of the material constants listed in Table 10.3, we obtain

$$\beta_x - \beta_y \approx -1.36 \times 10^6 \frac{b}{R} \left[\frac{b}{R} + 3.59 \frac{\delta L}{L} \right] \text{ rad/m} \qquad (10.31)$$

which is more complicated than the equation (10.29) for bending stress alone. In most cases, the effect due to the tensile strain $\delta L/L$ is much smaller than that due to the bending b/R, unless the bending radius R is very large.

Strong **electric fields** E_m (Figure 10.6d) also cause linear birefringence

$$\beta_x - \beta_y = 2\pi R_K E_m^2 \qquad (10.32)$$

This is the **Kerr effect** in fibers, as $\beta_x - \beta_y$ is proportional to E_m^2. The Kerr coefficient R_K is very small, as mentioned in Chapter 6.

For linear birefringence due to external lateral stress (Figure 10.6a) or an electric field (Figure 10.6d), the **fast axis** is along the direction of lateral stress or the electric field. For linear birefringence due to free bending and bending

Table 10.3 Selected material constants of silica fibers

	Symbol	Value	Units	Remarks
Young's modulus	Y	7.6×10^{10}	N/m²	
Poisson ratio	σ	0.17	—	
Index	n_{co}	1.46	—	@ 0.633 μm
Photoelastic constant	p_{11}	0.12	—	@ 0.633 μm
Photoelastic constant	p_{12}	0.27	—	@ 0.633 μm
Verdet constant	V	4.68×10^{-6}	rad/A	@ 0.633 μm

under tension, the fast axis is in the plane of coil and normal to the fiber axis (Figure 10.6b and c). In the figures, the direction of the fast axis is represented by a thick arrow.

When a fiber of length L is twisted by an angle Θ, a shearing strain is induced in the fiber. The shearing strain and the photoelastic effect cause the fiber to become circularly birefringent. The circular birefringence is proportional to Θ/L, as follows:

$$\beta_r - \beta_l = \frac{n_{co}^2(p_{11} - p_{12})}{2} \frac{\Theta}{L} \tag{10.33}$$

which numerically is

$$\beta_r - \beta_l \approx -0.16 \frac{\Theta}{L} \text{ rad/m} \tag{10.34}$$

Thus, the *circular retardation* caused by *twisting* is

$$\delta_c = (\beta_r - \beta_l)L = -0.16\Theta.$$

As demonstrated in Example 1, a circular retarder with a circular retardation δ_c rotates linearly polarized fields by an angle $-\delta_c/2$. A twisting of angle Θ will cause linearly polarized fields to rotate by an angle of 0.08Θ. As an example, a twist of 360 degrees will cause the linearly polarized field to rotate by 28.8 degrees. This is in qualitative agreement with the experimental results reported by Smith [7].

The linear and circular birefringent effects discussed in the preceding paragraphs, including the effect of twisting, are the same for waves propagating in either direction, and are referred to as **reciprocal effects.**

Faraday rotation, discussed in Chapter 6, also occurs in fibers, when they are subjected to a strong axial magnetic field. If we let the axial **magnetic field** be H_m, then the circular birefringence due to Faraday rotation is

$$\beta_r - \beta_l = -2VH_m \tag{10.35}$$

where V is the Verdet constant of the glass material. As listed in Table 10.3, the Verdet constant of fused silica is approximately 4.68×10^{-6} rad/A at $\lambda = 0.633$ µm. However, the circular birefringence due to Faraday rotation is **nonreciprocal.** This is a crucial difference between the circular birefringence effects of twisting and those of axial magnetic fields.

10.3.4 Polarization Beat Length

To quantify the linear birefringence in fibers, we introduce the **polarization beat length,** or simply the **beat length,** which is defined as

$$L_b = \frac{2\pi}{|\beta_x - \beta_y|} \tag{10.36}$$

For typical single-mode fibers, the polarization beat length ranges from a few centimeters to a few meters [6, 8]. For high-birefringent fibers, also known as **polarization-maintaining fibers,** the beat length may be as short as a few millimeters. In all cases, $|\beta_x - \beta_y|$ is much smaller than either β_x or β_y. If we take fibers with a polarization beat length of 10 cm as an example, $|\beta_x - \beta_y|$ is approximately 60 m^{-1}, and β_x and β_y are on the order of 10^7 m^{-1}. Clearly, $|\beta_x - \beta_y| \ll \beta_x$ and β_y. Although $|\beta_x - \beta_y|$ is small, this small difference must be compensated for if the phase or polarization information of the optical signals is to be retained.

10.3.5 Fiber Axis Rotation

Since fibers can be long and they usually are, twisting is inevitable, and twisting leads to circular birefringence and **axis rotation.** Axis rotation can also be caused by a change in the direction of an externally applied lateral stress. In addition, for fibers with built-in stress, the direction of thermally induced stress may vary with thermal fluctuations. In short, a change in the force direction, for *any* reason, leads directly to a rotation of the fast axis.

10.3.6 Jones Matrices: Birefringent Fibers

For fibers with a linear *or* circular birefringence, but not both, the Jones matrices (10.11) or (10.12) are adequate to describe the fiber birefringence. For fibers with linear *and* circular birefringence, (10.13) should be used. If there are a finite number of discrete axis rotations, (10.17) and (10.18) may be used. If the circular birefringence is small or absent, then (10.21) may be used to account for the linear birefringence and continuous axis rotation.

For most fibers, however, linear and circular birefringence and axis rotation are all present simultaneously, and are distributed over the entire fiber section. Kapron, Borrelli, and Keck [9] showed that the combined polarization effects in fibers can be represented by a linear retarder followed by a circular retarder. The Jones matrix can be expressed as

$$\mathbf{J} = \mathbf{R}_c(\delta_c) \, \mathbf{R}_l(\Theta, \delta_l) \qquad (10.37)$$

where δ_l is the linear retardation of the linear retarder, Θ is the orientation of the fast axis of the linear retarder, and δ_c is the circular retardation of the circular retarder. Matrix manipulation will show that the matrix elements of (10.37) are

$$J_{11} = \cos\frac{\delta_l}{2}\cos\frac{\delta_c}{2} + j\sin\frac{\delta_l}{2}\cos(2\Theta - \frac{\delta_c}{2}) \qquad (10.38)$$

$$J_{21} = -\cos\frac{\delta_l}{2}\sin\frac{\delta_c}{2} + j\sin\frac{\delta_l}{2}\sin(2\Theta - \frac{\delta_c}{2}) \qquad (10.39)$$

and $J_{12} = -J^*_{21}$ and $J_{22} = J^*_{11}$.

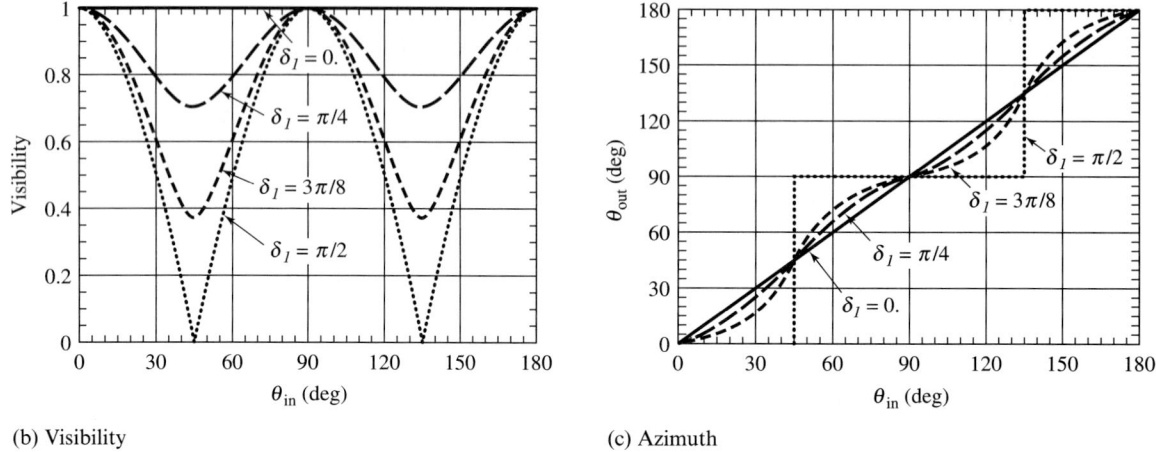

Figure 10.7 Evolution of the SOP in linearly birefringent fibers

To illustrate the effects of fiber birefringence on the evolution of the SOP in fibers, consider linearly polarized fields incident upon a fiber with a linear retardation δ_l (Figure 10.7a). Let the incident field be given by (10.6). The output is then

$$\begin{bmatrix} J_{11} & J_{12} \\ J_{21} & J_{22} \end{bmatrix} \begin{bmatrix} \cos\theta_{in} \\ \sin\theta_{in} \end{bmatrix}$$

where J_{ij} are given by (10.11).

Once the output Jones vector is known, the SOP can be deduced. We can apply the method discussed in connection with Figure 10.3 to measure the visibility and azimuth of the output fields. In Figure 10.7b and c, the visibility and azimuth (θ_{out}) of the output polarization ellipse are plotted as functions of the input azimuth (θ_{in}). When the input fields are aligned with the fast or slow axes of the fiber, corresponding to cases for which $\theta_{in} = 0°$, 90°, or 180°, etc., the output visibility is 1, which signifies a linearly polarized output. For all other input SOPs, the visibility is less than 1. In the special cases where δ_l is an odd multiple of $\pi/2$ and θ_{in} is 45 degrees, the visibility vanishes completely, indicating a circularly polarized output (Figure 10.7b). Although θ_{out} follows θ_{in}, as shown in Figure 10.7c, the curves are not straight lines. In fact, the relationship is highly nonlinear when δ_l is close to an odd multiple of $\pi/2$. For the special case of nonbirefringent fibers, $\delta_l = 0$, a linear input–output relationship exists, as shown by the solid line in Figure 10.7c.

10.4 SINGLE-MODE FIBER POLARIZATION COMPONENTS

In many fiber systems, fiber birefringence is regarded as an annoyance. On the other hand, several useful optical components take specific advantage of such birefringent properties. Examples of such components are discussed in the following subsections.

10.4.1 Single-Mode Fiber Polarizers

There are two basic types of fiber polarizers. In both, the cladding is partially removed by grinding and polishing, to gain access to the fields guided by the fiber core. In one type of polarizer, a thin metal film is deposited on the flat polished surface of the cladding (Figure 10.8). These polarizers are known as **metal film polarizers.** In the other type, an **anisotropic crystal** is placed on top of the polished surface (Figure 10.9).

Metal Film Polarizers A metal film polarizer is shown schematically in Figure 10.8a. In the absence of the metal film, there are two low-loss, nearly-degenerate polarization modes. However, in the presence of the metal layer, the two modes become lossy, but with quite different attenuation constants. Since the attenuation constants are different, one polarization mode is attenuated more than the other one. This is the basic operation of metal film polarizers.

Let the two modes be the x- and y-polarized modes. The electric fields of the x-polarized mode are normal to the metal film, while the y-polarized mode has electric field components parallel to the metal film. The boundary conditions require that the tangential electric field components vanish at a perfectly conducting boundary, or be small at a good conducting surface. Therefore, waves

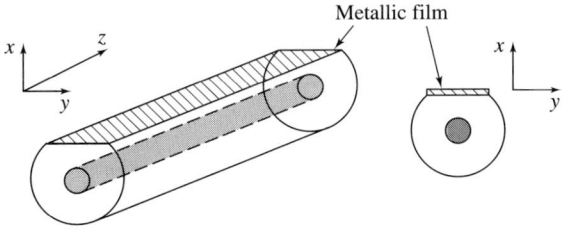

(a) Polished fiber with a thin metallic film

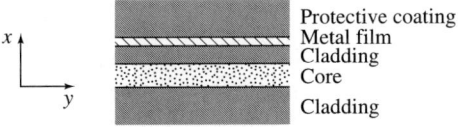

(b) Thin-film waveguide model

Figure 10.8 Fiber polarizer with a thin metallic film

with the electric fields parallel to the conducting surface are reflected or "repelled" by the conducting boundary. The boundary condition on the *normal* field component is different. The normal component of the electric flux density **D** is continuous at the boundary that is free of surface changes, and normal electric fields are therefore not repelled by that boundary. As a result, waves with electric fields normal to the conducting boundary penetrate more into the conductors and suffer more loss than waves with electric fields parallel to the conducting surface.

In Figure 10.8a, the *y*-polarized waves are the TE modes and the *x*-polarized waves are the TM modes. The TM mode is more lossy than the TE mode when both modes are far away from cutoff. A rough estimate of the loss characteristics can be made by modeling a polished fiber with a metal film as a five-layer thin-film waveguide structure, as shown in Figure 10.8b. Analysis of the five-layer waveguide structure confirms that $\alpha_{TM} > \alpha_{TE}$ when both modes are far from their cutoff conditions. The analysis also reveals that, while there is no cutoff for the lowest-order TM mode, there is a cutoff for the lowest-order TE mode. As the wavelength becomes longer, the TE mode is cutoff, and its attenuation constant increases drastically. In other words, α_{TE} may be larger that α_{TM} if the wavelength is sufficiently long.

To summarize, since the metal film is very effective in absorbing electric fields normal to the metallic boundary, $\alpha_{TM} > \alpha_{TE}$ at shorter wavelengths. At longer wavelengths, the situation is reversed: $\alpha_{TM} < \alpha_{TE}$, since the lowest TE mode may be below cutoff. In all cases, the operation of metal film polarizers is based on the fact that α_{TE} and α_{TM} are different.

In metal film polarizers, aluminum or gold films may be used. Since the index of Al has a large imaginary part at optical frequencies, Al is very effective in absorbing optical waves. The metal film is typically 1500Å thick or less [10]. In some cases, a dielectric layer is also deposited to protect the metallic film from moisture or other contaminants [11].

Anisotropic Superstrate Polarizers For the second type of fiber polarizer (Figure 10.9), the metal film is replaced by an anisotropic crystal. Recall that, if the index of the outside region is lower than that of the core, the waves are confined in the fiber core. When the outside index is larger than n_{co}, power leaks from the

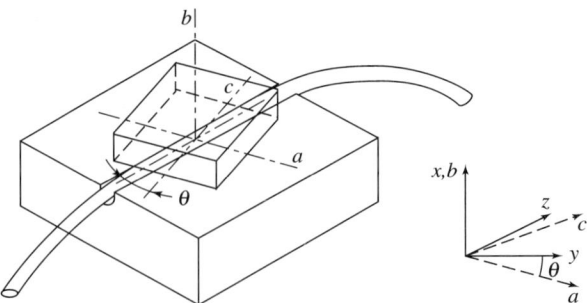

Figure 10.9 Fiber polarizer with an oriented crystal [12]

core into the outside region, and this constitutes loss for the guided mode. An anisotropic material may be chosen such that one polarization mode "experiences" a small index, while the other polarization mode "sees" a large index. As a result, one mode remains guided, while the other becomes lossy.

For this application, Bergh et al. [13] used a polished potassium pentaborate ($KB_5O_8 \cdot 4H_2O$) crystal, which is a biaxial material. Let the three principal directions be the *a, b,* and *c* axes, and note that $n_a = 1.49$, $n_b = 1.43$, and $n_c = 1.42$ at 0.633 μm. A $KB_5O_8 \cdot 4H_2O$ crystal is polished and mounted on the polished fiber, as shown in Figure 10.9. The polished surface is normal to the *b* axis. Since the *b* axis is in parallel with the *x* axis, the *x* polarized fields "see" an index of n_b that is smaller than the n_{co} of about 1.46. Thus, the *x*-polarized mode remains guided. For the *y*-polarized fields, the index may be calculated in the manner discussed in Chapter 6; that is,

$$\frac{1}{n^2} = \frac{\sin^2\theta}{n_c^2} + \frac{\cos^2\theta}{n_a^2} = \frac{\sin^2\theta}{1.42^2} + \frac{\cos^2\theta}{1.49^2}$$

where θ is the angle between the fiber axis and the crystal *c* axis. As θ changes from 90° to 0°, the index "seen" by the *y*-polarized fields changes from 1.42 to 1.49; that is, from a value smaller than n_{co} to a value larger than n_{co}. Therefore, the *y*-polarized fields change from guided to unguided modes. This is the basis of operation for polarizers with an anisotropic superstrate.

10.4.2 Fiber Retarders

As previously discussed, a stressed fiber may be used as a quarterwave or halfwave plate. To exert force on a fiber, simply clamp the fiber between the pole and armature of an electromagnetic relay, as shown in Figure 10.10a, and apply current to the relay. The direction of force is determined by the pole and armature surfaces. As the current to the relay coil is varied, the force acting on the fiber is varied. The linear retardation δ_l can thus be tuned electrically. The resulting **fiber squeezer** is probably the simplest fiber optic linear retarder with a variable linear retardation [13].

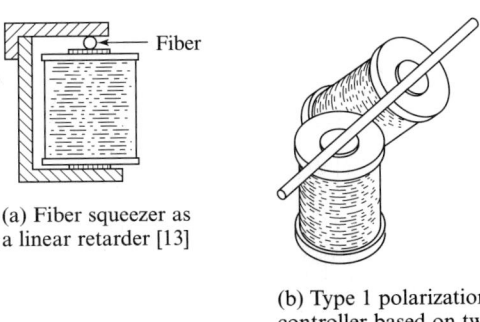

(a) Fiber squeezer as a linear retarder [13]

(b) Type 1 polarization controller based on two fiber squeezers [13]

Figure 10.10 Fiber optic type 1 polarization controller

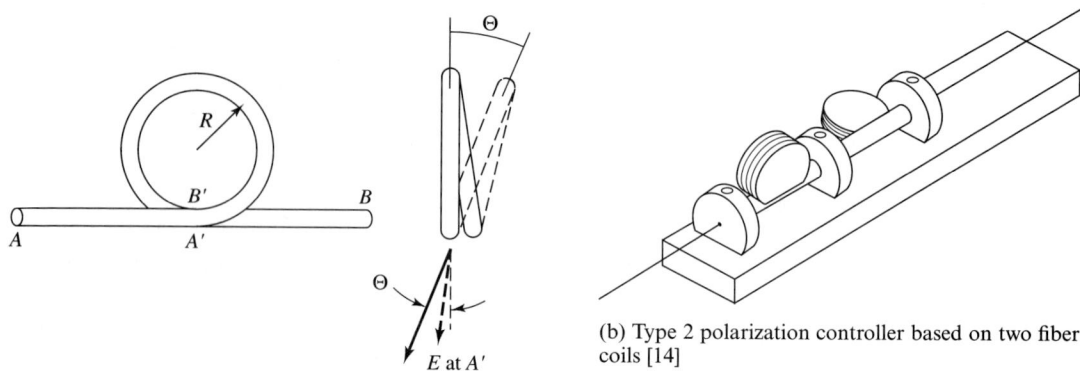

(a) Fiber coil as a linear retarder

(b) Type 2 polarization controller based on two fiber coils [14]

Figure 10.11 Fiber optic type 2 polarization controller

Bending may also be used to cause linear birefringence. With the help of (10.28) or (10.30), the radius and number of turns of the fiber coil may be selected to realize a specific linear retardation. These types of fiber optic linear retarders (Figure 10.11a) have been reported by Lefevre [14], who used fibers with a cladding radius of 40 μm and a drum with a radius of 0.85 cm to produce a linear birefringence of 29.67 rad/m. For a one-turn fiber coil, the fiber length is 5.34 cm, and the linear retardation is $\pi/2$. Therefore, a one-turn fiber coil may function as a quarterwave plate, and a two-turn coil, having a retardation of π, may be used as a halfwave plate [14].

10.4.3 Fiber Polarization Controllers

There are a number of designs of polarization controllers which can transform an arbitrary input SOP to fields linearly polarized in the x direction. According to Okoshi [15], these polarization controllers can be classified into three types (Figure 10.12). Types 1 and 2 each require two linearly birefringent elements, while type 3 uses two circular retardars and one linear retarder.

In **type 1 polarization controllers,** angles θ_1 and θ_2 of the fast axes are fixed, and the linear retardations, δ_{l1} and δ_{l2}, are tuned by any one of a number of means. A **type 2 controller** consists of a quarterwave plate and a halfwave plate, for which the linear retardations are fixed, $\delta_{l1} = \pi/2$ and $\delta_{l2} = \pi$. However, the orientations θ_1 and θ_2 of the two waveplates are varied to control the output polarization.

A **type 3 controller** has two circular retarders with variable retardations and a fixed linear retarder. The linear retarder is a quarterwave waveplate ($\delta_l = \pi/2$) with its fast axis aligned along the x axis ($\theta = 0$). Problems 8, 9, and 10 at the end of this chapter require that the reader prove that these controllers work as polarization controllers.

A type 1 controller can be created with two fiber squeezers for which the angles of force are offset by 45 degrees, as shown in Figure 10.10b [13]. Varying

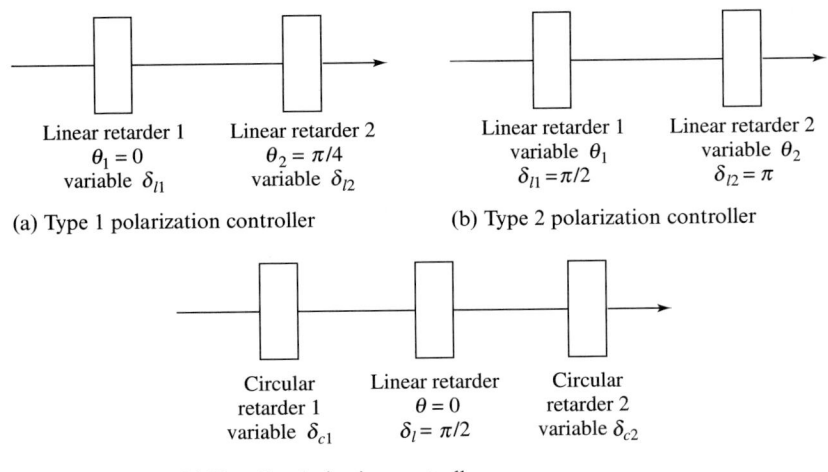

Figure 10.12 Three types of polarization controllers

the current to the electromagnetic relays results in δ_{l1} and δ_{l2} being varied.

To understand type 2 controllers, recall the functions of quarterwave and halfwave plates, as discussed in Examples 2 and 4. A quarterwave plate may be used to transform input fields with an arbitrary SOP to linearly polarized fields, with no loss in intensity. The transformation is accomplished by rotating the quarterwave plate such that the fast or slow axis is aligned with the major axis of the incoming polarization ellipse. Waves impinging on the quarterwave plate have a Jones vector of the form given in (10.5), and the two field components are out of phase by $\pm \pi/2$. This phase difference is compensated for by the quarterwave plate, which has a linear retardation of $\pi/2$. The output of the quarterwave plate is therefore linearly polarized.

To create a type 2 polarization controller, we need a means of rotating the fast axis. The fiber coils designed by Lefevre may be used for this application. In Figure 10.11a, four points are marked: A, A′, B′, and B. Points A and B are fixed to a stationary platform, and A′ and B′ are anchored to the fiber coil, which may be turned. When the plane of the fiber coil is turned by an angle Θ, points A′ and B′ also turn by the same angle. Since points A and B are fixed, fiber sections AA′ and B′B are twisted when the fiber coil is turned. Suppose the input field at point A is linearly polarized. Then, the field at A′ is also linearly polarized, but with the field direction rotated by $\xi\Theta$, where the constant ξ is given in (10.34). Also, when the fiber coil is turned by Θ, the direction of the fast axis at point A′ is turned. Consequently, at point A′, the angle between the linearly polarized field and the fast axis is $(1 - \xi)\Theta$. If the fiber coil behaves as a halfwave plate, the fields at point B′ must be in the direction of $2(1 - \xi)\Theta$, as explained in Example 2. Therefore, by cascading two fiber coils, one corresponding to a quarterwave waveplate and the other to a halfwave waveplate, as shown in Figure 10.11, we can produce a type 2 polarization controller [14].

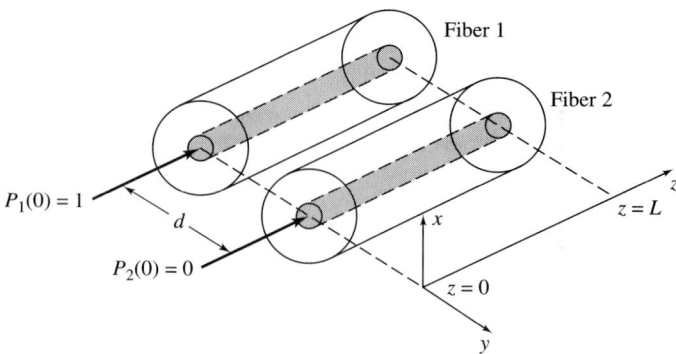

Figure 10.13 Fiber directional coupler

10.5 SINGLE-MODE FIBER DIRECTIONAL COUPLERS

A single-mode fiber directional coupler is formed by placing two single-mode fibers close to and parallel with each other, as shown in Figure 10.13. Although the geometry and construction of such couplers are similar to that of multimode fiber directional couplers (discussed in Chapter 9), their operation is actually more like that of the thin-film directional coupler (Chapter 7). This is because each single-mode fiber supports one and only one mode when the fiber separation d is large. Also, since each single-mode fiber supports only one mode, the coupled mode theory discussed in Chapter 7 is applicable to single-mode fiber directional couplers. According to the coupled mode theory, the necessary condition for a complete power transfer is that the two waveguides, or, in the case of fiber directional couplers, the two fibers, have the same propagation constants, $\beta_1 = \beta_2$. The power transfer is 100 percent when the phase-matched condition is met, and if κL is an odd multiple of $\pi/2$, where κ is the coupling constant.

Clearly, if $\beta_1 \neq \beta_2$, no significant power exchange between fibers is possible, even if the two fibers are very close. If the fibers are not very close, the coupling is weak, even if the two fibers are similar or identical. A small coupling constant would require a long interaction length, which is difficult to maintain. To increase κ, the fibers must be kept close, or ways must be found to "push" the fields out from the core. Typically, the spacing between cores is on the order of a few micrometers or less. To reduce the spacing, either the cladding is stripped off, or a preform with two closely spaced cores is used to create the fibers. Depending on the methods of fabrication, fiber directional couplers can be grouped into three types.

10.5.1 Twin Core Couplers

A **twin core coupler** is formed by using a preform with two cores [16]. Figure 10.14a depicts the cross section of a twin-core directional coupler.

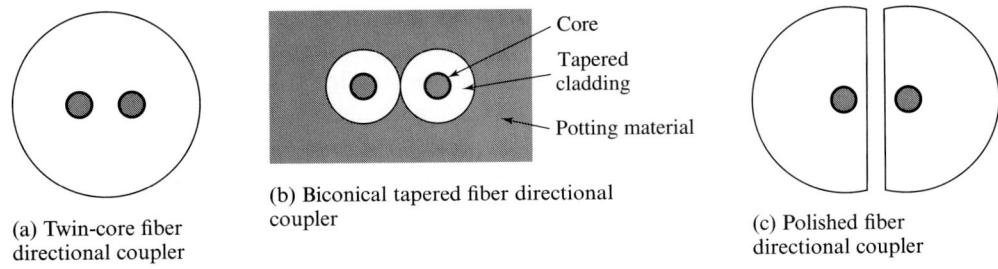

Figure 10.14 Cross sections of three types of fiber directional couplers

10.5.2 Biconical Tapered Structure Couplers

Couplers can also be made from existing fibers. The first step is to remove the outer jacket. The fibers are then etched with buffered hydrofloride ($NH_4F:HF = 4:1$) to thin the cladding region. At the same time, the two fibers are twisted and pulled together. Throughout the twisting, pulling, and etching processes, power coupled from one fiber to another is monitored. When the desired fraction of power transfer is obtained, or when the amount is just below the desired level by a few percent, etching is stopped. The fibers are then thoroughly rinsed. Potting material, glass gel, or some other other material with an index close to that of the cladding is used to hold the fibers in place and stabilize the coupler structure (Figure 10.14b).

Historically, the first single-mode fiber coupler was made in this manner [17]. It was known as a **bottle coupler,** since the etching and rinsing were done in a glass bottle, which also provided mechanical support for the coupler structure. When modern fiber couplers are made, there is one additional step: fiber sections in the interaction region are heated and stretched [18, 19]. The heating may be accomplished either with flames or electric heating elements.

Because of the heating and stretching, the core and cladding diameters are tapered. Since the core radius is reduced, the V value of a tapered fiber section is smaller and the percentage of power outside the core becomes larger. Consequently, for the same spacing d, the coupling between fibers is stronger. Also, because of the heating, the two fibers may be fused together, making the coupler structure mechanically stronger.

10.5.3 Polished Fiber Couplers

Another approach to gaining access to the core is to remove the cladding by mechanical polishing [20]. A fiber is secured on a quartz block that has grooves of known depth and curvature cut into its surface. The fiber is stretched and held in the curved grooves. The quartz block and fiber are ground and polished to within a few micrometers from the fiber core. A polished fiber coupler is then formed by holding two polished block–fiber combinations together (Figure

10.14c). An index-matching liquid may be used to increase the coupling between fibers and to facilitate mechanical adjustments. Also, the curvature of the slots on the quartz block is used to control the length and depth of the polished region, which determines the interaction length and the strength of interaction.

Since each fiber supports two orthogonal polarization modes, there are different propagation constants and coupling constants for the different polarizations. As a result, the fraction of power in each mode and the total power in each fiber depend on the input SOP. The polarization characteristics of these couplers can be analyzed using Jones calculus [21]. However, the analysis is too complicated to be presented here.

10.6 ERBIUM-DOPED FIBER AMPLIFIERS

As indicated in Chapter 8, attenuation and dispersion are two factors limiting the span of a fiber communications system. For single-mode fibers, the waveguide and material dispersion have opposite signs in the long-wavelength regime. Therefore, there are fibers that are designed to operate at, or near, the zero-dispersion wavelength where the two dispersion effects exactly cancel. If the effect of attenuation can also be compensated for by an amplification scheme, then the transmission span can be increased significantly, until other factors become dominant.

To compensate for the fiber loss, various optical amplifiers may be used. These include semiconductor optical amplifiers, Raman amplifiers, and rare-earth-doped optical fiber amplifiers [22]. As depicted in Figures 8.24a and b, loss in silica fibers is particularly low in the 1.2–1.3 μm and 1.5–1.6 μm bands. Therefore, there is considerable interest in optical amplifiers operating at these bands. Praseodymium-doped fluoride fibers pumped by InGaAs quantum-well lasers at 1.017 μm may be used to amplify signals at 1.3 μm. For signals at 1.55 μm, erbium-doped silica fibers may be used for amplification. Erbium-doped silica fibers may be pumped at wavelengths near 0.515, 0.65–0.68, 0.81, 0.98, or 1.48–1.49 μm. In the following discussions, we will concentrate mainly on **erbium-doped fiber amplifiers** (EDFAs), with semiconductor laser diodes emitting at 0.98 μm or 1.480 μm as the pump sources.

10.6.1 Erbium-Doped Fibers

Figure 10.15a is a schematic diagram of an EDFA. The pump and signal beams are fed through a fiber directional coupler. The two beams propagate in the same direction in the doped fiber. Optical isolators are used to reduce reflections and the attendant instabilities.

The key component of a fiber amplifier is the erbium-doped fiber. Erbium-doped fibers are similar to single-mode silica fibers: the core radius is on the order of 2.5 μm, and the NA is about 0.18. However, the fiber core is doped with Er^{3+} ions. The dopant concentration is typically in the 10^{+18} to 10^{+19} cm^{-3} range.

A simplified energy diagram of Er^{3+} ions in silica glass is shown in Figure

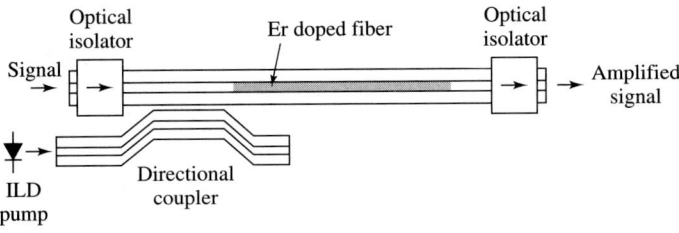

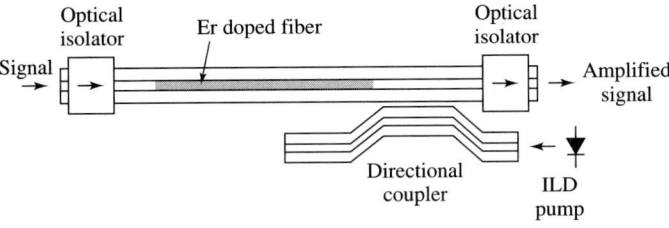

Figure 10.15 Schematic diagram of erbium-doped fiber amplifier

3.32. The "levels" are bands of finite bandwidths, because each energy level splits into multiplets in the presence of crystal fields. For example, the $^4I_{13/2}$ level splits into seven multiplets, and the $^4I_{15/2}$ level splits into eight multiplets [23]. In addition, because the glass is an amorphous material, the level splitting changes from site to site. Therefore, each group of multiplets coalesces into a band with a finite width.

As mentioned earlier, sources with various wavelengths may be used as the pump source. For example, optical beams with wavelengths of 0.98 μm and 1.490 μm can raise the states from the $^4I_{15/2}$ level to the $^4I_{11/2}$ multiplets and the upper multiplets of $^4I_{13/2}$, respectively. The advantages of EDFAs pumped at 0.980 or 1.490 μm include high gain, high saturated output power, polarization-independent gain, low noise, and low crosstalk [24]. EDFAs pumped at 0.980 μm are more efficient and have better noise performance, and may therefore be used as receiver preamplifiers. On the other hand, powerful ILDs at 1.490 μm are readily available, and they may be used as pump sources to drive EDFAs to a higher power level.

10.6.2 Rate Equations

Studies show that EDFAs pumped at 0.980 μm or 1.480 μm may be classified as *three-level* media. However fiber amplifiers differ from other three-level amplifiers, such as semiconductor optical amplifiers, in several respects. Since EDFAs are usually long (5 to 30 m), the photon densities and the densities of occupied states vary continuously in the longitudinal direction, and it is necessary to account for the spatial variation of the fields. In addition, the electric field intensity and the erbium ion concentration are not constant throughout the fiber core and both must be taken into account. To account for the nonuniform distribution of fields and the dopant concentration we introduce the *effective fiber cross sectional areas* at the pump and the signal frequencies. Finally, there is considerable noise, due to spontaneous emission generated everywhere in the doped fiber, and to amplification of the noise generated in preceding fiber sections [23]. However, noise is too complicated to be dealt with here, so we will ignore it in our discussions.

Consider an erbium-doped fiber extending from $z = 0$ to $z = L$, the input and output, respectively. To derive the differential equations describing the evolution of the signal and pump beams, consider an infinitesimal fiber section at z. We can use the same symbols and notations presented in Chapter 3 for three-level systems. For example, N_1, N_2, and N_3 are the densities of occupied states in levels E_1, E_2, and E_3, respectively. Also, $\Phi_p \delta f$ is the photon density of the pump beam with a frequency f_{31} corresponding to transitions between E_1 and E_3. Similarly $\Phi \delta f$ is the photon density of the signal beam with a frequency f_{21} corresponding to transitions between E_1 and E_2. Also assume that E_3-to-E_2 transitions are fast and nonradiative processes. All of these statements imply that the requirements in Chapter 3 for three-level systems are satisfied. The assumptions also imply that τ_{32} is much smaller than τ_{31}, τ_{21}, and $1/A_L$. Therefore, the expressions for N_1, N_2, etc., derived in Chapter 3 are applicable. Explicitly, from (3.28)–(3.30) we obtain

$$\frac{N_1 - N_3}{N_t} \approx \frac{A_L + \dfrac{1}{\tau_{21}}}{A_P + 2A_s + \dfrac{1}{\tau_{21}}} \tag{10.40}$$

$$\frac{N_2 - N_1}{N_t} \approx \frac{A_P - \dfrac{1}{\tau_{21}}}{A_P + 2A_s + \dfrac{1}{\tau_{21}}} \tag{10.41}$$

where $N_t = N_1 + N_2 + N_3$ is the total density of occupied states, $A_P = B_{13}(\Phi_p \delta f)$ and $A_L = B_{12}(\Phi \delta f)$. To use these expressions in this chapter, we note that these terms are all functions of z. To account for the variation of N_1, N_2, N_3, $\Phi_p \delta f$, and $\Phi \delta f$ as functions of z, recall that the change of photon density is due to absorption and stimulated emission. Therefore,

$$\frac{d}{dz}(\Phi_p \delta f) = \frac{d}{dt}(\Phi_p \delta f)\frac{dt}{dz} = \frac{n_{31}}{c}\frac{d}{dt}(\Phi_p \delta f) = \frac{n_{31}}{c} B_{13}(N_3 - N_1)\Phi_p \delta f \tag{10.42}$$

where n_{31} is the index at f_{31}. The situation is similar to that in the discussions on the absorption of electromagnetic waves by two-level systems (also in Chapter 3). Similarly, we obtain the expression for the rate of change of $\Phi \delta f$

$$\frac{d}{dz}(\Phi \delta f) = \frac{n_{21}}{c} B_{21}(N_2 - N_1)\Phi \delta f \qquad (10.43)$$

where n_{21} is the refractive index at f_{21}.

Since we are actually interested in optical power rather than photon density, we must now convert $\Phi_p \delta f$ and $\Phi \delta f$ to light intensities. Then, we can relate the light intensities to optical power. To begin the conversion, note that $\Phi_p \delta f$ relates to the photon flux density,

$$S_p = \frac{c}{n_{31}} \Phi_p \delta f$$

which in turn relates to the intensity of the pump beam

$$I_p = h f_{31} S_p = \frac{c h f_{31}}{n_{31}} \Phi_p \delta f$$

Similarly, we can develop the expression for the intensity of the signal beam

$$I_s = \frac{c h f_{21}}{n_{21}} \Phi \delta f$$

If the electric field distribution in fibers were uniform, then the optical power would simply be the product of the cross sectional area and the optical intensity. However, the field distribution is not uniform throughout the fiber cross section. Therefore, we introduce the **effective cross sectional area** A_p for the **pump beam** at the frequency f_{31}. In terms of this effective cross sectional area and the intensity of the pump beam, the pump power is

$$P_p = A_p I_p = A_p \frac{c h f_{31}}{n_{31}} \Phi_p \delta f$$

Similarly, we introduce the **effective cross sectional area** A_s for the **signal beam**, in terms of which the signal power P_s is

$$P_s = A_s I_s = A_s \frac{c h f_{21}}{n_{21}} \Phi \delta f$$

Then, in terms of P_p and P_s, (10.42) and (10.43) become

$$\frac{dP_p}{dz} = \sigma_p (N_3 - N_1) P_p \qquad (10.44)$$

$$\frac{dP_s}{dz} = \sigma_s (N_2 - N_1) P_s \qquad (10.45)$$

where

$$\sigma_p = n_{31} B_{31}/c$$
$$\sigma_s = n_{21} B_{21}/c$$

10.6.3 Normalized Variables

The amplification characteristics of EDFAs can be studied by examining (10.40), (10.41), (10.44), and (10.45). However, because the equations are too complicated to be studied analytically, numerical techniques may be used. To do so, first cast the variables in terms of normalized quantities. We will show shortly that signal amplification is possible if the pump power is greater than a threshold value, given by

$$P_{th} = A_P \frac{c}{n_{31}} \frac{hf_{31}}{B_{31}} \frac{1}{\tau_{21}} \tag{10.46}$$

This is the **threshold pump power.** We define the **normalized pump power** as

$$P_{pn} \equiv \frac{P_p}{P_{th}} = B_{31} \tau_{21} \Phi_p \delta f \tag{10.47}$$

We can then express $\Phi_p \delta f$ in terms of P_{pn}, as follows:

$$\Phi_p \delta f = \frac{P_{pn}}{B_{31} \tau_{21}} \tag{10.48}$$

Similarly, we define the **normalized signal power** P_{sn} as

$$P_{sn} \equiv \frac{\sigma_s}{\sigma_p} \frac{hf_{31}}{hf_{21}} \frac{P_s}{P_{th}} = B_{21} \tau_{21} \Phi \delta f \tag{10.49}$$

Then,

$$\Phi \delta f = \frac{P_{sn}}{B_{21} \tau_{21}} \tag{10.50}$$

We also label $z_n = z\sigma_p N_t$ as the **normalized distance.** By substituting (10.40) and (10.41) into (10.44) and (10.45) and using the normalized variables, we obtain

$$\frac{dP_{pn}}{dz_n} = -\frac{P_{sn} + 1}{P_{pn} + 2P_{sn} + 1} P_{pn} \tag{10.51}$$

$$\frac{dP_{sn}}{dz_n} = \frac{\sigma_s}{\sigma_p} \frac{P_{pn} - 1}{P_{pn} + 2P_{sn} + 1} P_{sn} \tag{10.52}$$

To study the evolution of P_{pn} and P_{sn}, we solve (10.51) and (10.52) numerically, subject to the initial conditions $P_{pn}(0)$ and $P_{sn}(0)$ [25]. Typical results are presented in Figures 10.16 and 10.17.

Figure 10.16 depicts the variations of P_{pn} and P_{sn} as functions of z_n for a fixed pump level, $[P_{pn}(0) = 10]$, and three signal levels $[P_{sn}(0) = 1 \times 10^{-3}, 1 \times 10^{-4},$ and $1 \times 10^{-5}]$ at the input. In the numerical calculation, we choose $\sigma_s/\sigma_p = 1.08$, which corresponds to EDFAs pumped at 0.980 μm [26]. In all cases, the signal P_{sn} increases to a maximum before it decreases. For a large input signal, the maximum is reached more quickly. Plots of the normalized pump power are

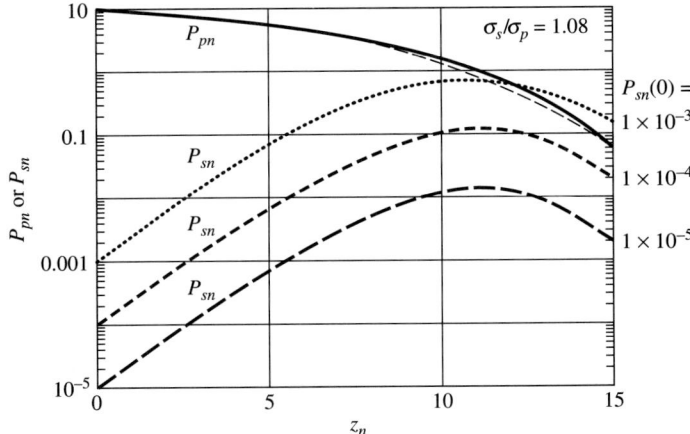

Figure 10.16 Typical variation of P_{pn} and P_{sn} of an erbium-doped fiber amplier pumped at 0.980 μm with $P_{pn}(0) = 10$

shown as the heavy solid line and the thin dash line. For small input signals, P_{pn} is essentially independent of the input signal level. For example, the two curves for P_{pn} with the initial condition of $P_{sn}(0) = 1 \times 10^{-4}$ and 1×10^{-5} are indistinguishable. In Figure 10.16, they are plotted as a single curve, that is the heavy solid line. The effect of $P_{sn}(0)$ on P_{pn} is discernible only when $P_{sn}(0)$ is large. The thin dash line depicts the variation of the normalized pump power with $P_{sn}(0)$ as large as 1×10^{-4}. Thus the change of P_{pn} is essentially independent of the signal power until the signal level increases to a value comparable to the pump level.

Figure 10.17 shows the effect of the pump level on the signal amplification. For the two sets of curves shown, the input signal is the same [$P_{sn}(0) = 1 \times 10^{-4}$]. However, the input pump levels are different [$P_{pn}(0) = 5$ and 10]. For a smaller pump input, the maximum signal level is reached sooner, and the maximum value is smaller, as expected.

10.6.4 Simple Amplification Characteristics

A qualitative understanding of the variation of signal power is possible by examining (10.52). In the region where P_{pn} is much greater than P_{sn} and 1, P_{sn} increases exponentially. Also, from (10.51) we see that P_{pn} decreases monotonically. As long as P_{pn} is greater than 1, P_{sn} keeps increasing. However, the rate of increase of P_{sn} drops as P_{pn} decreases. If the doped fiber is sufficiently long, there is a power level where P_{pn} becomes less than 1; then, P_{sn} decreases rather than increases. Physically this means that the amplified signal is reabsorbed by the doped fiber. Thus, there is an **optimal fiber length** for which the signal level is maximized for given pump and signal inputs.

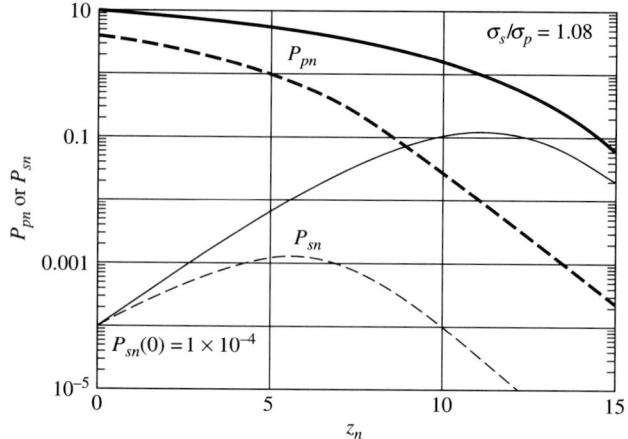

Figure 10.17 Variation of P_{pn} and P_{sn} for different input pump powers
Solid lines are for $P_{pn}(0) = 10$, and dashed lines are for $P_{pn}(0) = 5$.

The optimal length depends on the initial values $P_{pn}(0)$, $P_{sn}(0)$, and σ_p/σ_s. Note also that σ_p/σ_s is a function of the pump frequency. Mathematically, the optimal length, L_{op}, is given by the condition $P_{pn}(L_{op}) = 1$.

From (10.51) and (10.52), we also obtain

$$\frac{dP_{pn}}{dP_{sn}} = -\frac{\sigma_p}{\sigma_s} \frac{P_{sn} + 1}{P_{pn} - 1} \frac{P_{pn}}{P_{sn}} \tag{10.53}$$

which can be rearranged as

$$(1 - \frac{1}{P_{pn}})dP_{pn} = -\frac{\sigma_p}{\sigma_s}(1 + \frac{1}{P_{sn}})dP_{sn}$$

A simple integration leads to an expression relating the signal level at the output to the input and output pump power and the input signal level [26], as follows:

$$P_{sn}(L) - P_{sn}(0) + \frac{\sigma_s}{\sigma_p}[P_{pn}(L) - P_{pn}(0)] = -[ln\frac{P_{sn}(L)}{P_{sn}(0)} + \frac{\sigma_s}{\sigma_p} ln\frac{P_{pn}(0)}{P_{pn}(L)}] \tag{10.54}$$

Since the pump power at the input is stronger than the pump power at the output end of the fiber amplifier, that is, $P_{pn}(0) > P_{pn}(L)$, and that the signal after amplification is stronger than the input signal, that is, $P_{sn}(L) > P_{sn}(0)$. Therefore, the left-hand side of (10.54) is negative. We conclude that

$$P_{sn}(L) - P_{sn}(0) < \frac{\sigma_s}{\sigma_p}[P_{pn}(0) - P_{pn}(L)] \tag{10.55}$$

In terms of the unnormalized pump and signal power, this expression becomes

$$P_s(L) - P_s(0) < \frac{f_{21}}{f_{31}} [P_p(0) - P_p(L)] \quad (10.56)$$

Consequently, the maximum increase in signal power is the product of f_{21}/f_{31} and the net change in the pump power [26].

In the previous discussions, we have assumed that the signal and pump beams propagate in the same direction. This is the **copropagation scheme,** shown in Figure 10.15a. Amplification can also be achieved by having the signal and pump beams propagating in the opposite directions, as shown in Figure 10.15b. Studies also show that the maximum gain in the **counter propagating scheme** is *higher* than that in the copropagation scheme [27].

PROBLEMS

1. With reference to Figure 10.1, show that, by a rotation of coordinates and a shift of the time origin,

    $$\mathcal{E}(t) = \mathbf{a}_x E_{xo} \cos(\omega t + \phi_x) + \mathbf{a}_y E_{yo} \cos(\omega t + \phi_y)$$

    can be transformed to

    $$\mathcal{E}(t') = \mathbf{a}_\xi E_{mj} \cos\omega t' - \mathbf{a}_\eta E_{mn} \sin\omega t'$$

    where $\Delta\phi = \phi_y - \phi_x$, $\tan\nu = E_{yo}/E_{xo}$, and $\omega t + \phi_x = \omega t' + \delta$.

2. Refer to Problem 1. Show that $E_{mj}^2 + E_{mn}^2 = E_{xo}^2 + E_{yo}^2$.

3. Show that the major and minor axes of the polarization ellipse are along the x and y axes if $\Delta\phi$ is $\pm \pi/2$.

4. Show that circularly polarized fields are converted to linearly polarized fields by a quarterwave plate, without reducing light intensity.

5. Derive (10.18).

6. Show that the Jones matrix of a linear retarder with a linear retardation δ_l and a fast axis at θ relative to the x axis is

    $$\mathbf{R}_l(\theta, \delta_l) = \begin{bmatrix} \cos\frac{\delta_l}{2} + j\cos2\theta \sin\frac{\delta_l}{2} & j\sin2\phi\sin\frac{\delta_l}{2} \\ j\sin2\theta\sin\frac{\delta_l}{2} & \cos\frac{\delta_l}{2} - j\cos2\theta\sin\frac{\delta_l}{2} \end{bmatrix}$$

7. Consider an arrangement with three polarization components. The first and third components are linear polarizers with the fast axes normal to each other. To be specific, let $\theta_1 = 0°$ and $\theta_3 = 90°$. Waves with an arbitrary SOP are incident upon the arrangement. Calculate the output intensity if the second polarization component is one of the following (in all cases, ignore the effects due to reflection.):
    a. Absent.
    b. A linear polarizer with the fast axis in the direction θ_2.
    c. A right-hand circular polarizer.

8. Consider type 1 polarization controllers discussed in section 10.4.2. Show that any input SOP may be transformed to a field linearly polarized in the x direction with-

out loss of intensity. In particular, show that the output is linearly polarized in the x axis if $\delta_{l1} = \Delta\phi - (\pi/2)$, $\delta_{l2} = 2m\pi - 2v$ or $\delta_{l1} = \Delta\phi + (\pi/2)$, and $\delta_{l2} = 2v - 2m\pi$. Note that δ_{l1} and δ_{l2} are specified in Figure 10.12a, and v and $\Delta\phi$ are given in (10.4).

9. Consider type 2 polarization controllers discussed in section 10.4.2a. Show that with adjustments in the orientation of the wave plates, an arbitrary input SOP is transformed to linearly polarized fields in the x direction.

10. Consider type 3 polarization controllers discussed in section 10.4.2. Show that with adjustments in the circular retardation of the circular retarders, an arbitrary input SOP is transformed to linearly polarized fields in the x direction.

11. The figure shows an N-turn solenoid of length L. Let the solenoid current be I. A single-mode fiber is centered along the axis of the solenoid, to serve as a current sensor. A linearly-polarized wave impinges on one end of the fiber, and the SOP at the other end is monitored.

   a. Derive an expression relating NI to the angle of rotation of the linearly-polarized field.
   b. Suppose the maximum angle of rotation is 5 degrees, and a steady-state current of 20 A is to be measured. Specify the number of turns N and the wire size for safe, continuous operation. Estimate the solenoid length L. Justify your answers.

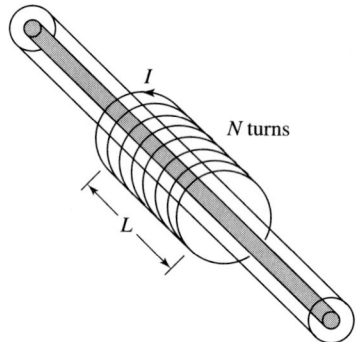

REFERENCES

1. Jones, R. C. "A new calculus for the treatment of optical systems, Part I, Description and discussion of the calculus." *J. Opt. Soc. Am.* 31, (1941), pp. 488–493. And Hurwits, H. Jr.; and R. C. Jones. "Part II, Proof of three general equivalent theorems." *J. Opt. Soc. Am.* 31, (1941), pp. 493–499.

2. Tabor, W. J.; and F. S. Chen. "Electromagnetic propagation through materials possessing both Faraday rotation and birefringence: experiment with Ytterbium orthoferrite." *J. Appl. Phys.* 40, (1969), pp. 2760–2765.

3. Barlow, A. J.; J. J. Ramskov-Hansen; and D. N. Payne. "Birefringence and polarization mode dispersion in spun single mode fibers." *Appl. Opt.* 20, (1981), pp. 2962–2968.

4. Tsao, C. Y. "Polarization parameters of plane waves in hybrid birefringent optical fibers." *J. Opt. Soc. Am. A* 4, (1987), pp. 1407–14127.
5. Kaminow, I. P. "Polarization in optical fibers." *IEEE J. of Quantum Electron.* QE-17, (1981), pp. 15–22.
6. Rashleigh, S. C. "Origins and control of polarization effects in single-mode fiber." *IEEE J. Lightwave Technol.* LT-1, (1983), pp. 312–331.
7. Smith, A. M. "Birefringence induced by bends and twists in single-mode optical fiber." *Appl. Opt.* 19, (1980), pp. 2606–2611.
8. Akers, F. I.; and R. E. Thompson. "Polarization-maintaining single-mode fibers." *Appl. Opt.* 21, (1982), pp. 1720–1721.
9. Kapron, F. P.; N. F. Borrelli; and D. B. Keck. "Birefringence in dielectric optical waveguides." *IEEE J. Quantum Electron.* QE-8, (1972), pp. 222–225.
10. Eickhoff, W. "In-line fiber-optic polarizer." *Electron. Lett.* 16, (1980), pp. 762–764.
11. Willsch, R. "High performance metal-clad fiber-optic polarizers." *Electron. Lett.* 26, (1990), pp. 1113–1115.
12. Bergh, R. A.; H. C. Lefevre; and H. J. Shaw. "Single-mode fiber-optic polarizer." *Opt. Lett.* 5, (1980), pp. 479–481.
13. Johnson, A. M. "An in-line fiber-optical polarization transformer." *Appl. Opt.* 18, (1979), pp. 1288–1289.
14. Lefevre, H. C. "Single-mode fiber fractional wave device and polarization controllers." *Electron. Lett.* 16, (1980), pp. 778–780.
15. Okoshi, A. T. "Polarization-State Control Schemes for Heterodyne or Homodyne Optical Fiber Communications." *J. Lightwave Technol.* 3, (1985), pp. 1232–1237.
16. Mertz, G.; J. R. Dunphy; W. W. Morey; and E. Snitzer. "Cross-talk fiber-optic temperature sensor." *Appl. Opt.* 22, (1983), pp. 464–477.
17. Sheem, S. K.; and T. G. Giallorenzi. "Single mode fiber optical power divider: encapsulated etching technique." *Opt. Lett.* 4, (1979), pp. 29–31.
18. Villarruel, C. A.; and R. P. Moeller. "Fused single mode fiber access couplers." *Electron. Lett.* 17, (1981), pp. 243–244.
19. Tran, D. C.; K. P. Koo; and S. K. Sheem. "Single-mode fiber directional couplers fabricated by twisting-etching techniques (stabilization)." *IEEE J. Quantum Electron.* QE-17, (1981), pp. 988–991.
20. Bergh, R. A., G. Kotler; and H. J. Shaw. "Single-mode fiber optic directional coupler." *Electron. Lett.* 16, (1980), pp. 260–261.
21. Chen, C. L.; and W. K. Burns. "Polarization characteristics of single-mode fiber couplers." *IEEE J. Quantum Electron.* QE-18, (1982), pp. 1589–1600.
22. Desurvire, E. "The golden age of optical fiber amplifiers." *Phys. Today,* 1994, pp. 20–27.
23. Desurvire, E.; and J. R. Simpson. "Amplification of spontaneous emission in erbium-doped single-mode fibers." *IEEE J. Lightwave Technol.* 7, (1989), pp. 835–845.
24. Li, T. Y. "The impact of optical amplifiers on long-distance lightwave telecommunications." *Proc. IEEE* 81, (1993), pp. 1568–1579.
25. Desurvire, E.; J. R. Simpson; and P. C. Becker. "High-gain erbium doped traveling-wave fiber amplifier." *Opt. Lett.* 12, (1987), pp. 888–890.

26. Peronic, M.; and M. Tamburrini. "Gain in erbium-doped fiber amplifiers: a simple analytic solution for the rate equations." *Opt. Lett.* 15, (1990), pp. 842–844.
27. Morkel, P. R.; and R. I. Laming. "Theoretical modeling of erbium-doped fiber amplifiers with excited-state absorption." *Opt. Lett.* 14, (1989), pp. 1062–1064.

JONES MATRIX FOR A LINEARLY BIREFRINGENT MEDIUM IN STRONG MAGNETIC FIELDS

APPENDIX E

APPENDIX OUTLINE

E.1 Introduction
E.2 Eigen Polarization Modes
E.3 Jones Matrix
References

E.1 INTRODUCTION

In this appendix, we derive the Jones matrix for a linearly birefringent medium under the influence of strong magnetic fields [E.1]. To anticipate the steps needed for this derivation, we will first review the derivations of the Jones matrices for linearly or circularly birefringent media, as presented in Chapter 10. The fields are expressed in terms of linearly or circularly polarized fields, which are eigen polarization modes for birefringent media. We will follow the same procedures here as in [E.2, E.3]

We begin with the time harmonic ($e^{+j\omega t}$) Maxwell equations

$$\nabla \times \mathbf{E} = -j\omega\mu_o \mathbf{H} \quad (E1)$$

$$\nabla \times \mathbf{H} = j\omega\varepsilon_o \overleftrightarrow{\varepsilon}_r \mathbf{E} \quad (E2)$$

where $\overleftrightarrow{\varepsilon}_r$ is the relative permittivity tensor of the medium

$$\overleftrightarrow{\varepsilon}_r = \begin{bmatrix} n_x^2 & j\gamma & 0 \\ -j\gamma & n_y^2 & 0 \\ 0 & 0 & n_z^2 \end{bmatrix} \quad (E3)$$

The off-diagonal terms $\pm j\gamma$ are the relative permittivity tensor components due to the magnetic fields in the z direction. If γ is zero and if n_x, n_y, and n_z are not identical, the medium is *linearly birefringent*. On the other hand, if $\gamma \neq 0$ and $n_x = n_y = n_z$, then (E3) is the relative permittivity tensor for *isotropic media subjected to magnetic fields*.

E.2 EIGEN POLARIZATION MODES

Consider transverse electromagnetic plane waves propagating along the direction of the strong magnetic fields, which is the z direction. The electric fields in the xy plane are of the form

$$\begin{bmatrix} E_x(z) \\ E_y(z) \end{bmatrix} = \begin{bmatrix} E_{xin} \\ E_{yin} \end{bmatrix} e^{-jknz} \quad (E4)$$

where $k = \omega\sqrt{\mu_o \varepsilon_o}$. The refractive index n and a relationship between E_{xin} and E_{yin} are to be determined. Substituting (E4) into (E1) and (E2) and using (E3), we obtain

$$\begin{bmatrix} n^2 - n_x^2 & -j\gamma \\ j\gamma & n^2 - n_y^2 \end{bmatrix} \begin{bmatrix} E_x(z) \\ E_y(z) \end{bmatrix} = 0 \quad (E5)$$

Nontrivial solutions for E_x and E_y exist if n satisfies the relation

$$(n^2 - n_x^2)(n^2 - n_y^2) - \gamma^2 = 0$$

There are two possible solutions labeled n_s and n_f for the *slow* and *fast* polarization modes, respectively:

$$n_{s,f}^2 = \frac{1}{2}\left[(n_x^2 + n_y^2) \pm \sqrt{(n_x^2 - n_y^2)^2 + 4\gamma^2}\right] \quad (E6)$$

By substituting (E6) into (E5), we have the eigen polarization modes in the column matrix form, as follows:

$$\mathbf{E}_s = \frac{A_s}{\sqrt{1 + \sigma^2}} \begin{bmatrix} j\sigma \\ 1 \end{bmatrix} e^{-jkn_s z} \quad (E7)$$

$$\mathbf{E}_f = \frac{A_f}{\sqrt{1 + \sigma^2}} \begin{bmatrix} 1 \\ j\sigma \end{bmatrix} e^{-jkn_f z} \quad (E8)$$

where A_s and A_f are the amplitude constants of the eigen polarization modes, and

$$\sigma = \frac{2\gamma}{n_y^2 - n_x^2 + \sqrt{(n_x^2 - n_y^2)^2 + 4\gamma^2}}$$

The eigen polarization modes can also be written in vector form,

$$\mathbf{E}_s(z) = \frac{j\sigma \mathbf{a}_x + \mathbf{a}_y}{\sqrt{1 + \sigma^2}} A_s e^{-jkn_s z} \quad (E9)$$

$$\mathbf{E}_f(z) = \frac{\mathbf{a}_x + j\sigma \mathbf{a}_y}{\sqrt{1 + \sigma^2}} A_f e^{-jkn_s z} \quad (E10)$$

For convenience, we define the basis vectors $\mathbf{a}_s$ and $\mathbf{a}_f$ in terms of $\mathbf{a}_x$ and $\mathbf{a}_y$, as follows:

$$\mathbf{a}_s = \frac{j\sigma \mathbf{a}_x + \mathbf{a}_y}{\sqrt{1+\sigma^2}}$$

$$\mathbf{a}_f = \frac{\mathbf{a}_x + j\sigma \mathbf{a}_y}{\sqrt{1+\sigma^2}}$$

Conversely, $\mathbf{a}_x$ and $\mathbf{a}_y$ can be written in terms of $\mathbf{a}_s$ and $\mathbf{a}_f$:

$$\mathbf{a}_x = \frac{-j\sigma \mathbf{a}_s + \mathbf{a}_f}{\sqrt{1+\sigma^2}}$$

$$\mathbf{a}_y = \frac{\mathbf{a}_s - j\sigma \mathbf{a}_f}{\sqrt{1+\sigma^2}}$$

Consider an arbitrary input field which may be expressed in terms of either the unit vectors $\mathbf{a}_x$ and $\mathbf{a}_y$ or the basis vectors $\mathbf{a}_s$ and $\mathbf{a}_f$:

$$\mathbf{E}(0) = \mathbf{E}_{in} = E_{xin}\mathbf{a}_x + E_{yin}\mathbf{a}_y = \frac{-j\sigma E_{xin} + E_{yin}}{\sqrt{1+\sigma^2}}\mathbf{a}_s + \frac{E_{xin} - j\sigma E_{yin}}{\sqrt{1+\sigma^2}}\mathbf{a}_f$$

At $z = L$, the output fields are

$$\mathbf{E}(L) = \mathbf{E}_{out} = \frac{-j\sigma E_{xin} + E_{yin}}{\sqrt{1+\sigma^2}}e^{-jkn_sL}\mathbf{a}_s + \frac{E_{xin} - j\sigma E_{yin}}{\sqrt{1+\sigma^2}}e^{-jkn_fL}\mathbf{a}_f$$

The x and y components of $\mathbf{E}_{out}$ are

$$E_{xout} = e^{-jk(n_f+n_s)L/2}\left[\left(\cos\frac{\delta}{2} + j\frac{1-\sigma^2}{1+\sigma^2}\sin\frac{\delta}{2}\right)E_{xin} + \frac{2\sigma}{1+\sigma^2}\sin\frac{\delta}{2}E_{yin}\right]$$

$$E_{yout} = e^{-jk(n_f+n_s)L/2}\left[-\frac{2\sigma}{1+\sigma^2}\sin\frac{\delta}{2}E_{xin} + \left(\cos\frac{\delta}{2} - j\frac{1-\sigma^2}{1+\sigma^2}\sin\frac{\delta}{2}\right)E_{yin}\right]$$

where $\delta = k(n_s - n_f)L$.

E.3 JONES MATRIX

By casting $\mathbf{E}_{in}$ and $\mathbf{E}_{out}$ in the form of Jones vectors, we immediately have the Jones matrix

$$\mathbf{J} = \begin{bmatrix} \cos\frac{\delta}{2} + j\frac{1-\sigma^2}{1+\sigma^2}\sin\frac{\delta}{2} & \frac{2\sigma}{1+\sigma^2}\sin\frac{\delta}{2} \\ -\frac{2\sigma}{1+\sigma^2}\sin\frac{\delta}{2} & \cos\frac{\delta}{2} - j\frac{1-\sigma^2}{1+\sigma^2}\sin\frac{\delta}{2} \end{bmatrix} \quad (E11)$$

This is the Jones matrix given in (10.13).

In the absence of strong magnetic fields, γ vanishes and the medium reverts back to being purely linearly birefringent. By setting γ to zero, we obtain, from

(E6), $n_s = n_y$ and $n_f = n_x$, assuming $n_y > n_x$ and $\delta = \delta_l = k(n_y - n_x)L$. Then (E11) reduces to

$$J = \begin{bmatrix} \cos\frac{\delta_l}{2} + j\sin\frac{\delta_l}{2} & 0 \\ 0 & \cos\frac{\delta_l}{2} - j\sin\frac{\delta_l}{2} \end{bmatrix} = \begin{bmatrix} e^{j\delta/2} & 0 \\ 0 & e^{-j\delta/2} \end{bmatrix}$$

which is the Jones matrix (10.11) for the medium without Faraday rotation.

If $n_x = n_y = n$ and $\gamma \neq 0$, then we have $\sigma = 1$ and $n_{sf}^2 = n^2 \pm \gamma$. This case corresponds to a simple isotropic medium subjected to a strong magnetic field. Since γ is much smaller than n, we can approximate n_{sf} as $\left(n \pm \frac{\gamma}{2n}\right)$. Then, $\delta \approx \delta_c = k\gamma L/n$, and (E11) becomes

$$J = \begin{bmatrix} \cos\frac{\delta_c}{2} & \sin\frac{\delta_c}{2} \\ -\sin\frac{\delta_c}{2} & \cos\frac{\delta_c}{2} \end{bmatrix}$$

which is the Jones matrix (10.12) for a circularly birefringent medium.

References

E.1 Tabor, W. J.; and F. S. Chen. "Electromagnetic propagation through materials possessing both Faraday rotation and birefringence: experiment with Ytterbium orthoferrite." *J. Appl. Phys.* 40, (1969), pp. 2760–2765.

E.2 Akers, F. I.; and R. E. Thompson. "Polarization-maintaining single-mode fibers." *Appl. Opt.* 21, (1982), pp. 1720–1721.

E.3 Kapron, F. P.; N. F. Borrelli; and D. B. Keck. "Birefringence in dielectric optical waveguides." *IEEE J. Quantum Electron.* QE-8, (1972), pp. 222–225.

JONES MATRIX FOR A LINEARLY BIREFRINGENT MEDIUM WITH A CONTINUOUSLY ROTATING AXIS

APPENDIX F

APPENDIX OUTLINE

F.1 Propagation
F.2 Axis Rotation
F.3 Coupled Mode Equations
F.4 Jones Matrix
References

In this appendix, we derive the Jones matrix for a linearly birefringent medium with a continuously rotating axis. Locally, the medium is linearly birefringent. However the fast and slow axes change continually. Instead of solving the Maxwell equations, as we did in Appendix E, we will set up and solve the coupled mode equations for the twisted, linearly birefringent medium. We will then obtain the Jones matrix for the medium with a rotating axis [F.1, F.2].

Assume that the fast and slow axes are along the x and y directions and have effective indices n_x and n_y. The fields in the transverse plane can then be written as

$$\mathbf{E}(z) = \mathbf{a}_x E_x(z) + \mathbf{a}_y E_y(z)$$

If the axes are not twisted, the two field components are not coupled. Each propagates independently. In the presence of axis twisting, however, the field components are coupled. Consider the fields at two neighboring points z and $z + \Delta z$. At these points, the field components are $E_x(z)$, $E_y(z)$, $E_x(z + \Delta z)$, and $E_y(z + \Delta z)$. $E_x(z)$ differs from $E_x(z + \Delta z)$ on two accounts: the propagation of the fields and the rotation of the axes. In an infinitesimal interval Δz, the two effects are vanishingly small. Therefore, we can consider each effect individually, independently of the other.

Appendix F Jones Matrix for a Linearly Birefringent Medium with a Continuously Rotating Axis

F.1 PROPAGATION

In the absence of axis rotation, the change of $E_x(z)$ is entirely due to the propagating nature of the fields, and $E_x(z)$ varies as $e^{-jkn_x z}$. Thus, the change in $E_x(z)$ due to *propagation* is

$$E_x(z+\Delta z) - E_x(z) \approx \frac{dE_x(z)}{dz}\Delta z = -jkn_x E_x(z)\Delta z \tag{F1}$$

Similarly,

$$E_y(z+\Delta z) - E_y(z) \approx \frac{dE_y(z)}{dz}\Delta z = -jkn_y E_y(z)\Delta z \tag{F2}$$

F.2 AXIS ROTATION

Next, consider the effects of axis rotation. For convenience, label the axes at z as the x' and y' axes, and at $(z + \Delta z)$ as the x'' and y'' axes. As shown in Figure F.1, the x'' and y'' axes are rotated with respect to the x' and y' axes by an angle $\Delta\Omega$. Therefore,

$$\mathbf{a}_{x'} = \mathbf{a}_{x''}\cos\Delta\Omega - \mathbf{a}_{y''}\sin\Delta\Omega$$
$$\mathbf{a}_{y'} = \mathbf{a}_{x''}\sin\Delta\Omega + \mathbf{a}_{y''}\cos\Delta\Omega$$

Since $\Delta\Omega$ is infinitesimally small, these equations can be approximated by

$$\mathbf{a}_{x'} \approx \mathbf{a}_{x''} - \mathbf{a}_{y''}\Delta\Omega \tag{F3}$$
$$\mathbf{a}_{y'} \approx \mathbf{a}_{x''}\Delta\Omega + \hat{\mathbf{a}}_{y''} \tag{F4}$$

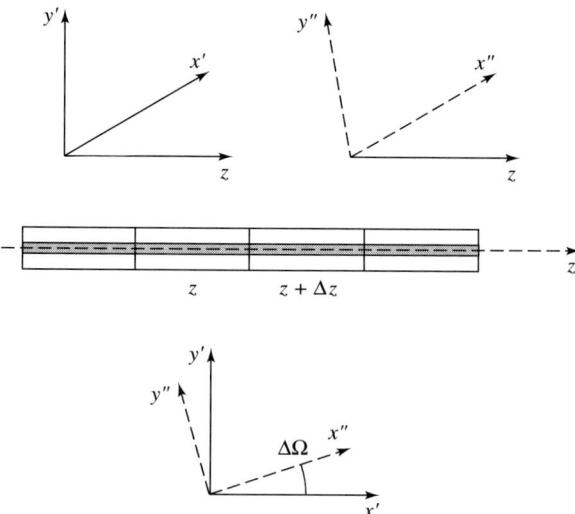

Figure F.1 Rotation of the coordinates

As noted earlier, we ignore the effects of propagation when the effects of twisting are considered. Consequently, $\mathbf{E}(z)$ must be identical to $\mathbf{E}(z + \Delta z)$; that is,

$$E_x(z)\mathbf{a}_{x'} + E_y(z)\mathbf{a}_{y'} = E_x(z + \Delta z)\mathbf{a}_{x''} + E_y(z + \Delta z)\mathbf{a}_{y''}$$

Using (F3) and (F4), we obtain

$$E_x(z)(\mathbf{a}_{x''} - \Delta\Omega\mathbf{a}_{y''}) + E_y(z)(\mathbf{a}_{y''} + \Delta\Omega\mathbf{a}_{x''}) \approx E_x(z + \Delta z)\mathbf{a}_{x''} + E_y(z + \Delta z)\mathbf{a}_{y''}$$

which requires that

$$E_x(z) + \Delta\Omega E_y(z) \approx E_x(z + \Delta z)$$
$$E_y(z) - \Delta\Omega E_x(z) \approx E_y(z + \Delta z)$$

These equations show that the two components are coupled when the axes are twisted. The coupling due to *axis rotation* is given by

$$E_x(z + \Delta z) - E_x(z) \approx \Delta\Omega E_y(z) \tag{F5}$$
$$E_y(z + \Delta z) - E_y(z) \approx -\Delta\Omega E_x(z) \tag{F6}$$

F.3 COUPLED MODE EQUATIONS

When *both axis rotation and propagation effects* are present, we have, from (F1), (F2), (F5) and (F6),

$$E_x(z + \Delta z) - E_x(z) \approx -jkn_x\Delta z E_x(z) + \Delta\Omega E_y(z)$$
$$E_y(z + \Delta z) - E_y(z) \approx -jkn_y\Delta z E_y(z) - \Delta\Omega E_x(z)$$

In the limit of $\Delta z \to 0$, we obtain the coupled mode equations

$$\frac{dE_x(z)}{dz} = -jkn_x E_x(z) + \xi E_y(z) \tag{F7}$$

$$\frac{dE_y(z)}{dz} = -jkn_y E_y(z) - \xi E_x(z) \tag{F8}$$

where $\xi = d\Omega/dz$ is the *rate of axis rotation per unit length*. We solve (F7) and (F8) under the assumption that n_x, n_y, and ξ are constants. By differentiating (F7) with respect to z, and using (F8), we obtain

$$\frac{d^2 E_x}{dz^2} + jk(n_x + n_y)\frac{dE_x}{dz} + (\xi^2 - k^2 n_x n_y)E_x = 0$$

A differential equation for E_y is similarly obtained. It is then possible to solve for E_x and E_y subject to the boundary conditions $E_{xin} = E_x(0)$ and $E_{yin} = E_y(0)$. Finally we obtain expressions for the output at $z = L$,

$$E_{xout} = E_x(L) = e^{-jk(n_x+n_y)L/2}\left[\left(\cos\frac{\delta_t}{2} + j\frac{\delta_l}{\delta_t}\sin\frac{\delta_t}{2}\right)E_{xin} + \frac{\delta_c}{\delta_t}\sin\frac{\delta_t}{2}E_{yin}\right]$$

$$E_{yout} = E_y(L) = e^{-jk(n_x+n_y)L/2}\left[-\frac{\delta_c}{\delta_t}\sin\frac{\delta_t}{2}E_{xin} + \left(\cos\frac{\delta_t}{2} - j\frac{\delta_l}{\delta_t}\sin\frac{\delta_t}{2}\right)E_{yin}\right]$$

where

$$\delta_l = k(n_y - n_x)L$$
$$\delta_r = 2\xi L$$
$$\delta_t = \sqrt{\delta_l^2 + \delta_r^2}$$

F.4 JONES MATRIX

We now have the Jones matrix for a linear birefringent medium with a *continuously twisted axis*:

$$\mathbf{J} = \begin{bmatrix} \cos\frac{\delta_t}{2} + j\frac{\delta_l}{\delta_t}\sin\frac{\delta_t}{2} & \frac{\delta_r}{\delta_t}\sin\frac{\delta_t}{2} \\ -\frac{\delta_r}{\delta_t}\sin\frac{\delta_t}{2} & \cos\frac{\delta_t}{2} - j\frac{\delta_l}{\delta_t}\sin\frac{\delta_t}{2} \end{bmatrix} \quad (F9)$$

This is precisely the same as (10.21). For comparison purposes, consider two special cases. First, if the axes are not twisted, then $\delta_r = 0$, $\delta_t = \delta_l$, and (F9) reduces to

$$\mathbf{J} = \begin{bmatrix} \cos\frac{\delta_l}{2} + j\sin\frac{\delta_l}{2} & 0 \\ 0 & \cos\frac{\delta_l}{2} - j\sin\frac{\delta_l}{2} \end{bmatrix} = \begin{bmatrix} e^{j\delta_l/2} & 0 \\ 0 & e^{-j\delta_l/2} \end{bmatrix}$$

which is the Jones matrix given in (10.11) for a linear retarder.

Second, for a twisted nonbirefringent medium, $\delta_l = 0$, and $\delta_t = \delta_r$. Then, (F9) reduces to

$$\mathbf{J} = \begin{bmatrix} \cos\frac{\delta_r}{2} & \sin\frac{\delta_r}{2} \\ -\sin\frac{\delta_r}{2} & \cos\frac{\delta_r}{2} \end{bmatrix}$$

which is the Jones matrix (10.12) for a circular birefringent medium or a circular retarder.

REFERENCES

F.1 Barlow, A. J.; J. J. Ramskov-Hansen; and D. N. Payne. "Birefringence and polarization mode dispersion in spun single mode fibers." *Appl. Opt.* 20, (1981), pp. 2962–2968.

F.2 Tsao, C. Y., "Polarization parameters of plane waves in hybrid birefringent optical fibers." *J. Opt. Soc. Am. A.* 4, (1987), pp. 1407–14127.

NAME INDEX

A

Abramowitz, M., 477 n, 483
Adams, M. J., 254 n, 477 n
Adler, R., 375 n
Agrawal, G. P., 528 n
Akers, F. I., 573 n, 578 n
Alfano, R. R., 163 n
Alferov, Z. I., 208, 253 n
Allard, F. A., 528 n
Anderson, L. K., 300 n, 301 n
Annovazzi-Lodi, V., 374 n
Arams, F. R., 301 n
Arecchi, F. T., 301 n
Arnaud, J. A., 82 n
Arnold, G., 254 n
Arrathoon, R., 162 n
Asakura, T., 82 n
Ashby, N., 162 n, 178 n
Aukerman, Y. W., 300 n
Auld, B. A., 390
Awwal, A. A. S., 83 n

B

Bachem, K. H., 253 n
Bagley, B. G., 478 n
Bailey, M. D., 529 n
Ballard, S. S., 477 n
Banerjee, P. P., 83 n, 163, 375
Barlow, A. J., 572 n, 583 n
Barnoski, M. K., 434 n, 479 n
Bartman, R. K., 21 n
Basch, E. E., 301 n, 479
Basit, A., 83 n
Beam, A. R., 390
Bean, J. C., 253 n
Becker, P. C., 573 n
Beisa, L., 374 n
Bell, T. E., 253 n
Bendow, B., 379, 529 n
Beni, G., 529 n
Bennett, G. A., 434 n

Bergh, R. A., 559, 573 n
Blakemore, J. S., 252 n
Boers, P. M., 245, 255 n
Bohr, Niels, 169, 171–172
Born, Max, 82 n, 162 n, 178 n, 473 n
Borrelli, N. F., 555, 573 n, 578 n
Botez, D., 253 n, 254 n
Bowers, J. E., 301 n
Brau, C. A., 163 n
Bridges, W. B., 163 n
Brouwer, W., 82 n
Browder, J. S., 477 n
Burch, J. M., 82 n
Burns, W. K., 255 n, 573 n
Burrus, C. A., Jr., 214, 253 n, 254 n, 255 n
Butler, J. K., 253 n
Byer, R. L., 163 n

C

Caird, J. A., 163 n
Campbell, J. C., 301 n
Carasso, M. G., 20 n
Carr, W. N., 254 n
Casey, H. C., Jr., 253 n
Casperson, L. W., 46, 47, 82 n
Chang, C. T., 478 n
Chang, I. C., 375 n
Chanin, D. J., 255 n
Chase, L. L., 163 n
Chen, C. L., 434 n, 573 n
Chen, F. S., 572 n, 578 n
Chen, H. Z., 253 n
Cheng, H., 253 n
Cheng, W. H., 255 n
Cheo, P. K., 434 n, 479
Cherin, A., 477 n
Chester, A. N., 163 n
Chynoweth, A. G., 478 n, 529 n
Claviere, B. de la, 82 n
Compaan, K., 20 n

Cook, B. D., 375 n
Cooper, G. R., 529 n
Cross, P. S., 163 n
Cuff, K. F., 301 n
Culshaw, B., 21 n

D

Dalven, R., 253 n
Danielsen, M., 245, 255 n
D'Asaro, L. A., 254 n
Davis, C. C., 163 n
Debye, P., 2, 20 n
Degnan, J. J., 163 n
DeLange, O. E., 300 n
DeMaria, J., 162 n
Dentai, A. G., 253 n, 254 n, 255 n
DePaula, R. P., 21 n
DePuydt, J. M., 253 n
Desurvire, E., 573 n
Dickson, L. D., 82 n
DiDomenico, M., Jr., 300 n
Ding, J., 253 n
Di Vita, E., 573 n
Di Vita, P., 529 n
Dixon, G. J., 163 n
Donati, S., 374 n
Dunphy, J. R., 573 n

E

Eberly, J. H., 83 n
Ebersole, J. F., 477 n
Eckhoff, W., 573 n
Eisenstein, G., 255 n
Emmons, R. B., 301 n
Enokihara, A., 20 n
Ettenberg, M., 254 n

F

Fan, T. Y., 163 n
Ferguson, A. I., 163 n

Field, R. L., Jr., 162 n, 162 n
Fisher, M. B., 300 n, 301 n
Florez, L. T., 253 n
Forrest, S. R., 301 n
Foy, P. W., 254 n
Franke, E. A., 82 n
Furuta, H., 434 n

G

Gale, H. G., 16, 21 n
Galus, W., 301 n
Garetz, B. A., 82 n
Gartner, W. W., 279, 301 n
Gauss, Carl Frederick, 28
Gayen, S. K., 163 n
Gerrard, A., 82 n
Ghaffari, A., 253 n
Ghausi, M. S., 529 n
Ghotak, A., 434 n
Giallorenzi, T. G., 573 n
Glass, A. M., 253 n
Gloge, D., 444, 449, 477 n, 488 n, 529 n
Golay, M. J. E., 301 n
Gooch, C. H., 254 n
Goodman, J. W., 83 n
Goodwin, D. W., 163 n
Goubau, G., 83 n
Gray, G. E., 477 n
Grudzien, M., 301 n
Guenther, B. D., 83 n, 375
Gunshor, R. L., 253 n
Guo, Y. L., 479

H

Hackwood, S., 529 n
Halbach, K., 82 n
Hale, P. G., 503, 529 n
Harana, M., 434 n
Harbison, J. P., 253 n
Harnagel, G. L., 163 n

585

Harris, S. E., 163 n, 178 n
Harvey, A. F., 162 n
Hasse, M. A., 253 n
Haus, H. A., 83, 434 n
Hawk, R. M., 529 n
Hawkes, J. F. B., 164, 375, 434, 479
Hawkins, S. R., 301 n
Hayashi, I., 254 n
Hayt, W. H., Jr., 82 n, 434, 478 n
Hecht, E., 83, 301 n, 529 n
Hecht, J., 163 n
Herczfeld, P. R., 21 n
Herman, R. M., 83
Herskowitz, G. J., 253 n
Herzberg, G., 162 n, 178 n
Herziger, G., 162 n
Hill, K. O., 528 n
Holden, W. D., 254 n
Hollberg, L., 163 n
Hondros, A., 2, 20 n
Horiguchi, M., 478 n
Hosaka, T., 478 n
Hubbard, W. M., 82 n
Hughes, R., 528 n
Hunsperger, R. G., 434 n
Hurd, F. K., 433 n, 477 n
Hurwits, H., Jr., 572 n
Hwang, C. J., 255 n

I

Iga, K. I., 253 n, 434, 479, 528 n
Ihaya, A., 434 n
Iksander, M. F., 449 n
Inagaki, N., 478 n
Inoue, K., 20 n
Ishikawa, R., 528 n
Izawa, T., 477 n, 478 n
Izutsu, M., 8, 20 n

J

Jackel, J. L., 466, 529 n
Jacobs, S. F., 301 n
Jefferies, J. A., 478 n
Jenkins, F. A., 83, 162 n, 163, 302 n, 374 n, 375, 529 n
Jeon, H., 253 n
Jeunhomme, J., 478 n
Jewell, J. L., 253 n
Johnson, A. M., 573 n
Johnson, D. C., 528 n
Jones, K. A., 163, 479
Jones, R. C., 572 n
Jones, W. B., Jr., 434, 479

K

Kaede, K., 528 n
Kaminow, I. P., 251, 255 n, 301 n, 374 n, 573 n
Kane, T. J., 163 n
Kapron, F. P., 555, 573 n, 578 n
Karim, M. A., 83 n, 302 n, 375, 479
Kasper, B. L., 301 n

Katzir, A., 253 n
Kawai, K., 528 n
Kawasaki, B. S., 528 n
Kayama, J., 20 n
Kazarinov, R. F., 208, 253 n
Kazorsky, L. G., 301 n
Keck, D. B., 478 n, 555, 573 n, 578 n
Keiser, G., 477 n
Kelly, P. L., 162 n
Kennard, E. M., 162 n, 178 n
Keyes, R. J., 300 n
Khostofian, J. M., 82 n
King, T. A., 163 n
Kino, G. S., 375
Kinoshita, S., 253 n
Kittel, C., 252 n
Kittle, C., 374 n
Klaiber, R. J., 478 n
Klein, W. R., 375
Kobayashi, K., 253 n, 528 n
Koechner, W., 163
Kogelnick, H., 44, 72, 82 n, 83 n, 434 n
Kokobun, Y., 528 n
Kompfner, R., 503, 529 n
Koo, K. P., 573 n
Korpel, A., 375
Kotler, G., 573 n
Koyama, F., 253 n
Kozlovsky, W. J., 163 n
Kramer, P., 20 n
Kraus, J. D., 374 n, 479
Krause, P. W., 300 n
Kressel, H., 253 n, 254 n
Kroemer, H., 208, 253 n
Krupke, W. F., 163 n
Kuizenga, D. J., 162 n
Kuppers, D., 478 n
Kurkjian, C. R., 478 n
Kurokawa, T., 528 n

L

Ladany, I., 253 n, 254 n
Lalor, M. J., 82 n
Lambo, W., 529 n
Laming, R. I., 574 n
Lass, H., 374 n
Lauritsen, T., 162 n, 178 n
Lee, T. P., 253 n, 254 n, 255 n
Lee, Y. H., 253 n
Lefevre, H. C., 573 n
Leune, K., 162 n
Li, T. Y., 44, 82 n, 573 n
Li, Y. F., 528 n
Librecht, Francois M., 375 n, 375 n
Lightstone, A., 528 n
Lin, C., 301 n
Lit, J. W. Y., 528 n
Liu, Y. S., 228, 254 n
Longurst, R. S., 529 n
Love, J. D., 477 n
Love, R., 478 n
Lydtin, H., 478 n

M

MacChesney, J. B., 478 n
Maiman, T. H., 162 n
Malcolm, G. P. A., 163 n
Mandeville, G. D., 82 n
Marcatili, E. A. J., 488 n, 529 n
Marcuse, D., 477 n, 478 n, 508, 529 n
Marton, L., 300 n
Matsushita, S., 528 n
McGillem, C. D., 529 n
McGlauchlin, L. D., 300 n
McMurty, B. J., 301 n
McQuistan, R. B., 300 n
Meijer, F., 478 n
Melchoir, H., 301 n
Mertz, G., 573 n
Michelson, A. A., 16, 21 n
Midwinter, J. E., 479
Miller, B. I., 254 n, 255 n
Miller, C. M., 529 n
Miller, G. D., 83 n
Miller, S. C., 162 n, 178 n
Miller, S. E., 301 n, 478 n, 529 n
Milonni, P. W., 83 n
Minemura, K., 528 n, 528 n
Minford, W. J., 21 n
Mitchell, J. W., 478 n
Mitra, S. S., 529 n
Miya, 478 n
Miyashita, T., 478
Moeller, R. P., 255 n, 573 n
Moore, C. E., 162 n
Morey, W. W., 573 n
Morkel, P. R., 574 n
Morkoc, H., 253 n
Morra, P., 528 n
Morrow, A. J., 478 n
Muoi, T. V., 301 n, 302 n
Murata, H., 478 n

N

Nakagome, H., 529 n
Nasiruddin, A. M., 83 n
Nelson, B. P., 478 n
Nelson, D. F., 374 n
Nemoto, S., 83 n
Neumann, E. G., 477 n
Nice, A., 529 n
Niezeki, N., 375 n
Nishihara, H., 20 n
Nishihara, N., 434 n
Nishimoto, H., 528 n
Noda, H., 434 n
Norton, P., 253 n
Nosu, K., 20 n, 528 n
Nuese, C. J., 253 n
Nurmikko, A. V., 253 n
Nussbaum, A., 164, 254 n, 302 n, 83, 375

O

O'Connor, P. B., 478 n
Ogawa, H., 21 n
Ohara, S., 529 n

Ohmori, Y., 478 n
Oikawa, M., 528 n
Okoshi, A. T., 477 n, 560, 573 n
Olson, H. M., 301 n
Orriols, G., 20 n
Osinski, M., 254 n
Osterberg, H., 477 n

P

Palais, J. C., 434, 479
Panish, M. B., 253 n, 254 n
Pankove, J. I., 252 n, 254 n
Paoli, T. L., 255 n
Pardo, J., 83
Paris, D. T., 433 n, 477 n
Partus, F. P., 478 n
Patel, B., 163 n
Payne, D. N., 572 n, 583 n
Payne, S. A., 163 n
Pearson, F., 16, 21 n
Pedrotti, F. L., 83, 162 n, 302 n, 375, 529 n
Pedrotti, L. S., 83, 162 n, 302 n, 375, 529 n
Peek, J. B. H., 20 n
Peronic, M., 574 n
Personick, S. D., 300, 301 n, 302 n
Peterson, G. E., 478 n
Petit, J. M., 529 n
Petricevic, V., 163 n
Phillips, M., 21 n
Phillips, R. A., 164, 254 n, 302 n, 375 n
Phillips, R. S., 83
Pi, F., 20 n
Pierce, J. R., 301 n
Piotrowski, J., 301 n
Poehler, T. O., 300 n
Poon, T. C., 83, 163, 375
Preier, H. M., 253 n
Presby, H. M., 478 n
Presley, R. J., 163 n, 254 n
Purohit, R. K., 252 n

Q–R

Qua, G. L., 301 n
Qui, J., 253 n
Radziemski, L. J., 162 n
Ramaswamy, V., 434 n
Ramos, S., 83, 434 n, 477 n
Ramponi, A. J., 163 n
Ramskov-Hansen, J. J., 572 n, 583 n
Rashleigh, S. C., 573 n
Rawson, E. G., 529 n
Reardon, A. C., 162 n
Reed, M. K., 163 n
Reich, S., 374 n
Richtmyer, F. K., 162 n, 178 n
Ridenour, L. N., 529 n
Ripper, J. E., 254 n, 255 n
Rosma, R., 253 n
Ross, D., 254 n
Ross, I. M., 2, 20 n
Rossi, U., 529 n

Rudolph, H. D., 477 n
Russer, P., 254 n

S

Saifi, M. A., 478 n
Saleh, B. E. A., 83, 164, 302 n, 375, 434
Scherer, A., 253 n
Schultz, P. C., 478 n
Schulz-Dubois, E. O., 301 n
Schwering, J. H., 83 n
Seib, D. H., 300 n
Self, S. A., 83 n
Sell, D. D., 253 n
Senior, J. M., 164, 302 n, 479
Shank, C. V., 163 n
Shankland, R. S., 21 n
Shannon, R. R., 82 n
Sharma, B. L., 252 n
Shaw, H. J., 573 n
Sheem, S. K., 573 n
Shimizu, N., 529 n
Shorthill, R. W., 21 n
Siegel, G. H., Jr., 479 n
Siegman, A. E., 82 n, 83 n, 162 n, 164
Simmons, P., 529 n
Simpson, J. R., 573 n
Sinjou, J. P., 20 n
Smith, A. M., 83 n, 573 n
Smith, D. A., 228, 254 n
Smith, P. W., 162 n
Smith, R. G., 300 n, 301 n
Snavely, B. B., 163 n
Snitzer, E., 477 n, 573 n
Snyder, A. W., 477 n
Sommerfeld, Arnold, 169, 171–172

Soref, R. A., 529 n
Staver, P. R., 163 n
Stegun, I. A., 477 n, 483
Stijns, E., 82 n
Stillman, G. E., 301 n
Stockman, M., 528 n
Stone, F. T., 21 n
Stulz, L. W., 255 n
Sudo, S., 477 n
Suematsu, Y., 253 n, 434, 479
Sueta, T., 20 n
Sugimoto, S., 528 n
Suhara, T., 20 n, 434 n
Sumski, S., 254 n
Suzaki, Y., 82 n
Svelto, O., 162 n, 164
Szarka, P., 528 n
Sze, S. M., 253 n

T

Tabor, W. J., 572 n, 578 n
Tachibana, A., 82 n
Takato, N., 528 n
Tambosso, T., 374 n
Tamburrini, M., 574 n
Tamir, T., 434 n
Teich, M. C., 83, 164, 375, 434
Terunma, Y., 478 n
Thiel, F. L., 529 n
Thompson, R. E., 573 n, 578 n
Tien, P. K., 434 n
Toba, H., 8, 20 n
Tomlinson, W. J., 528 n
Torge, R., 434 n
Tosco, F., 528 n
Tran, D. C., 479 n, 573 n
Tranter, W. H., 528 n
Treman, F. E., 529 n

Tridimas, Y., 82 n
Tsang, W. T., 301 n
Tsao, C. Y., 573 n
Tsuchiya, H., 529 n
Tynes, A. R., 478 n

U

Uchida, H., 528 n
Uchida, N., 375 n
Udd, E., 21 n
Ulrich, R., 434 n
Ura, S., 20 n

V

Vali, V., 21 n
Valley, G. E., Jr., 529 n
Van der Ziel, A., 301 n
VanDewoestine, R. V., 478 n
Van Duzer, T., 83, 434 n, 477 n
Van Etten, W., 529 n
Vannucci, R., 529 n
Vedak, D. A., 83 n
Verdeyn, 83, 164
Veselka, J. J., 529 n
Vezzoni, E., 528 n
Villarruel, C. A., 573 n
Vlaardingbrook, M. T., 245, 255 n

W

Wallman, H., 529 n
Wang, C. S., 255 n
Watanabe, R., 528 n
Watson, G. N., 483
Watson, H. A., 301 n
Weart, S. R., 21 n

Weber, H., 162 n
Whinnery, J. R., 83, 434 n, 477 n
White, H. E., 162 n, 163, 302 n, 374 n, 375, 529 n
White, K. I., 478 n
Wiggins, T. A., 83 n
Willsch, R., 573 n
Wilson, J., 164, 375, 434, 479
Witte, H. H., 528 n
Wittke, J. P., 254 n
Wolf, E., 82 n
Wolfe, R., 254 n
Wood, L. E., 477 n
Woolly, N. H., 82 n
Wynant, J. C., 82 n

X–Y

Xie, W., 253 n
Yariv, A., 83, 162 n, 163 n, 164, 253 n, 302 n, 375 n, 434 n, 479
Yeh, C., 477 n
Yeh, P., 375 n
Yoneyama, T., 21 n
Yoshida, A., 82 n
Youmans, B. R., 21 n
Ysao, C. Y., 583 n

Z

Zachos, T. H., 254 n
Zajac, A., 83, 301 n
Zayhowski, J. J., 162 n
Zhou, B., 163 n
Ziemmer, R. E., 528 n
Zory, P. S., 253 n

SUBJECT INDEX

A

Abbreviated indices, 389–390
ABCD law, 18, 25, 64–68
ABCD matrix/matrices, 18, 25–44
   factorization methods, 46–47
   lens-like and propagation-like, 47–49
   for mirrors, 37–38
   physical meaning of elements, 40–41
   propagation through a dielectric region of thickness d, 30–31
   for rays traveling in free-space region of thickness d, 29
   for spherical refractive surface, 31–32
   of thick lens, 35
Absorption, 102–103
   in semiconductors, 193–195
Absorption loss, 472–473
Acceptors, 189
Acoustic waves
   longitudinal, 354–357, 388
   transverse, 388–389
Acoustooptic effect, 18–19
   and elasticity, 351–357
   simple, 316–317
Acoustooptic modulators and deflectors, 357–370
   Bragg diffraction, 362–366
   material figure of merit, 369–370
   Parameter Q, 367
   Raman-Nath diffraction, 358–361
   resolvable spots, 368–369
   transition region diffraction, 367

Acoustooptic Q-switching, 127–128
Actinide ions, 97, 153–155
Active mode locking, 140
ADP/KDP crystals, 339–341
   and longitudinal EO modulators, 341–345
   and transverse EO modulators, 346–347
Air mode, 416
Alexandrite lasers, 157–158
Ampere, 95
Amplified modulation, of ILDs, 246–250
Amplifiers, erbium-doped, 564–571
Amplitude spontaneous emission, 250
Analyzer's axis, 314
Analyzers, 314
Angular magnification, 40
Angular misalignment, 522
Anisotropic media, 314
   dielectric properties, 328
   properties of wave propagation in, 330
Anisotropic superstrate polarizers, 558–559
Approximate characteristic equation, of weakly guided fibers, 485–486
Argon, electronic configuration, 178
Argon lasers, 144
Asymmetric stretch mode of CO_2 molecule, 146
Asymmetric waveguides, 414
Asymptotic expansions of Bessel functions, 482–483
Atoms
   helium, 175–176
   hydrogen, 170–173
   hydrogen-like, 173–175

   lithium, 176
   many-electron, 173–178
   neutral, 141
Attenuation, 270
Available thermal noise power, 286
Avalanche multiplication, 281
Avalanche multiplication gain factor, 282
Avalanche photodiodes, 275, 281–284
Axes rotation, 542–543, 555, 579–582
Azimuthal mode number, 442, 452
Azimuth of the ellipse, 310

B

Bandgaps, 185–187
   direct and indirect, 190–193
Band-to-band transition, 270
Bandwidth, 524
Bandwidth-length product, 460
Bandwidth-pulse width relationship, 525–526
Bandwidth-risetime product, 525
Bandwidth-risetime relationship, 524–525
Basis vectors, 313
Beam expanders, 71–72
Beam parameters, 54–55
   of optical cavities, 74–82
Beam radius, 52
   determination, 57–61
Beam splitter, 491
Beam waist, 52
Beat length; see Polarization beat length
Beryllium oxide laser tubes, 144
Bessel differential equations, 448
   and Bessel functions, 481–482

Bessel functions, 359, 443, 448–449
   first kind of, 481
   recurrence relations, 482
   second kind of, 489
Biaxial media, 330
   index ellipsoid, 336
Biconical tapered structure coupler, 563
Binary semiconductors, 189
Birefringence
   external applied stress, 551–554
   geometrical, 550
   internal or built-in stress, 550–551
   single-mode fiber, 550–556
Birefringent fibers, Jones matrices, 555–556
Birefringent media
   axes rotation, 542–543
   circularly, 540–541
   continuous axis rotation, 546–547
   linearly, 539–540
   in magnetic field, 541
Bit error rate, 491
Blackbody radiation, 99–100
   Planck's equation, 250–261
Bohr-Sommerfeld Theory, 170–172
Bohr's quantization rule, 170
Bohr theory for hydrogen atoms, 170–172
Bolometers, 264
Boltzmann's constant, 99, 186, 285
Boltzmann's distribution, 99–100
Boltzmann's equation, 88, 104, 107–113
Bosons, 172 n
Bottle coupler, 563
Bragg angle, 363

Subject Index

Bragg diffraction, 358, 362–366, 367
Breakdown voltage, 283
Brewster angle windows, 62
Broad area lasers, 209–210
Buried heterostructure laser, 213
Buried stripe waveguide, 433
Burrus LED, 230
Bus configuration, 482
bV diagram, 412–413

C

Candela, 95
Carbon dioxide lasers
 conventional, 147–148
 waveguide, 148–149
Carbon dioxide molecules, vibration modes, 146
Caustics, 486
CD-ROM readers, 3
Chalcogenide glasses, 438
Channeled substrate planar lasers, 213
Channel waveguides, 351
Characteristic equation
 of step index thin-film waveguides, 407–412
 of weakly guided fibers, 485–486
 of weakly guided step index fibers, 448–449
Charge carrier confinement, 220
Charge carrier transport, 270–271
Chloride glasses, 438
Cholesteric phase liquid crystal, 504
Chromatic dispersion; see Intramodal dispersion
Chromium-doped alexandrite, 157
Chromium-doped emerald, 157
Circularly birefringent fibers, 551, 555–556
Circularly birefringent media
 definition, 535
 Jones matrix, 540–541
Circularly polarized waves, 309–310
Circular retardation, 541, 554
Circular retarder, 541
Cladding region of fibers, 438–439
Clausius-Mosotti relation, 316
Cleaved coupled cavity, 215
Coherent detection, 292–294
 compared to incoherent detection, 294–295
 effect of beam diameters, 295–296
 effect of polarization states, 296
 effect of propagation directions, 296–297
 method, 261–262
 mode structures and wavefront curvature, 297–298

in single-mode fiber communication schemes, 534
Coherent light, 18
Coherent light source, 94
Coherent superposition of sources, 136
Cold junction, 263–264
Collimated beam, 74
Collimated range, 54
Collision broadening, 148
Commercially available lasers, 159–161
Communication system, local area networks, 490–491
Compact disk players, 2–5
Complex radius of curvature, 51, 55
Compression, 377
Concave lenses, 36
Concave spherical mirror, 42–43
Concentric resonators, 79
Conduction band, 185–186
Confocal resonators, 77–78
Connector loss
 extrinsic, 519–522
 intrinsic, 518–519
 normalized mismatch parameter, 522–523
Conservation of energy, 191–192
Conservation of momentum, 191–192
Contact noise, 288
Continuous axis rotation, 546–547
Converging lenses, 36
Convex lenses, 36
Convex spherical surface, 31–32
Core area mismatch, 519
Core electrons, 183–185
Core region, of fibers, 438–439
Corning 7059 glass, 423, 431–432
Cotton-Mouton effect, 307
Coulomb's law, 317
Coulomb force, 170
Counter-rotation matrix, 543
Coupled cavity lasers, 215
Coupled mode equation, 438, 581–582
Coupled mode theory, 427–431
Couplers
 biconical tapered structure, 563
 bottle coupler, 563
 polished fiber, 563–564
 twin core, 562
Coupling constant, 428
Coupling efficiency, 510–513
Coupling length, 430–431
Coupling ratio, 499
Cover modes, 416–417
Critical angle, 222, 402
Critical population inversion, 122
Critical pumping transition probability, 122
Crossed polarizers, 314

Crystal fields, 153–154
Crystal momenta, 191
Crystals
 liquid, 505
 with orthorhombic, monoclinic, or triclinic symmetry, 336
 trigonal, tetragonal, or hexagonal symmetry, 331
Current injection, 240
Current noise, 288
Current spreading, 212
Cutoff frequencies, 414
Cutoff thickness, 414

D

Dark current, 279, 288
Debye-Sears diffraction, 358
Deep Zn diffusion, 213
Deformation, types of, 377
Depletion layer, 204
Depletion region, 277
Depth of focus, 54
Detectivity, 263
Dielectric constant, 25
Dielectric slab waveguide, 403
Dielectric thin-film waveguides, 392
Dielectric waveguides, 426–427
Diffusion velocity of electrons, 270–271
Diodes
 light-emitting, 217–230
 superluminescent, 250–251
Dipole antennas, 475–476
Dipole moment, 475–476
Direct bandgap materials, 182
Direct bandgaps, 190–193
Direct detection method, 261; see also Incoherent detection method
Directional couplers
 coupling ratio, 499
 directional isolation, 499–500
 insertion loss/excess loss, 499
 multimode fiber, 496–500
 single-mode fiber, 562–564
Direct transition process, 193
Dispersion, 461
Dispersion relation, 395
Distributed Bragg reflectors (DBR), 215
Distributed feedback (DFB) lasers, 215
Diverging lenses, 36
Dominant mode, 443
Donors, 189
Dopants, 188
 losses due to, 472–473
Doping, 188
Doppler broadening, 148
Doppler effect, 110
Double crucible method of making glass, 467
Double degenerate bending mode of CO_2 molecules, 146

Double heterostructure diode, 206
Drift velocity of electrons, 270–271
Dye lasers, 149–153
 basic configurations, 151–153
 flowing dye jet, 151
 tunability, 150
Dynodes, 266

E

Edge-emitting ILDs, 217
Edge-emitting LEDs, 220
Effective active volume, 121–122
Effective cavity length, 121
Effective fiber cross-sectional areas, 566, 567
Effective focal length, 40
Effective index of refraction, 398
Effective mass of electrons, 187
Effective thickness of optical component, 40
Effects of wavefront curvature, in coherent detection, 297–298
Eigenfunctions, 172–173
Eigen polarization modes, 576–577
Eigenvalues, 172–173
Einstein relations, 104–105
Einstein's A-coefficient, 104
Einstein's B-coefficients, 102–103
Elastic constants, 383–385
Elastic equation of motion, 354, 385–386
Elasticity
 and acoustooptic effects, 351–357
 one-dimensional, 352–353
Elastic limit, 353, 383
Elastooptic effect, 357; see also Acoustooptic effect
Electric displacement, definition, 315
Electric polarizability, 315–316
Electromagentic theory, of charged particles, 170
Electron affinity, 203
Electron configuration, 99
 definition, 175
 of helium atoms, 175–176, 177
 of lithium, 176
 of neon and argon, 187
Electron density, time rate of change, 241
Electron emission efficiency, 266
Electron energy, 185–186
Electronic polarizability, 317
Electronic transitions, 473
Electron lifetime, 240
Electrons
 drift velocity and diffusion velocity, 270–271
 effective mass, 87
Electron spin, 169, 172
Electron spin quantum number, 172

Subject Index

Electron volts, 95
Electrooptical modulators, 341–351
  longitudinal EO modulators, 341–345
  versus magnetooptic modulators, 347–348
  transverse EO configuration, 346–347
Electrooptic crystals, 127
Electrooptic effect, 18–19
  in AKP/KDP crystals, 339–341
  linear, 337–341
  simple, 317–319
Electrooptic material figure of merit, 348–351
Electrowetting switches, 502, 503–505
Elemental semiconductors, 182
Ellipse, azimuth of, 310
Elliptically polarized waves, 310–312
Ellipticity, 310–311
Elongation deformation, 377
Embedded stripe waveguide, 433
Emerald lasers, 157–158
End fire coupling, 421
End pumping scheme, 157
End separation loss, 519–521
Energy bands, 185–187
Energy level, 97–99
  of acceptors, 189
  of donor states, 198
  ground state, 171
  many-electron atoms, 173–175
EO crystals; see Electrooptic crystals
Epitaxially grown materials, 213
Erbium-doped amplifiers
  normalized variables, 568–569
  rate equation, 565–567
  simple amplification characteristics, 569–571
Excess loss of directional coupler, 499
External applied stress or fields birefringence, 551–554
External photoelectric effect, 265
External quantum efficiency, 235–237
Extraordinary index, 332
Extrinsic loss
  angular misalignment, 525
  definition, 517
  end separation causing, 519–521
  Fresnel reflection, 521–522
  transverse offset, 521
Extrinsic semiconductors, 182, 183–189

F

Fabry-Perot cavities, 213
  definition, 79
Fabry-Perot interferometer, 13–14
Factorization methods, 46–47
Faraday effect, 307, 319–321
Faraday isolators, 327
Faraday rotation, 307, 322–323
  birefringence due to, 554
Far field region, 54
Fast axis, 553–554
Fermi-Dirac distribution function, 186
Fermi level, 185–186
Fermions, 172 n
Fiber axis rotation, 555
Fiber gyroscopes, 16–17
Fiber losses
  due to absorption, 472–473
  due to impurities, 473–474
  waveguide imperfections, 477
Fiber polarization controllers, 560–561
Fiber retarders, 559–560
Fibers; see also Optical fiber
  circularly birefringent, 551
  erbium-doped, 564–565
  linearly birefringent, 51
  polarization-maintaining, 555
  spun, 550–551
Fiber squeezer, 559
Field pattern, 49
Flowing dye jet, 151
Fluoride glasses, 438
Focal length
  effective, 40
  thin lens, 36
Form birefringence; see Geometrical birefringence
Forward biased junction, 204
Four-level laser system, 100–102
Franz-Keldysh effect, 307
Free charge carrier, 270
Free-charge carrier absorption, 307
Free-electron lasers, 142 n
Frequency chirping, 306
Frequency selection switch, 11, 12–13
Fresnel reflection, 521–522
Full width between half intensity, 224–225, 227, 228
Full width between half power points, 96, 138
Full width between nulls, 138
Fundamental Gaussian beams, 49–55, 297; see also Gaussian beams
  beam radius, 52
  complex radius of curvature, 55
  mathematical expression, 49–51
  radius of a curvature, 53–54
  Rayleigh range, 54–55
Fundamental mode, 443

Fused biconical taper fiber directional couplers, 496–500
FWHI; see Full width between half intensity

G

GaA1 As injection laser, 4
GaAs injection laser, 1
Gain-guided stripe geometry lasers, 211–212
Gain-guided structures, 212
Gas laser systems, 140–149
  ionized gas lasers, 142–146
  molecular gas lasers, 146–149
  neutral gas lasers, 141–142
Gaussian beams, 18, 24; see also Fundamental Gaussian beams
  definition, 52
  effects of spherical mirrors, 66–67
  fundamental, 49–55
  higher-order modes, 61–64
  transformations by thin lenses, 68–74
Gaussian distribution, 57
Gaussian function, 526
Gaussian optics, 28
Gaussian pulses, 525–526
Gem lasers, 157–158
Generalized film thickness, 406
Generalized frequency, 406, 443–444
Generalized guide index, 442–443
Geometrical birefringence, 550
Geometrical factor, 349, 443
Geometrical optics, 72–74
Glass, intrinsic scattering, 475–477
Glass cladding, 466
Glass fibers, 466–467; see also Fibers and Optical fibers
Glass receiving tubes, 142
Golay's pneumatic detectors, 264–265
Graded index fibers
  coupling efficiency, 512
  insertion loss, 512–513
  intermodal dispersion, 459–460
Graded index profile, of optical fibers, 439
Graded index thin-film waveguides, 418–421
Graded index waveguides, 403
Graded refractive index lenses, 504, 507, 512
  passive components based on, 492–495
Grating coupling, 421
GRIN; see Graded refractive index lenses
Grooved coupled cavity, 215
Ground states of inert gases, 178

Group III materials, 189
Group IV materials, 187
Group velocity, 389–399
Group V materials, 188–189
Guided waves, 392
  group velocity, 399–400
  parallel-plate waveguides, 393–397
  phase velocity, 399–400
  reflection by conducting boundaries, 393
Guide wavelength, 397
Gyroscopes, 13–16

H

Half-power beamwidth, 238
Half-power frequency, 524
Halfwave plates, 540
Half-wave voltage
  for longitudinal EO modulators, 342–343
  for transverse EO modulators, 347
Halogens, 141 n
Hankel functions, 481
He; see Helium
Helium atoms, 141–142, 175–176
  ground state electronic configuration, 177
Hemispherical domes, 223–224, 226
HeNe lasers, 101
Hermite polynomial, 49, 62
Heterojunction, 202, 204–205, 233
Hexagonal symmetry, 331
Higher-order modes, 414
High impedance amplifier circuit, 298–300
High-silica glasses, 467
Hole, 187
Homogeneous line broadening, 109–110
Homojunction, 202–204
Hooke's law, 353, 387
  compressed form, 389
Hot junction, 263–264
Huygen's principle, 66
Hybrid modes, 441–442
Hydrogen atoms, Bohr-Sommerfeld theory, 170–172
Hydrogen-like atoms, 173–175

I

II-IV semiconductors, 189, 200–201
III-IV semiconductors, 197–200
ILD; see semiconductor injection lasers
Impact ionization process, 281
Impurities, in fibers, 473–474
Impurity ionization process, 270
Incident angle, 28
Incident rays, 35

Subject Index

Incoherent detection method, 290–292
   compared to coherent detection, 294–295
   definition, 261
Index difference, 439
Index ellipsoid, 18–19, 308, 327–336
   of biaxial media, 336
   of isotropic media, 331
   of uniaxial media, 331–335
Index file parameter, 439
Index-guided stripe geometry lasers, 213
Index of refraction, 25, 198
Indirect bandgap materials, 182
Indirect bandgaps, 190–193
Indirect transitions, 194
Inert gases, ground states, 187
Information capacity, of optical fibers, 457–466
Inhomogeneous line broadening, 109–110
Injection lasers, 209
   basic parameters, 230–238
   broad-area lasers, 209–210
   single-frequency single mode, 213–215
   stripe geometry, 211–212
   surface-emitting semiconductor, 217
Insertion losses, 499, 512–513
   due to angular misalignment, 522
   due to core area mismatch, 519
   due to end separation, 519–521
   due to normalized mismatch parameters, 522–523
   due to numerical aperture mismatch, 519
Inside vapor deposition process, 471
Integrated etalon interference structure, 215
Integrated optic pickup unit, 5
Integrated optics, 19
Integrated optic star coupler, 500, 503
Integrated optic temperature sensors, 5–8
Integrating front end, 298
Intensity modulators, 427
Intermodal dispersion, 487–488
   definition, 457–458
   graded index fibers, 459–460
   step index fibers, 458–459
Internal current gain, 266
Internally striped structure, 212
Internal or built-in stress birefringence, 550–551
Internal photoelectric effect, 265
Internal quantum efficiency, 235–237
Intramodal dispersion, 458–466
Intrinsic conductors, 182

Intrinsic loss in fibers, definition, 517
Intrinsic semiconductors, 183–189
Inverse Stark effect, 308
Inverse Zeeman effect, 307
Inversion, 101
Ionic polarizability, 317–318
Ion implantation techniques, 211–212
Ionization level of atoms, 171
Ionized atoms, 141
Ionized gas lasers, 142–146
Ion lasers, 143
Irradiance, 52
   definition, 220
Isolators, 326–327
Isotropic liquids, 505
Isotropic media, 312
   elastic waves in, 387–389
   wave propagation, 327–328
Isotropic solids, 389–390
IV-VI semiconductors, 201–202

J

Johnson noise, 285
Johnson-Nyquist noise, 285
Jones calculus, 534, 535–536
   and birefringent media, 535–549
   definition, 535
Jones matrix, 535
   axis rotation, 542–543
   birefringent fibers, 555–556
   circularly birefringent media, 540–541
   definition, 537–538
   of isotropic media, 539
   linearly birefringent media, 539–540
   of linearly birefringent media in magnetic fields, 541, 575–578
   of linearly birefringent medium with continuously rotating axis, 579–582
   ordered product of, 538
   for polarizers, 542
   unitary, 547
Jones vectors, 535–537
Junction, semiconductor, 202–205

K

Kerr coefficient, 318
Kerr effect, 308, 318–319, 553
Kerr magnetooptic effect, 307
Knife edge method, 57–61
Kundt constant, 323

L

Laguerre polynomials, 49, 62–64
Lambert's cosine law, 225, 501
Lame's coefficients, 384

LANS (local area networks), 490–491
Lasers
   classification of, 100–102
   commercially available, 159–161
   dye lasers, 149–153
   gas laser systems, 140–149
   mode locking operation, 133–140
   optical cavities, 42
   Q-switched, 125–133
   rate equations with optical pumping, 110–125
   semiconductor injection, 182
   solid-state, 153–158
   steady-state output power, 123–124
Lasing stage, of Q-switching, 128–129
Lateral magnification, 40, 72, 513–514
Launching efficiency; see Coupling efficiency
Laws of conservation of energy, 191–192
Laws of conservation of momentum, 192
Lead selenide, 201
Lead sulphide, 201
Lead telluride, 201
LEDs; see Light emitting diodes
Left-hand circularly polarized waves, 310, 313, 321, 322
Lens/lenses
   convex or concave, 36
   and coupling efficiency, 513–517
   geometrical optics, 72–74
   lateral magnification, 40, 72, 513–514
   optical path lengths, 492–493
   positive or negative, 36
   simple, 32
   thick, 33–35
   thin, 35–36
Lens-like matrices, 47–49
LHCP; see Left-hand circularly polarized waves
Light emitting diodes, 183, 217–230
   compared with ILDs and SLDs, 251–252
   edge-emitting, 220, 226–227
   light surface-emitting, 219–226
   modulation characteristics, 228
   PI and spectral characteristics, 227–228
   power coupled into multimode fiber, 228–230
Light extraction efficiency of surface-emitting LEDs, 220–224
Linear birefringence in fiber, caused by bending, 560

Linear electrooptic effects, 337–341
   ADP/KDP crystals, 339–341
   linear EO coefficients, 337–339
Linearly birefringent fibers, 551
   caused by electrical fields, 554
   Jones matrix, 555–556
Linearly birefringent medium
   continuous axis rotation, 546–547, 579–582
   definition, 535
   in magnetic field, 541
Linearly polarized fields or waves, 309
Linearly polarized mode guided by weakly guided step index fibers, 444–457
   designation, 452–453
Linearly polarized waves, 309
Linear retardation, 540
Linear retarder, 540
Line broadening, 109–110
Liquid crystal phases, 505
Liquid crystal switches, 502, 505–508
Liquid dye lasers, 96
Liquid phase epitaxy technique, 213
Liquids, 390
Lithium atom, 176
Local numerical aperture, 421
Locking, 136; see also Mode locking operation
Longitudinal acoustic waves, 354–357, 388
Longitudinal EO modulator configuration, 341–345
   versus transverse EO modulators, 347–348
Longitudinal mode number, 80
Longitudinal modes, 214
Longitudinal mode separation, 81
Long-wavelength cutoff, 280
Lorentz force, 320
Loss; see Fiber loss
Low impedance amplifier circuits, 298–300
LS coupling, 176

M

Mach-Zehnder interferometer, 11–14
Magnetic quantum number, 172
Magnetic spin quantum number, 172
Magnetooptic amplitude modulators, 323–326
Magnetooptic effect, 18–19
   simple, 319–321
   types of, 307–308
Magnetooptic isolators, 326–327
Magnetooptic modulators, versus electrooptic modulators, 347–348

Many-electron atoms
  electron configuration, 173–175
  helium, 175–176
  lithium, 176
  term symbol, 176–178
Maser, 94
Matching length, 71
Material dispersion, 213
  of crystals and glasses, 465–466
  definition, 458
Material figure of merit
  for AO materials, 369–370
  for EO materials, 350–351
Maxwell's equations, 326–327, 408–409, 445–446, 575
Mechanical switches, 502, 503
Meniscus lens, 33
Meridional plane, 27
Meridional ray, 27
  of optical fibers, 440–441
Mesa isolated stripe, 213
Metal film polarizers, 557–558
Metal ion lasers, 143
Metallic waveguides, 392
Metals, use as reflectors, 392
Meter, 95
Michael Faraday Bicentennial conference, 2
Michelson interferometer, 5–8, 13–14
Microlasers, 217
Microns, 95
Minimum pumping transition probability, 114
Mirrors
  ABCD matrices, 42–44
  Brewster angle windows, 62
  optical cavities formed by, 42
  planar, 79
  planar or spherical, 36–39
Mixer plate, 500
Miya, 478 n
Mode designation
  step index fibers, 441–442
  weakly guided step index fibers, 452–453
Mode hoping, 306
Mode locking, 125, 133–140
  active or passive, 140
  definition, 133
Modified Bessel functions, 481–482
Modified chemical vapor deposition process, 470–471
Modulation
  characteristics of ILDs, 239–242
  electrooptic modulators, 341–351
Modulation bandwidth, 250
Modulation noise, 288–289
Modulus of rigidity, 384
Molecular gas lasers, 97
  conventional CO_2 lasers, 147–148
  types of, 146–147

Molecular rotational and vibrational states, 141
Molecular vibrational modes, 473
Mole fraction, 198
Monoclinic symmetry, 336
Monomode fibers, 453–454
Multilayer directional coupler, 431–432
Multimodal dispersion; see Intermodal dispersion
Multimode fiber components
  connector loss, 517–523
  optical switches compatible with, 502–508
  power coupled into fibers, 508–517
  taps and star couplers, 491–502
  weakly guided fibers, 485–488
Multimode fiber directional coupler, 496–500
Multimode fibers, coupling from LEDs to, 228–230
Multimode fiber switches
  electrowetting, 503–505
  liquid crystal, 506–508
  mechanical, 503
  types of, 502
Multiple quantum well lasers, 216
Multiplexing, 9–11
Multiplicity, 176–177

N

Narrow-bandgap semiconductor, 204
Ne; see Neon
Near field region, 54
Negative (or diverging) lenses, 36
Negative uniaxial materials, 332
Nematic axis, 505–506
Nematic liquid crystals, 506–507
Neodymium lasers, 155–157
Neon, 141–142
  electronic configuration, 178
Neutral atoms, 141
Neutral gas lasers, 141–142
Newton's second law of motion, 385–386
nN junction, 204
Noble gas ion lasers, 143
Noise
  in coherent detection, 291
  dark current, 288
  semiconductor noise, 288–289
  shot noise, 263, 286–287
  thermal noise types, 285–286
Noise equivalent power, 262–263, 289
Noise in signal, 262–263
Nonradiative recombination, 196
Nonradiative transition, 101, 111
Nonreciprocal effects, 554

Normalized cavity parameters, 76
Normalized detectivity D* and D**, 263
Normalized distance, 568
Normalized guide index, 406
Normalized Jones vector, 537
Normalized mismatch parameters, 522–523
Normalized pump power, 568–569
Normalized signal power, 568
Normal modes of wave propagation, 329
Numerical aperture, 229, 417–418
  local, 421
  of step index fiber, 440–441
Numerical aperture mismatch, 519

O

1 x 2 coupler, 491–492
One-dimensional elasticity, 352–353
1/f noise, 289
Optical axis, 26
Optical cavities, 42–44
  beam parameters, 74–82
  concentric resonators, 79
  confocal resonators, 77–78
  normalized cavity parameters, 76
  parallel planar mirror cavities, 79
  planar mirror/spherical mirror resonators, 79–80
  resonance frequency, 80–82
  stability condition, 77
  stable and unstable conditions, 43–44
  symmetric, 77
Optical components
  analyzers and polarizers, 314
  coupling between, 508–517
  effective thickness, 40
  GRIN lenses, 484–488
  sizes of pinholes, mirrors, and lenses, 56–57
  spherical mirrors, 66–67
  thin lenses, 68–74
Optical detectors
  comparison of, 284–285
  noise and noise equivalent power, 285–290
  photon detectors, 265–284
  thermal detectors, 263–265
Optical directional couplers, 426–433
  coupled mode theory, 427–431
  multilayer directional coupler example, 431–432
  waveguide directional coupler types, 432–433
Optical factor, 443
Optical fiber gyroscopes, 13–17
Optical fibers, 19

  absorption loss, 472–473
  fabrication process, 466–472
  index profiles, 439–440
  information capacity, 457–466
  intermodal dispersion, 457–460
  intramodal dispersion, 460–466
  intrinsic scattering, 475–477
  linearly polarized modes guided by weakly guided step index fibers, 444–457
  losses due to impurities, 473–474
  material components, 466–468
  single mode, 453–454
  step index, 440–444
  types of losses, 472
  types of pulse dispersion, 457–458
  waveguide imperfections, 477
  weakly guided fibers, 440
  weakly guiding, 446
  zero-dispersion wavelength, 466
Optical filters, 427
Optical frequency division multiplexing distribution system, 8–13
Optical gain, 124–125
Optical heterodyne detection method, 262, 294
Optical homodyne detection method, 294
Optical indicatrix, 329
Optical isolator, 9
Optically pumped lasers, 110–125
Optical planar waveguides, 392
Optical power, versus current characteristics, 227–228
Optical switches, 427
  electrowetting, 503–505
  liquid crystal, 505–508
  mechanical, 503
  types of, 502
Optical system synthesis, 44–48
Optic axis, of uniaxial media, 331
Optics, geometrical, 72–74
Optoelectronic integrated circuits (OEIC), 19
Orbital angular momentum quantum number, 171
Ordinary index, 332
Orthorhomic symmetry, 336
Outside vapor deposition process, 468
Overlap factor, 351

P

Parabolic index profile of weakly guided fibers, 486–487
Parallel planar mirror cavities, 79
Parallel-plate waveguides, 393–397

Parallel-plate waveguides—*Cont.*
  characteristic equation of TE modes guided by, 394–396
  cutoff thickness/frequencies, 414
  field distribution of TE modes guided by, 396–397
Parameter Q, 366–367
Paraxial approximation, 28, 31, 40–41
Passive components
  based on GRIN lenses, 492–495
  based on spherical reflectors, 495–496
  based on thin-film beam splitters, 491–492
Passive mode locking, 140
Pauli's exclusion principle, 172
PC detectors; *see* Photoconductive detectors
Perfect beam expanders, 71–72
Perfect polarizer, 542
Phase match condition, 429–430
Phase velocity, 397–400
Photoconductive detectors, 271–274
Photoconductive gain, 273–274
Photodetectors, 18
  circuit topology, 298–300
  thermal or photon, 261
Photoelastic constant, 317
Photoelastic effects, 357; *see also* Acoustooptic effects
Photoelastic tensor, 321
Photoemissive detectors, 265–268
Photoemissive effect, 265
Photomultiplier tubes, 265–268
Photon density
  rate of change, 120–122
  steady-state, 123–124
  time rate of charge, 130–131
Photon detectors, 265–285
  photoconductive detectors, 271–274
  photoemissive detectors, 265–268
  photovoltaic detectors, 274–284
Photon energy, 96
Photon flux intensity, 105–107
Photonics, 1–2
Photon lifetime, 241
Photons, 99
  absorption in semiconductors, 193–195
  spontaneous emission, 101, 104
  stimulated emission, 101, 103
Photon-to-electron conversion, 260
Photonics, 1–2
Photovoltaic detectors, 274–284
  avalanche photodiodes (APD), 275, 281–284
  PIN diodes, 275–280

PI characteristics, 206
PIN diode, 275–280
  long- or short-wavelength cutoff, 280
  quantum efficiency, 279
Pitch, of GRIN lenses, 493–494
Planar dielectric boundaries, reflection by, 400–403
Planar mirror, 36–39, 79
Planar mirror/spherical mirror resonators, 79–80
Planar thin-film waveguides, 351
Planar waveguide, 403
Planck's constant, 96
Planck's equation for blackbody radiation, 99–100, 104, 260–261
Plane of incidence, 400
Plane wave, 25, 312; *see also* Uniform plane wave
Plasma-impulse chemical vapor deposition process, 471–472
Plasma modified chemical vapor deposition process, 471–472
Plastic fibers, 438
Pn junction, 204
pN junction, 204
Pockels effect, 308, 318–319
Poisson distribution function, 287
Poisson ratio, 383
Polarization, 25, 296; *see also* States of polarization
  Eigen modes, 576–577
Polarization beat length, 554–555
Polarization controllers, 560–561
Polarization ellipse, 310
Polarization-maintaining fibers, 555
Polarized field, linearly, circularly, or elliptically, 536
Polarizers, 127, 314, 542
  anisotropic superstrate, 558–559
  metal film, 557–558
  single-mode fibers, 557–558
Polarizer's axis, 314
Polished fiber couplers, 563–564
Polymethyl methacrylate, 438
Population inversion, 107, 109
  critical, 122
  minimum pumping transition probability, 114
  necessary condition for, 113
  rate of change, 116
  steady-state, 123–124
Positive lenses, 36
Positive meniscus lens, 33
Positive uniaxial materials, 332
Power, coupled into multimode fibers, 508–517
Power budget, 526–528
Power combiner, 491
Power divider, 491
  with two tilted spherical surfaces, 495–496

Power efficiency, of ILDs, 237
Power extraction efficiency of surface-emitting LEDs, 220–224
Power-modulation bandwidth product of LEDs, 228
Power series expansions of Bessel function, 482–483
Poynting, vector, 328
pP junction, 204
Preform, 467
Preform forming techniques, 467
Pressure broadening, 148
Principal axes, 328
Principal axes of the ellipsoid, 329–330
Principal directions of wave propagation, 329
Principal mode number, 487
  for step index fibers, 454–455
Principal quantum number, 171
Principal values, of relative dielectric constant tensor, 328
Prism couplers, 421–424
Prism coupling, 421
Propagation, axis rotation due to, 580–581
Propagation constant, 397–400
Propagation-like matrices, 47–49
Pulse broadening, 483
  and information capacity, 523–524
  intermodal dispersion, 457–460
Pulse repetition rate, 458
Pulses, mode locking generation, 133–140
Pulse shape
  approximate width, 132
  Q-switched lasers, 129–130
Pulse width, 132–133
  definition, 138
Pumping, 95, 100
  critical transition probability, 122
  end pumping scheme, 157
  four-level laser systems, 116–119
  minimum transition probability, 114
  rate equations for lasers with, 110–125
  side pumping method, 157
  three-level laser systems, 111–116
  transition probabilities, 111–112
Pumping efficiency factor, 114–115
Pumping stage, of Q-switched lasers, 128–129
Pump power
  normalized, 568–571
  threshold, 568
Punchthrough voltage, 277–278
Pyroelectric detectors, 265
Pyroelectric effect, 265

Q

Q-spoiling, 125
Q-switched lasers, 125–133
  approximate pulse width, 132
  peak power, 130–131
  pulse shape, 129–130
  pumping stage or lasing stage, 128–129
  total energy level per pulse, 131–132
Q-switching, 125
Quantum efficiency
  definition, 266–268
  of ILDs, 234–237
  for PC detectors, 273
  for PIN diodes, 279
  of semiconductor detectors, 268–271
Quantum numbers, 97
  electron spin, 172
  electron spin angular momentum, 172
  magnetic, 172
  orbital angular momentum, 171
  principal, 171
Quantum well lasers, 216–217
Quaternary crystalline solid solutions, 197, 199–200
Quarterwave plates, 5, 540

R

Radial mode number, 442, 452, 486
Radiance, 516
Radiant intensity, 220–221
Radiant intensity patterns
  edge-emitting LEDs, 226–227
  surface-emitting LEDs, 224–226
Radiation
  by dipole antennas, 476
  of ILDs, 237–238
Radiative recombination, 196
Radiative transition, 101
Radius of a curvature, 53–54
  complex, 55
Raised strip waveguide, 433
Raman-Nath diffraction, 357–361, 364, 367
Rare earth ions, 153–155
Rate equation, for lasers with optical pumping, 110–125
Rate of radiative recombination, 196
Ray equation, 420
Rayleigh range, 54–55
Rayleigh scattering, 475–476
Rays, 24
  in homogeneous medium, 28
  multiple reflected, 43
  reflected by spherical and planar mirrors, 36–39
  refraction by a spherical surface, 31–33
  of step index fibers, 440–441

Ray transfer matrices, 25–44
Ray vectors, 25–44
Reciprocal effects, 554
Recurrence relationship for Bessel functions, 482
Reduced thickness, 31
Reference junction, 263
Reference plane, 26
Reflected angle, 28
Reflection, 269–270
  by planar and spherical mirrors, 36–39
  by planar dielectric boundaries, 400–403
Reflection coefficient, 402
Refracted angles, 28
Refraction, by a spherical surface, 31–33
Refractive index, second order tensor, 322
Relative dielectric constant tensor, 328
Relative dielectric constant, 25
Relative dielectric impermeability tensor, 329
Relative permeability, 25
Relative permittivity, 25
Resonators
  concentric, 79
  confocal, 77–78
  planar mirror/spherical mirror, 79–80
Retardation, 342
Reverse breakdown voltage, 281
RHCP; see Right-hand circularly polarized waves
Right-hand circularly polarized waves, 310, 313, 321, 322
Risetime, 524
Risetime bandwidth relationship, 524–525
Risetime budget, 526–528
Rodamine 6G, 150
Rotation deformation, 377–378
Rotation matrix, 543
Ruby lasers, 1, 94, 157
Russell-Saunders coupling scheme, 176
Rydberg constant, 170–171

S

Safety factor, 351
Sagnac interferometer, 13, 14–16
Saturable absorber, 128
Scattering process, 472
  fiber loss due to, 475–477
Schlottky barrier, 274
Schlottky's theorem, 287
Schrodinger equation, 172–172
Seebeck effect, 263–264
Selection rule, 178
Semiconducting bolometers, 264
Semiconductor detectors, quantum efficiency, 268–271
Semiconductor injection lasers, 97, 101, 182
  amplitude modulation, 246–250
  basic parameters, 230–238
  compared to LEDs and SLDs, 251–252
  dynamic characteristics, 239–240
  far field radiation pattern, 237–239
  internal and external quantum efficiencies, 234–237
  operation of, 230–232
  power efficiency, 237
  rate equations, 239–242
  spectral linewidth, 238
  steady-state PI characteristics, 242
  threshold current density, 233–234
  transient characteristics, 244–246
  wavelength of radiation, 233
Semiconductor luminescent diodes
  basic structures, 206–217
  broad area lasers, 209–210
  stripe geometry lasers, 211–213
Semiconductor noise, 288–289
Semiconductors
  absorption, 193–195
  attenuation, 270
  binary, 189
  direct/indirect bandgap, 191–193
  elemental, 187–189
  homojunctions and heterojunctions, 202–205
  intrinsic and extrinsic, 182, 183–189
  refractive index, 269–270
  ternary and quartenary, 197–202
Sharp state, 171
Shearing distortion, 377–378
Shearing strain, 378–379
Shear modulus, 384
Shear waves, 357
Short cavity laser, 214
Shot noise, 263, 286–287
  in coherent detection, 293–294
  in incoherent detection, 291
Shot noise limited signal-to-noise ratio, 291–292, 294
Side pumping method, 157
Signal-to-noise power ratio, 262
Signal-to-noise ratio, 294–295, 491
Silicate-based fibers, 438
Simple acoustooptic effect, 316–317
Simple electrooptic effect, 317–319
Simple magnetooptic effect, 319–321
Single-frequency single mode injection lasers, 213–215
Single heterostructure diode, 206
Single-mode double polarization fibers; see Single-mode fibers
Single-mode fiber birefringence, 550–556
  external applied stress or fields, 551–554
  geometrical, 550
  internal or built-in stress, 550–551
  polarization beat length, 554–555
Single-mode fiber directional couplers
  biconical tapered structure couplers, 563
  polished fiber couplers, 563–564
  twin core couplers, 562–563
Single-mode fiber polarizers, 557–558
Single-mode fibers, 20, 443, 453–454
  fiber retarders, 559–560
  polarization components, 557–561
  polarization controllers, 560–561
Single quantum well lasers, 216
SI system, 95
Skew ray, 27
  in optical fibers, 440
SLDs; see Superluminescent diodes
Small-signal gain, 125
Smetic liquid crystal, 505
Snell's law, 30, 31, 37, 400, 418
Solid-state lasers, 96
  alexandrite lasers, 157–158
  emerald lasers, 157–158
  neodymium lasers, 155–157
  ruby lasers, 157
  types of, 153–158
Soot, 468
Space-charge region, 204
Specific detectivity D* and D**, 263
Spectral linewidth, of ILDs, 238
Spectral width of light, 96
Spherical mirror, 36–39, 66–67
Spherical mirror cavities, resonance frequency, 80–82
Spherical mirror resonators, 79–80
Spherical reflectors, passive components based on, 495–496
Spherical waves, 49
Spin angular momentum quantum number, 172
Spontaneous emission lifetimes, 107, 111–112
Spontaneous emission of photons, 101, 104, 196, 241
Spot size, 52
Spun fibers, 550–551
Square law devices, 262
Stability condition for optical cavities, 76
Stability diagram, 44
Stable cavities, 43–44
Stable optical cavity, 74–77
Star couplers, 490–492
  fused biconical taper fiber directional couplers, 496–500
  and GRIN lenses, 492–495
  with mixer rod and plate, 500–502
  passive components based on spherical reflectors, 495–496
  transmission, 500
  transmission-reflection, 500
Stark effect, 308
States of polarization
  circularly polarized waves, 309–310
  definition, 308–309
  elliptically polarized waves, 310–312
  experimental determination, 547–548
  linearly polarized waves, 309
Steady-state population inversion, 123–124
Step index fiber
  bandwidth-length product, 460
  connector loss, 517–523
  coupling efficiency, 510–513
  generalized parameters, 442–444
  information capacity, 453
  intermodal dispersion, 458–459
  linearly polarized modes guided by weakly guided, 444–457
  numerical aperture, 229
  principal mode number, 454–455
  single-mode operation, 453–454
  total guided modes, 455–457
  traditional mode designation, 441–442
Step index profile, 403, 443
  of optical fibers, 439–440
Step index thin-film waveguides, 403–418
  characteristic equation, 403–406
  generalized parameters, 406–407
  modes guided by, 412–418
  numerical aperture, 417–418
Stimulated emission of photons, 101, 103
Strain, 353
Strain tensor, in three-dimensional objects, 377–379
Stress, 353, 379

Stress tensor, in three-dimensional objects, 379–383
Stripe geometry injection lasers, 211–213
Strip waveguides, 351
Substrate modes, 416–417
Superluminescent diodes, 182, 250–251
  compared to LEDs and ILDs, 251–252
Surface-emitting LEDs, 219–220
  power and light extraction efficiency, 220–224
  radiant intensity patterns, 224–226
Surface-emitting semiconductor injection lasers, 217
Surface plasma chemical vapor deposition process, 472
Symmetric confocal resonators, 77–78
Symmetric optical cavities, 77
Symmetric stretch modes of CO_2 molecules, 146

T

Tapered fiber directional couplers, 496–500
Taps, 490, 491
TE; see Transverse electric waves/modes
Tee coupler, 491
TEM; see Transverse electromagnetic waves
TEM modes, 64
Temperature sensors, 5–8
Tensile strain, 316–317
Tensor elements of relative dielectric impermeability tensor, 330
Term symbol, 99, 175–178
Ternary crystalline solid solutions, 197–198
Ternary II-VI semiconductors, 200–201
Tetragonal symmetry, 331
Thermal detectors, 263–265
  bolometers and thermistors, 264
  Golay's pneumatic detectors, 264–265
  pyroelectric, 265
  thermocouples and thermopiles, 263–264
Thermal equilibrium, distribution of occupied states, 107
Thermal equilibrium, 99
Thermal noise, 285–286
Thermistors, 264
Thermocouples, 263–264
Thermopiles, 263–264
Thevenin equivalent circuit, 349
Thick lenses, 33–35
Thin-film beam splitters, 491–492

Thin-film waveguides, 5, 19
  cover and substrate modes, 416–417
  general characteristics, 403–404
  graded index, 418–421
  guided mode fields, 414–416
  meriodional rays, 440
  number of guided modes, 414
  prism couplers, 421–424
  step index, 403–418
  three-dimensional, 433
  two-dimensional, 432
Thin lenses, 35–36
  transformation of Gaussian beams, 68–74
Three-dimensional objects
  elastic equations of motion, 385–386
  Hooke's law, 383–385
  strain tensor, 377–379
  stress tensor, 379–383
Three-dimensional waveguides, 433
Three-level laser system, 100–102
  necessary condition for population inversion, 113
  with optical pumping, 111–116
  pumping efficiency factor, 114–115
  steady-state gain, 124–125
Three-mirror folded cavity dye laser configuration, 151–152
Three-step process of making glass, 467
Threshold current density of ILDs, 206, 233–234
Threshold gain, 233
Threshold pump power, 568
Torr unit, 142
Total internal reflection, 222, 402–403
Traditional mode designation, 448, 452–453
Transimpedance amplifier circuit, 298–300
Transition metal ions, 154–155
Transition probabilities, 111
Transition region diffraction, 366–367
Transmission axis, 127, 314
Transmission-reflection star coupler, 500
Transmission star coupler, 500
Transverse acoustic waves, 357, 388–389
Transverse electric modes, 393, 395–397, 400–402, 441–442
  thin-film waveguides, 408–411
Transverse electromagnetic waves, 312, 442–443, 447
Transverse EO modulators, 341, 346–347
  versus longitudinal EO modulators, 347–348

Transverse junction stripe laser, 213
Transverse magnetic modes, 396, 400–402, 441–442
  thin-film waveguides, 411–412
Transverse magnification; see Lateral magnification
Transverse mode numbers, 81
Transverse mode separation, 82
Transverse offset, 521
Traveling wave tubes, 142
Triclinic symmetry, 336
Trigonal symmetry, 331
Trivalent rare earth ion, 97
Trivalent transition metal ions, 97
Turning points, 486
Turn-on time, 245
Twin core couplers, 562–563
Two-dimensional waveguides, 432
Two-level system rate equation, 107–109
Type one polarization controllers, 560–561
Type three polarization controllers, 560
Type two polarization controllers, 560–561

U

Uniaxial media, 330, 331–335
  optic axis, 331
Unidirectional device, 152
Unidirectional ring arrangement, 152–153
Uniform plane wave, 24, 25, 49, 312
United States National Research Council, 1
Units, of SI system, 95–97
Unpumped optical cavity, 120–121
Unstable optical cavity, 43–44, 74

V

Vacuum level, 203
Vacuum photodiodes, 265–268
Vacuum wavelength l [lambda], 25
Valence band, 185–186
Valence electrons, 183–185
Vapor deposition processes, 468–472
Vapor phase axial deposition process, 469–470
Vegard's law, 198
Verdet constant, 323
Vertical cavity surface-emitting lasers (VCSEL), 217
Vibration modes of CO_2 molecules, 146
Video long-play systems (VLP), 3–4
Visibility of the ellipse, 311, 548

Voigt effect, 307, 327
Voltage or current responsivity, 260–261
Vycor glass, 431–432

W

Wave attenuation, 105–107, 270
Wave equations, 407–412
Wavefront curvature, 297–298
Waveguide CO_2 lasers, 148–149
Waveguide dispersion, 458
Waveguide imperfections, 477
Waveguides, step index thin-film, 403–418
Wave-material interaction
  absorption, 102–103
  stimulated emission, 103
Wavenumber, 96
Wave propagation
  in biaxial media, 336
  in isotropic media, 327–328
  in uniaxial media, 331–335
Waves; see also Guided waves
  in anisotropic media, 314
  in isotropic media, 312
  plane/uniform plane, 312
  right- and left-handedness, 310 n
  states of polarization, 308–314
Wave vector k, 191
Weakly guided fibers, 440
  approximate characteristic equation, 485–486
  intermodal dispersion, 487–488
  intramodal dispersion, 458–466
  parabolic index profile, 486–487
Weakly guided step index fibers
  characteristic equation, 448–449
  linear mode designation, 452–453
  polarized modes guided by, 444–457
  principal mode number, 454–455
  single-mode operation, 453–454
  total guided modes, 455–457
  X-polarized fields, 450–451
Weakly guiding fibers, 19
Weierstrass spherical domes, 223–224, 226
White noise, 286
Wide-bandgap semiconductor, 204

Y–Z

YAG laser, 101, 154
Young's modulus, 353
Yttrium-aluminum garnet, 154
Zeeman effect, 307
Zero-dispersion wavelength, 466

Learning Resources
Brevard Community College
Cocoa, Florida